Handbook of Polymers

3rd Edition

George Wypych

ChemTec Publishing

Toronto 2022

Published by ChemTec Publishing
38 Earswick Drive, Toronto, Ontario M1E 1C6, Canada

© ChemTec Publishing, 2012, 2016, 2022
ISBN 978-1-927885-95-6 (hard copy)
ISBN 978-1-927885-96-3 (epub)

Cover design: Anita Wypych

All rights reserved. No part of this publication may be reproduced, stored or transmitted in any form or by any means without written permission of copyright owner. No responsibility is assumed by the Author and the Publisher for any injury or/and damage to persons or properties as a matter of products liability, negligence, use, or operation of any methods, product ideas, or instructions published or suggested in this book.

Library and Archives Canada Cataloguing in Publication

Title: Handbook of polymers / George Wypych.
Other titles: Polymers
Names: Wypych, George, author.
Description: 3rd edition. | Includes bibliographical references and index.
Identifiers: Canadiana (print) 20210194618
| Canadiana (ebook) 20210194642 | ISBN 9781927885956
  (hardcover) | ISBN 9781927885963 (PDF)
Subjects: LCSH: Polymers—Handbooks, manuals, etc.
| LCGFT: Handbooks and manuals.
Classification: LCC TP455.P58 W96 2022 | DDC 668.9—dc23

Printed in United States, United Kingdom, and Australia

Table of Contents

	Introduction	1
ABA	acrylonitrile-butadiene-acrylate copolymer	3
ABS	poly(acrylonitrile-co-butadiene-co-styrene)	5
AK	alkyd resin	13
ASA	poly(acrylonitrile-co-styrene-co-acrylate)	18
BIIR	bromobutyl rubber	23
BMI	polybismaleimide	26
BZ	polybenzoxazine	29
C	cellulose	32
CA	cellulose acetate	38
CAB	cellulose acetate butyrate	44
CAP	cellulose acetate propionate	48
CAPh	cellulose acetate phthalate	52
CAR	carrageenan	54
CB	cellulose butyrate	56
CEC	carboxylated ethylene copolymer	58
CHI	chitosan	60
CIIR	chlorobutyl rubber	64
CMC	carboxymethyl cellulose	67
CN	cellulose nitrate	70
COC	cyclic olefin copolymer	74
CPE	polyethylene, chlorinated	78
CPVC	poly(vinyl chloride), chlorinated	81
CR	polychloroprene	84
CSP	polyethylene, chlorosulfonated	89
CTA	cellulose triacetate	92
CY	cyanoacrylate	96
DAP	poly(diallyl phthalate)	99
E-RLPO	poly(ethyl acrylate-co-methyl methacrylate-co-triammonioethyl methacrylate chloride)	101
EAA	poly(ethylene-co-acrylic acid)	102
EAMM	poly(ethyl acrylate-co-methyl methacrylate)	105
EBAC	poly(ethylene-co-butyl acrylate)	107
EBCO	ethylene-n-butyl acrylate-carbon monoxide terpolymer	109
EC	ethyl cellulose	111
ECTFE	poly(ethylene-co-chlorotrifluoroethylene)	115
EEAC	poly(ethylene-co-ethyl acrylate)	119
EMA	poly(ethylene-co-methyl acrylate)	121
EMA-AA	poly(ethylene-co-methyl acrylate-co-acrylic acid)	124
ENBA	poly(ethylene-co-n-butyl acrylate)	126
EP	epoxy resin	128
EPDM	ethylene-propylene diene terpolymer	133
EPR	ethylene propylene rubber	138
ETFE	poly(ethylene-co-tetrafluoroethylene)	141
EVAC	ethylene-vinyl acetate copolymer	145
EVOH	ethylene-vinyl alcohol copolymer	150
FEP	fluorinated ethylene-propylene copolymer	154
FR	furan resin	158
GEL	gelatin	161
GT	gum tragacanth	164
HCP	hydroxypropyl cellulose	166
HDPE	high density polyethylene	169
HEC	hydroxyethyl cellulose	177
HPMC	hydroxypropyl methylcellulose	179
HPMM	poly(methacrylic acid-co-methyl methacrylate)	181

HANDBOOK OF POLYMERS 3rd Edition, Copyrights 2022; ChemTec Publishing

IIR	isobutylene-isoprene rubber	184
LCP	liquid crystalline polymers	188
LDPE	low density polyethylene	192
LLDPE	linear low density polyethylene	199
MABS	poly(methyl methacrylate-co-acrylonitrile-co-butadiene-co-styrene)	205
MBS	poly(styrene-co-butadiene-co-methyl methacrylate)	208
MC	methylcellulose	211
MF	melamine-formaldehyde resin	214
MP	melamine-phenolic resin	217
NBR	acrylonitrile-butadiene elastomer	219
PA-3	polyamide-3	222
PA-4,6	polyamide-4,6	224
PA-4,10	polyamide-4,10	228
PA-6	polyamide-6	230
PA-6,6	polyamide-6,6	236
PA-6,10	polyamide-6,10	242
PA-6,12	polyamide-6,12	246
PA-6,66	polyamide-6,66	250
PA-6I/6T	polyamide-6I/6T	253
PA-11	polyamide-11	256
PA-12	polyamide-12	261
PAA	poly(acrylic acid)	267
PAAm	polyacrylamide	270
PAC	polyacetylene	273
PAEK	polyaryletherketone	276
PAH	polyanhydride	278
PAI	poly(amide imide)	282
Palg	alginic acid	286
PAN	polyacrylonitrile	288
PANI	polyaniline	294
PAR	polyarylate	298
PARA	polyamide MXD6	301
PB	1,2-polybutylene	304
PBA	poly(p-benzamide)	308
PBAA	poly(butadiene-co-acrylonitrile-co-acrylic acid)	310
PBD,cis	*cis*-1,4-polybutadiene	312
PBD,trans	*trans*-1,4-polybutadiene	316
PBI	polybenzimidazole	319
PBMA	polybutylmethacrylate	323
PBN	poly(butylene 2,6-naphthalate)	325
PBT	poly(butylene terephthalate)	328
PC	polycarbonate	334
PCL	poly(ε-caprolactone)	342
PCS	polycarbodihydridosilane	345
PCT	poly(cyclohexylene terephthalate)	346
PCTFE	polychlorotrifluoroethylene	349
PCTG	poly(ethylene-co-1,4-cyclohexylenedimethylene terephthalate)	353
PDCPD	polydicyclopentadiene	356
PDL	polylysine	358
PDMS	polydimethylsiloxane	359
PDPD	poly(dicyclopentadiene-co-p-cresol)	364
PDS	polydioxanone	365
PE	polyethylene	368
PEA	poly(ethyl acrylate)	374
PEC	poly(ester carbonate)	377
PEDOT	poly(3,4-ethylenedioxythiophene)	381

PEEK	polyetheretherketone	385
PEF	poly(ethylene furanoate)	391
PEI	poly(ether imide)	393
PEK	polyetherketone	398
PEKK	polyetherketoneketone	401
PEM	poly(ethylene-co-methacrylic acid)	404
PEN	poly(ethylene 2,6-naphthalate)	406
PEO	poly(ethylene oxide)	411
PES	poly(ether sulfone)	415
PET	poly(ethylene terephthalate)	420
PEX	silane-crosslinkable polyethylene	427
PF	phenol-formaldehyde resin	430
PFA	perfluoroalkoxy resin	433
PFI	perfluorinated ionomer	437
PFPE	perfluoropolyether	440
PGA	poly(glycolic acid)	443
PHB	poly(3-hydroxybutyrate)	446
PHBV	poly(3-hydroxybutyrate-co-3-hydroxyvalerate)	450
PHEMA	poly(2-hydroxyethyl methacrylate)	452
PHSQ	polyhydridosilsesquioxane	455
PHT	polyhexahydrotriazine	456
PI	polyimide	457
PIB	polyisobutylene	463
PIP,cis	*cis*-polyisoprene	467
PIP,trans	*trans*-polyisoprene	471
PK	polyketone	474
PLA	poly(lactic acid)	479
PLGA	poly(DL-lactide-co-glycolide)	484
PLS	poly(L-serine)	485
PLT	poly(l-tyrosine)	486
PMA	poly(methyl acrylate)	487
PMAA	poly(methacrylic acid)	489
PMAN	polymethacrylonitrile	492
PMFS	polymethyltrifluoropropylsiloxane	498
PMMA	polymethylmethacrylate	497
PMP	polymethylpentene	503
PMPS	polymethylphenylsilylene	507
PMS	poly(p-methylstyrene)	509
PMSQ	polymethylsilsesquioxane	512
PN	polynorbornene	514
PNR	phthalonitrile resin	517
POE	very highly branched polyethylene	518
POM	polyoxymethylene	522
PP	polypropylene	529
PP,iso	polypropylene, isotactic	537
PP,syndio	polypropylene, syndiotactic	542
PPA	polyphthalamide	546
PPG	polypropylene glycol	549
PPMA	polypropylene, maleic anhydride modified	552
PPO	poly(phenylene oxide)	554
PPP	poly(1,4-phenylene)	558
PPS	poly(p-phenylene sulfide)	561
PPSE	poly(trimethylsilyl phosphate)	566
PPSQ	polyphenylsilsesquioxane	567
PPSU	poly(phenylene sulfone)	569
PPT	poly(propylene terephthalate)	572

PPTA	poly(p-phenylene terephthalamide)	574
PPTI	poly(m-phenylene isophthalamide)	578
PPV	poly(1,4-phenylene vinylene)	581
PPX	poly(p-xylylene)	583
PPy	polypyrrole	586
PR	proteins	589
PS	polystyrene	592
PS,iso	polystyrene, isotactic	600
PS,syndio	polystyrene, syndiotactic	603
PSM	polysilylenemethylene	607
PSMS	poly(styrene-co-α-methylstyrene)	608
PSR	polysulfide	610
PSU	polysulfone	613
PTFE	polytetrafluoroethylene	619
PTFE-AF	poly(tetrafluoroethylene-co-2,2-bis(trifluoromethyl)-4,5-difluoro-1,3-dioxole)	624
PTMG	poly(tetramethylene glycol)	626
PTT	poly(trimethylene terephthalate)	628
PU	polyurethane	632
PVAc	poly(vinyl acetate)	639
PVAl	poly(vinyl alcohol)	643
PVB	poly(vinyl butyrate)	648
PVC	poly(vinyl chloride)	651
PVCA	poly(vinyl chloride-co-vinyl acetate)	659
PVDC	poly(vinylidene chloride)	661
PVDF	poly(vinylidene fluoride)	664
PVDF-HFP	poly(vinylidene fluoride-co-hexafluoropropylene)	669
PVF	poly(vinyl fluoride)	673
PVK	poly(N-vinyl carbazole)	677
PVME	poly(vinyl methyl ether)	680
PVP	poly(N-vinyl pyrrolidone)	683
PZ	polyphosphazene	685
SAN	poly(styrene-co-acrylonitrile)	689
SBC	styrene-butadiene block copolymer	694
SBR	poly(styrene-co-butadiene)	698
SBS	styrene-butadiene-styrene triblock copolymer	702
SEBS	styrene-ethylene-butylene-styrene triblock copolymer	706
SIS	styrene-isoprene-styrene block copolymer	709
SMA	poly(styrene-co-maleic anhydride)	711
SMMA	poly(styrene-co-methylmethacrylate)	715
ST	starch	718
TPU	thermoplastic polyurethane	722
UF	urea-formaldehyde resin	727
UHMWPE	ultrahigh molecular weight polyethylene	730
ULDPE	ultralow density polyethylene	735
UP	unsaturated polyester	737
VE	vinyl ester resin	740
XG	xanthan gum	743

Introduction

Polymers selected for this edition of the Handbook of Polymers include primary polymeric materials used by plastics and other branches of the chemical industry as well as specialty polymers used in electronics, pharmaceutical, medical, and space fields. Extensive information is provided on biopolymers.

The data included in the Handbook of Polymers come from open literature (published articles, conference papers, and books), literature available from manufacturers of various grades of polymers, plastics, finished products, and patent literature. The above sources were searched, including the most recent literature. It can be seen from the references that a large portion of the data comes from information published in 2011-2021. This underscores one of the undertaking's significant goals: to provide readers with the most up-to-date information.

Frequently, data from different sources vary in a broad range, and they have to be reconciled. In such cases, values closest to their averages and values based on testing the most common grades of materials are selected to provide readers with information, which characterizes currently available products, focusing on the potential use of data to solve practical problems. In this verification process, many older data were rejected unless recently conducted studies have confirmed them.

The presentation of data for all polymers is based on a consistent pattern of data arrangement. However, depending on data availability, only data fields that contain actual values are included for each polymer. The entire scope of the data is divided into sections to make data comparison and search easy.

The following sections of data are included:
- General
- History
- Synthesis
- Structure
- Commercial polymers
- Physical properties
- Mechanical properties
- Chemical resistance
- Flammability
- Weather stability
- Biodegradation
- Toxicity
- Environmental impact
- Processing
- Blends
- Analysis

It can be anticipated from the above breakdown of information that the Handbook of Polymers contains information on all essential data used in practical applications, research, and legislation, providing such data are available for a particular material. In total, over 230 different types of data were searched for each polymer. The last number does not include special fields that might be added to characterize specialty polymers' performance in their applications.

In most cases, the information provided is self-explanatory, considering that each data field is composed of the parameter (or measured property), unit, value, and (in many cases) reference. In some cases, different values or a range of values are given. This indicates a disagreement in the published data that cannot be reconciled or that the data falls into a broader range because various grades differ in properties. Utmost care is taken that the specified range contains grades known from published data. If specific grades differ in properties, a set of separate ranges is given in some cases.

After some data, information is given in parenthesis to indicate additional characteristics of tested samples. The usual convention is that the first value given is for pure or typical material, followed by its different modifications (e.g., reinforcements with different fibers or different levels of crystallinity, structure, or different sample preparation conditions as to its temperature, state, etc.).

The range of molecular weights and related data (e.g., polymerization degree) requires additional explanation. In some cases, the number average molecular weight data do not correspond to mass average molecular data (as could be expected from a given range of polydispersities). This is because these data are given based on values found in literature without any attempts to reconcile them by means of calculation, which seems to be the correct approach because the data strictly reflect values found in the literature, not the results of any approximations, which will artificially compare sets of data for materials coming from different experimental or production conditions. This is in agreement with one essential goal of this collection – the authenticity of the data selected.

We hope that our thorough search results will be useful and that the data will be skillfully applied by users of this book to benefit their research and applications.

ABA acrylonitrile-butadiene-acrylate copolymer

PARAMETER	UNIT	VALUE	REFERENCES
GENERAL			
Common name	-	acrylonitrile-butadiene-acrylate copolymer	
Acronym	-	ABA	
PHYSICAL PROPERTIES			
Density at 20°C	g cm^{-3}	1.29-1.32	
Melting temperature, DSC	°C	340; 340-345 (30-40% glass fiber); 340 (30% carbon fiber)	
Thermal expansion coefficient, 23-80°C	°C^{-1}	0.45-0.47E-4; 0.16-0.17E-4 (30-40% glass fiber)	
Glass transition temperature	°C	150-158; 150-158 (30-40% glass fiber); 150 (30% carbon fiber)	
Heat deflection temperature at 1.8 MPa	°C	161-252; 213-286 (30-40% glass fiber); 267-276 (30% carbon fiber)	
Volume resistivity	ohm-m	2E16 (30-40% glass fiber)	
Electric strength K20/P50, d=0.60.8 mm	kV mm^{-1}	16 (30-40% glass fiber)	
MECHANICAL & RHEOLOGICAL PROPERTIES			
Tensile strength	MPa	84-93.8; 156-191 (30-40% glass fiber); 176-201 (30% carbon fiber)	
Tensile modulus	MPa	2,900-3,720; 9,900-15,200 (30-40% glass fiber); 18,800-22,100 (30% carbon fiber)	
Elongation	%	26-76; 1.8-2.9 (30-40% glass fiber); 1.5-2.0 (30% carbon fiber)	
Tensile yield strain	%	5.0-6.7	
Flexural strength	MPa	122-141; 234-253 (30-40% glass fiber); 259-317 (30% carbon fiber)	
Flexural modulus	MPa	3,100-3,720; 9,400-14,800 (30-40% glass fiber); 16,500-19,300 (30% carbon fiber)	
Compressive strength	MPa	228 (30-40% glass fiber)	
Izod impact strength, unnotched, 23°C	J m^{-1}	no break; 590-960 (30-40% glass fiber); 530 (30% carbon fiber)	
Izod impact strength, notched, 23°C	J m^{-1}	75-100; 53-110 (30-40% glass fiber)	
Shear strength	MPa	79 (30-40% glass fiber)	
Shrinkage	%	0.8-1.3; 0.3-1.3 (30-40% glass fiber); 0.1-0.5 (30% carbon fiber)	
Melt viscosity, shear rate=1000 s^{-1}	Pa s	410-450; 410-450 (30-40% glass fiber); 470 (30% carbon fiber)	
Melt index, 400°C/2.16 kg	g/10 min	1-5; 7-9 (30-40% glass fiber)	
CHEMICAL RESISTANCE			
Aromatic hydrocarbons	-	excellent	
Esters	-	excellent	
Halogenated hydrocarbons	-	excellent	
Ketones	-	excellent	
FLAMMABILITY			
UL 94 rating	-	V-0; V-0 or V-1 (30-40% glass fiber)	

HANDBOOK OF POLYMERS 3rd Edition, Copyrights 2022; ChemTec Publishing

ABA acrylonitrile-butadiene-acrylate copolymer

PARAMETER	UNIT	VALUE	REFERENCES
TOXICITY			
Carcinogenic effect	-	not listed by ACGIH, NIOSH, NTP	
PROCESSING			
Typical processing methods	-	extrusion blow molding, fiber spinning, film extrusion, injection blow molding, injection molding, machining, profile extrusion, thermoforming, wire and cable extrusion	
Preprocess drying: temperature/ time/residual moisture	°C/h/%	150/4/-; 149-175/2.5-4/- (30-40% glass fiber)	
Processing temperature	°C	354-382; 366-404 (30-40% glass fiber)	
Applications	-	aircraft, automotive, bearings, bushings, connectors, electrical/ electronics, film, fuel lines, gears, medical, oil/gas, semiconductors, seals	
Outstanding properties	-	ductile, high heat resistance, flame retardant	

ABS poly(acrylonitrile-co-butadiene-co-styrene)

PARAMETER	UNIT	VALUE	REFERENCES
GENERAL			
Common name	-	poly(acrylonitrile-co-butadiene-co-styrene)	
IUPAC name	-	buta-1,3-diene; prop-2-enenitrile; styrene	
CAS name	-	2-propenenitrile, polymer with 1,3-butadiene and ethenyl-benzene	
Acronym	-	ABS	
CAS number	-	9003-56-9	
RTECS number	-	AT6970000	
Linear formula	-	$[CH_2CH(CN)]_x(CH_2CH=CHCH_2)_y[CH_2CH(C_6H_5)]_z$	
HISTORY			
Details	-	ABS was patented in 1948 and introduced to commercial markets by the Borg-Warner Corporation in 1954	
SYNTHESIS			
Monomer(s) structure	-	$H_2C=CHC\equiv N$ $H_2C=CHCH=CH_2$	
Monomer(s) CAS number(s)	-	107-13-1 (acrylonitrile); 106-99-0 (butadiene); 100-42-5 (styrene)	
Monomer(s) molecular weight(s)	dalton, g/mol, amu	53.06; 54.09; 104.15	
Monomer(s) expected purity(ies)	%	variable	
Monomer ratio	-	acrylonitrile: 15-35%; butadiene: 5-30%; styrene: 40-60%	
SAN/BP		90-40/10-60	
Formulation example	-	H_2O (solvent) 17,070, emulsifier 3558, polybutadiene latex 6384, tertdodecyl mercaptan 70, $FeSO_4$ 1610, styrene 8565, acrylonitrile 4236, cumene hydroperoxide 53	Hu, K-H; Kao, C-S; Duh, Y-S, J. Hazardous Matrer., 159, 25-34, 2008.
Method of synthesis	-	the most frequently used are emulsion, mass, and suspension polymerizations; styrene and acrylonitrile are being grafted onto rubber by chemical grafting, chemical grafting blending, or physical mixing; chemical grafting blending is the most frequently used method and specifically emulsion grafting-bulk SAN blending is a method of choice	Huang, P; Tan, D; Luo, Y, J. Env. Sci., Technol., 3, 3, 148-58, 2010.
Temperature of polymerization	°C	62-75	
Heat of polymerization	J g^{-1}	styrene: 647; acrylonitrile: 2290; ABS: 890	
Number average molecular weight, M_n	dalton, g/mol, amu	30,000-200,000	
Mass average molecular weight, M_w	dalton, g/mol, amu	81,000-308,000	
Polydispersity, M_w/M_n	-	2.72-2.88	
STRUCTURE			
Domain size of rubber	nm	<1,000 (emulsion polymerization); 500-5,000 (mass polymerization)	Lucarini, M; Pedulli, G F; Motyakin, M V; Schlick, S, Prog. Polym. Sci., 28, 331–340, 2003.
Cis content	%	32.3-97.0 (polybutadiene); 1.5-51.6 (*trans* in polybutadiene)	Yu, Z; Li, Y; Zhao, Z; Wang, C; Yang, J; Zhang, C; Li, Z; Wang, Y, Polym. Eng. Sci., 49, 2249-56, 2009.

ABS poly(acrylonitrile-co-butadiene-co-styrene)

PARAMETER	UNIT	VALUE	REFERENCES
COMMERCIAL POLYMERS			
Some manufacturers	-	Ineos-Styrosolution; Daicel; Denka; Formosa; Sabic; LG Chem	
Trade names	-	Lustran, Terluran; Denka, Novodur; Tairilac; Cycolac	
PHYSICAL PROPERTIES			
Density at 20°C	g cm^{-3}	1.03-1.09; 0.93 (melt)	Terluran; Cevian; Daicel
Bulk density at 20°C	g cm^{-3}	0.6	
Refractive index, 20°C	-	1.540	
Transmittance	%	80-90	Cevian; Daicel
Haze	%	0.4-5	
Gloss, 60°, Gardner (ASTM D523)	%	85-95 (glossy); 1.8-6.6 (matt)	Arino, I; Kleist, U; Rigdahl, M, Polym. Eng. Sci., 45, 733-44, 2005.
Melting temperature, DSC	°C	220-260	Terluran; Karahaliou, E K; Tarantili, P A, Polym. Eng. Sci., 49, 2269-75, 2009.
Softening point	°C	105	
Heat deflection temperature	°C	76.5-79.2	Basurto, F C; Garcia-Lopez, D; Villarreal-Bastardo, N; Merino, J C; Pastor, J M, Composites: Part B, 47, 42-7, 2013.
Onset temperature of thermal degradation	°C	385-407	Li, Y; Zheng, Y; Liu, J; Shang, H, J. Appl. Polym. Sci., 115, 957-62, 2010.
Thermal expansion coefficient, 23-80°C	°C^{-1}	0.6-1.1E-4	Terluran; Cevian; Daicel
Thermal conductivity, melt	W m^{-1} K^{-1}	0.16	Terluran
Glass transition temperature	°C	102-107 (acrylonitrile-styrene mesophase) and -58 (butadiene component); 103 (DSC) and 121 (DMA); 112-115 (DMA); 108	Terluran; Xue, M-L; Yu, Y-L; Rhee, J M; Kim, N H; Lee, J H, Eur. Polym. J., 43, 9, 3826-37, 2007; Santos, R M; Botelho, G L; Machado, A V, J. Appl. Polym. Sci., 2005-14, 2010; Karahaliou, E K; Tarantili, P A, Polym. Eng. Sci., 49, 2269-75, 2009; Torrado, A R; Shemelya, C M; English, J D; Lin, Y; Wicker, R B; Roberson, D A, Addit. Manufac., 6, 16-29, 2015; Hadid, H, Mailand, B, Sundermann, T, Johnson, E, Sealy, M, Procedia Manuf, 34, 594-602, 2019.
Specific heat capacity	J K^{-1} kg^{-1}	1,390-2,130 (88°C); 2,300-2,400 (melt)	
Maximum service temperature	°C	62-80	
Heat deflection temperature at 0.45 MPa	°C	89-113	Terluran
Heat deflection temperature at 1.8 MPa	°C	67-109	Terluran; Cevian; Daicel; Denka
Vicat temperature VST/A/50	°C	90-112; 91-107	Terluran; Denka
Vicat temperature VST/B/50	°C	95-100	Terluran
Melting enthalpy peak	J g^{-1}	9.6	Karahaliou, E K; Tarantili, P A, Polym. Eng. Sci., 49, 2269-75, 2009.
Relative permittivity at 100 Hz	-	2.9; 2.74	Terluran; Malek, N A, Ramly, A M, Sidek, A, Mohamad, S, Ind. J. Elect. Eng. Comput. Sci., 6, 1, 116-23, 2017.

ABS poly(acrylonitrile-co-butadiene-co-styrene)

PARAMETER	UNIT	VALUE	REFERENCES
Relative permittivity at 1 MHz	-	2.7-3.4	
Dissipation factor at 100 Hz	E-4	48-160	Terluran
Dissipation factor at 1 MHz	E-4	79-140	Terluran
Volume resistivity	ohm-m	1E+13; 1E+1 (with 0.18 vol fraction of Ni coated mica)	Terluran; Kandasubramanian, B; Gilbert, M, Macromol. Symp., 211, 185-95, 2005.
Surface resistivity	ohm	1E+10 to 1E+16; 3E+17 (neat), 1E+9 (1 wt% MWCNTs)	Terluran, Denka; Park, K S, Youn, J R, Carbon, 50, 6, 2322-30, 2012.
Electric strength K20/P50, d=0.60.8 mm	kV mm^{-1}	37-41	Terluran
Comparative tracking index, CTI, test liquid A	-	600	Terluran
Comparative tracking index, CTIM, test liquid B	(-)-	225	Terluran
Shielding effectiveness	dB	-16-16.5; -17 (3D-printed components loaded with MWCTN); -49.6 (graphene-carbon nanotubes hybrid), -38.6 (MWCNT)	Kandasubramanian, B; Gilbert, M, Macromol. Symp., 211, 185-95, 2005; Schmitz, D P, Ecco, L G,. Dul, B, Pereira, E C L, Pegoretti, A, Mater. Today, Commun., 15, 70-80, 2018; Jyoti, J, Arya, A K, Polym. Testing, in press, 106839, 2020.
Coefficient of friction	ASTM D1894	0.21-0.28 (chrome steel); 0.40 (aluminum); 0.75-0.97 (ABS/ABS)	Maldonado, J E, Antec, 3431-35, 1998; Kumar, S, Roy, B S, Mater. Today, Proc., 26, 2, 2388-94, 2020.
Contact angle of water, 20°C	degree	80.9; 89.7; 90	Accu Dyne Test, Diversified Enterprizes; K. Fukuzawa, in Adhesion Science and Technology, H. Mizumachi, ed., International Adhesion Symposium, Yokohama, Japan, 1994; Rodríguez-Vidal, E, Sanz, C, Etxarri, J, Bejarano, A, Lebour, Y, Malet, R, Procedia CIRP, 74, 568-72, 2018.
Surface free energy	mJ m^{-2}	35-42	D.A. Markgraf, in Film Extrusion Manual, 2nd Ed., T.I. Butler, ed., TAPPI Press, Norcross, GA, 2005, p. 299.
Speed of sound	m s^{-1}	36.2-37.5	Alan R. Selfridge, IEEE Trans. Sonics Ultrasonics, SU-32, 3, 381-394, 1985.
Acoustic impedance		2.31-2.36	Alan R. Selfridge, IEEE Trans. Sonics Ultrasonics, SU-32, 3, 381-394, 1985.
Attenuation	dB cm^{-1}, 5 MHz	10.9-11.3	Alan R. Selfridge, IEEE Trans. Sonics Ultrasonics, SU-32, 3, 381-394, 1985.

MECHANICAL & RHEOLOGICAL PROPERTIES

PARAMETER	UNIT	VALUE	REFERENCES
Tensile strength	MPa	25-65	Li, J; Cai, C L, Current Appl. Phys., 11, 50, 2011; Lee, J-W; Lee, J-C; Pandey, J; Ahn, S-H; Kang, Y J, J. Compos. Mater., 44, 1701-16, 2010.
Tensile modulus	MPa	2000-3050	Denka; LG
Tensile stress at yield	MPa	37-61	Denka
Tensile stress at break	MPa	29-47	Denka
Tensile creep modulus, 1000 h, elongation 0.5 max	MPa	1250	
Elongation	%	8-20	Terluran; Karahaliou, E K; Tarantili, P A, Polym. Eng. Sci., 49, 2269-75, 2009.
Tensile yield strain	%	2.4-4	

ABS poly(acrylonitrile-co-butadiene-co-styrene)

PARAMETER	UNIT	VALUE	REFERENCES
Flexural strength	MPa	55-125	Jin, F-L; Lu, S-L; Song, Z-B; Pang, J-X; Zhang, L; Sun, J-D; Cai, X-P, Mater. Sci. Eng., A527, 3438-41, 2010; Terluran; Cevian; Daicel; Denka
Flexural modulus	MPa	1930-3000	Denka
Elastic modulus	MPa	1208-1939	Lee, J-W; Lee, J-C; Pandey, J; Ahn, S-H; Kang, Y J, J. Compos. Mater., 44, 1701-16, 2010; Karahaliou, E K; Tarantili, P A, Polym. Eng. Sci., 49, 2269-75, 2009.
Young modulus	MPa	1810-2390	Basurto, F C; Garcia-Lopez, D; Villarreal-Bastardo, N; Merino, J C; Pastor, J M, Composites: Part B, 47, 42-7, 2013.
Compressive strength	MPa	65-86; 120 (30% glass fiber)	
Charpy impact strength, unnotched, 23°C	kJ m^{-2}	120-190 to NB	Terluran; Cevian; Daicel
Charpy impact strength, unnotched, -30°C	kJ m^{-2}	80-140	Terluran; Cevian; Daicel
Charpy impact strength, notched, 23°C	kJ m^{-2}	5-40	Terluran; Cevian; Daicel; Denka
Charpy impact strength, notched, -30°C	kJ m^{-2}	2-33	Terluran; Cevian; Daicel;
Izod impact strength, notched, 23°C	J m^{-1}	30-470	Jin, F-L; Lu, S-L, Song, Z-B; Pang, J-X; Zhang, L; Sun, J-D; Cai, X-P, Mater. Sci. Eng., A527, 3438-41, 2010; Terluran; Cevian; Daicel
Izod impact strength, notched, -40°C	J m^{-1}	8-280	Terluran; Cevian; Daicel
Instrumental impact total energy, 23°C	J	60-68	
Shear modulus	MPa	700-1,50	
Rockwell hardness	-	101; 102-124; 95-118	(-); Jin, F-L; Lu, S-L, Song, Z-B; Pang, J-X; Zhang, L; Sun, J-D; Cai, X-P, Mater. Sci. Eng., A527, 3438-41, 2010; Denka
Ball indention hardness at 358 N/30 S (ISO 2039-1)	MPa	97	
Shrinkage	%	0.4-0.7; 0.72 (across the flow), 1.11 (along the flow); 0.4-0.7 (molding)	Terluran; Chang, T; Faison, E, Polym. Eng. Sci., 41, 5, 703-10, 2001; Denka
Melt viscosity, shear rate=1000 s^{-1}	Pa s	140-250	Xue, M-L; Yu, Y-L; Rhee, J M; Kim, N H; Lee, J H, Eur. Polym. J., 43, 9, 3826-37, 2007.
Melt volume flow rate (ISO 1133, procedure B), 220°C/10 kg	cm^3/10 min	2-34	Terluran
Pressure coefficient of melt viscosity, b	G Pa^{-1}	33.7	Aho, J; Syrjala, S, J. Appl. Polym. Sci., 117, 1076–84, 2010.
Melt index, 230°C/3.8 kg	g/10 min	1.5; 2.5-7.0; 18-34; 8-74	Karahaliou, E K; Tarantili, P A, Polym. Eng. Sci., 49, 2269-75, 2009; (-); Jin, F-L; Lu, S-L, Song, Z-B; Pang, J-X; Zhang, L; Sun, J-D; Cai, X-P, Mater. Sci. Eng., A527, 3438-41, 2010; Denka
Water absorption, equilibrium in water at 23°C	%	0.7-1.03	Terluran
Moisture absorption, equilibrium 23°C/50% RH	%	0.21-0.35	Terluran

ABS poly(acrylonitrile-co-butadiene-co-styrene)

PARAMETER	UNIT	VALUE	REFERENCES
CHEMICAL RESISTANCE			
Acid dilute/concentrated	-	no resistance to concentrated; good resistance to dilute	Terluran
Alcohols	-	limited resistance; insoluble	Terluran
Alkalis	-	good resistance to dilute	Terluran
Aliphatic hydrocarbons	-	limited resistance; insoluble	Terluran
Aromatic hydrocarbons	-	no resistance	Terluran,
Esters	-	no resistance	Terluran,
Greases & oils	-	limited resistance; insoluble: mineral oil	Terluran
Halogenated hydrocarbons	-	no resistance; soluble: dichloromethane	Terluran
Ketones	-	no resistance; soluble: acetone, methyl-ethyl ketone	Terluran
Other	-	resistant: water, salt solutions; soluble: dimethylformamide, tetrahydrofuran, toluene	Terluran
Good solvent	-	acetophenone, aniline, benzene, chlorobenzene, chloroform, dimethylformamide, dioxane, ethyl benzene	
Non-solvent	-	cyclohexane, diethanolamine, diethylene glycol, dipropylene glycol, petroleum ether	
Chemicals causing environmental stress cracking		nonionic surfactants	Kawaguchi, T; Nishimura, H; Kasahara, K; Kuriyama, T; Narisawa, I, Polym. Eng. Sci., 43, 2, 419-30, 2003.
Effect of EtOH sterilization (tensile strength retention)	%	105-110 (high gloss); 82-95 (low gloss)	Navarrete, L; Hermanson, N, Antec, 2807-18, 1996.
FLAMMABILITY			
Autoignition temperature	°C	>400; 466	Terluran, MSDS
Limiting oxygen index	% O$_2$	18.1-20.5; 23-35 (with flame retardants)	Yan, I; Zheng, Y; Liu, J; Shang, H, J. Appl. Polym. Sci., 115, 957-62, 2010; Hourston, D J, Shreir's Corrosion, Elsevier, 2010, Chapter 3.31, 2369-2386; Li, Y; Zheng, Y; Liu, J; Shang, H, J. Appl. Polym. Sci., 115, 957-62, 2010; Ren, Y-y; Chen, L; Zhang, Z-y; Wang, X-l; Yang, X-s; Kong, X-j; Yang, L, Polym. Deg. Stab., 109, 285-92, 2014.
Heat release	kW m^{-2}	1037; 602-796 (with organoclays); 243-268 (with flame retardant; 736	Du, X; Yu, H; Wang, Z; Tang, T, Polym. Deg. Stab., 95, 587-92, 2010; Yu, B; Liu, M; Lu, L; Dong, X; Gao, W; Tang, K, Fire Mater., 34, 251-61, 2010; Huang, G, Huo, S, Xu, X, Chen, W, Wang, H, Compos. Part B: Eng., 177, 107377, 2019.
NBS smoke chamber	Ds	800	Padey, D; Walling, J; Wood A, Polymers in Defence and Aerospace 2007, Rapra, 2007, paper 15.
Char, 554°C	%	0-0.6; 9.4; 0.43-2.89; 0.93	Yang, S; Castilleja, J R; Barrera, E V; Lozano, K, Polym. Deg. Stab., 83, 3, 383-88, 2004; Du, X; Yu, H; Wang, Z; Tang, T, Polym. Deg. Stab., 95, 587-92, 2010; Karahaliou, E K; Tarantili, P A, Polym. Eng. Sci., 49, 2269-75, 2009; Lyon, R E; Walters, R N, J. Anal. Appl. Pyrolysis, 71, 27-46, 2004; Huang, G, Huo, S, Xu, X, Chen, W, Wang, H, Compos. Part B: Eng., 177, 107377, 2019.

ABS poly(acrylonitrile-co-butadiene-co-styrene)

PARAMETER	UNIT	VALUE	REFERENCES
Heat of combustion	J g^{-1}	39,840; peak effective heat of combustion	Walters, R N; Hacket, S M; Lyon, R E, Fire Mater., 24, 5, 245-52, 2000; Huang, G, Huo, S, Xu, X, Chen, W, Wang, H, Compos. Part B: Eng., 177, 107377, 2019.
CO yield	mg g^{-1}	52	Rutkowski, J V, Levin, B C, Fire Mater., 10, 93-105, 1986.
UL 94 rating	-	HB	Denka
THERMAL STABILITY			
Activation energy under nitrogen	kJ mol^{-1}	134.5-242.4; 140-182	Yang, S; Castilleja, J R; Barrera, E V; Lozano, K, Polym. Deg. Stab., 83, 3, 383-88, 2004; Polli, H; Pontes, L A M; Araujo, A S; Barros, J M F; Fernandes, V J, J. Therm. Anal. Calorimetry, 95, 1, 131-34, 2009; Roussi, A T, Vouvoudi, E C, Achilias, D S, Thermochim. Acta, 690, 178705, 2020.
Activation energy under air	kJ mol^{-1}	156.3	Yang, S; Castilleja, J R; Barrera, E V; Lozano, K, Polym. Deg. Stab., 83, 3, 383-88, 2004.
Temperature of maximum degradation (air)	°C	428-445 (1st step); 554 (2nd step)	Yang, S; Castilleja, J R; Barrera, E V; Lozano, K, Polym. Deg. Stab., 83, 3, 383-88, 2004; Karahaliou, E K; Tarantili, P A, Polym. Eng. Sci., 49, 2269-75, 2009.
Weight loss	%	85.6 (1st step); 13.8 (2nd step)	Yang, S; Castilleja, J R; Barrera, E V; Lozano, K, Polym. Deg. Stab., 83, 3, 383-88, 2004.
Onset temperature of oxidation	°C	80 (isothermal test); 120 (dynamic scanning)	Duh, Y-S; Ho, T-C; Chen, J-R; Kao, C-S, Polymer, 51, 2, 171-84, 2010.
Heat of oxidation	J g^{-1}	2,800; 4,720 (polybutadiene)	
WEATHER STABILITY			
Activation wavelengths	nm	320, 385	
Activation energy for yellowing	kJ mol^{-1}	31	Pickett, J E, Kuvshinnikova, O, Sung, L-P, Ermi , B D, Polym. Deg. Stab., 181, 109330, 2020.
Depth of UV penetration	μm	110-150	Jouan, X; Gardette, J L, J. Polym. Sci., Polym. Chem., 29, 685, 1991; Bokria, J G; Schlick, S, Polymer, 43, 3239-46, 2002.
Products of degradation	-	hydroperoxides, carboxylic acids, anhydrides, gamma lactones, chain scission	Santos, R M; Botelho, G L; Machado, A V, J. Appl. Polym. Sci., 2005-14, 2010.

ABS poly(acrylonitrile-co-butadiene-co-styrene)

PARAMETER	UNIT	VALUE	REFERENCES
Stabilizers	-	UVA: 2-hydroxy-4-octyloxybenzophenone; 2-hydroxy-4-methoxybenzophenone; 2-(2H-benzotriazol-2-yl)-p-cresol; 2-(2H-benzotriazole-2-yl)-4,6-di-tert-pentylphenol; 2-(2H-benzotriazole-2-yl)-4-(1,1,3,3-tetraethylbutyl)phenol; 2,4-di-tert-butyl-6-(5-chloro-2H-benzotriazole-2-yl)-phenol; 2-[4,6-bis(2,4-dimethylphenyl)-1,3,5-triazin-2-yl]-5-(octyloxy)phenol; ethyl-2-cyano-3,3-diphenylacrylate; HAS: 1,3,5-triazine-2,4,6-triamine, N,N'''[1,2-ethane-diyl-bis[[[4,6-bis[butyl-(1,2,6,6-pentamethyl-4-piperidinyl)amino]-1,3,5-triazine-2-yl]imino]-3,1-propanediyl] bis[N',N''-dibutyl-N',N''-bis(1,2,2,6,6-pentamethyl-4-piperidinyl)-; bis(2,2,6,6-tetramethyl-4-piperidyl) sebacate; 2,2,6,6-tetramethyl-4-piperidinyl stearate; N,N'-bisformyl-N,N'-bis-(2,2,6,6-tetramethyl-4-piperidinyl)-hexamethylendiamine; alkenes, C20-24-.alpha.-, polymers with maleic anhydride, reaction products with 2,2,6,6-tetramethyl-4-piperidinamine; 1, 6-hexanediamine, N, N'-bis(2,2,6,6-tetramethyl-4-piperidinyl)-, polymers with 2,4-dichloro-6-(4-morpholinyl)-1,3,5-triazine; 1,6-hexanediamine, N,N'-bis(2,2,6,6-tetramethyl-4-piperidinyl)-, polymers with morpholine-2,4,6-trichloro-1,3,5-triazine reaction products, methylated; Phenolic antioxidants: ethylene-bis(oxyethylene)-bis(3-(5-tert-butyl-4-hydroxy-m-tolyl)-propionate); 2,6,-di-tert-butyl-4-(4,6-bis(octylthio)-1,3,5,-triazine-2-ylamino) phenol; pentaerythritol tetrakis(3-(3,5-di-tert-butyl-4-hydroxyphenyl)propionate); 2-(1,1-dimethylethyl)-6-[[3-(1,1-dimethylethyl)-2-hydroxy-5-methylphenyl] methyl-4-methylphenyl acrylate; isotridecyl-3-(3,5-di-tert-butyl-4-hydroxyphenyl)propionate; 2,2'-ethylidenebis (4,6-di-tert-butylphenol); 2,2'-methylenebis(4-ethyl-6-tertbutylphenol); 3,5-bis(1,1-dimethyethyl)-4-hydroxy-benzenepropanoic acid, C13-15 alkyl esters; phenol, 4-methyl-, reaction products with dicyclopentadiene and isobutene; Phosphite: trinonylphenol phosphite; isodecyl diphenyl phosphite	

BIODEGRADATION

PARAMETER	UNIT	VALUE	REFERENCES
Colonized products		bathroom fixtures, health care products, pipes	
Stabilizers	-	Microban, nanosilver	

TOXICITY

PARAMETER	UNIT	VALUE	REFERENCES
NFPA: Health, Flammability, Reactivity rating	-	1/1/0	
Carcinogenic effect	-	not listed by ACGIH, NIOSH, NTP	
MAK/TRK	mg m^{-3}	styrene: 86; acrylonitrile: 7; 1,3-butadiene: 11	
Oral rat, LD$_{50}$	mg kg^{-1}	>5,000	

ENVIRONMENTAL IMPACT

PARAMETER	UNIT	VALUE	REFERENCES
Cradle to grave non-renewable energy use	MJ/kg	92-95	

PROCESSING

PARAMETER	UNIT	VALUE	REFERENCES
Typical processing methods	-	calendering, casting, electroplating, extrusion, film lamination, injection molding, rotational molding, thermoforming, vacuum forming, vacuum metallization	Sarkar, K; Gomez, C; Zambrano, S; Ramirez, M; de Hoyos, E; Vasquez, H; Lozano, K, Mater. Today, 13, 11, 12-14, 2010.
Preprocess drying: temperature/time/residual moisture	°C/h/%	80-95/2-4/0.01	
Processing temperature	°C	190-275; 220-260 (injection molding)	

ABS poly(acrylonitrile-co-butadiene-co-styrene)

PARAMETER	UNIT	VALUE	REFERENCES
Processing pressure	MPa	5 (backpressure); 53 (holding pressure)	Ingnell, S; Kelist, U; Rigdahl, M, Polym. Eng. Sci., 50, 2114-21, 2010.
Additives used in final products	-	Fillers: antimony oxide, carbon black, glass beads, magnesium hydroxide, nickel or copper coated carbon fibers, talc; Plasticizers: hydrocarbon processing oil, phosphate esters (e.g., triphenyl phosphate, resorcinol bis(diphenyl phosphate), or oligomeric phosphate), long chain fatty acid esters, and aromatic sulfonamide; Antistatics: ethanol,2,2'-iminobis-,N-coco alkyl derivatives, glycerol monostearate, polyaniline, polyesteramide, sodium alkyl sufonate; Antiblocking: talc; Release: cetyl palmitate, fluorocarbon, methyl behenate, paraffin wax; Slip: bis-stearamide wax	
Applications	-	appliance (refrigerator liners, kitchen appliance housings, vacuum cleaners, power tools), automotive (instrument panels, consoles, door parts, knobs, trim, wheel covers, mirror and headlight housing, front radiator grilles), business machines (computers, discs, phones), furniture, hot tubs, lawn and garden equipment, luggage, lunch and tool boxes, medical applications, military, packaging, pipes and fittings, recreation (snowmobiles, boats, vehicles), toys	
Outstanding properties	-	combination of 3 monomers gives specific advantages: styrene gives rigidity, electrical properties, easy processability and surface gloss, butadiene improves low temperature toughness, and acrylonitrile improves ABS' chemical, weathering and heat resistance and increases tensile strength	Huang, P; Tan, D; Luo, Y, J. Env. Sic., Technol., 3, 3, 148-58, 2010.

BLENDS

PARAMETER	UNIT	VALUE	REFERENCES
Suitable polymer	-	chitosan, EPDM, ground rubber, PA6, PANI-EB, PC, PLA, PTT, PVC, SAN	Zhao, D, Yan, D, Fu, X, Zhang, N, Yang, G, Mater Lett., 274, 128013, 2020.
Compatibilizers	-	SBM	

ANALYSIS

PARAMETER	UNIT	VALUE	REFERENCES
FTIR (wavenumber-assignment)	cm^{-1}/-	hydroxy – 3460; carbonyl – 1646, 1718, 1722, 1730, 1785; C=N – 2237; C-O – 1450, 950; styrene – 700, 765, 1028, 1449, 1456-1495, 1582-1601; poly-1,2-butadiene – 910-911; poly-*trans*-1,4-butadiene – 966-967; C=C of 1,2 structures – 1640	Jouan, X; Gardette, J-L, J. Polym. Sci., Polym. Chem., 29, 685, 1991; Motyakin, M V; Schlick, S, Poly. Deg. Stab., 91, 7, 1462-70, 2006; Santos, R M; Botelho, G L; Machado, A V, J. Appl. Polym. Sci., 2005-14, 2010.

AK alkyd resin

PARAMETER	UNIT	VALUE	REFERENCES
GENERAL			
Common name	-	alkyd resin	
Acronym	-	AK	
CAS number	-	63148-69-6; 68333-62-0	
RTECS number	-	WZ6250000	
HISTORY			
Person to discover	-	Berzelius; Kienle	Hofland, A, Prog. Org. Coat., in press, 2011.
Date	-	1847, 1920s, 1976	
Details	-	Berzelius condensed glycerol tartrate; in 1920s, Kienle developed alkyd resins; in 1976 artist's alkyd paints were introduced by Winsor & Newton	Ploeger, R; Scalarone, D; Chiantore, O, J. Cultural Heritage, 9, 412-19, 2008.
SYNTHESIS			
Monomer(s) structure	-	polyol and dicarboxylic acid or anhydride; hyperbranched polyesters (di-trimethylolpropane and dimethylolpropionic acid = 1:4 to 1:12)	Jovičić, M, Radičević, R, Pavličević, J, Bera, O, Govedarica, D, Prog. Org. Coat., 148, 105832, 2020.
Monomer(s) molecular weight(s)	dalton, g/mol, amu	>1000; 714-1642	-; Jovičić, M, Radičević, R, Pavličević, J, Bera, O, Govedarica, D, Prog. Org. Coat., 148, 105832, 2020.
Oil or fatty acids contents	%	>70 (very long oil); 56-70 (long oil); 46-55 (medium oil); 35-45 (short oil); castor oil	Ploeger, R; Scalarone, D; Chiantore, O, J. Cultural Heritage, 9, 412-19, 2008.
Curative	-	melamine-formaldehyde resin; phthalic anhydride	Satdive, A, Mestry, S, Patil, D, Mhaske, S T, Prog. Org. Coat., 131, 165-75, 2019; Otabor, G O, Ifijen, I H, Mohammed, F U, Aigbodion, A I, Ikhuoria, E U, Heliyon, 5, 5, e01621, 2019.
Formulation example	wt%	glycerol – 25.9, oil – 33.3, phthalic anhydride – 40.8	Atimuttigul, V; Damrongsakkul, S; Tanthapanichakoon, W, Korean J. Chem. Eng., 23, 4, 672-77, 2006; Ikhuoria, E U; Maliki, M; Okieimen, F E; Aigbodion, A I; Obaze, E O; Bakare, I O, Prog. Org. Coat., 59, 134-37, 2007.
Method of synthesis	-	the mixture of oil, glycerol, and catalyst is heated to a required temperature and phthalic anhydride is added to accomplish esterification; the reaction lasted for more than 6 h and was discontinued when the acid number was below 12.	Atimuttigul, V; Damrongsakkul, S; Tanthapanichakoon, W, Korean J. Chem. Eng., 23, 4, 672-77, 2006; Jovičić, M, Radičević, R, Pavličević, J, Bera, O, Govedarica, D, Prog. Org. Coat., 148, 105832, 2020.
Temperature of polymerization	ºC	210-260; 140	Atimuttigul, V; Damrongsakkul, S; Tanthapanichakoon, W, Korean J. Chem. Eng., 23, 4, 672-77, 2006.
Time of polymerization	h	5-6	Atimuttigul, V; Damrongsakkul, S; Tanthapanichakoon, W, Korean J. Chem. Eng., 23, 4, 672-77, 2006.
Pressure of polymerization	Pa	atmospheric	Atimuttigul, V; Damrongsakkul, S; Tanthapanichakoon, W, Korean J. Chem. Eng., 23, 4, 672-77, 2006.

AK alkyd resin

PARAMETER	UNIT	VALUE	REFERENCES
Catalyst	-	LiOH; Mn and Co compounds (drying catalyst); methanesulfonic acid; modified iron(II) bispidine complex	Atimuttigul, V; Damrongsakkul, S; Tanthapanichakoon, W, Korean J. Chem. Eng., 23, 4, 672-77, 2006; Ikhuoria, E U; Maliki, M; Okieimen, F E; Aigbodion, A I; Obaze, E O; Bakare, I O, Prog. Org. Coat., 59, 134-37, 2007; Erich, S J F; Laven, J; Pel, L; Huinink, H P; Kopinga, K, Prog. Org. Coat., 55, 105-11, 2006; Křižan, M, Vinklárek, J, Erben, M, Růžičková, Z, Honzíček, J, Inorg. Chim. Acta, 486, 636-41, 2019.
Number average molecular weight, M_n	dalton, g/mol, amu	2,300-2,400; 3,754-6,611 (hyperbranched resins); 2550-4677 (hyperbranched resins)	Murillo, E A; Vallejo, P P; Lopez, B L, Prog. Org. Coat., 69, 235-40, 2010.
Mass average molecular weight, M_w	dalton, g/mol, amu	23,900-30,300; 8,125-19,537 (hyperbranched resins)	
Polydispersity, M_w/M_n	-	>10; 2.16-295 (hyperbranched resins); 1.94-2.58	Vallejo, P P; Lopez, B L; Murillo, E A, Prog. Org. Coat., 87, 213-21, 2015.

STRUCTURE

PARAMETER	UNIT	VALUE	REFERENCES
Crystalline structure	-	$\beta-MnO_2$ nanorods@graphene oxide hybrid's added to alkyd resin (amorphous) posses single crystallinity that can greatly enhance corrosion protection	Selim, M S, Hao, Z, Mo, P, Jiang, Y, Ou, H, Collods Surf. A: Physicochem. Eng Aspects, 601, 125057, 2020.
Cross-sectional surface area of chain	nm^2	0.34	Swarup, S; Nigam, A N, J. Appl. Polym. Sci., 39, 1727-31, 1990.
Entanglements		entanglements of the organic phase (matrix polymer) around inorganic matter (filler particles) cause increase in both T_g and crosslink density (increase strength but also increase of viscosity)	Salata, R R, Pellegrene, B, Soucek, M D, Prog. Org. Coat., 133, 340-9, 2019.
Number of carbon atoms per entanglement		440	Swarup, S; Nigam, A N, J. Appl. Polym. Sci., 39, 1727-31, 1990.

COMMERCIAL POLYMERS

PARAMETER	UNIT	VALUE	REFERENCES
Some manufacturers	-	Allnex; Mancuso, Polynt; Uniform Synthetics	
Trade names	-	Setyrene, Setal, Setaqua; Manrez, Resyd; Duramac, Rezimac; Unikyd	
Composition information	-	copolymer, long oil, waterborne; linseed, soya, TOFA; long oil, silicone-modified alkyd, and many other; linseed oil, castor oil	

PHYSICAL PROPERTIES

PARAMETER	UNIT	VALUE	REFERENCES
Density at 20°C	$g\ cm^{-3}$	1.10-1.25	
Color	-	yellow to white	
Refractive index, 20°C	-	1.42-1.62	Bora, M M, Deka, C, Tapadar, S A, Jha, D K, Kakati, D K, Prog. Org. Coat., 124, 71-9, 2018; Pathan, S, Ahmad, S, Prog. Org. Coat., 122, 189-98, 2018.
Gloss, 60°, Gardner (ASTM D523)	%	85-95 (coating); 85.2-90.9	Vallejo, P P; Lopez, B L; Murillo, E A, Prog. Org. Coat., 87, 213-21, 2015.
Odor	-	none	
Decomposition temperature	°C	150-250 (peroxide decomposition); 250-400 (oxidative decomposition); >400 (volatilization)	Lazzari, M; Chiantore, O, Polym. Degrad. Stab., 65, 303–13, 1999.

AK alkyd resin

PARAMETER	UNIT	VALUE	REFERENCES
Glass transition temperature	°C	8-10; 2 (uncrosslinked); 20-40 (naturally exposed for 25 years)	Erich, S J F; Adan, O C G; Pel, L; Huinink, H P; Kopinga, K, Chem. Mater., 18, 4500-4, 2006; Ploeger, R; Scalarone, D; Chiantore, O, Polym. Deg. Stab., 94, 2036-41, 2009.
Calorific value	MJ kg^{-1}	37.83 (*J. curcas.* oil)	Odetoye, T E, Ogunniyi, D S, Olatunji, G A, Ind. Crops Prod., 32, 3, 225-230, 2010.
Hansen solubility parameters, δ_D, δ_P, δ_H	MPa$^{0.5}$	20.42, 3.44, 4.56 (long oil); 18.50, 9.21, 4.91 (short oil)	
Relative permittivity at 100 Hz/1 MHz	-	3.5-5; 2.5-10	-; Sumi, V S, Arunima, S R, Deepa, M J, Sha, M A, Shibli, S M A, Mater. Chem. Phys., 247, 122881, 2020.
Speed of sound	m s^{-1} x 10^{-3}	1.29-1.35	
Diffusion coefficient of water vapor	cm^2 s^{-1} x10^{-12}	0.65-1.35	Gezici-Koç, O, Erich, S J F, Huinink, H P,. van der Ven, L G J, Adan, O C G, Prog. Org. Coat., 114, 135-44, 2018.

MECHANICAL & RHEOLOGICAL PROPERTIES

PARAMETER	UNIT	VALUE	REFERENCES
Tensile strength	MPa	0.7-23.2	Zafar, F, Ghosal, A, Sharmin, E, Chaturvedi, R, Nishat, N, Prog. Org. Coat., 131, 259-75, 2019.
Elongation	%	16-671	Zafar, F, Ghosal, A, Sharmin, E, Chaturvedi, R, Nishat, N, Prog. Org. Coat., 131, 259-75, 2019.
Young's modulus	MPa	2.3-720	Zafar, F, Ghosal, A, Sharmin, E, Chaturvedi, R, Nishat, N, Prog. Org. Coat., 131, 259-75, 2019.
Pencil hardness		2B-H	Bora, M M; Gogoi, P; Deka, D C; Kakati, D K, Ind. Crops Prod., 52, 721-8, 2014.

CHEMICAL RESISTANCE

PARAMETER	UNIT	VALUE	REFERENCES
Acid dilute/concentrated	-	poor	
Alcohols	-	very good	
Alkalis	-	poor-good	Huang, Q H; Liu, C; Chen, S; Bai, G; An, Q; Cao, J; Zheng, S; Liang, Y; Xiang, B, Prog. Org. Coat., 87, 189-96, 2015.
Aliphatic hydrocarbons	-	good	
Aromatic hydrocarbons	-	good	
Esters	-	good-fair	
Greases & oils	-	good-fair	
Halogenated hydrocarbons	-	fair-poor	
Good solvent	-	acids	
Non-solvent	-	carbon tetrachloride, methyl acetate, methanol	

FLAMMABILITY

PARAMETER	UNIT	VALUE	REFERENCES
Flash point	°C	212 (sunflower acid oil), 272 (refined sunflower oil); e.g., 26-32 (depends on solvent)	Chiplunkar, P P, Pratap, A P, Prog. Org. Coat., 93, 61-7, 2016; Bahadori, A, Essentials of Coating, Painting, and Lining for the Oil, Gas and Petrochemical Industries, GPP, 2015, pp. 499-581.

AK alkyd resin

PARAMETER	UNIT	VALUE	REFERENCES
Limiting oxygen index	% O_2	17.4-43.2 (depending on flame retarding additives)	Xu, L, Liu, X, An, Z, Y, R, Polym. Deg. Stab., 161, 114-120, 2019.
Peak heat release	kW m^{-2}	203-1233 (depending on flame retarding additives)	Xu, L, Liu, X, An, Z, Y, R, Polym. Deg. Stab., 161, 114-120, 2019.
Heat of combustion	kJ g^{-1}	69-80	Xu, L, Liu, X, An, Z, Y, R, Polym. Deg. Stab., 161, 114-120, 2019.

WEATHER STABILITY			
Activation wavelengths	nm	330	
Products of degradation	-	chalking, oxidation of double bonds	
Stabilizers	-	UVA: 2-hydroxy-4-methoxybenzophenone; 2,4-dihydroxy-benzophenone; 2-benzotriazol-2-yl-4,6-di-tert-butylphenol; 2-(2H-benzotriazole-2-yl)-4,6-di-tert-pentylphenol; N-(2-ethoxyphenyl)-N'-(4-isododecylphenyl)oxamide; HAS: decane-dioic acid, bis(2,2,6,6-tetramethyl-1-(octyloxy)-4-piperidinyl) ester, reaction products with 1,1-dimethylethylhydroperoxide and octane; 2,4-bis[N-butyl-N-(1-cyclohexyloxy-2,2,6,6-tetra-methylpiperidin-4-yl)amino]-6-(2-hydroxyethylamine)-1,3,5-triazine; bis(1,2,2,6,6-pentamethyl-4-piperidyl) sebacate and methyl 1,2,2,6,6-pentamethyl-4-piperidyl sebacate; 2-dodecyl-N-(2,2,6,6-tetramethyl-4-piperidinyl)succinimide; polymer of 2,2,4,4-tetramethyl-7-oxa-3,20-diaza-dispiro [5.1.11.2]-hene-icosan-21-on and epichlorohydrin; Screener: TiO2; Phosphite: phosphoric acid, (2,4-di-butyl-6-methylphenyl)ethylester	

BIODEGRADATION			
Colonized products		paints and coatings (triglycerides highly crosslinked and with nondegradable linkages are not biodegradable)	Shogren, R L; Petrovic, Z; Liu, Z; Erhan, S Z, J. Polym. Environ. 12, 3, 173-78, 2004.
Typical biodegradants	-	esterase action is responsible for the microbial degradation of alkyd resins	
Stabilizers	-	azole+iodopropargyl butylcarbamate, octylisothiazolinone, silver nanoparticles	

TOXICITY			
NFPA: Health, Flammability, Reactivity rating	-	1/2/0	
Carcinogenic effect	-	not listed by ACGIH, NIOSH, NTP	
Mutagenic effect	-	none known	
Oral rat, LD_{50}	mg kg^{-1}	>2000	
Skin rabbit, LD_{50}	mg kg^{-1}	non-irritant	

PROCESSING			
Typical processing methods	-	compounding/mixing, grinding, sand milling, molding	
Additives used in final products	-	Fillers: calcium carbonate, clay, glass fiber, iron oxides, litho-pone, mica, silica, titanium dioxide, zinc oxide	
Applications	-	adhesives, artist's paints, coatings, electrical applications, fibers, paints, pavement marking, printing inks, putties, var-nishes	

BLENDS			
Suitable polymers	-	acrylics, epoxy, melamine, melamine-formaldehyde	

AK alkyd resin

PARAMETER	UNIT	VALUE	REFERENCES
ANALYSIS			
FTIR (wavenumber-assignment)	cm^{-1}/-	O-H – 2500-3500, carbonyl – 1731-1701, C=C – 1648 (olefinic unsaturations), 1600-1500 (aromatic ring), C-O-H – 1406, C-O – 1275	Suarez, P A Z; Einloft, S; de Basso, N R; Fernandes, J A; da Motta, L; do Amaral, L C; Lima, D G, e-Polymers, 58, 1-8, 2008.
NMR (chemical shifts)	ppm	CH=CH – 5.30, CH$_2$OCOR – 4.21, CH$_3$, CH$_2$, CH – 0.5-3	Murillo, E A; Vallejo, P P; Lopez, B L, Prog. Org. Coat., 69, 235-40, 2010.

ASA poly(acrylonitrile-co-styrene-co-acrylate)

PARAMETER	UNIT	VALUE	REFERENCES
GENERAL			
Common name	-	poly(acrylonitrile-co-styrene-co-acrylate)	
IUPAC name	-	2-propenoic acid, butyl ester, polymer with ethenylbenzene and 2-propenenitrile	
CAS name	-	2-propenoic acid, butyl ester, polymer with ethenylbenzene and 2-propenenitrile	
Acronym	-	ASA	
CAS number	-	9003-54-7; 26299-47-8; 26716-29-0	
HISTORY			
Person to discover	-	Herbig and Salyer; Siebel and Otto	
Date	-	1964; 1965	
Details	-	first patent; refined product	
SYNTHESIS			
Monomer(s) structure	-		
Monomer(s) CAS number(s)	-	107-13-1; 100-42-5; 96-33-3	
Monomer(s) molecular weight(s)	dalton, g/mol, amu	104.15; 53.06; 86.09	
Monomer reactivity ratio		AN/ST=0.25/2	Badawy, S M; Dessouki, A M, J. Appl. Polym. Sci., 84, 268-75, 2002.
Method of synthesis	-	grafting rubber which is dispersed with a styrene acrylonitrile (SAN) phase	
Mass average molecular weight, M_w	dalton, g/mol, amu	60,000-200,000	
Rubber particle size	nm	100-1000	Hossain, M M; Moghbelli, E; Jahnke, E; Boeckmann, P; Guriyanova, S; Sander, R; Minkwitz, R; Sue, H-J, Polymer, 63, 71-81, 2015.
Morphology	-	random copolymer SAN matrix and grafted polybutyl-acrylate rubber particles with an average nominal diameter of 475 nm.	Liang, Y-L, Moghbelli, E, Sue, H-J, Minkwitz, R, Stark, R, Polymer, 53, 2, 604-12, 2012.
STRUCTURE			
Crystallinity	%	0	
COMMERCIAL POLYMERS			
Some manufacturers	-	Ineos-Styrolution; Sabic; LG Chem	
Trade names	-	Luran S; Geloy	
PHYSICAL PROPERTIES			
Density at 20ºC	g cm^{-3}	1.06-1.1; 1.18 (15% glass fiber)	
Bulk density at 20ºC	g cm^{-3}	0.6	
Refractive index, 20ºC	-	1.57-1.677	
Gloss, 60º, Gardner (ASTM D523)	%	93-94	
Odor	-	faint specific	
Melting temperature, DSC	ºC	180-200	

ASA poly(acrylonitrile-co-styrene-co-acrylate)

PARAMETER	UNIT	VALUE	REFERENCES
Softening point	°C	>85 to >100	
Decomposition temperature	°C	320; 395(TGA onset)	
Thermal expansion coefficient, -40 to 40°C	°C^{-1}	0.95-1.2E-4; 0.3E-4	
Thermal conductivity, melt	W m^{-1} K^{-1}	0.16-0.17	D M, de la Mata, M, Delgado, F J, Casal, V, Molina, S I, Mater Design, 191, 108577, 2020.
Glass transition temperature	°C	103-127; -46.5 to -50.9 (core) and 103.7 to 107.1 (shell)	Tolue, S; Moghbeli, M R; Ghafele-bashi, S M, Eur. Polym. J., 45, 714-20, 2009.
Specific heat capacity	J K^{-1} kg^{-1}	1,860-2,000 (melt)	
Long term service temperature	°C	-40 to 75	
Heat deflection temperature at 0.45 MPa	°C	84-106; 115 (15% glass fiber)	
Heat deflection temperature at 1.8 MPa	°C	75-103; 110 (15% glass fiber)	
Vicat temperature VST/A/50	°C	80-104; 115 (15% glass fiber)	
Vicat temperature VST/B/50	°C	82-98	
Melt flow index	g/10 min.	47	D M, de la Mata, M, Delgado, F J, Casal, V, Molina, S I, Mater Design, 191, 108577, 2020.
Dielectric loss factor at 1 kHz	-	0.02-0.05	
Relative permittivity at 100 Hz	-	3.5-3.9	
Relative permittivity at 1 MHz	-	3.2-3.5	
Dissipation factor at 100 Hz	E-4	90-110	
Dissipation factor at 1 MHz	E-4	240-340	
Volume resistivity	ohm-m	1E12	
Surface resistivity	ohm	1E13 to 1E15	
Electric strength K20/P50, d=0.60.8 mm	kV mm^{-1}	35	
Comparative tracking index	-	600	
Coefficient of friction	-	0.4-0.5	Hossain, M M; Moghbelli, E; Jahn-ke, E; Boeckmann, P; Guriyanova, S; Sander, R; Minkwitz, R; Sue, H-J, Polymer, 63, 71-81, 2015.
Permeability to nitrogen, 25oC	cm^3 m^{-2} d^{-1} day^{-1}	60-100	
Permeability to oxygen, 25oC	cm^3 m^{-2} d^{-1} day^{-1}	150-560	
Permeability to water vapor, 25oC	g m^{-2} day^{-1}	30-35	
Contact angle of water, 20oC	degree	93.7-97.0; 87	-; Qi, Y, Chen, T, Zhang, J, Appl. Surf. Sci., 435, 503-11, 2018.

MECHANICAL & RHEOLOGICAL PROPERTIES

PARAMETER	UNIT	VALUE	REFERENCES
Tensile strength	MPa	29-51; 110 (15% glass fiber); 30-52	Chang, M C O; Garrett, P D, Antec, 2588-93, 1996; Kumar, S R, Sridhar, S, Venkatramman, R, Venkatesan, M, Mater. today, Proc., in press, 2020.
Tensile modulus	MPa	1,870-2,600; 6,600 (15% glass fiber)	
Tensile stress at yield	MPa	38-56	Chang, M C O; Garrett, P D, Antec, 2588-93, 1996.

ASA poly(acrylonitrile-co-styrene-co-acrylate)

PARAMETER	UNIT	VALUE	REFERENCES
Tensile creep modulus, 1000 h, elongation 0.5 max	MPa	1,200-1,650	
Elongation	%	7-37	Chang, M C O; Garrett, P D, Antec, 2588-93, 1996.
Tensile yield strain	%	2.8-4.0; 2.5 (15% glass fiber)	
Tensile yield strength	MPa	55-57	Hossain, M M; Moghbelli, E; Jahnke, E; Boeckmann, P; Guriyanova, S; Sander, R; Minkwitz, R; Sue, H-J, Polymer, 63, 71-81, 2015.
Flexural strength	MPa	56-75; 40.8-62.7	-; Kumar, S R; Sridhar, S, Venkatramman, R, Venkatesan, M, Mater. today, Proc., in press, 2020.
Flexural modulus	MPa	1,880-2,570	
Compressive strength	MPa	35-65	
Compressive modulus	MPa	1,600-1,900	Hossain, M M; Moghbelli, E; Jahnke, E; Boeckmann, P; Guriyanova, S; Sander, R; Minkwitz, R; Sue, H-J, Polymer, 63, 71-81, 2015.
Young's modulus	MPa	1,900-2,600	Chang, M C O; Garrett, P D, Antec, 2588-93, 1996.
Charpy impact strength, unnotched, 23°C	kJ m^{-2}	160-270; 28 (15% glass fiber)	
Charpy impact strength, unnotched, -30°C	kJ m^{-2}	70-180; 17 (15% glass fiber)	
Charpy impact strength, notched, 23°C	kJ m^{-2}	10-40; 7 (15% glass fiber)	
Charpy impact strength, notched, -30°C	kJ m^{-2}	3-9; 6 (15% glass fiber)	
Izod impact strength, unnotched, 23°C	J m^{-1}	260	
Izod impact strength, notched, 23°C	J m^{-1}	65-210	
Izod impact strength, notched, -30°C	J m^{-1}	24-35	
Shear modulus	MPa	700-900	
Rockwell hardness	-	R100-103	
Shrinkage	%	0.45 (parallel); 0.9 (normal)	
Melt viscosity, shear rate=1000 s^{-1}	Pa s	150-200	
Melt volume flow rate (ISO 1133, procedure B), 220°C/10 kg	cm³/10 min	4-25	
Melt index, 220°C/10 kg	g/10 min	5.2-15	
Water absorption, equilibrium in water at 23°C	%	0.55-1.65; 1.42 (15% glass fiber)	
Moisture absorption, equilibrium 23°C/50% RH	%	0.15-0.35; 0.3 (15% glass fiber)	
CHEMICAL RESISTANCE			
Acid dilute/concentrated	-	resistant	
Alcohols	-	resistant-fair	
Alkalis	-	resistant	
Aliphatic hydrocarbons	-	resistant	
Aromatic hydrocarbons	-	non-resistant	

ASA poly(acrylonitrile-co-styrene-co-acrylate)

PARAMETER	UNIT	VALUE	REFERENCES
Esters	-	non-resistant	
Greases & oils	-	resistant	
Halogenated hydrocarbons	-	non-resistant	
Ketones	-	non-resistant	
Good solvent	-	chloroform, dichlorobenzene, diethyl ether, DMF, ethyl benzoate, ethyl chloride, mesityl oxide, methyl chloride, methyl propyl ketone, xylene	
Non-solvent	-	acetamide, ethylene glycol, glycerin, triethanolamine	

FLAMMABILITY

PARAMETER	UNIT	VALUE	REFERENCES
Flammability according to UL-94 standard; thickness 1.6/0.8 mm	class	HB; V-0 (PC/ASA)	Wagner, F, Peeters, J R, De Keyzer, J, Duflou, J R, Dewulf , W, Procedia CIRP, 90, 416-20, 2020.
Ignition temperature	^{o}C	>400	
Autoignition temperature	^{o}C	>400	
Limiting oxygen index	% O_2	19	
Char at 500°C	%	1.12	
Volatile products of combustion	-	CO, CO_2, cyanides, ammonia, acrylonitrile, styrene, nitrogen	

WEATHER STABILITY

PARAMETER	UNIT	VALUE	REFERENCES
Tensile strength retention	%	104 (40 months outdoors)	
Erosion rate (exposure for 10 years)	m year^{-1}	8.8E-10 (Florida) 12.0E-10 (Arizona)	
Stabilizers	-	UV absorbers: 2-(2H-benzotriazol-2-yl)-p-cresol; 2-(2H-benzotriazole-2-yl)-4,6-di-tert-pentylphenol; 2-(2H-benzotriazole-2-yl)-4-(1,1,3,3-tetraethylbutyl)phenol; 2-[4,6-bis(2,4-dimethylphenyl)-1,3,5-triazin-2-yl]-5-(octyloxy) phenol; HAS: 1,3,5-triazine-2,4,6-triamine, N,N'''[1,2-ethane-diyl-bis[[[4,6-bis[butyl(1,2,6,6-pentamethyl-4-piperidinyl) amino]-1,3,5-triazine-2-yl]imino]-3,1-propanediyl]-bis[N',N''-dibutyl-N',N''-bis(1,2,2,6,6-pentamethyl-4-piperidinyl)-; bis(2,2,6,6-tetramethyl-4-piperidyl) sebacate; 2,2,6,6-tetramethyl-4-piperidinyl stearate; 1, 6-hexanediamine, N, N'-bis(2,2,6,6-tetramethyl-4-piperidinyl)-, polymers with 2,4-dichloro-6-(4-morpholinyl)-1,3,5-triazine; 1,6-hexanedi-amine, N,N'-bis(2,2,6,6-tetramethyl-4-piperidinyl)-, polymers with morpholine-2,4,6-trichloro-1,3,5-triazine reaction products, methylated; Phenolic antioxidant: phenol, 4-methyl-, reaction products with dicyclopentadiene and isobutene; Screener: carbon black	

BIODEGRADATION

PARAMETER	UNIT	VALUE	REFERENCES
Stabilizers		silver compound is added to Luran S BX 13042 to impart its surface with germicidal effect	Anon., Plast. Addit. Compounding, Nov/Dec., p. 19, 2008.

TOXICITY

PARAMETER	UNIT	VALUE	REFERENCES
HMIS: Health, Flammability, Reactivity rating	-	1/1/0	
NFPA: Health, Flammability, Reactivity rating	-	0/0/0	
Carcinogenic effect	-	not listed by ACGIH, NIOSH, NTP	

ASA poly(acrylonitrile-co-styrene-co-acrylate)

PARAMETER	UNIT	VALUE	REFERENCES
PROCESSING			
Typical processing methods	-	extrusion, injection molding, thermoforming	
Preprocess drying: temperature/ time/residual moisture	oC/h/%	80-90/2-6/0.02-0.04	
Processing temperature	oC	225-280 (injection molding); 200-230 (pipe extrusion); 200-250 (sheet extrusion); 250/255/255 (extrusion)	-; -; -; Sánchez, D M, de la Mata, M, Delgado, F J, Casal, V, Molina, S I, Mater Design, 191, 108577, 2020.
Processing pressure	MPa	0.3-1 (back); 5-10 (injection)	
Additives used in final products	-	Fillers: carbon black, glass beads; release agents; thermal stabilizers	
Applications	-	exterior cable enclosures, impact modifier for PC, large screen displays, marine applications, mirrors for personal watercraft, pool accessories, profiles, recreational vehicle antennas, sheet outdoor furnishings, ski bindings, skylights, spas	
Outstanding properties	-	high service temperature, low thermal conductivity, weather resistant	
BLENDS			
Suitable polymers	-	AES, PBT, PC, PET, PMMA, PVC	
ANALYSIS			
FTIR (wavenumber-assignment)	cm^{-1}/-	C=O – 1733, C-O-C – 1169, CH$_3$ – 1387, 1456	Tomar, N; Maiti, S N, J. Appl. Polym. Sci., 113, 1657-63, 2009.

BIIR bromobutyl rubber

PARAMETER	UNIT	VALUE	REFERENCES
GENERAL			
Common name	-	bromobutyl rubber	
Acronym	-	BIIR	
CAS number	-	68441-14-5	
HISTORY			
Person to discover	-	R A Crawford and R T Morrissey	
Date	-	1954	
Details	-	BFGoodrich researchers obtained 3 patents for bromination of butyl rubber	
SYNTHESIS			
Isoprene contents	mol%	1.7-2	Xiong, X; Wang, J; Jia, H; Fang, E; Ding, L, Polym. Deg. Stab., 98, 2208-14, 2013.
Bromine contents	wt%	1.8-2.2	
Method of synthesis	-	the manufacture of the bromobutyl rubber is a two step process: the polymerization of isobutylene and isoprene to produce butyl rubber, followed by bromination to form bromobutyl rubber; a slurry of fine particles of butyl rubber dispersed in methyl chloride is formed in the reactor after Lewis acid initiation; bromine is added to the butyl solution in highly agitated reaction vessels	
Catalyst	-	aluminum trichloride, alkyl aluminum dichloride, boron trifluoride, tin tetrachloride, and titanium tetrachloride	
Mass average molecular weight, M_w	dalton, g/mol, amu	350,000-450,000	
Polydispersity, M_w/M_n	-	1.5	
STRUCTURE			
Trans content	%	50-60 (isoprenyl units)	
COMMERCIAL POLYMERS			
Some manufacturers	-	ExxonMobil; Lanxess	
Trade names	-	Bromobutyl rubber; Bromobutyl	
PHYSICAL PROPERTIES			
Density at 20°C	g cm^{-3}	0.92-0.93; 0.96-0.98 (compression molded)	-; He, F, Mensitieri, G, Lavorgna, M, Salzano de Luna, M, Scherillo, G, Compos. Part B: Eng, 116, 361-8, 2017.
Color	-	amber	
Odor	-	none to mild	
Degradation temperature (TGA)	°C	~300	Kim, H, Yarin, A L, Lee, M W, Compos. Part B: Eng., 182, 107598, 2020.
Activation energy of thermal decomposition	kJ mol^{-1}	213	Xiong, X; Wang, J; Jia, H; Fang, E; Ding, L, Polym. Deg. Stab., 98, 2208-14, 2013.
Thermal conductivity	W m^{-1} K^{-1}	0.125	Xiong, X; Wang, J; Jia, H; Fang, E; Ding, L, Polym. Deg. Stab., 98, 2208-14, 2013.
Long term service temperature	°C	316 (dry), 232 (wet)	

BIIR bromobutyl rubber

PARAMETER	UNIT	VALUE	REFERENCES
Vicat temperature VST/A/50	°C	85	
Permeability to oxygen, 27°C, O% RH	$cm^3_{ST-P}cm/s$ $atm\ cm^2$	2.07E-08	He, F, Mensitieri, G, Lavorgna, M, Salzano de Luna, M, Scherillo , G, Compos. Part B: Eng, 116, 361-8, 2017.
Permeability to carbon dioxide, 27°C, O% RH	$cm^3_{ST-P}cm/s$ $atm\ cm^2$	7.76E-08	He, F, Mensitieri, G, Lavorgna, M, Salzano de Luna, M, Scherillo , G, Compos. Part B: Eng, 116, 361-8, 2017.
Permeability to oxygen, 25°C	$cm^3\ mm$ $m^{-2}\ day^{-1}$ $mm\ Hg^{-1}$	0.71-0.78	

MECHANICAL & RHEOLOGICAL PROPERTIES

PARAMETER	UNIT	VALUE	REFERENCES
Tensile strength	MPa	9.3-14	
Elongation	%	400-840	
Tear strength	$kN\ m^{-1}$	54-59; 92-114 (peak load)	
Rebound, 23°C	%	9.8-10	
Fatigue to failure (ASTM 4482)	cycles at 136% strain	240,000-340,000	
Shore A hardness	-	47-50	
Mooney viscosity	-	28-64	

CHEMICAL RESISTANCE

PARAMETER	UNIT	VALUE	REFERENCES
Acid dilute/concentrated	-	good	
Alkalis	-	good	
Aliphatic hydrocarbons	-	poor	
Aromatic hydrocarbons	-	poor	
Halogenated hydrocarbons	-	poor	

FLAMMABILITY

PARAMETER	UNIT	VALUE	REFERENCES
Ignition temperature	°C	>210	
Autoignition temperature	°C	>300	
Volatile products of combustion	-	CO, CO_2	

TOXICITY

PARAMETER	UNIT	VALUE	REFERENCES
NFPA: Health, Flammability, Reactivity rating	-	1/1/0; 1/1/0 (HMIS)	
Carcinogenic effect	-	not listed by ACGIH, NIOSH, NTP	

PROCESSING

PARAMETER	UNIT	VALUE	REFERENCES
Typical processing methods	-	calendering, mixing, molding, vulcanization	
Processing temperature	°C	150 (vulcanization)	
Process time	min	20	
Additives used in final products	-	accelerator (MTBS); antidegradants (amine type), antioxidant; curing agents (ZnO, Zn stearate); fillers (carbon black and mineral fillers, such as silica, clays, talc, whiting), peroxide (e.g. dicumyl); release agent (metal stearates), retarder (MgO); plasticizers (petroleum based oils), sulfur; tackifying resins (phenolic, phenol-formaldehyde, phenol-acetylene, hydrocarbon resins; UV stabilizer (carbon black)	

BIIR bromobutyl rubber

PARAMETER	UNIT	VALUE	REFERENCES
Applications	-	automobile tires, conveyor belts, hoses, membranes, pharmaceutical stoppers, seals, protective clothing, tank lining, tire interliners	
Outstanding properties	-	fast cure, low gas transition temperature, low permeability to air, gases, moisture, processing safety	
BLENDS			
Suitable polymers	-	butyl rubber, chlorobutyl rubber, EPDM, SBR	

BMI polybismaleimide

PARAMETER	UNIT	VALUE	REFERENCES
GENERAL			
Common name	-	polybismaleimide	
IUPAC name	-	poly[N,N'-(1,4-phenylene)dimaleimide]	
CAS name	-	[1,1'-bi-1H-pyrrole]-2,2',5,5'-tetrone, homopolymer	
Acronym	-	BMI	
CAS number	-	62238-79-3, 26140-67-0	
SYNTHESIS			
Monomer(s) structure	-	$C_2H_2(CO)_2O$; diamine	
Monomer(s) CAS number(s)	-	108-31-6; large number of amines used	
Monomer(s) molecular weight(s)	dalton, g/mol, amu	98.06; from 100 to over 500	
Method of synthesis	-	maleic anhydride and diamines are reacted in the presence of catalyst such as triethylamine, these are further cured to form crosslinked polymers. Thermal curing is promoted by the presence of radical or ionic initiators. BMI can also be synthesized by Diels-Alder reaction (see ref.)	Jiang, B; Hao, J; Wang, W; Jiang, L; Cai, X, Eur. Polym. J., 37, 463-70, 2001.
Temperature of polymerization	°C	225-290	
Time of polymerization	h	0.5	
Catalyst	-	triethylamine	
Yield	%	93-97 (Diels-Alder)	
Activation energy of polymerization	kJ mol⁻¹	87.8-111.9	
COMMERCIAL POLYMERS			
Some manufacturers	-	Hexcel, Huntsman; Sumitomo	
Trade names	-	HexPly, Kerimid	
PHYSICAL PROPERTIES			
Density at 20°C	g cm⁻³	1.25-1.27	
Melting temperature, DSC	°C	90-360; 166-202 (naphthalene-containing)	
Storage temperature	°C	<0	
Shelf life	month	12 (at -18°C); 6 (at 4°C)	
Decomposition temperature	°C	400-430, 423	-: Xiong, X, Ma, X, Chen, P, Zhou, L, Ren, R, Liu, S, Reactive Funct. Polym., 129, 29-37, 2018.
Thermal expansion coefficient, 23-80°C	°C⁻¹	4.9-5.2E-5	
Glass transition temperature	°C	316-380; 225-232 (wet); 291-334 (naphthalene-containing); 260	Wang, C-S; Hwang, H-J, J. Appl. Polym. Sci., 60, 857-63, 1996; Xiong, X, Ma, X, Chen, P, Zhou, L, Ren, R, Liu, S, Reactive Funct. Polym., 129, 29-37, 2018.
Maximum service temperature	°C	232 (short term); 316 (structural integrity)	
Long term service temperature	°C	-75 to 204	
High temperature stability (special grades)		400-430	Kumar, D; Kaur, J, J. Macromol. Sci., Part A: Pure Appl. Chem., 29, 11, 267-275, 1992.
Dielectric constant at 100 Hz/1 MHz	-	3.09/3.4-3.7; 3.31 (47% glass fiber)	

BMI polybismaleimide

PARAMETER	UNIT	VALUE	REFERENCES
MECHANICAL & RHEOLOGICAL PROPERTIES			
Tensile strength	MPa	50-90; 418.5 (53% glass fiber); 744 (carbon fabric)	
Tensile modulus	MPa	3,500-4,500; 25,500 (53% glass fiber); 56,300 (carbon fabric)	
Elongation	%	3	
Flexural strength	MPa	637.8 (53% glass fiber); 917 (60% carbon fabric)	
Flexural modulus	MPa	31,100; 56,800 (carbon fabric)	
Compressive strength	MPa	480.6 (53% glass fiber); 889 (carbon fabric)	
Fracture toughness	MPa m(1/2)	0.46-0.97	
Strain energy release rate, G1C	kJ m-2	0.067	
Shear strength	MPa	96.5 (carbon fiber); 120 (carbon fabric)	
Shrinkage	%	0.007 (cure)	
Water absorption, equilibrium in water at 23°C	%	3.8-4.4	
Moisture absorption, equilibrium 23°C/50% RH	%	4.3	
CHEMICAL RESISTANCE			
Alcohols	-	poor	
Aromatic hydrocarbons	-	good	
Esters	-	poor	
Halogenated hydrocarbons	-	poor	
Ketones	-	poor	
Good solvent	-	methylethylketone, methylisobutylketone, dichloromethane, chloroform, tetrahydrofuran, DMF, NMP, and hot toluene	Xiong, X, Ma, X, Chen, P, Zhou, L, Ren, R, Liu, S, Reactive Funct. Polym., 129, 29-37, 2018.
Non-solvent	-	DMAC, ethanol, acetone	
FLAMMABILITY			
NBS smoke chamber	mg m^{-3}	0.025	
Burning rate (Flame spread index)		10	
Toxicity of smoke	HCN (ppm)	5-10	
Char at 500°C	%	7.6-18.5 (air); 43-71 (nitrogen); 43-44 (700°C)	Liu, Y-L; Chen, Y-J, Polymer, 45, 1797-1804, 2004; Surender, R; Mahendran, A; Than araichelvan, A; Alam, S; Vijayakumar, C T, Thermochim. Acta, 562, 11-21, 2013.
TOXICITY			
Oral rat, LD$_{50}$	mg kg^{-1}	>2,000	
Skin rabbit, LD$_{50}$	mg kg^{-1}	>5,400	
PROCESSING			
Typical processing methods	-	curing by free radical mechanism, prepreg preparation	
Processing temperature	°C	177-191; post cure at 232-246	
Processing pressure	kPa	586 (vacuum)	
Process time	h	6-4; post-cure time: 8	

BMI polybismaleimide

PARAMETER	UNIT	VALUE	REFERENCES
Applications	-	prepreg systems used in civil and military aircrafts, electrical boards, adhesives	
Outstanding properties	-	dimensional stability at high temperatures, high service temperature, low thermal conductivity	
BLENDS			
Suitable polymers	-	PEI, PEEK, PES, silicone	
ANALYSIS			
FTIR (wavenumber-assignment)	cm^{-1}/-	C=O – 1775-1780, 1710-1720; C-N-C – 1390-1400; C=C – 680-690	Wang, C-S; Hwang, H-J, J. Appl. Polym. Sci., 60, 857-63, 1996.

BZ polybenzoxazine

PARAMETER	UNIT	VALUE	REFERENCES
GENERAL			
Common name	-	polybenzoxazine	Ghosh, N N; Kiskan, B; Yagci, Y, Prog. Polym. Sci., 32, 1344-91, 2007.
IUPAC name	-	3-phenyl-3,4-dihydro-2H-1,3-benzoxazine	
Acronym	-	BZ	
HISTORY			
Person to discover	-	Holly, F W; Cope, A C	Ghosh, N N; Kiskan, B; Yagci, Y, Prog. Polym. Sci., 32, 1344-91, 2007.
Date	-	1944	
Details	-	condensation reaction of primary amines with formaldehyde and phenol	
SYNTHESIS			Ghosh, N N; Kiskan, B; Yagci, Y, Prog. Polym. Sci., 1344-91, 2007.
Monomer(s) structure	-		
Monomer(s) CAS number(s)	-	108-95-2; 50-00-0; 62-53-3	
Monomer(s) molecular weight(s)	dalton, g/mol, amu	94.11; 30.03; 93.13	
Method of synthesis	-	benzoxazine monomers are polymerized by solventless ring-opening polymerization; properties of polymer can be tailored by using monomers having different substitution groups which may provide different functionalities; polybenzoxazines are obtained using thermal or photoinitiated polymerization	Yagci, Y; Kiskan, B; Ghosh, N N, J. Polym. Sci., Part A; Polym. Chem., 47, 5565-76, 2009; Lu, H-C; Su, Y-C; Wang, C-F; Huang, C-F; Sheen, Y-C; Chang, F-C, Polymer, 49, 4852-60, 2008.
Temperature of polymerization	°C	101-110	Rimdusit, S; Tiptipakorn, S; Jubsilp, C; Takeichi, T, Reactive Functional Polym., 73, 369-80, 2013.
Time of polymerization	h	0.16-6	Rimdusit, S; Tiptipakorn, S; Jubsilp, C; Takeichi, T, Reactive Functional Polym., 73, 369-80, 2013.
Yield	%	77-87; 75-83	Takeichi, T; Kano, T; Aga, T, Polymer, 46, 12172-80, 2005; Rimdusit, S; Tiptipakorn, S; Jubsilp, C; Takeichi, T, Reactive Functional Polym., 73, 369-80, 2013.
Mass average molecular weight, M_w	dalton, g/mol, amu	6,000-10,000	
STRUCTURE			
Crystallinity	%	0	Kim, W-K; Mattice, W L, Computational Theoretical Polym. Sci., 8, 3/4, 353-61, 1998.
COMMERCIAL POLYMERS			
Some manufacturers	-	Henkel, Huntsman	
PHYSICAL PROPERTIES			
Density at 20°C	g cm^{-3}	1.1-1.19	Parkpoom, L; Wongkasemjit, S; Chaisuwan, T, Mater. Sci. Eng. A, 527, 77-84, 2009.

BZ polybenzoxazine

PARAMETER	UNIT	VALUE	REFERENCES
Decomposition temperature	°C	310-350	Rajput, A B; Ghosh, N N, Intl. J. Polym. Mater., 60, 1, 27-39, 2010.
Glass transition temperature	°C	146-247; 170-340	Takeichi, T; Kano, T; Agag, T, Polymer, 46, 26, 12172-80, 2005; Rimdusit, S; Tiptipakorn, S; Jubsilp, C; Takeichi, T, Reactive Functional Polym., 73, 369-80, 2013.
Maximum service temperature	°C	130-280	Rimdusit, S; Tiptipakorn, S; Jubsilp, C; Takeichi, T, Reactive Functional Polym., 73, 369-80, 2013.
Long term service temperature	°C	150-180	
Hildebrand solubility parameter	MPa$^{0.5}$	16.98	Kim, W-K; Mattice, W L, Computational Theoretical Polym. Sci., 8, 3/4, 353-61, 1998.
Dielectric constant at 100 Hz/1 MHz	-	3.6/3.5; 3.61	=; Hariharan, A, Prabunathan, P, Kumaravel, A, Manoj, M, Alagar, M, Polym. Testing, 86, 106443, 2020.
Dielectric loss factor at 1 kHz	-	0.0047	Hariharan, A, Prabunathan, P, Kumaravel, A, Manoj, M, Alagar, M, Polym. Testing, 86, 106443, 2020.
Dissipation factor at 1 MHz	E-4	60-110	
Diffusion coefficient of water vapor	cm^2 s^{-1} x10^2	3.6-4.9	
Contact angle of water, 20°C	degree	102-105; 139 (contains 0.05 silica)	Lu, H-C; Su, Y-C; Wang, C-F; Huang, C-F; Sheen, Y-C; Chang, F-C, Polymer, 49, 4852-60, 2008; Liu, J; Xin, Z; Zhou, C, Appl. Surf. Sci., 353, 1137-42, 2015.

MECHANICAL & RHEOLOGICAL PROPERTIES

PARAMETER	UNIT	VALUE	REFERENCES
Tensile strength	MPa	100-125	Rimdusit, S; Tiptipakorn, S; Jubsilp, C; Takeichi, T, Reactive Functional Polym., 73, 369-80, 2013.
Tensile modulus	MPa	2,000-5,300	
Elongation	%	1.6-4.1	
Flexural strength	MPa	132	
Flexural modulus	MPa	4,600	
Compressive strength	MPa	230; 5.2-12.4 (foam)	
Izod impact strength, notched, 23°C	J m^{-1}	18-31	
Shrinkage	%	0	Rimdusit, S; Tiptipakorn, S; Jubsilp, C; Takeichi, T, Reactive Functional Polym., 73, 369-80, 2013.
Water absorption, equilibrium in water at 23°C	%	1.3-1.9	

CHEMICAL RESISTANCE

PARAMETER	UNIT	VALUE	REFERENCES
Acid dilute/concentrated	-	very good	
Alkalis	-	very good	

FLAMMABILITY

PARAMETER	UNIT	VALUE	REFERENCES
Heat release	W g^{-1}	15 (peak)	

BZ polybenzoxazine

PARAMETER	UNIT	VALUE	REFERENCES
Char at 850°C	%	42, 60-64 (with high aromatic content); char yield can be increased to 74% at 800°C in siloxane-containing benzoxazines	Ohara, M, Yoshimoto, K, Kawauchi, T, Takeichi, T, Polymer, 202, 122668, 2020; Xu, J, Li, H, Zeng, K, Li, G, Zhao, X, Zhao, C, Thermochim. Acta, 671, 119-26, 2019.
Limiting oxygen index	% O_2	26.3-30.3	Zhu, Y, Su, J, Lin, R, Jiang, Y, Li, P, Thermochim. Acta, 683, 178465, 2020.
UL rating	-	V-1	Zhu, Y, Su, J, Lin, R, Jiang, Y, Li, P, Thermochim. Acta, 683, 178465, 2020.

PROCESSING

PARAMETER	UNIT	VALUE	REFERENCES
Typical processing methods	-	casting	
Processing temperature	°C	200-218; 160-230 (curing temperature)	Rimdusit, S; Tiptipakorn, S; Jubsilp, C; Takeichi, T, Reactive Functional Polym., 73, 369-80, 2013.
Process time	h	2-4	
Additives used in final products	-	Fillers: carbon fibers, clay, montmorillonite, titanium dioxide	
Applications	-	membranes, mold release agents in nanoimprint	
Outstanding properties	-	electrical performance, high glass transition temperature, near-zero shrinkage upon polymerization, very high char yield	

BLENDS

PARAMETER	UNIT	VALUE	REFERENCES
Suitable polymers	-	BM, PC, PCL, PEO, PI, POSS, PU, PVP, SPI, epoxy, polyacrylate, polyester, rubber	

C cellulose

PARAMETER	UNIT	VALUE	REFERENCES
GENERAL			
Common name	-	cellulose	
IUPAC name	-	cellulose; 2-(hydroxymethyl)-6-[4,5,6-trihydroxy-2-(hydroxymethyl)oxan-3-yl]oxy-oxane-3,4,5-triol	
Acronym	-	C	
CAS number	-	9004-34-6	
EC number	-	232-674-9	
RTECS number	-	FJ5691460	
HISTORY			
Person to discover	-	Emil Fischer	
Date	-	1891-1894	
Details	-	in 1902 Fischer received Nobel Price for establishing structure of carbohydrates, including cellulose	
SYNTHESIS			
Monomer(s) structure	-		
Monomer(s) CAS number(s)	-	50-99-7	
Monomer(s) molecular weight(s)	dalton, g/mol, amu	180.16	
Monomer ratio	-	100%	
Cellulose fractions	%	60.8-65.6 (holocellulose), 33.5-42.5 (cellulose), 19.9-27.3 (hemicellulose)	Xiao, M-Z, Chen, W-J, Cao, X-F, Chen, Y-Y, Sun, R-C, Carbohydrate Polym., 238, 116212, 2020.
Number-average molecular weight		36000-40000	Roig, F; Dantras, E; Dandurand, J; Lacabanne, C, J. Phys. D: Appl. Phys., 44, 045403, 2011.
Particle length	nm	700-1100 (nanofibrils)	Ladhar, A; Arous, M; Kaddami, H; Raihane, M; Kallel, A; Graca, M P F; Costa, L C, J. Mol. Liq., 209, 272-9, 2015.
Particle diameter	nm	8-12.5 (nanofibrils)	Ladhar, A; Arous, M; Kaddami, H; Raihane, M; Kallel, A; Graca, M P F; Costa, L C, J. Mol. Liq., 209, 272-9, 2015.
Cellulose content in some natural products	%	cotton – 94, hemp – 77, flax, kapok, sisal – 75, wood – 40-50, straw – 40-50	
Method of synthesis	-	bacterial cellulose can be biosynthesized by *Gluconacetobacter* sp. and some other bacteries	Pokalwar, S U; Mishra, M K; Manwar, A V, Recent Res. Sci. Technol., 2, 7, 14-19, 2010.
Mass average molecular weight, M_w	dalton, g/mol, amu	160,000-560,000	
Polymerization degree (number of monomer units)	-	300-1700 (wood); 800-10,000 (cotton), 1,000-3,000 (purified cotton); 200-600 (regenerated cellulose; e.g., rayon)	
STRUCTURE			
Crystallinity	%	40-60 (typical); 75 (cotton); 60 (wood pulp); 46-51 (switch-grass leaves); 35 (regenerated cellulose); 25 (viscose)	
Cell type (lattice)	-	triclinic/monoclinic (I); monoclinic (II)	

C cellulose

PARAMETER	UNIT	VALUE	REFERENCES
Cell dimensions	nm	a:b:c=0.835:0.70:1.03 (I); a:b:c=0.81:0.904:1.036 (II); a:b:c=1.025:0.778:1.034 (III); a:b:c=0.803:0.813:1.034 (allomorph IV/I); a:b:c=0.799:0.81:1.034 (allomorph IV/II)	Perez, S; Samain, D, Adv. Carbohydrate Chem. Biochem., 64, 25-116, 2010.
Unit cell angles	degree	γ= 84 (I); γ=117 (II); γ=122.4	
Number of chains per unit cell	-	2 (I); 2 (II)	
Crystallite size	nm	5-6 (cotton); 5.8-7 (wood pulps); 2-3.1 (viscose)	
Polymorphs	-	I, II (marine algae; occurs when form I is treated with NaOH), III (ammonia treatment of I and II gives III), and IV (heating of III generates IV)	
Chain conformation	-	gauche-gauche, gauche-trans, trans-gauche; P2/1 (I); P2/1 (II)	
Heat of crystallization	kJ kg^{-1}	105-134	
Lamellae thickness	nm	2302 middle lamellae in *Coccinia grandis* fiber; 2460-4070 (bark fibers of *Thespesia populnea*)	Jebadurai, S G, Raj, R E, Sreenivasan, V S, Binoj, J S, Carbohydrate Polym. 207, 675-83, 2019; Kathirselvam, M, Kumaravel, A, Arthanarieswaran, V P, Saravanakumar, S S, Int. J. Biol. Macromol., 129, 396-406, 2019.

COMMERCIAL POLYMERS

Some manufacturers	-	Sigma-Aldrich; GreenFiber, Insulmax, Applegate, Thermocell (insulation); Asahi Kasei, FMC, FrieslandCampina, Mingtai Chemical, and Tembec (microcrystalline cellulose); Weidmann (powder)	

PHYSICAL PROPERTIES

Density at 20°C	g cm^{-3}	1.54-1.57; 1.59-1.63 (crystalline); 1.482-1.489 (amorphous)	
Bulk density at 20°C	g cm^{-3}	0.3	
Color	-	white	
Refractive index, 20°C	-	1.534-1.618	
Birefringence	-	1.573-1.595/1.527-1.534	
Haze	%	4	
Gloss, 60°, Gardner (ASTM D523)	%	90	
Odor	-	none	
Melting temperature, DSC	°C	260-270 (decomp.)	
Thermal expansion coefficient, 23-80°C	°C^{-1}	0.2-1.6E-5	
Thermal conductivity, melt	W m^{-1} K^{-1}	0.054-0.13	
Glass transition temperature	°C	220-245; 237-277 (regenrated cellulose)	Okugawa, A, Sakaino, M, Yuguchi, Y, Yamane, C, Carbohydrate Polym., 231, 115663, 2020.
Specific heat capacity	J K^{-1} kg^{-1}	1364 (wood); 1318 (cotton)	
Calorific value	kJ kg^{-1}	29,000 (lignocellulose); 14,000 (wood)	Afra, E, Abyaz, A, Saraeyan, A, J. Cleaner, Prod., 278, 123543, 2021; Gamba, A, Charlier, M, Franssen, J-M, Fire Safety J., 117, 103213, 2020.
Activation energy of thermal degradation	kJ mol^{-1}	225-238 (pellets and powder); 120 (fiber)	Roig, F; Dantras, E; Dandurand, J; Lacabanne, C, J. Phys. D: Appl. Phys., 44, 045403, 2011.
Maximum service temperature	°C	225	
Hildebrand solubility parameter	MPa$^{0.5}$	18.03-32.02	
Surface tension	mN m^{-1}	36-42 (regenerated from pulp); 42 (regenerated from cotton)	

C cellulose

PARAMETER	UNIT	VALUE	REFERENCES
Relative permittivity at 100 Hz/1 MHz	-	3-7.5	
Dielectric loss factor at 1 kHz	-	0.02	
Surface resistivity	ohm	1E16 (pure cellulose); 2.4E7 (raw cotton); 1E4 (viscose)	
Electric strength K20/P50, d=0.60.8 mm	kV mm^{-1}	30-50	
Coefficient of friction	-	0.2 (dynamic); 0.25 (static)	
Permeability to nitrogen, 25°C	cm^3 cm cm^{-2} s^{-1} Pa^{-1} x 10^{-13}	0.02-0.06	
Permeability to oxygen, 25°C	cm^3 cm cm^{-2} s^{-1} Pa^{-1} x 10^{-13}	0.004-0.04	
Permeability to water vapor, 25°C	cm^3 cm cm^{-2} s^{-1} Pa^{-1} x 10^{-13}	20,000	
Speed of sound	m s^{-1}	4315	Koshani, R, van de Ven, T G M, J. Colloid Interface Sci., 563, 252-60,, 2020

MECHANICAL & RHEOLOGICAL PROPERTIES

PARAMETER	UNIT	VALUE	REFERENCES
Tensile strength	MPa	69-170 (regenerated); 50-120 (cellophane); 2.2-5.9 (pulp handsheets)	Spence, K L; Venditti, R A; Habibi, Y; Rojas, O J; Pawlak, J J, Bioresource Technol., 101, 5961-68, 2010.
Tensile modulus	MPa	3,000-5,000	
Tensile creep modulus, 1000 h, elongation 0.5 max	MPa	70-125	
Elongation	%	18-70 (film); 6-10 (fiber); 22-70 (regenerated)	
Young's modulus	MPa	137,000 (crystalline microfibril); 3,490-9,080 (microcrystalline)	Orts, W J; Imam, S H; Glenn, G M; Inglesby, M K; Guttman, M E; Nguyen, A; Revol, J-F, Antec, 2427-31, 2004; Roberts, R J; Rowe, R C; York, P, Int. J. Pharmaceutics, 105, 177-80, 1994.
Elastic modulus	MPa	15,850-35,45 (bamboo fibers depended on extraction method)	Wang, F, Shao, Z, Ind. Crops Prod., 152, 112521, 2020.
Compressive modulus	MPa	56,800	Prapavesis, A, Tojaga, V, Östlund, S, van Vuure, A W, Compos. Part A: Appl. Sci. Manuf, 135, 105930, 2020.
Tenacity (fiber) (standard atmosphere)	cN tex^{-1} (daN mm^{-2})	18-75 (25-125)	Fourne, F, Synthetic Fibers. Machines and Equipment Manufacture, Properties. Carl Hanser Verlag, 1999.
Tenacity (wet fiber, as % of dry strength)	%	40-110	Fourne, F, Synthetic Fibers. Machines and Equipment Manufacture, Properties. Carl Hanser Verlag, 1999.
Fineness of fiber (titer)	dtex	1.3-3.6	Fourne, F, Synthetic Fibers. Machines and Equipment Manufacture, Properties. Carl Hanser Verlag, 1999.

C cellulose

PARAMETER	UNIT	VALUE	REFERENCES
Length (elemental fiber)	mm	25-220	Fourne, F, Synthetic Fibers. Machines and Equipment Manufacture, Properties. Carl Hanser Verlag, 1999.
Poisson's ratio	-	0.30 (microcrystalline)	Roberts, R J; Rowe, R C; York, P, Int. J. Pharmaceutics, 105, 177-80, 1994.

CHEMICAL RESISTANCE			
Acid dilute/concentrated	-	good-poor	
Alcohols	-	good	
Alkalis	-	poor	
Aliphatic hydrocarbons	-	good	
Aromatic hydrocarbons	-	good	
Esters	-	good	
Greases & oils	-	good	
Halogenated hydrocarbons	-	good	
Ketones	-	good	
Good solvent	-	alkalies, calcium thiocyanate, sodium xantanate, phosphoric acid, sulfuric acid	
Non-solvent	-	diluted alkalies and acids, hydrocarbons, mineral oils, water, organic solvents	

FLAMMABILITY			
Ignition temperature	^oC	390-420	
Autoignition temperature	^oC	400-410	
Limiting oxygen index	$\% \ O_2$	18-20; 22.8-30.3 (treated cellulose)	Gaan, S; Rupper, P; Salimova, V; Heuberger, M, Polym. Deg. Stab., 94, 1125-34, 2009.
Heat release	$kW \ m^{-2}$	197; 130-190 (treated cellulose)	Gaan, S; Rupper, P; Salimova, V; Heuberger, M, Polym. Deg. Stab., 94, 1125-34, 2009.
Burning rate (Flame spread rate)	$mm \ min^{-1}$	195-399	Flisi, U, Polym. Deg. Stab., 30, 153-68, 1990.
Char at 500°C	%	7.8; 8-22.6 (treated cellulose)	
Heat of combustion	$J \ g^{-1}$	15,090-18,855	
Volatile products of combustion	-	carbon dioxide	Kim, N K, Dutta, S, Bhattacharyya, D, Compos. Sci., Technol. 162, 64-78, 2018.

WEATHER STABILITY			
Spectral sensitivity	nm	328 (rayon); 290-340 (without oxygen), 290-380 (with oxygen)	
Activation energy of photo-oxidation	$kJ \ mol^{-1}$	79 (bond scission)	Hill, D J T; Lee, T T; Darveniza, M; Saha, T, Polym. Deg. Stab., 48, 79, 1995.
Depth of UV penetration	μm	500-2500 (depth of lignin degradation)	
Important initiators and accelerators	-	nitroxyl radicals, ozone, thermal degradation	Biliuta, G; Fras, L; Strnad, S; Harabagiu, V; Coseri, S, J. Polym. Sci., Part A: Polym. Chem., 48, 4790-99, 2010.
Products of degradation	-	bond scission	

C cellulose

PARAMETER	UNIT	VALUE	REFERENCES
Stabilizers	-	UVA: 2-(2H-benzotriazol-2-yl)-p-cresol; phenol, 2-(5-chloro-2H-benzotriazole-2-yl)-6-(1,1-dimethylethyl)-4-methyl-; 2-(2H-benzotriazol-2-yl)-4,6-bis(1-methyl-1-phenylethyl)phenol; isopropenyl ethinyl trimethyl piperidol (cellulose diacetate), biphenyl cellulose (UV absorber fro paper), phenylbenzimidazole (reactive stabilizer for application in cellulosic textiles); Optical brighteners: 2,2'-(2,5-thiophenediyl)bis(5-tert-butyl-benzoxazole); Mixtures: an ortho-hydroxy tris-aryl-s-triazine compound+hindered hydroxybenzoate compound+hindered amine compound containing a 2,2,6,6-tetraalkylpiperidine or 2,2,6,6-tetraalkylpiperazinone radical	

BIODEGRADATION

PARAMETER	UNIT	VALUE	REFERENCES
Typical biodegradants	-	β-glucosidase	
Stabilizers	-	borate base supplemented with azole or thujaplicin; guar gum benzamide is water resistant biocide	Clausen, C A; Yang, V, Int. Biodeter. Biodeg., 59, 20-24, 2007Das, D; Ara, T; Dutta, S; Mukherjee, A, Bioresource Technol., 102, 5878-83, 2011.

TOXICITY

PARAMETER	UNIT	VALUE	REFERENCES
NFPA: Health, Flammability, Reactivity rating	-	0-2/1-2/0	
Carcinogenic effect	-	not listed by ACGIH, NIOSH, NTP	
TLV, ACGIH	mg m^{-3}	3 (respirable), 10 (total)	
NIOSH	mg m^{-3}	5 (respirable), 10 (total)	
OSHA	mg m^{-3}	5 (respirable), 15 (total)	
Oral rat, LD$_{50}$	mg kg^{-1}	>5,000	
Skin rabbit, LD$_{50}$	mg kg^{-1}	>2,000	

ENVIRONMENTAL IMPACT

PARAMETER	UNIT	VALUE	REFERENCES
Aquatic toxicity, *Daphnia magna*, LC$_{50}$, 48 h	mg l^{-1}	<1,000 to 111,000	Dave, G; Aspegren, P, Ecotoxicology Env. Safety, 73, 1629-32, 2010.
Biological oxygen demand, BOD$_5$	mg l^{-1}	148-163	Boroski, M; Rodriques, A C; Carcia, J C; Gerola, A P; Nozaki, J; Hioka, N, J. Hazardous Mater., 160, 135-41, 2008.
Chemical oxygen demand	mg O$_2$ g^{-1}	1193	Raposo, F; de la Rubia, M A; Borja, R; Alaiz, M, Talanta, 76, 448-53, 2008.
Theoretical oxygen demand	mg O$_2$ g^{-1}	1184	Raposo, F; de la Rubia, M A; Borja, R; Alaiz, M, Talanta, 76, 448-53, 2008.

PROCESSING

PARAMETER	UNIT	VALUE	REFERENCES
Typical processing methods	-	paper and pulp processing methods, chemical processing methods used to produce cellulose derivatives, spinning (rayon), compounding (adhesives and binders), extrusion (cellophane)	
Applications	-	conversion products (e.g., cellophane, rayon, etc.), derivatives (e.g., cellulose acetate, nitrocellulose, etc), fiber, medical (wound dressings, bandages), paper, reinforcement, textiles, thickeners, and many other	

C cellulose

PARAMETER	UNIT	VALUE	REFERENCES
Length (elemental fiber)	mm	25-220	Fourne, F, Synthetic Fibers. Machines and Equipment Manufacture, Properties. Carl Hanser Verlag, 1999.
Poisson's ratio	-	0.30 (microcrystalline)	Roberts, R J; Rowe, R C; York, P, Int. J. Pharmaceutics, 105, 177-80, 1994.

CHEMICAL RESISTANCE

PARAMETER	UNIT	VALUE	REFERENCES
Acid dilute/concentrated	-	good-poor	
Alcohols	-	good	
Alkalis	-	poor	
Aliphatic hydrocarbons	-	good	
Aromatic hydrocarbons	-	good	
Esters	-	good	
Greases & oils	-	good	
Halogenated hydrocarbons	-	good	
Ketones	-	good	
Good solvent	-	alkalies, calcium thiocyanate, sodium xantanate, phosphoric acid, sulfuric acid	
Non-solvent	-	diluted alkalies and acids, hydrocarbons, mineral oils, water, organic solvents	

FLAMMABILITY

PARAMETER	UNIT	VALUE	REFERENCES
Ignition temperature	$^{\circ}C$	390-420	
Autoignition temperature	$^{\circ}C$	400-410	
Limiting oxygen index	$\% \ O_2$	18-20; 22.8-30.3 (treated cellulose)	Gaan, S; Rupper, P; Salimova, V; Heuberger, M, Polym. Deg. Stab., 94, 1125-34, 2009.
Heat release	$kW \ m^{-2}$	197; 130-190 (treated cellulose)	Gaan, S; Rupper, P; Salimova, V; Heuberger, M, Polym. Deg. Stab., 94, 1125-34, 2009.
Burning rate (Flame spread rate)	$mm \ min^{-1}$	195-399	Flisi, U, Polym. Deg. Stab., 30, 153-68, 1990.
Char at 500°C	%	7.8; 8-22.6 (treated cellulose)	
Heat of combustion	$J \ g^{-1}$	15,090-18,855	
Volatile products of combustion	-	carbon dioxide	Kim, N K, Dutta, S, Bhattacharyya, D, Compos. Sci., Technol. 162, 64-78, 2018.

WEATHER STABILITY

PARAMETER	UNIT	VALUE	REFERENCES
Spectral sensitivity	nm	328 (rayon); 290-340 (without oxygen), 290-380 (with oxygen)	
Activation energy of photo-oxidation	$kJ \ mol^{-1}$	79 (bond scission)	Hill, D J T; Lee, T T; Darveniza, M; Saha, T, Polym. Deg. Stab., 48, 79, 1995.
Depth of UV penetration	μm	500-2500 (depth of lignin degradation)	
Important initiators and accelerators	-	nitroxyl radicals, ozone, thermal degradation	Biliuta, G; Fras, L; Strnad, S; Harabagiu, V; Coseri, S, J. Polym. Sci., Part A: Polym. Chem., 48, 4790-99, 2010.
Products of degradation	-	bond scission	

C cellulose

PARAMETER	UNIT	VALUE	REFERENCES
Stabilizers	-	UVA: 2-(2H-benzotriazol-2-yl)-p-cresol; phenol, 2-(5-chloro-2H-benzotriazole-2-yl)-6-(1,1-dimethylethyl)-4-methyl-; 2-(2H-benzotriazol-2-yl)-4,6-bis(1-methyl-1-phenylethyl)phenol; isopropenyl ethinyl trimethyl piperidol (cellulose diacetate), biphenyl cellulose (UV absorber fro paper), phenylbenzimidazole (reactive stabilizer for application in cellulosic textiles); Optical brighteners: 2,2'-(2,5-thiophenediyl)bis(5-tert-butyl-benzoxazole); Mixtures: an ortho-hydroxy tris-aryl-s-triazine compound+hindered hydroxybenzoate compound+hindered amine compound containing a 2,2,6,6-tetraalkylpiperidine or 2,2,6,6-tetraalkylpiperazinone radical	

BIODEGRADATION

PARAMETER	UNIT	VALUE	REFERENCES
Typical biodegradants	-	β-glucosidase	
Stabilizers	-	borate base supplemented with azole or thujaplicin; guar gum benzamide is water resistant biocide	Clausen, C A; Yang, V, Int. Biodeter. Biodeg., 59, 20-24, 2007Das, D; Ara, T; Dutta, S; Mukherjee, A, Bioresource Technol., 102, 5878-83, 2011.

TOXICITY

PARAMETER	UNIT	VALUE	REFERENCES
NFPA: Health, Flammability, Reactivity rating	-	0-2/1-2/0	
Carcinogenic effect	-	not listed by ACGIH, NIOSH, NTP	
TLV, ACGIH	mg m^{-3}	3 (respirable), 10 (total)	
NIOSH	mg m^{-3}	5 (respirable), 10 (total)	
OSHA	mg m^{-3}	5 (respirable), 15 (total)	
Oral rat, LD$_{50}$	mg kg^{-1}	>5,000	
Skin rabbit, LD$_{50}$	mg kg^{-1}	>2,000	

ENVIRONMENTAL IMPACT

PARAMETER	UNIT	VALUE	REFERENCES
Aquatic toxicity, *Daphnia magna*, LC$_{50}$, 48 h	mg l^{-1}	<1,000 to 111,000	Dave, G; Aspegren, P, Ecotoxicology Env. Safety, 73, 1629-32, 2010.
Biological oxygen demand, BOD$_5$	mg l^{-1}	148-163	Boroski, M; Rodriques, A C; Carcia, J C; Gerola, A P; Nozaki, J; Hioka, N, J. Hazardous Mater., 160, 135-41, 2008.
Chemical oxygen demand	mg O$_2$ g^{-1}	1193	Raposo, F; de la Rubia, M A; Borja, R; Alaiz, M, Talanta, 76, 448-53, 2008.
Theoretical oxygen demand	mg O$_2$ g^{-1}	1184	Raposo, F; de la Rubia, M A; Borja, R; Alaiz, M, Talanta, 76, 448-53, 2008.

PROCESSING

PARAMETER	UNIT	VALUE	REFERENCES
Typical processing methods	-	paper and pulp processing methods, chemical processing methods used to produce cellulose derivatives, spinning (rayon), compounding (adhesives and binders), extrusion (cellophane)	
Applications	-	conversion products (e.g., cellophane, rayon, etc.), derivatives (e.g., cellulose acetate, nitrocellulose, etc), fiber, medical (wound dressings, bandages), paper, reinforcement, textiles, thickeners, and many other	

C cellulose

PARAMETER	UNIT	VALUE	REFERENCES
ANALYSIS			
Raman (wavenumber-assignment)	cm^{-1}/-	C-H – 2800-3000; O-H – 1475, 1640; water – 1640	Fechner, P M; Wartewig, S; Füting, M; Heilmann, A; Neuber,t R H H; Kleinebudde, P, AAPS PharmaSci., 5, 4, art. 31, 2003.

CA cellulose acetate

PARAMETER	UNIT	VALUE	REFERENCES
GENERAL			
Common name	-	cellulose acetate	
Acronym	-	CA	
CAS number	-	9004-35-7	
HISTORY			
Person to discover	-	Paul Schuetzenberger; Camille & Henri Dreyfus	
Date	-	1865; 1904	
Details	-	Dreyfus brothers begun experimental work on the development of cellulose acetate in 1904. In 1910 they opened a factory capable to produce 3 tons of cellulose acetate per day, mainly used as base for motion picture film and lacquer also used by growing aircraft industry for fabric coatings for wings and fuselage covering	
SYNTHESIS			
Monomer(s) structure	-	$(CH_3CO)_2O$; cellulose	
Monomer(s) CAS number(s)	-	108-24-7; 9004-34-6	
Monomer(s) molecular weight(s)	dalton, g/mol, amu	102.09	
Acetyl content	%	32.0-60.9	
Hydroxyl content	%	3.5-8.7	
Formulation example	-	cellulose/acetic anhydride ratio = 3.43-10.29	Daud, W R W; Djuned, F M, Carbohydrate Polym., 132, 252-60, 2015.
Method of synthesis	-	cellulose derived from wood pulp is reacted with acetic anhydride in the presence of sulfuric acid, followed by the controlled partial hydrolysis in which sulfuric acid and some acetic acid groups are removed to achieve required degree of acetylation	
Temperature of polymerization	°C	0-5 (1 h), 30 (3 h)	Shaikh, H M; Pandare, K V; Nair, G; Varma, A J, Carbohydrate Polym., 76, 23-29, 2009.
Time of polymerization	h	4-6; 1.5-3	Daud, W R W; Djuned, F M, Carbohydrate Polym., 132, 252-60, 2015.
Number average molecular weight, M_n	dalton, g/mol, amu	30,000-125,000	
Mass average molecular weight, M_w	dalton, g/mol, amu	45,000-237,000	Fischer, S; Thuemmler, K; Volkert, B; Hettrich, K; Schmidt, I; Fischer, K, Macromol. Symp., 262, 89-96, 2008.
Polydispersity, M_w/M_n	-	1.47-3.25	Fischer, S; Thuemmler, K; Volkert, B; Hettrich, K; Schmidt, I; Fischer, K, Macromol. Symp., 262, 89-96, 2008.
Polymerization degree (number of monomer units)	-	175-360	
Molar volume at 298K	cm^3 mol^{-1}	246-264	Necula, A M; Olaru, N; Olaru, L; Homocianu, M; Ioan, S, J. Appl. Polym. Sci., 115, 1751-57, 2010.

38   **HANDBOOK OF POLYMERS** 3rd Edition, Copyrights 2022 ChemTec Publishing

CA cellulose acetate

PARAMETER	UNIT	VALUE	REFERENCES
STRUCTURE			
Crystallinity	%	12; 28 (bacterial cellulose based)	Sousa, M; Bras, A R; Veiga, H I M; Ferreira, F C; de Pinho, M N; Correela, N T; Dionision, M, J. Phys. Chem. B, 114, 10939-53, 2010; Barud, H S; de Araujo, A M; Santos, D B; de Assuncao, R M N; Meireles, C S; Cerquira, D A; Filho, G R; Ribeiro, C A; Messsaddeq, Y; Ribeiro, S J L, Themochim. Acta, 471, 61-69, 2008.
Cell type (lattice)	-	orthorhombic	
Cell dimensions	nm	a:b:c=0.594:1.143:1.046	Perez, S; Samain, D, Adv. Carbohydrate Chem. Biochem., 64, 25-116, 2010.
Unit cell angles	degree	γ=95.4	
Number of chains per unit cell	-	1	
Chain conformation	-	2_1 helix	
COMMERCIAL POLYMERS			
Some manufacturers	-	Eastman	
Trade names	-	Cellulose Acetate, Estron and Chromspun (yarns)	
PHYSICAL PROPERTIES			
Density at 20°C	g cm^{-3}	1.27-1.34; 1.375 (crystalline)	
Bulk density at 20°C	g cm^{-3}	0.22-0.32; 0.43 (tapped)	
Color	-	white flakes	
Refractive index, 20°C	-	1.46-1.49	
Birefringence	-	0.005	
Transmittance	%	>90	
Haze	%	4-8.5	
Gloss, 60°, Gardner (ASTM D523)	%	95	
Odor	-	odorless	
Melting temperature, DSC	°C	230-260	
Softening point	°C	190-229	
Decomposition temperature	°C	304	
Thermal expansion coefficient, 23-80°C	10^{-4} °C^{-1}	0.8-1.8	
Thermal conductivity, melt	W m^{-1} K^{-1}	0.17-0.33	
Glass transition temperature	°C	173-203; 197 (DMTA); 190 (DSC); 185-193 (film); 136-148 (bagasse cellulose)	Sousa, M; Bras, A R; Veiga, H I M; Ferreira, F C; de Pinho, M N; Correela, N T; Dionision, M, J. Phys. Chem. B, 114, 10939-53, 2010; Yuan, J; Dunn, D; Clipse, N M; Newton, R J, Pharm. Technol., 88-100, 2009.
Specific heat capacity	J K^{-1} kg^{-1}	1,260-1,670	
Activation energy of thermal degradation	kJ mol^{-1}	83.9	
Long term service temperature	°C	-20 to 70	
Heat deflection temperature at 0.45 MPa	°C	52-105	

CA cellulose acetate

PARAMETER	UNIT	VALUE	REFERENCES
Heat deflection temperature at 1.8 MPa	$^{\circ}C$	46-87	
Hansen solubility parameters, δ_D, δ_P, δ_H	$MPa^{0.5}$	17.1, 13.1, 9.4; 18.6, 12.73, 11.01	
Interaction radius		10.6	
Hildebrand solubility parameter	$MPa^{0.5}$	23.5-27.83	
Surface tension	$mN\ m^{-1}$	calc.=45.9	
Dielectric constant at 100 Hz/1 MHz	-	2.15-7/3.3-7	
Dissipation factor at 1 MHz	E-4	1500	
Volume resistivity	ohm-m	1E8 to 1E11 (varies with humidity)	
Electric strength K20/P50, d=0.60.8 mm	$kV\ mm^{-1}$	11-19	
Coefficient of friction	-	0.6 (dynamic); 0.7 (static)	
Permeability to nitrogen, 25°C	$cm^3\ cm$ $cm^{-2}\ s^{-1}$ $Pa^{-1}\ x$ 10^{-13}	0.2	
Permeability to oxygen, 25°C	$cm^3\ cm$ $cm^{-2}\ s^{-1}$ $Pa^{-1}\ x$ 10^{-13}	0.6	
Permeability to water vapor, 25°C	$cm^3\ cm$ $cm^{-2}\ s^{-1}$ $Pa^{-1}\ x$ 10^{-13}	4,000-5,000	
Contact angle of water, 20°C	degree	44	
Surface free energy	$mJ\ m^{-2}$	41.1	

MECHANICAL & RHEOLOGICAL PROPERTIES

PARAMETER	UNIT	VALUE	REFERENCES
Tensile strength	MPa	12-95; 100-140 (commercial films)	
Tensile modulus	MPa	2,900-4,000	
Tensile stress at yield	MPa	29.6-124	
Elongation	%	15-70	
Tensile yield strain	%	2-47	
Flexural strength	MPa	41-88	
Flexural modulus	MPa	1,000-2,700	
Compressive strength	MPa	29.76-52.99	
Young's modulus	MPa	2,400-4,100	
Charpy impact strength, un-notched, 23°C	$kJ\ m^{-2}$	NB	
Charpy impact strength, notched, 23°C	$kJ\ m^{-2}$	6-15	
Izod impact strength, unnotched, 23°C	$J\ m^{-1}$	100-450	
Izod impact strength, notched, 23°C	$J\ m^{-1}$	110-450	
Izod impact strength, notched, -40°C	$J\ m^{-1}$	53-64	
Shear modulus	MPa	862-1,474	

CA cellulose acetate

PARAMETER	UNIT	VALUE	REFERENCES
Tenacity (fiber) (standard atmosphere)	cN tex^{-1} (daN mm^{-2})	10-15 (13-20)	Fourne, F, Synthetic Fibers. Machines and Equipment Manufacture, Properties. Carl Hanser Verlag, 1999.
Tenacity (wet fiber, as % of dry strength)	%	50-80	Fourne, F, Synthetic Fibers. Machines and Equipment Manufacture, Properties. Carl Hanser Verlag, 1999.
Fineness of fiber (titer)	dtex	2-10	Fourne, F, Synthetic Fibers. Machines and Equipment Manufacture, Properties. Carl Hanser Verlag, 1999.
Length (elemental fiber)	mm	40-120	Fourne, F, Synthetic Fibers. Machines and Equipment Manufacture, Properties. Carl Hanser Verlag, 1999.
Abrasion resistance (ASTM D1044)	mg/1000 cycles	65	
Poisson's ratio	-	0.3827-0.3989	
Shore D hardness	-	73-82	
Rockwell hardness	-	R34-125	
Shrinkage	%	0.2-0.6	
Melt index, 230°C/3.8 kg	g/10 min	1.4-2.4	
Water absorption, equilibrium in water at 23°C	%	2-6.5	
Moisture absorption, equilibrium 23°C/50% RH	%	3.7-6.5; 2.3-2.6 (24 h)	

CHEMICAL RESISTANCE

Acid dilute/concentrated	-	poor	
Alcohols	-	poor	
Alkalis	-	poor	
Aliphatic hydrocarbons	-	poor	
Aromatic hydrocarbons	-	poor	
Esters	-	poor	
Greases & oils	-	good	
Halogenated hydrocarbons	-	poor	
Ketones	-	good	
Θ solvent, Θ-temp.=155, 37°C	-	acetone, butanone	
Good solvent	-	acetic acid, acetone, aniline, benzyl alchol, cyclohexanone, diethanolamine, formic acid, methyl acetate, phenols, pyridine	
Non-solvent	-	aliphatic esters, hydrocarbons, weak mineral acids	

FLAMMABILITY

Ignition temperature	$^{\circ}$C	304	
Autoignition temperature	$^{\circ}$C	475	
Limiting oxygen index	% O_2	18-19	
Minimum ignition energy	J	0.015	
Burning rate (Flame spread rate)	mm min^{-1}	12.7-50.8	
Char at 500°C	%	12	
Volatile products of combustion	-	CH_3COOH, CO, CO_2	

CA cellulose acetate

PARAMETER	UNIT	VALUE	REFERENCES
UL 94 rating	-	HB	
WEATHER STABILITY			
Stabilizers	-	acid scavenger, antioxidant	
Activation energy of hydrolysis	kJ mol^{-1}	121.8	Miller, R L; Stewart, M E, Antec, 2411-16, 1996.
TOXICITY			
NFPA: Health, Flammability, Reactivity rating	-	0-1/0-2/0; 1/1/0 (HMIS)	
Carcinogenic effect	-	not listed by ACGIH, NIOSH, NTP	
TLV, ACGIH	mg m^{-3}/ ppm	10	
MAK/TRK	mg m^{-3}/ ppm	3	
OSHA	mg m^{-3}/ ppm	15	
Oral rat, LD$_{50}$	mg kg^{-1}	>5,000	Thomas, W C; McGrath, L F; Baarson, K A; Auletta, C S; Daly, I W; McConnell, R F, Fd Chem. Toxic., 29, 7, 453-58, 1991.
PROCESSING			
Typical processing methods	-	extrusion, solution	
Preprocess drying: temperature/ time/residual moisture	°C/h/%	65-70/2-3	
Processing temperature	°C	193-210 (injection)	
Processing pressure	MPa	14.5 (injection), 0.45 (back)	
Process time	s	2-12 (cycle); 10-70 (cure)	
Additives used in final products	-	Plasticizers: acetyl triethyl citrate, di-(2-ethylhexyl) phthalate, diethyl phthalate, dimethyl phthalate, dimethyl sebacate, dioctyl sebacate, polyethylene glycol, polypropylene glycol, sulfolane, toluenesulfonamide derivatives, tri-(2-ethylhexyl) phosphate, triacetin, tributyl citrate, triethyl citrate, triphenyl phosphate; Antistatics: silver-doped vanadium pentoxide, vanadium pentoxide; Antiblocking: hydrogenated tallow amide, laponite, silica, talc; Release: magnesium stearate, sodium benzoate; Slip: silicone oil	
Applications	-	coatings (for glass, paper/paperboard), consumer electronics, electrical, fibers, films, food packaging, lacquers (for electric insulators, glass, paper, plastics, wire), membranes, pharmaceutical (osmotic drug delivery, excipient, tableting, task-masking), pressure-sensitive tape, sealants, wood sealers	
BLENDS			
Suitable polymers	-	PEEK, PSU, TPU, epoxy	
ANALYSIS			
FTIR (wavenumber-assignment)	cm^{-1}/-	major bands: C=O stretch of ester group – 1746-1755; asymmetric stretching of C-C-O of ester group – 1234-1237; asymmetric stretching of O-C-C bond attached to carbonyl – 915	Schilling, M; Bouchard, M; Khanjian, H; Learner, T; Phenix, A; Rivenc, R, Accounts Chem. Res., 43, 6, 888-96, 2010.
NMR (chemical shifts)	ppm	^{1}H and ^{13}C chemical shifts (comprehensive paper)	Kono, H; Hashimoto, H; Shimizu, Y, Carbohydrate Polym., 118, 91-100, 2015.

CA cellulose acetate

PARAMETER	UNIT	VALUE	REFERENCES
x-ray diffraction peaks	degree	8.2, 11.9, 16.8, 21.5 (bagasse cellulose based); 13 and 17	Shaikh, H M; Pandare, K V; Nair, G; Varma, A J, Carbohydrate Polym., 76, 23-29, 2009; Daud, W R W; Djuned, F M, Carbohydrate Polym., 132, 252-60, 2015.

CAB cellulose acetate butyrate

PARAMETER	UNIT	VALUE	REFERENCES
GENERAL			
Common name	-	cellulose acetate butyrate	
ACS name	-	cellulose, acetate butanoate	
Acronym	-	CAB	
CAS number	-	9004-36-8	
HISTORY			
Date	-	1935	
Details	-	introduced into the photographic film industry	Schilling, M; Bouchard, M; Khanjian, H; Learner, T; Phenix, A; Rivenc, R, Accounts Chem. Res., 43, 6, 888-96, 2010.
SYNTHESIS			
Monomer(s) structure	-	cellulose, butyric acid, acetic acid	
Monomer(s) CAS number(s)	-	9004-36-8; 107-92-6; 64-19-7	
Monomer(s) molecular weight(s)	dalton, g/mol, amu	typical of raw material; 88.1; 60.05	
Acetyl content	%	2-29.5	
Butyryl content	%	16.5-54	
Hydroxyl content	%	0.8-4.8	
Number average molecular weight, M_n	dalton, g/mol, amu	12,000-70,000	
Polydispersity, M_w/M_n	-	3.2-3.5	
STRUCTURE			
Crystallinity	%	0; amorphous	Suttiwijitpukdee, N; Sato, H; Zhang, J; Hashimoto, T; Ozaki, Y, Polymer, 52, 461-71, 2011.
COMMERCIAL POLYMERS			
Some manufacturers	-	Eastman	
Trade names	-	Cellulose Acetate Butyrate	
PHYSICAL PROPERTIES			
Density at 20°C	g cm^{-3}	1.16-1.26	
Bulk density at 20°C	g cm^{-3}	0.224-0.512; 0.256-0.612 (tapped)	
Color	-	white	
Refractive index, 20°C	-	1.4740-1.48	
Transmittance	%	90	
Haze	%	8.5	
Odor	-	slight, characteristic	
Melting temperature, DSC	°C	150-240	
Decomposition temperature	°C	313-350; 175-180 (plasticized)	
Thermal expansion coefficient, 23-80°C	10^{-4} °C^{-1}	1.2-1.7	
Thermal conductivity, melt	W m^{-1} K^{-1}	0.17-0.33	

44   **HANDBOOK OF POLYMERS** 3rd Edition, Copyrights 2022; ChemTec Publishing

CAB cellulose acetate butyrate

PARAMETER	UNIT	VALUE	REFERENCES
Glass transition temperature	°C	85-161; 113.5	Schilling, M; Bouchard, M; Khanjian, H; Learner, T; Phenix, A; Rivenc, R, Accounts Chem. Res., 43, 6, 888-96, 2010.
Maximum service temperature	°C	60-100	
Long term service temperature	°C	-40 to 60	
Heat deflection temperature at 0.45 MPa	°C	54-108	
Heat deflection temperature at 1.8 MPa	°C	43-94	
Vicat temperature VST/B/50	°C	65-102	
Hansen solubility parameters, δ_D, δ_P, δ_H	MPa$^{0.5}$	17.2, 13.8, 2.8	
Interaction radius		12.6	
Hildebrand solubility parameter	MPa$^{0.5}$	22.2	
Surface tension	mN m^{-1}	calc.=34.0	
Dielectric constant at 100 Hz/1 MHz	-	3.2-3.8	
Dissipation factor at 100 Hz		100-150	
Volume resistivity	ohm-m	1E9 to 1E13	
Surface resistivity	ohm	1E13 to 1E14	
Electric strength K20/P50, d=0.60.8 mm	kV mm^{-1}	10-98	
Permeability to oxygen, 25°C	cm^3 cm cm^{-2} s^{-1} Pa^{-1} x 10^{12}	0.356	
Contact angle of water, 20°C	degree	71.5-75	
Surface free energy	mJ m^{-2}	15-51	

MECHANICAL & RHEOLOGICAL PROPERTIES

Tensile strength	MPa	16-51	
Tensile modulus	MPa	300-2,000	
Tensile stress at yield	MPa	17-44	
Elongation	%	19-90	
Tensile yield strain	%	4-4.7	
Flexural strength	MPa	21-70	
Flexural modulus	MPa	620-2,400	
Elastic modulus	MPa	345-1,380	
Compressive strength	MPa	14-52	
Charpy impact strength, unnotched, 23°C	kJ m^{-2}	NB	
Charpy impact strength, notched, 23°C	kJ m^{-2}	10-30	
Charpy impact strength, notched, -30°C	kJ m^{-2}	7-8	
Izod impact strength, unnotched, 23°C	J m^{-1}	260	
Izod impact strength, notched, 23°C	J m^{-1}	80-530	

CAB cellulose acetate butyrate

PARAMETER	UNIT	VALUE	REFERENCES
Izod impact strength, notched, -40°C	J m⁻¹	100	
Rockwell hardness	-	R26-116	
Ball indention hardness at 358 N/30 S (ISO 2039-1)	MPa	30-65	
Shrinkage	%	0.3-0.9	
Melt volume flow rate (ISO 1133, procedure B), 210°C/2.16 kg	cm3/10 min	5-40	
Water absorption, equilibrium in water at 23°C	%	1.3-2.2; 1.4 (24 h)	
Moisture absorption, equilibrium 23°C/50% RH	%	0.8-1.2	

CHEMICAL RESISTANCE			
Acid dilute/concentrated	-	poor	
Alcohols	-	poor	
Alkalis	-	poor	
Aliphatic hydrocarbons	-	poor	
Aromatic hydrocarbons	-	poor	
Esters	-	good	
Greases & oils	-	poor	
Halogenated hydrocarbons	-	poor	
Ketones	-	poor	
Good solvent	-	acetone, amyl acetate, chloroform, cyclohexanone, dioxane, methanol (hot) nitromethane, tetrachloroethylene, toluene	
Non-solvent	-	aliphatic hydrocarbons, diethyl ether, ethanol, methanol	

FLAMMABILITY			
Limiting oxygen index	% O_2	17	
Volatile products of combustion	-	CO, CO_2, acetic acid	
UL 94 rating	-	HB	

TOXICITY			
NFPA: Health, Flammability, Reactivity rating	-	1/1/0; 1/1/0 (HMIS)	
Carcinogenic effect	-	not listed by ACGIH, NIOSH, NTP	
Reproductive toxicity	-	not reported	
TLV, ACGIH	mg m⁻³	3 (respirable), 10 (total)	
OSHA	mg m⁻³	5 (respirable), 15 (total)	
Oral rat, LD_{50}	mg kg⁻¹	>6,400	
Skin rabbit, LD_{50}	mg kg⁻¹	>1,000	

PROCESSING			
Typical processing methods	-	compression molding, extrusion, injection molding	
Preprocess drying: temperature/ time/residual moisture	°C/h/%	70/2	
Processing temperature	°C	129-199 (compression molding); 168-249 (injection molding)	

CAB cellulose acetate butyrate

PARAMETER	UNIT	VALUE	REFERENCES
Additives used in final products	-	Antistatic: hydroxyethyl cellulose; Antiblocking: silica; Release: fluorochemical, microcystalline wax, polyethylene wax, silicone; Slip: alumina, magnesium stearate, polyethylene wax, silica	
Applications	-	aircraft, automotive, coatings (for automotive plastics, cloth, leather, paper/paperboard, plastics, wood); lacquers (for automotive, paper, plastics, wood), nail care, panel for illuminated signs, printing inks, tool handles	
Outstanding properties	-	toughness, dimensional stability, resistance to extreme weather	

BLENDS

Suitable polymers	-	acrylics, alkyds, amino resins, isocyanate resins, PC, PMMA, polyesters	

ANALYSIS

FTIR (wavenumber-assignment)	cm^{-1}/-	major bands: C=O stretch of ester group – 1746; asymmetric stretching of C-C-O of ester group – 1234; asymmetric stretching of O-C-C bond attached to carbonyl	Schilling, M; Bouchard, M; Khanjian, H; Learner, T; Phenix, A; Rivenc, R, Accounts Chem. Res., 43, 6, 888-96, 2010.

CAP cellulose acetate propionate

PARAMETER	UNIT	VALUE	REFERENCES
GENERAL			
Common name	-	cellulose acetate propionate	
CAS name	-	cellulose, acetate propanoate	
Acronym	-	CAP	
CAS number	-	9004-39-1	
HISTORY			
Date	-	1924; 1931	
Details	-	introduced into the photographic film industry in 1924; in 1931 Celanese developed commercial product	Schilling, M; Bouchard, M; Khanjian, H; Learner, T; Phenix, A; Rivenc, R, Accounts Chem. Res., 43, 6, 888-96, 2010.
SYNTHESIS			
Monomer(s) structure	-	cellulose; propionic acid; acetic acid	
Monomer(s) CAS number(s)	-	9004-34-6; 79-09-4; 64-19-7	
Monomer(s) molecular weight(s)	dalton, g/mol, amu	typical for raw materials: 74.08; 60.05	
Acetyl content	%	0.6-2.5	
Propionyl content	%	42.5-89.7	
Hydroxyl content	%	1.7-10.0	
Method of synthesis	-	similar to other acetates, it is made with the addition of propionic acid in place of acetic anhydride	
Number average molecular weight, M_n	dalton, g/mol, amu	17,000-75,000	
Mass average molecular weight, M_w	dalton, g/mol, amu	25,000-247,000	
Polydispersity, M_w/M_n	-	3.07-3.31	
STRUCTURE			
Crystallinity	%	0, amorphous	
COMMERCIAL POLYMERS			
Some manufacturers	-	Eastman; Rotuba	
Trade names	-	Cellulose Acetate Propionate, Tenite; Auracel	
PHYSICAL PROPERTIES			
Density at 20°C	g cm^{-3}	1.17-1.25	
Bulk density at 20°C	g cm^{-3}	0.40	
Color	-	white	
Refractive index, 20°C	-	1.46-1.475	
Birefringence	-	5-8E-4 (depending on wavelength)	Yamaguchi, M; Masuzawa, K, Eur. Polym. J., 43, 3277-82, 2007; Nobukawa, S, Nakao, A, Songsurang, K, Pulkerd, P, Yamaguchi, M, Polymer, 111, 53-60, 2017.
Transmittance	%	90	
Haze	%	8.5	

CAP cellulose acetate propionate

PARAMETER	UNIT	VALUE	REFERENCES
Odor	-	none	
Melting temperature, DSC	^{o}C	184-210	
Thermal expansion coefficient, 23-80°C	10^{-4} $^{o}C^{-1}$	1.2-1.8	
Thermal conductivity, melt	W m^{-1} K^{-1}	0.16-0.36	
Glass transition temperature	^{o}C	128-159; 117.5	Schilling, M; Bouchard, M; Khanjian, H; Learner, T; Phenix, A; Rivenc, R, Accounts Chem. Res., 43, 6, 888-96, 2010.
Specific heat capacity	J K^{-1} kg^{-1}	1,200-1,900	
Maximum service temperature	^{o}C	60	
Heat deflection temperature at 0.45 MPa	^{o}C	76-96	
Heat deflection temperature at 1.8 MPa	^{o}C	67-90	
Vicat temperature VST/A/50	^{o}C	87-96	
Vicat temperature VST/B/50	^{o}C	94	
Dielectric constant at 100 Hz/1 MHz	-	3.55-4/3.3-3.8	
Dissipation factor at 1 MHz	E-4	80	
Volume resistivity	ohm-m	1E10	
Electric strength K20/P50, d=0.60.8 mm	kV mm^{-1}	11-19	
Contact angle of water, 20°C	degree	66	Amim, J; Kosaka, P M; Petri, D F S; Maia, F C B; Miranda, P B, J. Colloid Interface Sci., 332, 477-83, 2009.

MECHANICAL & RHEOLOGICAL PROPERTIES

Tensile strength	MPa	22-66	
Tensile modulus	MPa	1,000-2,200	
Tensile stress at yield	MPa	22-32	
Elongation	%	3-45	
Tensile yield strain	%	4	
Flexural strength	MPa	29-58	
Flexural modulus	MPa	1,100-1,750	
Compressive strength	MPa	58	
Izod impact strength, notched, 23°C	J m^{-1}	130-520	
Izod impact strength, notched, -40°C	J m^{-1}	85-120	
Abrasion resistance (ASTM D1044)	mg/1000 cycles	65	
Rockwell hardness	-	R40-96	
Shrinkage	%	0.2-0.6	
Water absorption, equilibrium in water at 23°C	%	1.8-2; 1.5 (24 h)	
Moisture absorption, equilibrium 23°C/50% RH	%	1.0	

CAP cellulose acetate propionate

PARAMETER	UNIT	VALUE	REFERENCES
CHEMICAL RESISTANCE			
Acid dilute/concentrated	-	good-fair	
Alcohols	-	poor	
Alkalis	-	good	
Aliphatic hydrocarbons	-	good	
Aromatic hydrocarbons	-	poor	
Esters	-	poor	
Greases & oils	-	good	
Halogenated hydrocarbons	-	poor	
Ketones	-	poor	
Good solvent	-	acetone, butyl acetate, cellosolve acetate, ethyl acetate, methyl alcohol, methyl ethyl ketone	
Non-solvent	-	ethylene glycol, heptane, turpentine, water	
FLAMMABILITY			
Autoignition temperature	^{o}C	432	
Limiting oxygen index	$\% \, O_2$	17-19	
Volatile products of combustion	-	CO, CO_2	
UL 94 rating	-	HB	
TOXICITY			
NFPA: Health, Flammability, Reactivity rating	-	1/1/0; 0/0/0 (HMIS)	
Carcinogenic effect	-	not listed by ACGIH, NIOSH, NTP	
Oral rat, LD_{50}	mg kg^{-1}	>6,400	
Skin rabbit, LD_{50}	mg kg^{-1}	>5,000	
PROCESSING			
Typical processing methods	-	mixing, injection molding	
Preprocess drying: temperature/ time/residual moisture	$^{o}C/h/\%$	65-70/2	
Processing temperature	^{o}C	168-195	
Processing pressure	MPa	8-10 (injection)	
Process time	s	8-12 (cycle time)	
Additives used in final products	-	Plasticizers: poly(1,3-butylene glycol adipate) (Drapex 429), polyester sebacate (Paraplex G-25), octyl adipate; Antistatic: hydroxyethyl cellulose; Antiblocking: silica; Release: fluoro-chemical, microcystalline wax, polyethylene wax, silicone; Slip: alumina, magnesium stearate, polyethylene wax, silica	
Applications	-	films, housewares, medical, membranes, nail care, ophthal-mic, printing inks	
Outstanding properties	-	fast solvent release, high melting point, solubility in ink solvents	
BLENDS			
Suitable polymers	-	acrylics, alkyds, amino resins, isocyanate resins, PHB	

CAP cellulose acetate propionate

PARAMETER	UNIT	VALUE	REFERENCES
ANALYSIS			
FTIR (wavenumber-assignment)	cm^{-1}/-	major bands: C=O stretch of ester group – 1746; asymmetric stretching of C-C-O of ester group – 1234; asymmetric stretching of O-C-C bond attached to carbonyl	Schilling, M; Bouchard, M; Khanjian, H; Learner, T; Phenix, A; Rivenc, R, Accounts Chem. Res., 43, 6, 888-96, 2010.

CAPh cellulose acetate phthalate

PARAMETER	UNIT	VALUE	REFERENCES
GENERAL			
Common name	-	cellulose acetate phthalate, cellacefate	
Acronym	-	CAPh	
CAS number	-	9004-38-0	
RTECS number	-	FJ5692000	
Linear formula	-	$C_{116}H_{116}O_{64}$	García-Casas, I, Montes, A, Pereyra, C, de la Ossa, E J M, Eur. J. Pharm. Sci., 100, 79-86, 2017.
SYNTHESIS			
Monomer(s) structure	-	phthalic anhydride; partial acetate ester of cellulose	
Monomer(s) CAS number(s)	-	85-44-9; 9004-35-7	
Monomer(s) molecular weight(s)	dalton, g/mol, amu	148.1; range	
Acetyl content	%	21.5-26	
Phthalyl content	%	30-36	
Method of synthesis	-	partially substituted cellulose acetate is reacted with phthalic anhydride in the presence of an organic solvent and a basic catalyst	
Catalyst	-	base	
Number average molecular weight, M_n	dalton, g/mol, amu	4,400-19,200	
Mass average molecular weight, M_w	dalton, g/mol, amu	2,500-65,900	
STRUCTURE			
Crystallinity	%	0, amorphous	
COMMERCIAL POLYMERS			
Some manufacturers	-	Eastman; FMC BioPolymer	
Trade names	-	Cellulose Acetate Phthalate; Aquacoat	
PHYSICAL PROPERTIES			
Bulk density at 20°C	g cm^{-3}	0.26	
Color	-	white to off-white	
Odor		odorless	
Melting temperature, DSC	°C	192	
pH solubility		above 6.2	
Glass transition temperature	°C	145.59, 175 (Eastman)	Bhat, K D; Jois, H S S, Procedia Mater. Sci., 5, 995-1004, 2014.
MECHANICAL & RHEOLOGICAL PROPERTIES			
Intrinsic viscosity, 25°C	dl g^{-1}	0.2-0.6	
Water absorption, equilibrium in water at 23°C	%	2.2-5	
CHEMICAL RESISTANCE			
Alcohols	-	poor	
Esters	-	poor	

CAPh cellulose acetate phthalate

PARAMETER	UNIT	VALUE	REFERENCES
Ketones	-	poor	
Good solvent	-	acetone:water=97:3, acetone:ethyl alcohol:50:50	
FLAMMABILITY			
Autoignition temperature	°C	416	
Residue on ignition	%	0.06	
Volatile products of combustion	-	CO, CO_2	
BIODEGRADATION			
Typical biodegradants	-	cellulase, esterase	
TOXICITY			
HMIS: Health, Flammability, Reactivity rating	-	1/1/0	
Carcinogenic effect	-	not listed by ACGIH, NIOSH, NTP	
Teratogenic effect	-	none	
Oral rat, LD_{50}	mg kg^{-1}	>5,000	
Skin rabbit, LD_{50}	mg kg^{-1}	>2,000	
NOAEL	ppm	>50,000	
PROCESSING			
Preprocess drying: temperature/ time/residual moisture	°C/h/%	mixing, spraying	
Additives used in final products	-	Plasticizers: diethyl phthalate, triethyl citrate, triacetin, dibutyl tartrate, glycerol, propylene glycol, tripropionin, triacetin citrate, acetylated monoglycerides	
Applications	-	delayed release, enteric coatings, pharmaceutical excipient, sustained release, tableting	
Outstanding properties	-	withstands prolonged contact with gastric fluids but dissolves readily in the mildly acidic to neutral environment of the small intestine	
BLENDS			
Suitable polymers	-	EC, PES, PVP	

CAR carrageenan

PARAMETER	UNIT	VALUE	REFERENCES
GENERAL			
Common name	-	carrageenan	
CAS name	-	carrageenan	
Acronym	-	CAR	
CAS number	-	9000-07-1	
EC number	-	232-524-2	
HISTORY			
Date	-	600 BC, 400 AD, 1930	
Details	-	Gigartina was first used in China, then in Ireland, and in 1930 the first industrial production begun	
SYNTHESIS			
Monomer(s) structure	-	carrageenan molecules are linear chains of alternating 3-O-substituted β-d-galactopyranosyl units and 4-O-substituted α-d-galactopyranosyl units	BeMiller, J N, Carbohydrate Chemistry for Food Scientists, 3rd Ed., AACC International, 2019, pp. 279-91.
Method of production	-	extraction from red seaweeds (*Chondrus crispus; Eucheuma cottonii; Eucheuma spinosum, Gigartina stellata* (red algae))	
Mass average molecular weight, M_w	dalton, g/mol, amu	20,000-913,000; 215,000	Bondu, S; Deslandes, E; Fabre, M S; Berthou, C; Yu, G, Carbohydrate Polym., 81, 448-60, 2010; Cosenza, V A; Navarro, D A; Pujol, C A; Damonte, E B; Stiortz, C A, Carbohydrate Polym., 128, 199-206, 2015.
STRUCTURE			
Cell type (lattice)	-	trigonal	
Cell dimensions	nm	a=b:c=1.373:1.328 (calcium salt, iota)	Chandrasekaran, R, Adv. Food Nutrition Res., 42, 131-210, 1998.
Polymorphs	-	κ (hazy gels), ι (clear gels), λ (no gel formation), δ, β, ω (families)	Janaswamy, S; Chandrasekaran, R, Carbohydrate Polym., 60, 499-505, 2005.
Chain conformation	-	double helix (ι and κ)	Janaswamy, S; Chandrasekaran, R, Carbohydrate Res., 343, 364-73, 2008.
COMMERCIAL POLYMERS			
Some manufacturers	-	Evonik, FMC BioPolymer, Kelco, Rhodia Food, Shemberg	
PHYSICAL PROPERTIES			
Density at 20°C	g cm^{-3}	1.3-1.48	
Color	-	yellowish to colorless	
Odor	-	slight marine	
Melting temperature, DSC	°C	50-70	
Setting point	°C	30-50	
pH		7-10; 9.4-10.5	Cosenza, V A; Navarro, D A; Pujol, C A; Damonte, E B; Stiortz, C A, Carbohydrate Polym., 128, 199-206, 2015.

CAR carrageenan

PARAMETER	UNIT	VALUE	REFERENCES
Glass transition temperature	°C	-7 (K salt)	Kasapis, S; Mitchell, J R, Int. J. Biological Macromol., 29, 315-21, 2001.

MECHANICAL & RHEOLOGICAL PROPERTIES

PARAMETER	UNIT	VALUE	REFERENCES
Tensile strength	MPa	3.52	Nogueira, L F B; Maniglia, B C; Pereira, L S; Tapia-Blacido, D R; Ramos, A P, Mater. Sci. Eng. C, in press, 2016.
Young's modulus	MPa	309	Nogueira, L F B; Maniglia, B C; Pereira, L S; Tapia-Blacido, D R; Ramos, A P, Mater. Sci. Eng. C, in press, 2016.
Water absorption, equilibrium in water at 23°C	%	75 (max); usually contains 8-10% water	

CHEMICAL RESISTANCE

PARAMETER	UNIT	VALUE	REFERENCES
Aliphatic hydrocarbons	-	good	
Aromatic hydrocarbons	-	good	
Esters	-	good	
Greases & oils	-	good	
Halogenated hydrocarbons	-	good	
Ketones	-	good	
Good solvent	-	hot water	
Non-solvent	-	diluted acids, organic solvents	

TOXICITY

PARAMETER	UNIT	VALUE	REFERENCES
NFPA: Health, Flammability, Reactivity rating	-	1/1/0	
Carcinogenic effect	-	not listed by ACGIH, NIOSH, NTP	
TLV, ACGIH	mg m^{-3}	10	
OSHA	mg m^{-3}	15	
Oral rat, LD$_{50}$	mg kg^{-1}	5,650 (Na salt)	

PROCESSING

PARAMETER	UNIT	VALUE	REFERENCES
Typical processing methods	-	compounding	
Applications	-	beer, cosmetics, diet sodas, excipient, shampoo, soy milk, thickening agents due to their pseudoplasticity, toothpaste, vegan alternative to gelatin, water-based drilling fluid	

CB cellulose butyrate

PARAMETER	UNIT	VALUE	REFERENCES
GENERAL			
Common name	-	cellulose butyrate	
CAS name	-	cellulose, butanoate	
Acronym	-	CB	
CAS number	-	9015-12-7	
SYNTHESIS			
Monomer(s) structure	-	cellulose; $CH_3CH_2CH_2CHOCl$; $(CH_3CH_2CH_2CO)_2O$	
Monomer(s) CAS number(s)	-	9004-34-6; 141-75-3; 106-31-0	
Monomer(s) molecular weight(s)	dalton, g/mol, amu	depends on raw material; 106.55; 158.19	
Method of synthesis	-	cellulose butyrate is produced by reaction of cellulose with butyric anhydride or butyryl chloride, and condensing agents such as $ZnCl_2$ or H_2SO_4; use of catalyst such as H_2SO_4 is very important in this synthesis because reaction between cellulose and anhydride is very slow	
Catalyst	-	H_2SO_4	
STRUCTURE			
Crystallinity	%	36-45	
Cell type (lattice)	-	orthorhombic	Zugenmaier, P J, Appl. Polym. Sci., Polym. Symp., 37, 223, 1983.
Cell dimensions	nm	a:b:c=3.13:2.56:1.036	Zugenmaier, P J, Appl. Polym. Sci., Polym. Symp., 37, 223, 1983.
Number of chains per unit cell	-	8	Zugenmaier, P J, Appl. Polym. Sci., Polym. Symp., 37, 223, 1983.
Chain conformation	-	2/1 helix	
COMMERCIAL POLYMERS			
Some manufacturers	-	Eastman	
PHYSICAL PROPERTIES			
Density at 20°C	g cm^{-3}	1.17; 1.19 (crystalline)	
Refractive index, 20°C	-	1.47-1.48	
Melting temperature, DSC	°C	115-178; 192	
Decomposition temperature	°C	315	
Glass transition temperature	°C	106-130; 81 (amorphous)	
Maximum service temperature	°C	80	
Long term service temperature	°C	50	
Hildebrand solubility parameter	MPa$^{0.5}$	17-24	
Speed of sound	m s^{-1}	35.7	
Acoustic impedance		2.56	
Attenuation	dB cm^{-1}, 5 MHz	21.9	
MECHANICAL & RHEOLOGICAL PROPERTIES			
Tensile strength	MPa	24-76	
Elongation	%	8-80	

CB cellulose butyrate

PARAMETER	UNIT	VALUE	REFERENCES
Water absorption, equilibrium in water at 23°C	%	0.9-2.4	
Moisture absorption, equilibrium 23°C/50% RH	%	0.2	
CHEMICAL RESISTANCE			
Acid dilute/concentrated	-	good/poor	
Alcohols	-	poor	
Alkalis	-	good	
Aliphatic hydrocarbons	-	good	
Aromatic hydrocarbons	-	poor	
Halogenated hydrocarbons	-	poor	
Ketones	-	poor	
Θ solvent, Θ-temp.=57°C	-	tetrachloroethane	
Good solvent	-	benzene, chloroform, cyclohexanone, tetrachloroethane	
Non-solvent	-	cyclohexane, diethyl ether, hexanol, methanol	
PROCESSING			
Preprocess drying: temperature/ time/residual moisture	°C/h/%	extrusion, thermoforming	
Additives used in final products	-	Plasticizers: octyl adipate, o-phenylphenol ethylene oxide adduct, N-toluene sulfonamide; Antiblocking: silica; Release: fluorochemical, microcystalline wax, polyethylene wax, silicone; Slip: alumina, magnesium stearate, polyethylene wax, silica	
Applications	-	corner guards, rocket fuels, sheets, tools, tubing	
BLENDS			
Suitable polymers	-	PHB	

CEC carboxylated ethylene copolymer

PARAMETER	UNIT	VALUE	REFERENCES
GENERAL			
Common name	-	carboxylated ethylene copolymer (salt) (ionomer), poly(ethylene-co-acrylic acid)	Laney, K A, B.Sc. Thesis, Princeton Uni., 2010.
Acronym	-	CEC	
CAS number	-	187410-30-6; 9078-96-0	
HISTORY			
Person to discover	-	Armitage, J B of DuPont	
Date	-	1961	
SYNTHESIS			
Monomer(s) structure	-	$CH_2=CH_2$; $CH_2=C(CH_3)COOH$	
Monomer(s) CAS number(s)	-	74-85-1; 79-41-4	
Monomer(s) molecular weight(s)	dalton, g/mol, amu	28.05; 86.06	
Methacrylic acid content	%	usually <15 mol% (range 5-70 wt%)	
Method of synthesis	-	copolymers are produced by high-temperature/high pressure free radical polymerization similar to the one used in production of LDPE. Carboxyl groups are completely or partially neutralized to form ionomers (mostly Na or Zn). Neutralization extent increase causes adequate increase in viscosity of material	
Polydispersity, M_w/M_n	-	10	
Degree of branching		2/100 carbons (short chain), 1/1000 carbons (long chain)	
STRUCTURE			
Crystallinity	%	30	
COMMERCIAL POLYMERS			
Some manufacturers	-	DOW	
Trade names	-	Surlyn	
PHYSICAL PROPERTIES			
Density at 20°C	g cm^{-3}	0.94-0.97; 1.01 (crystalline)	
Color	-	white	DOW
Odor	-	mild, like methacrylic acid	DOW
Melting point	°C	70-100	
Freeze point	°C	36-75	
Decomposition temperature	°C	>325	
Refractive index, 20°C	-	1.49	
Haze	%	3-7	
Thermal expansion coefficient, 23-80°C	°C^{-1}	6E-5	
Vicat temperature VST/A/50	°C	73	
Heat of fusion	kJ mol^{-1}	2.32	

CEC carboxylated ethylene copolymer

PARAMETER	UNIT	VALUE	REFERENCES
Melt flow index	g 10 min^{-1}	0.7-20	
Dielectric constant at 100 Hz/1 MHz	-	3.8/	

MECHANICAL & RHEOLOGICAL PROPERTIES

Tensile strength	MPa	15.8-37.2	
Elongation at break	%	285-770	
Hardness	Shore D	36-68	
Water absorption, equilibrium in water at 23°C	%	11-19	

PROCESSING

Typical processing methods	-	blown film extrusion, cast film, sheet/cast extrusion	
Applications	-	film, laminate, adhesive accelerator, sealant	
Maximum processing temperature	°C	285	DOW

BLENDS

Suitable polymers	-	EOP, HDPE, LDPE, LLDPE	Besser, K D; Gereke, J; Haubler, L; Leitner, B; Rapthel, I, US Patent, 2011/0077356, Dow, Mar. 31, 2001.

CHI chitosan

PARAMETER	UNIT	VALUE	REFERENCES
GENERAL			
Common name	-	chitosan, poly-D-glucosamine	
Acronym	-	CHI	
CAS number	-	9012-76-4; 1398-61-4 (chitin)	
EC number	-	222-311-2	
RTECS number	-	FM6300000	
HISTORY			
Person to discover	-	Bracconot; Ledderhose; Rammelberg	
Date	-	1811; 1878; 1930	
Details	-	Bracconot discovered chitin in 1811; Ledderhose, determined composition of chitin in 1878; and, in 1930, Rammelberg obtained chitosan from chitin	
SYNTHESIS			
Monomer(s) structure	-	D-glucosamine; 2-acet-amido 2-deoxy-β-d-glucose	Malar, C G, Seenuvasan, M, Kumar, K S, Kumar, M A, Microbial and Natural Macromolecules, Academic Press, 2021, pp. 73-86.
Monomer(s) CAS number(s)	-	3416-24-8; 7512-17-6	
Monomer(s) molecular weight(s)	dalton, g/mol, amu	161.156; 221.21	
Number of d-glucosamine units	-	<5000	Akpan, E I, Gbenebor, O P, Adeosu, S O, Cietus, O, Handbook of Chitin and Chitosan, Elsevier, 2020, pp. 131-64.
Method of synthesis	-	commercially produced by deacetylation of chitin (chitin is a structural element of shrimp and crab shells); deacetylation can be accomplished by treating chitin with an aqueous 40-45% NaOH for 4-5 h	
Molecular formula of chitin	-	structure is similar to cellulose but with 2-acetamido-2-deoxy-β-d-glucose; $(C_8H_{13}O_5N)_n$	Kumari, S, Kishor, R, Handbook of Chitin and Chitosan, Elsevier, 2020, pp. 1-33.
Degree of deacetylation	%	65-95; 90; 99	Hu, J; Wang, X; Xiao, Z; Bi, W, LTW Food Sci. Technol., 63, 519-26, 2015; Akpan, E I, Gbenebor, O P, Adeosu, S O, Cietus, O, Handbook of Chitin and Chitosan, Elsevier, 2020, pp. 131-64.
Number average molecular weight, M_n	dalton, g/mol, amu	33,700-99,400	
Mass average molecular weight, M_w	dalton, g/mol, amu	20,000-190,000 (low M_w chitosan); 190,000-375,000 (high M_w chitosan); 1,000,000-2,500,000 (chitin); 150,000	Hu, J; Wang, X; Xiao, Z; Bi, W, LTW Food Sci. Technol., 63, 519-26, 2015.
Polydispersity, M_w/M_n	-	3.3-8.1	
STRUCTURE			
Crystallinity	%	α-chitin – >90, β-chitin – 75-89, and γ-chitin – 68.6	Kumari, S, Kishor, R, Handbook of Chitin and Chitosan, Elsevier, 2020, pp. 1-33.
Cell type (lattice)	-	orthorhombic	Okuyama, K; Noguchi, K; Miyazawa, T; Yui, T; Ogawa, K, Macromolecules, 30, 19, 5849-55, 1997.
Cell dimensions	nm	a=0.895, b=1.697, c=1.037	Okuyama, K; Noguchi, K; Miyazawa, T; Yui, T; Ogawa, K, Macromolecules, 30, 19, 5849-55, 1997.

CHI chitosan

PARAMETER	UNIT	VALUE	REFERENCES
Number of chains per unit cell	-	4 (8 water molecules)	Okuyama, K; Noguchi, K; Miyazawa, T; Yui, T; Ogawa, K, Macromolecules, 30, 19, 5849-55, 1997.
Crystallite size	nm	1.04	Ogawa, K, J. Metals, Materials, Minerals, 15, 1, 1-5, 2005.
Polymorphs	-	α, β, γ	Dash, M; Chiellini, F; Ottenbritte, R M; Chiellini, E, Prog. Polym. Sci., 36, 981-1014, 2011.
Chain conformation	-	2-fold helix	

COMMERCIAL POLYMERS

Some manufacturers	-	BASF, Cognis	
Trade names	-	Chitopharm	

PHYSICAL PROPERTIES

Density at 20°C	g cm^{-3}	1.4-1.42	
Bulk density at 20°C	g cm^{-3}	0.4-0.68	
Color	-	off-white to gray	
Refractive index, 20°C	-	1.52-1.54	
Birefringence	-	0.012	
Melting temperature, DSC	°C	199-230	
Decomposition temperature	°C	313-317	
Glass transition temperature	°C	163-172	Martinez-Camacho, A P; Cortez-Rocha, M O; Ezquerra-Brauer, J M; Graciano-Verdugo, A Z; Rodriguez-Felix, F; Castillo-Ortega, M M; Yepiz-Gomez, M S; Plascencia-J; M, Carbohydrate Polym., 82, 305-15, 2010.
Volume resistivity	ohm-m	1.25E-7	
Permeability to water vapor, 25°C	g m^{-1} s^{-1} Pa^{-1} x 10^{11}	7.24	Pinotti, A; Garcia, M A; Martino, M N; Zaritzky, Food Hydrocolloids, 21, 66-72, 2007.

MECHANICAL & RHEOLOGICAL PROPERTIES

Tensile strength	MPa	6.7-150.2	Park, S Y; Marsh, K S; Rhim, J W, J. Foo Sci., Food Eng. Phys. Properties, 67, 1, 194-97, 2002.
Elongation	%	4.1-117.8	Park, S Y; Marsh, K S; Rhim, J W, J. Foo Sci., Food Eng. Phys. Properties, 67, 1, 194-97, 2002.
Young's modulus	MPa	32.6	
Tenacity (fiber)	cN tex^{-1}	10-15; 3-7 (wet)	Pillai, C K S; Paul, W; Sharma, C P, Prog. Polym. Sci., 34, 641-78, 2009.
Moisture absorption, equilibrium 23°C/50% RH	%	10	

CHEMICAL RESISTANCE

Acid dilute/concentrated	-	poor	
Alcohols	-	poor	
Alkalis	-	poor	
Esters	-	poor	
Halogenated hydrocarbons	-	poor	

CHI chitosan

PARAMETER	UNIT	VALUE	REFERENCES
Ketones	-	poor	
Good solvent	-	acetic acid, formic acid, concentrated mineral acids, water, soluble in acidic solutions of many organic and inorganic acids (pH < 6) because of protonation of its amino groups	Malar, C G, Seenuvasan, M, Kumar, K S, Kumar, M A, Microbial and Natural Macromolecules, Academic Press, 2021, pp. 73-86.

FLAMMABILITY

PARAMETER	UNIT	VALUE	REFERENCES
Autoignition temperature	ºC	>530	

WEATHER STABILITY

PARAMETER	UNIT	VALUE	REFERENCES
Activation wavelengths	nm	320	

BIODEGRADATION

PARAMETER	UNIT	VALUE	REFERENCES
Typical biodegradants	-	chitosan can be degraded by enzymes able to hydrolyze glucosamine–glucosamine, glucosamine and N-acetyl-glucosamine–N-acetylglucosamine; it can also be degraded by lysozyme	
Stabilizers	-	chitosan itself has antimicrobial properties	Kong, M; Chen, X G; Xing, K; Park, H J, Int. J. Food Microbiol., 144, 1, 51-63, 2010.

TOXICITY

PARAMETER	UNIT	VALUE	REFERENCES
NFPA: Health, Flammability, Reactivity rating	-	0-2/0-1/0-1	
Carcinogenic effect	-	not listed by ACGIH, NIOSH, NTP	
Oral mouse, LD$_{50}$	mg kg^{-1}	>16,000	

ENVIRONMENTAL IMPACT

PARAMETER	UNIT	VALUE	REFERENCES
Aquatic toxicity, *Daphnia magna*, LC$_{50}$, 48 h	% survival	100	Protech, Techn. Rep. TR01.1, Jul. 2004.
Aquatic toxicity, *Fathead minnow*, LC$_{50}$, 48 h	% survival	100	Protech, Techn. Rep. TR01.1, Jul. 2004.
Aquatic toxicity, *Rainbow trout*, LC$_{50}$, 48 h	% survival	100; LC$_{50}$: >10,000 mg/l	Protech, Techn. Rep. TR01.1, Jul. 2004.

PROCESSING

PARAMETER	UNIT	VALUE	REFERENCES
Typical processing methods	-	electrospinning, extrusion, hydrogel formation, precipitation, preparation (deminaralization, deproteinazation, decoloration, and deacetylation), spinning, spray drying	Youn, D K; No, H K; Prinyawiwat-kul, W, Carbohydrate Polym., 69, 707-12, 2007.
Applications	-	agriculture (biopeticide, seed treatment, plant growth enhance-ment), fibers, medical (wound treatments, artificial skin, hemo-static agent), pharmaceutical (drug delivery systems); textile industry, veterinary medicine, water filtration (helps to remove turbidity), beer clarification	Dash, M; Chiellini, F; Ottenbritte, R M; Chiellini, E, Prog. Polym. Sci., 36, 981-1014, 2011; Gassara, F; Antzak, C; Ajila, C M; Sarma, S J; Brar, S K; Verma, M, J. Food Eng., 166, 80-85, 2015.
Outstanding properties	-	accelerates wound healing, anti-itching effect, antimicrobial agent, moisturizing action; antioxidant activity, fat-binding, antimicrobial activity	Zou, P; Yang, X; Wang, J; Li, Y; Yu, H; Zhang, Y; Liu, G, Food Chem., 190, 1174-81, 2016.
Commercial forms	-	microspheres, fibers, hydrogels, membranes, foams, and nanoparticles	

BLENDS

PARAMETER	UNIT	VALUE	REFERENCES
Suitable polymers	-	C, CMC, PAM, PEG, PEO, PMMA, PVOH, PVP, polylysine	Lewandowska, K, J. Mol. Liquids, 209, 301-5, 2015.

CHI chitosan

PARAMETER	UNIT	VALUE	REFERENCES

ANALYSIS

PARAMETER	UNIT	VALUE	REFERENCES
FTIR (wavenumber-assignment)	cm^{-1}/-	N-H – 332-3349; amide – 1646-1648; chitins (α-, β-, and γ-chitin) can be distinguished by FTIR spectra. In the case of α-chitin, two separate peaks are observed at ~1662 and ~1630 (amide-I band) is associated with the occurrence of the inter-molecular hydrogen band –CO...HN and –CO...HOCH$_2$. In the case of β-chitin, a single band is observed at 1656 due to the hydrogen bond present between the amide group (–C=O) of the neighboring intrasheet chain. The –NH stretching bands at 3264 and 3107 are clearly observed in both α- and β-chitin. The bands at 703 and 750 represent the bending vibration of –OH groups and NH– groups present in the α-chitin. In the case of β-chitin, they are shifted to 682 and 710 cm. However, β-chitin shows a single band at approximately 1656, which is associated with intermolecular hydrogen bond of CO...HN. The β-chitin have the weaker inter- and intramolecular hydrogen bonding and a more loosely ordered structure compared to α-chitin. γ-Chitin shows two sharp peaks at 1660 and 1620 for amide-I band, which are also available in the α-chitin	Cardenas, G; Miranda, S P, J. Chilean Chem. Soc., 49, 4, 291-95, 2004; Kumari, S, Kishor, R, Handbook of Chitin and Chitosan, Elsevier, 2020, pp. 1-33.
x-ray diffraction peaks	degree	The α-chitin shows the major crystalline reflection peaks at 9.6 (020 plane) and 19.6 (110 plane), β-chitin at 9.1 (020 plane) and 20.3 (110 plane), and γ-chitin at 9.6 (020 plane) and 19.80 (110 plane)	Kumari, S, Kishor, R, Handbook of Chitin and Chitosan, Elsevier, 2020, pp. 1-33.

CIIR chlorobutyl rubber

PARAMETER	UNIT	VALUE	REFERENCES
GENERAL			
Common name	-	chlorobutyl rubber	
CAS name	-	butyl rubber, chlorinated	
Acronym	-	CIIR	
CAS number	-	68081-82-3	
HISTORY			
Person to discover	-	Baldwin, F P; Thomas, R M	Baldwin, F P; Thomas, R M, US Patent 2,926,718, Esso, Mar. 1, 1960
Date	-	1955, 1960	
Details	-	Esso researchers patented vulcanization of chlorinated butyl rubber	
SYNTHESIS			
Monomer ratio	-	0.8-2.5 mol% isoprene	
Chlorine contents	%	0.6-1.4	
Method of synthesis	-	the manufacture of the bromobutyl rubber is a two step process: the polymerization of isobutylene and isoprene to produce butyl rubber, followed by bromination to form bromobutyl rubber; a slurry of fine particles of butyl rubber dispersed in methyl chloride is formed in the reactor after Lewis acid initiation; bromine is added to the butyl solution in highly agitated reaction vessels	
Catalyst	-	aluminum trichloride, alkyl aluminum dichloride, boron trifluoride, tin tetrachloride, and titanium tetrachloride	
Mass average molecular weight, M_w	dalton, g/mol, amu	350,000-450,000	
STRUCTURE			
Trans content	%	50-60 (isoprenyl units)	
COMMERCIAL POLYMERS			
Some manufacturers	-	Exxon; Lanxess; Ravago	
Trade names	-	Chlorobutyl Rubber; Chlorobutyl; Ravaflex	
PHYSICAL PROPERTIES			
Density at 20°C	g cm^{-3}	0.92-0.93	
Color	-	amber	
Odor	-	mild	
Decomposition temperature	°C	>140; >170	
Storage temperature	°C	>500	
Glass transition temperature	°C	-73 to -39	
Permeability to oxygen, 25°C	cm^3 mm m^{-2} day^{-1} mm Hg^{-1}	0.958	

64 **HANDBOOK OF POLYMERS** 3rd Edition, Copyrights 2022; ChemTec Publishing

CIIR chlorobutyl rubber

PARAMETER	UNIT	VALUE	REFERENCES
MECHANICAL & RHEOLOGICAL PROPERTIES			
Tensile strength	MPa	9.2-20.6	
Tensile stress at yield	MPa	0.71-1.04	
Elongation	%	330-870	
Elastic modulus	MPa	5.1-9.7	
Tear strength	kN m^{-1}	42-56	
Rebound, 23°C	%	11.2	
Payne effect	Pa	4×10^6 (as measured by storage shear modulus)	Scagliusi, S R; Cardoso, E C L; Parra, D F; Lima, L F C P; Lugao, A B, Radiat. Phys. Chem., 84, 42-6, 2013.
Compression set, 24h 70°C	%	20-25	
Shore A hardness	-	52-69	
Mooney viscosity	-	38-55	
CHEMICAL RESISTANCE			
Alcohols	-	good	
Aliphatic hydrocarbons	-	poor	
Greases & oils	-	poor	
Halogenated hydrocarbons	-	poor	
Ketones	-	good	
FLAMMABILITY			
Autoignition temperature	°C	>210	
Volatile products of combustion	-	CO, CO_2, flammable hydrocarbons, HCl	
TOXICITY			
NFPA: Health, Flammability, Reactivity rating	-	1/1/0; 1/1/0 (HMIS)	
Carcinogenic effect	-	not listed by ACGIH, NIOSH, NTP	
OSHA	mg m^{-3}	5 (respiratory), 15 (total)	
Oral rat, LD$_{50}$	mg kg^{-1}	>2,000	
ENVIRONMENTAL IMPACT			
Aquatic toxicity, *Daphnia magna*, LC$_{50}$, 48 h	mg l^{-1}	125-2,100 (tires)	Wik, A; Dave, G, Chemosphere, 58, 645-51, 2005.
PROCESSING			
Typical processing methods	-	calendering, mixing, molding, vulcanization	
Additives used in final products	-	accelerator (MTBS); antioxidant; curing agents (ZnO, Zn stearate); peroxide (e.g. dicumyl); retarder (MgO); sulfur; tackifying resin (phenolic); UV absorber (carbon black)	
Applications	-	conveyor belts, curing bladders, hoses, membranes, pharmaceutical stoppers, seals, tank liners, tire innerlines, tire non-staining sidewalls	
Outstanding properties	-	fast cure, low permeability to air, gases, moisture, low gas transition temperature, processing safety, vibration damping	

CIIR chlorobutyl rubber

PARAMETER	UNIT	VALUE	REFERENCES
BLENDS			
Suitable polymers	-	NBR, NR, PA12	

CMC carboxymethyl cellulose

PARAMETER	UNIT	VALUE	REFERENCES
GENERAL			
Common name	-	carboxymethyl cellulose	
IUPAC name	-	acetic acid; 2,3,4,5,6-pentahydroxyhexanal	
ACS name	-	cellulose, carboxymethyl ether	
Acronym	-	CMC	
CAS number	-	9000-11-7	
RTECS number	-	FJ5700000	
HISTORY			
Person to discover	-	Payen, A; Lilienfield, L	
Date	-	1838;	
Details	-	Payen determined elemental composition of carboxymethyl cellulose in 1838; production technology patented in 192-1916	Varshney, V K, Gupta, P K, Naithani, S, Khullar, R, Bhatt, A, Soni, P L, Carbohydrate Polym., 63, 40-5, 2006.
SYNTHESIS			
Monomer(s) structure	-	chloroacetic acid	
Monomer(s) CAS number(s)	-	79-11-8	
Monomer(s) molecular weight(s)	dalton, g/mol, amu	94.50	
Degree of substitution	-	0.4-1.5 (theoretical maximum is 3 when all 3 groups in monomeric unit are substituted); 0.11-2.41	Yeasmin, S; Mondal, I H, Int. J. Biol. Macromol., 80, 725-31, 2015
Method of synthesis	-	carboxymethyl cellulose is obtained from reaction between cellulose and chloroacetic acid in the presence of alkalis which catalyze reaction	
Catalyst	-	alkalis	
Mass average molecular weight, M_w	dalton, g/mol, amu	80,000-560,000; 90,000-2,000,000	-; Kanikireddy, V, Varaprasad, K, Jayaramudu, T, Karthikeyan, C, Sadiku, R, Int. J. Biol. Macromol., 164, 963-75, 2020.
Polymerization degree (number of monomer units)	-	350-2,500	
STRUCTURE			
Crystallinity	%	80	Li, H; Wu, B; Mu, C; Lin, W, Carbohydrate Polym., 84, 881-86, 2011.
COMMERCIAL POLYMERS			
Some manufacturers	-	Dow	
Trade names	-	Walocel	
PHYSICAL PROPERTIES			
Density at 20°C	g cm^{-3}	1.05	
Decomposition temperature	°C	140-150	
pH	-	6.8	Yeasmin, S; Mondal, I H, Int. J. Biol. Macromol., 80, 725-31, 2015

CMC carboxymethyl cellulose

PARAMETER	UNIT	VALUE	REFERENCES
Permeability to water vapor, 25°C	g mm m^{-2} day^{-1} kPa^{-1}	1.8	Sayanjali, S; Ghanbarzadeh, B; Ghiassifar, S, LWT Food Sci. Technol., 44, 1133-38, 2011.

MECHANICAL & RHEOLOGICAL PROPERTIES

Tensile strength	MPa	17.6-17.8	Sayanjali, S; Ghanbarzadeh, B; Ghiassifar, S, LWT Food Sci. Technol., 44, 1133-38, 2011.
Elastic modulus	MPa	1,350	Ghanbarzadeh, B; Almasi, H, Int. J. Biol. Macromol., 48, 44-49, 2011.
Young's modulus	MPa	1227	Su, J-F; Huang, Z; Yuan, X-Y; Wang, X-Y; Li, M, Carbohydrate Polym., 79, 145-53, 2010.
Water absorption, equilibrium in water at 23°C	%	6.5; 10 max.	

CHEMICAL RESISTANCE

Acid dilute/concentrated	-	poor	
Alcohols	-	poor	
Alkalis	-	poor	
Aliphatic hydrocarbons	-	good	
Aromatic hydrocarbons	-	fair	
Esters	-	poor	
Greases & oils	-	fair	
Halogenated hydrocarbons	-	poor	
Ketones	-	poor	
Good solvent	-	alkalies, acetone, chloroform, esters, mixture of water and alcohols, pyridine, water	

FLAMMABILITY

Autoignition temperature	°C	287-370	

BIODEGRADATION

Typical biodegradants	-	bacteria which can produce cellulase	
Stabilizers	-	Carbosan; potassium sorbate	Sayanjali, S; Ghanbarzadeh, B; Ghiassifar, S, LWT Food Sci. Technol., 44, 1133-38, 2011.

TOXICITY

NFPA: Health, Flammability, Reactivity rating	-	0/2/0	
Carcinogenic effect	-	not listed by ACGIH, NIOSH, NTP	
Oral rat, LD$_{50}$	mg kg^{-1}	>5,000	
Skin rabbit, LD$_{50}$	mg kg^{-1}	>5,000	

PROCESSING

Typical processing methods	-	compounding	

CMC carboxymethyl cellulose

PARAMETER	UNIT	VALUE	REFERENCES
Applications	-	ceramics, cosmetics, fabric finishing, food products, flotation, leather pasting, paints, paper, pharmaceuticals, textile sizing, thickener and emulsion stabilizer in coatings, toothpaste, washing powders and liquids, well drilling, wound dressing	
BLENDS			
Suitable polymers	-	PEO, PR, PAM, PANI, carrageenan, PVAl	Miao, J; Zhang, R; Bai, R, J. Membrane Sci., 493, 654-63, 2015.
ANALYSIS			
FTIR (wavenumber-assignment)	cm^{-1}/-	3420 - OH stretching, 2925 - CH and CO stretching, 1640 - H-O-H, C=C, 896 - glucosidic linkages	Yeasmin, S; Mondal, I H, Int. J. Biol. Macromol., 80, 725-31, 2015.

CN cellulose nitrate

PARAMETER	UNIT	VALUE	REFERENCES
GENERAL			
Common name	-	cellulose nitrate, nitrocellulose	
CAS name	-	cellulose, nitrate	
Acronym	-	CN	
CAS number	-	9004-70-0	
RTECS number	-	QW0970000	
HISTORY			
Person to discover	-	Henri Braconnot; Alexander Parker	
Date	-	1832; 1855	
Details	-	Henri Braconnot discovered that nitric acid with starch or wood fibers produces explosive material; Alexander Parker invented celluloid	
SYNTHESIS			
Monomer(s) structure	-	cellulose; HNO_3	
Monomer(s) CAS number(s)	-	9004-34-6; 7697-37-2	
Monomer(s) molecular weight(s)	dalton, g/ mol, amu	depends on raw material; 63.012	
Nitrification degree	%	76-89 (lacquer grades); >87 (explosive grades)	
Nitrogen content	%	11.1-12.2 (non-explosive because their nitrogen content is less than 12.%) 12.55-13.42 (explosive)	Alinat, E; Delaunay, N; Archer, X; Gareil, P, Carbohydrate Polym., 128, 99-104, 2015; Alinat, E; Delaunay, N; Archer, X; Vial, J; Gareil, P, Forensic Sci. Int., 250, 68-76, 2015.
Method of synthesis	-	concentrated sulfuric acid and 70% nitric acid are mixed with cellulose at 0°C to produce nitrocellulose	
Nitrating agent diffusion	-	cellulose fibril interior is inaccessible because of low dielectric constant of cellulose. Cellulose microcrystallites nitration is controlled by fibrils untwisting. Nanofibril thickness is in the range of 3.5-9 nm (depending on the source of cellulose	Nikolsky, S N, Zlenko, D V, Melnikov, V P, Stovbun, S V, Carbohydrate Polym., 204, 232-7, 2019.
Temperature of polymerization	°C	0	
Free enthalpy of formation	kJ mol^{-1}	-753 (cellulose dinitrate); - 653 (cellulose trinitrate)	Burcat, A, Combustion Flame, 222, 181-5, 2020.
Yield	%	35-100	Adekunle, I M, E-J. Chem., 7, 3, 709-16, 2010.
Nitrogen content	%	12.1 to above 12.5 (explosives); 7-11 (plastics)	Jamal, S H, Rosla, N J, Shah, N A A, Noor, S A M, Yunus, W M Z W, Mater. Today, Proc., 29, 1, 185-9, 2020; Mazurek, J, Laganà, A, Dion, V, Etyemez, S, Schilling , M R, J. Cultural Heritage, 35, 263-70, 2019.
Mass average molecular weight, M$_w$	dalton, g/ mol, amu	125,000-150,00; 750,000-875,000 (dynamite) 20,000-312,000 (non-explosive) 69,000-200,000	Alinat, E; Delaunay, N; Archer, X; Gareil, P, Carbohydrate Polym., 128, 99-104, 2015; Alinat, E; Delaunay, N; Archer, X; Vial, J; Gareil, P, Forensic Sci. Int., 250, 68-76, 2015.
Polymerization degree (number of monomer units)	-	500-600; 3,000-5,000 (dynamite)	
Radius of gyration	nm	6.3-34.1	Alinat, E; Delaunay, N; Archer, X; Gareil, P, Carbohydrate Polym., 128, 99-104, 2015.

CN cellulose nitrate

PARAMETER	UNIT	VALUE	REFERENCES
STRUCTURE			
Cell type (lattice)	-	orthorhombic	Meader, D; Atkins, E; Happey, T, Polymer 19, 1371, 1978.
Cell dimensions	nm	a:b:c=1.22:2.54:0.90	Meader, D; Atkins, E; Happey, T, Polymer 19, 1371, 1978.
PHYSICAL PROPERTIES			
Density at 20°C	g cm^{-3}	1.35-1.40	
Color	-	white to yellow	
Refractive index, 20°C	-	1.49-1.51	
Odor	-	odorless	
Melting temperature, DSC	°C	142-170 (ignites); 169-170	Kim, D H, Reference Module in Biochemical Science from Encyclopedia of Toxicology, 3rd Ed., Elsevier, 2014, pp. 540-42.
Decomposition temperature	°C	>170	
Thermal expansion coefficient, 23-80°C	°C^{-1}	0.8-1.2E-4	
Thermal conductivity, melt	W m^{-1} K^{-1}	0.23	
Glass transition temperature	°C	53-66	
Heat of fusion	kJ mol^{-1}	3.8-6.3	
Long term service temperature	°C	-20 to 70	
Heat deflection temperature at 1.8 MPa	°C	60-71	
Hansen solubility parameters, δ_D, δ_P, δ_H	MPa$^{0.5}$	16.2, 14.1, 9.5; 15.4, 14.7, 8.8	
Interaction radius		10.7; 11.5	
Hildebrand solubility parameter	MPa$^{0.5}$	21.4-23.5	
Surface tension	mN m^{-1}	calc.=38.0	
Dielectric constant at 100 Hz/1 MHz	-	7/6	
Power factor	-	3-5	
Permeability to nitrogen, 25°C	cm^3 cm cm^{-2} s^{-1} Pa^{-1} x 10^{12}	0.0087	
Permeability to oxygen, 25°C	cm^3 cm cm^{-2} s^{-1} Pa^{-1} x 10^{12}	0.146	
Permeability to water vapor, 25°C	cm^3 cm cm^{-2} s^{-1} Pa^{-1} x 10^{12}	472	
Diffusion coefficient of nitrogen	cm^2 s^{-1} x10^6	0.0193	
Diffusion coefficient of oxygen	cm^2 s^{-1} x10^6	0.15	

CN cellulose nitrate

PARAMETER	UNIT	VALUE	REFERENCES
Diffusion coefficient of water vapor	$cm^2\ s^{-1}$ $x10^6$	0.0262	
Contact angle of water, 20°C	degree	54.7	
Surface free energy	$mJ\ m^{-2}$	42.7	

MECHANICAL & RHEOLOGICAL PROPERTIES			
Tensile strength	MPa	35-70	
Elongation	%	10-40	
Flexural strength	MPa	62-76	
Flexural modulus	MPa	1,300-1,500	
Compressive strength	MPa	14-55	
Izod impact strength, notched, 23°C	$J\ m^{-1}$	270-370	
Rockwell hardness	-	R95-115	
Water absorption, equilibrium in water at 23°C	%	0.6-2.0	
Moisture absorption, equilibrium 23°C/50% RH	%	1	

CHEMICAL RESISTANCE			
Acid dilute/concentrated	-	poor	
Alcohols	-	good	
Aromatic hydrocarbons	-	fair	
Esters	-	poor	
Halogenated hydrocarbons	-	poor	
Ketones	-	poor	
Good solvent	-	acetic acid (glacial), acetone, amyl acetate, ethylene glycol ethers	
Non-solvent	-	higher alcohols, higher carboxylic acids, higher ketones	

FLAMMABILITY			
Ignition temperature	°C	140	
Autoignition temperature	°C	140	
Flash point	°C	12.8	Kim, D H, Reference Module in Biochemical Science from Encyclopedia of Toxicology, 3rd Ed., Elsevier, 2014, pp. 540-42.
Volatile products of combustion	-	H_2O, CO, CO_2, NO_x	

WEATHER STABILITY			
Spectral sensitivity	nm	285	
Important initiators and accelerators	-	NO_x (autocatalytic process)	
Products of degradation	-	chain scission, radical formation	
Stabilizers	-	diphenylamine, 2-nitrodiphenylamine	

TOXICITY			
NFPA: Health, Flammability, Reactivity rating	-	2/3/3	

CN cellulose nitrate

PARAMETER	UNIT	VALUE	REFERENCES
Carcinogenic effect	-	not listed by ACGIH, NIOSH, NTP	Kim, D H, Reference Module in Biochemical Science from Encyclopedia of Toxicology, 3rd Ed., Elsevier, 2014, pp. 540-42.
Mutagenic effect	-	not listed	
TLV, ACGIH	mg m^{-3}/ ppm	n/a	
NIOSH	mg m^{-3}/ ppm	n/a	
MAK/TRK	mg m^{-3}/ ppm	n/a	
OSHA	mg m^{-3}/ ppm	n/a	
Oral rat, LD$_{50}$	mg kg^{-1}	>5,000; 8798-10,373 (no observable adverse effect)	Kim, D H, Reference Module in Biochemical Science from Encyclopedia of Toxicology, 3rd Ed., Elsevier, 2014, pp. 540-42.

ENVIRONMENTAL IMPACT

PARAMETER	UNIT	VALUE	REFERENCES
Aquatic toxicity, algae, LC$_{50}$, 48 h	mg l^{-1}	579	Kim, D H, Reference Module in Biochemical Science from Encyclopedia of Toxicology, 3rd Ed., Elsevier, 2014, pp. 540-42.

PROCESSING

PARAMETER	UNIT	VALUE	REFERENCES
Typical processing methods	-	blow molding, casting, compounding, compression molding, machining	
Processing temperature	ºC	85-93 (compression molding)	
Processing pressure	MPa	14-34 (compression molding)	
Additives used in final products	-	Plasticizers: 2-ethylhexyl diphenyl phosphate, acetyl tributyl citrate, acrylic resin (Acronal 700 L), aliphatic polyurethane, butyl benzyl phthalate, camphor (plasticizer of celluloid), castor oil, dibutyl phthalate, dimethyl phthalate, diisooctyl phthalate, epoxidized soybean oil, glyceryl triacetate, glyceryl tribenzoate, glyceryl tribenzoate, N-ethyl (o,p)-toluenesulfonamide, octyl diphenyl phosphate, sucrose acetate isobutyrate, tricresyl phosphate, triethylene glycol, urea resin; Antistatics: poly(3,4-ethylenedioxythiophene sulfonate), vanadium pentoxide; Slip: alumina, silica; Amines (stabilizers of gunpowder)	
Applications	-	celluloid, electroexplosive devices, explosives, lacquers	

BLENDS

PARAMETER	UNIT	VALUE	REFERENCES
Suitable polymers	-	CA, PEG, PMMA	

COC cyclic olefin copolymer

PARAMETER	UNIT	VALUE	REFERENCES
GENERAL			
Common name	-	cyclic olefin copolymer	
CAS name	-	bicyclo[2.2.1]hept-2-ene, polymer with ethene	
Acronym	-	COC	
CAS number	-	26007-43-2	
RTECS number	-	RC0190000	
SYNTHESIS			
Monomer(s) structure	-	$H_2C{=}CH_2$	
Monomer(s) CAS number(s)	-	498-66-8; 74-85-1	
Monomer(s) molecular weight(s)	dalton, g/mol, amu	94.17; 28.05	
Monomer(s) expected purity(ies)	%	99; 99.95	
Norbornene contents	%	20.2-82	
Temperature of polymerization	°C	80-120	
Time of polymerization	h	60	
Pressure of polymerization	kPa	120-240	Young, M-J; Chang, W-S; Ma, C-C M, Eur. Polym. J., 39, 165-71, 2003.
Catalyst	-	ethylenebis-(indenyl)zirconium dichloride; metallocene	
Initiation rate constant	mol s^{-1}	1.94E5	Young, M-J; Chang, W-S; Ma, C-C M, Eur. Polym. J., 39, 165-71, 2003.
Propagation rate constant	mol s^{-1}	2.804E5, 2.782E5	Young, M-J; Chang, W-S; Ma, C-C M, Eur. Polym. J., 39, 165-71, 2003.
Termination rate constant	mol s^{-1}	2.4E8	Young, M-J; Chang, W-S; Ma, C-C M, Eur. Polym. J., 39, 165-71, 2003.
Chain transfer rate constant	mol s^{-1}	4.159E3, 3.471E1	Young, M-J; Chang, W-S; Ma, C-C M, Eur. Polym. J., 39, 165-71, 2003.
Number average molecular weight, M_n	dalton, g/mol, amu	51,000-173,000	
Mass average molecular weight, M_w	dalton, g/mol, amu	41,000-188,000	
Polydispersity, M_w/M_n	-	1.5-4.1	
Molar volume at 298K	cm^3 mol^{-1}	51.1-67.5	Poulsen, L; Zebger, I; Klinger, M; Eldrup, M, Sommer-Larse, P; Ogilby, P R, Macromolecules, 36, 7189-98, 2003.
Van der Waals volume	cm^3 mol^{-1}	32.5-46.1	Poulsen, L; Zebger, I; Klinger, M; Eldrup, M, Sommer-Larse, P; Ogilby, P R, Macromolecules, 36, 7189-98, 2003.
STRUCTURE			
Crystallinity	%	amorphous	
Entanglement molecular weight	dalton, g/mol, amu	31,000	Blochowiak, M; Pakula, T; Butt, H-J; Bruch, M; Fluodas, G, J. Chem. Phys., 124, 134903,1-8, 2006.

COC cyclic olefin copolymer

PARAMETER	UNIT	VALUE	REFERENCES
COMMERCIAL POLYMERS			
Some manufacturers	-	Mitsui Chemical; Topas	
Trade names	-	Apel; Topas COC	
PHYSICAL PROPERTIES			
Density at 20°C	g cm^{-3}	1.00-1.08	
Bulk density at 20°C	g cm^{-3}	0.55-0.60	
Refractive index, 20°C	-	1.51-1.54, 1.53 (Topas)	
Birefringence	-	0.02	Oh, G K; Inoue, T, Rheol. Acta, 45, 116-23, 2005.
Transmittance	%	90-92, 91.4 (Topas)	
Haze	%	0.5-4	
Gloss, 60°, Gardner (ASTM D523)	%	>100	
Softening point	°C	60	
Decomposition temperature	°C	407-440	Liu, C; Yu, J; Sun, X; Zhang, J; He, J, Polym. Deg. Stab., 81, 197-205, 2003.
Thermal expansion coefficient, 23-80°C	°C^{-1}	0.6-0.7E-4	
Glass transition temperature	°C	62-177; 114-122 (metallocene catalyst); 96-125 (depending on annealing temperature; 179.4	Benavente, R; Scrivani, T; Cerrada, M L; Zamfirova, G; Perez, E; Perena, J M, J. Appl. Polym. Sci., 89, 3666-71, 2003; Tritto, I; Marestin, C; Boggioni, L; Sacchi, M C; Brintzinger, H-H; Ferro, D R, Macromolecules, 34, 5770-77, 2001; Nam, S, Jeong, Y J, Jang, J, Org. Electronics, 85, 105828, 2020.
Heat deflection temperature at 0.45 MPa	°C	75-150; 75-170 (Topas)	
Heat deflection temperature at 1.8 MPa	°C	60-125; 68-151 (Topas)	
Vicat temperature VST/B/50	°C	80-137	
Hansen solubility parameters, δ_D, δ_P, δ_H	MPa$^{0.5}$	18.0, 3.0, 2.0	Hansen, C M; Just, L, Ind. Eng. Chem. Res., 40, 21-25, 2001.
Molar volume	kmol m^{-3}	5.0	Hansen, C M; Just, L, Ind. Eng. Chem. Res., 40, 21-25, 2001.
Dielectric constant at 100 Hz/1 MHz	-	2.2 (Topas); 2.7	-; Nam, S, Jeong, Y J, Jang, J, Org. Electronics, 85, 105828, 2020.
Relative permittivity at 1-10 Hz	-	2.35	
Relative permittivity at 1 GHz	-	2.3	
Dissipation factor at 1 GHz		7E-5	
Volume resistivity	ohm-m	1E14	
Charge carrier mobility	cm^2 V^{-1}	1.075 (100 nm film)	Nam, S, Jeong, Y J, Jang, J, Org. Electronics, 85, 105828, 2020.
Comparative tracking index, CTI, test liquid A	-	>600	
Permeability to oxygen, 25°C	cm^3 cm m^{-2} day^{-1} bar^{-1}	1.7-4	
Permeability to water vapor, 25°C	g mm m^{-2} day^{-1}	0.200; 57% reduction with addition of 0.06 wt% graphene; 0.023-0.045 (Topas)	Lai, C-L; Fu, Y-J; Chen, J-T; Wang, D-M; Sun, Y-M; Huang, S-H; Hung, W-S; Hu, C-C; Lee, K-R, Carbon, 90, 85-93, 2015.

COC cyclic olefin copolymer

PARAMETER	UNIT	VALUE	REFERENCES
Diffusion coefficient of oxygen	$cm^2\ s^{-1}$ $\times 10^8$	2.2-5.8	Poulsen, L; Zebger, I; Klinger, M; Eldrup, M, Sommer-Larse, P; Ogilby, P R, Macromolecules, 36, 7189-98, 2003.
Contact angle of water, 20°C	degree	96.1	Nam, S, Jeong, Y J, Jang, J, Org. Electronics, 85, 105828, 2020.
Surface free energy	$mJ\ m^{-2}$	41.28	Nam, S, Jeong, Y J, Jang, J, Org. Electronics, 85, 105828, 2020.

MECHANICAL & RHEOLOGICAL PROPERTIES

PARAMETER	UNIT	VALUE	REFERENCES
Tensile strength	MPa	22-72, 46-63 (Topas)	
Tensile modulus	MPa	1260-3200	
Tensile stress at yield	MPa	37-60	
Elongation	%	1.1-100	
Tensile yield strain	%	6.5	
Flexural strength	MPa	90-110	
Flexural modulus	MPa	2,400-3,200	
Charpy impact strength, 23°C	$kJ\ m^{-2}$	13-20, 13-20 (Topas)	
Charpy impact strength, notched, 23°C	$kJ\ m^{-2}$	1.6-2.6; 1.8-2.6 (Topas)	
Izod impact strength, notched, 23°C	$J\ m^{-1}$	25-45	
Ball indention hardness at 358 N/30 S (ISO 2039-1)	MPa	130-184	
Shrinkage	%	0.1-0.7	
Melt viscosity, shear rate=1000 s^{-1}	Pa s	100-650	
Melt volume flow rate (ISO 1133, procedure B), 230°C/2.16 kg	$cm^3/10$ min	1-9 (extrusion grades); 14-48 (injection molding grades)	
Melt index, 260°C/2.16 kg	g/10 min	2-36; 48	Akin, D; Kosgoz, A; Durmus, A, Composites: Part A, 60, 44-51, 2014.
Water absorption, equilibrium in water at 23°C	%	<0.01	

CHEMICAL RESISTANCE

PARAMETER	UNIT	VALUE	REFERENCES
Acid dilute/concentrated	-	resistant	
Alcohols	-	resistant	
Alkalis	-	resistant	
Aliphatic hydrocarbons	-	non-resistant	
Aromatic hydrocarbons	-	non-resistant	
Esters	-	resistant	
Greases & oils	-	non-resistant	
Halogenated hydrocarbons	-	non-resistant	
Ketones	-	resistant (short chain)	

FLAMMABILITY

PARAMETER	UNIT	VALUE	REFERENCES
Autoignition temperature	°C	445	
UL 94 rating	-	HB	

COC cyclic olefin copolymer

PARAMETER	UNIT	VALUE	REFERENCES
WEATHER STABILITY			
Spectral sensitivity	nm	280-380	
Activation wavelengths	nm	267	
Activation energy of aging	kJ mol^{-1}	1,522	Huang, W-J; Chang, F-C, J. Polym. Res., 10, 195-200, 2003.
Products of degradation	-	chromophores, hydroperoxides, COOH	Pu, Q; Oyesanya, O; Thompson, B; Liu, S; Alvarez, J C, Langmuir, 23, 1577-83, 2007.
Stabilizers	-	antioxidants (e.g. Irganox 1010)	
TOXICITY			
NFPA: Health, Flammability, Reactivity rating	-	1/1/0	
TLV, ACGIH	mg m^{-3}	10; 3 (respirable fraction)	
OSHA	mg m^{-3}	5 (respirable dust); 15 (total dust)	
Oral rat, LD$_{50}$	mg kg^{-1}	3250	
PROCESSING			
Typical processing methods	-	injection molding, extrusion, thermoforming	
Processing temperature	°C	190-240	
Processing pressure	MPa	14 (extrusion); 50-110 (injection pressure)	
Process time	min	15 (injection molding)	
Additives used in final products	-	rheological additives, graphene, montmorillonite, carbon fillers, silica	Lai, C-L; Fu, Y-J; Chen, J-T; Wang, D-M; Sun, Y-M; Huang, S-H; Hung, W-S; Hu, C-C; Lee, K-R, Carbon, 90, 85-93, 2015.
Applications	-	bottles, contact lenses, cosmetics, dielectric films, electronics, film, healthcare, industrial parts, optical parts, pharmaceuticals, packaging, printer toner, sheet	
Outstanding properties	-	high clarity, outstanding moisture barrier, high heat distortion temperature; thickness of 100 nm film gives satisfactory dielectric layer for high-performance and reliable organic field-effect transistors	Nam, S, Jeong, Y J, Jang, J, Org. Electronics, 85, 105828, 2020.
BLENDS			
Suitable polymers	-	LLDPE, PC, POE, PP	
ANALYSIS			
FTIR (wavenumber-assignment)	cm^{-1}/-	C=O – 1850-1680; C=C – 1680-1590; C-O-C – 1400-1100	Nakade, K; Nagai, Y; Ohishi, F, Polym. Deg. Stab., 95, 2654-58, 2010.

CPE chlorinated polyethylene

PARAMETER	UNIT	VALUE	REFERENCES
GENERAL			
Common name	-	polyethylene, chlorinated	
ACS name	-	chlorinated polyethylene rubber	
Acronym	-	CPE	
CAS number	-	63231-66-3	
HISTORY			
Person to discover	-	Fawcett, E W; Gibson, R O; Perrin, M W	Fawcett, E W; Gibson, R O; Perrin, M W, US Patent 2,153,553, ICI, 1939.
Date	-	1939	
Details	-	chlorination in solution at elevated temperature	
SYNTHESIS			
Monomer(s) structure	-	polyethylene; chlorine	
Monomer(s) CAS number(s)	-	9002-88-4; 7782-50-5	
Monomer(s) molecular weight(s)	dalton, g/mol, amu	variable; 35.453	
Chlorine content	%	10-48, 36	Varma, A J; Deshpande, S V; Kondapalli, P, Polym. Deg. Stab., 63, 1-3, 1999; Mondal, S, Nayak, L, Rahaman, M, Aldalbahi, A, Das, N C, Compos. Part B: Eng., 109, 155-69, 2017.
Method of synthesis	-	chlorination in solution or powder form	Steenbakkers-Menting, H N A M, Chlorination of Ultrahigh Molecular eight Polyethylene, Diss. Techn. Uni. Eindhoven, 1995.
Temperature of polymerization	°C	20-130	
Time of polymerization	h	2-4	
Mass average molecular weight, M_w	dalton, g/mol, amu	96,500-120,000; 68,000 (resin for 3D printing)	Wady, P, Wasilewski, A, Brock, L, Edge, R, Vallés, C, Addit. Manuf., 31, 100907, 2020.
Number average molecular weight, M_n	dalton, g/mol, amu	21,000 (resin for 3D printing)	Wady, P, Wasilewski, A, Brock, L, Edge, R, Vallés, C, Addit. Manuf., 31, 100907, 2020.
Polydispersity, M_w/M_n	-	4.6-5.0	
STRUCTURE	-		
Crystallinity	%	2-5 (non-crystalline); 20-50 (semi-crystalline); 13.0-58.5 (CPE films having different chlorine content); 25% (25% Cl); 2% (36-42 Cl%); crystalline up to 30 wt% Cl if chlorination in solution and up to 50% if chlorination in powder form (blocky placement of chlorine atoms)	Stoeva, S, J. Appl. Polym. Sci., 101, 2602-13, 2006; Whiteley, M J; Pan, W-P, Thermochim. Acta., 166, 27-39, 1990; Steenbakkers-Menting, H N A M, Chlorination of Ultrahigh Molecular eight Polyethylene, Diss. Techn. Uni. Eindhoven, 1995.
Crystallite size	nm	7-16 (thickness)	
Lamellae thickness	nm	6.5-14.5	Steenbakkers-Menting, H N A M, Chlorination of Ultrahigh Molecular eight Polyethylene, Diss. Techn. Uni. Eindhoven, 1995.
COMMERCIAL POLYMERS			
Some manufacturers	-	Showa Denko, Dow	
Trade names	-	Elaslen, Tyrin	

CPE chlorinated polyethylene

PARAMETER	UNIT	VALUE	REFERENCES
PHYSICAL PROPERTIES			
Density at 25°C	g cm^{-3}	1.12-1.20	
Bulk density at 20°C	g cm^{-3}	0.39-0.55	
Color	-	off-white	
Odor	-	odorless	
Melting temperature, DSC	°C	108.9-123.3	Stoeva, S, J. Appl. Polym. Sci., 101, 2602-13, 2006.
Decomposition temperature	°C	215-239	Varma, A J; Deshpande, S V; Kondapalli, P, Polym. Deg. Stab., 63, 1-3, 1999.
Glass transition temperature	°C	-10 to 55	Stoeva, S, J. Appl. Polym. Sci., 101, 2602-13, 2006.
Heat of fusion	J g^{-1}	2	
Thermal conductivity	W m^{-1} K^{-1}	0.039	Zhang, Z X, Wang, C, Wang, S, Wen, S, Phule, A D, Radiat. Phys. Chem., 173, 108890, 2020.
Vicat temperature VST/A/50	°C	49.9-71.7	
Hildebrand solubility parameter	MPa$^{0.5}$	19.2 (44% Cl)	
Electric conductivity	S cm^{-1}	0.022 (with 15 wt% carbon nanofiber+Ketjen black)	Mondal, S, Ravindren, R, Bhawal, P, Shi, B, Das, N C, Compos. Part B: Eng., 197, 108071, 2020.
Electromagnetic interference shielding	dB	33 (with 15 wt% carbon nanofiber+Ketjen black), 24 (10 wt% carbon nanofibers), 37-39 (critical thickness of 2 mm with 15 wt% carbon nanofiber+Ketjen black); 42.4	Mondal, S, Ravindren, R, Bhawal, P, Shi, B, Das, N C, Compos. Part B: Eng., 197, 108071, 2020; Mondal, S, Ganguly, S, Das, P, Bhawal, P, Das , N C, Mater. Sci. Eng. B, 225, 140-9, 2017.
MECHANICAL & RHEOLOGICAL PROPERTIES			
Tensile strength	MPa	6.0-16.8; 3.1	Guo, Z; Ran, S; Fang, Z, Compos. Sci. Technol., 86, 157-63, 2013.
Elongation	%	550-1,000	
Elastic modulus	MPa	1.5	Guo, Z; Ran, S; Fang, Z, Compos. Sci. Technol., 86, 157-63, 2013.
Shore A hardness	-	47-70; 44	-; Mondal, S, Ravindren, R, Bhawal, P, Shi, B, Das, N C, Compos. Part B: Eng., 197, 108071, 2020.
Shore D hardness	-	46-48	
Brittleness temperature (ASTM D746)	°C	-55 to <-70	
Mooney viscosity	-	64-115	
Melt viscosity, shear rate=1000 s^{-1}	Pa s	800-2,900	
Melt index, 180°C/21.6 kg	g/10 min	0.1-25	
CHEMICAL RESISTANCE			
Alcohols	-	good	
Aromatic hydrocarbons	-	poor	
Esters	-	good	
Halogenated hydrocarbons	-	poor	
Ketones	-	good	
Good solvent	-	chlorobenzene, cyclohexanone, tetrachloroethylene, toluene, xylene	

CPE chlorinated polyethylene

PARAMETER	UNIT	VALUE	REFERENCES
Non-solvent	-	ketones, alcohols, esters	
FLAMMABILITY			
Limiting oxygen index	% O_2	29-33	
Volatile products of combustion	-	HCl, H_2, CH_4, CO_2, H_2O	Whiteley, M J; Pan, W-P, Thermochim. Acta., 166, 27-39, 1990.
TOXICITY			
Carcinogenic effect	-	not listed by ACGIH, NIOSH, NTP	
Oral rat, LD_{50}	mg kg^{-1}	>8,000; 920	
PROCESSING			
Typical processing methods	-	extrusion, mixing, molding, peroxide vulcanization	
Preprocess drying: temperature/time/residual moisture	ºC/h/%	80-85/4/-	
Processing temperature	ºC	145-165	
Additives used in final products	-	Fillers: calcium carbonate, carbon black, clay, silica, magnesium oxide (used as thermal stabilizer; typically 5-10 phr), titanium dioxide; Antistatics: polymers of ethylene oxide and epihalohydrin	
Applications	-	autoignition wire, automotive air ducts and hoses, car axle boots, EMI shielding composites, fiber optic cable, impact modification for PVC in pipe, power steering hose, roofing membranes, technical hoses, transmission oil cooler hose, vinyl siding, window profiles and FR ABS, wire and cable	
Outstanding properties	-	fire resistance, impact resistance, solvent resistance	
BLENDS			
Suitable polymers	-	ABS, ASA, ENR, EVAC, PMMA, PVC	Zhang, Z; Zhu, W; Zhang, J; Tian, T, Polym. Testing., 44, 23-9, 2015; Mao, Z, Zhang, X, Jiang, G, Zhang, J, Polym. Testing, 73, 21-30, 2019.
Compatibilizers	-	ENR	
ANALYSIS			
FTIR (wavenumber-assignment)	cm^{-1}/-	C-Cl – 660, 609; CH_2 – 1263, 1469	O'Keefe, J F, Rubber World, June 2004, 27-37.

CPVC chlorinated poly(vinyl chloride)

PARAMETER	UNIT	VALUE	REFERENCES
GENERAL			
Common name	-	poly(vinyl chloride), chlorinated	
CAS name	-	ethene, chloro-, homopolymer, chlorinated	
Acronym	-	CPVC	
CAS number	-	68648-82-8	
Linear formula		F01930000	
HISTORY			
Person to discover	-	Schoenburg, C of IG Farbenindustrie	Schoenburg, C, US Patent 1,982,765, IG Farbenindustrie, 1934.
Date	-	1934	
Details	-	product containing 64-68% chlorine was obtained	
SYNTHESIS			
Monomer(s) structure	-	PVC, chlorine	
Monomer(s) CAS number(s)	-	9002-86-2; 7782-50-5	
Monomer(s) molecular weight(s)	dalton, g/mol, amu	variable; 35.453	
Chlorine content	%	63-74	
Method of synthesis	-	chlorination is performed by free radical process; initiation occurs due to thermal or UV energy which decomposes chlorine gas to radicals	
Pressure of polymerization	Pa	pressure affects diffusion of chlorine and thus the rate of chlorination	Barriere, B; Glotin, M; Leibler, L, J. Polym. Sci., B: Polym. Phys., 38, 3201-9, 2000.
COMMERCIAL POLYMERS			
Some manufacturers	-	Avient, Sekisui	
Trade names	-	Geon, Durastream	
PHYSICAL PROPERTIES			
Density at 20°C	g cm^{-3}	1.47-1.56	
Bulk density at 20°C	g cm^{-3}	0.64-0.68	
Melting temperature, DSC	°C	199-212	
Thermal expansion coefficient, 23-80°C	10^{-4} °C^{-1}	0.8	
Thermal conductivity, melt	W m^{-1} K^{-1}	0.14-0.48	
Glass transition temperature	°C	103-135	Merah, N; Al-Qahtani, T; Khan, Z, Plast. Rubber Composites, 37, 8, 353-58, 2008.
Specific heat capacity	J K^{-1} kg^{-1}	900	
Long term service temperature	°C	80	
Heat deflection temperature at 1.8 MPa	°C	100-110	
Vicat temperature VST/B/50	°C	106-115	
Relative permittivity at 60 Hz	-	3.7	
Volume resistivity	ohm-m	3.4x10^{15}	

CPVC chlorinated poly(vinyl chloride)

PARAMETER	UNIT	VALUE	REFERENCES
Electric strength K20/P50, d=0.60.8 mm	kV mm^{-1}	49	

MECHANICAL & RHEOLOGICAL PROPERTIES

PARAMETER	UNIT	VALUE	REFERENCES
Tensile strength	MPa	50-80	
Tensile modulus	MPa	2,590-3,030	
Tensile stress at yield	MPa	49-56	
Elongation	%	20-40	
Flexural strength	MPa	92-108	
Flexural modulus	MPa	2,700-3,500	
Elastic modulus	MPa	2,900	Merah, N, J. Mater. Process. Technol., 191, 198-201, 2007.
Compressive strength	MPa	69-100	
Young's modulus	MPa	2,900-3,400	
Izod impact strength, notched, 23°C	J m^{-1}	80-450	
Shore D hardness	-	80-84	
Rockwell hardness	-	R110-119	
Water absorption, equilibrium in water at 23°C	%	0.03-0.4; 5-15 (pipes in hot water)	Barthelemy, E; Munier, C; Verdu, J, J. Mater. Sci. Lett., 20, 1143-45, 2001.

CHEMICAL RESISTANCE

PARAMETER	UNIT	VALUE	REFERENCES
Acid dilute/concentrated	-	good	
Alcohols	-	fair	
Alkalis	-	good	
Aliphatic hydrocarbons	-	poor	
Aromatic hydrocarbons	-	poor	
Esters	-	poor	
Greases & oils	-	good	
Halogenated hydrocarbons	-	poor	
Ketones	-	poor	
Good solvent	-	acetone, aromatic hydrocarbons, butyl acetate, chlorobenzene, chloroform, cyclohexanone, dioxane, DMF, DMSO, nitrobenzene, THF	
Non-solvent	-	aliphatic and cycloaliphatic hydrocarbons, carbon tetrachloride, methyl acetate, nitromethane, organic and inorganic acids	

FLAMMABILITY

PARAMETER	UNIT	VALUE	REFERENCES
Ignition temperature	°C	482	
Autoignition temperature	°C	>399	
Limiting oxygen index	% O$_2$	53-60	
Volatile products of combustion	-	HCl, CO, CO$_2$	
UL 94 rating	-	V-0	

WEATHER STABILITY

PARAMETER	UNIT	VALUE	REFERENCES
Important initiators and accelerators	-	24 month exposure in Saudi Arabia decreased tensile strength by 43% and in Florida by 26%	Merah, N, J. Mater. Process. Technol., 191, 198-201, 2007.

CPVC chlorinated poly(vinyl chloride)

PARAMETER	UNIT	VALUE	REFERENCES
TOXICITY			
NFPA: Health, Flammability, Reactivity rating	-	2/1/0	
Carcinogenic effect	-	not listed by ACGIH, NIOSH, NTP	
TLV, ACGIH	mg m^{-3}	10	
PROCESSING			
Typical processing methods	-	extrusion; injection molding; pipe extrusion	
Processing temperature	°C	190-210 (extrusion)	
Additives used in final products	-	Plasticizers: seldom used (e.g., 1,4-cyclohexane dimethanol dibenzoate (Benzoflex R 352)) and phthalates and phosphates; Release: ester of fatty acid, oxidized polyethylene	
Applications	-	fittings, industrial (ducts, pumps, scrubers, strainers, tanks, valves), photocatalysis, pipes	Su, Y, Wang, C, Zhang, Y, Yang, z, Dionysiou, D D, Mater. Res. Bull., 118, 110524, 2019.
Outstanding properties	-	frame resistance, thermal resistance	
BLENDS			
Suitable polymers	-	ABS polyester, PMMA, PVC, PVP, SAN	Kang, J S; Kim, K Y; Lee, Y M, J. Membrane Sci., 214, 311-21, 2003.

CR polychloroprene

PARAMETER	UNIT	VALUE	REFERENCES
GENERAL			
Common name	-	polychloroprene, neoprene	
CAS name	-	1,3-butadiene, 2-chloro-, homopolymer	
Acronym	-	CR	
CAS number	-	9010-98-4	
RTECS number	-	EI9640000	
HISTORY			
Person to discover	-	Wallace Carothers and Julius Arthur Nieuwland	
Date	-	April 17, 1930	
Details	-	polychloroprene was invented by DuPont scientists on April 17, 1930 after Dr. Elmer K. Bolton of DuPont laboratories attended a lecture by Fr. Julius Arthur Nieuwland, a professor of chemistry at the University of Notre Dame. Fr. Nieuwland's research was focused on acetylene chemistry and during the course of his work he produced divinyl acetylene, a jelly that firms into an elastic compound similar to rubber when passed over sulfur dichloride. After DuPont purchased the patent rights from the university, Wallace Carothers of DuPont took over commercial development of Nieuwland's discovery in collaboration with Nieuwland himself. DuPont focused on monovinyl acetylene and reacted the substance with hydrogen chloride gas, manufacturing chloroprene.	
SYNTHESIS			
Monomer(s) structure	-	chloroprene, C_4H_5Cl	
Monomer(s) CAS number(s)	-	126-99-8	
Monomer(s) molecular weight(s)	dalton, g/mol, amu	88.54	
Monomer ratio	-	100%	
Formulation example	-	n-dodecyl mercaptan (and sometimes xanthogen disulfide) is used as a chain transfer agent in linear grades; slow crystallizing grades are copolymerized with 2,3-dichloro-1,3-butadiene	
Method of synthesis	-	butadiene is converted into the monomer 2-chlorobutadiene-1,3 (chloroprene) via 3,4-dichlorobutene-1, and monomer is then polymerized by free radical emulsion polymerization using batch or continuous process. The polymerization is stopped at desired conversion by stopping agent. Finally the latex is freeze-coagulated to form a thin sheet. After washing and drying, it is shaped into a rope and chopped to chips or granules.	
Heat of polymerization	J g^{-1}	768	
Mass average molecular weight, M_w	dalton, g/mol, amu	140,000	Le Gac, P Y; Roux, G; Verdu, J; Davies, P; Fayolle, B, Polym. Deg. Stab., 109, 175-83, 2014.
Molar volume at 298K	cm^3 mol^{-1}	65.0 (crystalline)	
Van der Waals volume	cm^3 mol^{-1}	45.6 (crystalline)	
STRUCTURE			
Crystallinity	%	18-34	
Cell type (lattice)	-	monoclinic	

CR polychloroprene

PARAMETER	UNIT	VALUE	REFERENCES
Cell dimensions	nm	a:b:c=1.325:0.763:1.415 (macromer); 0.917:0.992:1.22 (crosslinked rubber)	
Crosslink density	mol kg^{-1}	0.18; 0746 10^4 mol cm^{-3} (186 monomers/chain)	Le Gac, P Y; Roux, G; Davies, P; Fayolle, B; Verdu, J, Polymer, 55, 2861-6, 2014; Le Gac, P-Y, Albouy, P-A, Petermann, D, Polymer, 142, 209-17, 2018.
Tacticity	%	trans: 70-90 (cis - 5-10)	
Rapid crystallization temperature	$^{\circ}$C	-5	

COMMERCIAL POLYMERS

Some manufacturers	-	BRP Manufacturing, DuPont, ExxonMobile, Bayer, EniChem, Sumitomo, Tosoh	
Trade names	-	Neoprene	

PHYSICAL PROPERTIES

Density at 20°C	g cm^{-3}	1.22-1.25	
Color	-	white, amber, gray	
Refractive index, 20°C	-	1.552-1.558	
Odor	-	odorless, mild	
Melting temperature, DSC	$^{\circ}$C	45-92; 70 (cis); 80-115 (trans)	
Decomposition temperature	$^{\circ}$C	>200	
Service temperature	$^{\circ}$C	-35 to 100	Eslami, H, C, Mekonnen, T H, Compos. Part C: Open Access, 1, 100009, 2020.
Thermal expansion coefficient, 23-80°C	$^{\circ}$C^{-1}	6E-4	
Thermal conductivity, 20°C	W m^{-1} K^{-1}	0.15-0.19	
Glass transition temperature	$^{\circ}$C	-25 to -46; -20 (cis)	
Specific heat capacity	J K^{-1} kg^{-1}	2200	
Heat of fusion	kJ mol^{-1}	8.37	
Hansen solubility parameters, δ_D, δ_P, δ_H	MPa$^{0.5}$	18.1, 4.3, 6.7	
Interaction radius		8.9	
Hildebrand solubility parameter	MPa$^{0.5}$	calc.=16.59-19.19; exp.=17.6-19.13	
Surface tension	mN m^{-1}	43.8	Wu, S, Adhesion, 5, 39, 1973.
Dielectric constant at 100 Hz/1 MHz	-	5-9	
DC conductivity	S cm^{-1}	2.84	Maya, M G, Soney, C G, Thomasukutty, J, Lekshmi, K, Sabu, T, Polym. Testing, 65, 256-63, 2018.
Surface resistivity	ohm	9E6 to 8.4E10 (antistatic)	
Permeability to nitrogen, 25°C	cm^3 cm cm^{-2} s^{-1} Pa^{-1} x 10^{12}	0.088	
Permeability to oxygen, 25°C	cm^3 cm cm^{-2} s^{-1} Pa^{-1} x 10^{12}	0.296	

CR polychloroprene

PARAMETER	UNIT	VALUE	REFERENCES
Permeability to water vapor, 25°C	cm^3 cm cm^{-2} s^{-1} Pa^{-1} x 10^{12}	68.3	
Diffusion coefficient of nitrogen	cm^2 s^{-1} $x10^6$	0.24	
Diffusion coefficient of oxygen	cm^2 s^{-1} $x10^6$	0.38	
Surface free energy	mJ m^{-2}	40.9	

MECHANICAL & RHEOLOGICAL PROPERTIES

PARAMETER	UNIT	VALUE	REFERENCES
Tensile strength	MPa	4.7-20.9	
Tensile stress at yield	MPa	0.57	
Tensile modulus	MPa	1.77 (can be increased with 1.5 wt% reduced graphene oxide to 3.97)	Maya, M G, Soney, C G, Thomasukutty, J, Lekshmi, K, Sabu, T, Polym. Testing, 65, 256-63, 2018.
Elongation	%	380-1086	
Tear strength	N mm^{-1}	8.8-50	
Young's modulus	MPa	1.3, 19.9 (3 wt% modified cellulose nanocrystals)	Eslami, H, C, Mekonnen, T H, Compos. Part C: Open Access, 1, 100009, 2020.
Compression set, 24h 70°C	%	10-32	
Shore A hardness	-	42-85	
Brittleness temperature (ASTM D746)	°C	-35 to -55	
Mooney viscosity	-	34-59	
Water absorption, equilibrium in water at 23°C	%	0.9	

CHEMICAL RESISTANCE

PARAMETER	UNIT	VALUE	REFERENCES
Acid dilute/concentrated	-	good	
Alcohols	-	good	
Alkalis	-	good	
Aliphatic hydrocarbons	-	good	
Aromatic hydrocarbons	-	fair to poor	
Esters	-	fair to poor	
Greases & oils	-	poor	
Ketones	-	fair to poor	
Θ solvent, Θ-temp.=	-	butanone, cyclohexane, trans-decalin	

FLAMMABILITY

PARAMETER	UNIT	VALUE	REFERENCES
Ignition temperature	°C	>260	
Limiting oxygen index	% O_2	28-47; 46-59 (with FR)	Hornsby, P R, Cusack, P A, Antec, 3310-12, 1998; Hornsby, P R; Mitchell, P A; Cusack, P A, Polym. Deg. Stab., 32, 299-312, 1991.
Heat release	kW m^{-2}	314	
NBS smoke chamber	Ds	800	Hornsby, P R; Mitchell, P A; Cusack, P A, Polym. Deg. Stab., 32, 299-312, 1991.

CR polychloroprene

PARAMETER	UNIT	VALUE	REFERENCES
Char at 700°C	%	24.53	Vijayan, D, Mathiazhagan, A, Joseph, R, Polymer, 132, 143-56, 2017.
WEATHER STABILITY			
Important initiators and accelerators	-	$FeCl_3$	Freitas, A R; Vidotti, G J; Rubira, A F; Muniz, E C, Polym. Deg. Stab., 87, 425-32, 2005.
Products of degradation	-	conjugated double bonds	
Stabilizers	-	UVA: dialkyl aryl substituted triazine; Screener: carbon black; Phenolic antioxidant: isotridecyl-3-(3,5-di-tert-butyl-4-hydroxy-phenyl) propionate; 2,2'-isobutylidenebis(2,4-dimethylphenol); phenol, 4-methyl-, reaction products with dicyclopentadiene and isobutene; Thiosynergist: 4,6-bis(dodecylthiomethyl)-o-cresol; Amine: nonylated diphenylamine	
TOXICITY			
NFPA: Health, Flammability, Reactivity rating	-	0/1/0	
Carcinogenic effect	-	not listed by ACGIH, NIOSH, NTP	
TLV, ACGIH	mg m^{-3}	2 (talc)	
OSHA	mg m^{-3}	3.3 (talc)	
Oral rat, LD_{50}	mg kg^{-1}	>5,000; >20,000	
PROCESSING			
Typical processing methods	-	calendering, compounding in solution, dip coating, extrusion, molding (compression, injection), sheeting, vulcanization	
Processing temperature	°C	50-100 (sheet calendering), 40-100 (extrusion)	
Additives used in final products	-	Fillers: carbon black, fumed silica, magnesium oxide, zinc oxide in EMI shielding field: montmorillonite, nickel and carbon black, silver, silver coated glass spheres, silver plated copper, silver plated aluminum, silver plated nickel; Other: acid aceptors (MgO, red lead), vulcanizing agent (ZnO), vulcanization accelerator (thioureas, sulfur-based), vulcanization retarder (MBTS, CBS, TMTD), antioxidant (octylated diphenylamine), antiozonant (diaryl-p-phenylene diamines with selected waxes up to 3 phr), plasticizers (aromatic or naphthenic process oils, mono esters, polyester, chlorinated waxes) processing aids (stearic acid, waxes, low molecular weight polyethylene, high-cis polybutadiene, special factices)	Neoprene, a guide to grades, compounding and processing neoprene rubber, DuPont, Oct. 2008.
Applications	-	adhesives, automotive gaskets, bitumen additive, cellular products, construction applications (bridge pads/seals, soil pipe gaskets, waterproof membranes, asphalt modification), CVJ boots and air springs, foamed wet suits, hose, foam, latex dipped goods (gloves, weather balloons, automotive), paper, and industrial binders (shoe board), molded and extruded goods, protective coatings, power transmission belts, sealants, seals, tear-resistant rubber, tubes and covers (auto and industrial), water-swellable rubber, wire and cable jacketing	
Outstanding properties	-	mechanical strength, ozone and weather resistance, low flammability, chemical resistance, good adhesion	
BLENDS			
Suitable polymers	-	NR, epoxidized polyisoprene	Freitas, A R; Gaffo, L; Rubira, A F; Muniz, E C, J. Molec. Liq., 190, 146-50, 2014.

CR polychloroprene

PARAMETER	UNIT	VALUE	REFERENCES
ANALYSIS			
FTIR (wavenumber-assignment)	cm^{-1}/-	C=O – 1725, C=C – 1695, 1660; CH$_2$ – 1444, 1431, C-Cl – 658, 602	O'Keefe, J F, Rubber World, June 2004, 27-37.
x-ray diffraction peaks	degree	see reference	Sathasivam, K; Haris, M; Mohan, S, Intl. J. Chem. Res., 2, 3, 1780-85, 2010.

CSP chlorosulfonated polyethylene

PARAMETER	UNIT	VALUE	REFERENCES
GENERAL			
Common name	-	chlorosulfonated polyethylene	
CAS name	-	chlorosulfonated polyethylene rubber	
Acronym	-	CSP	
CAS number	-	9008-08-6; 68037-39-8	
HISTORY			
Person to discover	-	McQueen, D M, DuPont	McQueen, D M, US Patent 2,212,786, DuPont, 1940.
Date	-	1940; 1951	
Details	-	polyethylene is dissolved or suspended in hot carbon tetrachloride and reacted with SO_2 and Cl_2; begining of marketing by DuPont under the name of Hypalon	
SYNTHESIS			
Monomer(s) structure	-	PE, SO_2, Cl_2	
Monomer(s) CAS number(s)	-	9002-88-4; 7446-09-5; 7782-50-5	
Monomer(s) molecular weight(s)	dalton, g/mol, amu	broad range; 64.07; 70.9	
Chlorine content	%	24-45	
Sulfur content	%	0.8-1.7	
Temperature of polymerization	°C	40-80	Zhao, R; Cheng, S; Shun, Y; Huang, Y, J. Appl. Polym. Sci., 81, 3582-88, 2001.
cure time	min	26.6	Malas, A; Das, C K; Composites Part B, 79, 639-48, 2015.
Mass average molecular weight, M_w	dalton, g/mol, amu	180,000-550,000	
STRUCTURE	-		
Crystallinity	%	16-21	Wang, Z; Ni, H; Bian, Y; Zhang, M; Zhang, H, J. Appl. Polym. Sci., 116, 2095-100, 2010.
Avrami constants, k/n	-	n=4	
COMMERCIAL POLYMERS			
Some manufacturers	-	Jiangxi, HongRun Chemical, Sundow Polymers, Tosoh	
Trade names	-	Acsium, HongRun, Sunpren, Toso	
PHYSICAL PROPERTIES			
Density at 20°C	g cm^{-3}	1.0-1.27	
Color	-	white to slightly yellow	
Odor	-	slight, ether-like	
Melting temperature, DSC	°C	87-140	
Decomposition temperature	°C	150-200	
Glass transition temperature	°C	7 to -27	
Activation energy of thermal degradation	kJ mol^{-1}	106-116 (N_2); 99-101 (air)	Sandelin, M J; Gedde, U W, Polym. Deg. Stab., 86, 331-38, 2004; Gilen, K T; Bernstein, R; Celina, M, Polym. Deg. Stab., 335-46, 2005.

CSP chlorosulfonated polyethylene

PARAMETER	UNIT	VALUE	REFERENCES
Hansen solubility parameters, δ_D, δ_P, δ_H	$MPa^{0.5}$	18.1, 3.4, 4.9; 18.2, 4.7, 2.0	
Molar volume	$kmol\ m^{-3}$	3.6; 5.0	
Permeability to oxygen, 25°C	$cm^3\ mm$ $m^{-2}\ day^{-1}$ atm^{-1}	60-78	

MECHANICAL & RHEOLOGICAL PROPERTIES

PARAMETER	UNIT	VALUE	REFERENCES
Tensile strength	MPa	2.5-31.7	
Elongation	%	260-820	
Tear strength	$kN\ m^{-1}$	2.5-44	
Abrasion resistance (ASTM D1044)	mg/1000 cycles	99-243	
Compression set, 22h 100°C	%	21-86	
Shore A hardness	-	40-75	
Shore D hardness	-	45-98	
Brittleness temperature (ASTM D746)	°C	8 to -44	
Mooney viscosity	-	28-94	

CHEMICAL RESISTANCE

PARAMETER	UNIT	VALUE	REFERENCES
Alcohols	-	very good	
Aromatic hydrocarbons	-	poor	
Greases & oils	-	good	
Halogenated hydrocarbons	-	fair-poor	
Ketones	-	poor	
Good solvent	-	chlorinated hydrocarbons, MEK, THF, toluene	

FLAMMABILITY

PARAMETER	UNIT	VALUE	REFERENCES
Limiting oxygen index	$\%\ O_2$	25; 26 (35 wt% Cl)	
Volatile products of combustion	-	HCl, CO, CO_2	

WEATHER STABILITY

PARAMETER	UNIT	VALUE	REFERENCES
Products of degradation	-	HCl, double bonds formation, crosslinking, yellowing	
Stabilizers	-	carbon black, tetrakis(methylene (3,5-di-tert-butyl-4-hydroxy-hydrocinnamate))methane,2 antiacids	

TOXICITY

PARAMETER	UNIT	VALUE	REFERENCES
Carcinogenic effect	-	IARC 2B, NTP X, ACGIH A2 (carbon tetrachloride present in concentration of 0.4%)	
Mutagenic effect	-	tests on bacterial or mammalian cell cultures did not show mutagenic effects	
Teratogenic effect	-	animal testing showed effects on embryo-foetal development at levels equal to or above those causing maternal toxicity	
TLV, ACGIH	ppm	2 (HCl)	
OSHA	ppm	5 (HCl)	

CSP chlorosulfonated polyethylene

PARAMETER	UNIT	VALUE	REFERENCES
Oral rat, LD$_{50}$	mg kg^{-1}	>20,000	
PROCESSING			
Typical processing methods	-	coating, extrusion, compounding, injection molding, vulcanization	
Additives used in final products	-	Acid acceptor (hydrotalcite, magnesia, calcium hydroxide); Fillers: aluminum powder, basic magnesium carbonate, calcium carbonate, carbon black, carbonyl iron powder, graphite, magnesium hydroxide, metal oxides (typically MgO) are used as curing agents, silica; Flame retardant (antimony oxide, hydrated alumina, halogenated hydrocarbons); Plasticizers: seldom used (e.g., polyethylene glycol, dioctyl sebacate, or dioctyl adipate used in small quantities as process oil); Processing aids (waxes, stearic acid, low MW PE, polyethylene glycol); Antistatics: glycerol monostearate, trineoalkoxy amino zirconate, trineoalkoxy sulfonyl zirconate; Release: fluorocarbon, polydimethylsiloxane, silicone coated paper; Vulcanizing agent, sulfur, TMTD, MBTS, NBC. DOTG, peroxide plus coagent, HVA-2 plus coagent)	
Applications	-	adhesives, automotive components (high-temperature timing belts, power steering pressure hose, gaskets, spark plugs), boots, coated fabrics, industrial effluent pit liners and lining for chemical processing equipment, industrial products (hose, rolls, seals, gaskets, diaphragms), inflatable boats, microwave absorbing rubber, pool liners, radiator and heater hoses, roofing, wire and cable	
Outstanding properties	-	resistance to ozone, heat, weather, oxygen and oils and high tensile and abrasion resistance	
BLENDS			
Suitable polymers	-	EPDM, EPR, HNBR, NR, PVC	
ANALYSIS			
FTIR (wavenumber-assignment)	cm^{-1}/-	SO$_2$ (asymmetric stretch) – 1369, SO$_2$ (symmetric stretch) – 1160	O'Keefe, J F, Rubber World, June 2004, 27-37.

CTA cellulose triacetate

PARAMETER	UNIT	VALUE	REFERENCES
GENERAL			
Common name	-	cellulose triacetate	
Acronym	-	CTA	
CAS number	-	9012-09-3	
HISTORY			
Person to discover	-	Walker, W H	Walker, W H, US Patent, 774,714, Nov. 8, 1904.
Date	-	1904; 1954	
Details	-	Walker patented production of cellulose acetate; first commercially produced by Celanese and Eastman Chemical	
SYNTHESIS			
Monomer(s) structure	-	cellulose; acetic acid	
Monomer(s) CAS number(s)	-	9004-34-6; 64-19-7	
Monomer(s) molecular weight(s)	dalton, g/mol, amu	depends on raw material; 60.05	
Acetyl content	%	42.5-43.6; 60.7-61.2 (combined)	
Hydroxy content	%	0.82 (min. 92% hydroxyl groups must be acetylated)	
Degree of substitution		0.89-2.84	El Nemr, A; Ragab, S; El Sikaily, A; Khaled, A, Carbohydrate Polym., 130, 41-8, 2015.
Method of synthesis	-	see cellulose acetate	
Catalyst	-	N-iodosuccinimide	El Nemr, A; Ragab, S; El Sikaily, A; Khaled, A, Carbohydrate Polym., 130, 41-8, 2015.
Number average molecular weight, M_n	dalton, g/mol, amu	30,000-350,000	Fredercik, T J; Godfrey, D A, US Patent 6,683,174, Eastman, 2004; Shimada, H; Nobucawa, S; Yamaguchi, M, Carbohydrate Polym., 120, 22-8, 2015.
Mass average molecular weight, M_w	dalton, g/mol, amu	14,000-408,000	Fredercik, T J; Godfrey, D A, US Patent 6,683,174, Eastman, 2004.
Polydispersity, M_w/M_n	-	3.32-6.11	
Polymerization degree (number of monomer units)	-	165-1670	
Radius of gyration	nm	7.01-7.55	Nair, P R M; Gohil, R M; Patel, K C; Patel, R D, Eur. Polym. J., 13, 273-76, 1977.
STRUCTURE	-		
Crystallinity	%	24-40; 42 (210ºC), 72.2 (250ºC), 99.5 (290ºC)	Hindeleh, A M; Johnson, D J, Polymer, 11, 12, 666-680, 1970.
Cell type (lattice)	-	monoclinic	
Cell dimensions	nm	a=0.5939, b=1.1431, c=1.046	Sikorski, P; Wada, M; Heux, L; Shintani, H; Stokke, B T, Macromolecules, 37, 12, 4547-53, 2004.
Unit cell angles	degree	γ=95.4	Sikorski, P; Wada, M; Heux, L; Shintani, H; Stokke, B T, Macromolecules, 37, 12, 4547-53, 2004.
Number of chains per unit cell	-	2	
Crystallite size	nm	10-20 molecules	

CTA cellulose triacetate

PARAMETER	UNIT	VALUE	REFERENCES
Polymorphs	-	I (one chain monoclinic), II, N	Zugenmaier, P, Macromol. Symp., 208, 81-166, 2004; Numata, Y; Kumagai, H; Kono, H; Erata, T; Takai, M, Sen'I Gakkaishi, 60, 3, 75-80, 2004.
Chain conformation	-	2/1 helix	Sikorski, P; Wada, M; Heux, L; Shintani, H; Stokke, B T, Macromolecules, 37, 12, 4547-53, 2004.

COMMERCIAL POLYMERS

Some manufacturers	-	Eastman	
Trade names	-	Cellulose Triacetate	

PHYSICAL PROPERTIES

Density at 20°C	g cm^{-3}	1.28-1.34	
Color	-	off white	
Refractive index, 20°C	-	1.472-1.475	
Birefringence	-	-0.003; 0.051 to -0.30 (elongation ratio 0 to 30%)	Hayakawa, D; Ueda, K, Carbohydrate Res., 402, 146-51, 2015.
Transmittance	%	89-93	
Haze	%	0.2-7.6	
Odor	-	odorless	
Melting temperature, DSC	°C	230	
Softening point	°C	190-205	
Decomposition temperature	°C	>240	
Fusion temperature	°C	260	
Thermal expansion coefficient, 23-80°C	10^{-4} °C^{-1}	1.-1.5	
Glass transition temperature	°C	120-195	
Specific heat capacity	J K^{-1} kg^{-1}	1,500	
Molar volume	kmol m^{-3}	3.6; 5.0	
Hildebrand solubility parameter	MPa$^{0.5}$	18.84-19.4	Shimada, H; Nobucawa, S; Yamaguchi, M, Carbohydrate Polym., 120, 22-8, 2015.
Dielectric constant at 100 Hz/1 MHz	-	3.0-4.5	
Dielectric loss factor at 1 kHz	-	0.01-0.02	
Volume resistivity	ohm-m	1E11 to 1E13	

MECHANICAL & RHEOLOGICAL PROPERTIES

Tensile strength	MPa	28-56	
Elongation	%	20-50	
Flexural strength	MPa	42-69	
Izod impact strength, notched, 23°C	J m^{-1}	1.1-4.5	
Intrinsic viscosity, 25°C	dl g^{-1}	1.7	
Water absorption, equilibrium in water at 23°C	%	2-7	
Moisture absorption, equilibrium 23°C/50% RH	%	3	

CTA cellulose triacetate

PARAMETER	UNIT	VALUE	REFERENCES
CHEMICAL RESISTANCE			
Acid dilute/concentrated	-	good/poor	
Aliphatic hydrocarbons	-	resistant	
Aromatic hydrocarbons	-	resistant	
Esters	-	non-resistant	
Greases & oils	-	resistant	
Ketones	-	non-resistant	
Θ solvent, Θ-temp.=27	-	acetone	
Good solvent	-	chloroform, dioxane, ethyl acetate, ethylene carbonate, methyl acetate, methylene chloride, THF, trichloroethane	
Non-solvent	-	aliphatic hydrocarbons, aliphatic ethers, chlorobenzene, dichloroethane, ethanol (absolute), MIBK, weak mineral acids	
FLAMMABILITY			
Ignition temperature	^{o}C	>540	
Limiting oxygen index	% O_2	18.4	
Heat of combustion	J g^{-1}	20,230	
THERMAL STABILITY			
Activation energy	kJ mol^{-1}	70.7-90.2	Ahmad, I R, Cane, C, Townsend, J H, Triana, C, Curran, K, Polym. Deg. Stab., 172, 109050, 2020.
Product of degradation	-	acetic acid	Ahmad, I R, Cane, C, Townsend, J H, Triana, C, Curran, K, Polym. Deg. Stab., 172, 109050, 2020.
BIODEGRADATION			
Typical biodegradants	-	bacterium *Sphingomonas paucimobilis*	Abrusci, C; Marquina, D; Santos, A; Del Amo, A; Corralees, T; Catalina, F, Int. Biodet. Biodeg., 63, 759-64, 2009.
TOXICITY			
NFPA: Health, Flammability, Reactivity rating	-	0/2/0; 1/1/0 (HMIS)	
Carcinogenic effect	-	not listed by ACGIH, NIOSH, NTP	
OSHA	mg m^{-3}	5 (respirable dust) 15 t(total dust)	
PROCESSING			
Additives used in final products	-	glycerin, triethyl citrate, dioctyl phthalate, diisodecyl adipate	Azadimanesh, F; Mohammadi, N, Carbohydrate Polym., 130, 316-24, 2015; Shimada, H; Nobucawa, S; Yamaguchi, M, Carbohydrate Polym., 120, 22-8, 2015.
Applications	-	clothing, coatings, consumer electronics, drug delivery, LCD displays, membranes, photographic films, protective film for polarizing plate	
Outstanding properties	-	low shrinkage, wrinkle resistant, optical clarity	

CTA cellulose triacetate

PARAMETER	UNIT	VALUE	REFERENCES
ANALYSIS			
FTIR (wavenumber-assignment)	cm^{-1}/-	C=O – 1735; C-O – 1216, 1029	
NMR (chemical shifts)	ppm	carbonyl carbon 170.9, 172.2; ring carbon 80.6, 76.3, 72.9, 62.7; methyl carbon 23.2, 22.3	

CY cyanoacrylate

PARAMETER	UNIT	VALUE	REFERENCES
GENERAL			
Common name	-	cyanoacrylate (super glue)	
ACS name	-	2-propenoic acid, 2-cyano-, ethyl ester, homopolymer	
Acronym	-	CY	
CAS number	-	123-31-9 (methyl); 7085-85-0 (ethyl); 25067-30-5 (2-propenoic); 6606-65-1 (butyl); 6701-17-3 (octyl)	
EC number	-	230-391-5 (ethyl)	
HISTORY			
Person to discover	-	Harry Coover and Fred Joyner	
Date	-	1942	
Details	-	discovered that after determination of refractive index of monomer the prisms of refractometer could not any longer be separated	
SYNTHESIS			
Monomer(s) structure	-	$CH_2=C(CN)C(O)OCH_3$; $CH_2=C(CN)C(O)OCH_2CH_3$; $CH_2=C(CN)C(O)O(CH_2)_3CH_3$	
Monomer(s) CAS number(s)	-	137-05-3; 7085-85-0; 6606-65-1	
Monomer(s) molecular weight(s)	dalton, g/mol, amu	111.1; 125.13; 153.18	
Method of synthesis	-	reaction of cyanoacrylic acid with formaldehyde in aqueous solution in the presence of basic condensation catalyst; polymerization is conducted in the presence of initiator (e.g., N,N'-dimethyl-p-toluidine), the reaction is spontaneous	
Initiation rate constant	s^{-1}	1E10	Zhou, Y; Bei, F; Ji, H; Yang, X; Lu, L; Wang, X, J. Molecular Structure, 737, 117-23, 2005.
Propagation rate constant	s^{-1}	100-500	Zhou, Y; Bei, F; Ji, H; Yang, X; Lu, L; Wang, X, J. Molecular Structure, 737, 117-23, 2005.
Number average molecular weight, M_n	dalton, g/mol, amu	852,000-1,112,000	Han, M G; Kim, S; Liu, S X, Polym. Deg. Stab., 93, 1243-51, 2008.
Mass average molecular weight, M_w	dalton, g/mol, amu	954,000-1,440,000	Han, M G; Kim, S; Liu, S X, Polym. Deg. Stab., 93, 1243-51, 2008.
Polydispersity, M_w/M_n	-	1.09-1.35	Han, M G; Kim, S; Liu, S X, Polym. Deg. Stab., 93, 1243-51, 2008.
STRUCTURE			
Crystallinity	%	amorphous	
COMMERCIAL POLYMERS			
Some manufacturers	-	Cyberbond; Elmer; Permabond	
Trade names	-	Apollo; Alpha; Permabond	
PHYSICAL PROPERTIES			
Density at 20°C	g cm^{-3}	1.05-1.08	
Refractive index, 20°C	-	1.483 (ethyl CY); 1.479 (butyl); 1.482 (hexyl)	Shankland, K; Whateley, T L, J. Colloid Interface Sci., 154, 1, 160-6, 1992.
Transmittance	%	80	

CY cyanoacrylate

PARAMETER	UNIT	VALUE	REFERENCES
Softening point	°C	150	
Decomposition temperature	°C	160; 300 (completely degraded)	Han, M G; Kim, S; Liu, S X, Polym. Deg. Stab., 93, 1243-51, 2008.
Activation energy of thermal degradation	kJ mol^{-1}	8.92	Han, M G; Kim, S; Liu, S X, Polym. Deg. Stab., 93, 1243-51, 2008.
Activation energy of curing	kJ mol^{-1}	458-933 (ethyl cyanoacrylate)	Stefanov, T, Ryan, B, Ivankovic, A, Murphy, N, Int. J. Adhesion Adhesives, 101, 102630, 2020.
Thermal expansion coefficient, 23-80°C	10^{-4} °C^{-1}	1.6	
Thermal conductivity, melt	W m^{-1} K^{-1}	0.1	
Glass transition temperature	°C	50-120; 74 (butyl cyanoacrylate); 131-6 (ethyl cyanoacrylate)	-; -; Stefanov, T, Ryan, B, Ivankovic, A, Murphy, N, Int. J. Adhesion Adhesives, 101, 102630, 2020.
Maximum service temperature	°C	80-120 (typical for adhesives)	
Long term service temperature	°C	-55 to 250; -55 to 80 (typical)	
Dielectric constant at 100 Hz/1 MHz	-	3.3	
Electric strength K20/P50, d=0.60.8 mm	kV mm^{-1}	24.6	
Permeability to water vapor, 25°C	g m^{-2} day^{-1}	1,800-2,100 (octyl, medical)	Zhang, S; Ruiz, R, World Patent 2010/008822, Adhezion Biomedical, 2010.

MECHANICAL & RHEOLOGICAL PROPERTIES

Tensile strength	MPa	22.8-42.0	
Shear strength	MPa	10-42; 10.3-22.1 (ethyl); 24.8 (methyl)	
Adhesive bond strength	MPa	25	
Failure load	kN	2310	
Melt viscosity, shear rate=1000 s^{-1}	mPa s	1-25,000	

CHEMICAL RESISTANCE

Acid dilute/concentrated	-	poor	
Alcohols	-	good	
Alkalis	-	poor	
Aliphatic hydrocarbons	-	good	
Aromatic hydrocarbons	-	good	
Greases & oils	-	good	
Halogenated hydrocarbons	-	good	

FLAMMABILITY

Ignition temperature	°C	77-93	
Autoignition temperature	°C	485	
Volatile products of combustion	-	CO, H_2O, CO_2	

BIODEGRADATION

Typical biodegradants	-	hydrolase	Williams, D F; Zhong, S P, Int. Biodet. Biodeg., 95-130, 1994.

CY cyanoacrylate

PARAMETER	UNIT	VALUE	REFERENCES
TOXICITY			
Carcinogenic effect	-	not listed by ACGIH, NIOSH, NTP	
Cytotoxicity		linear relationship with formation of formaldehyde (higher alkyl homologues, e.g., octyl, less toxic)	Park, D H; Kim, S B; Ahn, K-D; Kim, E Y; Kim, Y J; Han, D K, J. Appl. Polym. Sci., 89, 3272-78, 2003.
TLV, ACGIH	ppm	0.2 (methyl, ethyl)	
Oral rat, LD$_{50}$	mg kg^{-1}	>5,000; 30,000 (octyl)	
Skin rabbit, LD$_{50}$	mg kg^{-1}	>2,000	
PROCESSING			
Typical processing methods	-	compounding	
Additives used in final products	-	Plasticizers: acetyl tributyl citrate, dioctyl phthalate; Release: flujoroaliphatic polymer, silicone oil; Slip: cetyl palmitate, polydimethylsiloxane-trifluoropropylsiloxane; Inhibitor of degradation - acids	
Applications	-	drug delivery applications, fast curing glues (most frequently ethyl for general purpose glues); medical glue (most frequently butyl and octyl), endovascular use, tissue adhesive	Loffroy, R, Diagnostic Interventianal Imaging, in press, 2015.
Outstanding properties	-	fast cure, one component, low viscosity, modified with PLCL gives bioglue	Lim, J I; Kim, J H, Colloids Surf. B: Biointerfaces, 133, 19-23, 2015.
ANALYSIS			
FTIR (wavenumber-assignment)	cm^{-1}/-	CH$_2$ – 2991; C=O – 1750; C-O – 1254; CN – 2249	Zhou, Y; Bei, F; Ji, H; Yang, X; Lu, L; Wang, X, J. Molecular Structure, 737, 117-23, 2005.

DAP poly(diallyl phthalate)

PARAMETER	UNIT	VALUE	REFERENCES
GENERAL			
Common name	-	poly(diallyl phthalate)	
CAS name	-	1,2-benzenedicarboxylic acid, di-2-propenylester, homopolymer	
Acronym	-	DAP	
CAS number	-	25053-15-0	
HISTORY			
Person to discover	-	Dannenberg, H; Adelson, D E	Dannenberg, H; Adelson, D E, US Patent 2,294,286, Shell, Aug. 25, 1942.
Date	-	1942	
SYNTHESIS			
Monomer(s) structure	-	CH$_3$CH$_2$CH$_2$OH	
Monomer(s) CAS number(s)	-	85-44-9; 71-23-8	
Monomer(s) molecular weight(s)	dalton, g/mol, amu	148.1; 60.1	
Method of synthesis	-	obtained by polycondensation of phthalic anhydride and propylene alcohol	
Mass average molecular weight, M$_w$	dalton, g/mol, amu	65,000	
Polydispersity, M$_w$/M$_n$	-	5.9	
COMMERCIAL POLYMERS			
Some manufacturers	-	Cosmic; Rogers	
Trade names	-	DAP; DAP	
PHYSICAL PROPERTIES			
Density at 20°C	g cm^{-3}	1.22	
Color	-	white, off-white	
Refractive index, 20°C	-	1.572	
Odor	-	odorless	
Melting temperature, DSC	°C	80-110	
Softening point	°C	175	
Decomposition temperature	°C	260	Gu, A, Polym. Plast. Technol. Eng., 45, 8, 957-61, 2006.
Thermal expansion coefficient, 23-80°C	10^{-4} °C^{-1}	0.4	
Thermal conductivity, melt	W m^{-1} K^{-1}	0.6	
Glass transition temperature	°C	150-206	
Maximum service temperature	°C	177-204	
Long term service temperature	°C	150-180	
Heat deflection temperature at 0.45 MPa	°C	138-143	

DAP poly(diallyl phthalate)

PARAMETER	UNIT	VALUE	REFERENCES
Heat deflection temperature at 1.8 MPa	°C	143	
Dielectric constant at 1 kHz/1 MHz	-	3.4-3.5/3.1-3.6	
Relative permittivity at 1 MHz	-	5.2	
Dissipation factor at 1 kHz	E-4	80-160	
Dissipation factor at 1 MHz	E-4	120-210	
Volume resistivity	ohm-m	1E8	
Surface resistivity	ohm	1E10	
Electric strength K20/P50, d=0.60.8 mm	kV mm^{-1}	14	
Arc resistance	s	125	

MECHANICAL & RHEOLOGICAL PROPERTIES

Tensile strength	MPa	21-35	
Tensile stress at yield	MPa	29	
Flexural strength	MPa	70-76	
Compressive strength	MPa	150-200	
Young's modulus	MPa	10,000-15,000	
Izod impact strength, notched, 23°C	J m^{-1}	35-160	
Shrinkage	%	0.1-1.2	
Water absorption, 24h at 23°C	%	0.12-0.4	

FLAMMABILITY

UL 94 rating	-	HB	

TOXICITY

Carcinogenic effect	-	not listed by ACGIH, NIOSH, NTP	

PROCESSING

Typical processing methods	-	injection molding	
Processing temperature	°C	135-166	
Processing pressure	MPa	3.5-55 (injection)	
Additives used in final products	-	Fillers: mineral, glass fibers, polyamide fibers	
Applications	-	aviation, automotive, electronic, electrical, instrumentation industries, machinery, textile industry, production of transistors, resistors and tubes, computers, insulating materials	
Outstanding properties	-	dimensional stability, ease of molding, electrical properties	

BLENDS

Suitable polymers	-	PC, PMMA, PVAC, PVC (process aid)	

E-RLPO poly(ethyl acrylate-co-methyl methacrylate-co-triammonioethyl methacrylate chloride)

PARAMETER	UNIT	VALUE	REFERENCES
GENERAL			
Common name	-	poly(ethyl acrylate-co-methyl methacrylate-co-triammonioethyl methacrylate chloride)	
IUPAC name	-	poly(ethyl acrylate-co-methyl methacrylate-co-triammonioethyl methacrylate chloride)	
Acronym	-	E-RLPO	
CAS number	-	33434-24-1 (RL & RS)	
SYNTHESIS			
Monomer ratio (EA:MM:TAM)	-	1:2:0.2	
Mass average molecular weight, M_w	dalton, g/mol, amu	32,000-150,000	
Polydispersity, M_w/M_n	-	1.5	
COMMERCIAL POLYMERS			
Some manufacturers	-	Evonik	
Trade names	-	Eudragit RS & RL	
PHYSICAL PROPERTIES			
Density at 20°C	$g\ cm^{-3}$	0.816-0.836	
Color	-	milky white to light yellow	
Refractive index, 20°C	-	1.380-1.385	
Odor		characteristic, faint	
Film forming temperature (min)	°C	40-45	
Glass transition temperature	°C	50-70	
WEATHER STABILITY			
Spectral sensitivity	nm	196-210	de Oliveira, H P; Tavares, G F; Nogueiras, C; Rieymont, J, Int. J. Pharmaceutics, 55-61, 2009.
PROCESSING			
Typical processing methods	-	spraying, drying	
Processing temperature	°C	90-100 (drying air)	
Applications	-	MUPS, time controlled release coating, pH independent	Abdul, S; Chandewar, A V; Jaiswal, S B, J. Controlled Release, 147, 2-16, 2010.
Outstanding properties	-	low permeability, pH independent swelling, pH independednt drug release, highly flexible	
ANALYSIS			
FTIR (wavenumber-assignment)	$cm^{-1}/-$	ester groups – 1150-1190, 1240-1270; C=O – 173; CH – 1385, 1450, 1475, 2950-3000	

EAA poly(ethylene-co-acrylic acid)

PARAMETER	UNIT	VALUE	REFERENCES
GENERAL			
Common name	-	poly(ethylene-co-acrylic acid)	
ACS name	-	2-propenoic acid, polymer with ethene	
Acronym	-	EAA	
CAS number	-	9010-77-9	
SYNTHESIS			
Monomer(s) structure	-	$H_2C=CH_2$ $H_2C=CHCOH$ (with O double-bonded)	
Monomer(s) CAS number(s)	-	74-85-1; 79-10-7	
Monomer(s) molecular weight(s)	dalton, g/mol, amu	28.05; 72.06	
Acrylic acid content	%	5-38	
Aromacity	%	0-10 of aromatic protons	
Temperature of polymerization	°C	240-300	
Pressure of polymerization	MPa	200-300	
Number average molecular weight, M_n	dalton, g/mol, amu	280-160,000	Wiggins, K M; Bielawski, C W, Polym. Chem., 4, 2239-45, 2013.
Mass average molecular weight, M_w	dalton, g/mol, amu	450-86,000; 17,200	-; Ushiki, I, Yoshino, Y, Hayashi, S, Kihara, S-i, Takishima. S, J. Super-critical Fluids, 158, 104733, 2020.
Polydispersity, M_w/M_n	-	1.1-3.97	McAlduff, M; Reven, L, Macromolecules, 38, 3745-53, 2005.
STRUCTURE	-		
Crystallinity	%	8-37	Zhang, J; Chen, S; Su, J; Shi, X; Jin, J; Wang, X; Xu, Z, J. Therm. Anal. Calorim., 97, 959-67, 2009.
Peak crystallization temperature	°C	85-90	
Avrami constants, k/n	-	-/3-4	Zhang, J; Chen, S; Su, J; Shi, X; Jin, J; Wang, X; Xu, Z, J. Therm. Anal. Calorim., 97, 959-67, 2009.
COMMERCIAL POLYMERS			
Some manufacturers	-	Dow	
Trade names	-	Primacor; Nucrel	
PHYSICAL PROPERTIES			
Density at 20°C	g cm^{-3}	0.92-0.96	
Bulk density at 20°C	g cm^{-3}	0.5-0.6	
Color	-	clear to white to off-white to yellow	
Haze	%	3.7-4	
Gloss, 60°, Gardner (ASTM D523)	%	74-76	
Odor	-	acidic	
Melting temperature, DSC	°C	75-112	
Softening point	°C	92-140	
Thermal degradation	°C	325	
Glass transition temperature	°C	-28 to 86	
Vicat temperature VST/A/50	°C	40-90	

EAA poly(ethylene-co-acrylic acid)

PARAMETER	UNIT	VALUE	REFERENCES
pKa		4.25	Laney, K A, Elastic Modulus of Poly(ethylene-co-acrylic acid) Copolymers and Ionomers, Diss., Princeton, May 2010.
Coefficient of friction	-	0.3; 0.15-0.18 (with slip)	Luo, N; Janorkar, A V; Hirt, D E; Husson, S M; Schwark, D W, J. Appl. Polym. Sci., 97, 2242-48, 2005.
Permeability to oxygen, 25°C	cm^3 mm m^{-2} day^{-1} atm^{-1}	180-550	
Permeability to water vapor, 25°C	g mm m^{-2} atm^{-1} 24 h^{-1}	0.0.37-0.44	

MECHANICAL & RHEOLOGICAL PROPERTIES

PARAMETER	UNIT	VALUE	REFERENCES
Tensile strength	MPa	5.8-24	Li, D; Sur, G S, J. Ind. Eng. Chem., 20, 3122-7, 2014.
Tensile modulus	MPa	22-130	
Tensile stress at yield	MPa	7.2-10	
Elongation	%	390-640	
Flexural modulus	MPa	110	
Young's modulus	MPa	65-115	Valenza, A; Visco, A M; Acierno, D, Polym. Test., 21, 101-9, 2002.
Dart drop impact	g	410	
Elmendorf tear strength	g	270-730	
Shore D hardness	-	50-51	
Melt viscosity, shear rate=1000 s^{-1}	Pa s	0.93-7	
Melt index, 190°C/2.16 kg	g/10 min	1.5-1300	

CHEMICAL RESISTANCE

PARAMETER	UNIT	VALUE	REFERENCES
Alkalis	-	good	
Aliphatic hydrocarbons	-	poor	
Aromatic hydrocarbons	-	poor	

FLAMMABILITY

PARAMETER	UNIT	VALUE	REFERENCES
Ignition temperature	°C	>250 to 340	

WEATHER STABILITY

PARAMETER	UNIT	VALUE	REFERENCES
Stabilizers	-	silica coated ZnO particles shield polymer for UV radiation	Ramasamy, M; Kim, Y J; Gao, H; Yi, D K; An, J H, Mater. Res. Bull., 51, 85-91, 2014.

BIODEGRADATION

PARAMETER	UNIT	VALUE	REFERENCES
Stabilizers	-	benzoyl chloride	Matche, R S; Kulkarni, G; Raj B, J. Appl. Polym. Sci., 100, 3063-68, 2006.

TOXICITY

PARAMETER	UNIT	VALUE	REFERENCES
NFPA: Health, Flammability, Reactivity rating	-	1/1/0; 1/0/0 (HMIS)	

EAA poly(ethylene-co-acrylic acid)

PARAMETER	UNIT	VALUE	REFERENCES
Carcinogenic effect	-	not listed by ACGIH, NIOSH, NTP	
TLV, ACGIH	ppm	2 (acrylic acid)	
OSHA	ppm	10 (acrylic acid)	
Oral rat, LD_{50}	mg kg^{-1}	>2,350; >5,000	
Skin rabbit, LD_{50}	mg kg^{-1}	>2,000	
PROCESSING			
Typical processing methods	-	cast film, extrusion blown film, extrusion coating, lamination	
Processing temperature	°C	193-288 (extrusion); 305-325 (blown film)	
Additives used in final products	-	Slip: erucamide, grafted 12-aminododecamide	Luo, N; Janorkar, A V; Hirt, D E; Husson, S M; Schwark, D W, J. Appl. Polym. Sci., 97, 2242-48, 2005.
Applications	-	packaging multilayer films, resins for hot-melt adhesives, resins for pressure-sensitive adhesives; products: hot-melt packaging, curtain coating, bookbinding, glue stick, masking tapes, carpet tape, mounting tape, paper, strapping tapes, thermoplastic road marking	
Outstanding properties	-	adhesion, environmental stress cracking resistance, optical properties, strength	
BLENDS			
Suitable polymers	-	cellulose, PA6, PE, PP, starch	
ANALYSIS			
FTIR (wavenumber-assignment)	cm^{-1}/-	OH – 3500; C-H – 2925, 2850, 1450, 1465, 1375; C=O – 1710, 1230-1320; C-C – 940	Valenza, A; Visco, A M; Acierno, D, Polym. Test., 21, 101-9, 2002.

EAMM poly(ethyl acrylate-co-methyl methacrylate)

PARAMETER	UNIT	VALUE	REFERENCES
GENERAL			
Common name	-	poly(ethyl acrylate-co-methyl methacrylate)	
IUPAC name	-	poly(ethyl acrylate-co-methyl methacrylate)	
Acronym	-	EAMM	
CAS number	-	9010-88-2	
SYNTHESIS			
Monomer(s) structure	-		
Monomer(s) CAS number(s)	-	140-88-5; 80-62-6	
Monomer(s) molecular weight(s)	dalton, g/mol, amu	100.11; 100.11	
Monomer ratio	-	2:1	
Mass average molecular weight, M_w	dalton, g/mol, amu	750,000	
COMMERCIAL POLYMERS			
Some manufacturers	-	Evonik	
Trade names	-	Eudragit NE 30 D	
Composition information	-	30% aqueous dispersion	
PHYSICAL PROPERTIES			
Density at 20°C	g cm^{-3}	1.15-1.2	
Color	-	milky white	
Odor	-	characteristic, faint	
Melting temperature, DSC	°C	132-149	
Film forming temperature (min)	°C	5	
Decomposition temperature	°C	250	
Glass transition temperature	°C	-8.5	El-Malah, Y; Nazzal, S, Int. J. Pharmaceutics, 357, 219-27, 2008.
MECHANICAL & RHEOLOGICAL PROPERTIES			
Tensile strength	MPa	1	El-Malah, Y; Nazzal, S, Int. J. Pharmaceutics, 357, 219-27, 2008.
Elongation	%	900	El-Malah, Y; Nazzal, S, Int. J. Pharmaceutics, 357, 219-27, 2008.
CHEMICAL RESISTANCE			
Alcohols	-	good	
Aliphatic hydrocarbons	-	good	
FLAMMABILITY			
Ignition temperature	°C	>250	
Autoignition temperature	°C	>400	
Volatile products of combustion	-	CO, CO_2, methyl methacrylate	

EAMM poly(ethyl acrylate-co-methyl methacrylate)

PARAMETER	UNIT	VALUE	REFERENCES
WEATHER STABILITY			
Spectral sensitivity	nm	225, 275	Goepferich, A; Lee, G, J. Controlled Release, 18, 133-44, 1992.
TOXICITY			
Carcinogenic effect	-	not listed by ACGIH, NIOSH, NTP	
Oral rat, LD_{50}	mg kg^{-1}	>5,000	
Skin rabbit, LD_{50}	mg kg^{-1}	>5,000	
PROCESSING			
Typical processing methods	-	spraying, drying	
Processing temperature	°C	90-100 (drying air)	
Applications	-	time controlled release coating, pH independent	
Outstanding properties	-	low permeability, pH independent swelling, highly flexible	

EBAC poly(ethylene-co-butyl acrylate)

PARAMETER	UNIT	VALUE	REFERENCES
GENERAL			
Common name	-	poly(ethylene-co-butyl acrylate)	
IUPAC name	-	poly(ethylene-co-butyl acrylate)	
CAS name	-	2-propenoic acid, butyl ester, polymer with ethene	
Acronym	-	EBAC	
CAS number	-	25750-84-9	
Linear formula	-	$(CH_2CH_2)_x[CH_2CH[CO_2(CH_2)_3CH_3]]_y$	
HISTORY			
Date of discovery	-	1952	
SYNTHESIS			
Monomer(s) structure	-	$H_2C{=}CH_2$ $H_2C{=}CHCOCH_2CH_2CH_2CH_3$ (with O double-bonded above the C)	
Monomer(s) CAS number(s)	-	74-85-1; 141-32-2	
Monomer(s) molecular weight(s)	dalton, g/mol, amu	280.5; 128.2	
Butyl acrylate content	wt%	3-35; 70 (Lotryl 30BA02)	-; da Silva Ribeiro, S P, de Moura Estevão, L R, Veiga Nascimento, R S, Appl. Clay Sci., 143, 399-407, 2017.
COMMERCIAL POLYMERS			
Some manufacturers	-	Arkema; Borealis; DuPont; Repsol	
Trade names	-	Lotryl; Borealis PE FA; Elvaloy AC; Alcudia	
PHYSICAL PROPERTIES			
Density at 20°C	g cm^{-3}	0.923-0.944	
Color	-	white	
Transmittance	%	36-55	
Haze	%	10	
Gloss, 60°, Gardner (ASTM D523)	%	70	
Odor	-	characteristic acrylate	
Melting temperature, DSC	°C	78-99	
Decomposition temperature	°C	330	
Glass transition temperature	°C	-54	
Vicat temperature VST/A/50	°C	41-70	
Coefficient of friction	-	0.8	
MECHANICAL & RHEOLOGICAL PROPERTIES			
Tensile strength	MPa	6-25	
Tensile modulus	MPa	62	
Tensile stress at yield	MPa	4-8	
Elongation	%	350-710	
Tensile yield strain	%	15	
Flexural modulus	MPa	45	
Shore A hardness	-	75-90	

EBAC poly(ethylene-co-butyl acrylate)

PARAMETER	UNIT	VALUE	REFERENCES
Shore D hardness	-	32-34	
Ball indention hardness at 358 N/30 S (ISO 2039-1)	MPa	8	
Melt index, 190°C/2.16 kg	g/10 min	0.35-45	
FLAMMABILITY			
Ignition temperature	°C	430	
Limiting oxygen index	% O_2	19	da Silva Ribeiro, S P, de Moura Estevão, L R, Veiga Nascimento, R S, Appl. Clay Sci., 143, 399-407, 2017.
Volatile products of combustion	-	CO, H_2O, CO_2, organic acids, aldehydes, alcohols	
UL rating	-	V-0 (with ammonium polyphosphate and pentaerythritol)	da Silva Ribeiro, S P, de Moura Estevão, L R, Veiga Nascimento, R S, Appl. Clay Sci., 143, 399-407, 2017.
TOXICITY			
NFPA: Health, Flammability, Reactivity rating	-	0/0/0	
Carcinogenic effect	-	not listed by ACGIH, NIOSH, NTP	
OSHA	mg m^{-3}	5 (respirable), 15 (total)	
PROCESSING			
Typical processing methods	-	extrusion (blown film, cast film, coextrusion), coating	
Processing temperature	°C	160-285; 180 (extrusion)	
Additives used in final products	-	slip	
Applications	-	film, heavy-duty bags, foam, performance booster for other resins, packaging, profile, shrink wrap, wire & cable	
Outstanding properties	-	easy processing, compatible with LDPE	
BLENDS			
Suitable polymers	-	PA6	Balamurugan, G P; Maiti, S N, Eur. Polym. J., 43, 5, 1786-1805, 2007.
ANALYSIS			
NMR (chemical shifts)	ppm	1.27, 2.27. 4.02	

EBCO ethylene-n-butyl acrylate-carbon monoxide terpolymer

PARAMETER	UNIT	VALUE	REFERENCES
GENERAL			
Common name	-	ethylene-n-butyl acrylate-carbon monoxide terpolymer	
Acronym	-	EBCO	
SYNTHESIS			
Monomer(s) structure	-	$H_2C{=}CH_2$ $H_2C{=}CHCOCH_2CH_2CH_2CH_3$ CO	
Monomer(s) CAS number(s)	-	74-85-1; 141-32-2; 630-08-0	
Monomer(s) molecular weight(s)	dalton, g/mol, amu	280.5; 128.2; 28.01	
Monomer ratio	-	65/25/10	
COMMERCIAL POLYMERS			
Some manufacturers	-	DuPont	
Trade names	-	Elvaloy HP	
PHYSICAL PROPERTIES			
Density at 20°C	g cm^{-3}	0.96-1	
Color	-	clear	
Odor	-	mild acrylate-like	
Melting temperature, DSC	°C	59-63	
Decomposition temperature	°C	250	
Glass transition temperature	°C	-50 to -54	
Maximum service temperature	°C	235	
MECHANICAL & RHEOLOGICAL PROPERTIES			
Tensile strength	MPa	2.2-6.2	
Elongation	%	580-1213	
Shore A hardness	-	59-69	
Melt index, 190°C/2.16 kg	g/10 min	8-100	
FLAMMABILITY			
Ignition temperature	°C	430	
Volatile products of combustion	-	CO_2, H_2O, CO, organic acids, aldehydes, alcohols,	
TOXICITY			
OSHA	mg m^{-3}	5 (respirable), 15 (total)	
PROCESSING			
Typical processing methods	-	compounding	
Processing temperature	°C	<235	
Applications	-	non-migratory plasticizer and flexibilizer for numerous polymers	
Outstanding properties	-	modification of polymers	

EBCO ethylene-n-butyl acrylate-carbon monoxide terpolymer

PARAMETER	UNIT	VALUE	REFERENCES
BLENDS			
Compatible polymers	-	ABS, chrorinated polyolefin, PC, PVC	

EC ethyl cellulose

PARAMETER	UNIT	VALUE	REFERENCES
GENERAL			
Common name	-	ethyl cellulose	
CAS name	-	cellulose, ethyl ether	
Acronym	-	EC	
CAS number	-	9007-57-3	
RTECS number	-	FJ5950500	
SYNTHESIS			
Monomer(s) structure	-	cellulose, CH_3CH_2Cl; NaOH	
Monomer(s) CAS number(s)	-	9004-34-6; 75-00-3; 1310-73-2	
Monomer(s) molecular weight(s)	dalton, g/mol, amu	depends on source; 64.51; 40.00	
Ethoxy content	%	44.0-51.0	
Method of synthesis	-	cellulose is first reacted with sodium hydroxide solution, followed by reaction between alkali cellulose and chloroethylene	
Number average molecular weight, M_n	dalton, g/mol, amu	24,000-75,000; 9,530 (adhesive)	Gong, X, Cheng, Z, Gao, S, Zhang, D, Chu, F, Carbohydrate Polym., 250, 116846, 2020.
Mass average molecular weight, M_w	dalton, g/mol, amu	18,300-82,000	
Polydispersity, M_w/M_n	-	1.09-2.8	Sanchez, R; Franco, J M; Delgado, M A; Valencia, C; Gallegos, C, Carbohydrate Polym., 83, 151-58, 2011; Marucci, M; andersson, H; Hjartstam, J; Stevenson, G; Baderstedt, J; Standing, A; Larsson, A; von Corswant, C, Int. J. Pharm., 458, 218-23, 2013.
Molar volume at 298K	cm^3 mol^{-1}	calc.=220.5	
Van der Waals volume	cm^3 mol^{-1}	141.28	
STRUCTURE			
Crystallinity	%	close to 0	
Cell type (lattice)	-	orthorhombic	Zugenmaier, P, J. Appl. Polym. Sci., Polym. Symp., 37, 223, 1983.
Cell dimensions	nm	a:b:c=1.56:2.71:1.50	Zugenmaier, P, J. Appl. Polym. Sci., Polym. Symp., 37, 223, 1983.
Number of chains per unit cell	-	6	Zugenmaier, P, J. Appl. Polym. Sci., Polym. Symp., 37, 223, 1983.
Chain conformation	-	3/2	
Entanglement molecular weight	dalton, g/mol, amu	calc.=9126	
COMMERCIAL POLYMERS			
Some manufacturers	-	DOW	
Trade names	-	Ethocel	
PHYSICAL PROPERTIES			
Density at 20°C	g cm^{-3}	1.07-1.18	
Bulk density at 20°C	g cm^{-3}	0.29	

HANDBOOK OF POLYMERS 3rd Edition, Copyrights 2022; ChemTec Publishing

EC ethyl cellulose

PARAMETER	UNIT	VALUE	REFERENCES
Color	-	white to light tan	
Refractive index, 20°C	-	calc.=1.4745-1.489; exp.=1.479	
Melting temperature, DSC	°C	240-255	
Softening point	°C	152-180	
Thermal expansion coefficient, 23-80°C	10^{-4} °C^{-1}	1-2	
Thermal conductivity, melt	W m^{-1} K^{-1}	calc.=0.1721; exp.=0.159-0.293	
Glass transition temperature	°C	43-142	
Heat deflection temperature at 1.8 MPa	°C	46-88	
Hansen solubility parameters, δ_D, δ_P, δ_H	MPa$^{0.5}$	19.0, 5.6, 4.9	
Interaction radius		7.9	
Hildebrand solubility parameter	MPa$^{0.5}$	20.4-21.1	
Surface tension	mN m^{-1}	calc.=31.8-34.1	
Power factor	-	0.002-0.02	
Permeability to nitrogen, 25°C	cm^3 cm cm^{-2} s^{-1} Pa^{-1} x 10^{12}	0.332	
Permeability to oxygen, 25°C	cm^3 cm cm^{-2} s^{-1} Pa^{-1} x 10^{12}	1.1	
Permeability to water vapor, 25°C	cm^3 cm cm^{-2} s^{-1} Pa^{-1} x 10^{12}	670	
Diffusion coefficient of nitrogen	cm^2 s^{-1} x10^6	0.233	
Diffusion coefficient of oxygen	cm^2 s^{-1} x10^6	0.639	
Diffusion coefficient of water vapor	cm^2 s^{-1} x10^6	0.0286	
Contact angle of water, 20°C	degree	84.5	
Surface free energy	mJ m^{-2}	30.3	
MECHANICAL & RHEOLOGICAL PROPERTIES			
Tensile strength	MPa	14-62	Rowe, R C; Roberts, R J, J. Mater. Sci. Lett., 14, 6, 420-21, 1995.
Tensile stress at yield	MPa	30-45	Rowe, R C; Roberts, R J, J. Mater. Sci. Lett., 14, 6, 420-21, 1995.
Elongation	%	5-40	
Flexural strength	MPa	28-83	
Young's modulus	MPa	860-1,800	Rowe, R C; Roberts, R J, J. Mater. Sci. Lett., 14, 6, 420-21, 1995; Lee, S; Ko, K-H; Shin, J; Kim, N-K; Kim, Y-W; Kim, J-S, 121, 284-94, 2015.

EC ethyl cellulose

PARAMETER	UNIT	VALUE	REFERENCES
Izod impact strength, notched, 23°C	J m^{-1}	20	
Poisson's ratio	-	calc.=0.374	
Rockwell hardness	-	R50-115	
Shrinkage	%	0.5-0.9	
Water absorption, equilibrium in water at 23°C	%	0.8-1.8	
Moisture absorption, equilibrium 23°C/50% RH	%	2.0	

CHEMICAL RESISTANCE

Acid dilute/concentrated	-	resistant/non-resistant	
Alcohols	-	non-resistant	
Alkalis	-	resistant	
Aromatic hydrocarbons	-	non-resistant	
Esters	-	non-resistant	
Halogenated hydrocarbons	-	non-resistant	
Ketones	-	non-resistant	
Good solvent	-	acetic acid, formic acid, pyridine	
Non-solvent	-	ethanol	

FLAMMABILITY

Ignition temperature	°C	291	
Autoignition temperature	°C	296	
Minimum ignition energy	J	0.01	
Char at 500°C	%	0.4	

WEATHER STABILITY

UV/VIS transmittance	%	56	Lee, S; Ko, K-H; Shin, J; Kim, N-K; Kim, Y-W; Kim, J-S, 121, 284-94, 2015.

TOXICITY

NFPA: Health, Flammability, Reactivity rating	-	1-2/1/0	
Carcinogenic effect	-	not listed by ACGIH, NIOSH, NTP	
Oral rat, LD$_{50}$	mg kg^{-1}	>5,000	
Skin rabbit, LD$_{50}$	mg kg^{-1}	>5,000	

PROCESSING

Typical processing methods	-	compression molding, sheet extrusion, injection molding	
Processing temperature	°C	121-199 (compression molding), 177-260 (injection molding)	
Processing pressure	MPa	55-220	
Additives used in final products	-	Release: fluorosilicone, silicone coating; Slip: silicone oil	
Applications	-	pharmaceutical (microencapsulation, self-healing adhesives, sustained release, tablet coating, water insoluble films)	
Outstanding properties	-	adhesion, emulsifier	

EC ethyl cellulose

PARAMETER	UNIT	VALUE	REFERENCES
BLENDS			
Suitable polymers	-	MC, PHB, PVP	

ECTFE poly(ethylene-co-chlorotrifluoroethylene)

PARAMETER	UNIT	VALUE	REFERENCES
GENERAL			
Common name	-	poly(ethylene-co-chlorotrifluoroethylene)	
IUPAC name	-	1-chloro-1,2,2-trifluoroethene; ethene	
CAS name	-	ethene, chlorotrifluoro-, polymer with ethene	
Acronym	-	ECTFE	
CAS number	-	25101-45-5; 9044-11-5	
HISTORY			
Person to discover	-	Hanford, W E of DuPont	Hanford, W E, US Patent 2,392,378, Du Pont, Jan. 8, 1946.
Date	-	1946 (patent), 1970 (commercialization)	
Details	-	reactor was charged with both monomers (40 parts of chlorotrifluoroethylene and 10 parts of ethylene) and copolymerized in the presence of benzoyl peroxide	
SYNTHESIS			
Monomer(s) structure	-	$H_2C{=}CH_2 \qquad F_2C{=}CClF$	
Monomer(s) CAS number(s)	-	74-85-1; 79-38-9	
Monomer(s) molecular weight(s)	dalton, g/mol, amu	28.05; 116.469	
Monomer ratio	-	1	
Catalyst	-	tributyl boron	Ebnesajjad, S, Applied Plastics Engineering Handbook, 2nd Ed. Processing, Materials, and Applications, William Andrew, 2017, pp. 55-71.
Formulation example	-	water, monomers, solvent, initiator, chain transfer agent	
Temperature of polymerization	ºC	60-120	Ebnesajjad, S, Applied Plastics Engineering Handbook, 2nd Ed. Processing, Materials, and Applications, William Andrew, 2017, pp. 55-71.
Pressure of polymerization	MPa	5	Ebnesajjad, S, Applied Plastics Engineering Handbook, 2nd Ed. Processing, Materials, and Applications, William Andrew, 2017, pp. 55-71.
Time of polymerization	h	1.5	
Pressure of polymerization	MPa	1	
STRUCTURE			
Crystallinity	%	44-63	
Cell type (lattice)	-	hexagonal	Ebnesajjad, S, Applied Plastics Engineering Handbook, 2nd Ed. Processing, Materials, and Applications, William Andrew, 2017, pp. 55-71.
Rapid crystallization temperature	ºC	221-222	
Chain conformation	-	extended zigzag in which ethylene and CTFE alternate	Ebnesajjad, S, Applied Plastics Engineering Handbook, 2nd Ed. Processing, Materials, and Applications, William Andrew, 2017, pp. 55-71.

ECTFE poly(ethylene-co-chlorotrifluoroethylene)

PARAMETER	UNIT	VALUE	REFERENCES
COMMERCIAL POLYMERS			
Some manufacturers	-	Daaikin; Solvay	
Trade names	-	Neoflon; Halar	
PHYSICAL PROPERTIES			
Density at 20°C	g cm^{-3}	1.68-1.72	
Color	-	white	
Haze	%	3-5	
Refractive index, 20°C	-	1.4470	
Odor	-	none	
Melting temperature, DSC	°C	188-242	
Decomposition temperature	°C	>350; 405 (1% weight loss)	
Continuous use temperature	°C	165	
Crystallization point	°C	162-222	
Thermal expansion coefficient, 23-80°C	°C^{-1}	0.8-1.7E-4	
Thermal conductivity, 40°C	W m^{-1} K^{-1}	0.15-0.16	
Glass transition temperature	°C	63-85	
Specific heat capacity	J K^{-1} kg^{-1}	950-963 (23°C); 1,620 (melt)	
Heat of fusion	kJ kg^{-1}	28-42	
Crystallization heat	kJ kg^{-1}	5-40	
Maximum service temperature	°C	245-260	
Long term service temperature	°C	150-170	
Heat deflection temperature at 0.45 MPa	°C	62-116	
Heat deflection temperature at 1.8 MPa	°C	56-77	
Hansen solubility parameters, δ_D, δ_P, δ_H	MPa$^{0.5}$	16.8, 8.4, 7.8	
Interaction radius		2.7	
Dielectric constant at 1000 Hz/1 MHz	-	2.47-2.6/2.50-2.59	Ebnesajjad, S, Fluoroplastics. Vol. 2. Melt Processable Fluoroplastics, William Andrew, 2003.
Dielectric strength	kV mm^{-1}	220	Drobny, J G, Applications of Fluoropolymer Films, William Andrew, 2020, pp. 161-2.
Dissipation factor at 1000 Hz		0.0014-0.0017	Ebnesajjad, S, Fluoroplastics. Vol. 2. Melt Processable Fluoroplastics, William Andrew, 2003.
Dissipation factor at 1 MHz		0.0013-0.009	Ebnesajjad, S, Fluoroplastics. Vol. 2. Melt Processable Fluoroplastics, William Andrew, 2003.
Volume resistivity	ohm-m	5.5E13 to 1E14	
Surface resistivity	ohm	1E15	
Electric strength K20/P50	kV mm^{-1}	14-110	
Arc resistance	s	135	
Coefficient of friction	-	0.1-0.3 (static); 0.1-0.3 (dynamic)	

ECTFE poly(ethylene-co-chlorotrifluoroethylene)

PARAMETER	UNIT	VALUE	REFERENCES
Permeability to nitrogen, 25°C	cm³ mm m⁻² s⁻¹ atm 24 h	0.002	
Permeability to oxygen, 25°C	cm³ mm m⁻² s⁻¹ atm 24	0.001	
Permeability to water vapor, 25°C	cm³ mm m⁻² s⁻¹ atm 24	0.075	
Diffusion coefficient of water vapor	cm² s⁻¹ x10⁷	4.12 (20°C); 1.42 (90°C)	Hansen, C M, Prog. Org. Coat., 42, 167-78, 2001.
Contact angle of water, 20°C	degree	99 (adv) and 78 (rec)	Lee, S; Park, J-S; Lee, T R, Langmuir, 24, 4817-26, 2008.

MECHANICAL & RHEOLOGICAL PROPERTIES

PARAMETER	UNIT	VALUE	REFERENCES
Tensile strength	MPa	31-57	
Tensile modulus	MPa	1,375-1,800	
Tensile stress at yield	MPa	22-32	
Elongation	%	200-325	
Tensile yield strain	%	3-9	
Flexural strength	MPa	40-55	
Flexural modulus	MPa	1,400-1,800	
Elastic modulus	MPa	1650	
Izod impact strength, notched, 23°C	J m⁻¹	NB	
Izod impact strength, notched, -40°C	J m⁻¹	50-200	
Abrasion resistance (ASTM D1044)	mg/1000 cycles	5	
Shore D hardness	-	70-75	
Rockwell hardness	R	90-94	
Shrinkage	%	2.3-2.5	
Brittleness temperature (ASTM D746)	°C	-61 to <-76	
Melt index, 275°C/2.16 kg	g/10 min	1.5-60	
Water absorption, equilibrium in water at 23°C	%	<0.01 to <0.1	

CHEMICAL RESISTANCE

PARAMETER	UNIT	VALUE	REFERENCES
Acid dilute/concentrated	-	good	
Alcohols	-	good	
Alkalis	-	good	
Aliphatic hydrocarbons	-	good	
Aromatic hydrocarbons	-	good	
Esters	-	good	
Greases & oils	-	good	
Halogenated hydrocarbons	-	good	

ECTFE poly(ethylene-co-chlorotrifluoroethylene)

PARAMETER	UNIT	VALUE	REFERENCES
Ketones	-	good	
Good solvent	-	insoluble at room temperature	
FLAMMABILITY			
Ignition temperature	°C	none	
Autoignition temperature	°C	655	
Limiting oxygen index	% O_2	52-60	
Heat release	kW m^{-2}	74	
Volatile products of combustion	-	CDFA, TFA, DFA	
UL 94 rating	-	V-0	
WEATHER STABILITY			
Low earth orbit erosion yield	cm^3 atom^{-1} x 10^{-24}	1.79	Waters, D L; Banks, B A; De Groh, K K; Miller, S K R; Thorson, S D, High Performance Polym., 20, 512-22, 2008.
TOXICITY			
HMIS: Health, Flammability, Reactivity rating	-	1/0/0	
Carcinogenic effect	-	not listed by ACGIH, NIOSH, NTP	
TLV, ACGIH	mg m^{-3}	5 (respirable; 10 (total)	
OSHA	mg m^{-3}	5 (respirable); 15 (total)	
PROCESSING			
Typical processing methods	-	compression molding, electrostatic powder coating, fluidized bed coating, film/tube extrusion, flame and plasma spraying, injection molding monofilament extrusion, wire & cable injection molding	
Processing temperature	°C	250-280	
Applications	-	cable insulation, capacitors, coatings (agitators, centrifuges, electroplating equipment, exhaust hoods, filters, piping systems, reactors, semiconductor storage tanks, vessels), films, fibers, hollow fiber membranes, pipes, printed circuits, rods, sheet, solar collectors	
Outstanding properties	-	chemical and thermal resistance, purity, surface characteristics	

EEAC poly(ethylene-co-ethyl acrylate)

PARAMETER	UNIT	VALUE	REFERENCES
GENERAL			
Common name	-	poly(ethylene-co-ethyl acrylate)	
IUPAC name	-	ethene; ethyl prop-2-enoate	
Acronym	-	EEAC	
CAS number	-	9010-86-0	
HISTORY			
Person to discover	-	White, W G	White, W G, US Patent 2,953,551, Union Carbide, Sept. 20, 1960.
Date	-	1960	
Details	-	process of production	
SYNTHESIS			
Monomer(s) structure	-		

$$H_2C{=}CH_2 \qquad H_2C{=}CHCOCH_2CH_3$$
$$\overset{\displaystyle O}{\underset{\|}{}}$$

PARAMETER	UNIT	VALUE	REFERENCES
Monomer(s) CAS number(s)	-	74-85-1; 140-88-5	
Monomer(s) molecular weight(s)	dalton, g/mol, amu	28.05; 100.11	
Ethyl acrylate content	wt%	10-25	
STRUCTURE			
Crystallinity	%	22.8	Han, S H; Yeom, Y S; Ko, J G; Kang, H C; Yoon, HG, Compos. Sci. Technol., 117, 351-6, 2015.
Rapid crystallization temperature	°C	78-82	
COMMERCIAL POLYMERS			
Some manufacturers	-	SK Global; Dow; Japan Polychem Corp.	
Trade names	-	Lotader; Amplify, Elvaloy AC; Rexpearl	
PHYSICAL PROPERTIES			
Density at 20°C	g cm^{-3}	0.92-0.94	
Color	-	white	
Odor	-	characteristic acrylate	
Melting temperature, DSC	°C	92-112	Koulouri, E G; Gravalos, K G; Kallitsis, J K, Polymer, 37, 12, 2555-63, 1996.
Softening point	°C	116	
Glass transition temperature	°C	-33	
Heat deflection temperature at 0.45 MPa	°C	31-33	
Vicat temperature VST/A/50	°C	40-82	
MECHANICAL & RHEOLOGICAL PROPERTIES			
Tensile strength	MPa	6.0-24.0	
Tensile stress at yield	MPa	2.6-3.8	
Elongation	%	600-980	

EEAC poly(ethylene-co-ethyl acrylate)

PARAMETER	UNIT	VALUE	REFERENCES
Tensile yield strain	%	8-11	
Flexural modulus	MPa	24-77	
Shore A hardness	-	70-87	
Shore D hardness	-	19-37	
Brittleness temperature (ASTM D746)	°C	-64 to -76	
Intrinsic viscosity, 25°C	dl g^{-1}	0.78	
Melt index, 190°C/2.16 kg	g/10 min	1-21	
FLAMMABILITY			
Ignition temperature	°C	430	
Volatile products of combustion	-	CO, H_2O, CO_2, organic acids, aldehydes, alcohols	
TOXICITY			
Carcinogenic effect	-	not listed by ACGIH, NIOSH, NTP	
OSHA	mg m^{-3}	15 (total dust); 5 (respirable)	
PROCESSING			
Typical processing methods	-	extrusion (blown film, cast film, coextrusion)	
Processing temperature	°C	160-310; 310 (max)	
Additives used in final products	-	MWCNT	Han, S H; Yeom, Y S; Ko, J G; Kang, H C; Yoon, HG, Compos. Sci. Technol., 117, 351-6, 2015.
Applications	-	packaging, performance booster for other resins, profile, tubing, wire & cable	
Outstanding properties	-	easy processing, compatible with LDPE, thermal stability	
BLENDS			
Suitable polymers	-	LDPE, PA6, PBT, PET, PVOH	Han, S H; Yeom, Y S; Ko, J G; Kang, H C; Yoon, HG, Compos. Sci. Technol., 117, 351-6, 2015.

EMA poly(ethylene-co-methyl acrylate)

PARAMETER	UNIT	VALUE	REFERENCES
GENERAL			
Common name	-	poly(ethylene-co-methyl acrylate)	
CAS name	-	2-propenoic acid, methyl ester, polymer with ethene	
Acronym	-	EMA	
CAS number	-	25103-74-6	
SYNTHESIS			
Monomer(s) structure	-	$CH_2=CH_2$; $CH_2=CHCOOCH_3$	
Monomer(s) CAS number(s)	-	74-85-1; 96-33-3	
Monomer(s) molecular weight(s)	dalton, g/mol, amu	28.05; 86.04	
Methyl acrylate content	wt%	6.8-55	
Method of synthesis	-	in the presence of a mixture of initiators (e.g., peroxides and azo compounds), ethylene and methyl acrylate can be copolymerized via a free-radical mechanism	Kiparissides, C; Baltsas, A; Papadopoulos, S; Congalidis, J P; Richards, J R; Kelly, M B; Ye, Y, Ind. Eng. Chem. Res., 44, 2592-2605, 2005.
Mass average molecular weight, M_w	dalton, g/mol, amu	100,000-390,000	Albrecht, A; Bruell, R; Macko, T; Sinha, P; Pasch, H, Macromol. Chem. Phys., 209, 1909-19, 2008.
Polydispersity, M_w/M_n	-	3.7-8.7	
STRUCTURE			
Crystallinity	%	13.7	Ravindren, R, Mondal, S, Nath, K, Das, N C, Compos. Part B: Eng, 164, 559-69, 2019.
Chain conformation	-	planar zig-zag	
Surface organization		at low concentrations of MA (<20 wt%) surface is dominated by MA-depleted layer; at high MA concentrations surface contains EMA backbone	Galuska, A; Surf. Interface Anal., 24, 380-8, 1996.
COMMERCIAL POLYMERS			
Some manufacturers	-	SK Global; DOW; ExxonMobil; Westlake Polymers	
Trade names	-	Lotader, Lotryl; Elvaloy AC, Vamac; Optema; Emac	
PHYSICAL PROPERTIES			
Density at 20°C	g cm^{-3}	0.93-0.95	
Color	-	translucent to whitish	
Odor	-	ester-like	
Melting temperature, DSC	°C	33-101	
Crystallization temperature	°C	56.1; 64	Ravindren, R, Mondal, S, Nath, K, Das, N C, Compos. Part B: Eng, 164, 559-69, 2019; Remanan, S, Ghosh, S, Das, T K, Das, N C, Nano-Sturctures Nano-Objects, 23, 100487, 2020.
Degradation temperature	°C	>350; 330; >282	Mongal, N; Chakraborty, D; Bhattacharyya, R; Chaki, T K; Bhattacharta, P, J. Appl. Polym. Sci., 117, 75-83, 2010.

HANDBOOK OF POLYMERS 3rd Edition, Copyrights 2022; ChemTec Publishing

EMA poly(ethylene-co-methyl acrylate)

PARAMETER	UNIT	VALUE	REFERENCES
Glass transition temperature	°C	-29.8 to - 35.6; -36.6	Kanis, L A; Generoso, M; Meier, M M; Pires, A T N; Soldi, V. Eur. J. Phamaceutics Biopharmaceutics, 60, 383-90, 2005; Ravindren, R, Mondal, S, Nath, K, Das, N C, Compos. Part B: Eng, 164, 559-69, 2019.
Heat of fusion	J g^{-1}	19-45	
Vicat temperature VST/A/50	°C	43-70	
Dielectric constant at 100 Hz/1 MHz	-	4	
EMI shielding (1 mm)	dB	18.4 (15 wt% ketjen carbon black); 30 (5 wt% reduced graphene oxide); 53.5 (15 wt% copper nanowire in EMA/EOC blend)	Ravindren, R, Mondal, S, Nath, K, Das, N C, Compos. Part B: Eng, 164, 559-69, 2019.
Volume resistivity	ohm-m	1-2.5E11	
Contact angle of water, 20°C	degree	64-85	Kanis, L A; Generoso, M; Meier, M M; Pires, A T N; Soldi, V. Eur. J. Phamaceutics Biopharmaceutics, 60, 383-90, 2005.

MECHANICAL & RHEOLOGICAL PROPERTIES

Tensile strength	MPa	5-11; 19.7-24.2 (MD); 19.8-25.3 (TD)	
Elongation	%	380-850; 370-380 (MD); 610-670 (TD)	
Flexural modulus	MPa	37	
Elastic modulus	MPa	6.5-81	
Dart drop impact	g	370-520	
Elmendorf tear strength	g	30-45 (MD); 200-320 (TD)	
Shore A hardness	-	76-86	
Shore D hardness	-	24-34	
Melt index, 190°C/2.16 kg	g/10 min	0.5-110	

CHEMICAL RESISTANCE

Alcohols	-	poor	

FLAMMABILITY

Autoignition temperature	°C	>450	
Volatile products of combustion	-	CO_2, CO, alcohols, ketones, aldehydes, esters, acids, acrolein	

TOXICITY

HMIS: Health, Flammability, Reactivity rating	-	1/1/0	
Carcinogenic effect	-	not listed by ACGIH, NIOSH, NTP	
TLV, ACGIH	mg m^{-3}	10 (inhalable), 3 (respirable)	
OSHA	mg m^{-3}	15 (total dust); 5 (respirable)	

PROCESSING

Typical processing methods	-	coextrusion coating, extrusion (blown film, cast film, coextrusion); extrusion lamination	
Processing temperature	°C	165-310; 310 (max)	

EMA poly(ethylene-co-methyl acrylate)

PARAMETER	UNIT	VALUE	REFERENCES
Additives used in final products	-	Antiblock; Slip; Thermal stabilizer, MWCNT	Basuli, U; Panja, S; Chaki, T K; Chattopadhyay, Handbook of Polymer Nanocomposites. Volume B, Springer, 2015, 245-280.
Applications	-	disposable gloves, drug delivery membrane, EMI shielding, food packaging, heat seals, hospital drapes, performance booster for other resins, packaging, upholstery film	
Outstanding properties	-	easy processing, compatible with LDPE, printability, sealing	
BLENDS			
Suitable polymers	-	CR, EOC, PMA, PP, PVDF	Remanan, S, Ghosh, S, Das, T K, Das, N C, Nano-Sturctures Nano-Objects, 23, 100487, 2020.
ANALYSIS			
FTIR (wavenumber-assignment)	cm^{-1}/-	C=O – 1740; CH_3 – 1376	Albrecht, A; Bruell, R; Macko, T; Sinha, P; Pasch, H, Macromol. Chem. Phys., 209, 1909-19, 2008.

EMA-AA poly(ethylene-co-methyl acrylate-co-acrylic acid)

PARAMETER	UNIT	VALUE	REFERENCES
GENERAL			
Common name	-	poly(ethylene-co-methyl acrylate-co-acrylic acid)	
Synonym	-	2-propenoic acid, polymer with ethene and methyl 2-propenoate	
Acronym	-	EMA-AA	
CAS number	-	41525-41-1	
SYNTHESIS			
Monomer(s) structure	-	$H_2C{=}CH_2 \quad H_2C{=}CHCOCH_3 \quad H_2C{=}CHCOH$ (acrylate and acid groups with C=O)	
Monomer(s) CAS number(s)	-	74-85-1; 96-33-3; 79-10-7	
Monomer(s) molecular weight(s)	dalton, g/mol, amu	28.05; 86.04; 72.06	
Methyl acrylate content	wt%	6.5-24.0	
Acrylic acid content	wt%	5-6.5	
Number average molecular weight, M_n	dalton, g/mol, amu	16,000	
Mass average molecular weight, M_w	dalton, g/mol, amu	2,000-68,000	
STRUCTURE			
Crystallinity	%	8.48-27	Cerezo, F T; Preston, C M L; Shanks, R A, Composites Sci. Tech., 67, 79-91, 2007; Preston, C M L; Amarasinghe, G; Hopewell, J L; Shanks, R A; Mathys, Z, Polym. Deg. Stab., 84, 533-44, 2004.
Rapid crystallization temperature	°C	53-56	
COMMERCIAL POLYMERS			
Some manufacturers	-	SK Global; Dow; ExxonMobil	
Trade names	-	Lotader; Nucrel; Escor	
PHYSICAL PROPERTIES			
Density at 20°C	g cm^{-3}	0.92-0.94	
Color	-	clear to opaque, white to off-white	
Odor	-	odorless	
Melting temperature, DSC	°C	60-105	
Storage temperature	°C	20	
Vicat temperature VST/A/50	°C	40-87	
Acid number	mg KOH g^{-1}	45	
MECHANICAL & RHEOLOGICAL PROPERTIES			
Tensile strength	MPa	20-29 (MD); 22-28 (TD)	
Tensile stress at yield	MPa	10 (MD); 3.8 (TD)	
Elongation	%	200-370 (MD); 570-600 (TD)	
Flexural modulus	MPa	30	
Dart drop impact	g	200-600	

124　　**HANDBOOK OF POLYMERS** 3rd Edition, Copyrights 2022 ChemTec Publishing

EMA-AA poly(ethylene-co-methyl acrylate-co-acrylic acid)

PARAMETER	UNIT	VALUE	REFERENCES
Elmendorf tear strength	g	70-420 (MD); 610-1100 (TD)	
Puncture force	N	50	
Shore A hardness	-	64-80	
Shore D hardness	-	18-44	
Melt index, 190°C/2.16 kg	g/10 min	1.5-20	
FLAMMABILITY			
Ignition temperature	°C	316	
Heat release	kW m^{-2}	1,700	Preston, C M L; Amarasinghe, G; Hopewell, J L; Shanks, R A; Mathys, Z, Polym. Deg. Stab., 84, 533-44, 2004.
Heat of combustion	J g^{-1}	22,400	Preston, C M L; Amarasinghe, G; Hopewell, J L; Shanks, R A; Mathys, Z, Polym. Deg. Stab., 84, 533-44, 2004.
TOXICITY			
NFPA: Health, Flammability, Reactivity rating	-	1/1/0; 1/1/0 (HMIS)	
Carcinogenic effect	-	not listed by ACGIH, NIOSH, NTP	
Oral rat, LD$_{50}$	mg kg^{-1}	>5,000	
PROCESSING			
Typical processing methods	-	casting, coating, extrusion, laminating, molding	
Additives used in final products	-	antiblock; slip; thermal stabilizer	
Applications	-	adhesion promoter, compatibilizer, film, heat seal layer, lamination film, sealants, tie layers; wire & cable	
Outstanding properties	-	adhesion to polar and non-polar substrates, flexibility, thermal stability	

ENBA poly(ethylene-co-n-butyl acrylate)

PARAMETER	UNIT	VALUE	REFERENCES
GENERAL			
Common name	-	poly(ethylene-co-n-butyl acrylate)	
CAS name	-	2-propenoic acid, butyl ester, polymer with ethene	
Acronym	-	ENBA	
CAS number	-	25750-84-9	
SYNTHESIS			
Monomer(s) structure	-	$H_2C=CH_2 \quad H_2C=CHCOCH_2CH_2CH_2CH_3$ (with C=O)	
Monomer(s) CAS number(s)	-	74-85-1; 141-32-2	
Monomer(s) molecular weight(s)	dalton, g/mol, amu	28.05; 128.2	
Butyl acrylate content	%	0.18-32.5	
COMMERCIAL POLYMERS			
Some manufacturers	-	SK Global; BASF; DOW; ExxonMobil, Westlake Chemical	
Trade names	-	Lotryl; Lucalen; Elvaloy; EnBA; EBAC	
PHYSICAL PROPERTIES			
Density at 20°C	g cm^{-3}	0.90-0.94	
Color	-	colorless	
Haze	%	16	
Gloss, 60°, Gardner (ASTM D523)	%	60	
Odor		faint	
Melting temperature, DSC	°C	50-107	
Glass transition temperature	°C	-46 to -54	
Vicat temperature VST/A/50	°C	45-59	
Surface tension	mN m^{-1}	23.9	
Dielectric constant at 100 Hz/1 MHz	-	2.7/	
Dissipation factor at 100 Hz	E-4	10	
Volume resistivity	ohm-m	1E14	
Coefficient of friction	-	0.45 (itself, dynamic)	
MECHANICAL & RHEOLOGICAL PROPERTIES			
Tensile strength	MPa	1-24	
Tensile stress at yield	MPa	7-8	
Elongation	%	140-850	
Flexural modulus	MPa	5.4-7.2	
Shore A hardness	-	44-56	
Shore D hardness	-	40-56	
Brittleness temperature (ASTM D746)	°C	-50 to -73	
Melt viscosity, shear rate=0 s^{-1}	Pa s	3,122	Tinson, A; Takacs, E; Vlachopoulos, J, Antec, 870-4, 2004.
Melt index, 190°C/3.8 kg	g/10 min	0.45-900	

ENBA poly(ethylene-co-n-butyl acrylate)

PARAMETER	UNIT	VALUE	REFERENCES
Water absorption, equilibrium in water at 23°C	%	0.5	

FLAMMABILITY

Autoignition temperature	°C	>350	

PROCESSING

Typical processing methods	-	film extrusion, extrusion, injection molding	
Preprocess drying: temperature/ time/residual moisture	°C/h/%	60/5	
Processing temperature	°C	140-290	
Additives used in final products	-	Antiblock; Slip; Thermal stabilizer	
Applications	-	agricultural applications, automotive applications, cosmetics, film, hot melt adhesives and sealants, medical, wax blends	

BLENDS

Compatible polymers	-	polyolefins, polyesters, ionomers, PVC, many polar polymers	

EP epoxy resin

PARAMETER	UNIT	VALUE	REFERENCES
GENERAL			
Common name	-	epoxy resin	
CAS name	-	epoxy resin	
Acronym	-	EP	
CAS number	-	25036-25-3; 25068-38-6; 55818-57-0; 61788-97-4; 90598-46-2	
HISTORY			
Person to discover	-	Castan, P, licensed to Ciba	
Date	-	1936	
Details	-	bisphenol A based epoxy	
SYNTHESIS			
Monomer(s) structure	-	most frequently used epoxy monomer is a product of $C_{15}H_{16}O_2$ (bisphenol A); C_3H_5ClO (epichlorohydrin); oxirane groups can be generated from peroxidation of C=C bonds, most frequently oil or cycloaliphatic compounds are used	Pascault, J-P; Williams, R J J, Epoxy Polymers, Pascault, J-P; Williams, R J J, Eds., Wiley, 2010.
Monomer(s) CAS number(s)	-	80-05-7; 106-89-8	
Monomer(s) molecular weight(s)	dalton, g/mol, amu	228.29; 92.52	
Hardener(s)		polyamines (e.g., triethylenetetramine, $C_6H_{18}N_4$); anhydride (4-methylcyclohexane-1,2-dicarboxylic anhydride)	-; Hsu, Y-I, Huang, L, Asoh, T-A, Uyama, H, Polym. Deg. Stab., 178, 109213, 2020.
CAS number(s)		112-24-3	
Molecular weight(s)	dalton, g/mol, amu	146.23	
Epoxide percentage	%	7.7-8.3	
Method of synthesis	-	final resin is obtained from combination of epoxy monomer and hardener; properties depend on monomers and their proportions; epoxy polymers are produced by step or chain polymerizations or their combinations, leading to linear or crosslinked polymers	Pascault, J-P; Williams, R J J, Epoxy Polymers, Pascault, J-P; Williams, R J J, Eds., Wiley, 2010; Jin, F-L; Li, X; Park, S-J, J. Ind. Eng. Chem., 29, 1-11, 2015.
Temperature of polymerization	°C	120-160 (baking); 5-150 (adhesives)	
Time of polymerization	min	15-25 (baking); 5 min to over 24 h (adhesives)	
Activation energy of polymerization	kJ mol⁻¹	46(foaming); 37-80 (depending on conversion; energy increases with conversion increasing), 53-60 (with graphene oxide)	Mondy, L A; Rao, R R; Moffar, H; Adolf, D; Celina, M, Epoxy Polymers, Pascault, J-P; Williams, R J J, Eds., Wiley, 2010; Jouyandeh, M, Yarahmadi, E, Didehban, K, Ghiyasi, S, Ganjali M R, Prog. Org. Coat., 136, 105217, 2019.
Activation energy of gelation	kJ mol⁻¹	54-92	Osbaldiston, J R; Smith, W; Farquharson, S; Shaw, M T, Antec, 939-44, 1998.
Heat of polymerization	J g⁻¹	250 (foaming); 345-387	Mondy, L A; Rao, R R; Moffar, H; Adolf, D; Celina, M, Epoxy Polymers, Pascault, J-P; Williams, R J J, Eds., Wiley, 2010; Karami, Maryam Jouyandeh, M, Ali, J A, Z,, M R, Saeb, M R, Prog. Org. Coat., 136, 105218, 2019.
STRUCTURE			
Cross-section surface area of chain	nm²	0.39-0.44	Swarup, S; Nigam, A N, J. Appl. Polym. Sci., 39, 1727-31, 1990.

EP epoxy resin

PARAMETER	UNIT	VALUE	REFERENCES
Number of carbon atoms per entanglement		500-533	Swarup, S; Nigam, A N, J. Appl. Polym. Sci., 39, 1727-31, 1990.

COMMERCIAL POLYMERS

Some manufacturers	-	DOW, Hexion	
Trade names	-	D.E.R. (solid and liquid epoxy resins); D.E.H (curing agents); Epikote/Epicure, Epi-Rez, Epon, Eponex, Eponol	

PHYSICAL PROPERTIES

PARAMETER	UNIT	VALUE	REFERENCES
Density at 20°C	g cm^{-3}	1.15-1.3	
Refractive index, 20°C	-	1.51-1.58	
Melting temperature, DSC	°C	90-245	
Softening point	°C	80-90	
Thermal expansion coefficient, 23-80°C	°C^{-1} x 10^{-6}	17-67	Meijerink, J I; Eguchi, S; Ogata, M; Ishii, T; Amagi, S; Numata, S; Sashima, H, Polymer, 35, 1, 179-86, 1994.
Thermal conductivity, melt	W m^{-1} K^{-1}	0.15-0.25	
Glass transition temperature	°C	37-127 (thermoplastic); 130-246 (adhesives); 54.5-62 (commercial)	White, J E, Epoxy Polymers, Pascault, J-P; Williams, R J J, Eds., Wiley, 2010; Abuin, S P, Epoxy Polymers, Pascault, J-P; Williams, R J J, Eds., Wiley, 2010; Jin, F-L; Li, X; Park, S-J, J. Ind. Eng. Chem., 29, 1-11, 2015; Michels, J; Widmann, R; Czaderski, C; Allahvirdizadeh, R; Motavalli, M, Composites: Part B, 77, 484-93, 2015.
Maximum service temperature	°C	-260 to 350 (Duralco, adhesive)	Bhowmik, S; Benedictus, R; Poulis, H, Polymers in Defence and Aerospace 2007, Rapra, 2007, paper 8.
Heat deflection temperature at 0.45 MPa	°C	53-194	
Heat deflection temperature at 1.8 MPa	°C	46-187	
Hansen solubility parameters, δ_D, δ_P, δ_H	MPa$^{0.5}$	19.2, 10.9, 9.6; 20.36, 12.03, 11.48; 18.1, 11.4, 9.0	
Interaction radius		11.1; 9.1	
Hildebrand solubility parameter	MPa$^{0.5}$	22.0-27.1	
Surface tension	mN m^{-1}	36.6-51.6	Lemesle, C, Bellayer, S, Duquesne, S, Schuller, A-S, Jimenez , M, Appl. Surf. Sci., 536, 147687, 2021
Dielectric constant at 100 Hz/1 MHz	-	3.5-5.0/3.6	
Volume resistivity	ohm-m	1E14	
Electric strength K20/P50, d=0.60.8 mm	kV mm^{-1}	15	
Coefficient of friction	-	0.5-0.6	Larsen, T O; Andersen, T L; Thorning, B; Horsewell, A; Vigild, Wear, 265, 203-13, 2008.
Contact angle of water, 20°C	degree	54.0-87.5	
Surface free energy	mJ m^{-2}	45.3	
Speed of sound	m s^{-1} x 10^{-3}	1.01-1.08	

EP epoxy resin

PARAMETER	UNIT	VALUE	REFERENCES
MECHANICAL & RHEOLOGICAL PROPERTIES			
Tensile strength	MPa	27-200 (thermoplastic); 40-65 (adhesives)	White, J E, Epoxy Polymers, Pascault, J-P; Williams, R J J, Eds., Wiley, 2010; Abuin, S P, Epoxy Polymers, Pascault, J-P; Williams, R J J, Eds., Wiley, 2010.
Tensile modulus	MPa	850-4,800	Abuin, S P, Epoxy Polymers, Pascault, J-P; Williams, R J J, Eds., Wiley, 2010.
Tensile stress at yield	MPa	36.6-117.7 (thermoplastic)	White J E, Epoxy Polymers, Pascault, J-P; Williams, R J J, Eds., Wiley, 2010.
Elongation	%	1.3-705 (thermoplastic); 3-5 (adhesives)	White J E, Epoxy Polymers, Pascault, J-P; Williams, R J J, Eds., Wiley, 2010; Abuin, S P, Epoxy Polymers, Pascault, J-P; Williams, R J J, Eds., Wiley, 2010.
Tensile yield strain	%	4	
Flexural strength	MPa	74-325	
Flexural modulus	MPa	2,550-15,500	
Elastic modulus	MPa	2,700-4,100	Ruggiero, A; Merola, M; Carlone, P; Archoduolaki, V-M, Composites: Part B, 79, 595-603, 2015; Jin, F-L; Li, X; Park, S-J, J. Ind. Eng. Chem., 29, 1-11, 2015.
Compressive strength	MPa	116-404	
Compressive modulus	MPa	3,100-4,500	Hergenrother, P M; Thompson, C M; Smith, J G; Connell, J W; Hinkley, J A; Lyon, R E; Moulton, R, Polymer, 46, 5012-24, 2005.
Young's modulus	MPa	3,600-4,300	
Izod impact strength, notched, 23°C	J m^{-1}	25-1246 (thermoplastic)	White J E, Epoxy Polymers, Pascault, J-P; Williams, R J J, Eds., Wiley, 2010.
Shear strength	MPa	12-24	Abuin, S P, Epoxy Polymers, Pascault, J-P; Williams, R J J, Eds., Wiley, 2010.
Fracture toughness	MPa m$^{1/2}$	3.53-4.48	Jin, F-L; Li, X; Park, S-J, J. Ind. Eng. Chem., 29, 1-11, 2015.
Poisson's ratio	-	0.22-0.50 (decreases with crosslinking increase)	Liu, C, Ning, W, Tam, L-h, Yu , Z, J. Molec. Graphics Modelling, in press, 107757, 2020; Esmkhani, M, Shokrieh, M M,Taheri-Behrooz, F, Fatigue Life Prediction of Composites and Composite Structures, 2nd Ed., Woodhead Publishing, 2020, pp. 135-93.
Shore D hardness	-	62-95	
Shrinkage	%	0.001-0.13	
Intrinsic viscosity, 25°C	dl g^{-1}	0.4-0.94 (thermoplastic)	
Water absorption, equilibrium in water at 23°C	%	0.04-4.0 (thermoplastic); 2-5 (adhesives)	Abuin, S P, Epoxy Polymers, Pascault, J-P; Williams, R J J, Eds., Wiley, 2010.
CHEMICAL RESISTANCE			
Acid dilute/concentrated	-	fair-excellent	
Alcohols	-	excellent-good	
Alkalis	-	excellent	

EP epoxy resin

PARAMETER	UNIT	VALUE	REFERENCES
Aliphatic hydrocarbons	-	excellent-good	
Aromatic hydrocarbons	-	excellent	
Esters	-	good	
Greases & oils	-	good	
Halogenated hydrocarbons	-	excellent	
Ketones	-	poor	
FLAMMABILITY			
Ignition temperature	^{o}C	>120 to 249; 32 (solvent based)	
Limiting oxygen index	% O_2	18.3-19; 23	Kumar, S A; Denchev, Z, Prog. Org. Coat., 66, 1-7, 2009; Szolnoki, B; Bocz, K; Soti, P L; Bodzay, B; Zimonyi, E; Toldy, A; Morlin, B; Bujnowicz, K; Wladyka-Przybylak, M; Marosi, G, Polym. Deg. Stab., 119, 68-76, 2015.
Heat release	kW m^{-2}	51-97	Hergenrother, P M; Thompson, C M; Smith, J G; Connell, J W; Hinkley, J A; Lyon, R E; Moulton, R, Polymer, 46, 5012-24, 2005.
Char at 500°C	%	3.9-15.9; 25-44 (flame retarded)	Lyon, R E; Walters, R N, J. Anal. Appl. Pyrolysis, 71, 27-46, 2004; Hergenrother, P M; Thompson, C M; Smith, J G; Connell, J W; Hinkley, J A; Lyon, R E; Moulton, R, Polymer, 46, 5012-24, 2005.
WEATHER STABILITY			
Spectral sensitivity	nm	300-330	Wypych, G, Handbook of Materials Weathering, 6th Ed., ChemTec Publishing, 2018.
Important initiators and accelerators	-	alkaline products of corrosion, aromatic carbonyl groups, quinoic structures, hydroxide ions, double bonds	
Products of degradation	-	benzene, styrene, radicals, benzoic acid, benzaldehyde, benzophenone, water	
Stabilizers	-	UVA: 2,4-dihydroxybenzophenone; 2-(2H-benzotriazol-2-yl)-p-cresol; 2-benzotriazol-2-yl-4,6-di-tert-butylphenol; Screener: nano-ZnO; nano-silica-titania	
BIODEGRADATION			
Colonized products		coatings, marine coatings	
Typical biodegradants	-	fungi	Warscheid, T; Braams, J, Int. Biodet. Biodeg., 46, 343-68, 2000.
Stabilizers	-	silver-containing zeolite, coal tar, ferric benzoate, 4,5-dichloro-2-n-octyl-4-isothiazolin-3-one	
TOXICITY			
HMIS: Health, Flammability, Reactivity rating	-	2/1/0	
Carcinogenic effect	-	not listed by ACGIH, NIOSH, NTP	
Oral rat, LD$_{50}$	mg kg^{-1}	>2,000 to 5,800	
Skin rabbit, LD$_{50}$	mg kg^{-1}	>2,150	

EP epoxy resin

PARAMETER	UNIT	VALUE	REFERENCES
ENVIRONMENTAL IMPACT			
Aquatic toxicity, *Daphnia magna*, LC$_{50}$, 48 h	mg l^{-1}	1.4-19.6; 2,000-114,000 (EC50)	Lithner, Ph D Thesis, Univrsity of Gothenburg, 2011.
Aquatic toxicity, *Rainbow trout*, LC$_{50}$, 48 h	mg l^{-1}	1.5-2.4	
PROCESSING			
Typical processing methods	-	casting, coatings, compounding, dipping, infusion molding, *in situ* polymerization, lamination, pultrusion, sheet molding; spraying, transfer molding	Constantino, S; Waldvogel, U, Epoxy Polymers, Pascault, J-P; Williams, R J J, Eds., Wiley, 2010.
Additives used in final products	-	Plasticizers/flexibilizers: epoxidized oils, low molecular polyamides, polysulfidesdibutyl phthalate, condensation products of adipic acid and glycols, isodecyl pelargonate, cyclohexyl pyrrolidone; Other: diluents (glycide ether), modifiers, rheological additives, flame retardants; Antistatics: alkyl dipolyoxyethylene ethyl ammonium ethyl sulfate, carbon black, carbon monofiber, graphite, quaternary ammonium compound, silver-coated basalt, tin oxide; Release: calcium carbonate, carnauba wax, ceramic microspheres, ethylene bis stearoformamide, montan wax, silicone oil; Slip: carbon fiber, PTFE, sorbitan tristearate	
Applications	-	adhesives, encapsulating compounds, biosensors, bonding and adhesives, coatings, composites (building/construction, encapsulation, marine, electrical/electronics, aircraft, communication satellites, automotive, pipes, consumer products), electrical/electronics (printed circuit panels, conductive adhesives), flooring, fuel cells, repair mortar, semiconductor packaging, surface protective coatings (protective and decorative - automotive, metal cans, industrial flooring, anticorrosive paints), tooling and casting, wear resistant tools	
BLENDS			
Suitable polymers	-	PA, PBT, PC, PCL, PEO, PMMA, PP, PSU, PVP	Jin, H; Yang, B; Jin, F-L; Park, S-J, J. Ind. Eng. Chem., 25, 9-11, 2015.

EPDM ethylene-propylene diene terpolymer

PARAMETER	UNIT	VALUE	REFERENCES
GENERAL			
Common name	-	ethylene-propylene diene terpolymer	
CAS name	-	EPDM rubber	
Acronym	-	EPDM	
CAS number	-	25038-36-2; 308064-28-0	
HISTORY			
Person to discover	-	Ziegler, K; Natta, G	
Date	-	1951; 1962	
Details	-	discovery of catalyst essential for polymerization; commercial production	
SYNTHESIS			
Monomer(s) structure	-	$CH_2=CH_2$; $CH_3CH=CH_2$; diene (e.g., dicylopentadiene, ethylidene norbornene)	
Monomer(s) CAS number(s)	-	74-85-1; 115-07-1;	
Monomer(s) molecular weight(s)	dalton, g/mol, amu	28.05; 42.08;	
Monomer ratio (general)	-	ethylene – 50%, diene – 4%	
Ethylene content	wt%	42-85	
Ethylenenorbornene (vinylnorbornene) content	wt%	0.5-10.0	
Propylene content	wt%	10-53	
Method of synthesis	-	ENB is present in solution, catalyst is added and ethylene and propylene are bubbled through the solution	Bavarian, N; Baird, M C; Parent, J S, Macromol. Chem. Phys., 202, 3248-52, 2001.
Temperature of polymerization	°C	0	
Catalyst	-	Ziegler-Natta, metallocene	
Number average molecular weight, M_n	dalton, g/mol, amu	120,000-410,000	Kontos, E G, Antec, 2256-60, 1999; Young, H W; Brignac, S D; Kolbert, A C, Antec, 3429-4, 1997.
Mass average molecular weight, M_w	dalton, g/mol, amu	150,000-1,638,000	Kontos, E G, Antec, 2256-60, 1999; Young, H W; Brignac, S D; Kolbert, A C, Antec, 3429-4, 1997.
Polydispersity, M_w/M_n	-	1.6-6.3; 2-2.5 (typical)	Kontos, E G, Antec, 2256-60, 1999; Young, H W; Brignac, S D; Kolbert, A C, Antec, 3429-4, 1997; Snijders, E A; Boersma, A; van Baarle, B; Noordermeer, Polym. Deg. Stab., 89, 200-7, 2005.
Degree of branching	mol-ecule^{-1}	27-28	Mitra, S; Jorgensen, M; Pedersen, W B; Almdal, K; Banerjee, J. Appl. Polym.Sci., 113, 2962-72, 2009.
Crystallinity	%	13-21 (high ethylene content); 0 (low ethylene content; e.g., Et<70%)	Parikh, D R; Edmondson, M S; Smith, B W; Winter, J M; Castille, M J; Magee, J M; Patel, R M; Karajala, T P, Antec, 3434-9, 1997; Mitra, S; Jorgensen, M; Pedersen, W B; Almdal, K; Banerjee, J. Appl. Polym.Sci., 113, 2962-72, 2009.
Cell type (lattice)	-	orthorhombic; pseudo-hexagonal	
Cell dimensions	nm	a:b:c=0.788:0.497:0.254 (86 mol% Et, similar to PE); a:b:c=0.866:0.50:0.254 (75 mol%, stretched; crystallinity disappears on heating)	Bassi, I W; Corradini, P; Fagherazzi, G; Valvassori, A, Eur. Polym. J., 6, 709-18, 1970.

EPDM ethylene-propylene diene terpolymer

PARAMETER	UNIT	VALUE	REFERENCES
COMMERCIAL POLYMERS			
Some manufacturers	-	Crosspolimeri, Dow; ExxonMobil; Arlanxeo	
Trade names	-	Poligom; Nordel; Vistalon; Buna, Keltan	
PHYSICAL PROPERTIES			
Density at 20°C	g cm^{-3}	0.85-0.90	
Color	-	white to off-white	
Refractive index, 20°C	-	1.48 (vulcanized)	
Odor	-	none to mild	
Decomposition temperature	°C	>300	
Storage temperature	°C	<35	
Glass transition temperature	°C	-48 to -69	
Thermal conductivity	W m^{-1} K^{-1}	0.24; 0.29	Azizi, S, Momen, G, Ouellet-Plamondon, C, David, E, Polym. Testing, 84, 106281, 2020; Zirnstein, B, Schulze, D, Schartel, B, Thermochim. Acta, 673, 2019, 92-104, 2019.
Heat capacity	MJ m^{-3}	1.7	Zirnstein, B, Schulze, D, Schartel, B, Thermochim. Acta, 673, 2019, 92-104, 2019.
Long term service temperature	°C	-54 to 100	
Hildebrand solubility parameter	MPa$^{0.5}$	16.0-16.5	
Surface tension	mN m^{-1}	34.5	Wu, S, Polym. Eng. Sci., 27, 335, 1987.
Dielectric constant at 100 Hz/1 MHz	-	2.35	Canaud, C; Visconte, L L Y; Sens, M A; Nunes, R C R, Polym. Deg. Stab., 70, 259-62, 2000.
Volume resistivity	ohm-m	1E13 to 1E16	
Surface resistivity	ohm	1.5E16	
Electrical conductivity	S cm^{-1}	5.2E-16	Zirnstein, B, Schulze, D, Schartel, B, Thermochim. Acta, 673, 2019, 92-104, 2019.
Electric strength K20/P50, d=0.60.8 mm	kV mm^{-1}	27.4	
Coefficient of friction	-	1.5	Martinez, L; Nevshupa, R; Felhos, D; de Segovia, J L; Roman, E, Tribology Int., 2011 in press.
Diffusion coefficient of nitrogen	cm^2 s^{-1} x10^7	5.5	Rutherford, S W; Limmer, D T; Smith, M G; Honnell, K G, Polymer, 48, 6719-27, 2007.
Diffusion coefficient of oxygen	cm^2 s^{-1} x10^7	6.5	Rutherford, S W; Limmer, D T; Smith, M G; Honnell, K G, Polymer, 48, 6719-27, 2007.
Diffusion coefficient of water at 70°C	m^2 s^{-1}	1.2-3.0E-10	Lacuve, M, Colin, X, Perrin, L, Flandin, L, Tanzeghti, H, Polym. Deg. Stab., 168, 108949, 2019.
Contact angle of water, 20°C	degree	91-110	Martinez, L; Nevshupa, R; Felhos, D; de Segovia, J L; Roman, E, Tribology Int., 2011 in press.
Surface free energy	mJ m^{-2}	32.5	

EPDM ethylene-propylene diene terpolymer

PARAMETER	UNIT	VALUE	REFERENCES
MECHANICAL & RHEOLOGICAL PROPERTIES			
Tensile strength	MPa	8.8-33.12	Wang, F; Zhang, Y; Zhang, B B; Hong, R Y; Kumar, M R; Xie, C R, Composites: Part B, 2015, in press.
Tensile stress at yield	MPa	5-12.5	
Tensile creep modulus, 1000 h, elongation 0.5 max	MPa	13.7-16.9	
Elongation	%	250-760	
Flexural strength	MPa	18-23	
Flexural modulus	MPa	580-770	
Tear strength	kN m^{-1}	114-142	
Izod impact strength, notched, 23°C	J m^{-1}	540-680 to NB; 16,270-17,330	Wang, F; Zhang, Y; Zhang, B B; Hong, R Y; Kumar, M R; Xie, C R, Composites: Part B, 2015, in press.
Izod impact strength, notched, -30°C	J m^{-1}	91-96	
Compression set, 24h 70°C	%	50	
Shore A hardness	-	50-90	
Shrinkage	%	1.2-1.4	
Mooney viscosity	-	18-100	
Melt viscosity, shear rate=0 s^{-1}	MPa s	0.123-25	
Melt index, 190°C/2.16 kg	g/10 min	0.5-4.1	
CHEMICAL RESISTANCE			
Acid dilute/concentrated	-	fair-poor	
Alcohols	-	very good	
Alkalis	-	good	
Aliphatic hydrocarbons	-	poor	
Aromatic hydrocarbons	-	poor	
Esters	-	good	
Greases & oils	-	poor	
Halogenated hydrocarbons	-	poor	
Ketones	-	good	
FLAMMABILITY			
Limiting oxygen index	% O$_2$	16.9, 20.6	Hirsch, D B; Beeson, H D, Improved methods to determine flammability of aerospace materials, Halon Options Technical Working Conference, 2001; Zirnstein, B, Schulze, D, Schartel, B, Thermochim. Acta, 673, 2019, 92-104, 2019.
UL 94 rating	-	V-2; V-0 (flame retarded); HB	-; Zirnstein, B, Schulze, D, Schartel, B, Thermochim. Acta, 673, 2019, 92-104, 2019.
WEATHER STABILITY			
Excitation wavelengths	nm	274	
Emission wavelengths	nm	365	

EPDM ethylene-propylene diene terpolymer

PARAMETER	UNIT	VALUE	REFERENCES
Products of degradation	-	hydroperoxides, unsaturations and products of their degradation, crosslinks, chain scission, caboxylic acids, alcohols, aldehydes, and radicals	
Stabilizers	-	UVA: 2-hydroxy-4-octyloxybenzophenone; 2-(2H-benzotriazol-2-yl)-p-cresol; Screener: carbon black, titanium dioxide; HAS: 1,3,5-triazine-2,4,6-triamine, N,N'''[1,2-ethane-diyl-bis[[[4,6-bis[butyl(1,2,6,6-pentamethyl-4-piperidinyl)amino]-1,3,5-triazine-2-yl]imino]-3,1-propanediyl]bis[N',N''-dibutyl-N',N''-bis(1,2,2,6,6-pentamethyl-4-piperidinyl)-; bis(2,2,6,6-tetramethyl-4-piperidyl) sebacate Phenolic antioxidant: 2,6,-di-tert-butyl-4-(4,6-bis(octylthio)-1,3,5,-triazine-2-ylamino) phenol; pentaerythritol tetrakis(3-(3,5-di-tert-butyl-4-hydroxyphenyl)propionate); 2-(1,1-dimethylethyl)-6-[[3-(1,1-dimethylethyl)-2-hydroxy-5-methylphenyl] methyl-4-methylphenyl acrylate; 1,3,5-tris(3,5-di-tert-butyl-4-hydroxybenzyl)-1,3,5-triazine-2,4,6(1H,3H,5H)-trione; 2,2'-ethylidenebis (4,6-di-tert-butylphenol); Other: hydrotalcite; 2,2'-thiodiethylene bis[3-(3,5-ditert-butyl-4-hydroxyphenyl)propionate]; 4,4'-thiobis(2-t-butyl-5-methyl-phenol); 2,2'-thiobis(6-tert-butyl-4-methylphenol); octylated diphenylamine, nickel dibutyldithiocarbamate	

BIODEGRADATION

Colonized products		membranes	
Stabilizers	-	carbolic acid	

TOXICITY

NFPA: Health, Flammability, Reactivity rating	-	0-1/0-1/0; 1/1/0 (HMIS)	
Carcinogenic effect	-	not listed by ACGIH, NIOSH, NTP	

PROCESSING

Typical processing methods	-	calendering, coating, extrusion, molding	
Processing temperature	°C	175-225; 160 (vulcanization); 170 at 20 MPa pressure	Lacuve, M, Colin, X, Perrin, L, Flandin, L, Tanzeghti, H, Polym. Deg. Stab., 168, 108949, 2019.
Additives used in final products	-	Fillers: aluminum hydroxide, antimony trioxide, calcinated clay, calcium borate, calcium carbonate, graphene, huntite, hydromagnesite, magnesium carbonate, magnesium hydroxide, MWCNF, nanoclay, silica, talc, titanium dioxide, zinc oxide; Plasticizers: polyisobutylene, paraffin oil, dibutyl phthalate, dioctyl phthalated, vulcanized vegetable oil; Anti-static: polyaniline; Antiblocking: silica; Release: magnesium stearate, PTFE, siloxane; Slip: erucamide, fatty acid amide, graphite; Antioxidant: tetrakis[methylene 3-(30,50-di-tert-butyl-40-hydroxyphenyl)-propionate]methane	Khan, M A; Kumar, S S; Raghu, T S; Kotresh, T M; Sailaja, R R N, Mater. Today. Commun., 4, 50-62, 2015; Wang, F; Zhang, Y; Zhang, B B; Hong, R Y; Kumar, M R; Xie, C R, Composites: Part B, 2015, in press.
Applications	-	automotive, cable jacketing, hoses, innertubes for automobile and bicycle tires, pond liners, profiles, roofing	
Outstanding properties	-	flexibility, ozone resistance, service life	

BLENDS

Suitable polymers	-	ABS, NR, NBR, LDPE, PA6, PA66, PA12, PBT, PE, PP, PPy, PS, SAN, silicone	
Compatibilizers	-	EPDM-g-MAH	Moustafa, H; Darwish, Int. J. Adh. Adh., 61, 15-22, 2015.

EPDM ethylene-propylene diene terpolymer

PARAMETER	UNIT	VALUE	REFERENCES
ANALYSIS			
FTIR (wavenumber-assignment)	cm^{-1}/-	carbonyl – 1713; CH$_2$ – 1642	Snijders, E A; Boersma, A; van Baarle, B; Noordermeer, Polym. Deg. Stab., 89, 200-7, 2005.
Raman (wavenumber-assignment)	cm^{-1}/-	C=C – 1603; C-H – 1365	Zhao, Q; Li, X; Gao, J, Polym. Deg. Stab., 94, 339-43, 2009.

EPR ethylene-propylene rubber

PARAMETER	UNIT	VALUE	REFERENCES
GENERAL			
Common name	-	ethylene-propylene rubber	
CAS name	-	1-propene, polymer with ethene	
Acronym	-	EPR	
CAS number	-	9010-79-1; 61789-00-2	
Linear formula		$[CH_2CH(CH_3)]_X(CH_2CH_2)_Y$	
HISTORY			
Person to discover	-	Ziegler, K; Natta, G	
Date	-	1951; 1962	
Details	-	discovery of catalyst essential in polymerization; commercial production	
SYNTHESIS			
Monomer(s) structure	-	$CH_2=CH_2$; $CH_3CH=CH_2$	
Monomer(s) CAS number(s)	-	74-85-1; 115-07-1	
Monomer(s) molecular weight(s)	dalton, g/mol, amu	28.05; 42.08	
Ethylene content	%	45-78	
Method of synthesis	-	ethylene and propylene are copolymerized in the presence of catalyst solution (e.g., metallocene) and cocatalyst (e.g, MAO). Composites containing carbon nanotubes were prepared by gas-phase polymerization of the monomers in an autoclave loaded with carbon nanotubes under agitation.	Lu, L; Niu, H; Dong, J-Y; Zhao, X; Hu, X, J. Appl. Polym. Sci., 118, 3218-26, 2010; Shen, K-q, Li, H-y, Wang, Y, Carbon, 167, 932, 2020.
Temperature of polymerization	°C	40-100	
Heat of polymerization	J g^{-1}	55-188	
Mass average molecular weight, M_w	dalton, g/mol, amu	80,000-1,000,000; 180,000	Javadi, S, Panahi-Sarmad, M, Razzaghi-Kashani, M, Polymer, 145, 31-40, 2018.
Polydispersity, M_w/M_n	-	1.1-5.6	
STRUCTURE			
Crystallinity	%	20.1-43.8	van Reene, A J; Shebani, A N, Polym. Deg. Stab., 94, 1558-63, 2009.
Crystallite size	nm	1.5-2.2	
COMMERCIAL POLYMERS			
Some manufacturers	-	Lanxess	
Trade names	-	Buna	
PHYSICAL PROPERTIES			
Density at 20°C	g cm^{-3}	0.86-0.91	
Color	-	colorless, white, off-white	
Haze	%	5-10	
Odor		odorless	
Melting temperature, DSC	°C	120-170	
Degradation temperature	°C	250-300	

EPR ethylene-propylene rubber

PARAMETER	UNIT	VALUE	REFERENCES
Thermal conductivity, melt	W m^{-1} K^{-1}	0.3	
Thermal diffusivity	cm^2 s^{-1}	0.0012	
Glass transition temperature	°C	-35 to -69, -37.2	-; Javadi, S, Panahi-Sarmad, M, Razzaghi-Kashani, M, Polymer, 145, 31-40, 2018.
Specific heat capacity	J K^{-1} kg^{-1}	2800	
Maximum service temperature	°C	260 (without presence of oxygen)	
Long term service temperature	°C	-55 to 150	
Heat deflection temperature at 0.45 MPa	°C	68-102	
Relative permittivity at 100 Hz	-	3-4	
Volume resistivity	ohm-cm	10^{15}	
Cohesive energy density	J cm^{-3}	292.4	Javadi, S, Panahi-Sarmad, M, Razzaghi-Kashani, M, Polymer, 145, 31-40, 2018.
Contact angle of water, 20°C	degree	110, 24-63 (after plasma treatment)	Mrsic, I, Bäuerle, T, Ulitzsch, S, Lorenz, G, Chassé, T, Appl. Surf. Sci., 536, 147782, 2021.
Permeability to nitrogen, 25°C	cm^3 mm m^{-2} day^{-1} atm^{-1}	553	
Permeability to water vapor, 25°C	cm^3 cm cm^{-2} s^{-1} Pa^{-1} x 10^{12}	15.7	

MECHANICAL & RHEOLOGICAL PROPERTIES

Tensile strength	MPa	5.5-38	
Tensile stress at yield	MPa	19-35.2	
Elongation	%	200-730	
Tensile yield strain	%	5	
Flexural modulus	MPa	550-1,650	
Izod impact strength, notched, 23°C	J m^{-1}	37-480	
Shore A hardness	-	70-80	
Rockwell hardness	-	R80-106	
Intrinsic viscosity, 25°C	dl g^{-1}	0.89	
Mooney viscosity	-	25-69	
Melt index, 230°C/2.16 kg	g/10 min	0.5-35	

CHEMICAL RESISTANCE

Acid dilute/concentrated	-	fair/poor	
Alcohols	-	very good	
Alkalis	-	good	
Aliphatic hydrocarbons	-	poor	
Aromatic hydrocarbons	-	poor	
Esters	-	good	
Greases & oils	-	poor	
Halogenated hydrocarbons	-	poor	

EPR ethylene-propylene rubber

PARAMETER	UNIT	VALUE	REFERENCES
Ketones	-	good	

FLAMMABILITY

PARAMETER	UNIT	VALUE	REFERENCES
Ignition temperature	°C	250	
Volatile products of combustion	-	CO, CO_2, soot	

WEATHER STABILITY

PARAMETER	UNIT	VALUE	REFERENCES
Spectral sensitivity	nm	300-360	
Activation wavelengths	nm	300, 310	
Depth of UV penetration	μm	>3000; 1500	
Stabilizers	-	UVA: 2-hydroxy-4-octyloxybenzophenone; 2,2'-methylenebis(6-(2H-benzotriazol-2-yl)-4-1,1,3,3-tetrameth-ylbutyl)phenolHAS: 1,3,5-triazine-2,4,6-triamine, N,N'''[1,2-eth-ane-diyl-bis[[[4,6-bis[butyl(1,2,6,6-pentamethyl-4-piperidinyl)amino]-1,3,5-triazine-2-yl]imino]-3,1-propanediyl]bis[N',N''-di-butyl-N',N''-bis(1,2,2,6,6-pentamethyl-4-piperidinyl)-; Phenolic antioxidant: 1,3,5-tris(3,5-di-tert-butyl-4-hydroxybenzyl)-1,3,5-triazine-2,4,6(1H,3H, 5H)-trione	

TOXICITY

PARAMETER	UNIT	VALUE	REFERENCES
NFPA: Health, Flammability, Reactivity rating	-	1/0-1/0	
Carcinogenic effect	-	not listed by ACGIH, NIOSH, NTP	
Oral rat, LD_{50}	mg kg^{-1}	>5,000	
Skin rabbit, LD_{50}	mg kg^{-1}	>2,000	

PROCESSING

PARAMETER	UNIT	VALUE	REFERENCES
Typical processing methods	-	calendering, coating, extrusion, molding	
Additives used in final products	-	Fillers: aluminum hydroxide, antimony trioxide, calcinated clay, calcium borate, calcium carbonate, huntite, hydromagnesite, magnesium carbonate, magnesium hydroxide, nanoclay, silica, talc, titanium dioxide, zinc oxide; Release: fluoropolymer; Slip: erucamide, graphite	
Applications	-	automotive (including bumper, instrument panel), blending, cables, nuclear cable insulation, o-rings, roofing sheets, seals	
Outstanding properties	-	paintable, low temperature impact resistance	

BLENDS

PARAMETER	UNIT	VALUE	REFERENCES
Suitable polymers	-	PA6, PE, PP, PS	

ANALYSIS

PARAMETER	UNIT	VALUE	REFERENCES
FTIR (wavenumber-assignment)	cm^{-1}/-	carbonyl – 1712, 1737, 1780; aldehyde – 1735; hydroxyl – 3400; vinylidene – 888	

ETFE poly(ethylene-co-tetrafluoroethylene)

PARAMETER	UNIT	VALUE	REFERENCES
GENERAL			
Common name	-	poly(ethylene-co-tetrafluoroethylene)	
ACS name	-	ethene, 1,1,2,2-tetrafluoro-, polymer with ethene	
Acronym	-	ETFE	
CAS number	-	25038-71-5	
HISTORY			
Person to discover	-	Hanford, W E, Roland, J R; Joyce, R M; Sauer, J C	Ebnesajjad, S, Fluoroplastics. Vol. 2. Melt Processible Fluoroplastics, William Andrew, 2003.
Date	-	1949 (first patents); 1970 (commercialization)	
SYNTHESIS			
Monomer(s) structure	-	$CH_2=CH_2$; $CF_2=CF_2$	
Monomer(s) CAS number(s)	-	74-85-1; 116-14-3	
Monomer(s) molecular weight(s)	dalton, g/mol, amu	28.05; 100.02	
Monomer ratio	-	1	
Tetrafluoroethylene contents	%	39-71	Arai, K; Funaki, A; Phongtamrug, S; Tashiro, K, Polymer, 51, 4831-35, 2010.
Method of synthesis	-	suspension or emulsion polymerization	
Mass average molecular weight, M_w	dalton, g/mol, amu	500,000-1,200,000	
STRUCTURE			
Crystallinity	%	32.8-44.6 (depending on orientation)	Pieper, T; Heise, B; Wilke, W, Polymer, 30, 1768-75, 1989.
Cell type (lattice)	-	orthorhombic, monoclinic	
Cell dimensions	nm	a:b:c=0.96:0.925:0.5	
Unit cell angles	degree	$\gamma=96$	
Chain conformation	-	zig-zag	
COMMERCIAL POLYMERS			
Some manufacturers	-	3M; AGC; Daikin; Chemours	
Trade names	-	Dyneon; Fluon; Neoflon; Tefzel	
PHYSICAL PROPERTIES			
Density at 20°C	g cm^{-3}	1.67-1.78; 1.9 (crystalline)	
Bulk density at 20°C	g cm^{-3}	0.55-1	
Color	-	clear	
Refractive index, 25°C	-	1.403-1.42	
Transmittance	%	90-97 (visible); 92 (300 nm); 93 (350 nm); 94 (400-600 nm); 95 (700-800 nm)	Ebnesajjad, S, Fluoroplastics. Vol. 2. Melt Processible Fluoroplastics, William Andrew, 2003.
Odor	-	odorless	
Melting temperature, DSC	°C	225-280	Spencer, P, Polymers in Defence and Aerospace 2007, Rapra, 2007, paper 16.
Softening point	°C	200-300	

ETFE poly(ethylene-co-tetrafluoroethylene)

PARAMETER	UNIT	VALUE	REFERENCES
Decomposition temperature	ºC	>270; 356.5 (1%)	
Thermal expansion coefficient, 23-80ºC	ºC⁻¹	1.3-2.6E-4	
Thermal conductivity, melt	W m⁻¹ K⁻¹	0.174	Charbonneau, L, Moreno, A, Chemisana, D, Reitsma, F, Claria, F, , Renewable Sustainable Energy Rev., 82, Part 3, 2186-2201, 2018.
Glass transition temperature	ºC	60-110	
Specific heat capacity	J K⁻¹ kg⁻¹	250-380	
Service temperature range	ºC	-200 to 150	Charbonneau, L, Moreno, A, Chemisana, D, Reitsma, F, Claria, F, , Renewable Sustainable Energy Rev., 82, Part 3, 2186-2201, 2018.
Long term service temperature	ºC	-200 to 165	
Heat deflection temperature at 0.45 MPa	ºC	81-88	
Heat deflection temperature at 1.8 MPa	ºC	50-74; 204 (20% glass fiber)	
Transmittance	%	94-97	Charbonneau, L, Moreno, A, Chemisana, D, Reitsma, F, Claria, F, , Renewable Sustainable Energy Rev., 82, Part 3, 2186-2201, 2018.
Surface tension	mN m⁻¹	25.5	Becker, K, Int. Biodet. Biodeg., 41, 93-100, 1998.
Dielectric constant at 100 Hz/1 MHz	-	2.5-2.7/2.5-2.6	
Dissipation factor at 1000 Hz	E-4	8	
Volume resistivity	ohm-m	1E13 to 1E15	
Surface resistivity	ohm	1E15	
Electric strength K20/P50, d=0.60.8 mm	kV mm⁻¹	15-150	
Coefficient of friction	-	0.23 (ETFE/steel); 0.19-0.20 (20% glass fiber)	
Permeability to nitrogen, 25ºC	cm³ mm m⁻² day⁻¹ atm⁻¹	11.8	
Permeability to oxygen, 25ºC	cm³ mm m⁻² day⁻¹ atm⁻¹	39.4	
Contact angle of water, 20ºC	degree	93-94; 106; 108 (adv) and 84 (rec)	Kwok, D Y; Neuman, A W, Colloid Surf. A, 161, 49-62, 2000; Lee, S; Park, J-S; Lee, T R, Langmuir, 24, 4817-26, 2008.
Insulation value	W m⁻² K⁻¹	1.18-2.94 (depending on number of layers)	Hu, J, Chen, W, Zhao, B, Yang, D, Contruction Build. Mater., 131, 411-22, 2017.

MECHANICAL & RHEOLOGICAL PROPERTIES			
Tensile strength	MPa	33-64	
Tensile modulus	MPa	300-1,100	
Tensile stress at yield	MPa	25	
Elongation	%	150-650	
Tensile yield strain	%	4.5-23	
Flexural strength	MPa	38	
Flexural modulus	MPa	880-1380; 5,170 (20% glass fiber)	

ETFE poly(ethylene-co-tetrafluoroethylene)

PARAMETER	UNIT	VALUE	REFERENCES
Compressive strength	MPa	11-17.2	
Izod impact strength, unnotched, 23°C	J m⁻¹	480	
Poisson's ratio	-	0.43-0.45	Galliot, C; Luchsinger, R H, Polym. Testing, 30, 356-65, 2011; Hu, J; Chen, W; Zhao, B; Wang, K, Construction Bld. Mater., 75, 200-7, 2015.
Shore D hardness	-	63-75	
Shrinkage	%	0.2-4	
Melt index, 230°C/3.8 kg	g/10 min	2-50	Spencer, P, Polymers in Defence and Aerospace 2007, Rapra, 2007, paper 16.
Water absorption, 24h at 23°C	%	0.005-0.007	

CHEMICAL RESISTANCE

PARAMETER	UNIT	VALUE	REFERENCES
Acid dilute/concentrated	-	very good	
Alcohols	-	very good	
Alkalis	-	very good	
Aliphatic hydrocarbons	-	very good	
Aromatic hydrocarbons	-	very good	
Esters	-	very good	
Greases & oils	-	very good	
Halogenated hydrocarbons	-	very good	
Ketones	-	very good	
Non-solvent	-	not soluble in any solvent below 100°C	

FLAMMABILITY

PARAMETER	UNIT	VALUE	REFERENCES
Ignition temperature	°C	470	
Autoignition temperature	°C	510-515	
Limiting oxygen index	% O_2	30-31	
Heat release	kW m⁻²	16	
Heat of combustion	J g⁻¹	13,700	
Volatile products of combustion	-	HF, CO, CO_2, toxic fluorinated compounds	
UL 94 rating	-	V-0	

WEATHER STABILITY

PARAMETER	UNIT	VALUE	REFERENCES
Spectral sensitivity	nm	92-94% radiation in the range of 300-400 is transmitted	
Important initiators and accelerators	-	glass fibers decrease stability	
Stabilizers	-	not used	
Low earth orbit erosion yield	cm³ atom⁻¹ x 10⁻²⁴	0.961	Waters, D L; Banks, B A; De Groh, K K; Miller, S K R; Thorson, S D, High Performance Polym., 20, 512-22, 2008.

TOXICITY

PARAMETER	UNIT	VALUE	REFERENCES
NFPA: Health, Flammability, Reactivity rating	-	0/1/0 (HMIS)	
Carcinogenic effect	-	not listed by ACGIH, NIOSH, NTP	

ETFE poly(ethylene-co-tetrafluoroethylene)

PARAMETER	UNIT	VALUE	REFERENCES
PROCESSING			
Typical processing methods	-	extrusion, injection molding, wire coating	
Processing temperature	°C	<380	
Additives used in final products	-	Fillers: graphite, glass fiber, bronze powder	
Applications	-	aircraft insulated wires, Building insulation, components of valves and pumps, data transmission cable, filler in PTFE to reduce wear, lined pipes, release film for electronics, wire coating	
Outstanding properties	-	mechanical toughness, chemical resistance, radiation resistance	
BLENDS			
Suitable polymers	-	PA, PE, PMMA, PVDF	
ANALYSIS			
FTIR (wavenumber-assignment)	cm^{-1}/-	CH_2 – 2976; C=C – 1750, CH – 1454; CF_2 – 1000-1300	Chen, J; Asano, M; Yamaki, T; Yoshida, M, J. Membrane Sci., 269, 194-204, 2006.

EVAC ethylene-vinyl acetate copolymer

PARAMETER	UNIT	VALUE	REFERENCES
GENERAL			
Common name	-	ethylene-vinyl acetate copolymer	
ACS name	-	acetic acid ethenyl ester, polymer with ethene	
Acronym	-	EVAc	
CAS number	-	9003-20-7; 24937-78-8	
EC number	-	203-545-4	
RTECS number	-	AK0920000	
SYNTHESIS			
Monomer(s) structure	-	$CH_3COOCH=CH_2$; $CH_2=CH_2$	
Monomer(s) CAS number(s)	-	108-05-4; 74-85-1	
Monomer(s) molecular weight(s)	dalton, g/mol, amu	86.09; 28.05	
Vinyl acetate content	wt%	3-32 (packaging resins); 9-40 (industrial resins)	
Method of synthesis	-	both monomers are polymerized in the presence of initiator (details in ref.)	Lee, H-Y; Yang, T-H; Chien, I-L; Huang, H-P, Computers Chem. Eng., 33, 1371-78, 2009.
Temperature of polymerization	°C	164-187	
Catalyst	-	Ziegler-Natta, metallocene	
Number average molecular weight, M_n	dalton, g/mol, amu	16,000-42,000	McAlduff, M; Reven, L, Macromolecules, 38, 3745-53, 2005; Martin-Alfonso, J E; Franco, J M, Polym. Testing, 37, 78-85, 2014.
Mass average molecular weight, M_w	dalton, g/mol, amu	35,200-210,000	Martin-Alfonso, J E; Franco, J M, Polym. Testing, 37, 78-85, 2014.
Polydispersity, M_w/M_n	-	2.03-6	McAlduff, M; Reven, L, Macromolecules, 38, 3745-53, 2005.
STRUCTURE			
Crystallinity	%	5.9-60; 1053-12.31	Shi, X M; Zhang, J; Jin, J; Chen, S J, eXPRESS Polym. Lett., 2, 89, 623-29, 2008; Jin, J; Chen, S; Zhang, J, Polym. Deg. Stab., 95, 725-32, 2010; Martin-Alfonso, J E; Franco, J M, Polym. Testing, 37, 78-85, 2014; Carvalho de Oliveira, M C, Guimarães Soares, L, Machado Viana; M, Cardoso Diniz, A S A, de Freitas Cunha, V, Int. J. Adhesion Adhesives, 100, 102595, 2020.
Crystallite size	nm	9.1 (L_{110}); 4.8 (L_{200})	Chen, Y; Zou, H; Liang, M; Cao, Y, Thermochim Acta, 586, 1-8, 2014.
Crystallization rate coefficent	min^{-1}	1.957	Chen, Y; Zou, H; Liang, M; Cao, Y, Thermochim Acta, 586, 1-8, 2014.
Rapid crystallization temperature	°C	52-76	
COMMERCIAL POLYMERS			
Some manufacturers	-	DuPont; Exxon; LyondellBasel	
Trade names	-	Elvax; Escorene; Ultrathene	
PHYSICAL PROPERTIES			
Density at 20°C	g cm^{-3}	0.92-0.98	
Color	-	colorless to white	

EVAC ethylene-vinyl acetate copolymer

PARAMETER	UNIT	VALUE	REFERENCES
Refractive index, 20°C	-	1.467-1.498	
Haze	%	0.7-20	
Gloss, 60°, Gardner (ASTM D523)	%	34-100	
Odor	-	mild, ester-like	
Melting temperature, DSC	°C	58-112, 51/68	-; Carvalho de Oliveira, M C, Guimarães Soares, L, Machado Viana; M, Cardoso Diniz, A S A, de Freitas Cunha, V, Int. J. Adhesion Adhesives, 100, 102595, 2020.
Decomposition temperature	°C	221-240	
Fusion enthalpy	J g^{-1}	16.9-78.4	Martin-Alfonso, J E; Franco, J M, Polym. Testing, 37, 78-85, 2014.
Thermal expansion coefficient, 23-80°C	°C^{-1}	1.6-2.5E-4	
Thermal conductivity, melt	W m^{-1} K^{-1}	0.311-0.324; 0.26 (tubes)	Ghose, S; Watson, K A; Working, D C; Smith, J G; Lin, Y; Sun, Y-P, Polymers in Defence and Aerospace 2007, Rapra, 2007, paper 5.
Glass transition temperature	°C	-38 to -42	
Maximum service temperature	°C	<230	
Long term service temperature	°C	<204	
Vicat temperature VST/A/50	°C	36-86	
Hildebrand solubility parameter	MPa$^{0.5}$	17.0-19.2	
Volume resistivity	ohm-m	9.3E13	
Speed of sound	m s^{-1}	28-30	
Acoustic impedance	MRayl	1.60-1.69	

MECHANICAL & RHEOLOGICAL PROPERTIES

PARAMETER	UNIT	VALUE	REFERENCES
Tensile strength	MPa	2-41	Ghose, S; Watson, K A; Working, D C; Smith, J G; Lin, Y; Sun, Y-P, Polymers in Defence and Aerospace 2007, Rapra, 2007, paper 5.
Tensile modulus	MPa	10	Ghose, S; Watson, K A; Working, D C; Smith, J G; Lin, Y; Sun, Y-P, Polymers in Defence and Aerospace 2007, Rapra, 2007, paper 5.
Tensile stress at yield	MPa	4.6-7.4	
Elongation	%	300-860	Ghose, S; Watson, K A; Working, D C; Smith, J G; Lin, Y; Sun, Y-P, Polymers in Defence and Aerospace 2007, Rapra, 2007, paper 5.
Tensile yield strain	%	9-150	
Flexural modulus	MPa	28-121	
Elastic modulus	MPa	164-188	
Charpy impact strength, unnotched, 23°C	kJ m^{-2}	NB	
Izod impact strength, notched, 23°C	J m^{-1}	NB	
Elmendorf tear strength	g/10 min	80-260 (MD); 60-330 (TD)	
Dart drop impact	g/10 min	100-660	
Adhesive bond strength	MPa	1.5 (Al)	
Compression set	%	23 (23°C/24 h); 66 (50°C/6 h)	

EVAC ethylene-vinyl acetate copolymer

PARAMETER	UNIT	VALUE	REFERENCES
Shore A hardness	-	65-96	
Shore D hardness	-	15-43	
Brittleness temperature (ASTM D746)	°C	-70 to -85	
Melt viscosity, shear rate=92.5 s^{-1}	Pa s	1,000-2,800	Ghose, S; Watson, K A; Working, D C; Smith, J G; Lin, Y; Sun, Y-P, Polymers in Defence and Aerospace 2007, Rapra, 2007, paper 5.
Melt index, 190°C/2.16 kg	g/10 min	0.35-800	Dhamdhere, M; Deshpande, B; Patil, P; Hansen, M G, Antec, 2000.
Water absorption, equilibrium in water at 23°C	%	0.005-0.13	

CHEMICAL RESISTANCE

Acid dilute/concentrated	-	poor	
Alcohols	-	good	
Alkalis	-	poor	
Aliphatic hydrocarbons	-	poor	
Aromatic hydrocarbons	-	poor	
Halogenated hydrocarbons	-	poor	
Ketones	-	poor	
Good solvent	-	toluene, THF, MEK	

FLAMMABILITY

Ignition temperature	°C	260	
Autoignition temperature	°C	343-426	
Limiting oxygen index	% O$_2$	23; 18.5	Chang, M-K; Hwang, S-S; Liu, S-P, J. Ind. Eng. Chem., 20, 1596-1601, 2014; Du, J-Z, Jin, L, Zeng, H-Y, Shi, X-k, Zhou, E-g, Feng, B, Sheng, X, Appl. Clay Sci., 180, 105193, 2019.
Heat release	kW m^{-2}	1680; 810	Nyambo, C; Kandare, E; Wilkie, C A, Polym. Deg. Stab., 94, 513-20, 2009; Cavodeau, F; Sonnier, R; Otazaghine, B; Lopez-Cuesta, J-M; Delaite, C, Polym. Deg. Stab., 120, 23-31, 2015.
Total heat release	kJ g^{-1}	23.62	Cavodeau, F; Sonnier, R; Otazaghine, B; Lopez-Cuesta, J-M; Delaite, C, Polym. Deg. Stab., 120, 23-31, 2015.
Volatile products of combustion	-	CO_2, H_2O, acetic acid, vinyl acetate, CO, aldehydes, acrolein, alcohols, oxides of nitrogen	Hull, T R; Quinn, R E; Areri, I G; Purser, D A, Polym. Deg. Stab., 77, 235-42, 2002.
Residue at 600°C	%	1.73	Padhi, S, Jena, D P, Nayak, N C, Mater. Today: Proc., 30, 2, 355-9, 2020.
Char yield at 800°C	%	0.2	Du, J-Z, Jin, L, Zeng, H-Y, Shi, X-k, Zhou, E-g, Feng, B, Sheng, X, Appl. Clay Sci., 180, 105193, 2019.
UL rating	-	HB	Du, J-Z, Jin, L, Zeng, H-Y, Shi, X-k, Zhou, E-g, Feng, B, Sheng, X, Appl. Clay Sci., 180, 105193, 2019.

EVAC ethylene-vinyl acetate copolymer

PARAMETER	UNIT	VALUE	REFERENCES
WEATHER STABILITY			
Spectral sensitivity	nm	285, 335	Pern, F J, Solar Energy Mater. Solar Cells, 41/42, 587-615, 1996.
Excitation wavelengths	nm	350	Pern, F J, Solar Energy Mater. Solar Cells, 41/42, 587-615, 1996.
Emission wavelengths	nm	420	
Important initiators and accelerators	-	thermal processing	
Products of degradation	-	hydroperoxides, hydroxyl groups, polyene sequences, aldehyde, acetic acid	
Stabilizers	-	UVA: 2-hydroxy-4-octyloxybenzophenone; 2-(2H-benzo-triazol-2-yl)-6-dodecyl-4-methylphenol, branched & linear; propanedioic acid, [(4-methoxyphenyl)-methylene]-dimethyl ester; HAS: 1,3,5-triazine-2,4,6-triamine, N,N'''[1,2-ethane-diyl-bis[[[4,6-bis[butyl(1,2,6,6-pentamethyl-4-piperidinyl) amino]-1,3,5-triazine-2-yl]imino]-3,1-propanediyl]bis[N',N''-dibutyl-N',N''-bis(1,2,2,6,6-pentamethyl-4-piperidinyl)-; poly[[(6-[1,1,3,3-tetramethylbutyl)amino]-1,3,5-triazine-2,4-diyl] [2,2,6,6-tetramethyl-4-piperidinyl)imino]-1,6-hexanediyl[2,2,6,6-tetramethyl-4-piperidinyl)imino]]; 1,6-hexanediamine- N,N'-bis(2,2,6,6-tetramethyl-4-piperidinyl)-polymer with 2,4,6-trichlo-ro-1,3,5-triazine, reaction products with N-butyl-1-butanamine an N-butyl-2,2,6,6-tetramethyl-4-piperidinamine; butanedioic acid, dimethylester, polymer with 4-hydroxy-2,2,6,6-tetrameth-yl-1-piperidine ethanol; Phenolic antioxidant: pentaerythritol tetrakis(3-(3,5-di-tert-butyl-4-hydroxyphenyl)propionate); Amine: benzenamine, N-phenyl-, reaction products with 2,4,4-trimethylpentene; Optical brightener: 2,2'-(1,2-ethyl-enediyldi-4,1-phenylene)bisbenzoxazole, C.I.F.B. 367	
TOXICITY			
NFPA: Health, Flammability, Reactivity rating	-	0/1/0	
Carcinogenic effect	-	not listed by ACGIH, NIOSH, NTP	
Reproductive toxicity	-	not expected	
OSHA	mg m^{-3}	5 (respirable), 15 (total)	
Oral rat, LD$_{50}$	mg kg^{-1}	3,080	
Skin rabbit, LD$_{50}$	mg kg^{-1}	7,940	
PROCESSING			
Typical processing methods	-	Banbury mixer, coextrusion, cold feed extruders, extrusion, injection molding, mixing/compounding, reaction injection molding, two-roll mills	
Processing temperature	ᵒC	150-230	
Processing pressure	MPa	8-10 (injection); <2 (back pressure)	
Additives used in final products	-	Fillers: aluminum hydroxide, calcium carbonate, clay, carbon nanotubes, magnesium hydroxide, montmorillonite, red phosphorus, quartz, silica, wood fiber, zinc oxide, zinc powder; Plasticizers: EVAc is used as plasticizer in PVC and PLA therefore it seldom requires plasticization; Antistatics: 2-methyl-3-propyl benzothiazolium iodide, alkylether triethyl ammonium sulfate, organic amide; Antiblocking: fatty amide, laponite, silica; Release: methylstyryl silicone oil; Slip: eru-camide, oleamide, stearamide; Thermal stabilizer: BHT	

EVAC ethylene-vinyl acetate copolymer

PARAMETER	UNIT	VALUE	REFERENCES
Applications	-	asphalt modification, automotive wire, automotive ignition, baby products, cap liners, controlled release devices, encapsulant of photovoltaic cells, footwear, greenhouse film, hot-melt adhesives, hot-melt coatings, low-smoke cable, paints, semiconductor shields, slow burning candles, sporting goods, tubing, wall covering adhesives, wire and cable	
Outstanding properties	-	toughness, clarity, impact strength	

BLENDS

Suitable polymers	-	HDPE, LDPE, LLDPE, NBR, PA, PBT, PMMA, PP, PPy, PVC	

ANALYSIS

FTIR (wavenumber-assignment)	cm^{-1}/-	C=O – 1715, 1175-1163; C-O-C – 1160	Jin, J; Chen, S; Zhang, J, Polym. Deg. Stab., 95, 725-32, 2010.
Raman (wavenumber-assignment)	cm^{-1}/-	O-C=O – 629, 630; C=O – 1730-1740; C-H – 2800, 3000	Deshpande, B J; Dhamdhere, M S; Li, J; Hansen, M G, Antec, 1672-6, 1998.

EVOH ethylene-vinyl alcohol copolymer

PARAMETER	UNIT	VALUE	REFERENCES
GENERAL			
Common name	-	ethylene-vinyl alcohol copolymer	
CAS name	-	ethenol, polymer with ethene	
Acronym	-	EVOH	
CAS number	-	25067-34-9, 26221-27-2	
HISTORY			
Person to discover	-	Perrin, M W; Fawcett, E W; Paton, J G; Williams, E G, of ICI	Perrin, M W; Fawcett, E W; Paton, J G; Williams, E G, US Patent 2,200,429, ICI, May 14, 1940.
Date	-	1940	
Details	-	polymerization patented	
SYNTHESIS			
Monomer(s) structure	-	$CH_3COOCH=CH_2$; $CH_2=CH_2$	
Monomer(s) CAS number(s)	-	108-05-4; 74-85-1	
Monomer(s) molecular weight(s)	dalton, g/mol, amu	86.09; 28.05	
Ethylene content	mol%	27-48	
Hydroxyl group content	%	52-70	
Method of synthesis	-	EVOH is produced by a controlled hydrolysis of EVAc	
Mass average molecular weight, M_w	dalton, g/mol, amu	60,000	
STRUCTURE			
Crystallinity	%	35-67; 41.9	Franco-Urquiza, E; Santana, O O; Gámez-Pérez, J; Martínez, A B; Maspoch, M L, eXPRESS Polym. Lett., 4, 3, 153-60, 2010; Cejudo-Bastante, M J, Cejudo-Bastante, C, Cran, M J, Heredia, F J, Bigger, S W, Food Packaging Shelf Life, 24, 100502, 2020.
Cell type (lattice)	-	hexagonal, orthorhombic	
Number of chains per unit cell	-	2	
Chain conformation	-	planar-zigzag	
Crystallization enthalpy	$J\ g^{-1}$	69	Gimenez, E; Cabedo, L; Lagaron, J M; Gavara, R; Saura, J J, Antec, 2035-39, 2004.
Rapid crystallization temperature	°C	142-171; 155 (DSC)	
COMMERCIAL POLYMERS			
Some manufacturers	-	Kuraray; Noltex; Endovascular, Inc.; Chang Chun Petrochemical	Kolber, M; Shukla, P A; Kumar, A; Silberzweig, J. E., J. Vasc. Interv. Radiol., 26, 809-15, 2015.
Trade names	-	Eval; Soarnol, Onyx, Evasin	
Composition information	-	slip - 2100 ppm; antiblock - 6300-15000 ppm	
PHYSICAL PROPERTIES			
Density at 20°C	$g\ cm^{-3}$	1.12-1.2	
Bulk density at 20°C	$g\ cm^{-3}$	0.64-0.74	
Color	-	white to slightly yellowish	

EVOH ethylene-vinyl alcohol copolymer

PARAMETER	UNIT	VALUE	REFERENCES
Transmittance	%	60-75	
Haze	%	1.5-4	
Gloss, 60°, Gardner (ASTM D523)	%	82-95	
Odor	-	odorless	
Melting temperature, DSC	°C	155-205	Ethylene vinyl alcohol copolymer explained, EVAL (Kuraray).
Melting enthalpy	J g^{-1}	87.1	Muriel-Galet, V; Cran, M J; Bigger, S W; Hernandez-Munoz, P; Gavara, R, J. Food Eng., 149, 9-16, 2015.
Decomposition temperature	°C	>245	
Thermal expansion coefficient	°C^{-1}	5-8E-5 (below T_g); 1.1-1.3E-4 (above T_g)	Ethylene vinyl alcohol copolymer explained, EVAL (Kuraray).
Glass transition temperature	°C	44-72	Ethylene vinyl alcohol copolymer explained, EVAL (Kuraray).
Vicat temperature VST/A/50	°C	155-173	Ethylene vinyl alcohol copolymer explained, EVAL (Kuraray).
Surface tension	mN m^{-1}	30.6-35.5	
Surface resistivity	ohm	1.9-2.7E15	
Coefficient of friction	-	0.34 (metal); 0.66 (itself)	
Permeability to nitrogen, 25°C	cm^3 20 µm m^{-2} 24 h atm	0.017-0.13 (LDPE=3100)	Ethylene vinyl alcohol copolymer explained, EVAL (Kuraray).
Permeability to oxygen, 25°C	cm^3 20 µm m^{-2} 24 h atm	0.27-1.23 (LDPE=12,000)	Ethylene vinyl alcohol copolymer explained, EVAL (Kuraray).
Permeability to water vapor, 25°C	cm^3 mm^2 m^{-2} day^{-1}	5	Sunny, M C; Ramesh, P; Mohanan, P V; George, K E, Polym. Adv. Technol., 21, 621-31, 2010.
Contact angle of water, 20°C	degree	65; 63	Sunny, M C; Ramesh, P; Mohanan, P V; George, K E, Polym. Adv. Technol., 21, 621-31, 2010; Hutfles, J, Ravichandran, S A, Pellegrino, J, Colloids Surf. A: Physicochem. Eng. Aspects, 582, 123869, 2019.

MECHANICAL & RHEOLOGICAL PROPERTIES

PARAMETER	UNIT	VALUE	REFERENCES
Tensile strength	MPa	50-70	Ethylene vinyl alcohol copolymer explained, EVAL (Kuraray).
Tensile modulus	MPa	1,960-3,100	Ethylene vinyl alcohol copolymer explained, EVAL (Kuraray).
Tensile stress at yield	MPa	65-75	
Elongation	%	180-400	Ethylene vinyl alcohol copolymer explained, EVAL (Kuraray).
Young's modulus	MPa	2,160-3,600	Lambert, S; Chan, J; Hayashi, N; Takada, S; Michihata, K; Haneda, Y, Antec, 2411-5, 2001; Cabedo, L; Lagaron, J M; Cava, D; Saura, J J; Giminez, E, Polym. Testing, 25, 860-67, 2006.
Tear strength	N mm^{-1}	41	
Abrasion resistance (ASTM D1175)	mg/1000 cycles	1.2-2.2	Ethylene vinyl alcohol copolymer explained, EVAL (Kuraray).
Shore A hardness	-	85	
Rockwell hardness	-	M88-M100	

EVOH ethylene-vinyl alcohol copolymer

PARAMETER	UNIT	VALUE	REFERENCES
Brittleness temperature (ASTM D746)	°C	-76	
Melt viscosity, shear rate=1000 s^{-1}	Pa s	1400-2700 (190°C); 900-1700 (210°C); 600-1300 (230°C)	Ethylene vinyl alcohol copolymer explained, EVAL (Kuraray).
Melt index, 230°C/3.8 kg	g/10 min	1.6-5.5 (190°C); 1.8-13.0 (210°C); 6.2-22 (230°C)	Ethylene vinyl alcohol copolymer explained, EVAL (Kuraray).
Maximum melt temperature	°C	221	
Melt density	g cm^{-3} at 200°C	1.02-1.06	Ethylene vinyl alcohol copolymer explained, EVAL (Kuraray).

CHEMICAL RESISTANCE			
Acid dilute/concentrated	-	fair	
Alcohols	-	poor-good	
Alkalis	-	fair	
Aliphatic hydrocarbons	-	very good	
Aromatic hydrocarbons	-	very good	
Esters	-	very good	
Greases & oils	-	good	
Halogenated hydrocarbons	-	good	
Ketones	-	good	

FLAMMABILITY			
Ignition temperature	°C	>200	
Autoignition temperature	°C	500	
Heat release	kW m^{-2}	1,750	Matsuda, N; Shirasaka, H; Takayama, K; Ishikawa, T; Takeda, K, Polym. Deg. Stab., 79, 13-20, 2003.
Heat of combustion	J g^{-1}	30,000-32,000	
Volatile products of combustion	-	CO, CO_2, H2O, organic acids, aldehydes, alcohols	

BIODEGRADATION			
Typical biodegradants	-	*Salmonella*	
Stabilizers	-	chitosan	Fernandez-Saiz, P; Ocio, M J; Lagaron, J M, Carbohydrate Polym., 80, 874-84, 2010.

TOXICITY			
Carcinogenic effect	-	not listed by ACGIH, NIOSH, NTP	
Oral rat, LD$_{50}$	mg kg^{-1}	6000	
Skin rabbit, LD$_{50}$	mg kg^{-1}	4000	

ENVIRONMENTAL IMPACT			
Aquatic toxicity, *Daphnia magna*, LC$_{50}$, 48 h	mg l^{-1}	>8,300	
Aquatic toxicity, *Bluegill sunfish*, LC$_{50}$, 48 h	mg l^{-1}	>10,000	
Biological oxygen demand, BOD$_5$	-	0-5	

EVOH ethylene-vinyl alcohol copolymer

PARAMETER	UNIT	VALUE	REFERENCES
Chemical oxygen demand	mg O_2 g^{-1}	1800	

PROCESSING

PARAMETER	UNIT	VALUE	REFERENCES
Typical processing methods	-	mono- and multi-layer film extrusion, blow molding, pipe coextrusion, extrusion coating, co-injection molding, lamination	
Preprocess drying: temperature/time/residual moisture	°C/h/%	110/10/	
Processing temperature	°C	185-240; 170-220 (extrusion)	
Applications	-	food packaging, fuel tanks, tubes	
Outstanding properties	-	oxygen and odor barrier properties	

BLENDS

PARAMETER	UNIT	VALUE	REFERENCES
Suitable polymers	-	HDPE, PA, PMMA, PP, PVP	

FEP fluorinated ethylene-propylene copolymer

PARAMETER	UNIT	VALUE	REFERENCES
GENERAL			
Common name	-	fluorinated ethylene-propylene copolymer	
CAS name	-	1-propene, 1,1,2,3,3,3-hexafluoro-, polymer with 1,1,2,2-tetrafluoroethene	
Acronym	-	FEP	
CAS number	-	25067-11-2	
HISTORY			
Person to discover	-	Bro, M I and Sandt B W	
Date	-	1960 (commercialization)	
Details	-	polymerization in aqueous medium	
SYNTHESIS			
Monomer(s) structure	-	$F_2C{=}CF_2$ $\underset{CF_3CF}{\overset{CF_2}{\shortparallel}}$	
Monomer(s) CAS number(s)	-	116-14-3; 116-15-4	
Monomer(s) molecular weight(s)	dalton, g/ mol, amu	100.02; 150.03	
Tetrafluoroethylene content	%	15-50	Steward, C W; Wheland, R C; Anolick, C; Tattersall, T L, J. Vinyl Addit. Technol., 4, 4, 229-32, 1998.
Formulation example	-	water, monomers, surfactant, initiator, solvent	
Temperature of polymerization	°C	110-120; 350-380 (thermal polymerization)	Ebnesajjad, S, Fluoroplastics. Vol. 2. Melt Processible Fluoroplastics, William Andrew, 2003; Guerrero, G R; Sevilla, L; Soriano, C, Appl. Surf. Sci., 353, 686-92, 2015.
Pressure of polymerization	MPa	3.97-4.14	Ebnesajjad, S, Fluoroplastics. Vol. 2. Melt Processible Fluoroplastics, William Andrew, 2003.
Mass average molecular weight, M_w	dalton, g/ mol, amu	76,000-603,000	Kazatchkov, I B; Rosenbaum, E E; Hatzikiriakos, S G; Steward, C W, Antec, 2120-24, 1996.
Polydispersity, M_w/M_n	-	2.28-3.57	
STRUCTURE			
Crystallinity	%	65-75	Ebnesajjad, S, Applied Plastics Engineering Handbook, 2nd Ed., William Andrew, 2017, pp. 55-71.
Cell type (lattice)	-	pseudohexagonal	
Cell dimensions	nm	a:c=0.644:4.15	
Chain conformation	-	helical	
COMMERCIAL POLYMERS			
Some manufacturers	-	3M; DuPont	
Trade names	-	Dyneon	
PHYSICAL PROPERTIES			
Density at 20°C	g cm^{-3}	2.13-2.15	Ebnesajjad, S, Applied Plastics Engineering Handbook, 2nd Ed., William Andrew, 2017, pp. 55-71.
Color	-	transparent	

FEP fluorinated ethylene-propylene copolymer

PARAMETER	UNIT	VALUE	REFERENCES
Refractive index, 20°C	-	1.3380-1.344	
Odor	-	odorless	
Melting temperature, DSC	°C	250-280; 250	-; Hu, J, Li, Y, Li, C, Chen, W, Yang, D, Polym. Testing, 59, 362-70, 2017.
Thermal degradation	°C	230	
Decomposition temperature	°C	380	
Continuous working temperature	°C	less than 200	Ebnesajjad, S, Applied Plastics Engineering Handbook, 2nd Ed., William Andrew, 2017, pp. 55-71.
Thermal expansion coefficient, 23-80°C	°C^{-1}	0.8-2.8E-4	
Thermal conductivity, melt	W m^{-1} K^{-1}	0.19-0.24	
Glass transition temperature	°C	80, 112	-; Hu, J, Li, Y, Li, C, Chen, W, Yang, D, Polym. Testing, 59, 362-70, 2017.
Specific heat capacity	J K^{-1} kg^{-1}	1100-1200	
Maximum service temperature	°C	-267 to 205	
Long term service temperature	°C	150-200	
Heat deflection temperature at 0.45 MPa	°C	50-79	
Heat deflection temperature at 1.8 MPa	°C	48-70; 91 (20% glass fiber)	
Molar volume	kmol m^{-3}	19.0, 4.0, 3.0	
Interaction radius		4.0	
Hildebrand solubility parameter	MPa$^{0.5}$	19.6	
Surface tension	mN m^{-1}	20.5	Becker, K, Int. Biodet. Biodeg., 41, 93-100, 1998.
Dielectric constant at 100 Hz/1 MHz	-	2.1/2.1	
Dielectric loss factor at 1 kHz	-	0.0002	
Relative permittivity at 100 Hz	-	0.0002	
Dissipation factor at 1000 Hz	E-4	3-8	
Dissipation factor at 1 MHz	E-4	7	
Volume resistivity	ohm-m	1E15-1E18	
Surface resistivity	ohm	1E15-1E16	
Electric strength K20/P50, d=0.608 mm	kV mm^{-1}	19.7-20	
Surface arc resistance	s	165	
Coefficient of friction	-	0.05-0.67 (itself, static); 0.3 (FEP/steel); 0.11-0.12 (20% glass fiber)	
Permeability to nitrogen, 25°C	cm^3 m^{-2} day^{-1} atm^{-1}	5,000	
Permeability to oxygen, 25°C	cm^3 m^{-2} day^{-1} atm^{-1}	1,600	
Diffusion coefficient of water vapor	cm^2 s^{-1} x10^7	7.56 (20°C); 4.42 (90°C)	Hansen, C M, Prog. Org. Coat., 42, 167-78, 2001.
Contact angle of water, 20°C	degree	102-120; 118 (adv) and 98 (rec)	Lee, S; Park, J-S; Lee, T R, Langmuir, 24, 4817-26, 2008.

FEP fluorinated ethylene-propylene copolymer

PARAMETER	UNIT	VALUE	REFERENCES
Surface free energy	mJ m^{-2}	18.5	
Surface tension	mN m^{-1}	16 (lower than PTFE – 18)	Guerrero, G R; Sevilla, L; Soriano, C, Appl. Surf. Sci., 353, 686-92, 2015.

MECHANICAL & RHEOLOGICAL PROPERTIES			
Tensile strength	MPa	14-30; 12 (20% glass fiber)	
Tensile modulus	MPa	345-1,010	
Tensile stress at yield	MPa	20	
Elongation	%	150-300; 6 (20% glass fiber)	
Tensile yield strain	%	10	
Flexural strength	MPa	19-21; 28	Lee, M H, Kim, H Y, Oh, S M, Kim, B C, Bang, D, Han, J T, Woo, J S, Int. J. Hydrogen Energy, 43, 48, 21918-27, 2018.
Flexural modulus	MPa	600-797	
Compressive strength	MPa	15.2	
Izod impact strength, unnotched, 23°C	J m^{-1}	980 to NB	
Izod impact strength, notched, 23°C	J m^{-1}	570-680	
Abrasion resistance (ASTM D1044)	mg/1000 cycles	5-15	
Poisson's ratio	-	0.36 (100°C); 0.48 (23°C)	
Shore D hardness	-	55-66	
Rockwell hardness	-	R25-45	
Shrinkage	%	3.5-6.0	
Brittleness temperature (ASTM D746)	°C	-73	
Melt index, 200°C/15 kg	g/10 min	0.17-20	
Water absorption, equilibrium in water at 23°C	%	<0.01 to <0.03	

CHEMICAL RESISTANCE			
Acid dilute/concentrated	-	resistant	
Alcohols	-	resistant	
Alkalis	-	resistant	
Aliphatic hydrocarbons	-	resistant	
Aromatic hydrocarbons	-	resistant	
Esters	-	resistant	
Greases & oils	-	resistant	
Halogenated hydrocarbons	-	resistant	
Ketones	-	resistant	
Other	-	fluorine, molten alkali metals, and molten NaOH react with FEP	

FLAMMABILITY			
Ignition temperature	°C	530-550	
Autoignition temperature	°C	>500	
Limiting oxygen index	% O$_2$	>95	

FEP fluorinated ethylene-propylene copolymer

PARAMETER	UNIT	VALUE	REFERENCES
Heat of combustion	$J\ g^{-1}$	5,114-10,460	
Volatile products of combustion	-	carbonyl fluoride, CO, CO_2, HF, perfluoroisobutylene, toxic vapor	
UL 94 rating	-	V-0	

WEATHER STABILITY

Spectral sensitivity	nm	113-180	Dever, J A; McCracken, C A, High Performance Polym., 16, 289-301, 2004.
Stabilizers	-	not known to be used	
Crack growth rate in space	μm/year	30	De Groh, K K; Banks, B A; Dever, J A; Hodermarsky, J C, High Performance Polym., 16, 319-37, 2004.
Exposure in Florida		no change in tensile strength during 20 years of exposure	

TOXICITY

NFPA: Health, Flammability, Reactivity rating	-	1/0/0	
Carcinogenic effect	-	not listed by ACGIH, NIOSH, NTP	
OSHA	$mg\ m^{-3}$	5 (respirable), 15 (total dust)	
Oral rat, LD_{50}	$mg\ kg^{-1}$	>4,900 (inhalation of dust)	

PROCESSING

Typical processing methods	-	extrusion	
Preprocess drying: temperature/ time/residual moisture	^{o}C/h/%	121-149/4/	
Processing temperature	^{o}C	350-370; 310-330 (injection)	
Processing pressure	MPa	6.7-10.1; 0.17-0.34 (back)	

BLENDS

Suitable polymers	-	FEP, PANI, PF, PTFE	

ANALYSIS

FTIR (wavenumber-assignment)	cm^{-1}/-	COF – 1883, COOH – 1814, COOH (DIMER) – 1781, $CF+CF_2$ -1793, $COOCH_3$ – 1795, $CONH_2$ – 1768, CH_2OH – 3648	Ebnesajjad, S, Fluoroplastics. Vol. 2. Melt Processible Fluoroplastics, William Andrew, 2003.

FR furan resin

PARAMETER	UNIT	VALUE	REFERENCES
GENERAL			
Common name	-	furan resin	
IUPAC name	-	poly(furan-2,5-dimethylene)	
CAS name	-	2,5-furandione, polymer with 2-furanmethanol	
Acronym	-	FR	
CAS number	-	25054-13-1	
HISTORY			
Person to discover	-	Heberer, A J; Marshall, W R	Heberer, A J; Marshall, W R, US Patent 2,095,250, Glidden, Oct. 12, 1937.
Date	-	1937	
Details	-	synthetic coating composition	
SYNTHESIS			
Monomer(s) structure	-		
Monomer(s) CAS number(s)	-	98-00-0; 108-31-6	
Monomer(s) molecular weight(s)	dalton, g/mol, amu	98.10; 98.06	
Method of synthesis	-	furan can be obtained by electrochemical and chemical polymerization (details in references). The synthesis of furan resins proceeds in a pH range of 3 to 5, at a temperature range of 80 to 100°C. The condensation is stopped, when a desired viscosity value is reached, by neutralizing the liquid resin.	Gonzalez-Tejera, M J; Sanchez de la Blanca, E; Carrillo, I, Synthetic Metals, 158, 165-189, 2008; Rivero, G; Fasce, L A; Cere, S M; Manfredi, L B, Prog. Org. Coat., 77, 247-56, 2014; Fink, J K, Reactive Polymers: Fundamentals and Applications, 3rd Ed., William Andrew, 2018, pp. 287-301.
Temperature of polymerization	°C	135	
Time of polymerization	h	4	
Catalyst	-	p-toluene sulfonic acid monohydrate (cure catalyst)	Kandola, B K; Ebdon, J R; Chowdhury, K P, Polymers, 7, 298-315, 2015.
Polydispersity, M_w/M_n	-	2.33	de Vergara, U L; Sarrionandia, M; Gondra, K; Aurrekoetxea, J, Thermochim. Acta, 581, 92-9, 2014.
STRUCTURE	-		
Crystallinity	%	10	Gok, A; Can, H K; Sari, B; Talu, M, Mater. Lett., 59, 80-84, 2005.
COMMERCIAL POLYMERS			
Some manufacturers	-	Sika; Transfurans Chemicals	
Trade names	-	Asplit; Furolite	
PHYSICAL PROPERTIES			
Density at 20°C	g cm^{-3}	1.18-1.22	
Color	-	brown	
pH		4.5-4.9	

FR furan resin

PARAMETER	UNIT	VALUE	REFERENCES
Glass transition temperature	°C	252; 237	Rivero, G; Fasce, L A; Cere, S M; Manfredi, L B, Prog. Org. Coat., 77, 247-56, 2014; Asaro, L, Seoane, I T,. Fasce, L A, Cyras, V P, Manfredi, L B, Prog. Org. Coat., 133, 229-36, 2019.
Maximum service temperature	°C	200	
Contact angle of water, 20°C	degree	66.8; 81.1	Rivero, G; Fasce, L A; Cere, S M; Manfredi, L B, Prog. Org. Coat., 77, 247-56, 2014; Asaro, L, Seoane, I T,. Fasce, L A, Cyras, V P, Manfredi, L B, Prog. Org. Coat., 133, 229-36, 2019.

MECHANICAL & RHEOLOGICAL PROPERTIES

PARAMETER	UNIT	VALUE	REFERENCES
Tensile strength	MPa	40-160 (40-70% glass fiber)	
Flexural strength	MPa	50-200 (40-70% glass fiber); 365 (carbon fiber)	
Flexural modulus	MPa	6,000-10,000 (40-70% glass fiber); 25,000 (carbon fiber)	
Charpy impact strength, unnotched, 23°C	kJ m^{-2}	10-50 (40-70% glass fiber); 0.382	-; Wang, P, Deng, G, Zhu, H, Zhang, H, Wu, X, Compos. Part B: Eng., 168, 572-80, 2019.

CHEMICAL RESISTANCE

PARAMETER	UNIT	VALUE	REFERENCES
Acid dilute/concentrated	-	good/poor	
Alcohols	-	good	
Alkalis	-	good	
Aliphatic hydrocarbons	-	good	
Aromatic hydrocarbons	-	good	
Esters	-	good	
Greases & oils	-	good	
Halogenated hydrocarbons	-	good	
Ketones	-	good	
Good solvent	-	hot $HClO_4$	
Non-solvent	-	NMP, THF, DMR, $CHCl_3$, CH_3COOH, CCl_4	Li, X-G; Kang, Y; Huang, M-R, J. Comb. Chem., 8, 670-78, 2006.

FLAMMABILITY

PARAMETER	UNIT	VALUE	REFERENCES
Ignition temperature	°C	104; 65 (furfuryl alcohol)	
Time to ignition	s	98	Monti, M; Hoydonckx, H; Stappers, F; Camino, G, Eur. Polym. J., 67, 561-69, 2015.
Autoignition temperature	°C	490 (furfuryl alcohol)	
Limiting oxygen index	% O_2	22.7-23.1	Kandola, B K; Ebdon, J R; Chowdhury, K P, Polymers, 7, 298-315, 2015.
Peak heat release	kW m^{-2}	682	Monti, M; Hoydonckx, H; Stappers, F; Camino, G, Eur. Polym. J., 67, 561-69, 2015.
Total heat release	MJ m^{-2}	30.9-39	Kandola, B K; Ebdon, J R; Chowdhury, K P, Polymers, 7, 298-315, 2015.
Char at 500°C	%	38-44.2	Kandola, B K; Ebdon, J R; Chowdhury, K P, Polymers, 7, 298-315, 2015.

FR furan resin

PARAMETER	UNIT	VALUE	REFERENCES
Effective heat of combustion	MJ kg^{-1}	15.3	Monti, M; Hoydonckx, H; Stappers, F; Camino, G, Eur. Polym. J., 67, 561-69, 2015.

TOXICITY

NFPA: Health, Flammability, Reactivity rating	-	2/2/0	
Carcinogenic effect	-	not listed by ACGIH, NIOSH, NTP (furfural has been shown to cause cancer in laboratory animals)	
TLV, ACGIH	ppm	10 (furfuryl alcohol)	
NIOSH	ppm	10 (furfuryl alcohol)	
OSHA	ppm	50 (furfuryl alcohol)	
Oral rat, LD$_{50}$	mg kg^{-1}	177 (furfuryl alcohol)	
Skin rabbit, LD$_{50}$	mg kg^{-1}	400 (furfuryl alcohol)	

PROCESSING

Typical processing methods	-	BMC, compounding, curing, filament winding, molding, prepreg, pultrusion, RTM, SMC, spraying	
activation energy of curing	kJ mol^{-1}	78.24-79.83	de Vergara, U L; Sarrionandia, M; Gondra, K; Aurrekoetxea, J, Thermochim. Acta, 581, 92-9, 2014
Additives used in final products	-	Fillers: carbon fiber, glass fiber, mineral fillers, natural fibers;	
Applications	-	brake linings, composites, foundry, mortar cements, refractory materials, sand and soil consolidation, wood modification	
Outstanding properties	-	dimensional stability, hardness, resistance to fungal attack, stiffness	

BLENDS

Suitable polymers	-	PEDOT, unsaturate polyester	Kandola, B K; Ebdon, J R; Chowdhury, K P, Polymers, 7, 298-315, 2015.

ANALYSIS

NMR (chemical shifts)	ppm	3-3.4 (H_2O), 3.7 ($-CH_2-OH$), 3.9-4 ($-CH_2-$) and many more for ^1H and ^{13}C	de Vergara, U L; Sarrionandia, M; Gondra, K; Aurrekoetxea, J, Thermochim. Acta, 581, 92-9, 2014.

GEL gelatin

PARAMETER	UNIT	VALUE	REFERENCES
GENERAL			
Common name	-	gelatin	
CAS name	-	gelatins	
Acronym	-	GEL	
CAS number	-	9000-70-8	
EC number	-	232-554-6	
RTECS number	-	LX8580000	
HISTORY			
Date	-	1685	
Details	-	earliest known production of gelatin was reported in Holland	
SYNTHESIS			
Raw material for production	-	gelatin is obtained from collagen, which is insoluble fraction of connective tissues and bones. Collagen contains high concentrations of proline and hydroxyproline. The 27 types of collagen have already been identified. Gelatin can also be obtained from fish (skins, bones, and fins)	Karim, A A; Bhat, R, Food Hydrocolloids, 23, 563-76, 2009.
Method of production	-	gelatin is manufactured from collagen by hydrolysis. Properties of gelatin depend on the source of collagen. Gelatin of type B (alkali-treated precursor as distinct from Type A which is made from acid-treated precursor) is obtained from cattle bones, hides and pork skin. Non-protein components (fat, minerals, albuminoids) are removed by chemical treatment, which gives purified collagen, which is then hydrolyzed to gelatin.	
Composition of gelatin	%	protein – 84-90, mineral salts – 1-2, water – rest carbon – 50.5%, hydrogen – 6.8%, oxygen – 25.2%, nitrogen – 17%	-; Ali, O M, Hashem, Y, Bekhit, A A, Khattab, S N, Elzoghby, A O, Biopolymer Nanostructures for Food Encapsulation Purposes, Academic Press, 2019, pp. 189-216.
Number average molecular weight, M_n	dalton, g/mol, amu	15,000-400,000	Ali, O M, Hashem, Y, Bekhit, A A, Khattab, S N, Elzoghby, A O, Biopolymer Nanostructures for Food Encapsulation Purposes, Academic Press, 2019, pp. 189-216.
Mass average molecular weight, M_w	dalton, g/mol, amu	2,700-1,300,000 (gelatin); 100,000-300,000 (collagen)	
Polydispersity, M_w/M_n	-	1.6-6.1	
Radius of gyration	nm	13.8-25.7	
Isoelectric point	-	8-9 (cationic), 4-5 (anionic)	Ali, O M, Hashem, Y, Bekhit, A A, Khattab, S N, Elzoghby, A O, Biopolymer Nanostructures for Food Encapsulation Purposes, Academic Press, 2019, pp. 189-216.
STRUCTURE			
Polymorphs	-	α (1 chain), β (two α chains covalently crosslinked), γ (three α chains covalently crosslinked)	
Chain conformation	-	triple-helix (collagen)	
Chain length	nm	300 (collagen)	
Avrami constants, k/n	-	n=1	
PHYSICAL PROPERTIES			
Density at 20°C	g cm⁻³	1.2-1.35	
Color	-	colorless to slightly yellow	

HANDBOOK OF POLYMERS 3ʳᵈ Edition, Copyrights 2022; ChemTec Publishing

GEL gelatin

PARAMETER	UNIT	VALUE	REFERENCES
Moisture contents	%	8-13	Gelatin Handbook. Gelatin Manufacturers Institute of America, 2012
Refractive index, 20°C	-	1.54	
Odor	-	nearly odorless	
Melting temperature, DSC	°C	20.3-33.4	Tongdeesoontorn, W, Rawdkuen, S, Reference Module in Food Science, Elsevier, 2019, 1-15.
Decomposition temperature	°C	>100	
Glass transition temperature	°C	41-52	Gomez-Estaca, J; Montero, P; Fernandez-Martin, F; Gomez-Guillen, M C, J. Food Eng., 90, 480-86, 2009.

MECHANICAL & RHEOLOGICAL PROPERTIES

PARAMETER	UNIT	VALUE	REFERENCES
Tensile strength	MPa	3.4-10.5	Tongdeesoontorn, W, Rawdkuen, S, Reference Module in Food Science, Elsevier, 2019, 1-15.
Young's modulus	MPa	3,300	Rao, Y Q, Antec, 367-71, 2006.
Water absorption, equilibrium in water at 23°C	%	34 (bovine); 40 (fish)	Gomez-Estaca, J; Montero, P; Fernandez-Martin, F; Gomez-Guillen, M C, J. Food Eng., 90, 480-86, 2009.

CHEMICAL RESISTANCE

PARAMETER	UNIT	VALUE	REFERENCES
Acid dilute/concentrated	-	poor	
Alcohols	-	good	
Alkalis	-	poor	
Aliphatic hydrocarbons	-	good	
Aromatic hydrocarbons	-	good	
Esters	-	good	
Ketones	-	good	
Good solvent	-	acetic acid, DMSO, ethylene glycol, glycerol, water (warm)	
Non-solvent	-	acetone, ethanol, THF	

FLAMMABILITY

PARAMETER	UNIT	VALUE	REFERENCES
Autoignition temperature	°C	620	

WEATHER STABILITY

PARAMETER	UNIT	VALUE	REFERENCES
Spectral sensitivity	nm	290-320	

BIODEGRADATION

PARAMETER	UNIT	VALUE	REFERENCES
Typical biodegradants	-	*Bacillus, Staphylococcus*, and fungus from the genus *Alternaria* (e.g, *Aspergillus, Penicillium* and several others) (cinematographic films)	Abrusci, C; Martin-Gonzalez, A; Del Amo, A; Corrales, T; Catalina, F, Polym. Deg. Stab, 86, 283-91, 2004; Abrusci, C; Marquina, D; Del Amo, A; Corrales, T; Catalina, F, Int. Biodet. Biodeg., 58, 142-49, 2006.

PROCESSING

PARAMETER	UNIT	VALUE	REFERENCES
Applications	-	cosmetics, food, pharmaceuticals, photographic applications	

GEL gelatin

PARAMETER	UNIT	VALUE	REFERENCES
BLENDS			
Suitable polymers	-	chitosan (scaffolds), PET, PLA, PVA, starch (edible films)	Serra, I R; Fradique, R; Vallejo, M C S; Correia, T R; Miguel, S P; Correia, I J, Mater. Sci. Eng., 55, 592-604, 2015.
ANALYSIS			
Mass spectrometry		permits determination of gelatine source	Grundy, H H; Reece, P; Buckley, M; Solazzo, C M; Dowle, A A; Ashford, D; Charlton, A J; Wadsley, M K; Collins, M J, Food Chem., 190, 276-84, 2016.

GT gum tragacanth

PARAMETER	UNIT	VALUE	REFERENCES
GENERAL			
Common name	-	gum tragacanth	
CAS name	-	gum tragacanth	
Acronym	-	GT	
CAS number	-	9000-65-1	
EC number	-	232-552-5	
RTECS number	-	XW7750000	
HISTORY			
Date	-	3,000 BC	
Details	-	Egyptians used it as a binder in cosmetics and inks	
SYNTHESIS			
Monomer(s) structure	-	its polysaccharide component is composed of arabinose, xylose, glucose, fructose, galactose, thamnose, and galacturonic acid in different proportions depending on source (species and country), its protein component contains 18 aminoacids which also vary depending on species and location	Balaghi, S; Mohannadifar, M A; Zargaraan, A; Gavlighi, H A; Mohammadi, M, Food Hydrocolloids, in press 2011; Anderson, D M W, Bridgeman, M M E, Phytochemistry, 24, 10, 2301-4, 1985.
Source	-	dried exudate from stems and branches of *Astragalus genus*	Nejatian, M, Abbasi, S, Azarikia, F, Int. J. Biol. Macromol., 160, 846-60, 2020.
Mass average molecular weight, M_w	dalton, g/mol, amu	180,000-1,600,000	Mohammadifar, M A; Musavi, S M; Kiumarsi, A; Williams, P A, Int. J. Biol. Macromol., 38, 31-39, 2006.
Polydispersity, M_w/M_n	-	2.7	
STRUCTURE			
Molecule dimensions	nm	320-420 (length), 1.45-1.9 (width)	Gralen, N; Karrholm, M, J. Colloid Sci., 5, 1, 21-36, 1950.
PHYSICAL PROPERTIES			
Color	-	dull white to yellow; yellow (crude gum)	
Moisture	%	8.8-12.9	Farzi, M; YArmand, M S; Safari, M; Eman-Djomeh, Z; Mohammadifar, M A, Int. J. Biol. Macromol., 79, 433-9, 2015.
Odor	-	odorless	
Initial decomposition temperature	°C	252.3	Zohuriaan, M J; Shokrolahi, F, Polym. Test., 23, 575-79, 2004.
Surface tension	mN m^{-1}	52-64	Farzi, M; Yarmand, M S; Safari, M; Eman-Djomeh, Z; Mohammadifar, M A, Int. J. Biol. Macromol., 79, 433-9, 2015.
CHEMICAL RESISTANCE			
Acid dilute/concentrated	-	poor	
Alcohols	-	poor	
Alkalis	-	poor	
Aliphatic hydrocarbons	-	good	
Aromatic hydrocarbons	-	good	
Esters	-	good	
Greases & oils	-	good	

GT gum tragacanth

PARAMETER	UNIT	VALUE	REFERENCES
Halogenated hydrocarbons	-	good	
Ketones	-	good	
Good solvent	-	water, alcohol	
Non-solvent	-	acetone	
FLAMMABILITY			
Char at 500°C	%	19.5	Zohuriaan, M J; Shokrolahi, F, Polym. Test., 23, 575-79, 2004.
TOXICITY			
NFPA: Health, Flammability, Reactivity rating	-	1/1/0	
Carcinogenic effect	-	not listed by ACGIH, NIOSH, NTP	
Oral rat, LD_{50}	mg kg^{-1}	16,400; 10,200	
PROCESSING			
Applications	-	biomedical applications, coating for harvested fruits, thickener and stabilizer in food, pharmaceuticals, and cosmetics	Zare, E N, Makvandi, P, Tay, F R, Carbohydrate Polym., 212, 450-67, 2019.
BLENDS			
Compatible polymers	-	PCL (scaffolds), PVAl	Mohammadi, M R; Bahrami, S H, Mater. Sci. Eng. C, 48, 71-9, 2015.
ANALYSIS			
FTIR (wavenumber-assignment)	cm^{-1}/-	hydroxyl – 3330-3445, asymmetric and symmetric stretching vibrations of CH$_2$ groups – 2925-2940 and 2856-2857, carbonyl groups – 1640-2150, polyol carbonyl – 1040-1258, pyranose ring – 628	Nejatian, M, Abbasi, S, Azarikia, F, Int. J. Biol. Macromol., 160, 846-60, 2020.
Raman (wavenumber-assignment)	cm^{-1}/-	amide I – 1665, 1655; C=C – 1560, and more	Edwards, H G M; Falk, M J; Sibley, M G; Alvarez-Benedi, J; Rull, F, Spectrochim. Acta, Part A, 54, 903-20, 1998.

HPC hydroxypropyl cellulose

PARAMETER	UNIT	VALUE	REFERENCES
GENERAL			
Common name	-	hydroxypropyl cellulose	
CAS name	-	cellulose 2-hydroxypropyl ether	
Acronym	-	HPC	
CAS number	-	9004-64-2	
RTECS number	-	NF9050000	
HISTORY			
Person to discover	-	Hagedorn, M; Moeller, P	Hagedorn, M; Moeller, P, US Patent 1,994,038, IG Farben, Mar. 12, 1935.
Date	-	1935 (German application 1929)	
Details	-	technology of production patented	
SYNTHESIS			
Monomer(s) structure	-	cellulose, C_3H_6O	
Monomer(s) CAS number(s)	-	9004-34-6; 75-56-9	
Monomer(s) molecular weight(s)	dalton, g/mol, amu	160,000-560,000; 58.08	
Method of synthesis	-	cellulose is converted to alkali cellulose by reacting it with sodium hydroxide solution, subsequently, propylene oxide is used to obtain final product	
Maximum degree of substitution	%	26-66	Lopez-Velazquez, D; Hernandez-Sosa, A R; Perez, E, Carbohydrate Polym., 125, 224-31, 2015.
Mass average molecular weight, M_w	dalton, g/mol, amu	11,700-910,0000	
Radius of gyration	nm	33	Nilsson, S; Sundelof, L-O; Porsch, Carbohydrate Polym., 28, 265-75, 1995.
STRUCTURE			
Crystallinity	%	18.3-20.6	Matsuoo, M; Yanagida, N, Polymer, 32, 14, 2561-76, 1991.
Cell type (lattice)	-	tetragonal	Samuels, R J, J. Polym. Sci., A-2, 7, 1197, 1969.
Cell dimensions	nm	a:b:c=1.13:1.13:1.50	Samuels, R J, J. Polym. Sci., A-2, 7, 1197, 1969.
Number of chains per unit cell	-	2	Samuels, R J, J. Polym. Sci., A-2, 7, 1197, 1969.
Chain conformation	-	3/1 helix	Samuels, R J, J. Polym. Sci., A-2, 7, 1197, 1969.
COMMERCIAL POLYMERS			
Some manufacturers	-	Hercules	
Trade names	-	Klucel	
PHYSICAL PROPERTIES			
Density at 20°C	g cm^{-3}	1.17-1.21; 1.09 (amorphous); 2.05 (crystalline)	
Bulk density at 20°C	g cm^{-3}	0.29-0.40	
Color	-	white to off-white	
Refractive index, 20°C	-	1.3370	

HPC hydroxypropyl cellulose

PARAMETER	UNIT	VALUE	REFERENCES
Odor	-	odorless	
Melting temperature, DSC	°C	189-211	
Softening point	°C	130	
Glass transition temperature	°C	-4.2 to -7.0; 21-43	Lopez-Velazquez, D; Hernandez-Sosa, A R; Perez, E, Carbohydrate Polym., 125, 224-31, 2015.
Heat of fusion	kJ mol^{-1}	10.6	
Surface tension	mN m^{-1}	43.6	
Diffusion coefficient of water vapor	cm^2 s^{-1} x10^8	2.83	Yanagida, N; Matsuo, M, Polymer, 33, 5, 996-1005, 1992.

MECHANICAL & RHEOLOGICAL PROPERTIES

Tensile strength	MPa	14-24	
Tensile modulus	MPa	400-1,200	
Tensile stress at yield	MPa	16	
Elongation	%	31	
Elastic modulus	MPa	1,200 (dry); 300 (8% water)	Yakimets, I; Wellner, N; Smith, A C; Wilson, R H; Farhat, I; Mitchell, J, Mechanics Mater., 39, 500-12, 2007.
Young's modulus	MPa	700	Yanagida, N; Matsuo, M, Polymer, 33, 5, 996-1005, 1992.
Moisture absorption, equilibrium 23°C/50% RH	%	4	

CHEMICAL RESISTANCE

Acid dilute/concentrated	-	non-resistant	
Alcohols	-	soluble	
Aromatic hydrocarbons	-	insoluble	
Esters	-	insoluble	
Greases & oils	-	insoluble	
Halogenated hydrocarbons	-	soluble	
Ketones	-	soluble	
Good solvent	-	cellosolve, dioxane, ethanol, methanol, water	
Non-solvent	-	aliphatic hydrocarbons, benzene, carbon tetrachloride, toluene	

FLAMMABILITY

Autoignition temperature	°C	400	

TOXICITY

NFPA: Health, Flammability, Reactivity rating	-	2-1/1-0/0	
Carcinogenic effect	-	not listed by ACGIH, NIOSH, NTP	
OSHA	mg m^{-3}/ ppm	n/a	
Oral rat, LD$_{50}$	mg kg^{-1}	10,200	

PROCESSING

Typical processing methods	-	blow molding, compression molding, extrusion, injection molding	

HPC hydroxypropyl cellulose

PARAMETER	UNIT	VALUE	REFERENCES
Applications	-	adhesives, aerosols, coatings, cosmetics, encapsulation, extrusion (film and sheet), fibers, foods, paper, pharmaceuticals (controlled release matrix, film coating, tablet binder), textile printing	
BLENDS			
Suitable polymers	-	acrylics, PAA, PVDF	
ANALYSIS			
Raman (wavenumber-assignment)	cm^{-1}/-	OH stretching – 3600-3000, C-H stretching vibrations of CH_2 and CH_3 groups – 3000-2850, glycosidic linkeage – 1235-100	Talik, P, Moskal, P, Proniewicz, L M, Wesełucha-Birczyńska, A, J. Mol. Structure, 1210, 128062, 2020.

HDPE high density polyethylene

PARAMETER	UNIT	VALUE	REFERENCES

GENERAL

Common name	-	high density polyethylene; poly(ethylene-co-1-hexene); poly(ethylene-co-1-octene)	
IUPAC name	-	polyethene	
Acronym	-	HDPE	
CAS number	-	25213-02-9 (hexene); 26221-73.8 (octene)	
Linear formula		$\left[CH_2CH_2 \right]_n$	

HISTORY

Person to discover	-	Karl Ziegler and Gulio Natta; Paul Hogan and Robert Banks	
Date	-	1950, 1951, 1961	
Details	-	Ziegler developed catalyst which permits polymerization at low pressure and Natta explained mechanism; in 1951 Paul Hogan and Robert Banks discovered HDPE; in 1961 Phillips process (commercial process used today) was commercialized	

SYNTHESIS

Monomer(s) structure	-	$H_2C = CH_2$	
Monomer(s) CAS number(s)	-	74-85-1	
Monomer(s) molecular weight(s)	dalton, g/mol, amu	28.05	
Formulation example	-	selection of chromium/silica catalysts, Ziegler-Natta catalysts, or metallocene catalysts makes it possible to achieve low degree of branching required to obtain high density polyethylene (density >0.94), paraffin or cycloparaffin are used as diluents	
Method of synthesis	-	slurry polymerization or gas-solid polymerization	Wolf, C R; de Carmago Forte, M M; dos Santos, J H Z, Catalysis Today, 107-108, 451-57, 2005.
Temperature of polymerization	ºC	80-100 (ZN)	
Pressure of polymerization	MPa	0.1	
Catalyst	-	chromium/silica catalysts, Ziegler-Natta catalysts, or metallocene	
Heat of polymerization	kJ mol^{-1}	93.6	Kaminsky, W, Adv. Catalysis, 46, 89-159, 2001.
Mass average molecular weight, M_w	dalton, g/mol, amu	low – 1,000-100,000 (injection molding); medium – 100,000-180,000 (blow molding, film, pipe, sheet); high – 250,000-750,000 (large part blow molding, high strength thin films, pipe, sheet); extra high – 750,000-1,500,000 (extra large part blow molding)	
Polydispersity, M_w/M_n	-	1 to >10 (polydispersity increases the end-vinyl content and decreases the mechanical energy needed for processing); 11	Harlin, A, Vainio, T, Polym. Deg. Stab., 39, 1, 29-34, 1993; Mehrjerdi, A K, Bashir, T, Skrifvars, M, Heliyon, 6, 5, e04060, 2020.
Degree of branching		5/1000C	
Type of branching	-	1 side branching per 200 atoms	

STRUCTURE

Crystallinity	%	60-90	
Cell type (lattice)	-	orthorhombic	

HDPE high density polyethylene

PARAMETER	UNIT	VALUE	REFERENCES
Cell dimensions	nm	a:b:c=0.740-0.748:0.493-0.497:0.253-0.255	
Unit cell angles	degree	α:β:γ=90:90:90	
Number of chains per unit cell	-	4	
Crystal size	nm	16.69 (110), 15.2 (200)	Wang, S; Zhang, J; Liu, L; Yang, F; Zhang, Y, Solar Energy Mater. Solar Cells, 143, 120-7, 2015.
Spherulite morphology	-	banded	Tantry, S, Anantharaman, D, Thimmappa, B H S, Kamala-karan, R, Polym. Testing, 89, 106631, 2020.
Chain conformation	-	planar zig-zag; the disordered chain conformation may reduce anisotropy of spatial interaction between adjacent stiff chain segments in crystal and then conduce to a more uniform spatial distribution of chain segments, namely, the ideal hexagonal phase. The stretched polymer chains could organize into crystalline shish with extended chain conformation, whereas un-stretched, or coiled, polymer chains could crystallize into lamellae, initiated by the shish, and form crystalline kebabs having folded chain conformation.	Wang, Z, Liu, Y, Liu, C, Yang, J, Li, L, Polymer, 160, 170-80, 2019; Liu, Y, Gao, S, Hsiao, B S, Norman, A, Zhang, Y, Polymer, 153, 223-31, 2018.
Entanglement molecular weight	dalton, g/mol, amu	800-1100	
Lamellae thickness	nm	0.42-14.1 (film; depending on melt flow rate, melt extension, and annealing temperature)	Lee, S-Y; Park, S-Y; Song, H-S, Polymer, 47, 3540-47, 2006.
Rapid crystallization temperature	°C	114-120; 117 (bimodal)	

COMMERCIAL POLYMERS

Some manufacturers	-	DOW; ExxonMobil	
Trade names	-	Continuum (bimodal), Dowlex, HDPE , Unival; HDPE, Paxon	

PHYSICAL PROPERTIES

Density at 20°C	g cm^{-3}	0.94-0.965; 0.996-1.0 (crystalline)	
Color	-	white	
Refractive index, 20°C	-	1.54; 1.4261-1.4327 (amorphous); 1.520-1.582 (crystalline)	
Birefringence	-	0.01-0.012 (drawn filaments); 0.005-0.025 (film; depending on melt flow rate, melt extension, and annealing temperature)	Choi, C-H; White, J L, Intl. Polym. Proces., 13, 1, 78-87, 1998; Lee, S-Y; Park, S-Y; Song, H-S, Polymer, 47, 3540-47, 2006.
Haze	%	6	
Gloss, 60°, Gardner (ASTM D523)	%	68	
Odor	-	odorless	
Melting temperature, DSC	°C	125-135	
Melting enthalpy	J g^{-1}	185.5 (64.7% crystalline), 286.7 (100% crystalline)	Pelto, J, Verho, T, Ronkainen, H, Kaunisto, K, Karttunen, M, Polym. Testing, 77, 105897, 2019.
Decomposition temperature	°C	487	Santos, E, Rijo, B, Lemos, F, Lemos, M A N D A, Chem. Eng. J, 378, 122077, 2019.
Thermal conductivity, melt	W m^{-1} K^{-1}	0.52-0.55	
Glass transition temperature	°C	-118 to -133	

170   **HANDBOOK OF POLYMERS** 3rd Edition, Copyrights 2022; ChemTec Publishing

HDPE high density polyethylene

PARAMETER	UNIT	VALUE	REFERENCES
Specific heat capacity	MJ m^{-3} K^{-1}	1.66; 1.03-1.541 (titanium dioxide composite)	Wang, S; Zhang, J; Liu, L; Yang, F; Zhang, Y, Solar Energy Mater. Solar Cells, 143, 120-7, 2015.
Heat of fusion	J g^{-1}	240, 290 (fully crystalline PE)	Weingrill, H M, Resch-Fauster, K, Lucyshyn, T, Zauner, C, Polym. Testing, 76, 433-42, 2019; ZhongZ. Y. ZhongJ. W. Huang, J W, Sun, T, Luo, S N, Materialia, 6, 100274, 2019.
Maximum service temperature	°C	>220 (bimodal)	
Long term service temperature	°C	-50 to 82	
Heat deflection temperature at 0.45 MPa	°C	62-90; 68.3 (bimodal)	
Heat deflection temperature at 1.8 MPa	°C	44-65	
Vicat temperature VST/A/50	°C	122-129; 127-131 (bimodal)	
Hildebrand solubility parameter	MPa$^{0.5}$	calc.=16.0-16.8; exp.=17.1	
Surface tension	mN m^{-1}	26.0-28.8	
Dielectric constant at 1000 Hz/1 MHz	-	2.3-2.35/	
Dissipation factor at 1000 Hz	E-4	2	
Volume resistivity	ohm-m	1E13	
Electric strength K20/P50, d=0.60.8 mm	kV mm^{-1}	17-45	
Arc resistance	s	200-250	
Coefficient of friction	ASTM D1894	0.17 (chrome steel); 0.27-0.33 (aluminum)	Maldonado, J E, Antec, 3431-35, 1998.
Permeability to nitrogen, 25°C	cm^3 cm cm^{-2} s^{-1} Pa^{-1} x 10^{12}	0.011	
Permeability to oxygen, 25°C	cm^3 cm cm^{-2} s^{-1} Pa^{-1} x 10^{12}	0.03	
Permeability to water vapor, 25°C	cm^3 cm cm^{-2} s^{-1} Pa^{-1} x 10^{12}	0.9	
Diffusion coefficient of nitrogen	cm^2 s^{-1} x10^6	0.093	
Diffusion coefficient of oxygen	cm^2 s^{-1} x10^6	0.17	
Contact angle of water, 20°C	degree	80.5	Daniloska, V; Blazevska-Gilev, J; Dimova, V; Fajgar, R; Tomovska, R, Appl. Surface Sci., 256, 2276-83, 2010.
Speed of sound	m s^{-1}	40.5	
Acoustic impedance		2.33	

HDPE high density polyethylene

PARAMETER	UNIT	VALUE	REFERENCES
MECHANICAL & RHEOLOGICAL PROPERTIES			
Tensile strength	MPa	13.0-51.0; 17.9-29.0 (bimodal)	
Tensile modulus	MPa	500-1,100; 1,120-1,350 (bimodal)	
Tensile stress at yield	MPa	21.4-31.0; 23.5-24.8 (bimodal)	
Tensile creep modulus, 1000 h, elongation 0.5 max	MPa	6-11	
Elongation	%	250-1,200; 600-860 (bimodal)	
Tensile yield strain	%	8.7-15; 3.7-6.3 (bimodal)	
Flexural modulus	MPa	750-1,600; 621-1,680 (bimodal)	
Elastic modulus	MPa	700-1000	Abedini, A; Rahimlou, P; Asiabi, T; Ahmadi, S R; Azdast, T, J. Manuf. Proc., 19, 155-62, 2015.
Compressive strength	MPa	20	
Young's modulus	MPa	800-1005	
Fracture toughness	kJ m^{-2}	19.5-31 (23°C); 15-18 (-60°C)	Salazar, A; Rodriguez, J; Arbeiter, F; Pinter, G; Martinez, A B, Eng. Fracture Mech., 149, 190-213, 2015.
Izod impact strength, notched, 23°C	J m^{-1}	20-220; 490 (bimodal)	
Hydrostatic strength, 23°C	MPa	9-12.4 (bimodal)	
Resistance to rapid crack propagation, P_c, 0°C	bar	38.6-45.9 (bimodal)	
Resistance to rapid crack propagation, T_c	^{o}C	-2 to -17 (bimodal)	
Slow crack growth	h	4,000-15,000 (bimodal)	
Poisson's ratio	-	0.46	Salazar, A; Rodriguez, J; Arbeiter, F; Pinter, G; Martinez, A B, Eng. Fracture Mech., 149, 190-213, 2015.
Shore D hardness	-	40-69; 59 (bimodal)	
Shrinkage	%	1.5-4; 2.84 (across the flow), 1.98 (along the flow)	Chang, T; Faison, E, Polym. Eng. Sci., 41, 5, 703-10, 2001.
Brittleness temperature (ASTM D746)	^{o}C	-20 to -76.1; -60 to -75 (bimodal)	
Melt viscosity, shear rate=1000 s^{-1}	Pa s	2.57-1630 (133°C); 10.1-64.5 (192°C)	
Melt index, 190°C/2.16 kg	g/10 min	10-100 (low M_W); 0.6-10 (medium M_W); 0.06-0.15 (high M_W)	
Water absorption, equilibrium in water at 23°C	%	0.005-0.01	
CHEMICAL RESISTANCE			
Acid dilute/concentrated	-	very good	
Alcohols	-	good	
Alkalis	-	very good	
Aliphatic hydrocarbons	-	poor	
Aromatic hydrocarbons	-	poor	
Esters	-	poor	
Greases & oils	-	good to poor	
Halogenated hydrocarbons	-	poor	

172 **HANDBOOK OF POLYMERS** 3rd Edition, Copyrights 2022; ChemTec Publishing

HDPE high density polyethylene

PARAMETER	UNIT	VALUE	REFERENCES
Ketones	-	poor	
Θ solvent, Θ-temp.=127, >200, 162, 163	-	biphenyl, dibutyl phthalate, diphenyl ether, p-nonyl phenol	
Good solvent	-	1,2,4-trichlorobenzene, decalin, halogenated hydrocarbons, aliphatic ketones, xylene (all above 60°C)	
Non-solvent	-	most common solvents	
Effect of EtOH sterilization (tensile strength retention)	%	92-96	Navarrete, L; Hermanson, N, Antec, 2807-18, 1996.

FLAMMABILITY

PARAMETER	UNIT	VALUE	REFERENCES
Ignition temperature	°C	340-343	
Autoignition temperature	°C	350	
Limiting oxygen index	% O_2	17-18	
Heat of combustion	J g^{-1}	47,740	
Volatile products of combustion	-	CO, CO_2, aldehydes, benzene	
UL 94 rating	-	HB	

WEATHER STABILITY

PARAMETER	UNIT	VALUE	REFERENCES
Spectral sensitivity	nm	<300	Wypych, G, Handbook of Materials Weathering, 6th Ed., ChemTec Publishing, 2018.
Activation wavelengths	nm	300, 330-360	
Excitation wavelengths	nm	230, 265, 275, 290, 292	
Emission wavelengths	nm	295, 312, 330, 344, 358, 450	
Oxidation induction time	min	11	Ran, S; Zhao, L; Han, L; Guo, Z; Fang, Z, Thermochim. Acta, 612, 55-62, 2015.
Important initiators and accelerators	-	unsaturations, aromatic carbonyl compounds (deoxyanisoin, dibenzocycloheptadienone, flavone, 4-methoxybenzophe- none, 10-thioxanthone), hydrogen bound to tertiary carbon at branching points, aromatic amines, groups formed on oxida- tion (hydroperoxides, carbonyl, carboxyl, hydroxyl) substi- tuted benzophenones, complexes with ground-state oxygen, quinones (anthraquinone, 2-chloroanthraquinone, 2-tert-butyl- athraquinone, 1-methoxyanthraquinone, 2-ethylanthraquinone, 2-methylanthraquinone), transition metal compounds (Ni < Zn < Fe < Co), ferrocene derivatives, titanium dioxide (anatase), ferric stearate, polynuclear aromatic compounds (anthracene, phenanthrene, pyrene, naphthalene	
Products of degradation	-	free radicals, hydroperoxides, carbonyl groups, chain scission, crosslinking	

HDPE high density polyethylene

PARAMETER	UNIT	VALUE	REFERENCES
Stabilizers	-	UVA: 2-hydroxy-4-octyloxybenzophenone; phenol, 2-(5-chloro-2H-benzotriazole-2-yl)-6-(1,1-dimethylethyl)-4-methyl-; 2,2'-methylenebis(6-(2H-benzotriazol-2-yl)-4-1,1,3,3-tetramethylbutyl)phenol; 2,4-di-tert-butyl-6-(5-chloro-2H-benzotriazole-2-yl)-phenol; reaction product of methyl 3(3-(2H-benzotriazole-2-yl)-5-t-butyl-4-hydroxyphenyl propionate/PEG 300; 2-[4,6-bis(2,4-dimethylphenyl)-1,3,5-triazin-2-yl]-5-(octyloxy) phenol; Screener: titanium dioxide; zinc oxide; carbon black; Acid scavenger: hydrotalcite; Fiber: carbon nanotube; HAS: 1,3,5-triazine-2,4,6-triamine, N,N'''[1,2-ethane-diyl-bis[[[4,6-bis[butyl(1,2,6,6-pentamethyl-4-piperidinyl)amino]-1,3,5-triazine-2-yl]imino]-3,1-propanediyl] bis[N',N''-dibutyl-N',N''-bis(1,2,2,6,6-pentamethyl-4-piperidinyl)-; bis(1,2,2,6,6-pentamethyl-4-piperidyl)sebacate + methyl-1,2,2,6,6-pentamethyl-4-piperidyl sebacate; 2,2,6,6-tetramethyl-4-piperidinyl stearate; reaction products of N,N'-ethane-1,2-diylbis(1,3-propanediamine), cyclohexane, peroxidized 4-butylamino-2,2,6,6-tetramethylpiperidine and trichloro-1,3,5-triazine; poly[[(6-[1,1,3,3-tetramethylbutyl)amino]-1,3,5-triazine-2,4-diyl][2,2,6,6-tetramethyl-4-piperidinyl)imino]-1,6-hexanediyl[2,2,6,6-tetramethyl-4-piperidinyl)imino]]; 1,6-hexanediamine- N,N'-bis(2,2,6,6-tetramethyl-4-piperidinyl)-polymer with 2,4,6-trichloro-1,3,5-triazine, reaction products with N-butyl-1-butanamine an N-butyl-2,2,6,6-tetramethyl-4-piperidinamine; butanedioic acid, dimethylester, polymer with 4-hydroxy-2,2,6,6-tetramethyl-1-piperidine ethanol; alkenes, C20-24-.alpha.-, polymers with maleic anhydride, reaction products with 2,2,6,6-tetramethyl-4-piperidinamine; 1,6-hexanediamine, N,N'-bis(2,2,6,6-tetramethyl-4-piperidinyl)-, polymers with morpholine-2,4,6-trichloro-1,3,5-triazine reaction products, methylated; Phenolic antioxidant: 2,6,-di-tert-butyl-4-(4,6-bis(octylthio)-1,3,5,-triazine-2-ylamino) phenol; pentaerythritol tetrakis(3-(3,5-di-tert-butyl-4-hydroxyphenyl) propionate); octadecyl-3-(3,5-di-tert-butyl-4-hydroxyphenyl)-propionate; 3,3',3',5,5',5'-hexa-tert-butyl-a,a',a'-(mesitylene-2,4,6-triyl)tri-p-cresol; 2-(1,1-dimethylethyl)-6-[[3-(1,1-dimethylethyl)-2-hydroxy-5-methylphenyl] methyl-4-methylphenyl acrylate; 1,3,5-tris(3,5-di-tert-butyl-4-hydroxybenzyl)-1,3,5-triazine-2,4,6(1H,3H,5H)-trione; 3,4-dihydro-2,5,7,8-tetramethyl-2-(4,8,12-trimethyltridecyl)-2H-1-benzopyran-6-ol; 2',3-bis[[3-[3,5-di-tert-butyl-4-hydroxyphenyl]propionyl]]propionohydrazide; isotridecyl-3-(3,5-di-tert-butyl-4-hydroxyphenyl) propionate; 2,2'-ethylidenebis (4,6-di-tert-butylphenol); ethylene bis[3,3-bis[3-(1,1-dimethylethyl)-4-hydroxyphenyl]butanoate];1,3,5-tris(4-tert-butyl-3-hydroxy-2,6-dimethylbenzyl)-1,3,5-triazine-2,4,6-(1H,3H,5H)-trione; 2,2'-methylenebis(4-methyl-6-tertbutylphenol); 3,5-bis(1,1-dimethyethyl)-4-hydroxy-benzenepropanoic acid, C13-15 alkyl esters; 2,2'-isobutylidenebis(2,4-dimethylphenol); 1,1,3-tris(2'methyl-4'-hydroxy-5'tert-butylphenyl)butane; Phosphite: bis-(2,4-di-t-butylphenol) pentaerythritol diphosphite; tris (2,4-di-tert-butylphenyl)phosphite; trinonylphenol phosphite; distearyl pentaerythritol diphosphite; trilauryl trithiophosphite; Thiosynergist: didodecyl-3,3'-thiodipropionate; dioctadecyl 3,3'-thiodipropionate; 2,2'-thiodiethylene bis[3-(3,5-ditert-butyl-4-hydroxyphenyl)propionate]; 4,4'-thiobis(2-t-butyl-5-methylphenol); 2,2'-thiobis(6-tert-butyl-4-methylphenol); pentaerythritol tetrakis(b-laurylthiopropionate); Quencher: (2,2'-thiobis(4-tert-octyl-phenolato))-N-butylamine-nickel(II); Optical brightener: 2,2'-(2,5-thiophenediyl)bis(5-tert-butylbenzoxazole)	

HDPE high density polyethylene

PARAMETER	UNIT	VALUE	REFERENCES

BIODEGRADATION

Typical biodegradants	-	fungi, bacteria, *Actinomycetes* (3.5% mass loss per year)	Sudhakar, M; Doble, M; Sriyutha Murthy, P; Venkatesan, R, Int. Biodet. Biodeg., 61, 203-13, 2008; Ahamed, A, Vallam, P, Shiva Iyer, N, Veksha, A, Lisak, G,, J. Cleaner Prod., 278, 123956, 2021.

TOXICITY

NFPA: Health, Flammability, Reactivity rating	-	0/1/0	
Carcinogenic effect	-	not listed by ACGIH, NIOSH, NTP	
TLV, ACGIH	mg m^{-3}	3 (respirable), 10 (total)	
OSHA	mg m^{-3}	5 (respirable), 15 (total)	
Oral rat, LD$_{50}$	mg kg^{-1}	>7,950	
Skin rabbit, LD$_{50}$	mg kg^{-1}	>2,000	

ENVIRONMENTAL IMPACT

Aquatic toxicity, *Daphnia magna*, LC$_{50}$, 48 h	mg l^{-1}	17,000-24,000	Lithner, Ph D Thesis, University of Gothenburg, 2011.
Cradle to grave non-renewable energy use	MJ/kg	72-76	Harding, K G; Dennis, J S; von Blottnitz, H; Harrison, S T L, J. Biotechnol., 130, 57-66, 2007.
Cradle to pellet greenhouse gasses	kg CO_2 kg^{-1} resin	1.5-2.0	

PROCESSING

Typical processing methods	-	blow molding, blown film extrusion, cast film extrusion, extrusion, extrusion coating, injection molding, rotational molding	
Preprocess drying: temperature/time/residual moisture	°C/h/%	60-80/2	
Processing temperature	°C	193-227 (extrusion)	
Additives used in final products	-	Fillers: aluminum, barium sulfate, calcium carbonate, calcium sulfate whiskers, carbon black, diatomaceous earth, ferromagnetic powder, glass fiber, glass spheres, ground tire rubber, hollow silicates, hydrotalcite, kaolin, lignin, magnesium hydroxide, marble, mica, nickel fibers, red mud, sand, silica, soot, starch, superconductor (YBa2Cu3O7-x), talc, wollastonite, wood flour, zirconium silicate; Plasticizers: dioctyl phthalate, EPDM, EVA, glycerin, glyceryl tribenzoate, mineral oil, paraffin oil, polyethylene glycol, sunflower oil; Antistatics: carbon black, copper complex of polyacrylic acid, ethoxylated amines, fatty diethanol amines, glycerol monostearate, graphite, ionomer, lauric diethanolamide, polyethylene glycol, quaternary ammonium compound, trineoalkoxy zirconate; Antiblocking: diatomaceous earth, natural silica, siloxane spheres, synthetic silica, talc, zeolite; Release: stearyl erucamide; Slip: erucamide, ethylene bisoleamide, oleamide	
Applications	-	freezer bags, industrial pipes, mining (slurry and leachate pipes), natural gas distribution pipes, oil and gas production, portable water pipes, wire and cable	

HDPE high density polyethylene

PARAMETER	UNIT	VALUE	REFERENCES
Outstanding properties	-	impact toughness processability; bimodal PE: high temperature/high pressure performance, resistance to rapid crack propagation, slow crack growth, toughness	
BLENDS			
Suitable polymers	-	EPDM, EVA, EVOH, LDPE, LLDPE, NR, PA6, PA12, PET, PMMA, PP, PS, PTFE, PVC, SBS, SEBS, TPU, UHMWPE	
ANALYSIS			
FTIR (wavenumber-assignment)	cm^{-1}/-	C=O – 1715; CH_2 – 1474, 1464, 730, 720; $RCH=CH_2$ – 908	Stark, N M; Mantuana, L M, Polym. Deg. Stab., 86, 1-9, 2004.

176   **HANDBOOK OF POLYMERS** 3rd Edition, Copyrights 2022; ChemTec Publishing

HEC hydroxyethyl cellulose

PARAMETER	UNIT	VALUE	REFERENCES
GENERAL			
Common name	-	hydroxyethylcellulose	
CAS name	-	cellulose, 2-hydroxyethyl ether	
Acronym	-	HEC	
CAS number	-	9004-62-0	
HISTORY			
Person to discover	-	Hagedorn, M; Ziese, W; Reyle, B; Bauer, R	Hagedorn, M; Ziese, W; Reyle, B; Bauer, R; US Patent 1,876,920, IG Farben, Sept. 13, 1932.
Date	-	1932 (first application 1929)	
Details	-	patent for production of HEC	
SYNTHESIS			
Monomer(s) structure	-	cellulose, ethylene oxide	
Monomer(s) CAS number(s)	-	9004-34-6; 75-21-8	
Monomer(s) molecular weight(s)	dalton, g/mol, amu	160,000-560,000	
Method of synthesis	-	cellulose is reacted with ethylene oxide in the presence of sodium hydroxide	
Catalyst	-	NaOH	
Typical additives	%	1 (fumed silica as a flow aid)	
Mass average molecular weight, M_w	dalton, g/mol, amu	100,000-230,000	
Radius of gyration	nm	47, 41.5	Nilsson, S; Sundelof, L-O; Porsch, Carbohydrate Polym., 28, 265-75, 1995; Hörner, K D, Töpper, M, Ballauff, M, Langmuir, 13, 3, 551-8, 1997.
COMMERCIAL POLYMERS			
Some manufacturers	-	Ashland, DOW	
Trade names	-	Natrosol, Cellosize	
PHYSICAL PROPERTIES			
Density at 20°C	g cm^{-3}	1.3-1.4	
Bulk density at 20°C	g cm^{-3}	0.3-0.6	
Color	-	white to cream	
Refractive index, 20°C	-	1.5	
Odor	-	odorless	
Softening point	°C	135-140	
Decomposition temperature	°C	205-210	
Glass transition temperature	°C	120-125	
MECHANICAL & RHEOLOGICAL PROPERTIES			
Tensile strength	MPa	36-76	
Tensile modulus	MPa	30	
Elongation	%	6-7	
Elastic modulus	MPa	590	

HEC hydroxyethylcellulose

PARAMETER	UNIT	VALUE	REFERENCES
Charpy impact strength, unnotched, 23°C	kJ m^{-2}	15-99	
CHEMICAL RESISTANCE			
Acid dilute/concentrated	-	non-resistant	
Alcohols	-	insoluble	
Aromatic hydrocarbons	-	insoluble	
Esters	-	insoluble	
Greases & oils	-	insoluble	
Halogenated hydrocarbons	-	insoluble	
Ketones	-	insoluble	
Good solvent		water	
FLAMMABILITY			
Ignition temperature	°C	400	
Autoignition temperature	°C	420	
Volatile products of combustion	-	CO, CO_2, hydrocarbons	
TOXICITY			
NFPA: Health, Flammability, Reactivity rating	-	2/1/0	
Carcinogenic effect	-	not listed by ACGIH, NIOSH, NTP	
Mutagenic effect	-	not known	
Teratogenic effect	-	not known	
Reproductive toxicity	-	not known	
Oral rat, LD$_{50}$	mg kg^{-1}	>8,700	
ENVIRONMENTAL IMPACT			
Aquatic toxicity, *Daphnia magna*, LC$_{50}$, 48 h	mg l^{-1}	>100	
Aquatic toxicity, *Bluegill sunfish*, LC$_{50}$, 48 h	mg l^{-1}	>100	
Aquatic toxicity, *Fathead minnow*, LC$_{50}$, 48 h	mg l^{-1}	>100	
Aquatic toxicity, *Rainbow trout*, LC$_{50}$, 48 h	mg l^{-1}	>100	
Chemical oxygen demand	mg O_2/ mg	1.41	
PROCESSING			
Additives used in final products	-	graphene oxide (drug delivery)	Mianehrow, H; Mogadam, M H M; Sharif, F; Mazinani, S, Int. J. Pharm., 484, 1-2, 276-82, 2015.
Applications	-	agriculture (pesticides), building materials (retarder in gypsum and cement formulations), cosmetics, detergents, light-emitting diodes, paper (film forming), polymerization, removal of spilled oil, sizing agent (textiles), thickener of paints and coatings	Wu, C-L; Chen, Y, Org. Electronics, 25, 156-64, 2015.
Outstanding properties	-	thickening, pseudoplasticity, film forming	

HPMC hydroxypropyl methylcellulose

PARAMETER	UNIT	VALUE	REFERENCES
GENERAL			
Common name	-	hydroxypropyl methylcellulose	
Acronym	-	HPMC	
CAS number	-	9004-65-3	
RTECS number	-	NF9125000	
SYNTHESIS			
Monomer(s) structure	-	cellulose, C_3H_6O; CH_3OH	
Monomer(s) CAS number(s)	-	9004-34-6; 75-56-9; 67-56-1	
Monomer(s) molecular weight(s)	dalton, g/mol, amu	160,000-560,000; 58.08; 32.04	
Hydroxypropyl content	wt%	4-12	
Number average molecular weight, M_n	dalton, g/mol, amu	8,700-125,000	
Radius of gyration	nm	64	Nilsson, S; Sundelof, L-O; Porsch, Carbohydrate Polym., 28, 265-75, 1995.
COMMERCIAL POLYMERS			
Some manufacturers	-	DOW	
Trade names	-	Methocel	
PHYSICAL PROPERTIES			
Density at 20°C	g cm^{-3}	1.26-1.31	
Bulk density at 20°C	g cm^{-3}	0.3-0.7	
Color	-	white to slightly beige	
Odor		odorless	
Glass transition temperature	°C	150-190	
CHEMICAL RESISTANCE			
Acid dilute/concentrated	-	non-resistant	
Alcohols	-	insoluble	
Aromatic hydrocarbons	-	insoluble	
Esters	-	insoluble	
Greases & oils	-	insoluble	
Halogenated hydrocarbons	-	insoluble	
Ketones	-	insoluble	
Good solvent	-	water (cold)	
Non-solvent	-	ethanol, ether	
FLAMMABILITY			
Ignition temperature	°C	185	
TOXICITY			
NFPA: Health, Flammability, Reactivity rating	-	1/0/0	
Carcinogenic effect	-	not listed by ACGIH, NIOSH, NTP	
Oral rat, LD_{50}	mg kg^{-1}	10,000	

HPMC hydroxypropyl methylcellulose

PARAMETER	UNIT	VALUE	REFERENCES
PROCESSING			
Additives used in final products	-	nanoclay	Klangmuang, P; Sothornvit, R, LWT- Food Sci. Technol., 65, 222-7, 2016.
Applications	-	adhesives, cement and gypsum products, cosmetics, detergents and cleaners, food additive (emulsifier, thickening and suspending agent), paints and coatings, pharmaceutical (film coatings, ophthalmic preparations, stabilizing agent, tablet binder, viscosity increasing agent)	
BLENDS			
Compatible polymers	-	collagen, PLA, starch	Ding, C; Zhang, M; Li, G, Carbohydrate Polym., 119, 194-201, 2015.

HPMM poly(methacrylic acid-co-methyl methacrylate)

PARAMETER	UNIT	VALUE	REFERENCES
GENERAL			
Common name	-	poly(methacrylic acid-co-methyl methacrylate)	
ACS name	-	2-propenoic acid, 2-methyl-, polymer with methyl 2-methyl-2-propenoate	
Acronym	-	HPMM	
CAS number	-	25086-15-1	
HISTORY			
Person to discover	-	Brubaker, M M	Brubaker, M M; US Patent 2,244,704, DuPont, June 10, 1941.
Date	-	1941	
Details	-	HPMM synthesized for application as textile sizing material	
SYNTHESIS			
Monomer(s) structure	-		
Monomer(s) CAS number(s)	-	79-41-4 ; 80-62-6	
Monomer(s) molecular weight(s)	dalton, g/mol, amu	86.06 ; 100.12	
Monomer ratio	-	1:1 (Eudragit L100); 1:2 (Eudragit S100)	Dong, W; Bodmeier, R, Int. J. Pharmaceutis, 326, 128-38, 2006.
Methacrylic acid content	%	1.6-25	
Formulation example	-	MMA - 13.7, MAA - 1.3, potassium persulfate - 0.20, water - 285	Okubo, M; Inoue, M; Suzuki, T; Kouda, M, Colloid Polym. Sci., 282, 1150-54, 2004.
Method of synthesis	-	methacrylic acid and methyl methacrylate are polymerized in the presence of benzoyl peroxide	
Number average molecular weight, M_n	dalton, g/mol, amu	15,000-125,000	Smeets, A, Koekoekx, R, Ruelens, W, Smet, M, Van den Mooter, G, Int. J. Pharmaceutics, 574, 118885, 2020.
Mass average molecular weight, M_w	dalton, g/mol, amu	34,000-626,000	
Polydispersity, M_w/M_n	-	1.7-2.3	
STRUCTURE			
Crystallinity	%	0, amorphous	
COMMERCIAL POLYMERS			
Some manufacturers	-	Lucite; Evonic	
Trade names	-	Elvacite; Eudragit L & S	
PHYSICAL PROPERTIES			
Density at 20°C	g cm^{-3}	0.811-.852	
Color	-	white	
Refractive index, 20°C	-	1.380-1.395	
Decomposition temperature	°C	170-186	Lin, S-Y; nYu, H-L, J. Polym. Sci. A, 37, 2061-67, 1999.
Glass transition temperature	°C	105-140	

HPMM poly(methacrylic acid-co-methyl methacrylate)

PARAMETER	UNIT	VALUE	REFERENCES
Acceptor number	-	K_A: 128.2 (L100); 119.2 (S100)	Ohta, M; Buckton, G, Int. J. Pharmaceutics, 272, 121-128, 2004.
Donor number in chloroform	kJ mol^{-1}	1.76 (Eudragit L100); 2.75 (Eudragit S100)	Ohta, M; Buckton, G, Int. J. Pharmaceutics, 272, 121-128, 2004.

MECHANICAL & RHEOLOGICAL PROPERTIES

PARAMETER	UNIT	VALUE	REFERENCES
Tensile strength	MPa	15.6-25.4	Obara, S; McGinity, J W, Int. J. Pharmaceutics, 126, 1-10, 1995.
Elongation	%	1.0-2.3	
Melt viscosity, shear rate=1000 s^{-1}	mPa s	50-200	

CHEMICAL RESISTANCE

PARAMETER	UNIT	VALUE	REFERENCES
Alcohols	-	soluble	
Ketones	-	soluble	
Good solvent	-	ethanol (96%), isopropanol, acetone	Dong, W; Bodmeier, R, Int. J. Pharmaceutis, 326, 128-38, 2006.

FLAMMABILITY

PARAMETER	UNIT	VALUE	REFERENCES
Ignition temperature	°C	300	
Volatile products of combustion	-	toxic, irritant vapors	

WEATHER STABILITY

PARAMETER	UNIT	VALUE	REFERENCES
Degradation rate coefficient (365 nm)	mol g^{-1} min^{-1} x 1E7	3.57	Vinu, R; Madras, G, Polym. Deg. Stab., 93, 1440-49, 2008.

TOXICITY

PARAMETER	UNIT	VALUE	REFERENCES
HMIS: Health, Flammability, Reactivity rating	-	0/0/0	
Carcinogenic effect	-	not listed by ACGIH, NIOSH, NTP	
Mutagenic effect	-	no data	
Teratogenic effect	-	no data	
Reproductive toxicity	-	no data	
OSHA	mg m^{-3}	5 (respirable), 15 (total dust)	

PROCESSING

PARAMETER	UNIT	VALUE	REFERENCES
Typical processing methods	-	spraying, hot melt extrusion	Bruce, C; Fegely, K A; Rajabi-Siahboomi, A R; McGinity, J W, Int. J. Pharmaceutics, 341, 162-72, 2007.
Processing temperature	°C	60-90	
Processing pressure	MPa	7	
Additives used in final products	-	Plasticizer: triethyl acetate; Crystallization inhibitor	
Applications	-	pharmaceutical (controlled and transdermal delivery, site-specific peroral delivery)	Carelli, V; Di Colo, G; Nannipieri, E; Poli, B; Serafini, M F, Int. J. Pharmaceutics, 202, 103-12, 2000; Pinto, J F, Int. J. Pharmaceutics, 395, 44-52, 2010.
Outstanding properties	-	enteric (insoluble in gastric juice)	de Oliveira, H P; Albuquerque, J J F; Nogueiras, C; Rieumont, J, Int. J. Pharmaceutics, 366, 185-89, 2009.

HPMM poly(methacrylic acid-co-methyl methacrylate)

PARAMETER	UNIT	VALUE	REFERENCES
BLENDS			
Suitable polymers	-	POE, PVP	
ANALYSIS			
FTIR (wavenumber-assignment)	cm⁻¹/-	CH_2 – 1486, 1480; CH_3 – 1454, 1447, 1385; C-O – 1150; C=O – 11730	Vinu, R; Madras, G, Polym. Deg. Stab., 93, 1440-49, 2008.

IIR isobutylene-isoprene rubber

PARAMETER	UNIT	VALUE	REFERENCES
GENERAL			
Common name	-	butyl rubber, isobutylene-isoprene rubber	
CAS name	-	1,3-butadiene, 2-methyl-, polymer with 2-methyl-1-propene	
Acronym	-	IIR	
CAS number	-	9010-85-9; 308063-42-5	
HISTORY			
Person to discover	-	William J. Sparks and Robert M. Thomas	
Date	-	1937; 1943	
Details	-	butyl rubber was invented in 1937 by William J. Sparks and Robert M. Thomas of Standard Oil; it was commercialized in 1943	
SYNTHESIS			
Monomer(s) structure	-	$CH_2=C(CH_3)_2$ (isobutylene) and $CH_2=C(CH_3)CH=CH_2$ (isoprene)	
Monomer(s) CAS number(s)	-	115-11-7, 78-79-5	
Monomer(s) molecular weight(s)	dalton, g/mol, amu	56.11; 68.12	
Monomer ratio	-	isobutylene:isoprene=98:2	
Formulation example	-	calcium stearate, butylated hydroxy toluene, and epoxidized soybean oil are added to prevent dehydrohalogenation and oxidation during finishing and storage	
Method of synthesis	-	cationic polymerization of high purity isobutylene and isoprene is used to produce butyl rubber in the presence of complex systems of catalysts; polymerization is terminated by irreversible destruction of the propagating carbenium ion by the collapse of the ion pair, by hydrogen abstraction from the co-monomer, by formation of stable allylic carbenium ions, or by reaction with nucleophilic species such as alcohols or amines	
Catalyst	-	aluminum trichloride, alkyl aluminum dichloride, boron trifluoride, tin tetrachloride, and titanium tetrachloride are used as co-initiators and water, hydrochloric acid, organic acid are used as initiators	
Heat of polymerization	$J\ g^{-1}$	1042-1101 (isoprene)	Joshi, R M, Makromol. Chem., 55, 35, 1962.
Number average molecular weight, M_w	dalton, g/mol, amu	175,000	Xia, L, Li, C, Zhang, X, Wang, J, Wu, H, Guo, S, Polymer, 141, 70-8, 2018.
Mass average molecular weight, M_w	dalton, g/mol, amu	350,000-4,500,000	
Polydispersity, M_w/M_n	-	3-5	
Unsaturations	mol%	0.9-2.25	
STRUCTURE			
Cell type (lattice)	-	orthorhombic	
Cell dimensions	nm	a:b:c=0.694:1.196:1.863	
Chain conformation	-	2*8/3	
Entanglement molecular weight	dalton, g/mol, amu	710	
Rapid crystallization temperature	°C	-34	

IIR isobutylene-isoprene rubber

PARAMETER	UNIT	VALUE	REFERENCES
COMMERCIAL POLYMERS			
Some manufacturers	-	Exxon; Lanxess	
Trade names	-	Butyl Rubber; Butyl	
PHYSICAL PROPERTIES			
Density at 20°C	g cm^{-3}	0.917-0.94	
Color	-	white to off-white	
Refractive index, 20°C	-	1.5081-1.5092	
Odor	-	none to mild	
Melting temperature, DSC	°C	1.5	
Decomposition temperature	°C	>200	
Storage temperature	°C	<35	
Thermal expansion coefficient, 23-80°C	°C^{-1}	1.3-7.5E-4	
Thermal conductivity, melt	W m^{-1} K^{-1}	0.13; 0.25-0.40 (with different carbon blacks)	Wang, W; Lamba, R; Herd, C; Tandon, D; Edwards, C, Rubber World, 27-48, Sept., 2005.
Glass transition temperature	°C	-71 to -65	Xia, L, Li, C, Zhang, X, Wang, J, Wu, H, Guo, S, Polymer, 141, 70-8, 2018.
Specific heat capacity	J K^{-1} kg^{-1}	1950	
Maximum service temperature	°C	150	
Long term service temperature	°C	120	
Hildebrand solubility parameter	MPa$^{0.5}$	15.9-16.47	
Surface tension	mN m^{-1}	33.6	
Dielectric constant at 1 Hz/1 MHz	-	2.38	
Dielectric loss factor at 1 kHz	-	0.003	
Permeability to nitrogen, 25°C	cm^3 cm cm^{-2} s^{-1} Pa^{-1} x 10^{12}	0.0243	
Permeability to oxygen, 25°C	cm^3 cm cm^{-2} s^{-1} Pa^{-1} x 10^{12}	0.0977	
Permeability to water vapor, 25°C	cm^3 cm cm^{-2} s^{-1} Pa^{-1} x 10^{12}	36.7	
Diffusion coefficient of nitrogen	cm^2 s^{-1} x10^6	0.045	
Diffusion coefficient of oxygen	cm^2 s^{-1} x10^6	0.081	
MECHANICAL & RHEOLOGICAL PROPERTIES			
Tensile strength	MPa	8.4-19	
Tensile stress at yield	MPa	7-9	
Elongation	%	670-800	
Tear strength	kN m^{-1}	31-50	

IIR isobutylene-isoprene rubber

PARAMETER	UNIT	VALUE	REFERENCES
Young's modulus	MPa	80	Yan, W, Li, B, YYan, S, Wu, W, Li, Y, Results Phys., 14, 102385, 2019.
Rebound	%	12-15	
Poisson's ratio	-	0.43-0.49	
Compression set, 105°C/24 h	%	72	
Shore A hardness	-	36-90	
Brittleness temperature (ASTM D746)	°C	-60	
Mooney viscosity	-	32-75	
CHEMICAL RESISTANCE			
Acid dilute/concentrated	-	very good	
Alcohols	-	good	
Alkalis	-	very good	
Aliphatic hydrocarbons	-	poor	
Aromatic hydrocarbons	-	poor	
Esters	-	good	
Greases & oils	-	poor	
Halogenated hydrocarbons	-	poor	
Ketones	-	poor	
FLAMMABILITY			
Ignition temperature	°C	>300	
Autoignition temperature	°C	560	
Volatile products of combustion	-	CO, CO_2, smoke	
BIODEGRADATION			
Typical biodegradants	-	*Actinomycetes*	Morton, L H G; Surman, S B, Int. Biodet. Biodeg., 34, 3-4, 203-21, 1994.
TOXICITY			
NFPA: Health, Flammability, Reactivity rating	-	1/1/0; 1/1/0 (HMIS)	
Carcinogenic effect	-	not listed by ACGIH, NIOSH, NTP	
OSHA	mg m^{-3}	5 (respirable), 15 (total)	
PROCESSING			
Typical processing methods	-	mixing, molding, vulcanization	
Additives used in final products	-	Fillers: carbon black, clay, ferrites, silica, vermiculite, zinc oxide; in EMI shielding field: carbon black, nickel, silver, silver coated glass spheres, silver plated copper, silver plated aluminum, silver plated nickel; Antioxidants; Dicumyl peroxide	
Applications	-	adhesives, automotive vibration damping, belting, chewing gum, curing bladders, gas-metering diaphragms, innertubes for tires and bicycles, O-rings, protective clothing, roof coatings, shock and vibration products, sport ball bladders, steam hose, structural caulks and sealants, tire curing bladders, tire curing envelopes, water-barrier applications	

IIR isobutylene-isoprene rubber

PARAMETER	UNIT	VALUE	REFERENCES
Outstanding properties	-	impermeability, resistance to heat, resistance to ozone, energy absorption	
BLENDS			
Suitable polymers	-	bromobutyl rubber, chlorobutyl rubber, ionomers, natural rubber, EPDM, NBR, PA12, PP, PS, SBR	

LCP liquid crystalline polymers

PARAMETER	UNIT	VALUE	REFERENCES
GENERAL			
Common name	-	liquid crystalline polymers	
Acronym	-	LCP	
CAS number	-	31072-56-7 (Xydar); 144114-03-4 (Vectra C)	
HISTORY			
Person to discover	-	George-Luis LeClerc, Comte de Buffon; Friedrich Reinitze; Stephanie Louise Kwolek	
Date	-	1707-1788; 1888, 1960	
Details	-	first observation by LeClerc; Reinitzer described properties based on observations of cholesterol benzoate; DuPont prepared the first commercially available LCP material Kevlar, a polyaryl amide fiber, in the 1960s.	Lyu, X, Xiao, A, Shi, D, Li, Y, Zhou, Q-F, Polymer, 202, 122740, 2020.
SYNTHESIS			
Monomer(s) structure	-	three groups of monomers are involved: stiff units (e.g, phenyl, biphenyl, or naphthoic units), linking units (e.g., ether, ester, amide, etc.), and flexible spacer units (e.g., aliphatic or polyether chains); these monomers are used in Xydar: terephthalic acid; benzoic acid; p,p'-biphenol	Fink, J K, High Performance Polymers, William Andrew, 2008.
Monomer(s) CAS number(s)	-	100-21-0; 99-96-7; 92-88-6	
Monomer(s) molecular weight(s)	dalton, g/mol, amu	166.1308; 138.1207; 186.2066	
Method of synthesis	-	various routes can be used to link the three types of monomers involved, but polycondensation (transesterification) to form copolyesters and polyester amides is the most frequently used	Fink, J K, High Performance Polymers, William Andrew, 2008.
Temperature of polymerization	°C	80-100	Ji, L; Wu, Y; Ma, L; Yang, X, Composites: Part A, 72, 32-9, 2015.
Time of polymerization	h	12-24	Ji, L; Wu, Y; Ma, L; Yang, X, Composites: Part A, 72, 32-9, 2015.
Yield	%	24.2-60	Ji, L; Wu, Y; Ma, L; Yang, X, Composites: Part A, 72, 32-9, 2015.
Number average molecular weight, M_n	dalton, g/mol, amu	10,600-24,200	
Mass average molecular weight, M_w	dalton, g/mol, amu	12,000-47,400	
Polydispersity, M_w/M_n	-	1.8-2	
STRUCTURE			
Crystallinity	%	18-38	Kim, J Y; Kim, S H, Antec, 2942-46, 2004.
Crystallite size	nm	1-3.5 (spun fiber)	Kim, J Y; Kim, S H, Antec, 2942-46, 2004.
Cell type (lattice)	-	pseudoorthorhombic cell, the dimension of which are a = 7.87 Å, b = 5.18 Å, and c = 12.9 Å (fiber axis)	Pegoretti, A, Traina, M, Handbook of Properties of Textile and Technical Fibres, 2nd Ed., Woodhead Publishing, 2018, pp. 621-97.
COMMERCIAL POLYMERS			
Some manufacturers	-	America Quantum Leap Packaging; DuPont; Kuraray; Solvay; Saint-Gobain; Sumitomo; Celanese/Ticona	
Trade names	-	LCPh; Titan, X7G; Vecstar; Xydar; Ekonol; Sumikasuper; Vectra, Zenite	

LCP liquid crystalline polymers

PARAMETER	UNIT	VALUE	REFERENCES
PHYSICAL PROPERTIES			
Density at 20°C	g cm^{-3}	1.34-1.4; 1.5-1.81 (15-50% glass fiber)	
Bulk density at 20°C	g cm^{-3}	0.6	
Birefringence	-	0.03-0.07	Kim, J Y; Kim, S H, Antec, 2942-46, 2004.
Melting temperature, DSC	°C	221-370; 280-350 (15-50% glass fiber)	
Decomposition temperature	°C	350-400	
Thermal expansion coefficient, 23-80°C	°C^{-1}	3.6-7.6E-6; 2-10E-6 (20-45% glass fiber); 6E-05	Long, V K, Antec, 1570-76, 1998; Pegoretti, A, Traina, M, Handbook of Properties of Textile and Technical Fibres, 2nd Ed., Woodhead Publishing, 2018, pp. 621-97.
Thermal conductivity, melt	W m^{-1} K^{-1}	0.05-0.2	
Glass transition temperature	°C	95-136; 164-181 (crosslinked)	Hakemi, H, Polymer, 41, 6145-50, 2000; Igbal, M; Knijnenberg, A; Poulis, H; Dingemans, T J, Intl. J. Adhesion Adhesives, 30, 682-88, 2010.
Specific heat capacity	J K^{-1} kg^{-1}	1,000	
Heat of fusion	J g^{-1}	1.3-6.5	Sauer, B B; Kampert, W G; McLean, R S, Polymer, 44, 2721-38, 2003.
Maximum service temperature	°C	340-400	
Long term service temperature	°C	240	
Temperature index (50% tensile strength loss after 20,000 h/5000 h)	°C	130	
Heat deflection temperature at 0.45 MPa	°C	210; 273 (15-50% glass fiber)	
Heat deflection temperature at 1.8 MPa	°C	108-187; 230-340 (15-50% glass fiber); 145	-; -; McKeen, L W, Film Properties of Plastics and Elastomers, 4th Ed., William Andrew, 2017, pp. 105-45.
Vicat temperature VST/A/50	°C	128	
Vicat temperature VST/B/50	°C	145; 160-200 (15-50% glass fiber)	
Surface tension	mN m^{-1}	20-43	Gomes, L S; Demarquette, N R; Shimizu, R N; Kamal, M R, Antec, 3589-93, 2003.
Dielectric constant at 60 Hz/1 MHz	-	4.2/3.9-5	
Relative permittivity at 1 MHz	-	3	
Dissipation factor at 1 MHz	E-4	20	
Volume resistivity	ohm-m	1E13-1E14; 1E14 (15-50% glass fiber)	
Surface resistivity	ohm	1E14; 1E16 (15-50% glass fiber)	
Electric strength K20/P50, d=0.60.8 mm	kV mm^{-1}	39-47; 28-42 (15-50% glass fiber)	
Comparative tracking index	-	150-185	
Arc resistance	MV/m	3.9	

LCP liquid crystalline polymers

PARAMETER	UNIT	VALUE	REFERENCES
MECHANICAL & RHEOLOGICAL PROPERTIES			
Tensile strength	MPa	117-200; 430 (oriented); 125-200 (15-50% glass fiber); 130.5	Lusignea, R; Perdikoulias, J, Antec, 28-35, 1997; Chen, T, Mansfield, C D, Ju, L, Baird, D G, Compos. Part B: Eng, 200, 108316, 2020.
Tensile modulus	MPa	7,500-13,200; 41,000 (oriented); 12,00-17,500 (15-50% glass fiber); 17,500	Lusignea, R; Perdikoulias, J, Antec, 28-35, 1997; Chen, T, Mansfield, C D, Ju, L, Baird, D G, Compos. Part B: Eng, 200, 108316, 2020.
Tensile stress at yield	MPa	176; 140-155 (15-50% glass fiber)	
Elongation	%	1.3-4.4; 1.0-3.1 (15-50% glass fiber)	
Flexural strength	MPa	158-180; 200-280 (15-50% glass fiber)	
Flexural modulus	MPa	9,100-13,400; 12,400-20,000 (15-50% glass fiber)	
Compressive strength	MPa	70; 85-125 (15-50% glass fiber)	
Charpy impact strength, unnotched, 23°C	kJ m^{-2}	60-270; 19-48 (15-50% glass fiber)	
Charpy impact strength, notched, 23°C	kJ m^{-2}	46-95; 9-42 (15-50% glass fiber)	
Izod impact strength, unnotched, 23°C	J m^{-1}	250-430; 14-61 (15-50% glass fiber)	
Izod impact strength, notched, 23°C	J m^{-1}	60-96; 14-45 (15-50% glass fiber)	
Shear strength	MPa	8-13	Igbal, M; Knijnenberg, A; Poulis, H; Dingemans, T J, Intl. J. Adhesion Adhesives, 30, 682-88, 2010.
Rockwell hardness	-	M66-85 (15-50% glass fiber)	
Shrinkage	%	0.1-0.4; 0.1-0.6 (15-50% glass fiber)	
Melt viscosity, shear rate=1000 s^{-1}	Pa s	30-40	Guo, T; Harrison, G M; Ogale, A A, Antec, 1154-58, 2001.
Melt index, 230°C/3.8 kg	g/10 min	2	
Water absorption, equilibrium in water at 23°C	%	0.01; 0.005-0.02 (15-50% glass fiber)	
Moisture absorption, equilibrium 23°C/50% RH	%	0.03-0.04; 0.006-0.04 (15-50% glass fiber)	
CHEMICAL RESISTANCE			
Alcohols	-	resistant	
Aromatic hydrocarbons	-	resistant	
Esters	-	resistant	
Greases & oils	-	resistant	
Halogenated hydrocarbons	-	resistant	
FLAMMABILITY			
Ignition temperature	°C	93	
Autoignition temperature	°C	>540	
Char at 600°C	%	40	Wang, X, Bai, L, Tang, X, Gao, Y, Meng, F, Eur. Polym. J., 100, 146-52, 2018.
Volatile products of combustion	-	CO, CO_2, phenol	
UL 94 rating	-	V-2 to V-0	

LCP liquid crystalline polymers

PARAMETER	UNIT	VALUE	REFERENCES
WEATHER STABILITY			
Absorption wavelengths	nm	285-298, 309-310, 350-355	Marin, L; Perju, E; Damaceanu, M D, Eur. Polym. J., 47, 1284-99, 2011.
TOXICITY			
NFPA: Health, Flammability, Reactivity rating	-	1/1/0	
Carcinogenic effect	-	not listed by ACGIH, NIOSH, NTP	
TLV, ACGIH	mg m^{-3}	3 (respirable), 10 (total)	
OSHA	mg m^{-3}	5 (respirable), 15 (total)	
Oral rat, LD$_{50}$	mg kg^{-1}	>2,000	
PROCESSING			
Typical processing methods	-	extrusion, fiber spinning, injection molding, micromolding, rotational molding	
Preprocess drying: temperature/ time/residual moisture	°C/h/%	140-160/4-8/0.01	
Processing temperature	°C	270-360; 330-340 (injection molding)	
Processing pressure	MPa	35-70 (injection); 0.35 (back)	
Additives used in final products	-	Fillers: calcium carbonate, carbon black, glass fiber, graphene oxide, graphite, magnesium carbonate, mica, synthetic graphite, wollastonite	
Applications	-	adhesives, aerospace structures, audiovisual equipment, barrier films, bobbins, cameras, capsules for electronic devices, coatings, composites, connectors and sockets, electric motor components, fiber optic connectors, fuel cells bipolar plates, information storage devices, lamp sockets, LED, microwave cookware, precision molded components, printers and copiers parts, SMT components, sporting goods, under-bonnet automotive components, watches	
Outstanding properties	-	flow, heat deflection temperature, stiffness, strength, weather resistance	
BLENDS			
Suitable polymers	-	acrylics, fluorocarbon elastomers, PA6, PA6,6, PBT, PC, PDMS, PEEK, PE, PEI, PEN, PET, PP, PPO, PPS, PS; PSF, PVC, SEBS	Fink, J K, High Performance Polymers, William Andrew, 2008; DeMeuse, M T; Kiss, G, High Temperature Polymer Blends, Elsevier, 2014, 141-64.
Compatibilizers	-	maleic anhydride grafted PP, SEBS, epoxy	Fink, J K, High Performance Polymers, William Andrew, 2008.

LDPE low density polyethylene

PARAMETER	UNIT	VALUE	REFERENCES
GENERAL			
Common name	-	low-density polyethylene	
IUPAC name	-	polyethylene	
Acronym	-	LDPE	
CAS number	-	9002-88-4 (homopolymer)	
Formula		$\left[CH_2CH_2 \right]_n$	
HISTORY			
Person to discover	-	Fawcett, E; Perrin, M, both of ICI	
Date	-	1933; 1935; 1939	
Details	-	Fawcett produced PE by accident; Perrin developed sound technology; ICI begun production	
SYNTHESIS			
Monomer(s) structure	-	$H_2C = CH_2$	
Monomer(s) CAS number(s)	-	74-85-1	
Monomer(s) molecular weight(s)	dalton, g/mol, amu	28.05	
Monomer(s) expected purity(ies)	%	99.9	
Monomer ratio	-	100% ethylene or less	
Vinyl acetate content (only some grades)	wt%	3-6	
Formulation example	-	oxygen or an organic peroxide such as dibutyl, benzoyl, or diethyl peroxide used as initiator; benzene or chlorobenzene are used as solvents; tubular and autoclave reactors are used for synthesis	
Temperature of polymerization	°C	132-332	
Time of polymerization	h	100-300	
Pressure of polymerization	MPa	150-300	
Number average molecular weight, M_n	dalton, g/mol, amu	13,000-18,000	
Mass average molecular weight, M_w	dalton, g/mol, amu	69,000-411,000	
Polydispersity, M_w/M_n	-	4-30	
Degree of branching	methyl/1000 C	7.8-33	
Type of branching	-	methyl, ethyl, butyl, amyl, and longer	Zhu, H; Wang, Y; Zhang, X; Su, Y; Dong, X; Chen, Q; Zhao, Y; Geng, C; Zhu, S; Han, C C; Wang, D, Polymer, 48, 5098-106, 2007.
Unsaturations	% total	80 (vinylidene), 10 (vinyl), 10 (trans)	
STRUCTURE			
Crystallinity	%	28.8-60	
Cell type (lattice)	-	orthorhombic	
Cell dimensions	nm	a:b:c=0.738:0.493:0.253	
Crystallite thickness	nm	6.5-10.5	
Rapid crystallization temperature	°C	96-100	

LDPE low density polyethylene

PARAMETER	UNIT	VALUE	REFERENCES
Avrami constants, k/n	-	n=1.5-3	Grady, B P; Genetti, W B, Conductive Polymers and Plastics, Rupprecht, L, Ed., WilliamAndrew, Norwich, 1999.

COMMERCIAL POLYMERS

Some manufacturers	-	DOW; ExxonMobil	
Trade names	-	LDPE; LDPE	

PHYSICAL PROPERTIES

Density at 20°C	g cm^{-3}	0.915-0.929; 0.855 (amorphous); 1.0-1.014 (crystalline)	
Refractive index, 20°C	-	1.517-1.526	
Transmittance	%	90-91	
Haze	%	2.2-27	
Gloss, 60°, Gardner (ASTM D523)	%	33-87	
Melting temperature, DSC	°C	105-115; 102 (copolymer with VAc)	
Thermal expansion coefficient, 23-80°C	°C^{-1}	1-5.1E-4	
Thermal conductivity, melt	W m^{-1} K^{-1}	0.55	
Glass transition temperature	°C	-103 to -133	
Heat of fusion	kJ mol^{-1}	1.37-2.18	
Long term service temperature	°C	70	
Heat deflection temperature at 0.45 MPa	°C	37-48	
Heat deflection temperature at 1.8 MPa	°C	36-40	
Vicat temperature VST/A/50	°C	76-109	
Dielectric constant at 100 Hz/1 MHz	-	2.25-2.31	
Dissipation factor at 1000 Hz	E-4	2	
Volume resistivity	ohm-m	1E13	
Electric strength K20/P50, d=0.60.8 mm	kV mm^{-1}	16-28	
Arc resistance	s	135-160	
Power factor	-	0.0003	
Percolation threshold for MWCNT	wt%	0.85 (DC conductivity = 3.47E-07 S/cm)	Han, S H; Yeom, Y S; Ko, J G; Kang, H C; Yoon, H G, Compos. Sci. Technol., 117, 351-6, 2015.
Coefficient of friction	-	0.6 (itself, dynamic)	
Permeability to nitrogen, 25°C	cm^3 cm cm^{-2} s^{-1} Pa^{-1} x 10^{12}	0.073	
Permeability to oxygen, 25°C	cm^3 cm cm^{-2} s^{-1} Pa^{-1} x 10^{12}	0.22	

LDPE low density polyethylene

PARAMETER	UNIT	VALUE	REFERENCES
Permeability to water vapor, 25°C	cm³ cm cm⁻² s⁻¹ Pa⁻¹ x 10¹²	6.8	
Diffusion coefficient of nitrogen	cm² s⁻¹ x10⁶	0.32	
Diffusion coefficient of oxygen	cm² s⁻¹ x10⁶	0.46	
Speed of sound	m s⁻¹	32.5	
Acoustic impedance		1.79	
Attenuation	dB cm⁻¹, 5 MHz	2.4	

MECHANICAL & RHEOLOGICAL PROPERTIES			
Tensile strength	MPa	10-20; 25.4-27.0 (MD) and 14.5-17.3 (TD) for 0.025 mm thick film; 21.6-31.8 (MD) and 16.9-28.9 (TD) for 0.051 mm thick film; 22.1-26.7 (copolymer with VAc)	
Tensile modulus	MPa	130-348	
Tensile stress at yield	MPa	10.8-14.1 (MD) and 10.0-12.3 (TD) for 0.025 mm thick film; 11.8-13.6 (MD) and -11.6-13.7 (TD) for 0.051 mm thick film	
Elongation	%	130-270 (MD) and 490-570 (TD) for 0.025 mm thick film; 180-580 (MD) and 560-780 (TD) for 0.051 mm thick film; 600 (copolymer VAc)	
Flexural strength	MPa	7.5	Yildirir, E; Miskolczi, N; Onwudili, J A; Nemeth, K E; Williams, P T; Soja, J, Composites: Part B, 78, 393-400, 2015.
Flexural modulus	MPa	230-495	
Charpy impact strength, unnotched, 23°C	kJ m⁻²	18.2	Yildirir, E; Miskolczi, N; Onwudili, J A; Nemeth, K E; Williams, P T; Soja, J, Composites: Part B, 78, 393-400, 2015.
Izod impact strength, notched, 23°C	J m⁻¹	420 to NB	
Film puncture resistance	J cm⁻³	2.8-5.7	
Film toughness	J cm⁻³	55.1-82.5 (MD) and 63.7-109 (TD) for 0.025 mm thick fim; 64.5-216 (MD) and 66.9-211 (TD) for film thickness of 0.051 mm	
Dart drop impact	g	72-250; 120-1,200 (copolymer with VAc)	
Elmendorf tear strength	g	140-510 (MD) and 110-180 (TD) for 0.025 mm thick film; 300-560 (MD) and 180-470 (TD) for 0.051 mm thick film	
Puncture force	N	37	
Shore D hardness	-	41-50	
Shrinkage	%	2.4	
Brittleness temperature (ASTM D746)	°C	-34 to -60	
Melt viscosity, shear rate=0 s⁻¹ at 150°C	Pa s	54,500	Hertel, D; Valette, R; Muenstedt, H, J. Non-Newtonian Fluid Mech., 153, 82-94, 2008.
Pressure coefficient of melt viscosity, b	G Pa⁻¹	17.6	Aho, J; Syrjala, S, J. Appl. Polym. Sci., 117, 1076–84, 2010.
Melt index, 230°C/3.8 kg	g/10 min	0.25-55	
Water absorption, equilibrium in water at 23°C	%	0.005-0.015	

LDPE low density polyethylene

PARAMETER	UNIT	VALUE	REFERENCES
CHEMICAL RESISTANCE			
Acid dilute/concentrated	-	very good	
Alcohols	-	good	
Alkalis	-	very good	
Aliphatic hydrocarbons	-	poor	
Aromatic hydrocarbons	-	poor	
Esters	-	poor	
Greases & oils	-	good to poor	
Halogenated hydrocarbons	-	poor	
Ketones	-	poor	
Θ solvent, Θ-temp.=141-170	-	di-(2-ethylhexyl) adipate	
Good solvent	-	1,2,4-trichlorobenzene, decalin, halogenated hydrocarbons, aliphatic ketones, xylene (all above 80°C)	
Non-solvent	-	most common solvents	
FLAMMABILITY			
Ignition temperature	°C	340-343	
Autoignition temperature	°C	340-410	
Limiting oxygen index	% O_2	17.8	Hegazy, E-S A, Ghaffar, A M A, Ali, H E, Mater. Chem. Phys., 252, 123204, 2020.
Char at 500°C	%	0	Lyon, R E; Walters, R N, J. Anal. Appl. Pyrolysis, 71, 27-46, 2004.
Heat of combustion	$J\ g^{-1}$	47,740	
Volatile products of combustion	-	CO, CO_2, aldehydes, benzene	
UL 94 rating	-	HB	
WEATHER STABILITY			
Spectral sensitivity	nm	<300	Wypych, G, Handbook of Materials Weathering, 6th Ed., ChemTec Publishing, Toronto, 2018.
Activation wavelengths	nm	300, 330-360	
Excitation wavelengths	nm	230, 254, 265, 273, 278, 280, 300, 331	
Emission wavelengths	nm	275, 295, 335, 350, 378, 381, 391, 405, 416, 420, 435, 455, 470	
Depth of UV penetration	μm	<1500	
Important initiators and accelerators	-	unsaturations, aromatic carbonyl compounds (deoxyanisoin, dibenzocycloheptadienone, flavone, 4-methoxybenzophenone, 10-thioxanthone), hydrogen bound to tertiary carbon at branching points, aromatic amines, groups formed on oxidation (hydroperoxides, carbonyl, carboxyl, hydroxyl) substituted benzophenones, complexes with ground-state oxygen, quinones (anthraquinone, 2-chloroanthraquinone, 2-tert-butyl-athraquinone, 1-methoxyanthraquinone, 2-ethylanthraquinone, 2-methylanthraquinone), transition metal compounds (Ni < Zn < Fe < Co), ferrocene derivatives, titanium dioxide (anatase), ferric stearate, polynuclear aromatic compounds (anthracene, phenanthrene, pyrene, naphthalene	
Products of degradation	-	free radicals, hydroperoxides, carbonyl groups, chain scission, crosslinking	

LDPE low density polyethylene

PARAMETER	UNIT	VALUE	REFERENCES
Stabilizers	-	UVA: 2-hydroxy-4-octyloxybenzophenone; phenol, 2-(5-chloro-2H-benzotriazole-2-yl)-6-(1,1-dimethylethyl)-4-methyl-; 2,2'-methylenebis(6-(2H-benzotriazol-2-yl)-4-1,1,3,3-tetramethylbutyl)phenol; 2,4-di-tert-butyl-6-(5-chloro-2H-benzotriazole-2-yl)-phenol; reaction product of methyl 3(3-(2H-benzotriazole-2-yl)-5-t-butyl-4-hydroxyphenyl propionate/PEG 300; 2-[4,6-bis(2,4-dimethylphenyl)-1,3,5-triazin-2-yl]-5-(octyloxy) phenol; Screener: titanium dioxide; zinc oxide; carbon black; Acid scavenger: hydrotalcite; Fiber: carbon nanotube; HAS: 1,3,5-triazine-2,4,6-triamine, N,N'''[1,2-ethane-diyl-bis[[[4,6-bis[butyl(1,2,6,6-pentamethyl-4-piperidinyl)amino]-1,3,5-triazine-2-yl]imino]-3,1-propanediyl] bis[N',N''-dibutyl-N',N''-bis(1,2,2,6,6-pentamethyl-4-piperidinyl)-; bis(1,2,2,6,6-pentamethyl-4-piperidyl)sebacate + methyl-1,2,2,6,6-pentamethyl-4-piperidyl sebacate; 2,2,6,6-tetramethyl-4-piperidinyl stearate; reaction products of N,N'-ethane-1,2-diylbis(1,3-propanediamine), cyclohexane, peroxidized 4-butylamino-2,2,6,6-tetramethylpiperidine and trichloro-1,3,5-triazine; poly[[(6-[1,1,3,3-tetramethylbutyl)amino]-1,3,5-triazine-2,4-diyl][2,2,6,6-tetramethyl-4-piperidinyl)imino]-1,6-hexanediyl[2,2,6,6-tetramethyl-4-piperidinyl)imino]]; 1,6-hexanediamine- N,N'-bis(2,2,6,6-tetramethyl-4-piperidinyl)-polymer with 2,4,6-trichloro-1,3,5-triazine, reaction products with N-butyl-1-butanamine an N-butyl-2,2,6,6-tetramethyl-4-piperidinamine; butanedioic acid, dimethylester, polymer with 4-hydroxy-2,2,6,6-tetramethyl-1-piperidine ethanol; alkenes, C20-24-.alpha.-, polymers with maleic anhydride, reaction products with 2,2,6,6-tetramethyl-4-piperidinamine; 1,6-hexanediamine, N,N'-bis(2,2,6,6-tetramethyl-4-piperidinyl)-, polymers with morpholine-2,4,6-trichloro-1,3,5-triazine reaction products, methylated; Phenolic antioxidant: 2,6,-di-tert-butyl-4-(4,6-bis(octylthio)-1,3,5,-triazine-2-ylamino) phenol; pentaerythritol tetrakis(3-(3,5-di-tert-butyl-4-hydroxyphenyl) propionate); octadecyl-3-(3,5-di-tert-butyl-4-hydroxyphenyl)-propionate; 3,3',3',5,5',5'-hexa-tert-butyl-a,a',a'-(mesitylene-2,4,6-triyl)tri-p-cresol; 2-(1,1-dimethylethyl)-6-[[3-(1,1-dimethylethyl)-2-hydroxy-5-methylphenyl] methyl-4-methylphenyl acrylate; 1,3,5-tris(3,5-di-tert-butyl-4-hydroxybenzyl)-1,3,5-triazine-2,4,6(1H,3H,5H)-trione; 3,4-dihydro-2,5,7,8-tetramethyl-2-(4,8,12-trimethyltridecyl)-2H-1-benzopyran-6-ol; 2',3-bis[[3-[3,5-di-tert-butyl-4-hydroxyphenyl]propionyl]]propionohydrazide; isotridecyl-3-(3,5-di-tert-butyl-4-hydroxyphenyl) propionate; 2,2'-ethylidenebis (4,6-di-tert-butylphenol); ethylene bis[3,3-bis[3-(1,1-dimethylethyl)-4-hydroxyphenyl]butanoate];1,3,5-tris(4-tert-butyl-3-hydroxy-2,6-dimethylbenzyl)-1,3,5-triazine-2,4,6-(1H,3H,5H)-trione; 2,2'-methylenebis(4-methyl-6-tertbutylphenol); 3,5-bis(1,1-dimethyethyl)-4-hydroxy-benzenepropanoic acid, C13-15 alkyl esters; 2,2'-isobutylidenebis(2,4-dimethylphenol); 1,1,3-tris(2'methyl-4'-hydroxy-5'tert-butylphenyl)butane; Phosphite: bis-(2,4-di-t-butylphenol) pentaerythritol diphosphite; tris(2,4-di-tert-butylphenyl)phosphite; trinonylphenol phosphite; distearyl pentaerythritol diphosphite; trilauryl trithiophosphite; Thiosynergist: didodecyl-3,3'-thiodipropionate; dioctadecyl 3,3'-thiodipropionate; 2,2'-thiodiethylene bis[3-(3,5-ditert-butyl-4-hydroxyphenyl)propionate]; 4,4'-thiobis(2-t-butyl-5-methylphenol); 2,2'-thiobis(6-tert-butyl-4-methylphenol); pentaerythritol tetrakis(b-laurylthiopropionate); Quencher: (2,2'-thiobis(4-tert-octyl-phenolato))-N-butylamine-nickel(II); Optical brightener: 2,2'-(2,5-thiophenediyl)bis(5-tert-butylbenzoxazole)	

LDPE low density polyethylene

PARAMETER	UNIT	VALUE	REFERENCES
BIODEGRADATION			
Typical biodegradants	-	fungi, bacteria, Actinomycetes (10% mass loss per year)	Sudhakar, M; Doble, M; Sriyutha Murthy, P; Venkatesan, R, Int. Biodet. Biodeg., 61, 203-13, 2008.
TOXICITY			
NFPA: Health, Flammability, Reactivity rating	-	1/0/1	
Carcinogenic effect	-	not listed by ACGIH, NIOSH, NTP	
Mutagenic effect	-	not known	
Teratogenic effect	-	not known	
Reproductive toxicity	-	not known	
Oral rat, LD_{50}	mg kg^{-1}	>5,000	
Skin rabbit, LD_{50}	mg kg^{-1}	>2,000	
ENVIRONMENTAL IMPACT			
Biological oxygen demand, BOD_5	-	3.08	Psomiadou, E; Arvanitoyannis, I; Biliaderis, C G; Ogawa, H; Kawasaki, N, Carbohydrate Polym., 33, 227-42, 1997.
Cradle to grave non-renewable energy use	MJ/kg	81.8	Harding, K G; Dennis, J S; von Blottnitz, H; Harrison, S T L, J. Biotechnol., 130, 57-66, 2007.
Cradle to pellet greenhouse gasses	kg CO_2 kg^{-1} resin	2.0-2.2	
PROCESSING			
Typical processing methods	-	blown film extrusion, cast film extrusion, coating, coextrusion, extrusion, injection molding, molding, lamination, rotational molding	
Processing temperature	°C	199-232 (extrusion); 212 (blown film); 316-332 (coating)	
Additives used in final products	-	Fillers: aluminum, barium sulfate, calcium carbonate, calcium sulfate whiskers, carbon black, diatomaceous earth, ferromagnetic powder, glass fiber, glass spheres, ground tire rubber, hollow silicates, hydrotalcite, kaolin, lignin, magnesium hydroxide, marble, mica, nickel fibers, red mud, sand, silica, soot, starch, superconductor ($YBa_2Cu_3O_{7-x}$), talc, wollastonite, wood flour, zirconium silicate; Plasticizers: dioctyl phthalate, EPDM, EVA, glycerin, glyceryl tribenzoate, mineral oil, paraffin oil, polyethylene glycol, sunflower oil; Antistatics: carbon black, copper complex of polyacrylic acid, ethoxylated amines, fatty diethanol amines, glycerol monostearate, graphite, ionomer, lauric diethanolamide, polyethylene glycol, quaternary ammonium compound, trineoalkoxy zirconate; Antiblocking: diatomaceous earth, natural silica, siloxane spheres, synthetic silica, talc, zeolite; Release: stearyl erucamide; Slip: erucamide, ethylene bisoleamide, oleamide	
Applications	-	car covers, cling wrap, moisture barriers in construction, liners for tanks and ponds, sandwich bags, squeeze bottles	
BLENDS			
Suitable polymers	-	chitosan, EEA, EPDM, EVA, HDPE, HIPS, LLDPE, NBR, NR, PA6, PET, iPP, PP, PS, PVDF, SBR, starch	

LDPE low density polyethylene

PARAMETER	UNIT	VALUE	REFERENCES
Compatibilizers	-	LDPE-g-MA	Telen, L; Jansens, K J A; Verpoest, I; Delcour, J A; Van Puyvelde, P; Goderis, B, Ind. Crops Prod., 74, 824-38, 2015.

LLDPE linear low density polyethylene

PARAMETER	UNIT	VALUE	REFERENCES
GENERAL			
Common name	-	linear low density polyethylene; poly(ethylene-co-1-octene); poly(ethylene-co-1-butene); poly(ethylene-co-1-hexene)	
IUPAC name	-	polyethylene	
Acronym	-	LLDPE	
CAS number	-	26221-73-8 (octene); 25087-34-7 (butene); 25213-02-9 (hexene; metallocene)	
HISTORY			
Person to discover	-	DuPont Canada in Corunna, Ontario	
Date	-	1960	
Details	-	the world's first commercial LLDPE line	
Monomer(s) structure	-	$H_2C=CH_2$; $CH_3CH_2CH=CH_2$; $CH_3(CH_2)_3CH=CH_2$	
Monomer(s) CAS number(s)	-	74-85-1; 106-98-9; 592-41-6	
Monomer(s) molecular weight(s)	dalton, g/mol, amu	28.05; 56.11; 84.16	
Monomer(s) expected purity(ies)	%	99.9	
Method of synthesis	-	Ziegler or Philips catalysts are used in solution or gas phase reactions to obtain LLDPE. Octene copolymer with ethylene is obtained in solution process and butene and hexene are copolymerized with ethylene in gas phase reactors; also metallocene catalyst are in use	
Catalyst	-	Ziegler, Philips, metallocene	
Typical additives	ppm	3,000 (antiblock), antioxidant (e.g., Tuflin)	
Mass average molecular weight, M_w	dalton, g/mol, amu	94,000-208,000	
Polydispersity, M_w/M_n	-	1.6-35	
Degree of branching	mol%	2-4	
STRUCTURE			
Crystallinity	%	30-53; 46 (DSC); 52-59 (NMR); 37.2	Lu, J; Zhao, B; Sue, H-J, Metallocene Technology in Commercial Applications, Benedikt, G M, Ed., WilliamAndrew, Norwich, 1999; Panrong, T, Karbowiak, T, Harnkarnsujarit, N, J. Food. Eng., 284, 110057, 2020.
Cell type (lattice)	-	orthorhombic	
Cell dimensions	nm	a:b:c=0.748:0.497:0.257	
Spherulite size	nm	2,000-12,000	Ruksakulpiwat, Y, Antec, 582-6, 2001.
Spacing between crystallites	nm	8.0-36.9	
Lamellae thickness	nm	4.3-16.3	
Rapid crystallization temperature	°C	107-123 (injection molding grades)	
COMMERCIAL POLYMERS			
Some manufacturers	-	DOW; ExxonMobil; LyondellBasell	
Trade names	-	Aspun (fiber grades), Dowlex, Tuflin, LLDPE; LLDPE; Starflex	

LLDPE linear low density polyethylene

PARAMETER	UNIT	VALUE	REFERENCES
PHYSICAL PROPERTIES			
Density at 20°C	g cm^{-3}	0.905-0.942	
Bulk density at 20°C	g cm^{-3}	0.35-0.38	
Color	-	clear to white	
Refractive index, 20°C	-	1.49-1.52	
Haze	%	1-19	
Gloss, 60°, Gardner (ASTM D523)	%	47-92	
Odor	-	odorless to mild hydrocarbon odor	
Melting temperature, DSC	°C	120-136; 94 (injection molding grades); 123.4 (food film grade)	-; -; Khumkomgoola, A, Saneluk-sanaa, T, Harnkarnsujarit, N, Food Packaging Shelf Life, 26, 100557, 2020.
Thermal expansion coefficient, 23-80°C	°C^{-1}	1.6-5E-4	
Thermal conductivity, melt	W m^{-1} K^{-1}	0.55	
Glass transition temperature	°C	-110	
Heat of fusion	kJ mol^{-1}	1.37-2.18	
Maximum processing temperature	°C	300 (Booster)	
Heat deflection temperature at 0.45 MPa	°C	43-59	
Heat deflection temperature at 1.8 MPa	°C	38	
Vicat temperature VST/A/50	°C	94-123	
Surface tension	mN m^{-1}	22.4-24.0	Tinson, A; Takacs, E; Vlachopoulos, J, Antec, 870-74, 2004.
Dielectric constant at 100 Hz/1 MHz	-	2.3	
Coefficient of friction	-	0.6 (itself, dynamic)	
Contact angle of water, 20°C	degree	97.1-99.1	Jeon, H J; Kim, M N, Eur. Polym. J., 52, 146-53, 2014.
Speed of sound	m s^{-1}	717-1009	
MECHANICAL & RHEOLOGICAL PROPERTIES			
Tensile strength	MPa	25-45; 33-71 (MD) and 25-54.2 (TD) for 0.02-0.25 mm thick film; 30-55 (MD) and 29-53.6 (TD) for 0.051 mm thick film; 7.6-15.5 (injection molding grades)	
Tensile modulus	MPa	260-520	
Tensile stress at yield	MPa	8.1-32.4 (MD) and 7.7-29.0 (TD) for 0.02-0.025 mm thick film; 9.3-15.2 (MD) and 10.0-16.4 (TD) for 0.051 mm thick film; 11-750-830 (film grade); 16.5 (injection molding grades)	
Elongation	%	300-830 (MD); 610-890 (TD); 50-800 (injection molding grades)	
Tensile yield strain	%	2-18	
Flexural modulus	MPa	280-735; 310-700 (injection molding grades)	
Izod impact strength, notched, 23°C	J m^{-1}	54 to NB	
Dart drop impact (film thickness in mm)	g	43-470 (0.02); 130-610 (0.051)	
Film puncture resistance (film thickness in mm)	J cm-3	7.7-30.8 (0.02); 7.0-24.0 (0.051	

200   **HANDBOOK OF POLYMERS** 3rd Edition, Copyrights 2022; ChemTec Publishing

LLDPE linear low density polyethylene

PARAMETER	UNIT	VALUE	REFERENCES
Toughness	J cm-3	63-303 (MD) and 76-353 (TD) for 0.02-0.025 mm thick film; 84-346 (MD) and 87-361 (TD) for 0.051 mm thick film	
Elmendorf tear strength (film thickness in mm)	g	15-370 (MD) and 210-700 (TD) for 0.02-0.025 mm thick film; 470-950 (MD) and 1,100-1,300 (TD) for 0.051 mm thick film	
Puncture force	N	30-43	
Shore D hardness	-	55-56; 44-53 (injection molding grades)	
Shrinkage	%	2.0-2.5	
Brittleness temperature (ASTM D746)	°C	-20 to -100 (injection molding grades)	
Melt viscosity, shear rate=0 s^{-1}	kPa s	2.9 (190°C); 25.5 (150°C)	Tinson, A; Takacs, E; Vlachopoulos, J, Antec, 870-74, 2004; Hertel, D; Valette, R; Muenstedt, H, J. Non-Newtonian Fluid Mech., 153, 82-94, 2008.
Melt index	g/10 min	0.2-2000, 0.92	Tornuk, F; Hancer, M; Sagdic, O; Yetim, H, LTW - Food Sci. Technol., 64, 540-6, 2015; Khumkomgoola, A, Saneluksanaa, T, Harnkarnsujarit, N, Food Packaging Shelf Life, 26, 100557, 2020.
Water absorption	%	0.005-0.01	

CHEMICAL RESISTANCE

PARAMETER	UNIT	VALUE	REFERENCES
Acid dilute/concentrated	-	very good	
Alcohols	-	good	
Alkalis	-	very good	
Aliphatic hydrocarbons	-	poor	
Aromatic hydrocarbons	-	poor	
Esters	-	poor	
Greases & oils	-	good to poor	
Halogenated hydrocarbons	-	poor	
Ketones	-	poor	
Good solvent	-	cyclohexene, decalin, toluene, xylene	
Non-solvent	-	o-dichlorobenzene, 1,2-dichloropropane, methylene chloride	

FLAMMABILITY

PARAMETER	UNIT	VALUE	REFERENCES
Ignition temperature	°C	340-343	
Autoignition temperature	°C	350	
Limiting oxygen index	% O_2	19.5; 26.9	Zhou, R, Mu, J, Sun, X, Ding, Y, Jiang, J, Safety Sci., 131, 104849, 2020.
Heat of combustion	J g^{-1}	47,740	Zhou, R, Mu, J, Sun, X, Ding, Y, Jiang, J, Safety Sci., 131, 104849, 2020.
Peak of heat release rate	kW m^{-2}	510	Zhou, R, Mu, J, Sun, X, Ding, Y, Jiang, J, Safety Sci., 131, 104849, 2020.
Total heat release	MJ m^{-2}	80	Zhou, R, Mu, J, Sun, X, Ding, Y, Jiang, J, Safety Sci., 131, 104849, 2020.
Char	%	0	Zhou, R, Mu, J, Sun, X, Ding, Y, Jiang, J, Safety Sci., 131, 104849, 2020.
Volatile products of combustion	-	CO, CO_2, aldehydes, benzene	

LLDPE linear low density polyethylene

PARAMETER	UNIT	VALUE	REFERENCES
UL 94 rating	-	HB; V-0 (with aluminum diethyphosphinate, neopentyl glycol, and melamine)	Zhou, R, Mu, J, Sun, X, Ding, Y, Jiang, J, Safety Sci., 131, 104849, 2020.

WEATHER STABILITY			
Spectral sensitivity	nm	<300	
Activation wavelengths	nm	300, 330-360	
Excitation wavelengths	nm	230, 265, 275, 290, 292	
Emission wavelengths	nm	295, 312, 330, 344, 358, 450	
Important initiators and accelerators	-	unsaturations, aromatic carbonyl compounds (deoxyanisoin, dibenzocycloheptadienone, flavone, 4-methoxybenzophenone, 10-thioxanthone), hydrogen bound to tertiary carbon at branching points, aromatic amines, groups formed on oxidation (hydroperoxides, carbonyl, carboxyl, hydroxyl) substituted benzophenones, complexes with ground-state oxygen, quinones (anthraquinone, 2-chloroanthraquinone, 2-tert-butyl-athraquinone, 1-methoxyanthraquinone, 2-ethylanthraquinone, 2-methylanthraquinone), transition metal compounds (Ni < Zn < Fe < Co), ferrocene derivatives, titanium dioxide (anatase), ferric stearate, polynuclear aromatic compounds (anthracene, phenanthrene, pyrene, naphthalene	
Products of degradation	-	free radicals, hydroperoxides, carbonyl groups, chain scission, crosslinking	
Stabilizers	-	UVA: 2-hydroxy-4-octyloxybenzophenone; phenol, 2-(5-chloro-2H-benzotriazole-2-yl)-6-(1,1-dimethylethyl)-4-methyl-; 2,2'-methylenebis(6-(2H-benzotriazol-2-yl)-4-1,1,3,3-tetramethylbutyl)phenol; 2,4-di-tert-butyl-6-(5-chloro-2H-benzotriazol-2-yl)-phenol; reaction product of methyl 3(3-(2H-benzotriazole-2-yl)-5-t-butyl-4-hydroxyphenyl propionate/ PEG 300; 2-[4,6-bis(2,4-dimethylphenyl)-1,3,5-triazin-2-yl]-5-(octyloxy) phenol; Screener: titanium dioxide; zinc oxide; carbon black; Acid scavenger: hydrotalcite; Fiber: carbon nanotube; HAS: 1,3,5-triazine-2,4,6-triamine, N,N'''[1,2-ethane-diyl-bis[[[4,6-bis[butyl(1,2,6,6-pentamethyl-4-piperidinyl)amino]-1,3,5-triazine-2-yl]imino]-3,1-propanediyl] bis[N',N''-dibutyl-N',N''-bis(1,2,2,6,6-pentamethyl-4-piperidinyl)-; bis(1,2,2,6,6-pentamethyl-4-piperidyl)sebacate + methyl-1,2,2,6,6-pentamethyl-4-piperidyl sebacate; 2,2,6,6-tetramethyl-4-piperidinyl stearate; reaction products of N,N'-ethane-1,2-diylbis(1,3-propanediamine), cyclohexane, peroxidized 4-butylamino-2,2,6,6-tetramethylpiperidine and trichloro-1,3,5-triazine; poly[[(6-[1,1,3,3-tetramethylbutyl) amino]-1,3,5-triazine-2,4-diyl][2,2,6,6-tetramethyl-4-piperidinyl) imino]-1,6-hexanediyl[2,2,6,6-tetramethyl-4-piperidinyl)imino]]; 1,6-hexanediamine- N,N'-bis(2,2,6,6-tetramethyl-4-piperidinyl)-polymer with 2,4,6-trichloro-1,3,5-triazine, reaction products with N-butyl-1-butanamine an N-butyl-2,2,6,6-tetramethyl-4-piperidinamine; butanedioic acid, dimethylester, polymer with 4-hydroxy-2,2,6,6-tetramethyl-1-piperidine ethanol; alkenes, C20-24-.alpha.-, polymers with maleic anhydride, reaction products with 2,2,6,6-tetramethyl-4-piperidinamine; 1,6-hexanediamine, N,N'-bis(2,2,6,6-tetramethyl-4-piperidinyl)-,	

LLDPE linear low density polyethylene

PARAMETER	UNIT	VALUE	REFERENCES
Stabilizers (continuation)	-	polymers with morpholine-2,4,6-trichloro-1,3,5-triazine reaction products, methylated; Phenolic antioxidant: 2,6,-di-tert-butyl-4-(4,6-bis(octylthio)-1,3,5,-triazine-2-ylamino) phenol; pentaerythritol tetrakis(3-(3,5-di-tert-butyl-4-hydroxyphenyl) propionate); octadecyl-3-(3,5-di-tert-butyl-4-hydroxyphenyl)-propionate; 3,3',3',5,5',5'-hexa-tert-butyl-a,a',a'-(mesitylene-2,4,6-triyl)tri-p-cresol; 2-(1,1-dimethylethyl)-6-[[3-(1,1-dimethylethyl)-2-hydroxy-5-methylphenyl] methyl-4-methylphenyl acrylate; 1,3,5-tris(3,5-di-tert-butyl-4-hydroxybenzyl)-1,3,5-triazine-2,4,6(1H,3H,5H)-trione; 3,4-dihydro-2,5,7,8-tetramethyl-2-(4,8,12-trimethyltridecyl)-2H-1-benzopyran-6-ol; 2',3-bis[[3-[3,5-di-tert-butyl-4-hydroxy-phenyl]propionyl]]propionohydrazide; isotridecyl-3-(3,5-di-tert-butyl-4-hydroxyphenyl) propionate; 2,2'-ethylidenebis (4,6-di-tert-butylphenol); ethylene bis[3,3-bis[3-(1,1-dimethylethyl)-4-hydroxyphenyl]butanoate];1,3,5-tris(4-tert-butyl-3-hydroxy-2,6-dimethylbenzyl)-1,3,5-triazine-2,4,6-(1H,3H,5H)-trione; 2,2'-methylenebis(4-methyl-6-tertbutylphenol); 3,5-bis(1,1-dimethyethyl)-4-hydroxy-benzenepropanoic acid, C13-15 alkyl esters; 2,2'-isobutylidenebis(2,4-dimethylphenol); 1,1,3-tris(2'methyl-4'-hydroxy-5'tert-butylphenyl)butane; Phosphite: bis-(2,4-di-t-butylphenol) pentaerythritol diphosphite; tris (2,4-di-tert-butylphenyl)phosphite; trinonylphenol phosphite; distearyl pentaerythritol diphosphite; trilauryl trithiophosphite; Thiosynergist: didodecyl-3,3'-thiodipropionate; dioctadecyl 3,3'-thiodipropionate; 2,2'-thiodiethylene bis[3-(3,5-ditert-butyl-4-hydroxyphenyl)propionate]; 4,4'-thiobis(2-t-butyl-5-methylphenol); 2,2'-thiobis(6-tert-butyl-4-methylphenol); pentaerythritol tetrakis(b-laurylthiopropionate); Quencher: (2,2'-thiobis(4-tert-octyl-phenolato))-N-butylamine-nickel(II); Optical brightener: 2,2'-(2,5-thiophenediyl)bis(5-tert-butylbenzoxazole)	

BIODEGRADATION

PARAMETER	UNIT	VALUE	REFERENCES
Typical biodegradants	-	*Bacillus*	Abrusci, C; Pablos, J L; Corrales, T; Lopez-Marin, J; Marin, I; Catalina, F, Int. Biodet. Biodeg., 65, 451-59, 2011.

TOXICITY

PARAMETER	UNIT	VALUE	REFERENCES
NFPA: Health, Flammability, Reactivity rating	-	0/1/0	
Carcinogenic effect	-	not listed by ACGIH, NIOSH, NTP	
Mutagenic effect	-	not known	
Teratogenic effect	-	not known	
Reproductive toxicity	-	not known	
Oral rat, LD_{50}	mg kg^{-1}	>5,000	
Skin rabbit, LD_{50}	mg kg^{-1}	>2,000	

ENVIRONMENTAL IMPACT

PARAMETER	UNIT	VALUE	REFERENCES
Cradle to grave non-renewable energy use	MJ/kg	69-73	
Cradle to pellet greenhouse gasses	kg CO_2 kg^{-1} resin	1.8-2.0	

LLDPE linear low density polyethylene

PARAMETER	UNIT	VALUE	REFERENCES
PROCESSING			
Typical processing methods	-	blown film, blow molding, coating, extrusion, injection molding, slot cast extrusion	
Processing temperature	°C	180-327; 180-220 (extrusion of blown film); 200-240 (injection)	
Processing pressure	MPa	20-60 (holding)	
Additives used in final products	-	Antistatics: carbon black, copper complex of polyacrylic acid, ethoxylated amines, fatty diethanol amines, glycerol mono-stearate, graphite, ionomer, lauric diethanolamide, polyethylene glycol, quaternary ammonium compound, trineoalkoxy zirconate; Antiblocking: diatomaceous earth, natural silica, siloxane spheres, synthetic silica, talc, zeolite; Blowing agents; Release: stearyl erucamide; Slip: erucamide, ethylene bisoleamide, oleamide; Process aid; Thermal stabilizers; UV stabilizers	
Applications	-	automotive parts, agricultural film, bags, bottles, cast film, cling film, closures, diaper backsheet, drum liners, film, greenhouse film, hoses, lids, melt strength enhancer (Booster); overwrap film, parts of industrial containers, playground equipment, point of display cabinets, potable water tanks, toys, trash cans, tubing	
Outstanding properties	-	tear strength, toughness, processability, stiffness	
BLENDS			
Suitable polymers	-	EPDM, EVA, EVOH, LDPE, NR, PA6, PC, PLA, PMMA, PP, PVC, SBS, starch	
ANALYSIS			
x-ray diffraction peaks	degree	peaks at 21.4° and 23.6° correspond to the LLDPE characteristic lattice planes (110) and (200), respectively	Li, T, Sun, H, Lei, F, Li, D, Sun, D, Polymer, 172, 142-51, 2019.

MABS poly(methyl methacrylate-co-butadiene-co-styrene)

PARAMETER	UNIT	VALUE	REFERENCES
GENERAL			
Common name	-	poly(methyl methacrylate-co-acrylonitrile-co-butadiene-co-styrene), methyl methacrylate ABS	
CAS name	-	2-propenoic acid, 2-methyl-, methyl ester, polymer with 1,3-butadiene, ethenylbenzene and 2-propenenitrile	
Acronym	-	MABS	
CAS number	-	9010-94-0	
HISTORY			
Person to discover	-	D'Alello, G F	D'Alello, G F, US Patent 2,414,803, General Electric, Jan. 28, 1947.
Date	-	1947	
Details	-	polymerization patented	
SYNTHESIS			
Monomer(s) structure	-	$H_2C{=}CHC{\equiv}N$ $H_2C{=}CHCH{=}CH_2$ (phenyl–vinyl) $H_2C{=}CCOCH_3$ with CH_3 and $=O$	
Monomer(s) CAS number(s)	-	107-13-1; 106-99-0; 100-42-5; 80-62-6	
Monomer(s) molecular weight(s)	dalton, g/mol, amu	53.06; 54.09; 104.15; 100.12	
Core/shell ratio		60-70/40-30 (core - SBR (Bd/St=77/23), shell - St-MMA=50/50)	Dompas, D; Groeninckx, G; Isogawa, M; Hasegawa, T; Kadokura, M, Polymer, 35, 22, 4750-59, 1994.
Method of synthesis	-	core is synthesized by radical emulsion polymerization, followed by graft polymerization of shell	Dompas, D; Groeninckx, G; Isogawa, M; Hasegawa, T; Kadokura, M, Polymer, 35, 22, 4750-59, 1994.
Mass average molecular weight, M_w	dalton, g/mol, amu	100,000	
COMMERCIAL POLYMERS			
Some manufacturers	-	Ineos-Styrolution	
Trade names	-	Terlux	
PHYSICAL PROPERTIES			
Density at 20°C	g cm^{-3}	1.04-1.09	
Bulk density at 20°C	g cm^{-3}	0.6	
Color	-	colorless	
Refractive index, 20°C	-	1.54	
Odor	-	faint, specific	
Softening point	°C	100-150	
Decomposition temperature	°C	300	
Thermal expansion coefficient, 23-80°C	10^{-4} °C^{-1}	0.8-1.1E-4	
Thermal conductivity, melt	W m^{-1} K^{-1}	0.17	
Glass transition temperature	°C	107	
Maximum service temperature	°C	75	
Heat deflection temperature at 0.45 MPa	°C	93-94	

MABS poly(methyl methacrylate-co-butadiene-co-styrene)

PARAMETER	UNIT	VALUE	REFERENCES
Heat deflection temperature at 1.8 MPa	°C	87-90	
Vicat temperature	°C	87-93	
Vicat temperature VST/B/50	°C	91	
Dielectric constant at 100 Hz/1 MHz	-	2.9/2.8	
Relative permittivity at 100 Hz	-	2.9-3.0	
Relative permittivity at 1 MHz	-	2.8	
Dissipation factor at 100 Hz	E-4	160	
Dissipation factor at 1 MHz	E-4	130-140	
Volume resistivity	ohm-m	1E13	
Surface resistivity	ohm	1E15	
Electric strength K20/P50, d=0.60.8 mm	kV mm^{-1}	34	
Comparative tracking index	-	600	

MECHANICAL & RHEOLOGICAL PROPERTIES			
Tensile modulus	MPa	1,900-2,000	
Tensile stress at yield	MPa	42-48	
Tensile creep modulus, 1000 h, elongation 0.5 max	MPa	1,250	
Elongation	%	12-20	
Tensile yield strain	%	4	
Charpy impact strength, unnotched, 23°C	kJ m^{-2}	110-120	
Charpy impact strength, unnotched, -30°C	kJ m^{-2}	70-80	
Charpy impact strength, notched, 23°C	kJ m^{-2}	5	
Charpy impact strength, notched, -30°C	kJ m^{-2}	2	
Izod impact strength, notched, 23°C	J m^{-1}	100	
Ball indention hardness at 358 N/30 S (ISO 2039-1)	MPa	70	
Shrinkage	%	0.4-0.7	
Melt viscosity, shear rate=1000 s^{-1}	Pa s	220	
Melt volume flow rate (ISO 1133, procedure B), 220°C/10 kg	cm^3/10 min	2-8	
Water absorption, equilibrium in water at 23°C	%	0.7	
Moisture absorption, equilibrium 23°C/50% RH	%	0.35	

FLAMMABILITY			
Ignition temperature	°C	405	
Autoignition temperature	°C	>400	
Volatile products of combustion	-	CO, CO_2, HCN	

MABS poly(methyl methacrylate-co-butadiene-co-styrene)

PARAMETER	UNIT	VALUE	REFERENCES
UL 94 rating	-	HB	
WEATHER STABILITY			
Products of degradation	-	oxidation of C=C bonds to C=O and C-OH groups	
TOXICITY			
HMIS: Health, Flammability, Reactivity rating	-	1/1/0	
Carcinogenic effect	-	not listed by ACGIH, NIOSH, NTP	
PROCESSING			
Typical processing methods	-	injection molding	
Preprocess drying: temperature/ time/residual moisture	ºC/h/%	70/2/	
Processing temperature	ºC	230-260 (injection molding)	
Mold temperature	ºC	50-80	
Applications	-	computers, consumer electronics, household devices, peripherials, sport equipment, telecommunication	Dompas, D; Groeninckx, G; Isogawa, M; Hasegawa, T; Kadokura, M, Polymer, 35, 22, 4750-59, 1994.
Outstanding properties	-	brilliant visual effect	
BLENDS			
Suitable polymers	-	PA6, PMMA, PVC	

MBS poly(styrene-co-butadiene-co-methyl methacrylate)

PARAMETER	UNIT	VALUE	REFERENCES
GENERAL			
Common name	-	poly(styrene-co-butadiene-co-methyl methacrylate)	
IUPAC name	-	buta-1,3-diene; methyl 2-methylprop-2-enoate; styrene	
CAS name	-	2-propenoic acid, 2-methyl-methyl ester, polymer with 1,3-butadiene and ethenyl-benzene	
Acronym	-	MBS	
CAS number	-	25053-09-2	
HISTORY			
Person to discover	-	D'Alello, G F	
Date	-	1947	
Details	-	polymerization patented	
SYNTHESIS			
Monomer(s) structure	-	$H_2C=CCOCH_3$ $H_2C=CHCH=CH_2$ (structure) with CH_3 and O	
Monomer(s) CAS number(s)	-	80-62-6; 106-99-0; 100-42-5	
Monomer(s) molecular weight(s)	dalton, g/mol, amu	100.12; 54.09; 104.15	
Methyl methacrylate content	%	40-55	
Styrene content	%	25-40	
Butadiene content	%	15-25	
Method of synthesis	-	MBS consists of an elastomeric core and a glass shell. The elastomeric core is polybutadiene or styrene–butadiene rubber (SBR), and the shell is poly(methyl methacrylate) and polystyrene. The MBS copolymers are synthesized by emulsion polymerization method. In the preparation process PB polymer or SBR have to be synthesized first and then St and MMA are polymerized on rubber particles.	Zhou, C; Chen, M; Tan, Z Y; Sun, S L; Ao, Y H; Zhang, M Y; Yang, H D; Zhang, H X, Eur. Polym. J., 42, 1811-18, 2006.
Mass average molecular weight, M_w	dalton, g/mol, amu	100,000-150,000	
COMMERCIAL POLYMERS			
Some manufacturers	-	Arkema; Dow; Roehm; Kaneka	
Trade names	-	Clearstrength; Paraloid; Cyrolite; Kane Ace	
PHYSICAL PROPERTIES			
Density at 20°C	g cm^{-3}	1.05-1.11	
Bulk density at 20°C	g cm^{-3}	0.20-0.45	
Color	-	white	
Refractive index, 20°C	-	1.52-1.57	
Transmittance	%	84-91	
Haze	%	2.5-2.7	
Odor		pungent, sweet odor	
Melting temperature, DSC	°C	132-149	
Decomposition temperature	°C	>250	
Storage temperature	°C	<50	

MBS poly(styrene-co-butadiene-co-methyl methacrylate)

PARAMETER	UNIT	VALUE	REFERENCES
Thermal expansion coefficient, 23-80°C	10^{-4} °C^{-1}	0.9	
Glass transition temperature	°C	-77 to -30	Zhou, C; Chen, M; Tan, Z Y; Sun, S L; Ao, Y H; Zhang, M Y; Yang, H D; Zhang, H X, Eur. Polym. J., 42, 1811-18, 2006.
Heat deflection temperature at 0.45 MPa	°C	83	
Heat deflection temperature at 1.8 MPa	°C	72-85	
Vicat temperature VST/B/50	°C	95	
Surface resistivity	ohm	1E13	
Electric strength K20/P50, d=0.60.8 mm	kV mm^{-1}	1E13	

MECHANICAL & RHEOLOGICAL PROPERTIES

PARAMETER	UNIT	VALUE	REFERENCES
Tensile strength	MPa	34-42	
Tensile modulus	MPa	2,300	
Tensile stress at yield	MPa	44-59	
Elongation	%	11-70	
Tensile yield strain	%	3.4-25	
Flexural strength	MPa	59-88	
Flexural modulus	MPa	1,800-2,200	
Charpy impact strength, unnotched, 23°C	kJ m^{-2}	NB	
Charpy impact strength, notched, 23°C	kJ m^{-2}	9-12	
Izod impact strength, unnotched, 23°C	J m^{-1}	1,000	
Izod impact strength, notched, 23°C	J m^{-1}	39-130	
Rockwell hardness	-	M29-40; R76-85	
Melt index, 200°C/5 kg	g/10 min	0.5-6.2	
Water absorption, equilibrium in water at 23°C	%	0.3	

FLAMMABILITY

PARAMETER	UNIT	VALUE	REFERENCES
Flammability according to UL-94 standard; thickness 1.6/0.8 mm	class	HB	
Ignition temperature	°C	400	
Autoignition temperature	°C	470	
Limiting oxygen index	% O_2	18	
Volatile products of combustion	-	CO, CO_2, acrylates, hazardous organic products	

WEATHER STABILITY

PARAMETER	UNIT	VALUE	REFERENCES
Products of degradation	-	depolymerization of PMMA chain in presence of PVC	Chen, Q; Wang, J; Shen, J, Polym. Deg. Stab., 87, 527-533, 2005.

TOXICITY

PARAMETER	UNIT	VALUE	REFERENCES
NFPA: Health, Flammability, Reactivity rating	-	2/0/0; 2/0/0 (HMIS)	

MBS poly(styrene-co-butadiene-co-methyl methacrylate)

PARAMETER	UNIT	VALUE	REFERENCES
Carcinogenic effect	-	not listed by ACGIH, NIOSH, NTP	
TLV, ACGIH	mg m^{-3}/ ppm	not established	
OSHA	mg m^{-3}	5 (respirable), 15 (total dust)	
Oral rat, LD$_{50}$	mg kg^{-1}	>5,000	
Skin rabbit, LD$_{50}$	mg kg^{-1}	>5,000	
ENVIRONMENTAL IMPACT			
Aquatic toxicity, *Daphnia magna*, LC$_{50}$, 48 h	mg l^{-1}	>100	
Aquatic toxicity, *Bluegill sunfish*, LC$_{50}$, 48 h	mg l^{-1}	>100	
Aquatic toxicity, *Rainbow trout*, LC$_{50}$, 48 h	mg l^{-1}	>100	
PROCESSING			
Typical processing methods	-	extrusion, film extrusion, injection molding, profile extrusion, thermoforming	
Preprocess drying: temperature/ time/residual moisture	°C/h/%	71-79/2-3/-	
Processing temperature	°C	218-260	
Processing pressure	MPa	69-103 (injection); 0.17-0.69 (back)	
Additives used in final products	-	epoxidized soybean oil	
Applications	-	core-shell impact modifier (automotive parts, housings for electronic components, thermoset adhesives), electrical parts, sheet	
Outstanding properties	-	high impact efficiency, part rigidity, part surface finish	
BLENDS			
Suitable polymers	-	ABS, PBT, PC, PET, PVC, SAN	

MC methylcellulose

PARAMETER	UNIT	VALUE	REFERENCES
GENERAL			
Common name	-	methylcellulose	
Acronym	-	MC	
CAS number	-	9004-67-5	
EC number	-	232-674-9	
RTECS number	-	FJ5959000	
HISTORY			
Person to discover	-	Houghton, A A; Taylor, C M	Houghton, A A; Taylor, C M, US Patent 2,285,514, ICI, June 9, 1942.
Date	-	1942	
Details	-	low substituted methylcellulose patented	
SYNTHESIS			
Monomer(s) structure	-	cellulose; methyl chloride	
Monomer(s) CAS number(s)	-	9004-34-6; 74-87-3	
Monomer(s) molecular weight(s)	dalton, g/mol, amu	160,000-560,000; 50.49	
Degree of substitution	-	1.3-2.6 (3 is maximum), 1.6-1.9 (pharmaceutical formulations)	
Method of synthesis	-	cellulose is heated with sodium hydroxide solution and then reacted with methyl chloride	
Number average molecular weight, M_n	dalton, g/mol, amu	38,000-326,000	Ziembowicz, F I, de Freitas, D V, Bender, C R, dos Santos Salbego, P R, Villetti, M A, Carbohydrate Polym., 214, 174-85, 2019.
Mass average molecular weight, M_w	dalton, g/mol, amu	70,000-290,000	Ye, D; Farriol, X, Ind. Crops Products, 26, 54-62, 2007.
Radius of gyration	nm	67	Ziembowicz, F I, de Freitas, D V, Bender, C R, dos Santos Salbego, P R, Villetti, M A, Carbohydrate Polym., 214, 174-85, 2019.
STRUCTURE			
Chain width	nm	0.91 (cellulose 0.79)	Chandrasekaran, R, Adv. Food Nutrition Res., 42, 131-210, 1999.
COMMERCIAL POLYMERS			
Some manufacturers	-	DOW	
Trade names	-	Methocel	
PHYSICAL PROPERTIES			
Density at 20°C	g cm^{-3}	1.01	
Color	-	white to off-white	
Refractive index, 20°C	-	1.4970	
Transmittance	%	92.2	
Odor	-	odorless	
Melting temperature, DSC	°C	290-305	

HANDBOOK OF POLYMERS 3rd Edition, Copyrights 2022; ChemTec Publishing

MC methylcellulose

PARAMETER	UNIT	VALUE	REFERENCES
Gelation temperature	°C	48	
Glass transition temperature	°C	150-165	
Surface tension	mN m^{-1}	47-59	
Volume resistivity	ohm-m	2.6E+03	Xu, W; Xu, Q; Huang, Q; Tan, R; Shen, W; Song, W, Mater. Lett., 152, 173-6, 2015.
Permeability to water vapor, 25°C	g m^{-1} s^{-1} Pa^{-1} x 10^{11}	6.77	Pinotti, A; Garcia, M A; Martino, M N; Zaritzky, Food Hydrocolloids, 21, 66-72, 2007.
MECHANICAL & RHEOLOGICAL PROPERTIES			
Tensile strength	MPa	10-100	Turhan, K N; Sahbaz, F, J. Food Eng., 61, 3, 459-66, 2004; Xu, W; Xu, Q; Huang, Q; Tan, R; Shen, W; Song, W, Mater. Lett., 152, 173-6, 2015.
Elongation	%	14-97	Turhan, K N; Sahbaz, F, J. Food Eng., 61, 3, 459-66, 2004.
Young's modulus	MPa	3,200	Xu, W; Xu, Q; Huang, Q; Tan, R; Shen, W; Song, W, Mater. Lett., 152, 173-6, 2015.
Intrinsic viscosity, 25°C	ml g^{-1}	220-715	Ye, D; Farriol, X, Ind. Crops Products, 26, 54-62, 2007.
CHEMICAL RESISTANCE			
Alcohols	-	good	
Alkalis	-	poor	
Aliphatic hydrocarbons	-	poor	
Aromatic hydrocarbons	-	poor	
Esters	-	poor	
Halogenated hydrocarbons	-	good	
Ketones	-	poor	
Good solvent	-	alkalies, acetone, chloroform, cyclohexanone, esters, pyridine, water	
Non-solvent	-	diethyl ether, methanol, methylene chloride	
BIODEGRADATION			
Typical biodegradants	-	bacteria (e.g, *Escherichia, Pseudomonas, Staphylococcus, Lactobacillus, Bacillus*) and fungi (e.g., *Penicilium, Trichophyton*)	Tunc, S; Duman, O, LWT Food Sci. Technol., 44, 465-72, 2011.
Stabilizers	-	carvacrol	Tunc, S; Duman, O, LWT Food Sci. Technol., 44, 465-72, 2011.
TOXICITY			
NFPA: Health, Flammability, Reactivity rating	-	0/0/0	
Carcinogenic effect	-	not listed by ACGIH, NIOSH, NTP	
Oral rat, LD$_{50}$	mg kg^{-1}	>10,000	

MC methylcellulose

PARAMETER	UNIT	VALUE	REFERENCES
ENVIRONMENTAL IMPACT			
Aquatic toxicity, *Fathead minnow*, LC$_{50}$, 48 h	mg l^{-1}	>1,000	
PROCESSING			
Applications	-	artificial fiber (food supplement), artificial tears, food (emulsifier, thickener), glue, mortar, paper, pharmaceutical (controlled release, scaffolds, tablet coating, thickener)	Shen, H; Ma, Y; Luo, Y; Zhang, Z; Dai, J, Colloids Surf. B: Biointerfaces, 135, 332-8, 2015.
Outstanding properties	-	thickening, binding	
BLENDS			
Suitable polymers	-	chitosan, PEG, PPy, PVA, xanthan gum	
ANALYSIS			
FTIR (wavenumber-assignment)	cm^{-1}/-	ether group – 1066; C-O-C – 1100	Aziz, N A N; Idris, N K; Isa, M I N, Intl. J. Polym. Anal. Charact., 15, 319-327, 2010.
x-ray diffraction peaks	degree	sharp peak at 2θ = 7.8 and a broad peak centered at 2θ = 20.5	Łupina, K, Kowalczyk, D, Zięba, E, Kazimierczak, W, Wiącek, A E, Food Hydrocolloids, 96, 555-67, 2019.

MF melamine-formaldehyde resin

PARAMETER	UNIT	VALUE	REFERENCES
GENERAL			
Common name	-	melamine-formaldehyde resin	
ACS name	-	1,3,5-triazine-2,4,6-triamine, polymer with formaldehyde	
Acronym	-	MF	
CAS number	-	9003-08-1	
RTECS number	-	OS1140000	
HISTORY			
Person to discover	-	Talbot, W F	Talbot, W F, US Patent 2,260,239, Monsanto, Oct. 21, 1941.
Date	-	1941	
Details	-	patent for manufacture of melamine aldehyde condensation products	
SYNTHESIS			
Monomer(s) structure	-		
Monomer(s) CAS number(s)	-	108-78-1; 50-00-0	
Monomer(s) molecular weight(s)	dalton, g/mol, amu	126.12; 30.03	
Monomer ratio	-	1:1.5-3	Hansmann, C; Deka, M; Wimmer, R; Gindl, W, Holz Roh Werkstoff, 64, 198-203, 2006.
COMMERCIAL POLYMERS			
Some manufacturers	-	BASF	
Trade names	-	Saduren	
PHYSICAL PROPERTIES			
Density at 20°C	g cm^{-3}	1-1.14	
Bulk density at 20°C	g cm^{-3}	0.7	
Color	-	white	
pH	-	8.5	
Melting temperature, DSC	°C	90-101	
Decomposition temperature	°C	185	Devallencourt, C; Saiter, J M; Fafet, A; Ubrich, E, Thermochim. Acta, 259, 143-51, 1995.
Maximum service temperature	°C	150	
Dielectric constant at 100 Hz/1 MHz	-	4.7-10.9/7.9	
Volume resistivity	ohm-m	1E10 to 1E12	
Surface resistivity	ohm	1E11	
Electric strength K20/P50, d=0.60.8 mm	kV mm^{-1}	11-16	

MF melamine-formaldehyde resin

PARAMETER	UNIT	VALUE	REFERENCES
Arc resistance	s	183	
Diffusion coefficient of water vapor	$cm^2\ s^{-1}$ $\times 10^{10}$	115-1970	Smith, P M; Fisher, M M, Polymer, 25, 84-90, 1984.

MECHANICAL & RHEOLOGICAL PROPERTIES

PARAMETER	UNIT	VALUE	REFERENCES
Tensile strength	MPa	45-52	
Flexural strength	MPa	80	
Shrinkage	%	1.5-1.8; 0.6-0.8 (mold)	
Viscosity	mPa s	500-800	

CHEMICAL RESISTANCE

PARAMETER	UNIT	VALUE	REFERENCES
Alcohols	-	good	
Aliphatic hydrocarbons	-	very good	
Aromatic hydrocarbons	-	very good	
Esters	-	good	
Greases & oils	-	very good	
Ketones	-	good	

FLAMMABILITY

PARAMETER	UNIT	VALUE	REFERENCES
Ignition temperature	^oC	475-500	
Autoignition temperature	^oC	623-645	
Limiting oxygen index	$\%\ O_2$	30-60	
Volatile products of combustion	-	CO_2, CO, hydrocarbons, NO_x, ammonia, formaldehyde	Devallencourt, C; Saiter, J M; Fafet, A; Ubrich, E, Thermochim. Acta, 259, 143-51, 1995.
UL 94 rating	-	V-0	

WEATHER STABILITY

PARAMETER	UNIT	VALUE	REFERENCES
Spectral sensitivity	nm	300	Bauer, D R, Polym. Deg. Stab., 19, 97-112, 1987.
Emission wavelengths	nm	393	Bauer, D R, Polym. Deg. Stab., 19, 97-112, 1987.

BIODEGRADATION

PARAMETER	UNIT	VALUE	REFERENCES
Typical biodegradants	-	*Pseudomonas, Acinetobacter, Agrobacterium, Pseudaminobacter* and *Rhodococcus*	El-Sayed, W S; El-Baz, A; Othman, A M, Int. Biodet. Biodeg., 57, 75-81, 2006.

TOXICITY

PARAMETER	UNIT	VALUE	REFERENCES
NFPA: Health, Flammability, Reactivity rating	-	1-2/0-1/0	
Carcinogenic effect	-	not listed by ACGIH, NIOSH, NTP	
OSHA	$mg\ m^{-3}$	5 (respirable), 15 (total)	
Oral rat, LD_{50}	$mg\ kg^{-1}$	>10,000	
Skin rabbit, LD_{50}	$mg\ kg^{-1}$	>10,000	

PROCESSING

PARAMETER	UNIT	VALUE	REFERENCES
Typical processing methods	-	coating, compounding, molding	
Processing pressure	MPa	3.5 (molding)	

MF melamine-formaldehyde resin

PARAMETER	UNIT	VALUE	REFERENCES
Applications	-	cabinets, chipboard, construction materials (high pressure laminates), fiber glass sizing agent, furniture, insulation, kitchen utensils, paper laminates, water flocculation	
Outstanding properties	-	toughness, chemical resistance	

BLENDS			
Suitable polymers	-	PET, POM	

ANALYSIS			
FTIR (wavenumber-assignment)	cm^{-1}/-	3378 – stretching vibration of NH and NH$_2$, 2976 – the C–H stretching vibration of methylol group, 540 and 1471 – C=N stretching vibration in triazine rings and the C–H bending vibration in methylene, 1187 and 1049 – aliphatic C–N–C and C–O–C stretching vibrations, 841 – vibration of triazine ring	Cui, H, Chen, H, Guo, Z, Xu, J, Shen, J, Microporous Mesoporous Mater., 309, 110591, 2020.

MP melamine-phenolic resin

PARAMETER	UNIT	VALUE	REFERENCES
GENERAL			
Common name	-	melamine-phenolic resin	
CAS name	-	phenol, polymer with 1,3,5-triazine-2,4,6-triamine	
Acronym	-	MP	
CAS number	-	35484-57-2	
SYNTHESIS			
Monomer(s) structure	-		
Monomer(s) CAS number(s)	-	108-95-2; 108-78-1	
Monomer(s) molecular weight(s)	dalton, g/mol, amu	94.12; 126.12	
Method of synthesis	-	condensation of methylolated melamines with phenol in the presence of stoichiometric amounts of hydrochloric acid	
COMMERCIAL POLYMERS			
Some manufacturers	-	Momentive (Hexion)	
Trade names	-	Bakelite	
PHYSICAL PROPERTIES			
Density at 20°C	g cm^{-3}	1.3-1.5; 1.6-1.90 (mineral and glass filled)	
Bulk density at 20°C	g cm^{-3}	0.58	
Melting temperature, DSC	°C	50-70	Stark, W, Polym. Test., 29, 723-28, 2010.
Thermal conductivity, melt	W m^{-1} K^{-1}	0.58	
Maximum service temperature	°C	152-155	
Heat deflection temperature at 1.8 MPa	°C	135-194 (reinforced)	
Dielectric constant at 100 Hz/1 MHz	-	6.5-7/	
Relative permittivity at 100 Hz	-	7.5	
Dissipation factor at 100 Hz	E-4	80-540	
Volume resistivity	ohm-m	1E10 to 1.8E11	
Surface resistivity	ohm	1E11	
Electric strength K20/P50, d=0.60.8 mm	kV mm^{-1}	8.5-28.5	
Comparative tracking index, CTI, test liquid A	-	300	
Arc resistance	s	180-186	
MECHANICAL & RHEOLOGICAL PROPERTIES			
Tensile strength	MPa	41-57; 56-60 (reinforced)	
Tensile modulus	MPa	10,000-12,500 (reinforced)	
Elongation	%	0.64-0.81	

HANDBOOK OF POLYMERS 3ʳᵈ Edition, Copyrights 2022; ChemTec Publishing

MP melamine-phenolic resin

PARAMETER	UNIT	VALUE	REFERENCES
Flexural strength	MPa	41-91; 84-262 (reinforced)	Wu, C, Chen, Z, Wang, F, Hu, Y, Zhang, X, Compos. Part B: Eng., 162, 378-87, 2019.
Flexural modulus	MPa	8,900-13,500 (reinforced)	
Compressive strength	MPa	165-600 (reinforced)	
Izod impact strength, notched, 23°C	J m⁻¹	430	
Rockwell hardness	-	R71-120	
Ball indention hardness at 358 N/30 S (ISO 2039-1)	MPa	290 (reinforced)	
Shrinkage	%	0.1 (injection), 0.3 (post), 0.02 (compression molding shrinkage)	
Water absorption, equilibrium in water at 23°C	%	0.36	
CHEMICAL RESISTANCE			
Acid dilute/concentrated	-	poor	
Alcohols	-	resistant	
Alkalis	-	poor	
Aliphatic hydrocarbons	-	resistant	
Aromatic hydrocarbons	-	resistant	
Esters	-	resistant	
Greases & oils	-	resistant	
Halogenated hydrocarbons	-	resistant	
Ketones	-	resistant	
FLAMMABILITY			
Limiting oxygen index	% O_2	33	
UL 94 rating	-	V-0	
TOXICITY			
NFPA: Health, Flammability, Reactivity rating	-	1/1/0	
Carcinogenic effect	-	not listed by ACGIH, NIOSH, NTP	
PROCESSING			
Typical processing methods	-	compression molding, injection molding, powder molding	
Processing temperature	°C	160-170	
Processing pressure	MPa	15 (injection), 0.5-2 (back)	
Process time	min	0.6-0.66 (per 1 mm wall thickness)	
Applications	-	actuators, fan motors, fuel pumps, garden appliances, HVAC motors, household appliances, photoresists, power tools, universal motors, window lift motors, wiper motors	
ANALYSIS			
NMR (chemical shifts)	ppm	-NH-CH$_2$-oPhOH – 406-41.1; -NH-CH$_2$-pPhOH – 44.2; -N(CH$_2$)-oPhOH – 45.6-46.4; -N(CH$_2$)CH$_2$-pPhOH – 48.5-49.2	Maciejewski, M; Kedzieski, M; Bednarek, E; Rudnik, E, Polym. Bull., 48, 251-59, 2002.

NBR acrylonitrile-butadiene elastomer

PARAMETER	UNIT	VALUE	REFERENCES
GENERAL			
Common name	-	acrylonitrile-butadiene elastomer, nitrile rubber	
CAS name	-	2-propenenitrile, polymer with 1,3-butadiene; nitrile rubber	
Acronym	-	NBR	
CAS number	-	9003-18-3; 9005-98-5	
HISTORY			
Person to discover	-	Semon, W L	Semon, W L, US Patent 2,380,551, BF Goodrich, July 31, 1945.
Date	-	1941	
Details	-	patent for copolymerization of butadiene and acylonitrile in water emulsion in reactor composed of nickel, chromium, and iron	
SYNTHESIS			
Monomer(s) structure	-	$CH_2{=}CHCN$; $CH_2{=}CHCH{=}CH_2$	
Monomer(s) CAS number(s)	-	107-13-1; 106-99-0	
Monomer(s) molecular weight(s)	dalton, g/mol, amu	53.06; 54.09	
Acrylonitrile content	%	19-49	Vennemann, N, Kummerlöwe, C, Mertes, M, Kühnast, F, Siebert, A, Polym. Testing, 90, 106639, 2020.
Formulation example	-	AC – 32, butadiene – 68, water – 180, PMHP – 0.223, $FeSO_4$·H_2O – 0.0056, SFS – 0.12, Dresinate – 1.25, Tamol – 2.85, mercaptan – 0.42	Washington, I D; Duever, T A; Penlidis, A, J. Macromol. Sci. A, 47, 747-69, 2010.
Method of synthesis	-	NBR is produced by an emulsion polymerization. The water, emulsifier/soap, monomers (butadiene and acrylonitrile), radical generating activator, and other ingredients are introduced into the polymerization vessels. The emulsion process yields a polymer latex that is coagulated using calcium chloride or aluminum sulfate to form crumb rubber that is dried and compressed into bales.	Minari, R J; Gugliotta, L M; Vega, J R; Meira, G R, Computers Chem. Eng., 31, 1073-80, 2007.
Number average molecular weight, M_n	dalton, g/mol, amu	58,00-75,000	
Mass average molecular weight, M_w	dalton, g/mol, amu	199,000-600,000	
Polydispersity, M_w/M_n	-	2-6	
Crosslink density	10^{-4} mol cm^{-3}	3.4-3.7	Yang, S, Tian, J, Bian, X, Wu, Y, Compos. Sci. Technol., 186, 107909, 2020.
STRUCTURE			
Crystallinity	%	amorphous	
Tacticity	%	78 (trans in butadiene segments), 12 (cis), 10 (1,2-sites)	
COMMERCIAL POLYMERS			
Some manufacturers	-	Arlanxeo	
Trade names	-	Krynac, Baymod N	
PHYSICAL PROPERTIES			
Density at 20°C	g cm^{-3}	0.92-1.01	
Odor	-	slight rubbery	

NBR acrylonitrile-butadiene elastomer

PARAMETER	UNIT	VALUE	REFERENCES
Thermal decomposition	°C	>200 (begins with excessive hardening due to crosslinking)	
Storage temperature	°C	<35	
Glass transition temperature	°C	-60 to -10	
Specific heat capacity	J K^{-1} kg^{-1}	0.25	
Specific enthalpy of T$_g$	J g^{-1}	1.43	Ong, H T; Julkapli, N M; Hamid, S B A; Boondamnoen, O; Tai, M F, J. Magnetism Magnetic Mater., 395, 173-9, 2015.
Latent heat of crystallization	J g^{-1}	808.33	Ong, H T; Julkapli, N M; Hamid, S B A; Boondamnoen, O; Tai, M F, J. Magnetism Magnetic Mater., 395, 173-9, 2015.
Maximum service temperature	°C	-45 to 125	
Long term service temperature	°C	-40 to 108	
Hansen solubility parameters, δ_D, δ_P, δ_H	MPa$^{0.5}$	17.2, 8.6, 4.3	Zhu, L; Cheung, C S; Zhang, W G; Huang, Z, Fuel, 158, 288-92, 2015.
Hildebrand solubility parameter	MPa$^{0.5}$	17.90-21.38	
Permeability to nitrogen, 25°C	cm^3 cm cm^{-2} s^{-1} Pa^{-1} x 10^{12}	0.0177-0.189 (decreases with acrylonitrile concentration increasing)	
Permeability to oxygen, 25°C	cm^3 cm cm^{-2} s^{-1} Pa^{-1} x 10^{12}	0.0721-1.44 (decreases with acrylonitrile concentration increasing)	
Diffusion coefficient of nitrogen	cm^2 s^{-1} x10^6	0.25-0.064	
Diffusion coefficient of oxygen	cm^2 s^{-1} x10^6	0.79-13.6	

MECHANICAL & RHEOLOGICAL PROPERTIES

PARAMETER	UNIT	VALUE	REFERENCES
Tensile strength	MPa	3.1-6.5 (pure rubber)	
Tensile modulus	MPa	20.1-29.4	
Elongation	%	300-650	
Young's modulus	MPa	2-5	
Tear strength	kN m^{-1}	42-65	
Compression set	%	9 (1 day at 23°C); 25-35 (1 day at 70°C); 12-54 (70 h at 100°C)	Cook, S; Patel, J; Tinker, A J, 1680-84, 2000.
Shore A hardness	-	25-95	
Brittleness temperature (ASTM D746)	°C	-28 to -55	
Mooney viscosity	-	30-120	

CHEMICAL RESISTANCE

PARAMETER	UNIT	VALUE	REFERENCES
Acid dilute/concentrated	-	very good	
Alcohols	-	good	
Alkalis	-	very good	
Aliphatic hydrocarbons	-	good	
Aromatic hydrocarbons	-	poor	
Esters	-	poor	
Greases & oils	-	good	

NBR acrylonitrile-butadiene elastomer

PARAMETER	UNIT	VALUE	REFERENCES
Halogenated hydrocarbons	-	poor	
Ketones	-	poor	
Θ solvent	-	butanone/isopropanol, cyclohexane/MEK=1/1	
Lifetime in aviation kerosene	day/temp	11,113/20, 6467//25, 3831/30, 2309/35, 1414/40	Xiong, Y; Chen, G; Guo, S; Li, G, J. Ind. Eng. Chem., 19, 1611-16, 2013.

FLAMMABILITY			
Ignition temperature	°C	>300	
Limiting oxygen index	% O$_2$	29-31 (different FRs)	Moon, S C; Jo, B W; Farris, R J, Polym. Compos. 30, 1732-42, 2009.
Heat release	MJ m^{-2}	2.6-17.9 (different FRs)	Moon, S C; Jo, B W; Farris, R J, Polym. Compos. 30, 1732-42, 2009.
Volatile products of combustion	-	CO, CO$_2$, HCN, hydrocarbons, soot	

WEATHER STABILITY			
Spectral sensitivity	nm	262 (isolated C=C)	Sreeja, R; Najidha, S; Jayan, S R; Predeep, P; Mazur, M; Sharma, P D, Polymer, 617-23, 2006.

TOXICITY			
NFPA: Health, Flammability, Reactivity rating	-	1/1/0	
Carcinogenic effect	-	not listed by ACGIH, NIOSH, NTP	

PROCESSING			
Typical processing methods	-	coating, molding, vulcanization	
Additives used in final products	-	Fillers: calcium carbonate, carbon black, cellulose fibers, graphite, kaolin, montmorillonite, talc, zinc oxide; Plasticizers: dibenzyl ether, dioctyl adipate, dioctyl phthalate, fatty acid ester, and polyglycol ether; Antistatics: conductive carbon black, high styrene resin; Release: silicone resin surface coating; Slip: crosslinking of nitrile rubber, stearic acid; Dusting agent: calcium stearate, silica, PVC (powder grades); Crosslinker: sulfur	
Applications	-	aerospace (airplane components, cockpit display components, fighter jet components, fighter pilot headgear, guided missiles electrical connectors, etc.), automotive hoses, belt covers, electronics (buttons, connectors, keypads, power supply gaskets, satellite, etc.), footwear, gaskets, hose jackets, industrial hoses, medical (angioplasty balloons, blood pumps, dialysis, insulin pumps, needle-less syringes, etc.), o-rings, precision diaphragms, printing rolls, polymer modification, seals, tires	

BLENDS			
Suitable polymers	-	BR, EPDM, epoxy, EVA, LDPE, PA, PANI, PEDOT, PP, PSU, PVC, SBR, UHMWPE	

ANALYSIS			
FTIR (wavenumber-assignment)	cm^{-1}/-	2260-2236 – cyano; 1540-1530 – C=C	Sreeja, R; Najidha, S; Jayan, S R; Predeep, P; Mazur, M; Sharma, P D, Polymer, 617-23, 2006.

PA-3 polyamide-3

PARAMETER	UNIT	VALUE	REFERENCES
GENERAL			
Common name	-	polyamide-3, poly(imino-1-oxotrimethylene), poly(β-alanine)	
ACS name	-	poly[imino(1-oxo-1,3-propanediyl)]	
Acronym	-	PA-3	
CAS number	-	24937-14-2	
RTECS number	-	DG0875000	
SYNTHESIS			
Monomer(s) structure	-	$CH_2=CHC(O)NH_2$	
Monomer(s) CAS number(s)	-	79-06-1	
Monomer(s) molecular weight(s)	dalton, g/mol, amu	71.08	
Method of synthesis	-	hydrogen transfer polymerization of acrylamide in the presence of anionic catalyst; vinyl polymerization is side reaction, which can be prevented if tert-BuOK catalyst is finely dispersed in polymerization solvents	Masamoto, J, Memoirs of Fukui University of Technology, 3, 1, 291, 2003.
Temperature of polymerization	°C	80-200	
Catalyst	-	t-BuONa	
Yield	%	98-99	
Mass average molecular weight, M_w	dalton, g/mol, amu	90,000-140,000	
STRUCTURE			
Crystallinity	%	38	Wolfe, E; Stoll, B, Colloid Polym. Sci,. 258, 300, 1980.
Cell type (lattice)	-	monoclinic (form I); orthorhombic (form II)	
Cell dimensions	nm	a:b:c=0.933:0.478:0.873	Masamoto, J, Memoirs of Fukui University of Technology, 3, 1, 291, 2003.
Unit cell angles	degree	β=60	
Spacing between crystallites	nm	0.378	
Polymorphs	-	I, II	
Chain conformation	-	planar zigzag	
COMMERCIAL POLYMERS			
Some manufacturers	-	Croda	
Trade names	-	OleoCraft	
PHYSICAL PROPERTIES			
Density at 20°C	g cm^{-3}	1.33 (theoretical=1.39)	
Melting temperature, DSC	°C	330-340 (decomposition)	
Glass transition temperature	°C	111	
Dielectric constant at 100 Hz/1 MHz	-	4.7/	
MECHANICAL & RHEOLOGICAL PROPERTIES			
Elongation	%	10-20	
Flexural modulus	MPa	19,900 (fiber glass-reinforced)	

PA-3 polyamide-3

PARAMETER	UNIT	VALUE	REFERENCES
Young's modulus	GPa	374-414 (theoretical)	Peeters, A; van Alsenoy, C; Bartha, F; Bogar, F; Zhang, M-L; van Doren, V E, Int. J. Quantum Chem. 87, 303-10, 2002.
Hildebrand solubility parameter	MPa$^{0.5}$	28.2-35.4	
Water absorption, equilibrium in water at 23°C	%	9 (similar to silk)	

CHEMICAL RESISTANCE

Acid dilute/concentrated	-	poor	
Alcohols	-	good	
Aliphatic hydrocarbons	-	good	
Aromatic hydrocarbons	-	good	
Halogenated hydrocarbons	-	poor	
Good solvent	-	dichloroacetic acid, formic acid, phenol, sulfuric acid, trifluoroethanol	
Non-solvent	-	chloroform, butanol, methanol, water	

TOXICITY

Carcinogenic effect	-	not listed by ACGIH, NIOSH, NTP	

PROCESSING

Typical processing methods	-	because polyamide 3 melts with decomposition, melt spinning cannot be used but wet or dry spinning is used; wet spinning is performed from formic acid solution	Masamoto, J, Memoirs of Fukui University of Technology, 3, 1, 291, 2003.
Applications	-	fiber for production of ropes, POM stabilizer (formaldehyde scavenger), emolient in cosmetics	
Outstanding properties	-	thermal stability, fiber strength	

ANALYSIS

FTIR (wavenumber-assignment)	cm^{-1}/-	C=O – 1640; amide II – 1530; CH$_2$ – 1420, 1360; amide III – 1283, 1220; amide IV – 1183, 1100, 1040, 960	Morgenstern, U; Berger, W, Makromol. Chem., 193, 2561-69, 1992.
Raman (wavenumber-assignment)	cm^{-1}/-	amide 1 – 1630; N-H – 1227; C-C – 1170	Hendra, P J; Maddams, W F; Royaud, I A M; Willis, H A; Zichy, V, Spectrochim. Acta, 64A, 5, 747-56, 1990.
x-ray diffraction peaks	degree	22.0 and 23.5	

PA-4,6 polyamide-4,6

PARAMETER	UNIT	VALUE	REFERENCES
GENERAL			
Common name	-	polyamide-4,6; poly(iminotetramethyleneiminoadipoyl)	
CAS name	-	poly[imino-1,4-butanediylimino(1,6-dioxo-1,6-hexanediyl)]	
Acronym	-	PA-4,6	
CAS number	-	50327-22-5	
HISTORY			
Person to discover	-	Wallace Hume Carothers	Carothers, W H, US Patent 2,130,948, DuPont, Sept. 20, 1938.
Date	-	1938	
Details	-	first patent (first application filed in 1931)	
SYNTHESIS			
Monomer(s) structure	-	$H_2N(CH_2)_4NH_2$ $HOC(CH_2)_4COH$ (with O double-bonds)	
Monomer(s) CAS number(s)	-	110-60-1; 124-04-9	
Monomer(s) molecular weight(s)	dalton, g/mol, amu	88.15; 146.14	
CH_2/CONH ratio		4	
Method of synthesis	-	polyamide-4,6 is a product of polycondensation of tetramethylene-diamine and adipic acid. Due to the very high melting temperature the production of polyamide-4,6 was very difficult. As late as 1985 a process was invented to produce polyamide-4,6 on a commercially attractive scale in two step process including precondensation and solid-state post-condensation	
Number average molecular weight, M_n	dalton, g/mol, amu	6,100-24,400	
Mass average molecular weight, M_w	dalton, g/mol, amu	14,000-55,000	
Polydispersity, M_w/M_n	-	1.95-2.3	
STRUCTURE			
Crystallinity	%	25-70; 49 (dry); 44 (wet)	Zhang, Q; Zhang, Z; Zhang, H; Mo, Z, J. Polym. Sci. B, 40, 1784-93, 2002; Extrand, C W, J. Colloid Interface Sci., 248, 136-42, 2002; Adriaensens, P; Pollaris, A; Carlleer, R; Vanderzande, D; Gelan, J; Litvinov, V M; Tijssen, J, Polymer, 42, 7943-52, 2001.
Cell type (lattice)	-	monoclinic	
Cell dimensions	nm	a:b:c=0.96:1.48:0.806	Franco, L; Puiggali, Polymer, 40, 3255-59.
Unit cell angles	degree	β=67	Franco, L; Puiggali, Polymer, 40, 3255-59.
Polymorphs	-	α (the most stable), γ	
Rapid crystallization temperature	°C	260-265	
Avrami constants, k/n	-	n=4.1-4.2	
COMMERCIAL POLYMERS			
Some manufacturers	-	Cool Polymers; DSM; ERJKS	
Trade names	-	CoolPoly; Stanyl; Ertalon	

PA-4,6 polyamide-4,6

PARAMETER	UNIT	VALUE	REFERENCES
PHYSICAL PROPERTIES			
Density at 20°C	g cm^{-3}	1.17-1.18; 0.978 (melt); 1.29-1.52 (15-60% glass fiber); 1.1-1.55 (15-60% glass fiber; melt)	
Melting temperature, DSC	°C	290-295; 295 (15-60% glass fiber)	
Thermal decomposition	°C	>350	
Thermal expansion coefficient, 23-80°C	10^{-4} °C^{-1}	0.7-1.1; 0.2-0.6 (15-60% glass fiber)	
Thermal conductivity, melt	W m^{-1} K^{-1}	0.252 (melt); 0.267- 0.43 (15-60% glass fiber, melt); 1.2 (special conductive polymer)	
Glass transition temperature	°C	43-80; 75 (15-60% glass fiber)	
Specific heat capacity	J K^{-1} kg^{-1}	2,800 (melt); 1,450-2,540 (15-60% glass fiber, melt)	
Heat of fusion	kJ mol^{-1}	15.1	
Maximum service temperature	°C	100	
Temperature index (50% tensile strength loss after 20,000 h/5000 h)	°C	130/152; 153/177 (15-60% glass fiber)	
Heat deflection temperature at 0.45 MPa	°C	220-285; 290 (15-60% glass fiber)	
Heat deflection temperature at 1.8 MPa	°C	160-190; 275-290 (15-60% glass fiber)	
Vicat temperature VST/A/50	°C	290; 290 (15-60% glass fiber)	
Dielectric constant at 100 Hz/1 MHz	-	3.83/	
Relative permittivity at 100 Hz	-	3.9; 22 (conditioned)	
Relative permittivity at 1 MHz	-	3.6; 4.5 (conditioned)	
Dissipation factor at 100 Hz	E-4	70; 8,700 (conditioned)	
Dissipation factor at 1 MHz	E-4	260; 1,200 (conditioned)	
Volume resistivity	ohm-m	5E12; 1E7 (conditioned)	
Surface resistivity	ohm	8E15; 1E13 (conditioned)	
Electric strength K20/P50, d=0.60.8 mm	kV mm^{-1}	25; 15 (conditioned); 30 (15-60% glass fiber)	
Comparative tracking index, CTI, test liquid A	-	400; 175 (15-60% glass fiber)	
Coefficient of friction	-	0.08-0.38	Gordon, D H; Kukureka, S N, Wear, 267, 669-78, 2009.
Contact angle of water, 20°C	degree	57.6/32.8 (ascending/receding)	
MECHANICAL & RHEOLOGICAL PROPERTIES			
Tensile strength	MPa	55-100; 140-255 (15-60% glass fiber)	
Tensile modulus	MPa	3,300-3,900; 6,100-20,000 (15-60% glass fiber)	
Tensile stress at yield	MPa	40-100	
Tensile creep modulus, 1000 h, elongation 0.5 max	MPa	550; 10,500 (15-60% glass fiber)	
Elongation	%	7; 2-4 (15-60% glass fiber)	
Tensile yield strain	%	10	
Flexural strength	MPa	140-150; 225 (15-60% glass fiber)	
Flexural modulus	MPa	3,000-3,800; 7,500-17,000 (15-60% glass fiber)	
Young's modulus	MPa	932	
Charpy impact strength, unnotched, 23°C	kJ m^{-2}	6.5-10; 40-90 (15-60% glass fiber)	

PA-4,6 polyamide-4,6

PARAMETER	UNIT	VALUE	REFERENCES
Charpy impact strength, unnotched, -30°C	kJ m^{-2}	4; 40-65 (15-60% glass fiber)	
Charpy impact strength, notched, 23°C	kJ m^{-2}	6-18 (15-60% glass fiber)	
Charpy impact strength, notched, -30°C	kJ m^{-2}	6-18 (15-60% glass fiber)	
Izod impact strength, notched, 23°C	J m^{-1}	10; 6-18 (15-60% glass fiber)	
Izod impact strength, notched, -30°C	J m^{-1}	4; 6-18 (15-60% glass fiber)	
Crack growth velocity	x 10^{-6} m s^{-1}	800	Rajesh, J J; Bijwe, J, Tribology lett., 18, 3, 331-40, 2005.
Fracture energy	x 10^{4} J m^{-2}	9.29	Rajesh, J J; Bijwe, J, Tribology lett., 18, 3, 331-40, 2005.
Ductility factor	mm	10.81	Rajesh, J J; Bijwe, J, Tribology lett., 18, 3, 331-40, 2005.
Stress necessary to cause spontaneous fracture	MPa	197.13	Rajesh, J J; Bijwe, J, Tribology lett., 18, 3, 331-40, 2005.
Poisson's ratio	-	0.37	
Shrinkage	%	0.3-2; 0.4-1.5 (15-60% glass fiber)	
Viscosity number	ml g^{-1}	175; 75-150 (15-60% glass fiber)	
Water absorption, equilibrium in water at 23°C	%	9.5-13.5; 5.4-11.5 (15-60% glass fiber)	Adriaensens, P; Pollaris, A; Rulkens, R; Litvinov, V M; Gelan, J, Polymer, 45, 2465-73, 2004; Adriaensens, P; Pollaris, A; Carlleer, R; Vanderzande, D; Gelan, J; Litvinov, V M; Tijssen, J, Polymer, 42, 7943-52, 2001.
Moisture absorption, equilibrium 23°C/50% RH	%	2.6-3.7; 1.4-3.15 (15-60% glass fiber)	

CHEMICAL RESISTANCE

PARAMETER	UNIT	VALUE	REFERENCES
Acid dilute/concentrated	-	poor	
Alcohols	-	good	
Alkalis	-	fair-poor	
Aliphatic hydrocarbons	-	good	
Aromatic hydrocarbons	-	good	
Esters	-	good	
Greases & oils	-	good	
Halogenated hydrocarbons	-	fair-poor	
Ketones	-	good	
Good solvent	-	formic acid, hexafluoroisopropanol, sulfuric acid	

FLAMMABILITY

PARAMETER	UNIT	VALUE	REFERENCES
Autoignition temperature	°C	420	
Limiting oxygen index	% O$_2$	22-27	
Volatile products of combustion	-	CO_2, H_2O, CO, NH_3, NO_x, HNC, cyclopentanone	
UL 94 rating	-	V-2; HB to V-0 (15-60% glass fiber)	

PA-4,6 polyamide-4,6

PARAMETER	UNIT	VALUE	REFERENCES
TOXICITY			
Carcinogenic effect	-	not listed by ACGIH, NIOSH, NTP	
TLV, ACGIH	mg m^{-3}	3 (respiratory), 10 (total)	
OSHA	mg m^{-3}	5 (respiratory), 15 (total)	
PROCESSING			
Typical processing methods	-	extrusion, injection molding	
Preprocess drying: temperature/ time/residual moisture	^{o}C/h/%	80-95/2-24/0.05	
Processing temperature	^{o}C	315-320; 305-315 (15-60% glass fiber)	
Processing pressure	MPa	5.2-10.4 (injection), 1.7-5.2 (hold), 0.35 (back)	
Additives used in final products	-	flame retardants; fillers (glass fiber); heat stabilizers; impact modifiers, lubricants; release agents	
Applications	-	automotive (ABS controllers, alternator parts, chain tensioners, clutch components, inlet manifolds, oil filters, radiator end caps, sensors and switches, etc.)	
BLENDS			
Suitable polymers	-	poly(phenylene oxide)	Ran, J, Xie, H, Lai, X, Li, H, Zeng, X, Tribology Int., 128, 204-13, 2018.

PA-4,10 polyamide-4,10

PARAMETER	UNIT	VALUE	REFERENCES
GENERAL			
Common name	-	poly(tetramethylene sebacamide), polyamide-4,10	
Acronym	-	PA-4,10	
CAS number	-	26247-06-3	
SYNTHESIS			
Monomer(s) structure	-	$H_2N(CH_2)_4NH_2$ $HOC(CH_2)_8COH$ (with two C=O groups)	
Monomer(s) CAS number(s)	-	110-60-1; 111-20-6	
Monomer(s) molecular weight(s)	dalton, g/mol, amu	88.15; 202.25	
Method of synthesis	-	in the first step a low molecular weight prepolymers with M_n=1000–2000 g mol^{-1} are synthesized to obtain a concentrated aqueous solution of prepolymers; after isolation of the prepolymers, the solid state post-condensation is performed at a temperature of about 35ºC below the corresponding T_m until the desired molecular weight is reached	Koning, C; Teuwen, L; de Jong, R; Janssen, G; Coussens, B, High Perform. Polym., 11, 387-94, 1999.
Number average molecular weight, M_n	dalton, g/mol, amu	17,200	
Chain-end groups	meq g^{-1}	NH_2 – 0.02; COOH – 0.058	Koning, C; Teuwen, L; de Jong, R; Janssen, G; Coussens, B, High Perform. Polym., 11, 387-94, 1999.
STRUCTURE			
Cell type (lattice)	-	triclinc	
Cell dimensions	nm	a:b:c=0.49:0.532:1.98 (α); a:b:c=0.49:0.800:1.98 (β)	Jones, N A; Atkins, E D T; Hill, M J; Cooper, S J; Franco, L, Polymer, 38, 11, 2689-99, 1997.
Unit cell angles	degree	α:β:γ=49:77:63 (α); α:β:γ=90:77:66 (β)	Jones, N A; Atkins, E D T; Hill, M J; Cooper, S J; Franco, L, Polymer, 38, 11, 2689-99, 1997.
Polymorphs	-	α, β	Jones, N A; Atkins, E D T; Hill, M J; Cooper, S J; Franco, L, Polymer, 38, 11, 2689-99, 1997.
Rapid crystallization temperature	ºC	210	
Avrami constants, k/n	-	n=2.0-3.3	
COMMERCIAL POLYMERS			
Some manufacturers	-	DSM	
Trade names	-	EcoPaXX	
PHYSICAL PROPERTIES			
Density at 20ºC	g cm^{-3}	1.09 (dry); 1.34-1.52 (30-50% glass fiber, dry)	
Odor	-		
Melting temperature, DSC	ºC	249-250 (dry); 250 (30-50% glass fiber, dry)	
Heat deflection temperature at 1.8 MPa	ºC	77 (dry); 215-220 (30-50% glass fiber, dry)	
Water absorption (50% RH)	%	2	Biron, M, Industrial Applications of Renewable Plastics, William Andrew, 2017, pp. 155-369.

PA-4,10 polyamide-4,10

PARAMETER	UNIT	VALUE	REFERENCES
MECHANICAL & RHEOLOGICAL PROPERTIES			
Tensile strength	MPa	135-220 (30-50% glass fiber, dry); 115 (30-50% glass fiber, conditioned)	
Tensile modulus	MPa	3,100-3,160 (dry); 1,700-1,840 (conditioned); 9,500-16,000 (30-50% glass fiber, dry); 7,000 (30-50% glass fiber, conditioned)	
Tensile stress at yield	MPa	82-85 (dry); 61 (conditioned)	
Elongation	%	10-16 (dry); 32 (conditioned); 2.5-4 (30-50% glass fiber, dry); 5.8 (30-50% glass fiber, conditioned)	
Tensile yield strain	%	5 (dry); 3 (30-50% glass fiber, dry)	
Charpy impact strength, unnotched, 23°C	kJ m^{-2}	NB; 40-80 (30-50% glass fiber, dry); 80 (30-50% glass fiber, conditioned)	
Charpy impact strength, unnotched, -30°C	kJ m^{-2}	NB; 60 (30-50% glass fiber, dry)	
Charpy impact strength, notched, 23°C	kJ m^{-2}	5-7 (dry); 13 (conditioned); 11-15 (30-50% glass fiber, dry); 15 (30-50% glass fiber, conditioned)	
Charpy impact strength, notched, -30°C	kJ m^{-2}	4 (dry); 9-12 (30-50% glass fiber, dry)	
Shrinkage	%	1.2-1.3; 0.4-1.1 (30-50% glass fiber, dry)	Biron, M, Industrial Applications of Renewable Plastics, William Andrew, 2017, pp. 155-369.
Moisture absorption, equilibrium 23°C/50% RH	%	2-2.24; 1.5 (30-50% glass fiber, dry)	
CHEMICAL RESISTANCE			
Greases & oils	-	good	
Other	-	hot water, salt	
FLAMMABILITY			
UL 94 rating	-	V-0 (30-50% glass fiber, dry)	
PROCESSING			
Typical processing methods	-	injection molding	
Preprocess drying: temperature/ time/residual moisture	°C/h/%	80/4-8/0.1	
Processing temperature	°C	240-275	
Additives used in final products	-	release agent	
BLENDS			
Compatible polymers	-	PLA	

PA-6 polyamide-6

PARAMETER	UNIT	VALUE	REFERENCES
GENERAL			
Common name	-	polyamide-6, nylon 6, poly(ε-caprolactam)	
IUPAC name	-	poly[imino(1-oxohexane-1,6-diyl)]	
CAS name	-	poly[imino(1-oxo-1,6-hexanediyl)]	
Acronym	-	PA-6	
CAS number	-	25038-54-4	
RTECS number	-	TQ9800000	
HISTORY			
Person to discover	-	Paul Schlack	Schlack, P, US Patent 2,241,321, IG Farben, May 6, 1941.
Date	-	1941	
Details	-	preparation of polyamides	
SYNTHESIS			
Monomer(s) structure	-		
Monomer(s) CAS number(s)	-	105-60-2	
Monomer(s) molecular weight(s)	dalton, g/mol, amu	113.2	
CH_2/CONH ratio		5	
Method of synthesis	-	caprolactam is melted and polymerized in the presence of catalyst and dulling agent	
Temperature of polymerization	°C	190-275	
Time of polymerization	h	24+22	
Catalyst	-	AH-salt	
Heat of polymerization	$J\ g^{-1}$	137-146	Riedel, O; Wittmer, P, Makromol. Chem., 97, 1, 1966.
Number average molecular weight, M_n	dalton, g/mol, amu	20,800-48,100	Robert, E C; Bruessau, R; Dubois, J; Jacques, B; Meijerink, N; Nguye, T Q; Niehaus, D E; Tobisch, W A, Pure Appl. Chem., 76, 11, 2009-25, 2004; Farina, H; Yuan, C M; Ortenzi, M; Di Silvestro, G, Macromol. Symp., 218, 51-60, 2004.
Mass average molecular weight, M_w	dalton, g/mol, amu	17,000-94,000	Farina, H; Yuan, C M; Ortenzi, M; Di Silvestro, G, Macromol. Symp., 218, 51-60, 2004.
Polydispersity, M_w/M_n	-	1.7-2.4	
Polymerization degree (number of monomer units)	-	130-250	
Molar volume at 298K	$cm^3\ mol^{-1}$	calc.=103.7; 92.0 (crystalline); 104.4 (amorphous); exp.=104.4	
Van der Waals volume	$cm^3\ mol^{-1}$	68.06; 64.2 (crystalline); 64.2 (amorphous)	
Chain-end groups	micro-equivalent g^{-1}	COOH – 28.8; NH_2 – 34.5; COOH – 0; NH_2 – 0.84	Fornes, T D; Paul, D R, Macromolecules, 37, 7698-7709, 2004; Davis, R D; Jarrett, W L; Mathias, L J, Polymer, 42, 2621-26, 2001.

230 **HANDBOOK OF POLYMERS** 3rd Edition, Copyrights 2022; ChemTec Publishing

PA-6 polyamide-6

PARAMETER	UNIT	VALUE	REFERENCES
STRUCTURE			
Crystallinity	%	24.4-50; 42 (dry); 34 (wet)	Extrand, C W, J. Colloid Interface Sci., 248, 136-42, 2002; Yebra-Rodriguez, A; Alvarez-Lloret, P; Cardell, C; Rrodriquez-Navarro, A B, Appl. Clay Sci., 43, 91-97, 2009; Lai, D Li, Y, Wang, C, Liu, Y, Yang, J, Mater. Today, 25, 101578, 2020.
Cell type (lattice)	-	monoclinic	Holmes, D R; Bunna, C W; Smith, C J, J. Polym. Sci., 17, 159, 1955.
Cell dimensions	nm	a:b:c=0.956:0.801:1.724	Holmes, D R; Bunna, C W; Smith, C J, J. Polym. Sci., 17, 159, 1955.
Unit cell angles	degree	γ=67.5	Holmes, D R; Bunna, C W; Smith, C J, J. Polym. Sci., 17, 159, 1955.
Number of chains per unit cell	-	8	Holmes, D R; Bunna, C W; Smith, C J, J. Polym. Sci., 17, 159, 1955.
Crystallite size	nm	23.3	Rajesh, J J; Bijwe, J, Wear, 661-68, 2005.
Cis content	%	1.64	Davis, R D; Jarrett, W L; Mathias, L J, Polymer, 42, 2621-26, 2001.
Entanglement molecular weight	dalton, g/mol, amu	calc.=1988-2490	
Crystallization temperature	ºC	191.8-193.2	Faghihi, M; Shojaei, A; Bagheri, R, Composites: Part B, 78, 50-64, 2015.
Melting point of crystallites	ºC	220 (α-form), 212 (γ-form)	Lai, D Li, Y, Wang, C, Liu, Y, Yang, J, Mater. Today, 25, 101578, 2020.
Rapid crystallization temperature	ºC	173-180	Adriaensens, P; Pollaris, A; Carlleer, R; Vanderzande, D; Gelan, J; Litvinov, V M; Tijssen, J, Polymer, 42, 7943-52, 2001.
COMMERCIAL POLYMERS			
Some manufacturers	-	BASF, DSM; EMS; DuPont; Kolon Plastics; Eurotec; Zig Shen	
Trade names	-	Ultramid B; Aculon; Grilon; Zytel; Kopa: Tecomid; Zisamide	
PHYSICAL PROPERTIES			
Density at 20ºC	g cm^{-3}	1.06-1.16; 1.23-1.56 (12-50% glass fiber, dry); 0.96-1.07 (melt); 1.34 (50% glass fiber)	
Bulk density at 20ºC	g cm^{-3}	0.5-0.8	
Color	-	colorless to white	
Refractive index, 20ºC	-	calc.=1.513-1.530; exp.=1.53	
Birefringence	-	1.580/1.582	
Transmittance	%	85	
Odor	-	odorless	
Melting temperature, DSC	ºC	220-260; 218.6-220	Faghihi, M; Shojaei, A; Bagheri, R, Composites: Part B, 78, 50-64, 2015.
Heat of melting	J g^{-1}	53.8-62.1	Faghihi, M; Shojaei, A; Bagheri, R, Composites: Part B, 78, 50-64, 2015.
Thermal decomposition temperature	ºC	>300	
Activation energy of thermal degradation	kJ mol^{-1}	162	Herrera, M; Matuschek, G; Kettrup, A, Chemosphere, 42, 601-7, 2001.
Thermal expansion coefficient, 23-80ºC	ºC^{-1}	0.83-0.93E-4 (dry); 0.12-0.78E-4 (12-50% glass fiber, dry)	

PA-6 polyamide-6

PARAMETER	UNIT	VALUE	REFERENCES
Thermal conductivity, melt	W m^{-1} K^{-1}	0.1968-0.23; 0.25-0.5	-; Ghahramani, N, Rahmati, M, Int. J. Heat Mass Transfer, 154, 119487, 2020.
Glass transition temperature	°C	calc.=49-61; exp.=50-75 (dry); 3-20 (exposed to 50% RH); -22 to -32 (exposed to 100% RH)	
Specific heat capacity	J K^{-1} kg^{-1}	1,400; 2,680-2,730 (melt); 1,950 (50% glass fiber)	
Heat of fusion	kJ kg^{-1}	188-280	
Heat deflection temperature at 0.45 MPa	°C	117-190 (dry); 217-250 (12-50% glass fiber, dry)	
Heat deflection temperature at 1.8 MPa	°C	42-65 (dry); 190-250 (12-50% glass fiber, dry)	
Vicat temperature VST/A/50	°C	195-204 (dry); 210-250 (12-50% glass fiber, conditioned)	
Hansen solubility parameters, δ_D, δ_P, δ_H	MPa$^{0.5}$	17.0, 3.4, 10.6	
Interaction radius		5.1	
Hildebrand solubility parameter	MPa$^{0.5}$	21.5-32; calc.=21.5; exp.=20.3	Ghahramani, N, Rahmati, M, Int. J. Heat Mass Transfer, 154, 119487, 2020.
Surface tension	mN m^{-1}	calc.=33.9-47.6; exp.=40.0-47.0	
Dielectric constant at 100 Hz/1 MHz	-	-/3.8	
Relative permittivity at 100 Hz	-	3.2-4.1 (dry); 3.6-14 (12-50% glass fiber, dry)	
Relative permittivity at 1 MHz	-	3.0-3.5 (dry); 4.5-7 (conditioned); 3.4-3.8 (12-50% glass fiber, dry)	
Dissipation factor at 100 Hz	E-4	50-100 (dry); 100-250 (12-50% glass fiber, dry); 1,700-3,000 (12-50% glass fiber, conditioned)	
Dissipation factor at 1 MHz	E-4	150-310 (dry); 1,200-3,000 (conditioned); 0.02-250 (12-50% glass fiber, dry); 1,700-2,400 (12-50% glass fiber, conditioned)	
Volume resistivity	ohm-m	1E13 (dry), 2E9 (saturated at 50% RH, 20°C), 4E6 (saturated at 100% RH, 20°C)	
Surface resistivity	ohm	1E15 (dry); 1E10 to 1E14 (conditioned)	
Electric strength K20/P50, d=0.60.8 mm	kV mm^{-1}	25-460; 20 (conditioned)	
Comparative tracking index	-	600 (conditioned); 550 (12-50% glass fiber, conditioned)	
Power factor	-	0.02-0.06	
Coefficient of friction	ASTM D1894	0.32-0.43 (static), 0.16-0.8 (dynamic); 0.26 (chrom steel), 0.44-0.47 (aluminum)	Maldonado, J E, Antec, 3431-35, 1998.
Diffusion coefficient of water vapor	cm^2 s^{-1} x10^6	0.004/0.055	
Contact angle of water, 20°C	degree	62.6-65.5; 69.2/42.4 (ascending/receding)	
Surface free energy	mJ m^{-2}	45.4	

MECHANICAL & RHEOLOGICAL PROPERTIES

PARAMETER	UNIT	VALUE	REFERENCES
Tensile strength	MPa	74-106; 130-240 (12-50% glass fiber, dry); 70-180 (12-50% glass fiber, conditioned)	Jia, N; Kagan, V A, Antec, 1706-12, 1998.
Tensile modulus	MPa	780-3,800 (dry); 570-1,200 (conditioned); 5,800-16,800 (12-50% ; 2,640-12,500 (12-50% glass fiber, conditioned)	
Tensile stress at yield	MPa	36-95 (dry); 32-55 (conditioned)	
Tensile creep modulus, 1000 h, elongation 0.5 max	MPa	700-1,100 (conditioned); 2,100-7,800 (12-50% glass fiber, conditioned)	
Elongation	%	10-160 (dry); 327 (conditioned); 2.5-4 (12-50% glass fiber, dry); 3.5-18 (12-50% glass fiber, conditioned)	

PA-6 polyamide-6

PARAMETER	UNIT	VALUE	REFERENCES
Tensile yield strain	%	3.5-7 (dry); 15-25 (conditioned)	
Flexural strength	MPa	100 (dry)	
Flexural modulus	MPa	2,600 (dry)	
Elastic modulus	MPa	3,080-3,330; 9,860-10,820 (33% glass fiber); 14,110-16,440 (50% glass fiber)	Jia, N; Kagan, V A, Antec, 1706-12, 1998.
Compressive strength	MPa	55	
Young's modulus	GPa	235-337 (theoretical); 270-377 (experimental)	Peeters, A; van Alsenoy, C; Bartha, F; Bogar, F; Zhang, M-L; van Doren, V E, Int. J. Quantum Chem. 87, 303-10, 2002.
Charpy impact strength, unnotched, 23°C	kJ m^{-2}	no break to 50-250 (dry); no break (conditioned); 40-95 (12-50% glass fiber, dry); 100-105 (12-50% glass fiber, conditioned)	
Charpy impact strength, unnotched, -30°C	kJ m^{-2}	51-200 to no break (dry); 55-90 (12-50% glass fiber, dry)	
Charpy impact strength, notched, 23°C	kJ m^{-2}	3.5-82 (dry); 35-150 to no break (conditioned); 6.5-18 (12-50% glass fiber, dry); 25 (12-50% glass fiber, conditioned)	
Charpy impact strength, notched, -30°C	kJ m^{-2}	3-23 (dry); 4-22 (conditioned); 8 (12-50% glass fiber, dry)	
Izod impact strength, unnotched, 23°C	J m^{-1}	1,700-2,000	Lin, H; Isayev, A I, Antec, 1518-22, 2003.
Crack growth velocity	x 10^{-6} m s^{-1}	527	Rajesh, J J; Bijwe, J, Tribology lett., 18, 3, 331-40, 2005.
Fracture energy	x 10^4 J m^{-2}	5.37	Rajesh, J J; Bijwe, J, Tribology lett., 18, 3, 331-40, 2005.
Ductility factor	mm	17.9	Rajesh, J J; Bijwe, J, Tribology lett., 18, 3, 331-40, 2005.
Stress necessary to cause spontaneous fracture	MPa	97.40	Rajesh, J J; Bijwe, J, Tribology lett., 18, 3, 331-40, 2005.
Tenacity (fiber) (standard atmosphere)	cN tex^{-1} (daN mm^{-2})	40-90 (45-100)	Fourne, F, Synthetic Fibers. Machines and Equipment Manufacture, Properties. Carl Hanser Verlag, 1999.
Tenacity (wet fiber, as % of dry strength)	%	80-90	Fourne, F, Synthetic Fibers. Machines and Equipment Manufacture, Properties. Carl Hanser Verlag, 1999.
Fineness of fiber (titer)	dtex	1.4-300	Fourne, F, Synthetic Fibers. Machines and Equipment Manufacture, Properties. Carl Hanser Verlag, 1999.
Length (elemental fiber)	mm	38-300, filament	Fourne, F, Synthetic Fibers. Machines and Equipment Manufacture, Properties. Carl Hanser Verlag, 1999.
Poisson's ratio	-	calc.=0.440; exp.=0.33	
Shrinkage	%	0.87-1.4; 0.33-0.82 (12-50% glass fiber)	
Viscosity number	ml g^{-1}	140-270; 130-170 (12-50% glass fiber, dry)	
Melt viscosity, shear rate=1000 s^{-1}	Pa s	220-1,070	
Melt volume flow rate (ISO 1133, procedure B), 275°C/5 kg	cm^3/10 min	40-185; 20-50 (15-50% glass fiber)	
Melt index, 190°C/2.16 kg	g/10 min	7.7-36.4	Faghihi, M; Shojaei, A; Bagheri, R, Composites: Part B, 78, 50-64, 2015.

PA-6 polyamide-6

PARAMETER	UNIT	VALUE	REFERENCES
Water absorption, equilibrium in water at 23°C	%	7.1-10; 4-8.1 (12-50% glass fiber, dry)	
Moisture absorption, equilibrium 20-90°C/saturation	%	2.0-3.3; 1.2-2.3 (12-50% glass fiber, dry)	

CHEMICAL RESISTANCE			
Acid dilute/concentrated	-	good/poor	
Alcohols	-	good	
Alkalis	-	dilute - good, concentrated - poor	
Aliphatic hydrocarbons	-	resistant	
Aromatic hydrocarbons	-	resistant	
Greases & oils	-	resistant	
Ketones	-	resistant	
Good solvent	-	acetic acid, chlorophenol, m-cresol, ethylene carbonate, formic acid, phosphoric acid, sulfuric acid, trichloroacetic acid	
Non-solvent	-	alcohols, chloroform, DMF, esters, ethers, hydrocarbons, ketones	

FLAMMABILITY			
Flammability according to UL-94 standard; thickness 1.6/0.8 mm	class	V-2 or HB; HB (12-50% glass fiber, dry)	
Ignition temperature	°C	400	
Autoignition temperature	°C	424	
Limiting oxygen index	$\% \, O_2$	20-27; 22.5 (15-50% glass fiber); 27.2 (5 wt% furan-phospha-mide)	
Peak of heat release rate	$kW \, m^{-2}$	989	Sun, J, Qian, L, Li, J, Polymer, 210, 122994, 2020.
Total heat release	$MJ \, m^{-2}$	128.7	Sun, J, Qian, L, Li, J, Polymer, 210, 122994, 2020.
Char at 500°C	%	0	Lyon, R E; Walters, R N, J. Anal. Appl. Pyrolysis, 71, 27-46, 2004.
Heat of combustion	$J \, g^{-1}$	31,400	
Volatile products of combustion	-	CO, HCN, caprolactam	
UL rating	-	NR, V-0 (5 wt% furan-phosphamide)	Sun, J, Qian, L, Li, J, Polymer, 210, 122994, 2020.

WEATHER STABILITY			
Spectral sensitivity	nm	290-310, 340-460	
Activation wavelengths	nm	300, 313, 334 (radical formation); 320 (yellowing)	
Excitation wavelengths	nm	282, 300, 310	
Emission wavelengths	nm	380, 420, 455, 460, 470	
Important initiators and accelerators	-	thermal degradation	
Products of degradation	-	CO, HCN, caprolactam, amines, carbon monoxide, hydrogen, hydrocarbons, crosslinks, carbon dioxide, acids, aldehydes, ketones, water, ammonia, hydroperoxides, pyrrole, ethylene	

PA-6 polyamide-6

PARAMETER	UNIT	VALUE	REFERENCES
BIODEGRADATION			
Typical biodegradants	-	bacterial hydrolases; filametous fungus (*Phanerochaete Chrysosporium*)	Klun, U; Ffriedrich, J; Krzan, A, Polym. Deg. Stab., 79, 99-104, 2003.
TOXICITY			
HMIS: Health, Flammability, Re-activity rating	-	0-1/0-1/0	
Carcinogenic effect	-	not listed by ACGIH, IARC, or NTP	
PROCESSING			
Typical processing methods	-	extrusion, injection molding	
Preprocess drying: temperature/time/residual moisture	°C/h/%	80/4-10/0.15	
Processing temperature	°C	250-280 (injection molding); 245-250 (film extrusion); 270-290 (profile extrusion)	
Processing pressure	MPa	3.5-12.5 (injection pressure)	
Applications	-	safety helmet parts, washers, gears, engine and motor parts, chutes; plugs, receptacles, covers, weed trimer components, clips, fasteners, flanges, key housing, flexible tubing, film, fans, hand-help power tools	
Outstanding properties	-	temperature resistance, chemical resistance to greases and oils	
BLENDS			
Suitable polymers	-	ABS, chitosan, HDPE, LDPE, NR, PA66, PANI, PBT, PC, PET, PP, PPO, PS, PVDF	
ANALYSIS			
FTIR (wavenumber-assignment)	cm^{-1}/-	C=O – 1715; CH_2 – 1463	Dong, W; Gijsman, P, Polym. Deg. Stab., 95, 1054-62, 2010.
Raman (wavenumber-assignment)	cm^{-1}/-	amide 1 – 1636; N-H – 1233; C-C – 1063/1076	Hendra, P J; Maddams, W F; Royaud, I A M; Willis, H A; Zichy, V, Spectrochim. Acta, 64A, 5, 747-56, 1990.
NMR (chemical shifts)	ppm	see ref.	Davis, R D; Jarrett, W L; Mathias, L J, Polymer, 42, 2621-26, 2001.
x-ray diffraction peaks	degree	21.3, 21.5 (γ polymorph) and 20, 23 (α polymorph)	Lai, D Li, Y, Wang, C, Liu, Y, Yang, J, Mater. Today, 25, 101578, 2020.

PA-6,6 polyamide-6,6

PARAMETER	UNIT	VALUE	REFERENCES
GENERAL			
Common name	-	polyamide-6,6, nylon-6,6, poly(iminoadipoyliminohexamethylene), poly(hexamethylene adipamide)	
IUPAC name	-	poly[N,N'-(hexane-1,6-diyl)adipamide]	
CAS name	-	poly[imino(1,6-dioxo-1,6-hexanediyl)imino-1,6-hexanediyl]	
Acronym	-	PA-6,6	
CAS number	-	32131-17-2	
RTECS number	-	DG0875000	
HISTORY			
Person to discover	-	Wallace Hume Carothers	
Date	-	February 28, 1935	
Details	-	deliberate research in DuPont resulted in synthesis of small amount of viscous mass which was capable of forming fibers	
SYNTHESIS			
Monomer(s) structure	-	$H_2N(CH_2)_6NH_2$ $\quad$ $HOC(CH_2)_4COH$ (each carbonyl with $=O$)	
Monomer(s) CAS number(s)	-	124-09-4; 124-04-9	
Monomer(s) molecular weight(s)	dalton, g/mol, amu	116.21; 146.14	
Monomer ratio	-	0.79; 0.79:1	
$CH_2/CONH$ ratio		5	
Number average molecular weight, M_n	dalton, g/mol, amu	17,500-18,040	
Mass average molecular weight, M_w	dalton, g/mol, amu	16,000-30,000	
Polydispersity, M_w/M_n	-	1.7-2.1	
Molar volume at 298K	cm^3 mol^{-1}	calc.=183 (crystalline); 211.5 (amorphous); exp.=193-208	
Van der Waals volume	cm^3 mol^{-1}	calc.=128.3 (crystalline); 128.3 (amorphous)	
Chain-end groups	-	NH_2 – 0.6; COOH – 0.8	Davis, R D; Jarrett, W L; Mathias, L J, Polymer, 42, 2621-26, 2001.
STRUCTURE			
Crystallinity	%	32-65 (DMA, X-ray); 43 (dry); 39 (wet)	Extrand, C W, J. Colloid Interface Sci., 248, 136-42, 2002.
Cell type (lattice)	-	monoclinic, triclinic (α); pseudohexagonal (γ)	
Cell dimensions	nm	a:b:c=0.914:0.484:1.668 (monoclinic); 0.49:0.54:1.72 (triclinic)	Jones, N A; Atkins, E D T; Hill, M J; Cooper, S J; Franco, L, Polymer, 38, 11, 2689-99, 1997.
Unit cell angles	degree	β: 67 (monoclinic); α:β:γ=48.5:77:63.5 (triclinic)	Jones, N A; Atkins, E D T; Hill, M J; Cooper, S J; Franco, L, Polymer, 38, 11, 2689-99, 1997.
Number of chains per unit cell	-	1 or 2	
Crystallite size	nm	15.6	Rajesh, J J; Bijwe, J, Wear, 661-68, 2005.

PA-6,6 polyamide-6,6

PARAMETER	UNIT	VALUE	REFERENCES
Spacing between crystallites	nm	0.37-0.65	Extrand, C W, J. Colloid Interface Sci., 248, 136-42, 2002; Elzein, T; Brogly, M; Castelein, G; Schultz, J, J. Polym. Sci. B, 40, 1464-76, 2002; Sengupta, R; Tikku, V K; Somani, A K; Chaki, T K; Bhowmick, A K, Radiat. Phys. Chem., 72, 625-33, 2005.
Polymorphs	-	α, γ	
Cis content	%	1.1	Davis, R D; Jarrett, W L; Mathias, L J, Polymer, 42, 2621-26, 2001.
Lamellae thickness	nm	17	Elzein, T; Brogly, M; Castelein, G; Schultz, J, J. Polym. Sci. B, 40, 1464-76, 2002.
Rapid crystallization temperature	°C	230	Adriaensens, P; Pollaris, A; Carlleer, R; Vanderzande, D; Gelan, J; Litvinov, V M; Tijssen, J, Polymer, 42, 7943-52, 2001.

COMMERCIAL POLYMERS

PARAMETER	UNIT	VALUE	REFERENCES
Some manufacturers	-	BASF; DSM; DuPont; EMS	
Trade names	-	Ultramid A; Akulon; Zytel; Grilon	

PHYSICAL PROPERTIES

PARAMETER	UNIT	VALUE	REFERENCES
Density at 20°C	g cm^{-3}	1.05-1.14; 0.969 (melt); 1.213-1.500 (13-43% glass fiber, dry)	
Bulk density at 20°C	g cm^{-3}	0.5-0.8	
Color	-	white	
Refractive index, 20°C	-	1.565-1.568	
Birefringence	-	1.582/1.519-1.53	
Odor		odorless	
Melting temperature, DSC	°C	257-270; 262 (13-43% glass fiber, dry)	
Decomposition temperature	°C	340; 369	-; Morimune-Moriya, S, Yada, S, Kuroki, N, Ito, S, Nishino, T, Compos. Sci. Technol., 199, 108356, 2020.
Activation energy of thermal degradation	kJ mol^{-1}	91	
Thermal expansion coefficient, -40 to 160°C	10^{-4} °C^{-1}	1; 0.09-1.58 (13-43% glass fiber, dry)	
Thermal conductivity, melt	W m^{-1} K^{-1}	0.23-0.343	Benaarbia, A; Chrysochoos, A; Robert, G, Polym. Testing, 34, 155-67, 2014.
Glass transition temperature	°C	56-70 (dry); 35 (50% RH); -15 (100% RH)	
Specific heat capacity	J K^{-1} kg^{-1}	1638-1,700; 2,750 (melt)	
Heat of fusion	kJ kg^{-1}	192-196	
Temperature index (50% tensile strength loss after 20,000 h/5000 h)	°C	65-85; 125-130 (13-43% glass fiber, dry)	Padey, D; Walling, J; Wood A, Polymers in Defence and Aerospace 2007, Rapra, 2007, paper 15.
Heat deflection temperature at 0.45 MPa	°C	200-230; 258-264 (13-43% glass fiber, dry)	
Heat deflection temperature at 1.8 MPa	°C	70-86; 238-256 (13-43% glass fiber, dry)	
Vicat temperature VST/A/50	°C	255 (13-43% glass fiber, dry)	
Enthalpy of melting	J g^{-1}	83	Cerrutti, P; Carfagna, C, Polym. Deg. Stab., 95, 2405-12, 2010.

PA-6,6 polyamide-6,6

PARAMETER	UNIT	VALUE	REFERENCES
Hansen solubility parameters, δ_D, δ_P, δ_H	MPa$^{0.5}$	17.2, 9.9, 16.5; 18.6, 5.1, 12.3	
Interaction radius		4.4	
Hildebrand solubility parameter	MPa$^{0.5}$	exp.=22.87-25.8	
Surface tension	mN m^{-1}	calc.=46.5; exp.=36-44	
Dielectric constant at 100 Hz/1 MHz	-	4.0/3.6	
Relative permittivity at 100 Hz	-	3.2-4.3 (dry); 10.3-15 (conditioned); 3.9-4.5 (13-43% glass fiber, dry)	
Relative permittivity at 1 MHz	-	3-3.6 (dry); 4.2-4.3 (conditioned); 3.2-4.1 (13-43% glass fiber, dry)	
Dissipation factor at 100 Hz	E-4	60-150 (dry); 2,000-2,400 (conditioned); 100 (13-43% glass fiber, dry)	
Dissipation factor at 1 MHz	E-4	170-240 (dry); 750-1,200 (conditioned); 145 (13-43% glass fiber, dry)	
Volume resistivity	ohm-m	1E13 (dry); 3E9-1E11 (saturated at 50% RH, 20ºC); 6E9 (saturated at 100% RH, 20ºC); 1E14 (13-43% glass fiber, dry); 1E10 (13-43% glass fiber, conditioned)	
Surface resistivity	ohm	1E14 (conditioned); 1E12 (13-43% glass fiber, dry)	
Electric strength K20/P50, d=0.60.8 mm	kV mm^{-1}	30-30.5; 25 (conditioned); 25-27 (13-43% glass fiber, dry)	
Comparative tracking index, CTI, test liquid A	-	>600; >600 (13-43% glass fiber, dry)	
Comparative tracking index, CTIM, test liquid B	-	>600 (13-43% glass fiber, dry)	
Arc resistance	MV/m	145 (13-43% glass fiber, dry)	
Power factor	-	0.04	
Percolation threshold for MWCNT	wt%	0.040.05	Krause, B; Boldt, R; Haussler, L; Poetschke, P, Compos. Sci. Tecnol., 114, 119-25, 2015.
Coefficient of friction	-	0.12-0.48 (depending on applied normal load and sliding velocity); 0.098-0.32 (20 wt% glass fiber)	Ravi, N, Shanmugam, M, Bheemappa, S, Gowripalan, N, J. Manuf. Proc., 58, 1052-63, 2020.
Diffusion coefficient of nitrogen	cm^2 s^{-1} x10^6	0.0002	
Diffusion coefficient of water vapor	cm^2 s^{-1} x10^6	0.002 (20ºC); 0.035 (60ºC); 0.35 (100ºC)	
Contact angle of water, 20ºC	degree	67.6-72; 68.5/40.9 (ascending/receding)	
Surface free energy	mJ m^{-2}	44.3	
Speed of sound	m s^{-1}	43.3-46.1	
Acoustic impedance		2.90-3.15	
Attenuation	dB cm^{-1}, 5 MHz	2.9-16.0	
MECHANICAL & RHEOLOGICAL PROPERTIES			
Tensile strength	MPa	70-88 (dry); 120-235 (13-43% glass fiber, dry); 75-167 (13-43% glass fiber, conditioned)	
Tensile modulus	MPa	3,000-3,600 (dry); 1,200-1,800 (conditioned); 5,500-14,000 (13-43% glass fiber, dry); 3,500-11,000 (13-43% glass fiber, conditioned)	
Tensile stress at yield	MPa	82-95; 55-60 (conditioned)	

PA-6,6 polyamide-6,6

PARAMETER	UNIT	VALUE	REFERENCES
Tensile creep modulus, 1000 h, elongation 0.5 max	MPa	7,960 (13-43% glass fiber, conditioned)	
Elongation	%	10-45 (dry); >50 to >100 (conditioned); 3 (13-43% glass fiber, dry); 4-13 (13-43% glass fiber, conditioned)	
Tensile yield strain	%	4-5 (dry); 20-27 (conditioned)	
Flexural strength	MPa	125; 160-340 (13-43% glass fiber, dry); 100-260 (13-43% glass fiber, conditioned); 226-263 (50% glass fiber)	Teixeira, D; Giovanela, M; Gonella, LB; Crespo, J S, Mater Design, 85, 695-706, 2015.
Flexural modulus	MPa	2,800 (dry); 1,070-1,200 (conditioned); 4,800-12,000 (13-43% glass fiber, dry); 2,900-8,900 (13-43% glass fiber, conditioned)	
Young's modulus	GPa	2.8	Sasayama, T; Okabe, T; Aoyagi, Y; Nishikawa, M, Composites: Part A, 52, 45-54, 2013.
Charpy impact strength, unnotched, 23°C	kJ m^{-2}	no break; 40-100 (13-43% glass fiber, dry); 70-105 (13-43% glass fiber, conditioned)	
Charpy impact strength, unnotched, -30°C	kJ m^{-2}	no break to 400; 40-85 (13-43% glass fiber, dry); 45-105 (13-43% glass fiber, conditioned)	
Charpy impact strength, notched, 23°C	kJ m^{-2}	4.9-6 (dry); 12-20 (conditioned); 4-16 (13-43% glass fiber, dry); 5-19 (13-43% glass fiber, conditioned)	
Charpy impact strength, notched, -30°C	kJ m^{-2}	3.8-6 (dry); 3-6 (conditioned); 4.0-12 (13-43% glass fiber, dry); 6-12 (13-43% glass fiber, conditioned)	
Crack growth velocity	x 10^{-6} m s^{-1}	508	Rajesh, J J; Bijwe, J, Tribology lett., 18, 3, 331-40, 2005.
Fracture energy	x 10^4 J m^{-2}	6.52	Rajesh, J J; Bijwe, J, Tribology lett., 18, 3, 331-40, 2005.
Ductility factor	mm	15.14	Rajesh, J J; Bijwe, J, Tribology lett., 18, 3, 331-40, 2005.
Specific wear rate	10^{-6} mm^3 N^{-1} m^{-1}	0.63-2.26	Ravi, N, Shanmugam, M, Bheemappa, S, Gowripalan, N, J. Manuf. Proc., 58, 1052-63, 2020.
Stress necessary to cause spontaneous fracture	MPa	129.16	Rajesh, J J; Bijwe, J, Tribology lett., 18, 3, 331-40, 2005.
Poisson's ratio	-	0.3-0.5	
Rockwell hardness	-	M105 (13-43% glass fiber, dry); M90 (13-43% glass fiber, conditioned); 125R (13-43% glass fiber, dry); 118 (13-43% glass fiber, conditioned)	
Ball indention hardness at 358 N/30 S (ISO 2039-1)	MPa	295 (13-43% glass fiber, dry); 218 (13-43% glass fiber, conditioned)	
Shrinkage	%	0.95-1.6; 0.2-1.7 (13-43% glass fiber, dry)	
Brittleness temperature (ASTM D746)	°C	-80 to -100 (dry); -65 to -85 (50% RH)	
Water absorption, equilibrium in water at 23°C	%	8.5-9.0; 4.7-7.6 (13-43% glass fiber, dry)	
Moisture absorption, equilibrium 23°C/50% RH	%	2.3-3.4; 1.4-2.2 (13-43% glass fiber, dry)	

CHEMICAL RESISTANCE

Acid dilute/concentrated	-	good/poor	
Alcohols	-	good	
Alkalis	-	good/poor	
Aliphatic hydrocarbons	-	good	
Aromatic hydrocarbons	-	fair	
Esters	-	good	

PA-6,6 polyamide-6,6

PARAMETER	UNIT	VALUE	REFERENCES
Greases & oils	-	good	
Halogenated hydrocarbons	-	poor	
Ketones	-	good	
Θ solvent, Θ-temp.=20°C	-	carbon tetrachloride/m-cresol/cyclohexane	
Good solvent	-	acetic acid, benzyl alcohol, chloroacetic acid, DMSO, formamide, formic acid, HCl, HF, H_3PO_3, H_2SO_4, methanol, phenol, sulfur dioxide, trichloroethanol, trifluoroethanol	
Non-solvent	-	aliphatic alcohols, aliphatic esters, aliphatic ketones, chloroform, diethyl ether, hydrocarbons	
Effect of EtOH sterilization (tensile strength retention)	%	95-105	Navarrete, L; Hermanson, N, Antec, 2807-18, 1996.
FLAMMABILITY			
Ignition temperature	°C	420	
Autoignition temperature	°C	530	
Limiting oxygen index	% O_2	20, 28-31; 21.5-24 (13-43% glass fiber, dry)	Morimune-Moriya, S, Yada, S, Kuroki, N, Ito, S, Nishino, T, Compos. Sci. Technol., 199, 108356, 2020.
Heat release	kW m^{-2}	328	Braun, U; Scchartel, B; Fichera, M A; Jaeger, C, Polym. Deg. Stab., 92, 1528-45, 2007.
Peak of heat release rate	kW m^{-2}	481	Morimune-Moriya, S, Yada, S, Kuroki, N, Ito, S, Nishino, T, Compos. Sci. Technol., 199, 108356, 2020.
Heat of combustion	J g^{-1}	30,900; 12,000	Walters, R N; Hacket, S M; Lyon, R E, Fire Mater., 24, 5, 245-52, 2000; Morimune-Moriya, S, Yada, S, Kuroki, N, Ito, S, Nishino, T, Compos. Sci. Technol., 199, 108356, 2020.
Volatile products of combustion	-	H_2O, CO_2, cyclopentanone, hexylamine, hexamethylene diamine	
Residue at 700°C	%	4.1 (air), 4.4 (nitrogen)	Morimune-Moriya, S, Yada, S, Kuroki, N, Ito, S, Nishino, T, Compos. Sci. Technol., 199, 108356, 2020.
UL rating	-	HB to V-2; HB (13-43% glass fiber)	
WEATHER STABILITY			
Excitation wavelengths	nm	295, 310, 315	
Emission wavelengths	nm	423, 450, 460, 465	
Important initiators and accelerators	-	chlorinated water, conjugated carboxylic unsaturations, products of photooxidation, titanium dioxide, zinc oxide, derivatives of anthraquinone, oxyethylsulfonic red and yellow, monochlorotriazine red and yellow, copper compounds	
Products of degradation	-	amines, carbon monoxide, hydrogen, hydrocarbons, crosslinks (photolysis); amines, carbon monoxide, carbon dioxide, acids, aldehydes, ketones, water, ammonia, hydroperoxides, pyrrole, ethylene (photooxidation)	
BIODEGRADATION			
Typical biodegradants	-	cutinase from *Fusarium solani pisi*	Araujo, R; Silva, C; O'Neill, A; Micaelo, N; Guebitz, G; Soares, C M; Casal, M; Cavaco-Paulo, A, J. Biotechnol., 128, 849-57, 2007.

PA-6,6 polyamide-6,6

PARAMETER	UNIT	VALUE	REFERENCES
TOXICITY			
HMIS: Health, Flammability, Reactivity rating	-	1/1/0	
Carcinogenic effect	-	not listed by ACGIH, IARC, or NTP	
Oral rat, LD$_{50}$	mg kg^{-1}	>10,000	
Skin rabbit, LD$_{50}$	mg kg^{-1}	not a skin irritant	
PROCESSING			
Typical processing methods	-	injection molding	
Preprocess drying: temperature/ time/residual moisture	°C/h/%	60-80/2-4/0.04-0.2	
Processing temperature	°C	280-305 (injection molding)	
Processing pressure	MPa	35-125 (injection pressure); 0-0.35 (back pressure)	
Applications	-	aerospace, automotive (fuel systems, under-the-hood applications such as shrouds and ducts), composite structures, stock shapes, waste water treatment	
Outstanding properties	-	mechanical strength, drawing behavior	
BLENDS			
Suitable polymers	-	HDPE, LCP, PA6, PBT, PC, PET, PP, SEBS	
ANALYSIS			
FTIR (wavenumber-assignment)	cm^{-1}/-	amides – 1635; conjugated carbonyls – 1700; aldehydes – 1725; isolated carboxylic acids – 1760	Yoshioka, Y; Tashiro, K; Ramesh, C, J. Polym. Sci. B, 41, 1294-1307, 2003; Cerruti, P; Lavorgna, M; Carfagna, C; Nicolais, L, Polymer, 46, 4571-83, 2005.
NMR (chemical shifts)	ppm	see ref.	Davis, R D; Jarrett, W L; Mathias, L J, Polymer, 42, 2621-26, 2001.
x-ray diffraction peaks	degree	13.5, 20.2, 22.3, 23.8 13.2°, 20.3° and 21.8° correspond to the (002), (100) and (010)/(110) planes, respectively, and the peak associated to the amorphous halo at around 2θ = 21.8°	Sengupta, R; Tikku, V K; Somani, A K; Chaki, T K; Bhowmick, A K, Radiat. Phys. Chem., 72, 625-33, 2005; Morimune-Moriya, S, Yada, S, Kuroki, N, Ito, S, Nishino, T, Compos. Sci. Technol., 199, 108356, 2020.

PA-6,10 polyamide-6,10

PARAMETER	UNIT	VALUE	REFERENCES
GENERAL			
Common name	-	polyamide-6,10, nylon-6,10; poly(iminohexamethyleneimino-sebacoyl), poly(hexamethylene sebacamide)	
CAS name	-	poly[imino-1,6-hexanediylimino(1,10-dioxo-1,10-decanediyl)]	
Acronym	-	PA-6,10	
CAS number	-	9008-66-6	
HISTORY			
Person to discover	-	Carothers, W H, 1937. Austin, P R	Carothers, W H, US Patent 2,071,250, DuPont, Feb. 16, 1937. Austin, P R, US Patent 2,244,183, DuPont, June 3, 1941.
Date	-	1937; 1941	
Details	-	Carothers patented polymerization; Austin patented plasticization	
SYNTHESIS			
Monomer(s) structure	-	$H_2N(CH_2)_6NH_2 \quad HOC(CH_2)_8COH$ (with two C=O groups)	
Monomer(s) CAS number(s)	-	124-09-4; 111-20-6	
Monomer(s) molecular weight(s)	dalton, g/mol, amu	116.21; 202.25	
Monomer ratio	-	0.575 (0.575:1)	
Number average molecular weight, M_n	dalton, g/mol, amu	18,900-22,140	
Polydispersity, M_w/M_n	-	2.0	
Molar volume at 298K	cm^3 mol^{-1}	calc.=238 (crystalline); 271.5 (amorphous)	
Van der Waals volume	cm^3 mol^{-1}	calc.=169.2 (crystalline); 169.2 (amorphous)	
Chain-end groups	meq g^{-1}	NH_2 – 0.021, COOH – 0.085; NH_2 – 0.21, COOH – 0.42	Koning, C; Teuwen, L; de Jong, R; Janssen, G; Coussens, B, High Perform. Polym., 11, 387-94, 1999; Davis, R D; Jarrett, W L; Mathias, L J, Polymer, 42, 2621-26, 2001.
STRUCTURE			
Crystallinity	%	26.5-45; 34 (dry); 31 (wet)	Extrand, C W, J. Colloid Interface Sci., 248, 136-42, 2002.
Cell type (lattice)	-	triclinic	Jones, N A; Atkins, E D T; Hill, M J; Cooper, S J; Franco, L, Polymer, 38, 11, 2689-99, 1997.
Cell dimensions	nm	a:b:c=0.49:0.53:2.23 (α); 0.49:0.799:2.23 (β); a:b:c=0.495:0.54:2.24	Jones, N A; Atkins, E D T; Hill, M J; Cooper, S J; Franco, L, Polymer, 38, 11, 2689-99, 1997; Ruehle, D A; Perbix, C; Castaneda, M; Dorgan, J R; Mittal, V; Halley, P; Martin, D, Polymer, 54, 6961-70, 2013.
Unit cell angles	degree	$\alpha{:}\beta{:}\gamma$=49:77:64 (α); 90:77:66 (β); $\alpha{:}\beta{:}\gamma$=49:76.5:63.5	Jones, N A; Atkins, E D T; Hill, M J; Cooper, S J; Franco, L, Polymer, 38, 11, 2689-99, 1997; Ruehle, D A; Perbix, C; Castaneda, M; Dorgan, J R; Mittal, V; Halley, P; Martin, D, Polymer, 54, 6961-70, 2013.

PA-6,10 polyamide-6,10

PARAMETER	UNIT	VALUE	REFERENCES
Number of chains per unit cell	-	1/2	Jones, N A; Atkins, E D T; Hill, M J; Cooper, S J; Franco, L, Polymer, 38, 11, 2689-99, 1997.
Polymorphs	-	α, β	Jones, N A; Atkins, E D T; Hill, M J; Cooper, S J; Franco, L, Polymer, 38, 11, 2689-99, 1997.
Cis content	%	1.1	Davis, R D; Jarrett, W L; Mathias, L J, Polymer, 42, 2621-26, 2001.
Lamellae thickness	nm	18	Elzein, T; Brogly, M; Castelein, G; Schultz, J, J. Polym. Sci. B, 40, 1464-76, 2002.
Rapid crystallization temperature	°C	179	
Avrami constants, k/n	-	n=2.5-3.1	

COMMERCIAL POLYMERS

Some manufacturers	-	BASF; EMS-Grivory	
Trade names	-	Ultramid; Grilamid 2S	

PHYSICAL PROPERTIES

Density at 20°C	g cm^{-3}	1.03-1.10	
Bulk density at 20°C	g cm^{-3}	0.62-0.65	
Refractive index, 20°C	-	1.52-1.57	
Melting temperature, DSC	°C	215-230	
Decomposition temperature	°C	350	
Thermal expansion coefficient, 23-80°C	10^{-4} °C^{-1}	1.1-2 (dry, parallel); 1.1-1.5 (dry, normal)	
Thermal conductivity, melt	W m^{-1} K^{-1}	0.23; 0.35 (moist)	
Glass transition temperature	°C	48, 65-70 (dry); 40 (50% RH); 10 (100% RH)	
Heat of fusion	kJ kg^{-1}	201-215	
Maximum service temperature	°C	110-130	
Long term service temperature	°C	160	
Heat deflection temperature at 0.45 MPa	°C	115-150 (dry)	
Heat deflection temperature at 1.8 MPa	°C	50-65 (dry)	
Surface tension	mN m^{-1}	37	
Dielectric constant at 100 Hz/1 MHz	-	3.9/3.3 (dry), 6.5/3.5 (at 65% RH)	
Relative permittivity at 100 Hz	-	0.04	
Relative permittivity at 1 MHz	-	0.03	
Volume resistivity	ohm-m	1E13 (dry); 2E10 (saturated at 50% RH, 20°C); 3E8 (saturated at 100% RH, 20°C)	
Surface resistivity	ohm	>1E15 (conditioned)	
Electric strength K20/P50, d=0.60.8 mm	kV mm^{-1}	36-38 (dry); 34-44 (conditioned)	
Comparative tracking index	-	600 (conditioned)	
Power factor	-	0.02	
Contact angle of water, 20°C	degree	71; 73.8/49.1 (asc/rec)	
Surface free energy	mJ m^{-2}	40.5	

PA-6,10 polyamide-6,10

PARAMETER	UNIT	VALUE	REFERENCES
MECHANICAL & RHEOLOGICAL PROPERTIES			
Tensile strength	MPa	40-66 (dry); 40-60 (conditioned)	
Tensile modulus	MPa	750-3,160 (dry); 450-1,500 (conditioned)	
Tensile stress at yield	MPa	60-66 (dry); 45-52 (conditioned)	
Elongation	%	37 (dry); 140 (conditioned)	
Tensile yield strain	%	4.5-5 (dry); 18-20 (conditioned)	
Charpy impact strength, unnotched, 23°C	kJ m^{-2}	NB (dry); NB (conditioned)	
Charpy impact strength, unnotched, -30°C	kJ m^{-2}	NB to 300 (dry); NB (conditioned)	
Charpy impact strength, notched, 23°C	kJ m^{-2}	NB to 5-8 (dry; NB to 10-18 (conditioned)	
Charpy impact strength, notched, -30°C	kJ m^{-2}	3-15 (dry); 8-15 (conditioned)	
Poisson's ratio	-	0.3-0.4	
Ball indention hardness at 358 N/30 S (ISO 2039-1)	MPa	50-120 (dry); 35-70 (conditioned)	
Shrinkage	%	0.9-2.3	
Brittleness temperature (ASTM D746)	°C	-90 (dry); -62 (50% RH)	
Viscosity number	ml g^{-1}	148-150	
Intrinsic viscosity, 25°C	dl g^{-1}	1.147	
Melt viscosity, shear rate=1000 s^{-1}	Pa s	27	
Melt volume flow rate (ISO 1133, procedure B), 275°C/5 kg	cm^3/10 min	120	
Water absorption, equilibrium in water at 23°C	%	3-3.3	
Moisture absorption, equilibrium 23°C/50% RH	%	0.5-1.7	
CHEMICAL RESISTANCE			
Acid dilute/concentrated	-	fair-poor	
Alcohols	-	fair	
Alkalis	-	good-fair	
Aliphatic hydrocarbons	-	good	
Aromatic hydrocarbons	-	good	
Esters	-	good-fair	
Greases & oils	-	good	
Halogenated hydrocarbons	-	good-poor	
Ketones	-	fair	
Good solvent	-	m-cresol, phenol, sulfolane (hot), trichloroethanol	
FLAMMABILITY			
Ignition temperature	°C	>350	
Autoignition temperature	°C	415	
Limiting oxygen index	% O$_2$	21-24	Levchik, S V; Costa, L; Camino, G, Polym. Deg. Stab., 43, 43-54, 1994.

244 **HANDBOOK OF POLYMERS** 3rd Edition, Copyrights 2022; ChemTec Publishing

PA-6,10 polyamide-6,10

PARAMETER	UNIT	VALUE	REFERENCES
Volatile products of combustion	-	CO_2, NH_3, CO, hydrocarbons	Levchik, S V; Costa, L; Camino, G, Polym. Deg. Stab., 43, 43-54, 1994.
UL 94 rating	-	HB	

PROCESSING

PARAMETER	UNIT	VALUE	REFERENCES
Typical processing methods	-	injection molding	
Preprocess drying: temperature/ time/residual moisture	°C/h/%	80/4/0.15	
Processing temperature	°C	250-270 (injection molding)	
Processing pressure	MPa	0.17-0.35 (back)	
Applications	-	automotive (air intake systems, compressed air systems, hydraulic systems, fuel systems, powertrain and chassis), electrical & electronics (electrical appliances, electrical equipment, cables & tubes, connectors), industry & consumer goods (housewares, hydraulics and pneumatics, mechanical engineering, sports & leisure, tools & accessories)	
Outstanding properties	-	low water absorption, low melting temperature, bio-based	

BLENDS

PARAMETER	UNIT	VALUE	REFERENCES
Suitable polymers	-	ABS, PA-11, PP, SMA	Ruehle, D A; Perbix, C; Castaneda, M; Dorgan, J R; Mittal, V; Halley, P; Martin, D, Polymer, 54, 6961-70, 2013.

PA-6,12 polyamide-6,12

PARAMETER	UNIT	VALUE	REFERENCES
GENERAL			
Common name	-	polyamide-6,12, nylon-6,12; poly(iminohexamethyleneimino-dodecanedioyl)	
CAS name	-	poly[imino-1,6-hexanediylimino(1,12-dioxo-1,12-dodecanedi-yl)]	
Acronym	-	PA-6,12	
CAS number	-	24936-74-1	
HISTORY			
Person to discover	-	Carothers, W H, 1937. Peterson, W R, 1939	Carothers, W H, US Patent 2,071,250, DuPont, Feb. 16, 1937. Peterson, W R, US Patent 2,174,527, DuPont, Oct. 3, 1939.
Date	-	1937; 1939	
Details	-	Carothers proposed basic synthesis; Peterson proposed process improvement	
SYNTHESIS			
Monomer(s) structure	-	$H_2N(CH_2)_6NH_2$ $HOC(CH_2)_{10}COH$ (with O double bonds)	
Monomer(s) CAS number(s)	-	124-09-4; 143-07-7	
Monomer(s) molecular weight(s)	dalton, g/mol, amu	116.21; 232.32	
Monomer ratio	-	0.5 (0.5:1)	
CH_2/CONH ratio		8	
Number average molecular weight, M_n	dalton, g/mol, amu	16,400	
Chain-end groups	meq g^{-1}	NH_2 – 0.036; COOH – 0.086	Koning, C; Teuwen, L; de Jong, R; Janssen, G; Coussens, B, High Perform. Polym., 11, 387-94, 1999.
STRUCTURE			
Crystallinity	%	7.7-28.4	Rhee, S; White, J L, Antec, 1690-94, 2000; Jones, N A; Atkins, E D T; Hill, M J; Cooper, S J; Franco, L, Polymer, 38, 11, 2689-99, 1997.
Cell type (lattice)	-	triclinic (α), pseudohexagonal (γ)	Jones, N A; Atkins, E D T; Hill, M J; Cooper, S J; Franco, L, Polymer, 38, 11, 2689-99, 1997; Menchaca, C; Manoun, B; Martinez-Barrera, G; Castado, V M; Lopez-Valdivia, H, J. Phys. Chem. Solids, 67, 2111-2118, 2006.
Cell dimensions	nm	a:b:c=0.49:0.533:2.48 (α); a:b:c=0.49:0.802:2.48 (γ)	Jones, N A; Atkins, E D T; Hill, M J; Cooper, S J; Franco, L, Polymer, 38, 11, 2689-99, 1997.
Unit cell angles	degree	α:β:γ=49:77:63.5 (α); α:β:γ=90:77:66.5 (β)	Jones, N A; Atkins, E D T; Hill, M J; Cooper, S J; Franco, L, Polymer, 38, 11, 2689-99, 1997.
Number of chains per unit cell	-	1/2	
Crystallite size	nm	20.4	Rajesh, J J; Bijwe, J, Wear, 661-68, 2005.
Polymorphs	-	α, γ	Menchaca, C; Manoun, B; Martinez-Barrera, G; Castado, V M; Lopez-Valdivia, H, J. Phys. Chem. Solids, 67, 2111-2118, 2006.

PA-6,12 polyamide-6,12

PARAMETER	UNIT	VALUE	REFERENCES
Lamellae thickness	nm	23	Elzein, T; Brogly, M; Castelein, G; Schultz, J, J. Polym. Sci. B, 40, 1464-76, 2002.
Rapid crystallization temperature	$^{\circ}$C	181	
Avrami constants, k/n	-	n=2.4-2.6	

PHYSICAL PROPERTIES

Density at 20°C	g cm^{-3}	1.05-1.07; 1.32-1.42 (33-43% glass fiber, dry)	
Melting temperature, DSC	$^{\circ}$C	215-218; 218 (33-43% glass fiber, dry)	
Decomposition temperature	$^{\circ}$C	291	
Activation energy of thermal degradation	kJ mol^{-1}	164	Herrera, M; Matuschek, G; Kettrup, A, Chemosphere, 42, 601-7, 2001.
Thermal expansion coefficient, 23-80°C	$^{\circ}$C^{-1}	9E-5-1.2E-4; 0.16-1.39E-4 (33-43% glass fiber, dry)	
Thermal conductivity, melt	W m^{-1} K^{-1}	0.22-0.25	
Glass transition temperature	$^{\circ}$C	54-62 (dry); 42 (100% RH)	
Specific heat capacity	J K^{-1} kg^{-1}	1260	
Heat of fusion	kJ kg^{-1}	95	
Temperature index (50% tensile strength loss after 20,000 h/5000 h)	$^{\circ}$C	65; 120 (33-43% glass fiber, dry)	
Heat deflection temperature at 0.45 MPa	$^{\circ}$C	135-160 (dry); 216-217 (33-43% glass fiber, dry)	
Heat deflection temperature at 1.8 MPa	$^{\circ}$C	62-80 (dry); 200-203 (33-43% glass fiber, dry)	
Vicat temperature VST/A/50	$^{\circ}$C	181 (dry)	
Surface tension	mN m^{-1}	25-31	
Relative permittivity at 100 Hz	-	3.6 (dry); 6 (conditioned); 4.1 (33-43% glass fiber, dry)	
Relative permittivity at 1 MHz	-	3.2 (dry); 4 (conditioned); 3.6-3.8 (33-43% glass fiber, dry)	
Dissipation factor at 100 Hz	E-4	140 (dry); 1,500 (conditioned); 135 (33-43% glass fiber, dry)	
Dissipation factor at 1 MHz	E-4	165 (dry); 1,000 (conditioned); 150-200 (33-43% glass fiber, dry)	
Volume resistivity	ohm-m	1E13 (dry); 1E11-1E12 (saturation at 50% RH, 20°C); 1E11 (saturation at 100% RH, 20°C); 1E13 (33-43% glass fiber, dry)	
Surface resistivity	ohm	1E12 (dry); 1E12 (33-43% glass fiber)	
Electric strength K20/P50, d=0.60.8 mm	kV mm^{-1}	27 (33-43% glass fiber)	
Comparative tracking index, CTI, test liquid A	-	>600; >600 (33-43% glass fiber, dry)	
Arc resistance	MV/m	145 (33-43% glass fiber, dry)	
Coefficient of friction	-	0.35-0.55	
Surface free energy	mJ m^{-2}	67.0	

MECHANICAL & RHEOLOGICAL PROPERTIES

Tensile strength	MPa	37; 168-200 (33-43% glass fiber, dry); 140-165 (33-43% glass fiber, conditioned)	
Tensile modulus	MPa	2,400-2,570 (dry); 1,500-1,680 (conditioned); 9,500-12,500 (33-43% glass fiber, dry); 7,90011,500 (33-43% glass fiber, conditioned)	

PA-6,12 polyamide-6,12

PARAMETER	UNIT	VALUE	REFERENCES
Tensile stress at yield	MPa	62-63 (dry); 52-55 (conditioned)	
Elongation	%	7-35 (dry); 33 (conditioned); 3-3.2 (33-43% glass fiber, dry); 3.2-5 (33-43% glass fiber, conditioned)	
Tensile yield strain	%	4.3-4.5 (dry); 19 (conditioned)	
Flexural modulus	MPa	2,150 (dry); 8,200-11,000 (33-43% glass fiber, dry); 7,000 (33% glass fiber, conditioned)	
Young's modulus	MPa	672	
Charpy impact strength, unnotched, 23°C	kJ m^{-2}	no break; 80-100 (33-43% glass fiber, dry)	
Charpy impact strength, unnotched, -30°C	kJ m^{-2}	no break; 60-85 (33-43% glass fiber, dry)	
Charpy impact strength, notched, 23°C	kJ m^{-2}	4.2-5 (dry); 8 (conditioned); 13-17 (33-43% glass fiber, dry); 12 (33-43% glass fiber, conditioned)	
Charpy impact strength, notched, -30°C	kJ m^{-2}	4.2 (dry); 4 (conditioned); 11-17 (33-43% glass fiber, dry); 10 (33-43% glass fiber, conditioned)	
Izod impact strength, notched, 23°C	J m^{-1}	40	
Crack growth velocity	x 10^{-6} m s^{-1}	2077	Rajesh, J J; Bijwe, J, Tribology lett., 18, 3, 331-40, 2005.
Fracture energy	x 10^4 J m^{-2}	3.78	Rajesh, J J; Bijwe, J, Tribology lett., 18, 3, 331-40, 2005.
Ductility factor	mm	14.52	Rajesh, J J; Bijwe, J, Tribology lett., 18, 3, 331-40, 2005.
Stress necessary to cause spontaneous fracture	MPa	110.95	Rajesh, J J; Bijwe, J, Tribology lett., 18, 3, 331-40, 2005.
Poisson's ratio	-	0.3-0.4; 0.39-0.42 (33-43% glass fiber, dry)	
Rockwell hardness	-	R114; R118 (33-43% glass fiber, dry)	
Shrinkage	%	1.1-1.5; 0.3-0.8 (33-43% glass fiber)	
Brittleness temperature (ASTM D746)	°C	-109	
Intrinsic viscosity, 25°C	dl g^{-1}	1.45	
Water absorption, equilibrium in water at 23°C	%	1.3-3.0; 1.7 (33-43% glass fiber)	
Moisture absorption, equilibrium 23°C/50% RH	%	1.3	
CHEMICAL RESISTANCE			
Acid dilute/concentrated	-	fair-poor	
Alcohols	-	good-fair	
Alkalis	-	good	
Aliphatic hydrocarbons	-	good	
Aromatic hydrocarbons	-	good	
Greases & oils	-	good	
Halogenated hydrocarbons	-	poor	
Ketones	-	good	
Good solvent	-	chloral hydrate, m-cresol, fluorinated alcohols, formic acid, mineral acids, phenols, trichloroethanol	

PA-6,12 polyamide-6,12

PARAMETER	UNIT	VALUE	REFERENCES
FLAMMABILITY			
Ignition temperature	°C	420-430	
Autoignition temperature	°C	420-445	
Limiting oxygen index	% O_2	25-28; 23 (33-43% glass fiber)	
UL rating	-	HB; HB (33-43% glass fiber, dry)	
WEATHER STABILITY			
Products of degradation	-	CO,CO_2, H_2O, NO_x, caprolactam	
TOXICITY			
NFPA: Health, Flammability, Reactivity rating	-	1/1/0	
Carcinogenic effect	-	not listed by ACGIH, NIOSH, NTP	
PROCESSING			
Typical processing methods	-	blow molding, extrusion, injection molding	
Preprocess drying: temperature/ time/residual moisture	°C/h/%	80/2-6/<0.05; 80/2-4/<0.15 (33-43% glass fiber)	
Processing temperature	°C	230-290 (injection molding); 230-240 (extrusion); 280-300 (33-43% glass fiber)	
ANALYSIS			
FTIR (wavenumber-assignment)	cm⁻¹/-	N-H – 1539, 689, more in refs.	Yoshioka, Y; Tashiro, K; Ramesh, C, J. Polym. Sci. B, 41, 1294-1307, 2003; Rusu, G; Rusu, E, Mater. Design, 31, 4601-10, 2010.
Raman (wavenumber-assignment)	cm⁻¹/-	C=O – 1634; CH_2 – 2884, 1439; C-C – 1129	Olivares, M; Mondragon, M A; Vazquez-Polo, G; Martinez, E; Castano, V M, Intern. J. Polymeric Mater., 40, 213-18, 1998.

PA-6,66 polyamide-6,66

PARAMETER	UNIT	VALUE	REFERENCES
GENERAL			
Common name	-	polyamide-6,66, nylon-6,66, poly(hexamethylene adipamide-co-caprolactam)	
Acronym	-	PA-6,66	
CAS number	-	24993-04-2	
HISTORY			
Person to discover	-	Owens, J K; Scroggie, A G, 1940. Owens, J K; Scroggie, A G, 1941	Owens, J K; Scroggie, A G, US Patent 2,201,741, DuPont, May 21, 1940; Joyce, R M; Ritter, D M, US Patent 2,251,519, DuPont, Aug. 5, 1941.
Date	-	1940; 1941	
Details	-	PA-6,66 with improved resistance to carbon arc; catalytic conversion of monomers	
SYNTHESIS			
Monomer(s) structure	-		
Monomer(s) CAS number(s)	-	628-94-4; 105-60-2	
Monomer(s) molecular weight(s)	dalton, g/mol, amu	144.17; 113.16	
Monomer ratio	-	4:1	
STRUCTURE			
Crystallinity	%	29	
Cell type (lattice)	-	pseudohexagonal	
Spacing between crystalline planes	nm	0.37-0.44	Men, Y; Rieger, J, Eur. Polym. J., 40, 2629-35, 2004.
Polymorphs	-	α, γ (also called phases); hydrogen bonds formed between antiparallel chains in α phase and parallel chains in γ phase	Men, Y; Rieger, J, Eur. Polym. J., 40, 2629-35, 2004.
COMMERCIAL POLYMERS			
Some manufacturers	-	BASF, EMS	
Trade names	-	Ultramid; Grilon	
PHYSICAL PROPERTIES			
Density at 20°C	g cm^{-3}	1.12-1.15; 1.2-1.72 (15-50% glass fiber)	
Bulk density at 20°C	g cm^{-3}	0.7	
Melting temperature, DSC	°C	189-199	
Thermal expansion coefficient, 23-80°C	10^{-4} °C^{-1}	0.6-0.8 (parallel); 0.9-1.2 (normal); 0.2 (15-50% glass fiber, parallel); 1.1 (15-50% glass fiber, normal)	
Glass transition temperature	°C	42 (dry); -35 (water saturated)	Men, Y; Rieger, J, Eur. Polym. J., 40, 2629-35, 2004.
Maximum service temperature	°C	80-120; 90-130 (15-50% glass fiber)	
Long term service temperature	°C	180-220; 180-200 (15-50% glass fiber)	
Heat deflection temperature at 0.45 MPa	°C	200-220	

PA-6,66 polyamide-6,66

PARAMETER	UNIT	VALUE
Heat deflection temperature at 1.8 MPa	°C	55-85; 215-230 (15-50% glass fiber)
Dielectric constant at 100 Hz/1 MHz	-	-/3.6 (dry); -/6 (conditioned)
Relative permittivity at 100 Hz	-	3 (dry); 8 (conditioned)
Relative permittivity at 1 MHz	-	3 (dry); 4 (conditioned)
Dissipation factor at 100 Hz	E-4	50 (dry); 1,500 (conditioned)
Dissipation factor at 1 MHz	E-4	150-200 (dry); 700-3,000 (conditioned)
Volume resistivity	ohm-m	1E11 to 1E12 (dry); 1E9 to 1E11 (conditioned)
Surface resistivity	ohm	1E10 to 1E12 (conditioned)
Electric strength K20/P50, d=0.60.8 mm	kV mm^{-1}	26-32 (dry); 25-28 (conditioned); 30-34 (15-50% glass fiber, dry); 27-30 (15-50% glass fiber, conditioned)
Comparative tracking index, CTI, test liquid A	-	600 (dry); 475-600 (conditioned)
Permeability to oxygen, 25°C	cm^3 m^{-2} day^{-1} bar^{-1}	14
Permeability to water vapor, 25°C	g m^{-2} day^{-1}	25

MECHANICAL & RHEOLOGICAL PROPERTIES

PARAMETER	UNIT	VALUE
Tensile strength	MPa	45-75 (dry); 50 (conditioned); 110-220 (15-50% glass fiber, dry); 65-155 (15-50% glass fiber, conditioned)
Tensile modulus	MPa	2,200-3,600 (dry); 600-1,600 (conditioned); 5,600-20,000 (15-50% glass fiber, dry); 2,900-13,500 (15-50% glass fiber, conditioned)
Tensile stress at yield	MPa	70-90 (dry); 40-55 (conditioned)
Elongation	%	5-25 (dry); >50 (conditioned); 2-4 (15-50% glass fiber, dry); 2.5-10 (15-50% glass fiber, conditioned)
Tensile yield strain	%	4-5 (dry); 15-18 (conditioned)
Flexural modulus	MPa	3,000 (dry)
Charpy impact strength, unnotched, 23°C	kJ m^{-2}	NB to 75-80 (dry); NB to 100 (conditioned); 75-100 (15-50% glass fiber, dry); 90 (15-50% glass fiber, conditioned)
Charpy impact strength, unnotched, -30°C	kJ m^{-2}	NB to 70-80 (dry); NB to 60 (conditioned); 60-80 (15-50% glass fiber, dry); 70-80 (15-50% glass fiber, conditioned)
Charpy impact strength, notched, 23°C	kJ m^{-2}	4-9 (dry); 10-40 (conditioned); 12-15 (15-50% glass fiber, dry); 17-25 (15-50% glass fiber, conditioned)
Charpy impact strength, notched, -30°C	kJ m^{-2}	3-6 (dry); 3-7 (conditioned); 5-12 (15-50% glass fiber, dry); 5-12 (15-50% glass fiber, conditioned)
Izod impact strength, notched, 23°C	J m^{-1}	4.5
Ball indention hardness at 358 N/30 S (ISO 2039-1)	MPa	135-145 (dry); 45-80 (conditioned); 160-180 (15-50% glass fiber, dry); 75-95 (15-50% glass fiber, conditioned)
Shrinkage	%	0.7-1.2 (parallel); 0.8-1.4 (normal); 0.1 (15-50% glass fiber, parallel); 0.3-0.7 (15-50% glass fiber, normal)
Viscosity number	ml g^{-1}	195
Melt volume flow rate (ISO 1133, procedure B), 275°C/5 kg	cm^3/10 min	140
Water absorption, equilibrium in water at 23°C	%	5-10.5; 13.7 (saturated film); 5-8 (15-50% glass fiber)
Moisture absorption, equilibrium 23°C/50% RH	%	2-3.2; 1.1-3 (15-50% glass fiber)

PA-6,66 polyamide-6,66

PARAMETER	UNIT	VALUE	REFERENCES
CHEMICAL RESISTANCE			
Acid dilute/concentrated	-	poor	
Alcohols	-	good	
Alkalis	-	good	
Aliphatic hydrocarbons	-	good	
Aromatic hydrocarbons	-	good	
Greases & oils	-	good	
FLAMMABILITY			
Limiting oxygen index	% O$_2$	35	
UL 94 rating	-	HB to V-0	
TOXICITY			
Carcinogenic effect	-	not listed by ACGIH, NIOSH, NTP	
Skin rabbit, LD$_{50}$	mg kg^{-1}	moderate irritant	
PROCESSING			
Typical processing methods	-	blown film, cast film, injection molding	
Preprocess drying: temperature/ time/residual moisture	ºC/h/%	80/2-4/0.15	
Processing temperature	ºC	240-285	
Processing pressure	MPa	3.5-10.5 (injection and packing pressure)	
Additives used in final products	-	flame retardant, nucleating agent	
Applications	-	automotive (air intake systems, electrical, electronics, interior, lighting, powertrain and chassis), electrical appliances and equipment, cables & tubes, connectors, energy distribution, lighting, industry and consumer goods (housewares, mechanical engineering, power transmission, sports & leisure, tools & accessories), medical packaging	
Outstanding properties	-	low friction, wear resistance	
BLENDS			
Suitable polymers	-	EVA, EVOH, PA-6,10	
ANALYSIS			
x-ray diffraction peaks	degree	20.7, 22.9 (dry)	Men, Y; Rieger, J, Eur. Polym. J., 40, 2629-35, 2004.

252 **HANDBOOK OF POLYMERS** 3rd Edition, Copyrights 2022; ChemTec Publishing

PA-6I,6T polyamide-6I/6T

PARAMETER	UNIT	VALUE	REFERENCES
GENERAL			
Common name	-	polyamide-6I/6T, copolymer of 1,6-hexamethylene diamine and isophthalic acid (6I) (70 wt %) and terephthalic acid (6T) (30 wt %)	
Acronym	-	PA-6I,6T	
CAS number	-	25750-23-6	
HISTORY			
Person to discover	-	Schlack, P	Schlack, P, US Patent 2,356,702, Alien Property Custodian, Aug. 22, 1944.
Date	-	1944	
Details	-	production of synthetic linear condensation polyamides	
SYNTHESIS			
Monomer(s) structure	-		
Monomer(s) CAS number(s)	-	124-09-4; 121-91-5; 100-21-0	
Monomer(s) molecular weight(s)	dalton, g/mol, amu	116.21; 166.13; 166.13	
Method of synthesis	-	manufactured by the condensation of hexamethylenediamine, terephthalic acid, and isophthalic acid such that 65 to 80 percent of the polymer units are derived from hexamethylene isophthalamide	
STRUCTURE			
Crystallinity	%	close to amorphous	
COMMERCIAL POLYMERS			
Some manufacturers	-	DuPont; EMS	
Trade names	-	Selar; Grivory	
PHYSICAL PROPERTIES			
Density at 20°C	g cm^{-3}	1.06-1.19	
Odor	-	odorless	
Melting temperature, DSC	°C	125-140	
Decomposition temperature	°C	340	
Thermal expansion coefficient, 23-80°C	10^{-4} °C^{-1}	0.6; 0.1-0.15 (20-60% glass fiber, parallel); 0.9-1 (20-60% glass fiber, normal)	
Glass transition temperature	°C	125-127	
Maximum service temperature	°C	70; 100-120 (20-60% glass fiber)	
Long term service temperature	°C	220 (20-60% glass fiber)	
Heat deflection temperature at 0.45 MPa	°C	115	
Heat deflection temperature at 1.8 MPa	°C	105; 230-235 (20-60% glass fiber)	
Volume resistivity	ohm-m	1E11 to 1E12; 1E12 (20-60% glass fiber)	
Surface resistivity	ohm	1-1.2E12; 1E13 (20-60% glass fiber)	

PA-6I,6T polyamide-6I/6T

PARAMETER	UNIT	VALUE	REFERENCES
Electric strength K20/P50, d=0.60.8 mm	kV mm^{-1}	27-35; (20-60% glass fiber)	
Comparative tracking index, CTI, test liquid A	-	575-600 (20-60% glass fiber)	
Permeability to oxygen, 25°C	cm^3 m^{-2} s^{-1} bar^{-1} 24 h^{-1}	10-30	
Permeability to water vapor, 25°C	g m^{-2} 24 h^{-1}	7	

MECHANICAL & RHEOLOGICAL PROPERTIES

Tensile strength	MPa	85; 145-260 (20-60% glass fiber)	
Tensile modulus	MPa	3,000; 8,200-22,000 (20-60% glass fiber)	
Tensile stress at yield	MPa	100; 2-4 (20-60% glass fiber)	
Elongation	%	50-300	
Tensile yield strain	%	5	
Tear strength	N m^{-1}	50	
Charpy impact strength, unnotched, 23°C	kJ m^{-2}	NB to 50-80; 50-90 (20-60% glass fiber)	
Charpy impact strength, unnotched, -30°C	kJ m^{-2}	35-80 (20-60% glass fiber)	
Charpy impact strength, notched, 23°C	kJ m^{-2}	6.9-11; 7-14 (20-60% glass fiber)	
Charpy impact strength, notched, -30°C	kJ m^{-2}	2-8; 6-13 (20-60% glass fiber)	
Ball indention hardness at 358 N/30 S (ISO 2039-1)	MPa	145; 225-315 (20-60% glass fiber)	
Shrinkage	%	0.3-0.5; 0.1-0.8 (20-60% glass fiber)	
Intrinsic viscosity, 25°C	dl g^{-1}	0.72-0.82	
Melt volume flow rate (ISO 1133, procedure B), 275°C/5 kg	cm^3/10 min	25	
Melt index, 230°C/3.8 kg	g/10 min	12-100	
Water absorption, equilibrium in water at 23°C	%	7; 3.5-5 (20-60% glass fiber)	
Moisture absorption, equilibrium 23°C/50% RH	%	2; 1.2-1.5 (20-60% glass fiber)	

CHEMICAL RESISTANCE

Acid dilute/concentrated	-	not resistant	
Alcohols	-	resistant to higher alcohols	
Alkalis	-	resistant (dilute	
Aliphatic hydrocarbons	-	resistant	
Aromatic hydrocarbons	-	resistant	

FLAMMABILITY

Volatile products of combustion	-	aldehydes, ammonia, CO, CO_2, oxides of nitrogen	
UL 94 rating	-	V-2; HB (20-60% glass fiber)	

PA-6I,6T polyamide-6I/6T

PARAMETER	UNIT	VALUE	REFERENCES
TOXICITY			
Carcinogenic effect	-	not listed by ACGIH, NIOSH, NTP	
PROCESSING			
Typical processing methods	-	extrusion, coextrusion, injection molding, blow molding	
Processing temperature	ºC	240-250; 310 (max)	
Applications	-	appliance components, automotive parts, blown containers, cast film, cosmetic packaging, flexible and rigid packaging, paper coatings, tubing	
Outstanding properties	-	transparency, barrier properties to gases water and solvents	
BLENDS			
Suitable polymers	-	other PA	

PA-11 polyamide-11

PARAMETER	UNIT	VALUE	REFERENCES
GENERAL			
Common name	-	polyamide-11, nylon-11; poly(imino-1-oxoundecamethylene)	
IUPAC name	-	poly[imino(1-oxoundecane-1,11-diyl)]	
CAS name	-	poly[imino(1-oxo-1,11-undecanediyl)]	
Acronym	-	PA-11	
CAS number	-	25035-04-5	
HISTORY			
Person to discover	-	Carothers, W H	Carothers, W H, US Patent 2,071,250, DuPont, Feb. 16, 1937.
Date	-	1937	
Details	-	patent for linear condensation polymers including PA11	
SYNTHESIS			
Monomer(s) structure	-	$H_2N(CH_2)_{10}COOH$	
Monomer(s) CAS number(s)	-	2432-99-7	
Monomer(s) molecular weight(s)	dalton, g/mol, amu	201.31	
Monomer ratio	-	100%	
CH_2/CONH ratio		10	
Method of synthesis	-	condensation polymerization reaction	
Temperature of polymerization	°C	240 (star-shaped)	Martino, L; Basilissi, L; Farina, H; Ortenzi, M A; Zini, E; Di Silvestro, G; Scandola, M, Eur. Polym. J., 59, 69-77, 2014.
Time of polymerization	h	4 (star-shaped)	Martino, L; Basilissi, L; Farina, H; Ortenzi, M A; Zini, E; Di Silvestro, G; Scandola, M, Eur. Polym. J., 59, 69-77, 2014.
Number average molecular weight, M_n	dalton, g/mol, amu	16,800-42,355	Robert, E C; Bruessau, R; Dubois, J; Jacques, B; Meijerink, N; Nguye, T Q; Niehaus, D E; Tobisch, W A, Pure Appl. Chem., 76, 11, 2009-25, 2004.
Mass average molecular weight, M_w	dalton, g/mol, amu	28,800-88,800	Robert, E C; Bruessau, R; Dubois, J; Jacques, B; Meijerink, N; Nguye, T Q; Niehaus, D E; Tobisch, W A, Pure Appl. Chem., 76, 11, 2009-25, 2004.
Polydispersity, M_w/M_n	-	1.72-2.5	Robert, E C; Bruessau, R; Dubois, J; Jacques, B; Meijerink, N; Nguye, T Q; Niehaus, D E; Tobisch, W A, Pure Appl. Chem., 76, 11, 2009-25, 2004.
Molar volume at 298K	cm^3 mol^{-1}	181.5 (amorphous)	
Van der Waals volume	cm^3 mol^{-1}	115.3 (amorphous)	

PA-11 polyamide-11

PARAMETER	UNIT	VALUE	REFERENCES
STRUCTURE			
Crystallinity	%	16.8-36; a critical crystallinity χ_c of 35.4% is used as a criterion for ductile-brittle-fracture transition	Apgar, G, Nylon Plastics Handbook, Kohan, M I, Ed., Hanser, Munich, 1995; Fruebig, P; Kremmer, A; Gerhard-Multhaupt, R; Spanoudaki, A; Pissis, P, J. Chem. Phys., 125, 214701,1-8, 2006; Mancic, L; Osman, R F M; Costa, A M L M; dèAlmeida, J R M; Marinkovic, B A; Rizzo, F C, Mater. Design, 83, 459-67, 2015; Scherer, B, Kottenstedde, I L, Bremser, W, Matysik, F-M, Polym. Testing, 91, 106786, 2020.
Cell type (lattice)	-	triclinic (α), monoclinic (β), hexagonal (γ, δ, δ')	Apgar, G, Nylon Plastics Handbook, Kohan, M I, Ed., Hanser, Munich, 1995.
Cell dimensions	nm	a:b:c=0.49:0.54:1.49 (α); a:b:c=0.98:1.5:0.80 (β); a:b:c=0.95:2.94:0.45 (γ)	Apgar, G, Nylon Plastics Handbook, Kohan, M I, Ed., Hanser, Munich, 1995; Zhang, Q; Mo, Z; Zhang, H; Liu, S; Cheng, S Z D, Polymer, 42, 5543-47, 2001.
Unit cell angles	degree	α:β:γ=49:77:63 (α); β=65 (β); β=118.5 (γ)	Apgar, G, Nylon Plastics Handbook, Kohan, M I, Ed., Hanser, Munich, 1995; Zhang, Q; Mo, Z; Zhang, H; Liu, S; Cheng, S Z D, Polymer, 42, 5543-47, 2001.
Crystallite size	nm	20.4	Rajesh, J J; Bijwe, J, Wear, 661-68, 2005.
Spacing between crystallites	nm	0.37-0.44 (intersheet distance)	
Polymorphs	-	α (triclinic), β (monoclinic), γ, δ, δ' (hexagonal)	Zhang, Q; Mo, Z; Zhang, H; Liu, S; Cheng, S Z D, Polymer, 42, 5543-47, 2001.
Peak crystallization temperature	^{o}C	142	Jariyavidyanont, K, Janke, A, Androsch, R, Polymer, 184, 121864, 2020.
COMMERCIAL POLYMERS			
Some manufacturers	-	Arkema	
Trade names	-	Rilsan	
PHYSICAL PROPERTIES			
Density at 20°C	g cm^{-3}	1.026-1.06; 1.15 (crystalline); 1.01 (amorphous)	Apgar, G, Nylon Plastics Handbook, Kohan, M I, Ed., Hanser, Munich, 1995.
Melting temperature, DSC	^{o}C	176-198	
Decomposition temperature	^{o}C	240-270; The thermal decomposition by TGA of neat PA11 starts at 420°C under nitrogen and 410°C in the air.	Oliveira, M J; Botelho, G, Polym. Deg. Stab., 93, 139-46, 2008; Benobeidallah, B, Benhamida, A, Kaci, M, Lopez-Cuesta, J-M, Appl. Clay Sci., 198, 105837, 2020.
Onset degradation temperature	^{o}C	390.7 (nitrogen), 395.2 (air)	Filippone, G; Carroccio, S C; Curcuruto, G; Passaglia, E; Cambarotti, C; Dintcheva, N T, Polymer, 73, 102-10, 2015.
Thermal expansion coefficient, 23-80°C	^{o}C^{-1}	8.5E-5	
Thermal conductivity, melt	W m^{-1} K^{-1}	0.267-0.29	Boudenne, A; Ibos, L; Gehin, E; Candau, Y, J. Phys. D: Appl. Phys., 37, 132-39, 2004.
Glass transition temperature	^{o}C	35-46	
Specific heat capacity	J K^{-1} kg^{-1}	1753	

HANDBOOK OF POLYMERS 3rd Edition, Copyrights 2022; ChemTec Publishing

PA-11 polyamide-11

PARAMETER	UNIT	VALUE	REFERENCES
Hansen solubility parameters, δ_D, δ_P, δ_H	MPa$^{0.5}$	17.0, 4.4, 10.6	
Interaction radius		5.1	
Hildebrand solubility parameter	MPa$^{0.5}$	calc.=19.2	
Dielectric constant at 100 Hz/1 MHz	-	3.9/3.1	
Volume resistivity	ohm-m	1E12	
Coefficient of friction	-	0.1-0.3	
Contact angle of water, 20°C	degree	82	

MECHANICAL & RHEOLOGICAL PROPERTIES

PARAMETER	UNIT	VALUE	REFERENCES
Tensile strength	MPa	37-69	
Tensile modulus	MPa	1,300	
Tensile stress at yield	MPa	31-41	
Elongation	%	310	
Tensile yield strain	%	18-24	
Flexural modulus	MPa	290-1,150	
Young's modulus	MPa	344; 365 (theoretical)	Peeters, A; van Alsenoy, C; Bartha, F; Bogar, F; Zhang, M-L; van Doren, V E, Int. J. Quantum Chem. 87, 303-10, 2002.
Charpy impact strength, notched, 23°C	kJ m^{-2}	5-15	
Charpy impact strength, notched, -30°C	kJ m^{-2}	5-13	
Izod impact strength, unnotched, 23°C	J m^{-1}	116	
Crack growth velocity	x 10^{-6} m s^{-1}	344	Rajesh, J J; Bijwe, J, Tribology lett., 18, 3, 331-40, 2005.
Fracture energy	x 10^4 J m^{-2}	5.89	Rajesh, J J; Bijwe, J, Tribology lett., 18, 3, 331-40, 2005.
Ductility factor	mm	17.60	Rajesh, J J; Bijwe, J, Tribology lett., 18, 3, 331-40, 2005.
Stress necessary to cause spontaneous fracture	MPa	66.7	Rajesh, J J; Bijwe, J, Tribology lett., 18, 3, 331-40, 2005.
Shear strength	MPa	35-42	
Tenacity (fiber) (standard atmosphere)	cN tex^{-1} (daN mm^{-2})	45-68 (47-70)	Fourne, F, Synthetic Fibers. Machines and Equipment Manufacture, Properties. Carl Hanser Verlag, 1999.
Tenacity (wet fiber, as % of dry strength)	%	100	Fourne, F, Synthetic Fibers. Machines and Equipment Manufacture, Properties. Carl Hanser Verlag, 1999.
Fineness of fiber (titer)	dtex	3-7	Fourne, F, Synthetic Fibers. Machines and Equipment Manufacture, Properties. Carl Hanser Verlag, 1999.
Length (elemental fiber)	mm	continuous filament	Fourne, F, Synthetic Fibers. Machines and Equipment Manufacture, Properties. Carl Hanser Verlag, 1999.
Shore D hardness	-	64-75	

PA-11 polyamide-11

PARAMETER	UNIT	VALUE	REFERENCES
Rockwell hardness	-	R106	
Viscosity number	ml g^{-1}	151	
Melt viscosity, shear rate=0 s^{-1}	Pa s	2,260	
Moisture absorption, equilibrium 23°C/50% RH	%	0.9-1.1	

CHEMICAL RESISTANCE

Acid dilute/concentrated	-	good-poor	
Alcohols	-	fair-poor	
Alkalis	-	good	
Aliphatic hydrocarbons	-	good	
Aromatic hydrocarbons	-	good	
Esters	-	good	
Greases & oils	-	good	
Halogenated hydrocarbons	-	good-poor	
Ketones	-	good	
Good solvent	-	higher primary alcohols, DMF, DMSO, hexafluoropropanol, formic acid/dichloromethane	

FLAMMABILITY

Ignition temperature	°C	400	
Autoignition temperature	°C	440	
Limiting oxygen index	% O$_2$	20-27	
Volatile products of combustion	-	CO_2, CO, unsaturated hydrocarbons, methane	Levchik, S V; Costa, L; Camino, G, Polym. Deg. Stab., 36, 31-41, 1992.
Peak of heat release rate	kW m^{-2}	1054	Benobeidallah, B, Benhamida, A, Kaci, M, Lopez-Cuesta, J-M, Appl. Clay Sci., 198, 105837, 2020.
Total heat release	MJ m^{-2}	140	Benobeidallah, B, Benhamida, A, Kaci, M, Lopez-Cuesta, J-M, Appl. Clay Sci., 198, 105837, 2020.
Char at 500°C	%	0	Benobeidallah, B, Benhamida, A, Kaci, M, Lopez-Cuesta, J-M, Appl. Clay Sci., 198, 105837, 2020.

WEATHER STABILITY

Spectral sensitivity	nm	385 (increases on heating)	Oliveira, M J; Botelho, G, Polym. Deg. Stab., 93, 139-46, 2008.

BIODEGRADATION

Stabilizers	-	silver	

PROCESSING

Typical processing methods	-	electrospinning, extrusion, injection molding, rotational molding, spinning	
Processing temperature	°C	209-222 (extrusion); 218-235 (injection molding)	
Applications	-	aeronautics, automotive, bearings, break lines, bushings, flexible pipe, food contact, fuel tanks, insulators, marine, medical, natural gas, oil, skis, ski boots, tennis rackets, transport, wire & cable	

PA-11 polyamide-11

PARAMETER	UNIT	VALUE	REFERENCES
Outstanding properties	-	low moisture absorption, monomer is made by thermal cracking of ricinoleic acid from renewable source (castor oil)	
BLENDS			
Suitable polymers	-	PA6,6, PA6,10, PANI, PE, polyepichlorohydrin, PSU, PVDF, starch	
Compatibilizers	-	EPDM-MAH	
ANALYSIS			
FTIR (wavenumber-assignment)	cm^{-1}/-	amide – 1640, 1543; C-C – 1126	Yu, H H, Mater. Chem. Phys., 56, 289-93, 1998.
Raman (wavenumber-assignment)	cm^{-1}/-	amide 1 – 1640; C-C – 1063/1107	Hendra, P J; Maddams, W F; Royaud, I A M; Willis, H A; Zichy, V, Spectrochim. Acta, 64A, 5, 747-56, 1990.
NMR (chemical shifts)	ppm	see ref.	Davis, R D; Jarrett, W L; Mathias, L J, Polymer, 42, 2621-26, 2001.

PA-12 polyamide-12

PARAMETER	UNIT	VALUE	REFERENCES
GENERAL			
Common name	-	polyamide-12, nylon-12, poly(imino-1-oxodecamethylene), polydodecanolactam	
IUPAC name	-	poly[imino(1-oxododecane-1,12-diyl)]	
CAS name	-	poly[imino(1-oxo-1,12-dodecanediyl)]	
Acronym	-	PA-12	
CAS number	-	24937-16-4	
HISTORY			
Person to discover	-	Schaaf, S; Griehl, W	Schaaf, S; Griehl, W, US Patent 3,564,599, Inventa AG, Feb. 16, 1971.
Date	-	1971	
SYNTHESIS			
Monomer(s) structure	-	lauryl lactam	
Monomer(s) CAS number(s)	-	947-04-6	
Monomer(s) molecular weight(s)	dalton, g/mol, amu	197.32	
Monomer(s) expected purity(ies)	%	99.9 min	
Monomer ratio	-	100%	
CH_2/CONH ratio		11	
Method of synthesis	-	hydrolytic polycondensation of dodecanolactam at 300-330°C in the presence of phosphoric acid	
Catalyst	-	phosphoric acid	
Number average molecular weight, M_n	dalton, g/mol, amu	23,400-44,100	Robert, E C; Bruessau, R; Dubois, J; Jacques, B; Meijerink, N; Nguye, T Q; Niehaus, D E; Tobisch, W A, Pure Appl. Chem., 76, 11, 2009-25, 2004.
Mass average molecular weight, M_w	dalton, g/mol, amu	42,400-144,300	Robert, E C; Bruessau, R; Dubois, J; Jacques, B; Meijerink, N; Nguye, T Q; Niehaus, D E; Tobisch, W A, Pure Appl. Chem., 76, 11, 2009-25, 2004.
Polydispersity, M_w/M_n	-	1.54-3.5	Robert, E C; Bruessau, R; Dubois, J; Jacques, B; Meijerink, N; Nguye, T Q; Niehaus, D E; Tobisch, W A, Pure Appl. Chem., 76, 11, 2009-25, 2004.
Molar volume at 298K	cm^3 mol^{-1}	171 (crystalline); 199.3 (amorphous)	
Van der Waals volume	cm^3 mol^{-1}	115.3 (crystalline); 125.1 (amorphous)	

HANDBOOK OF POLYMERS FOR ELECTRONICS / Copyrights 2021; ChemTec Publishing

PA-12 polyamide-12

PARAMETER	UNIT	VALUE	REFERENCES
STRUCTURE			
Crystallinity	%	30-52; 18.3-29.1 (non-isothermal crystallization); 45 (dry); 28 (wet)	McFerran, N L A; Armstrong, C G; McNally, T, J. Appl. Polym. Sci., 110, 1043-58, 2008; Urman, K; Otaigbe, J, Antec, 2063-66, 2004.
Cell type (lattice)	-	pseudohexagonal; monoclinic hexagonal; triclinic	Gogolewski, S; Czerniawska, K; Gasiorek. M, Coll. Polym. Sci,. 258, 1130, 1980; McFerran, N L A; Armstrong, C G; McNally, T, J. Appl. Polym. Sci., 110, 1043-58, 2008.
Cell dimensions	nm	a:b:c=0.479:3.19:0.958; a:b:c=0.958:3.19:0.479; a:b:c=0.91:0.53:3.18	Gogolewski, S; Czerniawska, K; Gasiorek. M, Coll. Polym. Sci,. 258, 1130, 1980; McFerran, N L A; Armstrong, C G; McNally, T, J. Appl. Polym. Sci., 110, 1043-58, 2008; Dosiere, M, Polymer, 34, 15, 3160-67, 1993.
Unit cell angles	degree	β=120; α=γ=90, β=120	Gogolewski, S; Czerniawska, K; Gasiorek. M, Coll. Polym. Sci,. 258, 1130, 1980.
Number of chains per unit cell	-	4	Gogolewski, S; Czerniawska, K; Gasiorek. M, Coll. Polym. Sci,. 258, 1130, 1980.
Crystallite size	nm	10.3-12.6	Rajesh, J J; Bijwe, J, Wear, 661-68, 2005.
Spacing between crystallites	nm	0.42-0.479	
Polymorphs	-	α; γ (the most stable, monoclinic)	
Avrami constants, k/n	-	n=2.05-2.55; n=2.3-2.9 (non-isothermal crystallization)	McFerran, N L A; Armstrong, C G; McNally, T, J. Appl. Polym. Sci., 110, 1043-58, 2008.
Activation energy for crystallization	kJ mol^{-1}	345.5	
Peak crystallization temperature	^{o}C	147-148	Yang, F, Jiang, T, Lalier, G, Bartolone, J, Chen, X, J. Manuf. Proc., 57, 828-46, 2020.
Crystallization enthalpy	J g^{-1}	50.15, 46.78-48.42 (agend powder)	Yang, F, Jiang, T, Lalier, G, Bartolone, J, Chen, X, J. Manuf. Proc., 57, 828-46, 2020.
COMMERCIAL POLYMERS			
Some manufacturers	-	EMS-Grivory, Arkema, Evonik	
Trade names	-	Grilamid, Rilsan, Vestosint	
PHYSICAL PROPERTIES			
Density at 20°C	g cm^{-3}	1.01-1.03; 1.11 (crystalline)	
Refractive index, 20°C	-	1.52-1.53	
Melting temperature, DSC	^{o}C	174-185, 171-183 (powder for laser sintering)	-; Yang, F, Jiang, T, Lalier, G, Bartolone, J, Chen, X, J. Manuf. Proc., 57, 828-46, 2020.
Activation energy of thermal degradation	kJ mol^{-1}	2208	Herrera, M; Matuschek, G; Kettrup, A, Chemosphere, 42, 601-7, 2001.
Melting enthalpy	J g^{-1}	98-105, 36-41 (remelting)	Yang, F, Jiang, T, Lalier, G, Bartolone, J, Chen, X, J. Manuf. Proc., 57, 828-46, 2020.
Fusion heat	J g^{-1}	58	Jun g, H S; Choi, M C; Chang, Y-W; Kang, P-H; Hong, S C, Eur. Polym. J., 66, 367-75, 2015.

PA-12 polyamide-12

PARAMETER	UNIT	VALUE	REFERENCES
Thermal expansion coefficient, 23-80°C	10^{-4} °C^{-1}	1.1-1.61; 0.2-1 (15-65% glass fiber, dry)	Bai, J; Goodridge, R D; Hague, R J M; Song, M; Okamoto, M, Polym. Testing, 36, 95-100, 2014.
Thermal conductivity, 10-100°C	W m^{-1} K^{-1}	0.24	
Glass transition temperature	°C	55 (dry); 45 (equilibrated at 50% RH)	
Specific heat capacity	J K^{-1} kg^{-1}	2,000-2,900	
Heat of fusion	kJ kg^{-1}	65-70	
Maximum service temperature	°C	140-150; 150-160 (15-65% glass fiber, dry); 180 (blend with EPDM)	Jun g, H S; Choi, M C; Chang, Y-W; Kang, P-H; Hong, S C, Eur. Polym. J., 66, 367-75, 2015.
Long term service temperature	°C	85-110; 90-120 (15-65% glass fiber, dry)	
Heat deflection temperature at 0.45 MPa	°C	110-130	
Heat deflection temperature at 1.8 MPa	°C	45-50; 150-160 (15-65% glass fiber, dry)	
Vicat temperature VST/A/50	°C	170	
Vicat temperature VST/B/50	°C	140	
Hansen solubility parameters, δ_D, δ_P, δ_H	MPa$^{0.5}$	18.5, 8.1, 9.1	
Interaction radius		6.3	
Hildebrand solubility parameter	MPa$^{0.5}$	calc.=19.0; exp.=22.2	
Surface tension	mN m^{-1}	calc.=35.8	
Dielectric constant at 100 Hz/1 MHz	-	-/3.00 (at 50% RH)	
Relative permittivity at 100 Hz	-	3.8	
Relative permittivity at 1 MHz	-	3.8 (dry), 6.7 (wet)	
Dissipation factor at 1 MHz	E-4	500 (dry); 170 (wet)	
Volume resistivity	ohm-m	1E11 to 1E12 (conditioned)	
Surface resistivity	ohm	1E13 (at 50% RH)	
Electric strength K20/P50, d=0.60.8 mm	kV mm^{-1}	27-34 (dry); 32 (conditioned); 35-45 (15-65% glass fiber, dry)	
Comparative tracking index, CTI, test liquid A	-	600 (conditioned); 600 (15-65% glass fiber, dry)	
Contact angle of water, 20°C	degree	72.4; 77/56.2 (asc/rec)	
Surface free energy	mJ m^{-2}	39.3	

MECHANICAL & RHEOLOGICAL PROPERTIES

Tensile strength	MPa	50-70 (dry); 45-50 (conditioned); 80-190 (15-65% glass fiber, dry); 70-170 (15-65% glass fiber, conditioned)	
Tensile modulus	MPa	1,400-1,600 (dry); 1,100 (conditioned); 3,500-20,000 (15-65% glass fiber, dry); 3,000-18,500 (15-65% glass fiber, conditioned)	
Tensile stress at yield	MPa	41-46	
Tensile creep modulus, 1000 h, elongation 0.5 max	MPa	45 (dry); 40-45 (conditioned)	
Elongation	%	>50 (dry); >50 (conditioned); 3-8 (15-65% glass fiber, dry); 3-10 (15-65% glass fiber, conditioned)	
Tensile yield strain	%	5-6 (dry); 12-15 (conditioned)	

PA-12 polyamide-12

PARAMETER	UNIT	VALUE	REFERENCES
Flexural modulus	MPa	360-1,260; 1160 (bone replacement)	Abdullah, A M, Mohamad, D, Rahim, T N A T, Akil, H M, Rajion, Z A, Mater. Sci. Eng.: C, 99, 719-25, 2019.
Young's modulus	MPa	460-1,900	
Charpy impact strength, unnotched, 23°C	kJ m^{-2}	NB (dry); NB (conditioned); 65-100 (15-65% glass fiber, dry); 60-90 (15-65% glass fiber, conditioned)	
Charpy impact strength, unnotched, -30°C	kJ m^{-2}	NB (dry); NB (conditioned); 65-100 (15-65% glass fiber, dry); 60-90 (15-65% glass fiber, conditioned)	
Charpy impact strength, notched, 23°C	kJ m^{-2}	6-25 (dry); 6 (conditioned); 12-15 (15-65% glass fiber, dry); 12-15 (15-65% glass fiber, conditioned)	
Charpy impact strength, notched, -30°C	kJ m^{-2}	4-8 (dry); 5-7 (conditioned); 10 (15-65% glass fiber, dry); 10 (15-65% glass fiber, conditioned)	
Izod impact strength, unnotched, 23°C	J m^{-1}	160	
Crack growth velocity	x 10^{-6} m s^{-1}	913-962	Rajesh, J J; Bijwe, J, Tribology lett., 18, 3, 331-40, 2005.
Fracture energy	x 10^4 J m^{-2}	3.9-4.43	Rajesh, J J; Bijwe, J, Tribology lett., 18, 3, 331-40, 2005.
Ductility factor	mm	14.21-14.76	Rajesh, J J; Bijwe, J, Tribology lett., 18, 3, 331-40, 2005.
Stress necessary to cause spontaneous fracture	MPa	96.98-98.24	Rajesh, J J; Bijwe, J, Tribology lett., 18, 3, 331-40, 2005.
Abrasion resistance (ASTM D1044)	mg/100 cycles	14	
Shore D hardness	-	61-79	
Ball indention hardness at 358 N/30 S (ISO 2039-1)	MPa	75 (dry); 70 (conditioned)	
Shrinkage	%	0.5-1.1; 0.1-0.7 (15-65% glass fiber, dry)	
Melt volume flow rate (ISO 1133, procedure B), 275°C/5 kg	cm^3/10 min	20-36	
Melt index, 235°C/2.16 kg	g/10 min	52.5	Jung, H S; Choi, M C; Chang, Y-W; Kang, P-H; Hong, S C, Eur. Polym. J., 66, 367-75, 2015.
Water absorption, equilibrium in water at 23°C	%	1.3-1.5; 0.8-1.1 (15-65% glass fiber, dry)	
Moisture absorption, equilibrium 23°C/50% RH	%	0.4-0.8; 0.4-0.8 (15-65% glass fiber, dry)	

CHEMICAL RESISTANCE			
Acid dilute/concentrated	-	good/poor	
Alcohols	-	poor	
Alkalis	-	good/poor	
Aliphatic hydrocarbons	-	good	
Aromatic hydrocarbons	-	good	
Esters	-	good-fair	
Greases & oils	-	good	
Halogenated hydrocarbons	-	fair-poor	
Ketones	-	good	
Good solvent	-	cresol, formic acid/dichloromethane	

PA-12 polyamide-12

PARAMETER	UNIT	VALUE	REFERENCES
FLAMMABILITY			
Limiting oxygen index	% O_2	21-22.5	
Volatile products of combustion	-	CO_2, CO, ethylene, propylene	Levchik, S V; Costa, L; Camino, G, Polym. Deg. Stab., 36, 31-41, 1992.
Peak of heat release rate	kW m^{-2}	1580	Batistella, M, Regazzi, A, Pucci, M F, Lopez-Cuesta, J-M, Ayme, F, Polym. Deg. Stab., 181, 109318, 2020.
Total heat release	MJ m^{-2}	151	Batistella, M, Regazzi, A, Pucci, M F, Lopez-Cuesta, J-M, Ayme, F, Polym. Deg. Stab., 181, 109318, 2020.
UL 94 rating	-	HB; HB (15-65% glass fiber, dry)	
TOXICITY			
NFPA: Health, Flammability, Reactivity rating	-	1/1/0	
Carcinogenic effect	-	not listed by ACGIH, NIOSH, NTP	
PROCESSING			
Typical processing methods	-	electrospinning, extrusion, injection molding, laser sintering, spinning	Yang, F, Jiang, T, Lalier, G, Barto-lone, J, Chen, X, J. Manuf. Proc., 57, 828-46, 2020.
Preprocess drying: temperature/ time/residual moisture	°C/h/%	80/3-5/0.1	
Processing temperature	°C	240-270	
Processing pressure	MPa	0-1 (back)	
Additives used in final products	-	MWCNT, exfoliated graphite, graphene	Karevan, M; Eshraghi, S; Gerhard, R; Das, S; Kalaitzidou, K, Carbon, 64, 122-31, 2013.
Applications	-	automotive (compressed air systems, hydraulic systems, electrical, lighting, cooling and climate control, fuel systems, powertrain chassis), connectors, electrical equipment, film, industry & consumer goods (housewares, hydraulics & pneu-matics, mechanical engineering, medical devices, sanitary, water and gas supply, sports & leisure, tools & accessories)	
Outstanding properties	-	low environmental stress cracking, low moisture absorption	
BLENDS			
Suitable polymers	-	EPDM, EPR, HDPE, PA6, PBT, PET, PP, SBM, SEBS	
ANALYSIS			
FTIR (wavenumber-assignment)	cm^{-1}/-	−NH stretching vibration − 3290, Fermi resonance of the v(NH−) stretching − 3095, amide-I (mostly of the v(CO=) stretches) −1637, amide III (C−N stretching + C=O in-plane bending) − 1267, splitting of amide II (CH_2 wagging or CH_2 twisting) − 1194, skeletal motion involving CONH (am, γ) − 1159, skeletal motion involving CONH − 1062, CONH in-plane − 948, CH_2 rocking − 710, amide VI (N−H out-of-plane bend) − 621	Yang, F, Jiang, T, Lalier, G, Barto-lone, J, Chen, X, J. Manuf. Proc., 57, 828-46, 2020.
Raman (wavenumber-assignment)	cm^{-1}/-	amide − 1636; C-C − 1063/1107	Hendra, P J; Maddams, W F; Royaud, I A M; Willis, H A; Zichy, V, Spectrochim. Acta, 64A, 5, 747-56, 1990.

HANDBOOK OF POLYMERS FOR ELECTRONICS / Copyrights 2021; ChemTec Publishing

PA-12 polyamide-12

PARAMETER	UNIT	VALUE	REFERENCES
NMR (chemical shifts)	ppm	see ref.	Davis, R D; Jarrett, W L; Mathias, L J, Polymer, 42, 2621-26, 2001.

PAA poly(acrylic acid)

PARAMETER	UNIT	VALUE	REFERENCES
GENERAL			
Common name	-	poly(acrylic acid)	
IUPAC name	-	poly(acrylic acid)	
CAS name	-	2-propenoic acid, homopolymer	
Acronym	-	PAA	
CAS number	-	9003-01-4	
RTECS number	-	AT4680000	
HISTORY			
Person to discover	-	Fikentscher, H; Wappes, H; Eifflaender, L; Schoeller, C; Schneevoigt, A	Fikentscher, H; Wappes, H; Eifflaender, L; Schoeller, C; Schneevoigt, A, US Patent 1,976,679, IG Farben, Oct. 9, 1934.
Date	-	1934	
Details	-	production of dispersions using PAA	
SYNTHESIS			
Monomer(s) structure	-	$H_2C=CHCOOH$	
Monomer(s) CAS number(s)	-	79-10-7	
Monomer(s) molecular weight(s)	dalton, g/mol, amu	72.06	
Monomer ratio	-	100%	
Heat of polymerization	$J\ g^{-1}$	1033-1075	McCurdy, K G; Laidler, K J, Ca. J. Chem., 42, 818, 1964.
Number average molecular weight, M_n	dalton, g/mol, amu	45,000-3,000,000	
Molar volume at 298K	$cm^3\ mol^{-1}$	53.3	
Van der Waals volume	$cm^3\ mol^{-1}$	36.48	
STRUCTURE			
Spacing between crystallites	nm	0.287-0.479 (crystalline, isotactic)	Miller, M L; O'Donnell, K; Skogman, J, J. Coll. Sci., 17, 649-59, 1962.
Tacticity	%	atactic (can be produced in isotactic form from its butyl ester)	Miller, M L; O'Donnell, K; Skogman, J, J. Coll. Sci., 17, 649-59, 1962.
Entanglement molecular weight	dalton, g/mol, amu	calc.=4785	
COMMERCIAL POLYMERS			
Some manufacturers	-	Lubrizol	
Trade names	-	Carbopol	
PHYSICAL PROPERTIES			
Density at 20°C	$g\ cm^{-3}$	1.22-1.44	
Color	-	white	
Refractive index, 20°C	-	calc.=1.4905-1.5112; exp.=1.492-1.527	
Odor	-	acetic	
Melting temperature, DSC	°C	179	

PAA poly(acrylic acid)

PARAMETER	UNIT	VALUE	REFERENCES
Thermal conductivity, melt	W m^{-1} K^{-1}	0.1888	
Glass transition temperature	°C	calc.=70-101; exp.=105-106	
Surface tension	mN m^{-1}	54.7-61.7	
MECHANICAL & RHEOLOGICAL PROPERTIES			
Tensile strength	MPa	0.084	Ponchel, G; Touchard, F; Ducheme, D; Peppas, N A, J. Controlled Release, 5, 129-141, 1987.
Tensile modulus	MPa	1.09	Ponchel, G; Touchard, F; Ducheme, D; Peppas, N A, J. Controlled Release, 5, 129-141, 1987.
Poisson's ratio	-	0.400	
CHEMICAL RESISTANCE			
Alcohols	-	poor	
Alkalis	-	poor	
Aliphatic hydrocarbons	-	good	
Aromatic hydrocarbons	-	good	
Ketones	-	good	
Θ solvent, Θ-temp.=	-	dioxane	
Good solvent	-	DMF, dioxane, ethanol, methanol, water	
Non-solvent	-	acetone, aliphatic hydrocarbons, benzene, diethyl ether	
FLAMMABILITY			
Autoignition temperature	°C	520	
Volatile products of combustion	-	CO, CO_2	
TOXICITY			
NFPA: Health, Flammability, Reactivity rating	-	1/1/0	
Carcinogenic effect	-	not listed by ACGIH, NIOSH, NTP	
TLV, ACGIH	mg m^{-3}	1	
NIOSH	mg m^{-3}	1	
MAK/TRK	mg m^{-3}	0.05 (Netherlands)	
Oral rat, LD$_{50}$	mg kg^{-1}	2,500	
ENVIRONMENTAL IMPACT			
Aquatic toxicity, *Daphnia magna*, LC$_{50}$, 48 h	mg l^{-1}	168-280	
Aquatic toxicity, *Bluegill sunfish*, LC$_{50}$, 48 h	mg l^{-1}	580-2000	
PROCESSING			
Typical processing methods	-	compounding	
Additives used in final products	-	Filler: calcium carbonate, kaolin, metal oxide	

PAA poly(acrylic acid)

PARAMETER	UNIT	VALUE	REFERENCES
Applications	-	bioadhesives, binder for ceramic, contact activation of blood coagulation, controlled drug release, dental cements, diapers, deodorants, dispersants for pigments and fillers, hydraulic fluids, hydrogels, ion exchange resins, paper, pharmaceutical (viscosity modifier in creams an gels), polyelectrolytes, rheology modifiers, suspending agents, thickeners, toothpaste	
BLENDS			
Suitable polymers	-	cellulose, chitosan, CMC, CR, PA6, PANI, PE, PEG, PVAc, PVAI, PVDF, poly(propylene carbonate), polyurethane, polyvinylpyrrolidone	

PAAm polyacrylamide

PARAMETER	UNIT	VALUE	REFERENCES
GENERAL			
Common name	-	polyacrylamide	
IUPAC name	-	polyacrylamide	
CAS name	-	2-propenamide, homopolymer	
Acronym	-	PAAm	
CAS number	-	9003-05-8	
RTECS number	-	AS3700000	
HISTORY			
Person to discover	-	Ornstein & Davis; Harper, Bashaw, Atkins	
Date	-	1959; 1966	
Details	-	first use of gel for electrophoresis; soil hydrators patented by DOW	
SYNTHESIS			
Monomer(s) structure	-	$H_2C=CHC(O)NH_2$	
Monomer(s) CAS number(s)	-	79-06-1	
Monomer(s) molecular weight(s)	dalton, g/mol, amu	69.08	
Monomer(s) expected purity(ies)	%	100%	
Monomer ratio	-	amine:aldehyde:acrylic acid:isocyanide=1.2:1:1.2:1	
Method of synthesis	-	a mixture of amine, aldehyde and methanol was stirred at room temperature for 30 min. Acrylic acid and isocyanide were added and reaction conducted for 24 hrs, after which methanol was removed	Sehlinger, A; Ochsenreither, K; Bartnick, N; Meier, M A R, Eur. Polym. J., 65, 313-24, 2015.
Heat of polymerization	$J\ g^{-1}$	1146	Joshi, R M, Makromol. Chem., 55, 35, 1962.
Mass average molecular weight, M_w	dalton, g/mol, amu	>5,000,000; 8,000,000-15,000,000	Bessaies-Bey, H; Baumann, R; Schmitz, M; Radler, M; Roussel, N, Cement Concrete Res., 76, 98-106, 2015.
Polymerization degree (number of monomer units)	-	>150,000	Sojka, R E; Bjorneberg, D L; Entry, J A; Lentz, R D; Orts, W J, Adv. Agronomy, 92, 75-162, 2007.
Molar volume at 298K	$cm^3\ mol^{-1}$	calc.=56.5; exp.=54.6	
Van der Waals volume	$cm^3\ mol^{-1}$	38.15	
STRUCTURE			
Entanglement molecular weight	dalton, g/mol, amu	calc.=4847	
PHYSICAL PROPERTIES			
Color	-	white to off-white	
Refractive index, 20°C	-	calc.=1.5207-1.5252; exp.=1.52	
Odor	-	odorless	
Melting temperature, DSC	°C	246	
Softening point	°C	208	
Decomposition temperature	°C	160	

PAAm polyacrylamide

PARAMETER	UNIT	VALUE	REFERENCES
Thermal conductivity, melt	W m^{-1} K^{-1}	calc.=0.1863	
Glass transition temperature	°C	calc.=93-148; exp.=153-165	
Surface tension	mN m^{-1}	calc.=50.7-52.3	
Dielectric constant at 100 Hz/1 MHz	-	-/5	

MECHANICAL & RHEOLOGICAL PROPERTIES

Tensile strength	MPa	66.2; 0.04-0.08 (5% gel)	Abdurrahmanoglu, S; Can, V; Okay, O, Polymer, 50, 5449-55, 2009.
Elongation	%	214-265	Abdurrahmanoglu, S; Can, V; Okay, O, Polymer, 50, 5449-55, 2009.
Elastic modulus	MPa	0.0181 (8% gel)	Gautreau, Z; Griffin, J; Peterson, T; Thongpradit, P, Characterizing Viscoelastic Properties of Polyacrylamide Gels, Worcester Polytechnic Institute, 2006.
Young's modulus	MPa	0.031-0.035 (8% gel)	Gautreau, Z; Griffin, J; Peterson, T; Thongpradit, P, Characterizing Viscoelastic Properties of Polyacrylamide Gels, Worcester Polytechnic Institute, 2006.
Poisson's ratio	-	calc.=0.399; exp.=0.45	Gautreau, Z; Griffin, J; Peterson, T; Thongpradit, P, Characterizing Viscoelastic Properties of Polyacrylamide Gels, Worcester Polytechnic Institute, 2006.
Ball indention hardness at 358 N/30 S (ISO 2039-1)	MPa	0.00541-0.00973 (5% gel)	Gautreau, Z; Griffin, J; Peterson, T; Thongpradit, P, Characterizing Viscoelastic Properties of Polyacrylamide Gels, Worcester Polytechnic Institute, 2006.
Swelling	%	6,800	Ortega-Gudino, P; Sanchez-Diaz, J C; Becerra, F; Martinez-Ruvalcaba, A; Gonzalez-Alvarez, A, Antec, 1479-82, 2007.
Water absorption, equilibrium in water at 23°C	%	15	

CHEMICAL RESISTANCE

Alcohols	-	good	
Aliphatic hydrocarbons	-	good	
Aromatic hydrocarbons	-	good	
Esters	-	good	
Θ solvent	-	methanol/water=2/3	
Good solvent	-	ethylene glycol, morpholine, water	
Non-solvent	-	alcohols, diethyl ether, DMF, esters, hydrocarbons, THF	

FLAMMABILITY

Ignition temperature	°C	>200	
Autoignition temperature	°C	>400	
Char at 500°C	%	8.3	Lyon, R E; Walters, R N, J. Anal. Appl. Pyrolysis, 71, 27-46, 2004.

PAAm polyacrylamide

PARAMETER	UNIT	VALUE	REFERENCES
WEATHER STABILITY			
Spectral sensitivity	nm	325	Sojka, R E; Bjorneberg, D L; Entry, J A; Lentz, R D; Orts, W J, Adv. Agronomy, 92, 75-162, 2007.
BIODEGRADATION			
Typical biodegradants	-	*Bacillus cereus, Bacillus flexu, Pseudomonas stutzeri, Rhodococcus spp., Xanthomonas spp.*; lignin oxidizing enzymes produced by fungi and abiotic processes (biodegradation)	Wen, Q; Chen, Z; Zhao, Y; Zhang, H; Feng, Y, J. Hazardous Mater., 175, 955-59, 2010; Nyyssölä, A, Ahlgren, J, Int. Biodeterioration Biodegrad., 139, 24-33, 2019.
TOXICITY			
NFPA: Health, Flammability, Reactivity rating	-	1/0/0	
Carcinogenic effect	-	not listed by ACGIH, NIOSH, NTP	
Oral rat, LD$_{50}$	mg kg^{-1}	>5,000	Sojka, R E; Bjorneberg, D L; Entry, J A; Lentz, R D; Orts, W J, Adv. Agronomy, 92, 75-162, 2007.
Skin rabbit, LD$_{50}$	mg kg^{-1}	>5,000	Sojka, R E; Bjorneberg, D L; Entry, J A; Lentz, R D; Orts, W J, Adv. Agronomy, 92, 75-162, 2007.
ENVIRONMENTAL IMPACT			
Aquatic toxicity, *Daphnia magna*, LC$_{50}$, 48 h	mg l^{-1}	150-230	Acharya, K; Schulman, C; Young, M H, Water Air Soil Pollut., 212, 309-17, 2010.
PROCESSING			
Typical processing methods	-	gel synthesis and modification	
Additives used in final products	-	alumina, graphene, starch, titanium dioxide	
Applications	-	drilling fluids, excipient, flocculation and dewatering of oil sands tailings, gel for chemical analysis, irrigation water treatment for erosion reduction, sizing agent, soft tissue filler, soil hydration, water separation membranes	
Outstanding properties	-	flocculating properties	
BLENDS			
Suitable polymers	-	alginate, chitosan, CR, HDPE, PANI, polydopamine, PS	
ANALYSIS			
FTIR (wavenumber-assignment)	cm^{-1}/-	NH$_2$ – 3352, 3180, 1353, 1282, 991, 708; C=O – 1675, 490; C=C – 1650; C-N – 1430	Murugan, R; Mohan, S; Bigotto, A, J. Korean Phys. Soc., 32, 4, 505-12, 1998.
Raman (wavenumber-assignment)	cm^{-1}/-	NH$_2$ – 3342, 3163, 1350, 1280, 990, 708; C=O – 1685, 490; C=C – 1639; C-N – 1432	Murugan, R; Mohan, S; Bigotto, A, J. Korean Phys. Soc., 32, 4, 505-12, 1998.

PAC polyacetylene

PARAMETER	UNIT	VALUE	REFERENCES
GENERAL			
Common name	-	polyacetylene	
IUPAC name	-	poly(ethene-1,2-diyl); polyethyne	
CAS name	-	ethyne, homopolymer	
Acronym	-	PAC	
CAS number	-	25067-58-7	
HISTORY			
Person to discover	-	Natta, G, Mazzanti, G, Corradini, P; Ito, T, Shirakawa, H, and Ikeda, S	Shirakawa, H, Rev. Mod. Phys., 73, 713-18, 2001.
Date	-	1958; 1967	
Details	-	first synthesis; synthesized PAC film	
SYNTHESIS			
Monomer(s) structure	-	$HC\equiv CH$	
Monomer(s) CAS number(s)	-	74-86-2	
Monomer(s) molecular weight(s)	dalton, g/mol, amu	26.04	
Monomer ratio	-	100%	
Method of synthesis	-	the most common method of synthesis is ring opening metathesis polymerization of molecules such as cyclooctatetraene; simple method of synthesis of cis isomer involves blowing acetylene onto the stationary surface of Ziegler catalyst	
Temperature of polymerization	ºC	-78	
Natural polyacetylene	-	polyacetylene can be found in red ginseng and other species (plants, fungi, marine organisms, and animals)	Wang, B-Y, Yang, X-Q, Hu, M, Shi, L-J, Ding, Z-T, J. Ginseng Res., in press, 2019.
Catalyst	-	Ziegler-Natta; $Pd(OAc)_2$	Huber, J; Mecking, S, Angew. Chem. Int. Ed., 45, 6314-17, 2006.
Number average molecular weight, M_n	dalton, g/mol, amu	21,500-286,100	
Polydispersity, M_w/M_n	-	1.25-1.46	
STRUCTURE			
Crystallinity	%	80	Saxena, V; Malhotra, B D, Handbook of polymers in Elecronics, Ed. Malhotra, B D, Rapra, 2002.
Cell type (lattice)	-	orthorhombic, hexagonal	
Cell dimensions	nm	a:b:c=0.720-0.741:0.406-0.492:0.245-0.260 (*trans*, othorhombic); 0.761-0.768:0.430-0.446:0.436-0.447 (*cis-transoid*, orthorhombic); 0.512:0.512:0.484 (*cis-cisoid*, hexagonal)	Shrikawa, H, Synthetic Metals, 125, 3-10, 2002.
Fibril diameter	nm	20-100	Shrikawa, H, Synthetic Metals, 125, 3-10, 2002.
Tacticity	%	70-95 (*cis*); 100 (*trans*) at 150ºC; 98.1 (*cis*) at -78ºC	Shrikawa, H, Synthetic Metals, 125, 3-10, 2002.
Cis content	%	depends on polymerization temperature; *cis*, which is insulator-like state, can be converted to *trans* by heating	Skanderi, Z; Djebaili, A; Bouzaher, Y; Belloum, M; Abadie, M J M, Composites, Part A, 36, 497-501, 2005.
Chain conformation	-	helix; polyacetylene exists in four conformations; i.e., *cis-transoid*, *cis-cisoid*, *tran-cisoid* and *trans-transoid*. The most stable configuration at low temperature is *cis-transoid* conformation	Akagi, K; Mori, T, Chem. Record, 8, 395-406, 2008; Zhang, C, Liu, L, Okamoto, Y, TrAC Trends Anal. Chem., 123, 115762, 2020.

HANDBOOK OF POLYMERS 3ʳᵈ Edition, Copyrights 2022; ChemTec Publishing

PAC polyacetylene

PARAMETER	UNIT	VALUE	REFERENCES
Space group		Pnam	Martens, J H F; Pichler, K; Marseglia, E A; Friend, R H; Cramail, H; Khosravi, E; Parker, D; Feast, W J; Polymer, 35, 2, 403-14, 1994.

PHYSICAL PROPERTIES

PARAMETER	UNIT	VALUE	REFERENCES
Density at 20°C	g cm^{-3}	1.0-1.23	
Refractive index, 20°C	-	1.7-3	
Isomerization temperature of cis-isomer	°C	0 (beginning), 100 (complete isomerization to trans)	
Glass transition temperature	°C	200	Fink, J K, High Performance Polymers, William Andrew, 2008.
Surface tension	mN m^{-1}	51 (cis); 52 (trans)	Schonhorn, H; Baker, G L; Bates, F S, J. Polym. Sci., Polym. Phys. Ed., 23, 1555, 1985.
Volume resistivity	ohm-m	1E2 (trans-rich); 2.4E6 (cis 80%)	Shrikawa, H, Synthetic Metals, 125, 3-10, 2002.
Contact angle of water, 20°C	degree	72	
Surface free energy	mJ m^{-2}	51.5	
Optical absorption edge	eV	1.4 (trans), 2.0 (cis)	Saxena, V; Malhotra, B D, Handbook of polymers in Elecronics, Ed. Malhotra, B D, Rapra, 2002.

MECHANICAL & RHEOLOGICAL PROPERTIES

PARAMETER	UNIT	VALUE	REFERENCES
Tensile strength	MPa	900	
Tensile modulus	MPa	50,000	
Young's modulus	MPa	25,000-30,000	

CHEMICAL RESISTANCE

PARAMETER	UNIT	VALUE	REFERENCES
Acid dilute/concentrated	-	good	
Alcohols	-	good	
Alkalis	-	good	
Aliphatic hydrocarbons	-	good	
Aromatic hydrocarbons	-	good	
Esters	-	good	
Greases & oils	-	good	
Halogenated hydrocarbons	-	good	
Ketones	-	good	
Good solvent	-	aniline, DMF, isopropylamine	
Non-solvent	-	acetone, carbon tetrachloride, methanol	

BIODEGRADATION

PARAMETER	UNIT	VALUE	REFERENCES
Stabilizers	-	some polyacetylene derivatives have insecticidal properties especially in the presence of UV	Haouas, D; Guido, F; Monia, B H-K; Habib, B H M, Ind. Crops Products, in press, 2011.

TOXICITY

PARAMETER	UNIT	VALUE	REFERENCES
Carcinogenic effect	-	not listed by ACGIH, NIOSH, NTP	

PROCESSING

PARAMETER	UNIT	VALUE	REFERENCES
Typical processing methods	-	printing using dispersion	

PAC polyacetylene

PARAMETER	UNIT	VALUE	REFERENCES
Additives used in final products	-	Antistatics: carbon black, various doping systems	
Applications	-	antistatics, environmental sensing devices, membranes, nano-electronic devices, rechargeable batteries, semiconductor devices, solar cells	
Outstanding properties	-	Marine polyacetylenes constitute a very structurally diverse and useful class of compounds with important biological activities such as antifungal, antibiotic, anticancer, antitumor, anti-HIV, anti-inflammatory, and antimicrobial properties.	Legrave, N; Elsebai, M F; Mehiri, M; Amade, P, Studies in Natural Products Chemistry, Chapter 8, 251-95, Elsevier, 2015.
BLENDS			
Suitable polymers	-	PANI, SBS	
ANALYSIS			
Raman (wavenumber-assignment)	cm^{-1}/-	*trans*-PAC – 1150 and 1450	Oshiro, T; Yamazato, M; Higa, A; Toguchi, M, Jpn J. Appl. Phys., 46, 2, 756-60, 2007.

PAEK polyaryletherketone

PARAMETER	UNIT	VALUE	REFERENCES
GENERAL			
Common name	-	Polyaryletherketone	
Acronym	-	PAEK	
Well-known members of the group	-	polyether ether ketone (PEEK), polyether ketone (PEK), PEEK ketone (PEEKK), PEK ether ketone ketone (PEKEKK), PEK ketone (PEKK)	
HISTORY			
Person to discover	-	Bonner, W H (DuPont); Goodman, J E (ICI)	
Date	-	1962, 1964	
COMMERCIAL POLYMERS			
Some manufacturers	-	Solvay	
Trade names	-	AvaSpire	
PHYSICAL PROPERTIES			
Density at 20°C	g cm^{-3}	1.29-1.32	
Melting temperature, DSC	°C	340; 340-345 (30-40% glass fiber); 340 (30% carbon fiber)	
Thermal expansion coefficient, 23-80°C	°C^{-1}	0.45-0.47E-4; 0.16-0.17E-4 (30-40% glass fiber)	
Thermal conductivity, melt	W m^{-1} K^{-1}	0.2	
Glass transition temperature	°C	150-158; 150-158 (30-40% glass fiber); 150 (30% carbon fiber)	
Specific heat capacity	J K^{-1} kg^{-1}	1450	
Maximum service temperature	°C	350	
Long term service temperature	°C	250	
Heat deflection temperature at 1.8 MPa	°C	161-252; 213-286 (30-40% glass fiber); 267-276 (30% carbon fiber)	
Dielectric constant at 100 Hz/1 MHz	-	3.88/4.00 (40% glass fiber)	
Relative permittivity at 1 MHz	-	3.1	
Dissipation factor at 1 MHz	E-4	40	
Volume resistivity	ohm-m	6.2E+17; 2E16 (30-40% glass fiber)	
Surface resistivity	ohm	1.9E+17	
Electric strength K20/P50, d=0.60.8 mm	kV mm^{-1}	16 (30-40% glass fiber)	
MECHANICAL & RHEOLOGICAL PROPERTIES			
Tensile strength	MPa	84-93.8; 156-191 (30-40% glass fiber); 176-201 (30% carbon fiber)	
Tensile modulus	MPa	2,900-3,720; 9,900-15,200 (30-40% glass fiber); 18,800-22,100 (30% carbon fiber)	
Tensile stress at yield	MPa	84-87	
Elongation	%	26-76; 1.8-2.9 (30-40% glass fiber); 1.5-2.0 (30% carbon fiber)	
Tensile yield strain	%	5.0-6.7	
Flexural strength	MPa	122-141; 234-253 (30-40% glass fiber); 259-317 (30% carbon fiber)	

276   **HANDBOOK OF POLYMERS FOR ELECTRONICS** / Copyrights 2021; ChemTec Publishing

PAEK polyaryletherketone

PARAMETER	UNIT	VALUE	REFERENCES
Flexural modulus	MPa	3,100-3,720; 9,400-14,800 (30-40% glass fiber); 16,500-19,300 (30% carbon fiber)	
Compressive strength	MPa	228 (30-40% glass fiber)	
Young's modulus	MPa	4100	
Izod impact strength, unnotched, 23°C	J m^{-1}	no break; 590-960 (30-40% glass fiber); 530 (30% carbon fiber)	
Izod impact strength, notched, 23°C	J m^{-1}	75-100; 53-110 (30-40% glass fiber)	
Shear strength	MPa	79 (30-40% glass fiber)	
Rockwell hardness	-	M93	
Shrinkage	%	0.8-1.3; 0.3-1.3 (30-40% glass fiber); 0.1-0.5 (30% carbon fiber)	
Melt viscosity, shear rate=1000 s^{-1}	Pa s	410-450; 410-450 (30-40% glass fiber); 470 (30% carbon fiber)	
Melt index, 400°C/2.16 kg	g/10 min	1-5; 7-9 (30-40% glass fiber)	

CHEMICAL RESISTANCE

Aromatic hydrocarbons	-	excellent	
Esters	-	excellent	
Halogenated hydrocarbons	-	excellent	
Ketones	-	excellent	

FLAMMABILITY

UL 94 rating	-	V-0; V-0 or V-1 (30-40% glass fiber)	

TOXICITY

Carcinogenic effect	-	not listed by ACGIH, NIOSH, NTP	

PROCESSING

Typical processing methods	-	extrusion blow molding, fiber spinning, film extrusion, injection blow molding, injection molding, machining, profile extrusion, thermoforming, wire and cable extrusion	
Preprocess drying: temperature/time/residual moisture	°C/h/%	150/4/-; 149-175/2.5-4/- (30-40% glass fiber)	
Processing temperature	°C	354-382; 366-404 (30-40% glass fiber)	
Applications	-	aircraft, automotive, bearings, bushings, connectors, electrical/electronics, film, fuel lines, gears, medical, membranes, oil/gas, semiconductors, seals	
Outstanding properties	-	ductile, high heat resistance, flame retardant	

PAH polyanhydride

PARAMETER	UNIT	VALUE	REFERENCES
GENERAL			
Common name	-	polyanhydride	
IUPAC name	-	e.g., poly(oxydecanedioyl)	
CAS name	-	poly[oxy(1,6-dioxo-1,6-hexanediyl)] (adipic); poly[oxy(1,10-dioxo-1,10-decanediyl)] (sebacic); poly[oxy(1,9-dioxo-1,9-nonanediyl)] and 2,10-oxecanedione, homopolymer (azelaic)	
Acronym	-	PAH	
CAS number	-	26913-47-3 (sebacic polyanhydride); 26968-29-6 (adipic polyanhydride); 26968-31-0 and 27306-28-1 (azelaic polyanhydride)	
HISTORY			
Person to discover	-	Bucher and Slade; Carothers & Hill; Rosen, Wnek, Linhardt, Langer	Jain, J P; Chirkara, D; Kumar, N, Expert Opin. Drug Deliv., 5, 8, 889-907, 2008; Carothers, W H, US Patent 2,071,250, DuPont, Feb. 16, 1937.
Date	-	1909; 1930-32; 1983	
Details	-	first report on synthesis of aromatic polyanhydride (1909); aliphatic polyanhydrides were reported in 1930-1932; in 1983 polyanhydrides were studied for drug delivery	
SYNTHESIS			
Monomer(s) structure	-	acids: 5-(p-carboxyphenoxy) valeric, 8-(p-carboxyphenoxy) octanoic, 1,3-bis(p-carboxyphenoxy) propane,1,6-bis(p-carboxyphenoxy) hexane, 1,6-bis(o-carboxyphenoxy) hexane, adipic, azelaic, dodecanedicarboxylic, dodecanedioic, fumaric, isophthalic, p-carboxyphenoxy acetic, pimelic, sebacic, suberic, terephthalic; other: erucic acid dimer, ricinoleic acid maleate, ricinoleic acid succinate, 12-hydroxystearic acid succinate; photopolymerizable monomers, e.g., methacrylated sebacic acid	Goepferich, A; Tessmar, J, Adv. Drug Delivery Rev., 54, 911-32, 2002; Jain, J P; Chirkara, D; Kumar, N, Expert Opin. Drug Deliv., 5, 8, 889-907, 2008.
Monomer ratio	-	20/80 to 80/20	Kipper, M J; Hou, S-S; Seifert, S; Thiyagarajan, P; Schmidt-Rohr, K; Narashimhan, B, Macromolecules, 38, 8468-72, 2005.
Method of synthesis	-	polyanhydrides can be prepared by melt condensation polymerization in which dicarboxylic acid monomer reacts with excess of acetic anhydride	
Temperature of polymerization	°C	150-200	
Catalyst	-	cadmium acetate	
Propagation rate constant	s^{-1}	0.002-0.077	Young, J S; Gonzalea, K D; Anseth, K S, Biomaterials, 21, 1181-88, 2000.
Number average molecular weight, M_n	dalton, g/mol, amu	10,000-15,000 (microspheres)	
Mass average molecular weight, M_w	dalton, g/mol, amu	1,200-5,000 (commercial); 19,000-75,000 (micropsheres); 10,000-65,000 (experimental)	
Polydispersity, M_w/M_n	-	2-4.9 (microspheres); 1.2-2.2 (experimental)	
Radius of gyration	nm	1.51-2.27 (sebacic)	Kipper, M J; Hou, S-S; Seifert, S; Thiyagarajan, P; Schmidt-Rohr, K; Narashimhan, B, Macromolecules, 38, 8468-72, 2005.

HANDBOOK OF POLYMERS 3rd Edition, Copyrights 2022; ChemTec Publishing

PAH polyanhydride

PARAMETER	UNIT	VALUE	REFERENCES
STRUCTURE			
Crystallinity	%	40-66	Goepferich, A; Tessmar, J, Adv. Drug Delivery Rev., 54, 911-31, 2002; Mathiowitz, E; Amato, C; Dor, P; Langer, R, Polymer, 31, 547-55, 1990.
COMMERCIAL POLYMERS			
Some manufacturers	-	Chevron Phillips; MGI Pharma	
Trade names	-	PA-18; Gliadel	
SYNTHESIS			
Methods of synthesis	-	melt polycondensation, solution polymerization, dehydrative coupling, ring-opening polymerization, etc.	Ghadi, R, Muntimadugu, E, Domb, A J, Khan, W, Zhang, X, Synthetic biodegradable medical polymer: Polyanhydrides. Science and Principles of Biodegradable and Bioresorbable Medical Polymers. Woodhead Publishing, 2017, pp. 153-88,
PHYSICAL PROPERTIES			
Density at 20°C	g cm^{-3}	0.97-1.07	
Color	-	white	
Melting temperature, DSC	°C	50-90 (aliphatic); >100 (up to 240) (aromatic)	
Erosion rate		70% in 48 h (short chain aliphatic); 20% in 48 (long chain aliphatic); 5% in 17 days (aromatic)	Goepferich, A; Tessmar, J, Adv. Drug Delivery Rev., 54, 911-31, 2002.
Decomposition temperature	°C	195-322	
Storage temperature	°C	-12	deRonde, B M; Carbone, A L; Uhrich, K, Polym. Deg. Stab., 95, 1778-82, 2010.
Glass transition temperature	°C	41-65	Jaszcz, K; Lukaszczyk, J, Reactive Functional Polym., 70, 630-38, 2010.
Long term service temperature	°C	150	
Heat deflection temperature at 0.45 MPa	°C	39	
Contact angle of water, 20°C	degree	69-71.5	
MECHANICAL & RHEOLOGICAL PROPERTIES			
Tensile strength	MPa	35 (azelaic polyanhydride)	
Tensile modulus	MPa	640-1,400 (crosslinked); 800-2,100 (tricarballylic acid for orthopedic applications)	Muggli, D S; Burkoth, A K; Anseth, K S, J. Biomed. Mater. Res., 46, 271-78, 1999; Young, J S; Gonzalea, K D; Anseth, K S, Biomaterials, 21, 1181-88, 2000.
Elongation	%	14.9 (azelaic polyanhydride)	
Compressive strength	MPa	0.0018-0.121; 32-40 (crosslinked)	Muggli, D S; Burkoth, A K; Anseth, K S, J. Biomed. Mater. Res., 46, 271-78, 1999.
Young's modulus	MPa	1.3	Gunatillake, P; Mayadunne, R; Adhikari, R, Biotech. Ann. Rev., 12, 301-47, 2006.
Shore D hardness	-	65-75 (azelaic polyanhydride)	

PAH polyanhydride

PARAMETER	UNIT	VALUE	REFERENCES
Intrinsic viscosity, 25°C	dl g^{-1}	0.3	
Melt viscosity, shear rate=1000 s^{-1}, 60°C	Pa s	800-1,000	
CHEMICAL RESISTANCE			
Alcohols	-	poor (reaction of esterification)	
Alkalis	-	poor	
Aromatic hydrocarbons	-	poor	
Esters	-	poor	
Halogenated hydrocarbons	-	poor	
Ketones	-	poor	
Good solvent	-	acetone, benzene, carbon tetrachloride, 1,2-dichloroethane, ethyl acetate, MIK, sodium and potassium hydroxides aqueous solutions	
Non-solvent	-	ethanol, methanol	
FLAMMABILITY			
Ignition temperature	°C	104-302	
BIODEGRADATION			
Typical biodegradants	-	hydrolysis of anhydride linkage causes surface erosion degradation	Williams, D F; Zhong, S P, Int. Biodet. Biodeg., 34, 2, 95-130, 1994; Lucas, N; Bienaime, C; Belloy, C; Queneudec, M; Silvestre, F; Nava-Saucedo, J-E, Chemosphere, 73, 429-42, 2008.
Stabilizers	-	generally not required in this biocompatible polymer	
TOXICITY			
NFPA: Health, Flammability, Reactivity rating	-	1-3/0-1/0	
Carcinogenic effect	-	not listed by ACGIH, NIOSH, NTP	
Oral rat, LD$_{50}$	mg kg^{-1}	>8,000	
Skin rabbit, LD$_{50}$	mg kg^{-1}	>2,000	
ENVIRONMENTAL IMPACT			
Aquatic toxicity, *Daphnia magna*, LC$_{50}$, 48 h	mg l^{-1}	>100	
PROCESSING			
Typical processing methods	-	coating, compounding, forming, spraying	
Applications	-	adjuvants, bone replacement, cancer vaccines, controlled drug delivery, corrosion protection, chemotherapy, curing agent, implants (e.g., antibiotic delivery), microspheres (protein delivery), paper, vaccine delivery	
Outstanding properties	-	biocompatible, bioerodible, easily metabolized	

PAH polyanhydride

PARAMETER	UNIT	VALUE	REFERENCES
BLENDS			
Suitable polymers	-	PEG, PLA	
ANALYSIS			
FTIR (wavenumber-assignment)	cm^{-1}/-	C=C – 1640; anhydride double peak – 1810 and 1740	Young, J S; Gonzalea, K D; Anseth, K S, Biomaterials, 21, 1181-88, 2000.
Raman (wavenumber-assignment)	cm^{-1}/-	PSA – 1739, 1803; P(CPP) – 1712, 1764	Kumar, N; Langer, R S; Domb, A J; Adv. Drug Delivery Rev., 54, 889-910, 2002.
x-ray diffraction peaks	degree	17, 18-28 (four peaks)	Gopferich, A, Biomaterials, 18, 397-403, 1997.

PAI poly(amide imide)

PARAMETER	UNIT	VALUE	REFERENCES
GENERAL			
Common name	-	poly(amide imide)	
ACS name	-	5-isobenzofurancarboxylic acid, 1,3-dihydro-1,3-dioxo-, polymer with 1,3-benzenediamine and 4,4'-oxybis[benzenamine] (Torlon)	
Acronym	-	PAI	
CAS number	-	42955-03-3; 914797-27-6	
Formula (Torlon)			
The most relevant sources of additional data			Torlon, Polyamide-imide, Design guide, Solvay Advanced Polymers, 2003, T-50246
HISTORY			
Person to discover	-	George, N J	George, N J, US Patent 3,554,984, Jan. 12, 1971.
Date	-	1971 (filed 1965)	
Details	-	PAI resin obtained from trimellitic anhydride and diamines	
SYNTHESIS			
Monomer(s) structure	-		Robertson, G P; Guiver, M D; Yoshikawa, M; Brownstein, S, Polymer, 45, 1111-17, 2004.
Monomer(s) CAS number(s)	-	1204-28-0; 101-80-4; 2479-46-1	
Monomer(s) molecular weight(s)	dalton, g/mol, amu	210.57; 200.24; 292.34	
Monomer ratio	-	1:0.7:0.3	Robertson, G P; Guiver, M D; Yoshikawa, M; Brownstein, S, Polymer, 45, 1111-17, 2004.
Amide content	%	10.52-16.25	Bai, L, Zhai, L, He, M, Wang, C, Mo, S, Fan, L, Polymer, 141, 155-64, 2019.
Formulation example	-	the above set of monomers is used in the production of Torlon 4000	Robertson, G P; Guiver, M D; Yoshikawa, M; Brownstein, S, Polymer, 45, 1111-17, 2004
Method of synthesis	-	polymer can be obtained by reacting the trimellitic anhydride chloride with m-phenylenediamine and 4,4'-oxydianiline, followed by dehydratation; other methods include isocyanate route and direct polymerization	Robertson, G P; Guiver, M D; Yoshikawa, M; Brownstein, S, Polymer, 45, 1111-17, 2004; Fink, J K, High Performance Polymers, William Andrew, 2008.
Yield	%	>95	Liaw, D-J; Chen, W-H, Polym. Deg. Stab., 91, 8, 1731-39, 2006.
Number average molecular weight, M_n	dalton, g/mol, amu	18,400-86,000	
Mass average molecular weight, M_w	dalton, g/mol, amu	20,100-220,000	
Polydispersity, M_w/M_n	-	1.74-2.56	Liaw, D-J; Chen, W-H, Polym. Deg. Stab., 91, 8, 1731-39, 2006.
Molecular cross-sectional area, calculated	$cm^2 \times 10^{-16}$	18.6	

282   **HANDBOOK OF POLYMERS** 3rd Edition, Copyrights 2022; ChemTec Publishing

PAI poly(amide imide)

PARAMETER	UNIT	VALUE	REFERENCES
Radius of gyration	nm	1.49-3.02	Chen, Y; Liu, Q L; Zhu, A M; Zhang, Q G; Wu, J W, J. Membrane Sci., 348, 204-12, 2010.

COMMERCIAL POLYMERS

Some manufacturers	-	Solvay	
Trade names	-	Torlon	

PHYSICAL PROPERTIES

PARAMETER	UNIT	VALUE	REFERENCES
Density at 20°C	g cm^{-3}	1.38-1.42; 1.61 (30% glass fiber); 1.48-1.49 (30% carbon fiber); 1.47 (graphite)	
pH	-	4.5	Wolff, M F H; Antonyuk, S; Heinrich, S; Schneider, G A, Particuology, 17, 92-96, 2014
Refractive index, 20°C	-	1.656	Bryce, R M; Nguyen, H T; Clement, T; Haugen, C J; Tykwinski, R R; DeCorby, R G; McMullin, J N, Thin Solid Films, 458, 233-36, 2004.
Melting temperature, DSC	°C	357	
Thermal expansion coefficient, 23-80°C	°C^{-1}	1.7-3.1E-5; 0.16E-4 (30% glass fiber); 5E-4 (30% carbon fiber); 1.4E-5 (graphite)	
Thermal conductivity, melt	W m^{-1} K^{-1}	0.259; 0.36 (30% glass fiber); 0.518 (30% carbon fiber); 0.533 (graphite)	
Glass transition temperature	°C	206-326; 342-412	Liaw, D-J; Hsu, P-N; Chen, W-H; Lin, S-L, Macromolecules, 35, 12, 4669-76, 2002; Chen, Y; Liu, Q L; Zhu, A M; Zhang, Q G; Wu, J W, J. Membrane Sci., 348, 204-12, 2010; Bai, L, Zhai, L, He, M, Wang, C, Mo, S, Fan, L, Polymer, 141, 155-64, 2019.
Specific heat capacity	J K^{-1} kg^{-1}	1013; 959 (30% glass fiber); 963 (30% carbon fiber); 1005 (graphite)	
Maximum service temperature	°C	400	
Long term service temperature	°C	-150 to 260	
Temperature index (50% tensile strength loss after 20,000 h/5000 h)	°C	200	Padey, D; Walling, J; Wood A, Polymers in Defence and Aerospace 2007, Rapra, 2007, paper 15.
Heat deflection temperature at 1.8 MPa	°C	278; 282 (30% glass fiber); 282 (30% carbon fiber); 279 (graphite)	
Zeta potential	mV	30	Wolff, M F H; Antonyuk, S; Heinrich, S; Schneider, G A, Particuology, 17, 92-96, 2014
Dielectric constant at 100 Hz/1 MHz	-	4.2/3.9; 4.4/4.2 (30% glass fiber); 6.0/5.4 (graphite)	
Dissipation factor at 1000 Hz		0.026; 0.022 (30% glass fiber); 0.037 (graphite)	
Dissipation factor at 1 MHz		0.031; 0.05 (30% glass fiber); 0.042 (graphite)	
Volume resistivity	ohm-m	2E15, (30% glass fiber); 8E13 (graphite)	
Surface resistivity	ohm	5E18; 1E18 (30% glass fiber); 8E17 (graphite)	
Electric strength K20/P50, d=0.60.8 mm	kV mm^{-1}	23; 33 (30% glass fiber)	

PAI poly(amide imide)

PARAMETER	UNIT	VALUE	REFERENCES
MECHANICAL & RHEOLOGICAL PROPERTIES			
Tensile strength	MPa	147-192; 205-221 (30% glass fiber); 203 (30% carbon fiber); 163 (graphite); 280 (based on the diamine N,N'-(1,4-phenylene)bis(4-aminobenzamide)0	Bai, L, Zhai, L, He, M, Wang, C, Mo, S, Fan, L, Polymer, 141, 155-64, 2019.
Tensile modulus	MPa	4,480-4,900; 9,360-14,500 (30% glass fiber); 22,220 (30% carbon fiber); 6,555 (graphite)	
Elongation	%	15-35; 2.3-7 (30% glass fiber)	
Flexural strength	MPa	196-241; 333 (30% glass fiber); 350 (30% carbon fiber); 215 (graphite)	
Flexural modulus	MPa	3620-5030; 11,730 (30% glass fiber); 16,560 (30% carbon fiber); 6,900 (graphite)	
Compressive strength	MPa	117-221; 264 (30% glass fiber); 255 (30% carbon fiber); 166 (graphite)	
Izod impact strength, unnotched, 23°C	J m^{-1}	1068-1100; 507-530 (30% glass fiber); 320-342 (30% carbon fiber); 406 (graphite)	
Izod impact strength, notched, 23°C	J m^{-1}	138-144; 80 (30% glass fiber); 48 (30% carbon fiber); 64 (graphite)	
Poisson's ratio	-	0.45; 0.43 (30% glass fiber); 0.39 (30% carbon fiber); 0.39 (graphite)	
Rockwell hardness	E	86-127; 94 (30% glass fiber); 72 (graphite)	
Shrinkage	%	0.6-0.85; 0.1-0.25 (30% glass fiber); 0.0-0.15 (30% carbon fiber)	
Intrinsic viscosity, 25°C	dl g^{-1}	0.83-1.51	Liaw, D-J; Chen, W-H, Polym. Deg. Stab., 91, 8, 1731-39, 2006.
Water absorption, equilibrium in water at 23°C	%	0.33; 0.24 (30% glass fiber); 0.26 (30% carbon fiber); 0.28 (graphite)	
Moisture absorption, equilibrium 23°C/50% RH	%	1.6-2.5	
CHEMICAL RESISTANCE			
Acid dilute/concentrated	-	excellent to poor	
Alcohols	-	excellent to poor	
Alkalis	-	poor	
Aliphatic hydrocarbons	-	excellent	
Aromatic hydrocarbons	-	excellent	
Esters	-	excellent	
Greases & oils	-	excellent	
Halogenated hydrocarbons	-	excellent	
Ketones	-	excellent	
Non-solvent	-	hot NH$_4$OH	
FLAMMABILITY			
Ignition temperature	°C	570	
Autoignition temperature	°C	620	
Limiting oxygen index	% O$_2$	39.5-45; 51 (30% glass fiber); 52 (30% carbon fiber); 44 (graphite)	Abdolmaleki, A; Mallakpour, S; Rostami, M, Prog. Org. Coat., 80, 71-6, 2015.

PAI poly(amide imide)

PARAMETER	UNIT	VALUE	REFERENCES
NBS smoke chamber, minimum light transmittance	%, smoldering/flaming	92/6; 96/56 (30% glass fiber); 95/28 (30% carbon fiber)	
Char at 500°C	%	53.6-54.9	Lyon, R E; Walters, R N, J. Anal. Appl. Pyrolysis, 71, 27-46, 2004.
Heat of combustion	$J\ g^{-1}$	24,970	Walters, R N; Hacket, S M; Lyon, R E, Fire Mater., 24, 5, 245-52, 2000.
Volatile products of combustion	-	CO_2, CO	
UL 94 rating	-	94 V-0	

WEATHER STABILITY

PARAMETER	UNIT	VALUE	REFERENCES
UV absorption maximum	nm	296-324	Behniafar, H; Beit-Saeed, A; Hadian, A, Polym. Deg. Stab., 94, 1991-98, 2009.
Weather-O-Meter exposure	10,000 h in carbon arc	tensile strength - 93% retention, elongation - 100% retention	

TOXICITY

PARAMETER	UNIT	VALUE	REFERENCES
NFPA: Health, Flammability, Reactivity rating	-	0-1/1/0	
Carcinogenic effect	-	not listed by ACGIH, NIOSH, NTP	

PROCESSING

PARAMETER	UNIT	VALUE	REFERENCES
Typical processing methods	-	coating, compression molding, extrusion, injection molding, lamination, machining	
Preprocess drying: temperature/time/residual moisture	°C/h/%	177/3/0.05	
Processing temperature	°C	304-371 (injection molding)	
Processing pressure	MPa	6.89 (back pressure)	
Additives used in final products	-	Fillers: carbon fiber, glass fiber, graphene, graphite, molybdenium dioxide, MWCNT, PTFE powder (0.5%), silica, talc, TiO_2 (3%), zinc oxide	
Applications	-	automotive and aircraft parts, bonding tapes, bushings, business equipment, chip-on-film, compressor valve plates, electrical, electronic bearings, fasteners, gears, hollow fiber membranes, magnet wire coating, membranes, metal compressors in aerospace applications, optoelectronics, piston rings and seals, plastic engine, pump housings, space shuttle, thermal transfer sheet for printers, wear pads	
Outstanding properties	-	performance temperature, high strength, wear resistance, thermal stability, low expansion coefficient, and resistance to automotive and aviation fluids	

BLENDS

PARAMETER	UNIT	VALUE	REFERENCES
Suitable polymers	-	PAEK, PI, PVAI, PVP	

ANALYSIS

PARAMETER	UNIT	VALUE	REFERENCES
FTIR (wavenumber-assignment)	cm^{-1}/-	imide − 1778; C=O − 1717; imide ring − 1109, 725	Setiawan, L; Wang, R; Li, K; Fane, A G, J. Membrane Sci., 369, 196-205, 2011.
NMR (chemical shifts)	ppm	benzene ring − 8.11-8.25 and 8.50-8.70	Robertson, G P; Guiver, M D; Yoshikawa, M; Brownstein, S, Polymer, 45, 1111-17, 2004.

Palg alginic acid

PARAMETER	UNIT	VALUE	REFERENCES
GENERAL			
Common name	-	alginic acid; poly(1,4-α,L-guluronic-co-1,4-β-D-mannuronic acid)	
Acronym	-	Palg	
CAS number	-	9005-32-7	
EC number	-	232-680-1	
RTECS number	-	AZ577500	
Linear formula	-	$(C_6H_8O_6)_n$	
SYNTHESIS			
Monomer(s) structure	-	β-D-mannuronic acid and α-L-guluronic acid	
Monomer ratio	-	G:M=39/61	Tolentino, A; Alla, A; Martinez de Ilarduya, A; Munoz Guerra, S, Carbohydrate Polym., in press, 2011.
Method of synthesis	-	natural polysaccharide; alginates are extracted from brown algae by firstly converting them to the Na salt by addition of Na_2CO_3. Na alginate is water soluble and extracted from the remaining residue. The soluble Na alginate is then precipitated by addition of an acid which forms alginic acid as a soft gel. The alginic acid is then removed by extraction with alcohol	Ross, A B; Hall, C; Anastasakis, K; Westwood, A; Jones, J M; Crewe, R J, J. Anal. Appl. Pyrolysis, 91, 344-51, 2011.
Mass average molecular weight, M_w	dalton, g/mol, amu	32,000-240,000	Tolentino, A; Alla, A; Martinez de Ilarduya, A; Munoz Guerra, S, Carbohydrate Polym., in press, 2011.
STRUCTURE			
Fibril diameter	nm	6.0	Robitzer, M; Di Renzo, F; Quignard, F, Microporous Mesoporous Mater., 140, 9-16, 2011.
Lamellar spacing	nm	3.5-4.2	
Conformation	-	twofold helical conformation. The conformations of the two molecules are stabilized by intra-molecular hydrogen bonds	Guo, X, Wang, Y, Qin, Y, Shen, P, Peng, Q, Int. J. Biol. Macromol., 162, 618-28, 2020.
PHYSICAL PROPERTIES			
Density at 20°C	g cm⁻³	1.79	Robitzer, M; Di Renzo, F; Quignard, F, Microporous Mesoporous Mater., 140, 9-16, 2011.
Color	-	white to light yellow	
pH	-	1.5-3.5	
Refractive index, 20°C	-	1.595-1.69	
MECHANICAL & RHEOLOGICAL PROPERTIES			
Tensile strength	MPa	200 (fiber)	Sa, V; V; Kornev, K G, Carbon, 49, 1859-68, 2011.
Tensile modulus	MPa	3,620 (fiber)	Sa, V; V; Kornev, K G, Carbon, 49, 1859-68, 2011.
Elongation	%	16	
FLAMMABILITY			
Volatile products of combustion	-	CO_2, CO, furfural, phenol, and more	Ross, A B; Hall, C; Anastasakis, K; Westwood, A; Jones, J M; Crewe, R J, J. Anal. Appl. Pyrolysis, 91, 344-51, 2011.

Palg alginic acid

PARAMETER	UNIT	VALUE	REFERENCES
TOXICITY			
Carcinogenic effect	-	not listed by ACGIH, NIOSH, NTP	
PROCESSING			
Typical processing methods	-	compounding, spraying, spinning	
Applications	-	dental health, drug delivery, excipient, hydrocolloids, wound dressing	
Outstanding properties	-	biocompatibility, swelling, antianaphylaxis, immunomodulatory activity, antioxidant activity, anti-inflammatory effect	Guo, X, Wang, Y, Qin, Y, Shen, P, Peng, Q, Int. J. Biol. Macromol., 162, 618-28, 2020.
BLENDS			
Suitable polymers	-	polyethylenimine, polylysine; poly(galacturonic acid); PVOH	
ANALYSIS			
FTIR (wavenumber-assignment)	cm^{-1}/-	C-O – 947.9; C-H – 878.1; mannuronic acid residue – 817.1	Gomez-Ordonez, E; Ruperez, P, Food Hydrocolloids, 25, 1514-20, 2011.

PAN polyacrylonitrile

PARAMETER	UNIT	VALUE	REFERENCES
GENERAL			
Common name	-	polyacrylonitrile	
IUPAC name	-	polyacrylonitrile	
ACS name	-	2-propenenitrile, homopolymer	
Acronym	-	PAN	
CAS number	-	25014-41-9; 63908-52-1	
EC number	-	not available	
RTECS number	-	AT6977900	
Linear formula	-	$(C_3H_3N)_n$	
HISTORY			
Person to discover	-	Herbert Rein (1920, first synthesis); 1942, spinning from dimethylformamide (DuPont chemist Ray Houtz in 1942, used dimethylacetamide for production of spinning solution, which was the base of Orlon production in DuPont)	
Date	-	1920	
Details	-	The name Orlon® has been trademarked by the DuPont company, discovered by a scientist (Ray C. Houtz) working with rayon. Production of the trademarked material began in 1950.	
SYNTHESIS			
Monomer(s) structure	-	CH_2=CHCN	
Monomer(s) CAS number(s)	-	107-13-1	
Monomer(s) molecular weight(s)	dalton, g/mol, amu	53.06	
Monomer(s) expected purity(ies)	%	min 99	
Monomer content	%	85-100; 100% – pure PAN; 100-85% – Dralon, Orlon, Acrilan; <85 – modacrylics	
Comonomers		acrylic acid methylester, itaconic acid, methacrylic acid ethylester, styrene, vinyl acetate, vinyl chloride	
Method of synthesis	-	chain growth polymerization (the chain-growth reaction occurs in several steps, including initiation, propagation, and termination). Acrylic fiber is commercially produced by free radical polymerization, initiated by a redox system. Industrial production of polyacrylonitrile is a variant of aqueous dispersion polymerization, which takes place in homogenous phase under isothermal conditions; anionic polymerization	Long, T E; McGrath, J E; Turner, S R, Encyclopedia of Physical Science and Technology, Elsevier, 2004, 751-74; Atasoy, I; Yuceer, M; Berber, R, Comp. Aided Chem. Eng., 21, 1617-22, 2006; Shi, X, Jiang, J, Chin. Chem. Lett., 30, 2, 473-9, 2019.
Temperature of polymerization	°C	50 (emulsion polymerization)	
Yield	%	>90% (emulsion polymerization)	
Activation energy of polymerization	J g^{-1}	426 (water), 292 (DMF)	Dainton, F S; Eaton, R S, J. Polym. Sci., 39, 313, 1959; Rabel, W, Ueberreiter, Ber. Bunsenges., 67, 710, 1963
Heat of polymerization	J g^{-1}	1450	Joshi, R M, J. Polym. Sci., 56, 313, 1962.
Number average molecular weight, M_n	dalton, g/mol, amu	30,000-40,000	
Mass average molecular weight, M_w	dalton, g/mol, amu	30,000-230,000; 80,000-100,000 (fiber); 150,000; 1,020,000-1,230,000	Long, T E; McGrath, J E; Turner, S R, Encyclopedia of Physical Science and Technology, Elsevier, 2004, 751-74; Cai, T, Ni, T, Yang, Y, Bi, E, Xue, S, Polymer, 196, 122486, 2020; Shi, X, Jiang, J, Chin. Chem. Lett., 30, 2, 473-9, 2019.

PAN polyacrylonitrile

PARAMETER	UNIT	VALUE	REFERENCES
Polydispersity, M_w/M_n	-	1.1-3.5; 1.9-2.2	-; Shi, X, Jiang, J, Chin. Chem. Lett., 30, 2, 473-9, 2019.
Polymerization degree (number of monomer units)	-	700	
Molar volume at 298K	cm^3 mol^{-1}	calc.=45.1-47.7; 41.5 (crystalline); exp.=41.8-44.8	
Van der Waals volume	cm^3 mol^{-1}	32.2; 30.7 (crystalline)	
Radius of gyration	nm	1.6-2.3	Karanikas, S; Economou, I G, Eur. Polym. J., 47, 735-45, 2011.
End-to-end distance of unperturbed polymer chain	nm	3.1-5.4	Karanikas, S; Economou, I G, Eur. Polym. J., 47, 735-45, 2011.

STRUCTURE

PARAMETER	UNIT	VALUE	REFERENCES
Crystallinity	%	18.5-45; 60 (fiber)	Qian, B; Lin, W, J. Polym. Eng., 15, 327, 1995; Esrafilzadeh, D; Morshed, M; Tavanai, H, Synthetic Metals, 159, 267-72, 2009; Jung, B; Yoon, J K; Kim, B; Rhee, H-W, J. Membrane Sci., 246, 67-76, 2005.
Crystalline structure	-	orthorhombic; pseudo-hexagonal; hexagonal	
Cell dimensions	nm	a:b:c=1.055:0.58:0.508; a:b:c=2.148:1.155:0.7096; a:b=1.036:0.598 (hexagonal)	Kobayashi, H J, J. Polym. Sci., B1, 209, 1963; Colvin, B G; Storr, P, Eur. Polym. J., 10, 337, 1974; Allen, R A; Ward, I M; Bashir, Z, Polymer, 35, 10, 2063-71, 1994
Unit cell angles	degree	$\alpha:\beta:\gamma$=90:90:90; $\alpha:\beta:\gamma$=90:90:90	
Crystallite size	nm	4.35-7.8; 30.3 (fiber)	Qian, B; Lin, W, J. Polym. Eng., 15, 327, 1995; Esrafilzadeh, D; Morshed, M; Tavanai, H, Synthetic Metals, 159, 267-72, 2009.
Polymorphs	-	hexagonal, orthorhombic	
Tacticity	%	isotactic: 25-29, heterotactic: 47-51, and syndiotactic: 22-27 in radical polymerization; syndiotactic (mainly orthorhombic), isotactic (tetragonal)	
Chain conformation	-	helix	
Dipole moment	Debye	3.5 (can be controlled by molecular conformation of polyacrylonitrile by using electric field polarization technology)	Cai, T, Ni, T, Yang, Y, Bi, E, Xue, S, Polymer, 196, 122486, 2020.
Entanglement molecular weight	dalton, g/mol, amu	calculated: 1,412-3,486	
Lamellae thickness	nm	11-13	Gohil, R M; Patel, K C; Patel, R D, Polymer, 15, 402-6, 1974.
Heat of crystallization	$kJ\ kg^{-1}$	22.8	
Rapid crystallization temperature	°C	110-130	
Avrami constants, k/n	-	1.634-1.648	Esrafilzadeh, D; Morshed, M; Tavanai, H, Synthetic Metals, 159, 267-72, 2009.

COMMERCIAL POLYMERS

PARAMETER	UNIT	VALUE	REFERENCES
Some manufacturers	-	Lenzing Plastics, Montefibre	
Trade names	-	Dolanit, Leacril	

PHYSICAL PROPERTIES

PARAMETER	UNIT	VALUE	REFERENCES
Density at 20°C	$g\ cm^{-3}$	1.184 (1.04-1.31), calculated: 1.112-1.177	
Color	-	white	

PAN polyacrylonitrile

PARAMETER	UNIT	VALUE	REFERENCES
Refractive index, 20°C	-	1.514-1.52	
Molar polarizability	cm^3 x 10^{-25}	calculated: 19.14-22.51	
Odor	-	odorless	
Melting temperature, DSC	°C	317 (297-341)	
Temperatures of carbonization	°C	20-300 - stretching and oxidation; 1000 - carbonization; 1500-3000 ordering and orientation	Rahaman, M S A; Ismail, A F; Mustafa, A, Polym. Deg. Stab., 92, 1421-32, 2007.
Thermal expansion coefficient, 23-80°C	$°C^{-1}$	0.3-2E-4	
Thermal conductivity, melt	$W\ m^{-1}\ K^{-1}$	0.272	Knappe, W; Lohe, P; Wutschig, R, Angew. Makromol. Chem., 7, 181-193, 1969.
Glass transition temperature	°C	84-85 (43.8-105); calculated: 89-148; 100	-; -; Agdebola, T A, Agboola, O, Fayomi, O S I, Results Eng., 7, 100144, 2020.
Specific heat capacity	$J\ K^{-1}\ mol^{-1}$	0.0688	
Heat of fusion	$kJ\ g^{-1}$	0.042-5.021	
Maximum service temperature	°C	140	
Continuous operation temperature	°C	130	Agdebola, T A, Agboola, O, Fayomi, O S I, Results Eng., 7, 100144, 2020.
Decomposition temperature	°C	175	Agdebola, T A, Agboola, O, Fayomi, O S I, Results Eng., 7, 100144, 2020.
Hansen solubility parameters, δ_D, δ_P, δ_H	$MPa^{0.5}$	21.7, 14.1, 9.1	
Interaction radius		10.9	
Hildebrand solubility parameter	$MPa^{0.5}$	calc.=26.2; exp.=25.27-27.4	
Surface tension	$mN\ m^{-1}$	calc.= 42.5-61.1; exp.= 44-50	Lee, H-L, J Appl. Polym. Sci., 12, 719, 1968; Wu, S, Polymer Interface and Adhesion, Marcel Dekker, New York, 1982, p. 87.
Dielectric constant	-	calc.=3.01-3.99; exp.=2.87-4.00/4.2	
Dissipation factor at 100 Hz		0.113	
Dissipation factor at 1 MHz		0.033	
Volume resistivity	ohm-m	1E11	
Surface resistivity	ohm	5.5E+7	
Electric strength K20/P50, d=0.60.8 mm	$kV\ mm^{-1}$	3.6E+8 to 4E+9	
Permeability to oxygen, 25°C	$cm^3\ cm\ cm^{-2}\ s^{-1}\ Pa^{-1}$ x 10^{12}	0.000015-0.00041	
Permeability to water vapor, 25°C	$cm^3\ cm\ cm^{-2}\ s^{-1}\ Pa^{-1}$ x 10^{12}	23-49	
Surface free energy	$mJ\ m^{-2}$	46.8	

HANDBOOK OF POLYMERS 3rd Edition, Copyrights 2022; ChemTec Publishing

PAN polyacrylonitrile

PARAMETER	UNIT	VALUE	REFERENCES
MECHANICAL & RHEOLOGICAL PROPERTIES			
Tensile strength	MPa	19-72; 133-480 (fiber); 115-117 (film)	Lenzing Plastics; Hu, Y; Sun, T; Wang, H; Wu, D, J. Appl. Polym. Sci., 114, 3668-72, 2009.
Tensile modulus	MPa	3,700	
Elongation	%	15-35	Lenzing Plastics, Montefibre
Elastic modulus	MPa	5,670 (fiber)	Lenzing Plastics
Tenacity (fiber) (standard atmosphere)	cN tex^{-1} (daN mm^{-2})	35-45 (40-55)	Fourne, F, Synthetic Fibers. Machines and Equipment Manufacture, Properties. Carl Hanser Verlag, 1999.
Tenacity (wet fiber, as % of dry strength)	%	80-95	Fourne, F, Synthetic Fibers. Machines and Equipment Manufacture, Properties. Carl Hanser Verlag, 1999.
Fineness of fiber (titer)	dtex	0.6-25	Fourne, F, Synthetic Fibers. Machines and Equipment Manufacture, Properties. Carl Hanser Verlag, 1999.
Length (elemental fiber)	mm	38-200	Fourne, F, Synthetic Fibers. Machines and Equipment Manufacture, Properties. Carl Hanser Verlag, 1999.
Shrinkage	%	14-22; 5.1 (fiber in boiling water)	Wu, G; Lu, C; Wu, X; Zhang, S; He, F; Ling, L, J. Appl. Polym. Sci., 94, 1705-9, 2004.
Melt viscosity, shear rate=1000 s^{-1}	Pa s	170	
Water absorption, equilibrium in water at 23°C	%	1.0-1.5	
CHEMICAL RESISTANCE			
Acid dilute	-	good	
Alcohols	-	very good	
Alkalis (weak)	-	good	
Aliphatic hydrocarbons	-	good	
Aromatic hydrocarbons	-	good	
Esters	-	poor	
Halogenated hydrocarbons	-	poor	
Θ solvent	-	ethyl carbonate	
Good solvent	-	γ-butyrolactone, chloroacetonitrile, dioxanone, DMA, DMF, DMSO, dimethyl phosphite, dimethyl sulfone, ethylene carbonate, nitric acid, sulfuric acid	Iovleva, M M; Smirnova, V N; Budnitskii, G A, Fibre Chem., 33, 4, 262-64, 2001.
Non-solvent	-	acetonitrile, alcohols, chlorinated hydrocarbons, diethyl ether, formamide, hydrocarbons, ketones, methanol	
FLAMMABILITY			
Ignition temperature	$^{\circ}$C	480	
Autoignition temperature	$^{\circ}$C	560	
Limiting oxygen index	% O$_2$	18-18.2	Crook, V; Ebdon, J; Hunt, B; Joesph, P; Wyman, P, Polym. Deg. Stab., 95, 2260-68, 2010.
Peak of heat release rate	kW m^{-2}	587.8	Kil, H-S, Lee, S, Compos. Part B: Eng., 178, 107458, 2019.
Total heat release	MJ m^{-2}	35.4	Kil, H-S, Lee, S, Compos. Part B: Eng., 178, 107458, 2019.

PAN polyacrylonitrile

PARAMETER	UNIT	VALUE	REFERENCES
Char at 500°C	%	31.5; 53 (nitrogen)	Yang, C Q; He, Q; Lyon, R E; Hu, Y, Polym. Deg. Stab., 95, 108-15, 2010; Crook, V; Ebdon, J; Hunt, B; Joesph, P; Wyman, P, Polym. Deg. Stab., 95, 2260-68, 2010.
Volatile products of combustion	-	HCN, NH_3, H, CO, H_2O	
WEATHER STABILITY			
Spectral sensitivity	nm	270, 310	
Important initiators and accelerators	-	CO, CO_2, NH_3, H_2O	
Products of degradation	-	polyenes, imides, hydroperoxides, lactones, amides, CO, CO_2, H_2O, NH_3	
Stabilizers	-	benzophenone, benzotriazole, and benzoates	
BIODEGRADATION			
Typical biodegradants	-	considerable resistance	
TOXICITY			
NFPA: Health, Flammability, Reactivity rating	-	2/0/0	
Carcinogenic effect	-	not listed by ACGIH, NIOSH, NTP	
TLV, ACGIH	mg m^{-3}	3	
MAK/TRK	ppm	3 (Germany)	
OSHA	ppm	2	
Oral rat, LD_{50}	mg kg^{-1}	11,000; 3,000	
ENVIRONMENTAL IMPACT			
Aquatic toxicity, *Daphnia magna*, LC_{50}, 48 h	mg l^{-1}	1,000-185,000	Lithner, Ph D Thesis, University of Gothenburg, 2011.
PROCESSING			
Processing temperature	°C	160-175 (filament extrusion)	
Additives used in final products	-	Plasticizers: dibutyl phthalate, propylene carbonate, ethylene carbonate, dimethylformamide, dimethylsulfoxide, glycerin, polyethylene glycol, polypropylene glycol, vegetable and mineral oils, tributyl phosphate; Antistatics: alkylolamine salts of branched alkylbenzenesulfonic acids, polyaniline, polymeric quaternary ammonium salt	
Applications	-	concrete, composites, dielectric material, fibers, production of carbon fiber, clothing, awnings, beach umbrellas, boat covers, car tops, lithium-sulfur battery, piezoelectric films	
BLENDS			
Suitable polymers	-	gelatin, methylcellulose, PANI, PMHS, PU, PVDF-HFP, poly-indole	
ANALYSIS			
FTIR (wavenumber-assignment)	cm^{-1}/-	CN – 2240, NH_2 – 3240, NH – 3500	
NMR (chemical shifts)	ppm	rr – 27.31; mr – 26.88; mm – 26.40	Katsuraya, K; Hatanaka, K; Matsuzaki, K; Minagawa, M, Polymer, 42, 6323-26, 2001.

PAN polyacrylonitrile

PARAMETER	UNIT	VALUE	REFERENCES
x-ray diffraction peaks	degree	17 (sharp), 20-30 (amorphous), 30 (weak)	Jung, B; Yoon, J K; Kim, B; Rhee, H-W, J. Membrane Sci., 246, 67-76, 2005.

PANI polyaniline

PARAMETER	UNIT	VALUE	REFERENCES
GENERAL			
Common name	-	polyaniline	
IUPAC name	-	poly(imino-1,4-phenylene)	
ACS name	-	benzenamine, homopolymer	
Acronym	-	PANI	
CAS number	-	25233-30-1	
HISTORY			
Person to discover	-	C J Fritzsche	
Date	-	1841	
Details	-	treating indigo with potassium hydroxide obtained oil which he named aniline; he then oxidized it to PANI	
SYNTHESIS			
Monomer(s) structure	-		
Monomer(s) CAS number(s)	-	62-53-3	
Monomer(s) molecular weight(s)	dalton, g/mol, amu	93.13	
Monomer(s) expected purity(ies)	%	99.5	Zhou, Z; Wang, J; Wang, Z; Zhang, F, Mater. Lett., 65, 2311-14, 2011.
Method of synthesis	-	p-toluenesulfonic acid protonated aniline is used to make anilinium complexes; slow addition of ammonium peroxydisulfate caused formation of polyaniline in the micelles and grew to needle-like aggregates potentially useful as conductive fillers	Jung, W-H; Kim, D-Y; Lee, Y-M; McCarthy, S P, Antec, 1786-90, 2006.
Temperature of polymerization	°C	18 (M_w=29,700); 0 (M_w=122,000); -35 (M_w=166,000)	Yilmaz, F; Kucukyavuz, Z, Polym. Int., 59, 552-56, 2010.
Number average molecular weight, M_n	dalton, g/mol, amu	25,000-127,000	
Mass average molecular weight, M_w	dalton, g/mol, amu	5,000-440,000	
Polydispersity, M_w/M_n	-	2.55-3.46	
Radius of gyration	nm	20-40	
STRUCTURE	-		
Crystallinity	%	30-50	Saxena, V; Malhotra, B D, Handbook of polymers in Elecronics, Ed. Malhotra, B D, Rapra, 2002.
Cell type (lattice)	-	orthorhombic	Pouget, J P; Josefowicz, M E; Epstein, A J; Tang, X; MacDiarmid, A G, Macromolecules, 24, 779, 1991.
Cell dimensions	nm	a:b:c=0.765:0.575:1.02	Pouget, J P; Josefowicz, M E; Epstein, A J; Tang, X; MacDiarmid, A G, Macromolecules, 24, 779, 1991.
Crystallite size	nm	2.8-7.2	

294   **HANDBOOK OF POLYMERS** 3rd Edition, Copyrights 2022; ChemTec Publishing

PANI polyaniline

PARAMETER	UNIT	VALUE	REFERENCES
COMMERCIAL POLYMERS			
Some manufacturers	-	Enthone	
Trade names	-	Panipol	
Composition information	-	water dispersion; NMP, Solvesso, toluene, and xylene solutions, melt processable grade, and masterbatches	
PHYSICAL PROPERTIES			
Density at 20°C	g cm^{-3}	1.36-1.4; 1.245 (emeraldine base)	
Color	-	dark green to black (emeraldine base)	
Refractive index, 20°C	-	1.85	
Transmittance	%	97	Bae, S; Lee, J U; Park, H-s; Jung, E H; Jung, J W; Jo, W H, Solar En. Mater. Solar Cells, 130, 599-604, 2014.
Melting temperature, DSC	°C	385	
Glass transition temperature	°C	100-190	
Long term service temperature	°C	>300	
Hansen solubility parameters, δ_D, δ_P, δ_H	MPa$^{0.5}$	17.4, 8.1, 10.7 (emeraldine base), 17.0, 8.9, 13.7 (emeraldine salt), 21.1, 5.6, 7.3 (leucoemeraldine base)	Shacklette, L W; Han, C C, Mat. Res. Soc. Symp. Proc., 328, 157, 1994.
Hildebrand solubility parameter	MPa$^{0.5}$	22.2 (emeraldine base), 23.6 (emeraldine salt), 23.25 (leucoemeraldine base)	Shacklette, L W; Han, C C, Mat. Res. Soc. Symp. Proc., 328, 157, 1994.
Electric conductivity	Siemens cm^{-1}	6-300; 100,000 (theoretically possible)	MacDiarmid, A G; Zhou, Y; Feng, J; Furst, G T; Shedlow, Antec, 1563-67, 1999.
Volume resistivity	ohm-m	2.5E-3	Saxena, V; Malhotra, B D, Handbook of polymers in Elecronics, Ed. Malhotra, B D, Rapra, 2002.
Surface resistivity	ohm	>1000	
Optical absorption edge	eV	1.6	Saxena, V; Malhotra, B D, Handbook of polymers in Elecronics, Ed. Malhotra, B D, Rapra, 2002.
Contact angle of water, 20°C	degree	74	Trân, T H, Debarnot, D, Richaud, E, Polym. Testing, 81, 106187, 2020.
MECHANICAL & RHEOLOGICAL PROPERTIES			
Tensile strength	MPa	40-168 (non crosslinked); 60-430 (crosslinked)	Oh, E J; Min, Y; Wiesinger, J M; Manohar, S K; Scherr, E M; Prest, P J; MacDiarmid, A G; Epstein, A J, Synthetic Metals, 55-57, 977-82, 1993.
Tensile stress at yield	MPa	120	
Young's modulus	MPa	900 (PANI HCl); 1,300 (PANI base); 1,260-1,750 (fibers)	Valentova, H; Stejskal, J, Synthetic Metals, 160, 832-34, 2010; Wang, H-L; Zhu, Y; Valdez, J A; Mattes, B R, Conductive Polymers and Plastics in Industrial Applications, Rupprecht, L, Ed., WilliamAndrew, 1999.
Intrinsic viscosity, 25°C	dl g^{-1}	0.42-1.42	Yilmaz, F; Kucukyavuz, Z, Polym. Int., 59, 552-56, 2010.
Charpy impact, notched	J cm^{-2}	900	

PANI polyaniline

PARAMETER	UNIT	VALUE	REFERENCES
CHEMICAL RESISTANCE			
Alcohols	-	good	
Aliphatic hydrocarbons	-	good	
Aromatic hydrocarbons	-	good	
Esters	-	good	
Halogenated hydrocarbons	-	good	
Ketones	-	good	
Good solvent	-	DMAC, DMF, DMSO, NMP (emeraldine base); DMPU (N,N'-dimethyl propylene urea)	
FLAMMABILITY			
Ignition temperature	°C	535	
Volatile products of combustion	-	CO, CO_2, NO_x	
WEATHER STABILITY			
Spectral sensitivity	nm	290, 325, 350, 400; 380-400 (photobleaching); 331 (emeraldine base)	Teo, C H; Rahman, F, Appl. Phys. A, 99, 311-16, 2010; Wang, H-L; Zhu, Y; Valdez, J A; Mattes, B R, Conductive Polymers and Plastics in Industrial Applications, Rupprecht, L, Ed., WilliamAndrew, 1999.
Activation wavelengths	nm	280	Lakshmi, G B V S; Dhillon, A; Siddiqui, A M; Zulfequar, M; Avasthi, D K, Eur. Polym. J., 45, 2873-77, 2009.
Stabilizers	-	Tinuvin 213	Teo, C H; Rahman, F, Appl. Phys. A, 99, 311-16, 2010.
TOXICITY			
NFPA: Health, Flammability, Reactivity rating	-	0/0/0-1	
Carcinogenic effect	-	not listed by ACGIH, NIOSH, NTP	
PROCESSING			
Typical processing methods	-	compounding, melt processing (PE, PP, PS, SEBS, EMA, EPDM/PP), *in situ* polymerization, spin coating, surface grafting	Perento, J, Polymers in Electronics 2007, Rapra, 2007, paper 5.
Additives used in final products	-	Fillers: carbon black, montmorillonite, multiwalled carbon nanotubes, silica; Plasticizers: hydroquinone, resorcinol, tert-butyl hydroquinone, 4-hexyl resorcinol, bisphenol-A, sulfonic acids, phosphonic acids (phenyl phosphonic acid), and aliphatic diesters of phosphoric acid (diphenyl, dioctyl and dibutyl); UV absorber: Tinuvin 213	
Applications	-	corrosion protection, EMI shielding, filler to make plastic conductive, materials having electric conductivity, montmorillonite-PANI coated, protection against static electricity, synthesis of carbon black coated by PANI (Eonomer), synthesis of hybrid filler (multiwalled carbon nanotubes+PANI particles); plastics: addition of 8-15 wt% lowers surface resistivity of plastics to 1E4-1E9 ohm; coatings and inks, solar cells, supercapacitors	Perento, J, Polymers in Electronics 2007, Rapra, 2007, paper 5.
Outstanding properties	-	chemical resistance, electrical conductivity	Perento, J, Polymers in Electronics 2007, Rapra, 2007, paper 5.

PANI polyaniline

PARAMETER	UNIT	VALUE	REFERENCES
BLENDS			
Suitable polymers	-	PA, PCL, PEO, PET, PMMA, PPy, PS, PVAl, PVC, PVDF	
ANALYSIS			
FTIR (wavenumber-assignment)	cm⁻¹/-	H-H – 3389, 3270; C-H – 3024; C-N – 1297, 1310	Ragachev, A A; Yarmolenko, M A; Xiaohong, J; Shen, R; Luchnikov, P A; Rogachev, A V, Appl. Surf. Sci., 351, 811-8, 2015.
Raman (wavenumber-assignment)	cm⁻¹/-	C-C – 1597-1603, 1563-1566; C=N – 1489	Annapoorni, M J S, Synthetic Metals, 160, 1727-32, 2010.
x-ray diffraction peaks	degree	8.9, 14.8, 20.8, 25.3, 26.7	Zhou, Z; Wang, J; Wang, Z; Zhang, F, Mater. Lett., 65, 2311-14, 2011.

PAR polyarylate

PARAMETER	UNIT	VALUE	REFERENCES
GENERAL			
Common name	-	polyarylate	
IUPAC name		poly[oxycarbonylphenylenecarbonyloxy-1,4-phenylene(1-methylethylidene)-1,4-phenylene]	
CAS name	-	poly[oxycarbonylphenylenecarbonyloxy-1,4-phenylene(1-methylethylidene)-1,4-phenylene]	
Acronym	-	PAR	
CAS number	-	39281-59-9	
HISTORY			
Person to discover	-	Korshak, V V; Vinogradova, A V; Salazkin, S N; Bereza, S V	Korshak, V V; Vinogradova, A V; Salazkin, S N; Bereza, S V, US Patent 3,480,597, Nov. 25, 1969.
Date	-	1969	
Details	-	method of producing polyarylates	
SYNTHESIS			
Monomer(s) structure	-		
Monomer(s) CAS number(s)	-	80-05-7; 100-21-0	
Monomer(s) molecular weight(s)	dalton, g/mol, amu	228.29; 166.13	
Monomer ratio	-	1:0.5:0.5 (Ardel, U-polymer)	
Method of synthesis	-	polyarylate is the polyester derived from bisphenol-A and a mixture of isophthalic and terephthalic acids by acidolysis, phenolysis, alcoholysis and esterolysis	Han, X; Padia, A B; Hall, H K, J. Polym. Sci. A, 37, 2891-97, 1999.
Temperature of polymerization	°C	260	
Time of polymerization	h	108	
Yield	%	100	
Mass average molecular weight, M_w	dalton, g/mol, amu	81,000-148,000	
COMMERCIAL POLYMERS			
Some manufacturers	-	Westlake; Unitika; Kuraray	
Trade names	-	Ardel; U-polymer; Vectran (fiber)	
PHYSICAL PROPERTIES			
Density at 20°C	g cm^{-3}	1.2-1.21	
Refractive index, 20°C	-	1.64	
Transmittance	%	87	
Haze	%	2.3	
Cutoff wavelength	nm	294 (polyarylates containing cyclohexane group)	Zhang, Y, Yan, G-m, Zhang, G, Liu, S-l, Yang, J, Polymer, 186, 122047, 2020.
Melting temperature, DSC	°C	228-252	
Thermal expansion coefficient, 23-80°C	10^{-4} °C^{-1}	0.61	
Thermal conductivity, melt	W m^{-1} K^{-1}	0.18	

PAR polyarylate

PARAMETER	UNIT	VALUE	REFERENCES
Glass transition temperature	°C	155-196	Serin, M; Sakar, D; Cankurtaran, O; Karaman Yilmaz, F, J. Optoelectronics Adv. Mater., 6, 1, 283-88, 2004.
Specific heat capacity	J K^{-1} kg^{-1}	1200	
Long term service temperature	°C	150	
Heat deflection temperature at 0.45 MPa	°C	175-180	
Heat deflection temperature at 1.8 MPa	°C	160-174	
Hildebrand solubility parameter	MPa$^{0.5}$	10.8-11.6	Serin, M; Sakar, D; Cankurtaran, O; Karaman, F, J. Optoelectronics Adv. Mater., 7, 3, 1533-38, 2005.
Dielectric constant at 100 Hz/1 MHz	-	1.9-3.3/3.0	
Relative permittivity at 100 Hz	-	2.6-3.3	
Dissipation factor at 1000 Hz	E-4	40-100	
Volume resistivity	ohm-m	2E12-2E15	
Surface resistivity	ohm	2E12	
Electric strength K20/P50, d=0.60.8 mm	kV mm^{-1}	31-106	Noniewicz, K; Brzozowski, Z K; Zadrozna, I, J. Appl. Polym. Sci., 60, 1071-82, 1996.
Arc resistance	s	127	

MECHANICAL & RHEOLOGICAL PROPERTIES

Tensile strength	MPa	48-75	
Tensile modulus	MPa	2,070-2,100	
Tensile stress at yield	MPa	69	
Elongation	%	25-65	
Tensile yield strain	%	8.4	
Flexural strength	MPa	76-83	
Flexural modulus	MPa	2,100-2,140	
Compressive strength	MPa	81-84	
Izod impact strength, notched, 23°C	J m^{-1}	260-288	
Rockwell hardness	-	R105-125	
Shrinkage	%	0.8-1.0	
Water absorption, 24h at 23°C	%	0.26	

CHEMICAL RESISTANCE

Alcohols	-	good	
Aliphatic hydrocarbons	-	good	
Aromatic hydrocarbons	-	good	
Halogenated hydrocarbons	-	poor	
Other	-	poor	
Θ solvent, Θ-temp.=25	-	dioxane:cyclohexane=68:32	
Good solvent	-	tetrachloroethane	
Non-solvent	-	THF, n-heptane	

PAR polyarylate

PARAMETER	UNIT	VALUE	REFERENCES
FLAMMABILITY			
Limiting oxygen index	% O_2	34-36	
Heat release	kJ g^{-1}	4-19	Zhang, H; Westmoreland, P R; Farrus, R J; Coughlin, E B; Plichta, A; Brzozowski, Z K, Polymer, 43, 5463-72, 2002.
Char at 500°C	%	27-41	Zhang, H; Westmoreland, P R; Farrus, R J; Coughlin, E B; Plichta, A; Brzozowski, Z K, Polymer, 43, 5463-72, 2002.
UL 94 rating	-	V-0	
WEATHER STABILITY			
Spectral sensitivity	nm	320-340	
Activation wavelengths	nm	343, 355, 370	
Important initiators and accelerators	-	30 weeks of WOM exposure without changes	Cooney, J D, Polym. Eng. Sci., 22, 8, 492-98, 1982.
Products of degradation	-	photo-Fries rearrangement producing UV absorber which protects against further degradation	
Stabilizers	-	uses own produced UV absorber. When exposed to UV light, it undergoes a molecular rearrangement resulting in the formation of a protective layer essentially serving as a UV stabilizer.	
PROCESSING			
Preprocess drying: temperature/ time/residual moisture	°C/h/%	120-140/6-8/	
Processing temperature	°C	310-360	
Processing pressure	MPa	0.981 (back)	
Applications	-	appliance parts, automotive, photolithographic emulsions, semiconductor components, snap-lock connectors, solar energy components	
Outstanding properties	-	UV stability, transparency, good electrical properties	

PARA polyamide MXD6

PARAMETER	UNIT	VALUE	REFERENCES
GENERAL			
Common name	-	polyamide MXD6, nylon MXD6, polyarylamide, hexanedioic acid polymer with 1,3-benzenedimethanamine	
ACS name	-	poly[iminomethylene-1,3-phenylenemethyleneimino(1,6-dioxo-1,6-hexanediyl)]	
Acronym	-	PARA	
CAS number	-	25805-74-7 (MXD6); 902465-02-5 (IXEF 300); 1008793-20-1 IXEF 2060)	
HISTORY			
Person to discover	-	Caldwell J R; Gilkey, R	Caldwell J R; Gilkey, R, US Patent 2,916,476, Eastman Kodak, Dec. 8, 1959.
Date	-	1959	
Details	-	polyamides of xylidenediamine and aliphatic dibasic acids	
SYNTHESIS			
Monomer(s) structure	-	H_2N —◯— NH_2 $HOOC(CH_2)_4COOH$	
Monomer(s) CAS number(s)	-	1477-55-0; 124-04-9	
Monomer(s) molecular weight(s)	dalton, g/mol, amu	136.20; 146.14	
Monomer ratio	-	0.93 (0.93:1)	
Method of synthesis	-	polycondensation of m-xylenediamine with adipic acid	
STRUCTURE			
Crystallinity	%	35; 6, 26-33 (water-crystallized)	-; Messin, T, Marais, S, Follain, N, Chappey, C, Sollogoub, C, Eur. Polym. J., 152-60, 2019.
Cell type (lattice)	-	triclinic	Ota, T; Yamashita, M; Yoshizaki, O; Nagai, E, J. Polym. Sci., A-2, 4, 959, 1966.
Cell dimensions	nm	a:b:c=1.201:0.483:2.98	Ota, T; Yamashita, M; Yoshizaki, O; Nagai, E, J. Polym. Sci., A-2, 4, 959, 1966.
Unit cell angles	degree	$\alpha{:}\beta{:}\gamma$=75:26:65	Ota, T; Yamashita, M; Yoshizaki, O; Nagai, E, J. Polym. Sci., A-2, 4, 959, 1966.
Number of chains per unit cell	-	2	
COMMERCIAL POLYMERS			
Some manufacturers	-	Mitsubishi Chemical; Solvay	
Trade names	-	Reny; Ixef	
PHYSICAL PROPERTIES			
Density at 20°C	g cm^{-3}	1.20-1.23; 1.19 (amorphous), 1.25 (crystalline); 1.43-1.77 (30-60% glass fiber); 1.34 (30% carbon fiber)	
Bulk density at 20°C	g cm^{-3}	0.6-0.8	
Refractive index, 20°C	-	1.581-1.586	
Haze	%	3.1	
Melting temperature, DSC	°C	234-240, 280 (30-60% glass fiber); 280 (30% carbon fiber)	

PARA polyamide MXD6

PARAMETER	UNIT	VALUE	REFERENCES
Thermal expansion coefficient, 23-80°C	°C^{-1}	5E-5; 1.4-1.8E-5	
Thermal conductivity, melt	W m^{-1} K^{-1}	0.38; 0.55 (50% glass fiber)	
Glass transition temperature	°C	75-85, 50	-; Messin, T, Marais, S, Follain, N, Chappey, C, Sollogoub, C, Eur. Polym. J., 152-60, 2019.
Specific heat capacity	J K^{-1} kg^{-1}	2510	
Heat of fusion	kJ mol^{-1}	37	
Heat deflection temperature at 0.45 MPa	°C	96; 237-238 (30-50% glass fiber)	
Heat deflection temperature at 1.8 MPa	°C	224-245 (30-50% glass fiber)	
Dielectric constant at 100 Hz/1 MHz	-	3.9-4.5 (30-60% glass fiber)	
Relative permittivity at 100 Hz	-	5	
Relative permittivity at 1 MHz	-	4-5	
Dissipation factor at 100 Hz	E-4	70 (30-50% glass fiber)	
Dissipation factor at 1 MHz	E-4	80-90 (30-50% glass fiber)	
Volume resistivity	ohm-m	2E13 (30-60% glass fiber)	
Surface resistivity	ohm	1E14	
Electric strength K20/P50, d=0.60.8 mm	kV mm^{-1}	24-31 (30-60% glass fiber)	
Comparative tracking index	V	>400 to 600 (30-60% glass fiber)	
Coefficient of friction	-	0.36-0.53	

MECHANICAL & RHEOLOGICAL PROPERTIES			
Tensile strength	MPa	99; 180-280 (30-60% glass fiber); 250 (30% carbon fiber)	
Tensile modulus	MPa	4,700-4,800; 11,500-24,000 (30-60% glass fiber)	
Elongation	%	1.5-2.3	
Tensile yield strain	%	1.8-2.0 (50-60% glass fiber); 1.3 (30% carbon fiber)	
Flexural strength	MPa	160; 280-400 (30-60% glass fiber)	
Flexural modulus	MPa	4,400; 11,000-21,000 (30-60% glass fiber); 2,300 (30% carbon fiber)	
Charpy impact strength, 23°C	kJ m^{-2}	35-72	
Charpy impact strength, notched, 23°C	kJ m^{-2}	6.3-11.8	
Izod impact strength, 23°C	J m^{-1}	460-850 (30-60% glass fiber); 450 (30% carbon fiber)	
Izod impact strength, notched, 23°C	J m^{-1}	20; 70-120 (30-60% glass fiber); 60 (30% carbon fiber)	
Poisson's ratio	-	0.35 (50% glass fiber)	
Rockwell hardness	-	M108	
Shrinkage	%	0.1-0.5 (30-60% glass fiber); 0.03-0.1 (30% carbon fiber)	
Melt volume flow rate (ISO 1133, procedure B), 275°C/2.16 kg	cm^3/10 min	3-31 (30-50% glass fiber)	
Melt index, 275°C/3.8 kg	g/10 min	6-41 (30-50% glass fiber)	
Water absorption, 24h at 23°C	%	0.10-0.20 (30-60% glass fiber); 0.22 (30% carbon fiber)	
Moisture absorption, equilibrium 23°C/50% RH	%	0.7-1.5	

PARA polyamide MXD6

PARAMETER	UNIT	VALUE	REFERENCES
CHEMICAL RESISTANCE			
Acid dilute/concentrated	-	non-resistant	
Aliphatic hydrocarbons	-	excellent	
Aromatic hydrocarbons	-	excellent	
Esters	-	excellent	
Greases & oils	-	excellent	
Halogenated hydrocarbons	-	good	
Ketones	-	excellent	
Good solvent	-	m-cresol, formic acid, hexafluoroisopropanol, phenol/ethanol=4/1, sulfuric acid, trifluoroacetic acid	
Non-solvent	-	n-butanol, n-heptane, water	
FLAMMABILITY			
Ignition temperature	°C	331	
Autoignition temperature	°C	385	
Limiting oxygen index	% O_2	25 (30-60% glass fiber)	
UL 94 rating	-	HB	
TOXICITY			
NFPA: Health, Flammability, Reactivity rating	-	0/1/0	
Oral rat, LD_{50}	mg kg^{-1}	>5000	
PROCESSING			
Typical processing methods	-	injection molding	
Preprocess drying: temperature/ time/residual moisture	°C/h/%	80-120/3-12/0.3	
Processing temperature	°C	250-280 (injection molding); 255-265 (film casting)	
Processing pressure	MPa	20-150 (injection)	
Additives used in final products	-	lubricants, release agents, color master batches, blowing agents, UV stabilizers	
Applications	-	automotive (mirror housings, door handles, headlamp surrounds, cam covers and clutch parts); food and water contact; small appliances (shaver heads, electric iron parts, sewing machine components and vacuum cleaner motor supports); electronics (phone housings, induction motor supports, safety switches, DVD disk supports and internal moving parts); healthcare products resistant to γ radiation	
Outstanding properties	-	gas barrier properties, recyclability, rigidity, resistance to mechanical stresses	
BLENDS			
Suitable polymers	-	PA6, PC, PET, PPTA	
ANALYSIS			
FTIR (wavenumber-assignment)	cm^{-1}/-	N-H – 3280, 1640, 1550; C-N – 1355, 1255	Seif, S; Cakmak, M, Polymer, 51, 3762-73, 2010.

PB 1,2-polybutylene

PARAMETER	UNIT	VALUE	REFERENCES
GENERAL			
Common name	-	1,2-polybutylene	
IUPAC name	-	poly(1-ethylethylene), poly(but-1-ene)	
CAS name	-	1-butene, homopolymer	
Acronym	-	PB	
CAS number	-	9003-28-5	
HISTORY			
Person to discover	-	Edwards, R W; Francis, A W; Eichenbaum, R; Ringelman, R E; Wu, W C L, 1968. Natta, G; Pino, P; Mazzanti, G, 1969. Klingensmith, G B; Higgins, T L, 1986.	Edwards, R W; Francis, A W; Eichenbaum, R; Ringelman, R E; Wu, W C L, US Patent 3,362,940, Mobil Oil Corp., Jan 9, 1968. Natta, G; Pino, P; Mazzanti, G, US Patent 3,435,017, Montecatini Edison, Mar. 25, 1969. Klingensmith, G B; Higgins, T L, US Statuatory Invention Registration H179, Dec. 2, 1986.
Date	-	1968; 1969; 1986	
Details	-	Edwards et al. patented stereoregular PB; Natta et al. patented isotactic PB; Klingensmith et al. patented elastomeric PB	
SYNTHESIS			
Monomer(s) structure	-	$H_2C=CHCH_2CH_3$	
Monomer(s) CAS number(s)	-	106-98-9	
Monomer(s) molecular weight(s)	dalton, g/mol, amu	56.1	
Method of synthesis	-	addition polymerization; polymerization occurs on contact of monomer with Zigler-Natta catalyst	
Temperature of polymerization	ºC	120	
Pressure of polymerization	MPa	7-15	
Catalyst	-	Ziegler-Natta, nickel-based	
Number average molecular weight, M_n	dalton, g/mol, amu	20,000-300,000	
Mass average molecular weight, M_w	dalton, g/mol, amu	85,000-2,200,000; 1,000,000-3,000,000	-; Fujii, T, Matsui, Y, Hirabayashi, H, Igawa, K, Yamada, K, Polym. Deg. Stab., 167, 1-9, 2019.
Polydispersity, M_w/M_n	-	4-12	
Molar volume at 298K	cm^3 mol^{-1}	calc.=64.1; 59.5 (crystalline); 65.2 (amorphous); exp.=65.2	
Van der Waals volume	cm^3 mol^{-1}	calc.=40.9-41.1 (crystalline); 40.9 (amorphous)	
Radius of gyration	nm	13.7-16.3	
STRUCTURE			
Crystallinity	%	45-55 (conventional); 47-58 (form I); 38 (form II)	Klingensmith, G B; Higgins, T L, US Statuatory Invention Registration H179, Dec. 2, 1986; Abedi, S; Sharifi-Sanjani, N, J. Appl. Polym. Sci., 78, 2533-39, 2000; Maring, D; Meurer, B; Weill, J. Polym. Sci. B, 33, 1235-47, 1995.
Cell type (lattice)	-	hexagonal, tetragonal, orthorhombic	
Cell dimensions	nm	a:b:c=1.77:1.77:0.65 (hexagonal); 1.542:1.542:2.05 (tetragonal); 1.238:0.892:0.745 (orthorhombic)	

PB 1,2-polybutylene

PARAMETER	UNIT	VALUE	REFERENCES
Number of monomers per unit cell	-	18 (hexagonal); 44 (tetragonal)	Winkel, A K; Miles, M J, Polymer, 41, 2313-17, 2000.
Polymorphs	-	I (hexagonal; three-fold helix); I' (hexagonal); II (tetragonal; four-fold helix); III (orthorhombic; 2/1 helix)	Natta, G; Corradini, P; Bassi, I W, Rend. Accad. Naz. Lincei, 19, 404, 1955; Kim, K-W; Lee, K-H; Park, J-H; Lee, D-R; Ko, J-A; Kim, H-Y, Fibers Polym., 10, 5, 667-72, 2009.
Tacticity	%	84.5-99.5 isotactic	Klingensmith, G B; Higgins, T L, US Statuatory Invention Registration H179, Dec. 2, 1986.
Chain conformation	-	I (3/1 helix, twinned); I' (3/1 helix, untwinned); II (four 11/3 or 40/11 helix); III (2/1 helix)	Kim, K-W; Lee, K-H; Park, J-H; Lee, D-R; Ko, J-A; Kim, H-Y, Fibers Polym., 10, 5, 667-72, 2009.
Entanglement molecular weight	dalton, g/mol, amu	calc.=4,344	
Lamellae thickness	nm	27-31	Samon, J M; Schultz, J M; Hsiao, B S, Macromolecules, 34, 2008-11, 2001.
Half-time of crystallization	min	form II: at 90, 95, and 100°C – 6, 28, and 70 min, respectively	Li, Y-K, Zhang, X-X, Sun, Z-Y, Polymer, 188, 122137, 2020.
Heat of crystallization	kJ kg^{-1}	33.1-38.9	
Rapid crystallization temperature	°C	72	Maring, D; Meurer, B; Weill, J. Polym. Sci. B, 33, 1235-47, 1995.
Avrami constants, k/n	-	7.1E-2 to 1.98E-3/2.43-2.69	Maring, D; Meurer, B; Weill, J. Polym. Sci. B, 33, 1235-47, 1995.

COMMERCIAL POLYMERS

Some manufacturers	-	LyondellBasell	
Trade names	-	PB	

PHYSICAL PROPERTIES

Density at 20°C	g cm^{-3}	0.87-0.92; 0.95 (crystalline)	
Refractive index, 20°C	-	calc.=1.4669-1.5012; exp.=1.5125-1.5246	
Odor	-	odorless	
Melting temperature, DSC	°C	113-126; 97-142 (isotactic); 50 (syndiotactic); 120-135 (form I); 90-100 (form I'); 110-120 (form II); 90-100 (form III)	Shieh, Y-T; Lee, M-S; Chen, S-A, Polymer, 42, 4439-48, 2001.
Decomposition temperature	°C	300-440 (form I)	
Thermal expansion coefficient, 23-80°C	°C^{-1}	1.1-6.7E-4	
Thermal conductivity, melt	W m^{-1} K^{-1}	0.1344-0.22	
Glass transition temperature	°C	calc.=-23.0 to -60.0; exp.=-17.0 to -45.0; -20.5 (form I); -26.9 (form II)	Maring, D; Meurer, B; Weill, J. Polym. Sci. B, 33, 1235-47, 1995.
Specific heat capacity	J K^{-1} kg^{-1}	2,150-2,600	
Heat of fusion	kJ mol^{-1}	63-6.5	
Long term service temperature	°C	70 (lifetime of 50 years)	
Vicat temperature VST/A/50	°C	99	
Surface tension	mN m^{-1}	calc.=34.3-36.1	
Dielectric constant at 100 Hz/1 MHz	-	2.53	
Dissipation factor at 100 Hz	E-4	0.0005	
Dissipation factor at 1 MHz	E-4	0.0005	

PB 1,2-polybutylene

PARAMETER	UNIT	VALUE	REFERENCES
Permeability to oxygen, 25°C	cm^3 mm m^{-2} s^{-1} atm^{-1} day^{-1}	160	
Permeability to water vapor, 25°C	g m^{-2} 24h	29	

MECHANICAL & RHEOLOGICAL PROPERTIES

Tensile strength	MPa	27-45	
Tensile modulus	MPa	150-295	
Tensile stress at yield	MPa	12-15	
Elongation	%	200-400	
Flexural modulus	MPa	250-450	
Elastic modulus	MPa	290-295	
Poisson's ratio	-	calc.=0.393; exp.=0.47	
Shrinkage	%	2.5-5	
Brittleness temperature (ASTM D746)	°C	-18 to -20	
Intrinsic viscosity, 25°C	dl g^{-1}	0.0178-0.039	
Melt index, 190°C/10 kg	g/10 min	1-30	
Water absorption, 24h at 23°C	%	<0.03	

CHEMICAL RESISTANCE

Acid dilute/concentrated	-	good	
Alcohols	-	good	
Alkalis	-	good	
Aliphatic hydrocarbons	-	good	
Aromatic hydrocarbons	-	poor	
Esters	-	poor	
Greases & oils	-	poor	
Halogenated hydrocarbons	-	good	
Ketones	-	poor	
Θ solvent, Θ-temp.=86	-	anisole	
Good solvent	-	benzene, chloroform, chlorobenzene, decalin, toluene	
Non-solvent	-	organic solvents	

FLAMMABILITY

Ignition temperature	°C	>121	
Volatile products of combustion	-	CO, CO$_2$, toxic fumes	

TOXICITY

NFPA: Health, Flammability, Reactivity rating	-	1/1/2	
Carcinogenic effect	-	not listed by ACGIH, NIOSH, NTP	

PROCESSING

Typical processing methods	-	blown film, casting, extrusion, injection molding, rotational molding, spinning, thermoforming	

PB 1,2-polybutylene

PARAMETER	UNIT	VALUE	REFERENCES
Additives used in final products	-	Plasticizers: dioctyl adipate, phthalate, maleate, dibutyl phthalate, isodecyl pelargonate, oleyl nitrile, mineral oil, polybutene; Slip; Antiblock	
Applications	-	film, flexible piping, hot-melt adhesives, membranes, packaging, seals, shoe soles, wire & cable	
Outstanding properties	-	creep resistance, semi-crystalline	

BLENDS

Suitable polymers	-	PE, PS, SBS	

ANALYSIS

FTIR (wavenumber-assignment)	cm^{-1}/-	form I: 925 and 810; form I': 925 and 792; form II: 900; form III: 900 and 810	Abedi, S; Sharifi-Sanjani, N, J. Appl. Polym. Sci., 78, 2533-39, 2000.
x-ray diffraction peaks	degree	form I: 9.8, 17.1, 20; form III: 12.1, 17.0, 18.5; 9.9°, 17.5°and 20.5°represents the lattice plane (200), (220) and (213) of form I, and 11.9°, 16.6°and 18.3°, corresponding to the lattice plane (200), (220) and (213) respectively, represents the characteristic peaks of form II	Kim, K-W; Lee, K-H; Park, J-H; Lee, D-R; Ko, J-A; Kim, H-Y, Fibers Polym., 10, 5, 667-72, 2009; Li, Y-K, Zhang, X-X, Sun, Z-Y, Polymer, 188, 122137, 2020.

PBA poly(p-benzamide)

PARAMETER	UNIT	VALUE	REFERENCES
GENERAL			
Common name	-	poly(p-benzamide)	
CAS name	-	poly(imino-1,4-phenylenecarbonyl); benzamide, homopolymer	
Acronym	-	PBA	
CAS number	-	24991-08-0; 55738-52-8	
HISTORY			
Person to discover	-	Kwolek, S L	US Patent 3,600,350, DuPont, Aug 17, 1971.
Date	-	1966	
Details	-	it may be obtained by the low temperature solution polymerization of p-aminobenzoyl halide salts	
SYNTHESIS			
Monomer(s) structure	-		
Monomer(s) CAS number(s)	-	150-13-0	
Monomer(s) molecular weight(s)	dalton, g/mol, amu	137.14	
Mass average molecular weight, M_w	dalton, g/mol, amu	12,000-23,500	
Polydispersity, M_w/M_n	-	1.10-1.18	Takagi, K; Nobuke, K; Nishikawa, Y; Yamakado, R, Polym. J., 45, 1171-76, 2013.
Van der Waals volume	cm^3 mol^{-1}	calc.=76.6	
STRUCTURE			
Cell type (lattice)	-	orthorhombic	
Cell dimensions	nm	a:b:c=0.775:0.530:1.287	Takahashi, Y; Ozaki, Y; Takase, M; Krigbaum, W R, J. Polym. Sci., Part B: Polym. Phys., 31, 9, 1135-43, 1993.
Number of chains per unit cell	-	1	
Interplanar spacing	nm	0.379-0.439	Shi, H; Zhao, Y; Zhang, X; Zhou, Y; Xu, Y; Zhou, S; Wang, D; Han, C C; Xu, D, Polymer, 45, 6299-6307, 2004.
Chain conformation	-	TCTC	
Avrami constants, k/n	-	1.01-1.10	Lin, J; Xi, S; Wu, H; Li, S, Eur. Polym. J., 33, 10-12, 1601-5, 1997.
PHYSICAL PROPERTIES			
Density at 20°C	g cm^{-3}	1.48	
Melting temperature, DSC	°C	>500	
Glass transition temperature	°C	230	
MECHANICAL & RHEOLOGICAL PROPERTIES			
Intrinsic viscosity, 25°C	dl g^{-1}	1.18-2.07	

PBA poly(p-benzamide)

PARAMETER	UNIT	VALUE	REFERENCES
CHEMICAL RESISTANCE			
Acid dilute/concentrated	-	poor	
Alcohols	-	good	
Aliphatic hydrocarbons	-	good	
Aromatic hydrocarbons	-	good	
Esters	-	good	
Greases & oils	-	good	
Good solvent	-	acetylpyrrolidone, diethylacetamide, dimethylacetamide, tetramethylurea	
Non-solvent	-	ethylene glycol, glycerol, water	
ANALYSIS			
x-ray diffraction peaks	degree	20.2, 23.4	Shi, H; Zhao, Y; Zhang, X; Zhou, Y; Xu, Y; Zhou, S; Wang, D; Han, C C; Xu, D, Polymer, 45, 6299-6307, 2004.

PBAA poly(butadiene-co-acrylonitrile-co-acrylic acid)

PARAMETER	UNIT	VALUE	REFERENCES
GENERAL			
Common name	-	poly(butadiene-co-acrylonitrile-co-acrylic acid)	
CAS name	-	2-propenoic acid, polymer with 1,3-butadiene and 2-propenenitrile	
Acronym	-	PBAA	
CAS number	-	25265-19-4	
RTECS number	-	RW9085000	
SYNTHESIS			
Monomer(s) structure	-	$H_2C{=}CHCH{=}CH_2$ $H_2C{=}CHC{\equiv}N$ $H_2C{=}CHCOOH$	
Monomer(s) CAS number(s)	-	106-99-0; 107-13-1; 79-10-7	
Monomer(s) molecular weight(s)	dalton, g/mol, amu	54.1; 53.06; 72.06	
Acrylonitrile content	%	Bu/Ac/Aa=62-83/13-32/4-6	Athey, R D, Prog. Org. Coat., 7, 289-329, 1979.
Mass average molecular weight, M_w	dalton, g/mol, amu	140,000-400,000	Pospisil, J; Laudat, J; Fahnrich, J; Havranek, A; Nespurek, S, Mol. Cryst. Liq. Cryst., 229, 195-201, 1993.
PHYSICAL PROPERTIES			
Density at 20°C	g cm^{-3}	0.936	
Refractive index, 20°C	-	1.5200	
Melting temperature, DSC	°C	25	
Decomposition temperature	°C	>100; 100-350 (70% mass lost)	Sell, T; Vyazovkin, S; Wight, C A, Combustion Flame, 119, 1-2, 174-81, 1999.
Activation energy for decomposition	kJ mol^{-1}	260	Sell, T; Vyazovkin, S; Wight, C A, Combustion Flame, 119, 1-2, 174-81, 1999.
Glass transition temperature	°C	-41	Pospisil, J; Laudat, J; Fahnrich, J; Havranek, A; Nespurek, S, Mol. Cryst. Liq. Cryst., 229, 195-201, 1993.
FLAMMABILITY			
Ignition temperature	°C	>110	
WEATHER STABILITY			
Activation energy of photo-oxidation	kJ mol^{-1}	70	
TOXICITY			
NFPA: Health, Flammability, Reactivity rating	-	1/1/0	
Carcinogenic effect	-	not listed by ACGIH, NIOSH, NTP	
Oral rat, LD_{50}	mg kg^{-1}	10,200	
Skin rabbit, LD_{50}	mg kg^{-1}	>2,000	

310   **HANDBOOK OF POLYMERS** 3rd Edition, Copyrights 2022; ChemTec Publishing

PBAA poly(butadiene-co-acrylonitrile-co-acrylic acid)

PARAMETER	UNIT	VALUE	REFERENCES
PROCESSING			
Applications	-	booster propellant (binder), epoxy modifier, drilling mud, solid rocket propellant, toughening agent in dental compositions	Mante, F K; Wadenya, R O; Bienstock, D A; Mendelson, J; LaFleur, E E, Dental Mater., 26, 164-8, 2010.
BLENDS			
Suitable polymers	-	PA-6, PANI, PVC	
ANALYSIS			
FTIR (wavenumber-assignment)	cm^{-1}/-	CN – 2240; C=N – 1650	Pospisil, J; Laudat, J; Fahnrich, J; Havranek, A; Nespurek, S, Mol. Cryst. Liq. Cryst., 229, 195-201, 1993.

PBD, *cis* cis-1,4-polybutadiene

PARAMETER	UNIT	VALUE	REFERENCES
GENERAL			
Common name	-	*cis*-1,4-polybutadiene	
CAS name	-	butadiene rubber, of cis-1,4-configuration	
Acronym	-	*cis*-PBD	
CAS number	-	9003-17-2	
HISTORY			
Person to discover	-	Smith, D R; Zielinski, R P	Smith, D R; Zielinski, R P, US Patent 3,976,630, Phillips Petroleum, Aug. 24, 1976.
Date	-	1976	
Details	-	emulsion polymerization	
SYNTHESIS			
Monomer(s) structure	-	$H_2C=CHCH=CH_2$	
Monomer(s) CAS number(s)	-	106-99-0	
Monomer(s) molecular weight(s)	dalton, g/mol, amu	54.09	
Monomer(s) expected purity(ies)	%	min. 90 wt% (typical 99.5 wt%)	
Monomer ratio	-	100%	
Formulation example	-	see in the method of synthesis	
Method of synthesis	-	a dry solution of butadiene in hexane (35 wt%) is first charged under nitrogen into reactor, followed by the addition of dry hexane. The reaction medium was heated up to the stated reaction temperature (50-100ºC) and catalyst was added. The monomer concentration in the reaction medium varies from 0.56 to 1.05 mol/l. Polymerization is terminated by adding a hexane solution of 2,6-di-trans-butyl-4-methmethylphenol (BHT) @ 50 wt% and stabilized by adding a hexane solution of TNPP @ 10 wt%. The polymer is coagulated in hot water under vigorous agitation and dried in an oven at 60ºC.	Pires, N M T; Coutinho, F M B; Costa, M A S, Eur. Polym. J, 40, 2599-2603, 2004.
Temperature of polymerization	ºC	70-8	Pires, N M T; Coutinho, F M B; Costa, M A S, Eur. Polym. J, 40, 2599-2603, 2004.
Time of polymerization	h	5	Pires, N M T; Coutinho, F M B; Costa, M A S, Eur. Polym. J, 40, 2599-2603, 2004.
Polymerization yield of *cis*-PBD	%	96.2	Gao, B, Li, D, Duan, Q, Polyhedron, 169, 287-90, 2019.
Pressure of polymerization	Pa	atmospheric (typical) or higher	
Optimum polymerization conditions	-	monomer concentration 22%, reaction time 3.3 h and reaction temperature 42ºC for the specified catalyst system	Maiti, M; Srivastava, V K; Shewale, S; Jasra, R V; Chavda, A; Modi, S, Chem. Eng. Sci., 107, 256-65, 2014.
Catalyst	-	triethylaluminum+titanium tetraiodate; also neodynium catalyst for ultrahigh cis-PB; cobalt complexes; optimum catalyst formulation was observed to be 1,3-butadiene/neodymium octanoate, diethylaluminum chloride/neodymium octanoate and triethylaluminum/neodymium octanoate molar ratio of 902.88, 1.74 and 27.51, respectively	Schonema, D P; Stachowiak, R W, US Patent 4,020,255, Goodyear Tire and Rubber Company, 1977; Rodriguez Garraza, A L; Sorichetti, P; Marzocca, A J; Matteo, C L; Monti, G A, Polym. Test., in press, 2011; Gong, D; Wang, B; Cai, H; Zhang, X; Jiang, L, J. Organometallic Chem., 696, 1584-90, 2011; Maiti, M; Srivastava, V K; Shewale, S; Jasra, R V; Chavda, A; Modi, S, Chem. Eng. Sci., 107, 256-65, 2014.

PBD, *cis* *cis*-1,4-polybutadiene

PARAMETER	UNIT	VALUE	REFERENCES
Yield	%	above 95% of *cis*-isomer	Schonema, D P; Stachowiak, R W, US Patent 4,020,255, Goodyear Tire and Rubber Company, 1977.
Heat of polymerization	J g^{-1}	1350-1442	Roberts, D E, J. Res. Natl. Bur. Std, 44, 221-7, 1950.
Number average molecular weight, M_n	dalton, g/mol, amu	5,000-152,000	
Mass average molecular weight, M_w	dalton, g/mol, amu	56,500-817,000	Kariyo, S; Gainaru, C; Schick, H; Brodin, A; Novikov, V N; Roessler, E A, Phys. Rev. Lett., 97, 207803,1-4, 2006.
Polydispersity, M_w/M_n	-	2.0-3.8; 1.2-22.1	Gao, B, Li, D, Duan, Q, Polyhedron, 169, 287-90, 2019.
Molar volume at 298K	cm^3 mol^{-1}	calc.=59.8; 53.5 (crystalline); 60.7 (amorphous); exp.=60.7	
Van der Waals volume	cm^3 mol^{-1}	37.1 (crystalline); 37.5 (amorphous); exp.=37.84	

STRUCTURE

Crystallinity	%	18-24	Di Lorenzo, M L, Polymer, 50, 578-84, 2009.
Cell type (lattice)	-	monoclinic	
Cell dimensions	nm	a:b:c=0.853:0.816:1.266	
Unit cell angles	degree	β=83.33	
Tacticity	%	*cis*: 95-98.4%	Di Lorenzo, M L, J. Appl. Polym. Sci., 116, 1408-13, 2010.
Dielectric moment	Debye	0.33 (*cis*); 0.5 (vinyl)	
Entanglement molecular weight	dalton, g/mol, amu	calc.=1,581; 1,844-2,000	Kariyo, S; Gainaru, C; Schick, H; Brodin, A; Novikov, V N; Roessler, E A, Phys. Rev. Lett., 97, 207803,1-4, 2006.
Isothermal crystallization temperature	°C	-26	

COMMERCIAL POLYMERS

Some manufacturers	-	Lanxess	
Trade names	-	Buna	

PHYSICAL PROPERTIES

Density at 20°C	g cm^{-3}	0.890-0.915	
Color	-	colorless to white	
Refractive index, 20°C	-	calc.=1.502-1.5188; exp.=1.516-1.605	
Molar polarizability	cm^3 x 10^{-25}	71.4	
Odor	-	rubber-like	
Melting temperature, DSC	°C	1	
Thermal expansion coefficient, 23-80°C	°C^{-1}	1.5-6.7E-4	
Thermal conductivity, melt	W m^{-1} K^{-1}	calc.=0.1788; exp.=0.22	
Glass transition temperature	°C	calc.=-72 to -103.0; exp.=-99 to -106; -102 (*cis*) and -83 (*trans*)	Di Lorenzo, M L, Polymer, 50, 578-84, 2009.

PBD, *cis* cis-1,4-polybutadiene

PARAMETER	UNIT	VALUE	REFERENCES
Specific heat capacity	J K^{-1} kg^{-1}	1,850-1,950	Di Lorenzo, M L, Polymer, 50, 578-84, 2009.
Heat of fusion	kJ mol^{-1}	2.51	
Hansen solubility parameters, δ_D, δ_P, δ_H	MPa$^{0.5}$	17.5, 2.3, 3.4; 17.3, 2.25, 3.42	
Interaction radius		6.5	
Hildebrand solubility parameter	MPa$^{0.5}$	16.2-18.0	
Surface tension	mN m^{-1}	calc.=33.5-39.8; exp.=32.0	
Dielectric constant at 50 Hz/1 MHz	-	2.3/-	
Permeability to nitrogen, 25°C	cm^3 cm cm^{-2} s^{-1} Pa^{-1} x 10^{12}	1.44	
Diffusion coefficient of nitrogen	cm^2 s^{-1} x10^6	1.5	Gestoso, P; Meunier, M, Molecular Simulations, 34, 10-15, 1135-41, 2008.
Diffusion coefficient of oxygen	cm^2 s^{-1} x10^6	1.1-3.0	Gestoso, P; Meunier, M, Molecular Simulations, 34, 10-15, 1135-41, 2008.
Contact angle of water, 20°C	degree	95-97	
Surface free energy	mJ m^{-2}	45.9	

MECHANICAL & RHEOLOGICAL PROPERTIES

PARAMETER	UNIT	VALUE	
Tensile strength	MPa	16.2-20	
Tensile stress at yield	MPa	8.9-9.3	
Elongation	%	450-620	
Tear strength	kN m^{-1}	36.3-69.6	
Abrasion resistance (ASTM D1044)	mg/1000 cycles	41-44	
Poisson's ratio	-	calc.=0.432	
Compression set, 22h 70°C	%	13-15	
Shore A hardness	-	64-88	
Mooney viscosity	-	33-55	

CHEMICAL RESISTANCE

PARAMETER	UNIT	VALUE	
Acid dilute/concentrated	-	not resistant	
Alcohols	-	resistant	
Alkalis	-	not resistant	
Aromatic hydrocarbons	-	not resistant	
Esters	-	not resistant	
Halogenated hydrocarbons	-	not resistant	
Θ solvent, Θ-temp.=10.3, -1, 35.5, 144°C	-	diethyl ketone, n-heptane, n-propyl acetate, propylene oxide	
Good solvent	-	higher ketones, higher aliphatic esters, hydrocarbons, THF	
Non-solvent	-	alcohol, dilute acids, dilute alkalies, nitromethane, propionitrile, water	

PBD, *cis* cis-1,4-polybutadiene

PARAMETER	UNIT	VALUE	REFERENCES
FLAMMABILITY			
Ignition temperature	°C	200	
WEATHER STABILITY			
Products of degradation	-	hydroperoxides, carbonyls, H_2O, chain scissions, crosslinks, carboxyl groups	
Stabilizers	-	UVA: 2(2'-hydroxy-5-methylphenyl)benzotriazole; Screener: carbon black; HAS: bis(1,2,2,6,6-pentamethyl-4-piperidinyl)-2-n-butyl-2-(3,5-di-tert-butyl-4-hydroxy-benzyl) malonate; Phenolic antioxidants: phenol, 4-methyl-, reaction products with dicyclopentadiene and isobutene; 2,6,-di-tert-butyl-4-(4,6-bis(octylthio)-1,3,5,-triazine-2-ylamino) phenol; 2-(1,1-dimethylethyl)-6-[[3-(1,1-dimethylethyl)-2-hydroxy-5-methylphenyl] methyl-4-methylphenyl acrylate; isotridecyl-3-(3,5-di-tert-butyl-4-hydroxyphenyl) propionate; 3,5-bis(1,1-dimethyethyl)-4-hydroxy-benzenepropanoic acid, C13-15 alkyl esters; 2,2'-isobutylidenebis(2,4-dimethylphenol); Amine: nonylated diphenylamine; Thiosynergist: 2,2'-thiobis(6-tert-butyl-4-methylphenol); 4,6-bis(octylthiomethyl)-o-cresol	
BIODEGRADATION			
Typical biodegradants	-	horseradish peroxidase	Enoki, M; Kaita, S; Wakatsuki, Y; Doi, Y; Iwata, T, Polym. Deg. Stab., 84, 321-26, 2004.
TOXICITY			
NFPA: Health, Flammability, Reactivity rating	-	1/1/0	
Carcinogenic effect	-	not listed by ACGIH, NIOSH, NTP	
PROCESSING			
Typical processing methods	-	blow molding, extrusion, injection molding, mixing, molding, vulcanization	
Additives used in final products	-	Fillers: carbon black, china clay, fly ash, mica, nano-calcium carbonate, nano-magnesium hydroxide, zinc oxide; Plasticizers: chlorinated paraffins, dioctyl sebacate, dibuthyl phthalate, dioctyl phthalate, paraffinic, aromatic, or naphthenic mineral oils, polyisobutylene; Antiblocking: diatomaceous earth; Release: liquid polybutadiene; Slip: erucamide+stearamide	
Applications	-	buffer springs, conveyor belts, golf balls, hoses, industrial flooring, modification of other polymers (e.g., HIPS and ABS), rubber pads for ballastless track, rubberized cloth, seals and gaskets, tires	
Outstanding properties	-	resiliency	
BLENDS			
Suitable polymers	-	NR, PE, PS	
ANALYSIS			
FTIR (wavenumber-assignment)	cm^{-1}/-	725 (*cis*-1,4), 910 (vinyl-1,2), and 965 (*trans*-1,4)	

PBD, *trans* *trans*-1,4-polybutadiene

PARAMETER	UNIT	VALUE	REFERENCES
GENERAL			
Common name	-	*trans*-1,4-polybutadiene	
Acronym	-	*trans*-PBD	
CAS number	-	9003-17-2	
HISTORY			
Person to discover	-	Carlson, E J; Horne, S E	Carlson, E J; Horne, S E, US Patent 3,657,209, Goodrich-Gulf Chemicals, Apr. 18, 1972.
Date	-	1972 (filed 1955)	
Details	-	all-*trans* polymerization	
SYNTHESIS			
Monomer(s) structure	-	$H_2C=CHCH=CH_2$	
Monomer(s) CAS number(s)	-	106-99-0	
Monomer(s) molecular weight(s)	dalton, g/mol, amu	54.09	
Method of synthesis	-	polymerization with application of Ziegler-Natta catalyst system; anionic polymerization	
Temperature of polymerization	°C	20	
Catalyst	-	$TiCl_4$+triisobutyl aluminum; iron bisiminopyridyl complexes; lithium catalyst	Gong, D; Jia, X; Wang, B; Wang, F; Zhang, C; Zhang, X; Jiang, L; Dong, W, Inorg. Chim. Acta, 373, 47-53, 2001; Rodriguez Garraza, A L; Sorichetti, P; Marzocca, A J; Matteo, C L; Monti, G A, Polym. Test., in press, 2011.
Yield	%	over 90% *trans*; up to 99% *trans*; mixture of isomers	
Number average molecular weight, M_n	dalton, g/mol, amu	68,000-290,000	
Mass average molecular weight, M_w	dalton, g/mol, amu	250,000-580,00	
Polydispersity, M_w/M_n	-	1.07-3.7	
Molar volume at 298K	cm^3 mol^{-1}	calc.=59.8; exp.=60.7	
Van der Waals volume	cm^3 mol^{-1}	54.1	
STRUCTURE	-		
Crystallinity	%	11-55.8; amorphous (less than 65 *trans*)	Benvenuta-Tapia, J J; Tenorio-Lopez, J A; Herrera-Najera, R; Rios-Guerrero, L, Polym. Eng. Sci., 49, 1-10, 2009; Yang, X; Cai, J; Kong, X; Dong, W; Li, G; Ling, W; Zhou, E, Eur. Polym. J., 37, 763-69, 2001.
Cell type (lattice)	-	monoclinic; hexagonal	Yang, X; Cai, J; Kong, X; Dong, W; Li, G; Ling, W; Zhou, E, Eur. Polym. J., 37, 763-69, 2001.
Cell dimensions	nm	a:b:c=0.863:0.911:0.483; a:b:c=0495:0.495:0.466	Yang, X; Cai, J; Kong, X; Dong, W; Li, G; Ling, W; Zhou, E, Eur. Polym. J., 37, 763-69, 2001.
Unit cell angles	degree	β=114; γ=120	Yang, X; Cai, J; Kong, X; Dong, W; Li, G; Ling, W; Zhou, E, Eur. Polym. J., 37, 763-69, 2001.
Number of chains per unit cell	-	4	

PBD, *trans* trans-1,4-polybutadiene

PARAMETER	UNIT	VALUE	REFERENCES
Trans content	%	40-80 (elastomer); >90 (thermoplastic resin)	Benvenuta-Tapia, J J; Tenorio-Lopez, J A; Herrera-Najera, R; Rios-Guerrero, L, Polym. Eng. Sci., 49, 1-10, 2009.
Dielectric moment	Debye	0 (*trans*)	
Chain conformation	-	1/0	
Entanglement molecular weight	dalton, g/mol, amu	calc.=1,581	
Lamellae thickness	nm	4-16	Yang, X; Cai, J; Kong, X; Dong, W; Li, G; Ling, W; Zhou, E, Eur. Polym. J., 37, 763-69, 2001.

PHYSICAL PROPERTIES

Density at 20°C	g cm^{-3}	0.91; 1.04 (crystalline)	
Color	-	white to light yellow	
Refractive index, 20°C	-	calc.=1.502-1.5188; exp.=1.516-1.605	
Melting temperature, DSC	°C	80-145	
Thermal conductivity	W m^{-1} K^{-1}	calc.=0.1788; exp.=0.22; 0.08-0.15	Katagiri, K, Totani, T, Isono, T, Goto, R, Sasaki, K, J. Energy Storage, 31, 101636, 2020.
Glass transition temperature	°C	calc.=-72 to -103.0; ; -102 (*cis*) and -72 to -87 (*trans*)	Di Lorenzo, M L, Polymer, 50, 578-84, 2009.
Heat of fusion	kJ mol^{-1}	4.18	
Maximum service temperature	°C	-101 to 93	
Enthalpy of melting	J g^{-1}	19.4-34.6	Yang, X; Cai, J; Kong, X; Dong, W; Li, G; Ling, W; Zhou, E, Eur. Polym. J., 37, 763-69, 2001.
Hildebrand solubility parameter	MPa$^{0.5}$	calc.=14.65-17.15; exp.=17.09-17.6	
Surface tension	mN m^{-1}	calc.=33.5-39.8; exp.=32.0	
Dielectric constant at 100 Hz/1 MHz	-	/2.8	
Dielectric loss factor at 1 kHz	-	0.002	
Diffusion coefficient of nitrogen	cm^2 s^{-1} x10^6	0.5	
Diffusion coefficient of oxygen	cm^2 s^{-1} x10^6	0.7	
Contact angle of water, 20°C	degree	96	

MECHANICAL & RHEOLOGICAL PROPERTIES

Tensile strength	MPa	250-400 (draw ratio - 10-20)	
Elongation	%	400	
Young's modulus	MPa	15,000-20,000 (draw ratio - 10-20)	
Poisson's ratio	-	calc.=0.432	
Intrinsic viscosity, 25°C	dl g^{-1}	2.1	
Mooney viscosity	-	50-68	

CHEMICAL RESISTANCE

Acid dilute/concentrated	-	good/poor	
Alcohols	-	good	
Alkalis	-	good/poor	
Aliphatic hydrocarbons	-	poor	

PBD, *trans* trans-1,4-polybutadiene

PARAMETER	UNIT	VALUE	REFERENCES
Aromatic hydrocarbons	-	poor	
Esters	-	poor	
Ketones	-	poor	
Theta solvent, Theta-temp.=212, 240, 146°C	-	diethyl ketone, ethyl propyl ketone, propylene oxide	
Good solvent	-	higher ketones, higher aliphatic esters, hydrocarbons, THF	
Non-solvent	-	alcohol, dilute acids, dilute alkalies, nitromethane, proponitrile, water	
FLAMMABILITY			
Ignition temperature	°C	200	
Autoignition temperature	°C	350	
BIODEGRADATION			
Typical biodegradants	-	horseradish peroxidase	Enoki, M; Kaita, S; Wakatsuki, Y; Doi, Y; Iwata, T, Polym. Deg. Stab., 84, 321-26, 2004.
TOXICITY			
NFPA: Health, Flammability, Reactivity rating	-	1/1/0	
Carcinogenic effect	-	not listed by ACGIH, NIOSH, NTP	
PROCESSING			
Applications	-	golf ball covers	
ANALYSIS			
FTIR (wavenumber-assignment)	cm^{-1}/-	725 (*cis*-1,4), 910 (vinyl-1,2), and 965 (*trans*-1,4)	Pires, N M T; Coutinho, F M B; Costa, M A S, Eur. Polym. J, 40, 2599-2603, 2004.
Raman (wavenumber-assignment)	cm^{-1}/-	C=C – 1666	Pathak, A; Saxena, V; Tandon, P; Gupta, V D, Polymer, 5154-60, 2006.
NMR (chemical shifts)	ppm	H NMR: =CH-- – 5.6; =CH$_2$ – 4.9; CH$_3$ – 0.85; C NMR: CH= – 142.7; CH$_2$= – 114.5	Hung, N Q; Sanglar, C; Grenier-Loustalot, M F; Huoung, P V; Cuong, H N, Polym. Deg. Stab., 96, 1255-60, 2011.

PBI polybenzimidazole

PARAMETER	UNIT	VALUE	REFERENCES
GENERAL			
Common name	-	polybenzimidazole	
CAS name	-	poly[(1,5-dihydrobenzo[1,2-d:4,5-d']diimidazole-2,6-diyl)-1,3-phenylene] (27233-57-4); poly[(1,5-dihydrobenzo[1,2-d:4,5-d']diimidazole-2,6-diyl)-1,4-phenylene] (32075-68-6); poly(1H-benzimidazole-2,5-diyl) (32109-42-5); poly(1H-benzimidazole-2,5-diylsulfonyl-1H-benzimidazole-5,2-diyl-1,4-phenyleneoxy-1,4-phenylene) (928655-57-6); poly[(1,5-dihydrobenzo[1,2-d:4,5-d']diimidazole-2,6-diyl)-1,2-phenylene] (1073316-19-4)	
Acronym	-	PBI	
CAS number	-	27233-57-4; 32075-68-6; 32109-42-5; 928655-57-6; 1073316-19-4	
Formula			
HISTORY			
Person to discover	-	Marvel C S; Vogel, H A	Marvel C S; Vogel, H A, US Patent 3,174,947, University of Illinois, Mar. 23, 1965.
Date	-	patent 1965 (filed in 1962)	
Details	-	synthesis	
SYNTHESIS			
Monomer(s) structure	-		
Monomer(s) CAS number(s)	-	91-95-2; 744-45-6	
Monomer(s) molecular weight(s)	dalton, g/mol, amu	214.27; 318.32	
Monomer ratio	-	1:1.49	
Method of synthesis	-	condensation of 3,3'4,4'-tetraaminobiphenyl and diphenyl isophthalate in suitable solvent	
Temperature of polymerization	°C	100-290	Marsano, E; Azzurri, F; Corsini, P, Macromol. Symp., 234, 33-41, 2006.
Time of polymerization	h	1-2.5	Marsano, E; Azzurri, F; Corsini, P, Macromol. Symp., 234, 33-41, 2006.
Yield	%	93-100	
Crosslinker	-	$K_2S_2O_8$	Zhao, B, Shi, G M, Wang, K Y, Lai, J-Y, Chung, T-S, Sepration Purification Technol., 255, 117702, 2021.
Number average molecular weight, M_n	dalton, g/mol, amu	2,500-32,700	
Mass average molecular weight, M_w	dalton, g/mol, amu	19,600-55,900	Ohishi, T; Sugi, R; Yokoyama, A; Yokozawa, T, J. Polym. Sci. a, 44, 4990-5003, 2006.
Polydispersity, M_w/M_n	-	1.08-5.4	

HANDBOOK OF POLYMERS 3rd Edition, Copyrights 2022; ChemTec Publishing

PBI polybenzimidazole

PARAMETER	UNIT	VALUE	REFERENCES
STRUCTURE			
Cell type (lattice)	-	monoclinic; triclinic	
Cell dimensions	nm	a:b:c=0.9921:1.868:1.422 (monoclinic); a:b:c=1.070:1.199:1.371 (triclinic)	Tanatani, A; Yokoyama, A; Azumaya, I; Takakura, Y; Mitsui, C; Shiro, M; Uchiyama, M; Muranaka, A; Kobayashi, N; Yokozawa, T, JACS, 127, 8553-61, 2005.
Unit cell angles	degree	α:β:γ=101.4; α:β:γ=87.72:87.36:85.13	Tanatani, A; Yokoyama, A; Azumaya, I; Takakura, Y; Mitsui, C; Shiro, M; Uchiyama, M; Muranaka, A; Kobayashi, N; Yokozawa, T, JACS, 127, 8553-61, 2005.
Crystallite size	nm	9.5-12.4	
COMMERCIAL POLYMERS			
Some manufacturers	-	PBI Performance Products	
Trade names	-	Celazole	
PHYSICAL PROPERTIES			
Density at 20°C	g cm^{-3}	1.30-1.43	
Color	-	black	
Odor	-	odorless	
Melting temperature, DSC	°C	300	
Thermal expansion coefficient, 23-80°C	10^{-4} °C^{-1}	0.23-0.25	
Thermal conductivity, melt	W m^{-1} K^{-1}	0.038 (fiber)	
Glass transition temperature	°C	420-510; 399 (amorphous)	MacKnight, W J; Kantor, S W; Zhu, H, Antec, 1594-98, 1996.
Specific heat capacity	J K^{-1} kg^{-1}	1,300	
Maximum service temperature	°C	-196 to 500; 760 (short burst)	Bhowmik, S; Benedictus, R; Poulis, H, Polymers in Defence and Aerospace 2007, Rapra, 2007, paper 8.
Long term service temperature	°C	260-315	
Heat deflection temperature at 1.8 MPa	°C	435	
Dielectric constant at 100 Hz/1 MHz	-	5.4/3.2	
Relative permittivity at 100 Hz	-	3.3	
Relative permittivity at 1 MHz	-	3.2	
Dissipation factor at 100 Hz	E-4	10	
Dissipation factor at 1 MHz	E-4	30	
Volume resistivity	ohm-m	1E11	
Surface resistivity	ohm	1E10-2E15	
Electric strength K20/P50, d=0.60.8 mm	kV mm^{-1}	21-23	
Arc resistance	s	185	
Proton conductivity, <180°C	S cm^{-1}	0.122	Duan, C; Luo, H; Li, J; Liu, C, Polymer, 201, 122555, 2020.
Coefficient of friction	-	0.19-0.27	

PBI polybenzimidazole

PARAMETER	UNIT	VALUE	REFERENCES
MECHANICAL & RHEOLOGICAL PROPERTIES			
Tensile strength	MPa	94-160	
Tensile modulus	MPa	5,900	
Tensile stress at yield	MPa	74.4	Ngamsantivongsa, P; Lin, H-L; Yu, T L, , J. Membrane Sci., 491, 10-21, 2015.
Elongation	%	3-8; 9-27 (fiber)	
Flexural strength	MPa	220	
Flexural modulus	MPa	6,500	
Elastic modulus	MPa	5,800	
Compressive strength	MPa	344-400	
Young's modulus	MPa	2,300	Ngamsantivongsa, P; Lin, H-L; Yu, T L, , J. Membrane Sci., 491, 10-21, 2015.
Charpy impact strength, notched, 23°C	kJ m^{-2}	3.5	
Izod impact strength, unnotched, 23°C	J m^{-1}	590	
Izod impact strength, notched, 23°C	J m^{-1}	30	
Poisson's ratio	-	0.34	
Shore D hardness	-	94	
Rockwell hardness	-	M125	
Ball indention hardness at 358 N/30 S (ISO 2039-1)	MPa	375	
Shrinkage	%	1	
Intrinsic viscosity, 25°C	dl g^{-1}	0.39-0.51	
Water absorption, 24h at 23°C	%	0.4-0.5; 5-15 (saturation)	
CHEMICAL RESISTANCE			
Acid dilute/concentrated	-	poor	
Alcohols	-	good	
Alkalis	-	poor	
Aliphatic hydrocarbons	-	good	
Aromatic hydrocarbons	-	good	
Esters	-	good	
Greases & oils	-	good	
Halogenated hydrocarbons	-	good/poor	
Ketones	-	good	
Good solvent	-	96% H_2SO_4, N,N-dimethylacetimide	
FLAMMABILITY			
Ignition temperature	°C	>540	
Autoignition temperature	°C	>540	
Limiting oxygen index	% O_2	41-58	

PBI polybenzimidazole

PARAMETER	UNIT	VALUE	REFERENCES
Char at 500°C	%	67.5	Lyon, R E; Walters, R N, J. Anal. Appl. Pyrolysis, 71, 27-46, 2004.
Heat of combustion	J g^{-1}	31,650	Walters, R N; Hacket, S M; Lyon, R E, Fire Mater., 24, 5, 245-52, 2000.
UL 94 rating	-	V-0	

WEATHER STABILITY			
Spectral sensitivity	nm	260-300	Tanatani, A; Yokoyama, A; Azumaya, I; Takakura, Y; Mitsui, C; Shiro, M; Uchiyama, M; Muranaka, A; Kobayashi, N; Yokozawa, T, JACS, 127, 8553-61, 2005.

TOXICITY			
HMIS: Health, Flammability, Reactivity rating	-	0/1/0	
Carcinogenic effect	-	not listed by ACGIH, NIOSH, NTP	

PROCESSING			
Typical processing methods	-	compression molding, melt spinning	
Additives used in final products	-	graphene oxide	
Applications	-	ball valve seats, clamp rings, electrical connectors, fiber, fuel cell membranes, hollow fiber membranes, insulator bushings, protective apparel (e.g., firefighter coats and suits)	
Outstanding properties	-	temperature stability (204°C) and chemical resistance, high heat deflection (427°C), with continuous service capability (399°C) in inert environment	

BLENDS			
Suitable polymers	-	aramid, PAES, PEI, PES, PI, PVF, poly(aminophosphonate ester)	Zhu, W-P; Sun, S-P; Gao, J; Fu, F-J; Chung, T-S, J. Membrane Sci.,456, 117-27, 2014.

ANALYSIS			
FTIR (wavenumber-assignment)	cm^{-1}/-	N-H – 3283; C=O – 1653	
Raman (wavenumber-assignment)	cm^{-1}/-	1563 and 1618 correspond to the stretching vibrations of C=N and C=C bonds, 1451 – stretching vibration of benzimidazole ring, 1281 and 1115 – in-plane vibration of C–H and stretching vibration of C–C skeletal	Dey, B, Ahmad, M W, AlMezeni, A, Sarkhel, G, Bag, D S, Choudhury, A, Compos. Commun., 17, 87-96, 2020.
x-ray diffraction peaks	degree	17.84, 18.76, 20.44-20.47, 22.76-23.67, 25.29, 28.06	Kobashi, K; Kobayashi, K; Yasuda, H; Arimachi, K; Uchida, T; Wakabayashi, K; Yamazaki, S; Kimura, K, Macromolecules, 42, 6128-35, 2009.

PBMA polybutylmethacrylate

PARAMETER	UNIT	VALUE	REFERENCES
GENERAL			
Common name	-	polybutylmethacrylate	
CAS name	-	2-propenoic acid, 2-methyl-, butyl ester, homopolymer	
Acronym	-	PBMA	
CAS number	-	9003-63-8	
EC number	-	202-615-1	
SYNTHESIS			
Monomer(s) structure	-	$CH_3(CH_2)_3OCCH_3$ with O double-bonded above and CH_2 below	
Monomer(s) CAS number(s)	-	97-88-1	
Monomer(s) molecular weight(s)	dalton, g/ mol, amu	142.2	
Monomer(s) expected purity(ies)	%	99.6	
Monomer ratio	-	100%	
Method of synthesis	-	suspension polymerization	Zhao, J, Han, P, Quan, Q, Shan, Y, Zhan, T, Wang, J, Xiao, C, Prog. Org. Coat., 115, 181-7, 2018.
Temperature of polymerization	°C	60	
Catalyst	-	1,1,2,2-tetramethyl-1-benzyl-2-n-propylethylene-1,2-diammonium bromide chloride	Vajjiravel, M; Umapathy, M J, Int. J. Polym. Mater., 59, 647-62, 2010.
Mass average molecular weight, M_w	dalton, g/ mol, amu	47,000-723,000	
Polydispersity, M_w/M_n	-	2-4.3	
Molar volume at 298K	cm^3 mol^{-1}	calc.=135.0-139.1 (amorphous); exp.=134.8	
Van der Waals volume	cm^3 mol^{-1}	calc.=86.8 (amorphous); exp.=87.34	
STRUCTURE			
Entanglement molecular weight	dalton, g/ mol, amu	calc.=18,533	
COMMERCIAL POLYMERS			
Some manufacturers	-	Lucite	
Trade names	-	Elvacite	
PHYSICAL PROPERTIES			
Density at 20°C	g cm^{-3}	1.03-1.07	
Color	-	white	
Refractive index, 20°C	-	calc.=1.4761-1.4888; exp.=1.43-1.483	
Odor	-	methacrylate	
Thermal conductivity, melt	W m^{-1} K^{-1}	calc.=0.1474	
Glass transition temperature	°C	calc.=0-19; exp.=15-50	
Hansen solubility parameters, δ_D, δ_P, δ_H	$MPa^{0.5}$	18.3, 4.9, 6.3; 16.0, 6.2, 6.6	

PBMA polybutylmethacrylate

PARAMETER	UNIT	VALUE	REFERENCES
Interaction radius		9.4; 9.5	
Hildebrand solubility parameter	$MPa^{0.5}$	calc.=18.41; exp.=18.4-20.0	
Surface tension	$mN\ m^{-1}$	calc.=27.2-37.1; exp.=31.2	Wu, S, J. Phys. Chem., 74, 632, 1970.
Contact angle of water, 20°C	degree	91; 108.1	
Surface free energy	$mJ\ m^{-2}$	30.9; 33.1	

MECHANICAL & RHEOLOGICAL PROPERTIES			
Tensile strength	MPa	1.38-14.1	
Tensile stress at yield	MPa	2.5	
Elongation	%	150-290	
Poisson's ratio	-	calc.=0.340	

CHEMICAL RESISTANCE			
Alcohols	-	poor	
Aliphatic hydrocarbons	-	poor	
Aromatic hydrocarbons	-	poor	
Θ solvent, Θ-temp.=10.7, 84.8, 45, 23°C	-	isobutanol, n-decane, ethanol, isopropanol	
Good solvent	-	carbon tetrachloride, cyclohexane, gasoline, hexane, turpentine, and hot ethanol,	
Non-solvent	-	ethanol (cold) formic acid	

FLAMMABILITY			
Ignition temperature	°C	300	
Char at 500°C	%	0	Lyon, R E; Walters, R N, J. Anal. Appl. Pyrolysis, 71, 27-46, 2004.
Volatile products of combustion	-	CO_2, CO, CH_4, C_2H_8, butyl methacrylate, MeOH, EtOH	Czech, Z; Pelech, Z, J. Therm. Anal. Calorim., 101, 309-13, 2010.

TOXICITY			
HMIS: Health, Flammability, Reactivity rating	-	0/0/0	
Carcinogenic effect	-	not listed by ACGIH, NIOSH, NTP	
OSHA	$mg\ m^{-3}$	5 (respirable); 15 (total dust)	

PROCESSING			
Typical processing methods	-	coating, spraying	
Additives used in final products	-	Plasticizers: phthalates, adipates, citrates; Antistatics: vanadium pentoxide, vinyl benzene quaternary ammonium polymer	
Applications	-	art conservation, clear coatings, plastic coatings, printing inks	
Outstanding properties	-	flexibility, UV resistance	

BLENDS			
Suitable polymers	-	NBR, PE, PMMA, PS, PVAc	Piunova, V A; Hogen-Esch, T E, Polymer, 69, 58-65, 2015.

PBN poly(butylene 2,6-naphthalate)

PARAMETER	UNIT	VALUE	REFERENCES
GENERAL			
Common name	-	poly(butylene 2,6-naphthalate), poly(butylene 2,6-naphthalenedicarboxylate)	
CAS name	-	poly(oxy-1,4-butanediyloxycarbonyl-2,6-naphthalenediylcarbonyl)	
Acronym	-	PBN	
CAS number	-	28779-82-0	
Formula			
SYNTHESIS			
Monomer(s) structure	-		
Monomer(s) CAS number(s)	-	1141-38-4; 110-63-4	
Monomer(s) molecular weight(s)	dalton, g/mol, amu	216.2; 90.12	
Monomer(s) expected purity(ies)	%	99.5	
Monomer ratio	-	1	
Method of synthesis	-	reaction of 2,6-naphthalene dicarboxylic acid with the 1,4-butanediol	
Catalyst	-	$Ti(OBu)_4$	
Number average molecular weight, M_n	dalton, g/mol, amu	23,000	
Polydispersity, M_w/M_n	-	2.1	
STRUCTURE	-		
Crystallinity	%	50-53	Papageorgiou, G Z; Karayannidis, G P; Bikiaris, D N; Stergiou, A; Litsardakis, G; Makridis, S S, J. Polym. Sci., Part B: Polym. Phys., 42, 843-60, 2004.
Cell type (lattice)	-	crystallization at slow cooling rate lower than 10 K/min (0.167 K/s), or at temperatures higher than 200°C causes formation of triclinic β'-crystals	Androsch, R, Soccio, M, Lotti, N, Jehnichen, D, Schick, C, Polymer, 158, 77-82, 2018.
Cell dimensions	nm	a:b:c=0.487:0.622:1.436 (α); a:b:c=0.455:0.643:1.531 (β)	Konishi, T; Nishida, K; Matsuba, G; Kanaya, T, Macromolecules, 41, 3157-61, 2008.
Unit cell angles	degree	α:β:γ=100.78:126.90:97.93 (α); α:β:γ=110.1:121.1:100.6 (β)	Konishi, T; Nishida, K; Matsuba, G; Kanaya, T, Macromolecules, 41, 3157-61, 2008.
Number of chains per unit cell	-	1	
Polymorphs	-	α, β	Konishi, T; Nishida, K; Matsuba, G; Kanaya, T, Macromolecules, 41, 3157-61, 2008.
Lamellae thickness	nm	10-15	Ding, Q, Janke, A, Schick, C, Androsch, R, Polymer, 194, 122404, 2020.
Rapid crystallization temperature	°C	212	Papageorgiou, G Z; Karayannidis, G P; Bikiaris, D N; Stergiou, A; Litsardakis, G; Makridis, S S, J. Polym. Sci., Part B: Polym. Phys., 42, 843-60, 2004.

PBN poly(butylene 2,6-naphthalate)

PARAMETER	UNIT	VALUE	REFERENCES
Enthalpy of crystallization	µJ	2.6	Androsch, R, Soccio, M, Lotti, N, Jehnichen, D, Schick, C, Polymer, 158, 77-82, 2018.
Avrami constants, k/n	-	0.112763/2.18-2.38	Papageorgiou, G Z; Karayannidis, G P; Bikiaris, D N; Stergiou, A; Litsardakis, G; Makridis, S S, J. Polym. Sci., Part B: Polym. Phys., 42, 843-60, 2004.

COMMERCIAL POLYMERS

Some manufacturers	-	Teijin Chemicals	
Trade names	-	PBN	

PHYSICAL PROPERTIES

Density at 20°C	g cm^{-3}	1.31; 1.36 (theoretical density of α form); 1.39 (theoretical density of β form)	Ju, M-Y; Huang, J-M; Chang, F-C, Polymer, 43, 2065-74, 2002.
Color	-	white	
Melting temperature, DSC	°C	the equilibrium melting temperature of the α-phase of 261°C is about 20K lower than that of β'-crystals	Androsch, R, Soccio, M, Lotti, N, Jehnichen, D, Schick, C, Polymer, 158, 77-82, 2018.
Thermal expansion coefficient, 23-80°C	10^{-4} °C^{-1}	0.75-0.94	
Glass transition temperature	°C	41-82	Konishi, T; Nishida, K; Matsuba, G; Kanaya, T, Macromolecules, 41, 3157-61, 2008.
Heat deflection temperature at 1.8 MPa	°C	77	
Dielectric constant at 100 Hz/1 MHz	-	3.6/3.4	
Relative permittivity at 100 Hz	-	3.6	
Dissipation factor at 60 Hz	E-4	42	
Dissipation factor at 1 MHz	E-4	23	
Volume resistivity	ohm-m	1E14	
Electric strength K20/P50, d=0.60.8 mm	kV mm^{-1}	42	
Breakdown voltage	MV/m	42	
Arc resistance	s	84	

MECHANICAL & RHEOLOGICAL PROPERTIES

Tensile strength	MPa	65	
Elongation	%	87-100	
Flexural strength	MPa	81	
Flexural modulus	MPa	1920	
Charpy impact strength, unnotched, 23°C	kJ m^{-2}	34	
Izod impact strength, unnotched, 23°C	J m^{-1}	NB	
Izod impact strength, notched, 23°C	J m^{-1}	34	
Rockwell hardness	-	M102	
Shrinkage	%	1.2-1.3	
Intrinsic viscosity, 25°C	dl g^{-1}	0.83-1.40	

PBN poly(butylene 2,6-naphthalate)

PARAMETER	UNIT	VALUE	REFERENCES
Melt volume flow rate (ISO 1133, procedure B), 220°C/10 kg	cm³/10 min	10	
Water absorption, 24h at 23°C	%	0.1	

CHEMICAL RESISTANCE

Acid dilute/concentrated	-	good/excellent	
Alkalis	-	good	
Good solvent	-	1,1,1,3,3,3-hexafluoro-2-propanol, chloroform/1,1,1,3,3,3-hexafluoro-2-propanol	

FLAMMABILITY

UL 94 rating	-	94HB	

WEATHER STABILITY

Spectral sensitivity	nm	330-340	Scheirs, J; Gardette, J-L, Polym. Deg. Stab., 56, 351-56, 1997.
Products of degradation	-	extensive yellowing compared with PBT, PET	

PROCESSING

Processing temperature	°C	270	
Applications	-	fuel hoses, medical devices, semiconductor wafer carrier, wire & cable	
Outstanding properties	-	fuel resistant and impermeable, high heat distortion, solder resistant	

BLENDS

Suitable polymers	-	PEN	

ANALYSIS

FTIR (wavenumber-assignment)	cm⁻¹/-	C=O – 1718 (α), 1713 (β); CH$_2$ *trans* – 1502; CH$_2$ gauche – 1447/1449, see more in ref.	Ju, M-Y; Huang, J-M; Chang, F-C, Polymer, 43, 2065-74, 2002.
x-ray diffraction peaks	degree	12.88, 16.64, 25.86	Ju, M-Y; Huang, J-M; Chang, F-C, Polymer, 43, 2065-74, 2002.

PBT poly(butylene terephthalate)

PARAMETER	UNIT	VALUE	REFERENCES
GENERAL			
Common name	-	poly(butylene terephthalate)	
ACS name	-	poly(oxy-1,4-butanediyloxycarbonyl-1,4-phenylenecarbonyl) (24968-12-5); 1,4-benzenedicarboxylic acid, polymer with 1,4-butanediol (26062-94-2)	
Acronym	-	PBT	
CAS number	-	24968-12-5; 26062-94-2	
Linear formula		$\left[OC-\bigcirc-COO(CH_2)_4O \right]_n$	
HISTORY			
Person to discover	-	Chipman, G R; Henk, M G; De Boer, J A; Blaha, E W	Chipman, G R; Henk, M G; De Boer, J A; Blaha, E W, US Patent 4,014,858, Standard Oil, Mar. 29, 1977.
Date	-	1977	
Details	-	reaction of terephthalic acid and butane diol using tetravalent tin catalyst having the organo-to-tin linkage	
SYNTHESIS			
Monomer(s) structure	-	$HOCH_2CH_2CH_2CH_2OH$ $HOOC-\bigcirc-COOH$	
Monomer(s) CAS number(s)	-	107-88-0; 100-21-0	
Monomer(s) molecular weight(s)	dalton, g/mol, amu	90.121; 166.13	
Monomer ratio	-	1:1.84	
Temperature of polymerization	°C	220	
Catalyst	-	tetravalent tin	
Number average molecular weight, M_n	dalton, g/mol, amu	5,600-34,100	
Mass average molecular weight, M_w	dalton, g/mol, amu	26,000-125,000	
Polydispersity, M_w/M_n	-	2.0-3.18	
STRUCTURE	-		
Crystallinity	%	27.8-42.7; 18-22 (single-bubble film); 20-47 (double-bubble film and biaxially stretched film)	Forouhashad, M; Saligheh, O; Arasteh, R, Farsani, R E, J. Macromol. Sci. B, 49, 833-42, 2010; Song, K; White, J L, Polym. Eng. Sci., 38, 3, 505-15, 1998.
Unit cell angles	degree	$\alpha{:}\beta{:}\gamma{=}99.7{:}115.2{:}110.8$	
Number of chains per unit cell	-	1	
Crystallite size	nm	1.16 (length; α); 1.3 (length, β)	
Polymorphs	-	α, β	
Chain conformation	-	GTG (α); TTT (β)	Konishi, T; Miyamoto, Y, Macromolecules, 43, 375-83, 2010.
Rapid crystallization temperature	°C	n=2.35-2.60	Al-Mulla, A; Mathew, J; Shanks, R, J. Polym. Sci. B, 45, 1344-53, 2007.

328   **HANDBOOK OF POLYMERS** 3rd Edition, Copyrights 2022; ChemTec Publishing

PBT poly(butylene terephthalate)

PARAMETER	UNIT	VALUE	REFERENCES
Cyrstallization half-time	s	4196	Arai, S, Tsunoda, S, Yamaguchi, A, Ougizawa, T, Optic Laser Technol., 117, 94-104, 2019.

COMMERCIAL POLYMERS

Some manufacturers	-	BASF, DuPont; Epsan; Mitsubishi Chemical; Sabic	
Trade names	-	Ultradur; Crastin; Epimix; Novaduran; Valox	

PHYSICAL PROPERTIES

PARAMETER	UNIT	VALUE	REFERENCES
Density at 20°C	g cm^{-3}	1.24-1.34; 1.404 (crystalline; α form); 1.283 (β form); 1.37 (10% glass fiber, GF); 1.45-1.47 (20% GF); 1.53-1.55 (30% GF); 1.73 (50% GF)	Shibaya, M; Ohkubo, N; Ishihara, H; Yamashita, K; Yoshihara, N; Nonomura, C, Antec, 1658-62, 2003.
Bulk density at 20°C	g cm^{-3}	0.60-0.90	
Color	-	white	
Odor	-	slight	
Refractive index, 20°C	°C	220-230; 220-250 (10-50% glass fiber)	
Melting temperature	°C	207.5-225	Wang, C; Fang, C-Y; Wang, C-Y, Polymer, 72, 21-9, 2015; Arai, S, Tsunoda, S, Yamaguchi, A, Ougizawa, T, Optic Laser Technol., 117, 94-104, 2019.
Softening point	°C	50 (amorphous fraction), 220 (crystalline fraction)	
Decomposition temperature	°C	288	
Thermal expansion coefficient, -40 to 160°C	°C^{-1}	0.9E-4 to 8.1E-5; 0.2-1.2E-4 (10-50% glass fiber)	
Thermal conductivity, melt	W m^{-1} K^{-1}	0.25-0.27; 0.23-0.36 (10-50% glass fiber); 0.18 (melt, 10-50% glass fiber)	
Glass transition temperature	°C	31-60; 46 (DSC)	
Specific heat capacity	J K^{-1} kg^{-1}	1,500; 1,400-1,700 (10-50% glass fiber); 1,990 (melt, 10-50% glass fiber)	
Heat of fusion	kJ mol^{-1}	21.2	
Maximum service temperature	°C	200; 210 (10-50% glass fiber)	
Temperature index (50% tensile strength loss after 20,000 h/5000 h)	°C	120/140; 120-160 (10-50% glass fiber)	
Heat deflection temperature at 0.45 MPa	°C	130-165; 215-222 (10-50% glass fiber)	
Heat deflection temperature at 1.8 MPa	°C	50-65; 168-215 (10-50% glass fiber)	
Vicat temperature VST/A/50	°C	175; 205-215 (10-50% glass fiber)	
Acid constant	-	K_a=-0.27 to -0.49; K_b=-0.96 to -0.97	Santos, J M R C A; Guthrie, J T, J Chromatog. A, 1379, 92-9, 2015.
Surface energy	mJ m^{-2}	34.9 (total); 30 (dispersive); 4.9 (polar)	Karsli, N G; Ozkan, C; Aytac, A; Deniz, V, Mater. Design, 87, 318-23, 2015.
Dielectric constant at 100 Hz/1 MHz	-	3.2-3.4/3.2-3.3; 3.6-4.7/3.6-4.5 (10-50% glass fiber)	
Relative permittivity at 100 Hz	-	3.3-3.8; 3.6-4.1 (10-50% glass fiber)	
Relative permittivity at 1 MHz	-	3.1-3.3; 3.5-3.8 (10-50% glass fiber)	
Dissipation factor at 100 Hz	E-4	10-20; 12-38 (10-50% glass fiber)	
Dissipation factor at 1 MHz	E-4	200-219; 130-200 (10-50% glass fiber)	
Volume resistivity	ohm-m	1-5E13 to 4E14; 1-4.7E14 (10-50% glass fiber)	

PBT poly(butylene terephthalate)

PARAMETER	UNIT	VALUE	REFERENCES
Surface resistivity	ohm	1E12 to 5E15; 1E13-1E15 (10-50% glass fiber)	
Electric strength K20/P50, d=0.60.8 mm	kV mm^{-1}	15-140; 16-100 (10-50% glass fiber)	
Comparative tracking index, CTI, test liquid A	-	475-600; 225-500 (10-50% glass fiber)	
Comparative tracking index, CTIM, test liquid B	-	375-450; 125-150 (10-50% glass fiber)	
Coefficient of friction	-	0.3-0.6	
Contact angle of water, 20°C	degree	81	Karsli, N G; Ozkan, C; Aytac, A; Deniz, V, Mater. Design, 87, 318-23, 2015.
Surface free energy	mJ m^{-2}	46.3-47.8	

MECHANICAL & RHEOLOGICAL PROPERTIES

PARAMETER	UNIT	VALUE	REFERENCES
Tensile strength	MPa	50-57; 90-165 (10-50% glass fiber)	
Tensile modulus	MPa	2,000-2,600; 4,500-9,800 (10-30% glass fiber)	
Tensile stress at yield	MPa	47-59; 90-165 (10-50% glass fiber)	
Tensile creep modulus, 1000 h, elongation 0.5 max	MPa	1,100-1,800; 2,500-11,600 (10-50% glass fiber)	
Elongation	%	35-300; 1.5-4.5 (10-50% glass fiber)	
Tensile yield strain	%	3.7-8; 1.7-4.0 (10-50% glass fiber)	
Flexural strength	MPa	62-85; 140-230 (10-50% glass fiber)	
Flexural modulus	MPa	2,000-2,400; 5,200-14,900 (15-50% glass fiber)	
Elastic modulus	MPa	2,200-2,600; 4,500-18,700 (10-50% glass fiber)	
Young's modulus	MPa	1906	Soudmand, B H, Shelesh-Nezhad, K, Hassanifard, S, Theoretical Appl. Mech., 108, 102662, 2020.
Charpy impact strength, unnotched, 23°C	kJ m^{-2}	190-290 to NB; 26-67 (10-50% glass fiber)	
Charpy impact strength, unnotched, -30°C	kJ m^{-2}	38-65 (10-50% glass fiber)	
Charpy impact strength, notched, 23°C	kJ m^{-2}	4-9; <5 to 11 (10-50% glass fiber)	
Charpy impact strength, notched, -30°C	kJ m^{-2}	4; 62 (10-50% glass fiber)	
Izod impact strength, unnotched, 23°C	J m^{-1}	1,600	
Izod impact strength, notched, 23°C	J m^{-1}	53	
Tenacity (fiber) (standard atmosphere)	cN tex^{-1} (daN mm^{-2})	25-30	Fourne, F, Synthetic Fibers. Machines and Equipment Manufacture, Properties. Carl Hanser Verlag, 1999.
Tenacity (wet fiber, as % of dry strength)	%	95-100	Fourne, F, Synthetic Fibers. Machines and Equipment Manufacture, Properties. Carl Hanser Verlag, 1999.
Fineness of fiber (titer)	dtex	3-20	Fourne, F, Synthetic Fibers. Machines and Equipment Manufacture, Properties. Carl Hanser Verlag, 1999.
Length (elemental fiber)	mm	38-200, filament	Fourne, F, Synthetic Fibers. Machines and Equipment Manufacture, Properties. Carl Hanser Verlag, 1999.

PBT poly(butylene terephthalate)

PARAMETER	UNIT	VALUE	REFERENCES
Rockwell hardness	M	72	
Ball indention hardness at 358 N/30 S (ISO 2039-1)	MPa	130-139; 155 (10-50% glass fiber)	
Shrinkage	%	0.9-2.2; 0.25-0.42 parallel to melt flow, 0.67-1.25 perpendicular to melt flow (glass fiber reinforced)	
Warpage	mm	0.3-1.0	
Viscosity number	ml g^{-1}	107-160; 90-115 (10-50% glass fiber)	
Melt viscosity, shear rate=1000 s^{-1}	Pa s	150-600	
Melt volume flow rate (ISO 1133, procedure B), 220°C/10 kg	cm^3/10 min	9-50; 6-24 (10-50 glass fiber)	
Melt index, 250°C/2.16 kg	g/10 min	18-50	
Water absorption, equilibrium in water at 23°C	%	0.1-0.5	
Moisture absorption, equilibrium 23°C/50% RH	%	0.2-0.25; 0.1-0.2 (10-50% glass fiber)	

CHEMICAL RESISTANCE

Acid dilute/concentrated	-	good/poor	
Alcohols	-	good	
Alkalis	-	poor	
Aliphatic hydrocarbons	-	good	
Aromatic hydrocarbons	-	good	
Esters	-	good-fair	
Greases & oils	-	very good	
Halogenated hydrocarbons	-	poor	
Ketones	-	poor	
Good solvent	-	m-cresol, o-dichlorobenzene, dichlormethane, phenol/tetra-chloroethane, trifluoroacetic acid	
Non-solvent	-	carbon tetrachloride	

FLAMMABILITY

Ignition temperature	°C	93	
Autoignition temperature	°C	>350	
Limiting oxygen index	% O$_2$	22-22.2; 19-20 (10-50% glass fiber)	
Burning rate (Flame spread rate)	mm min^{-1}	<100	
Glow wire test	°C	750-960	
Char at 500°C	%	1.5	Lyon, R E; Walters, R N, J. Anal. Appl. Pyrolysis, 71, 27-46, 2004.
Heat of combustion	J g^{-1}	27,910	Walters, R N; Hacket, S M; Lyon, R E, Fire Mater., 24, 5, 245-52, 2000.
Total heat release	kJ g^{-1}	22.5	Arai, S, Tsunoda, S, Yamaguchi, A, Ougizawa, T, Optic Laser Technol., 117, 94-104, 2019.
Volatile products of combustion	-	CO, CO$_2$, hydrocarbon fragments, tetrahydrofuran	Dzieciol, M, Intern. J. Environ. Anal. Chem., 89, 8-12, 881-89, 2009.
UL 94 rating	-	94HB; V-0 (10 wt% polypentabromobenzyl acrylate and 5 wt% antimony trioxide)	Arai, S, Tsunoda, S, Yamaguchi, A, Ougizawa, T, Optic Laser Technol., 117, 94-104, 2019.

PBT poly(butylene terephthalate)

PARAMETER	UNIT	VALUE	REFERENCES
WEATHER STABILITY			
Spectral sensitivity	nm	290-325	
Activation wavelengths	nm	305, 325	
Excitation wavelengths	nm	340, 383, 395	
Emission wavelengths	nm	440-450	
Activation energy for yellowing	kJ mol^{-1}	16	Pickett, J E, Kuvshinnikova, O, Sung, L-P, Ermi, B D, Polym. Deg. Stab., 166, 135-44, 2019.
Important initiators and accelerators	-	ferrocene, cobalt octoates and naphthenates, compounds containing aromatic keto-ester groups	
Products of degradation	-	radicals, crosslinks, hydroperoxides, hydroxyl and carbonyl groups, CO, CO_2	
Stabilizers	-	UVA: 2-hydroxy-4-octyloxybenzophenone; 2-hydroxy-4-methoxybenzophenone; 2,4-dihydroxybenzophenone; 2,2',4,4'-tetrahydroxybenzophenone; 2,2'-dihydroxy-4-methoxybenzophenone; 2-(2H-benzotriazol-2-yl)-p-cresol; 2-benzotriazol-2-yl-4,6-di-tert-butylphenol; 2-(2H-benzotriazole-2-yl)-4,6-di-tert-pentylphenol; 2-(2H-benzotriazole-2-yl)-4-(1,1,3,3-tetraethylbutyl)phenol; 2-(2H-benzotriazol-2-yl)-4,6-bis(1-methyl-1-phenylethyl) phenol; 2,2'-methylenebis(6-(2H-benzotriazol-2-yl)-4-1,1,3,3-tetramethylbutyl)phenol; 2-(2H-benzotriazol-2-yl)-6-dodecyl-4-methylphenol, branched & linear; 2-(2H-benzotriazol-2-yl)-4,6-bis(1-methyl-1-phenylethyl) phenol; 2,4-di-tert-butyl-6-(5-chloro-2H-benzotriazole-2-yl)-phenol; 2-(3-sec-butyl-5-tert-butyl-2-hydroxyphenyl) benzotriazole; 2-[4-[(2-hydroxy-3-(2'-ethyl)hexyl)oxy]-2-hydroxyphenyl]-4,6-bis(2,4-dimethylphenyl)-1,3,5-triazine; 2-(4,6-diphenyl-1,3,5-triazin-2-yl)-5-hexyloxy-phenol; 2-[4,6-bis(2,4-dimethylphenyl)-1,3,5-triazin-2-yl]-5-(octyloxy) phenol; (2-ethylhexyl)-2-cyano-3,3-diphenylacrylate; 1,3-bis-[(2'-cyano-3',3'-diphenylacryloyl)oxy]-2,2-bis-{[(2'-cyano-3',3'-diphenylacryloyl)oxy]methyl}-propane; propane-dioic acid, [(4-methoxyphenyl)-methylene]-dimethyl ester; 2,2'-(1,4-phenylene)bis[4H-3,1-benzoxazin-4-one]; Screener: carbon black, zinc oxide; Acid scavenger: hydrotalcite; HAS: 1,3,5-triazine-2,4,6-triamine, N,N'''[1,2-ethane-diyl-bis[[[4,6-bis[butyl(1,2,6,6-pentamethyl-4-piperidinyl)amino]-1,3,5-triazine-2-yl]imino]-3,1-propanediyl]bis[N',N''-dibutyl-N',N''-bis(1,2,2,6,6-pentamethyl-4-piperidinyl)-; decanedioic acid, bis(2,2,6,6-tetramethyl-1-(octyloxy)-4-piperidinyl)ester, reaction products with 1,1-dimethylethylhydroperoxide and octane; bis(1,2,2,6,6-pentamethyl-4-piperidyl)sebacate + methyl-1,2,2,6,6-pentamethyl-4-piperidyl sebacate; alkenes, C20-24-. alpha.-, polymers with maleic anhydride, reaction products with 2,2,6,6-tetramethyl-4-piperidinamine; 1, 6-hexanediamine, N, N'-bis(2,2,6,6-tetramethyl-4-piperidinyl)-, polymers with 2,4-dichloro-6-(4-morpholinyl)-1,3,5-triazine; 1,6-hexanedi-amine, N,N'-bis(2,2,6,6-tetramethyl-4-piperidinyl)-, polymers with morpholine-2,4,6-trichloro-1,3,5-triazine reaction products, methylated; Phenolic antioxidants: ethylene-bis(oxyethylene)-bis(3-(5-tert-butyl-4-hydroxy-m-tolyl)-propionate); N,N'-hexane-1,6-diylbis(3-(3,5-di-tert-butyl-4-hydroxy-phenylpropionamide)); 3,3',3',5, 5',5'-hexa-tert-butyl-a,a',a'-(mesitylene-2,4,6-triyl)tri-p-cresol; 1,3,5-tris(3,5-di-tert-butyl-4-hydroxybenzyl)-1,3,5-triazine-2,4,6(1H,3H,5H)- trione; Phosphite: bis-(2,4-di-t-butylphenol) pentaerythritol diphosphite; tris (2,4-di-tert-butylphenyl)phosphite; phos-phoric acid, (2,4-di-butyl-6-methylphenyl)ethylester; distearyl pentaerythritol diphosphite; Optical brightener: 2,2'-(2,5-thio-phenediyl)bis(5-tert-butylbenzoxazole)	

PBT poly(butylene terephthalate)

PARAMETER	UNIT	VALUE	REFERENCES
Results of exposure	-	3-year exposure in Central Europe caused only slight color change; 90% tensile strength is retained after 3,600 h in Xenotest 1200	Ultradur Brochure KT/K, F204, BASF 2010.

TOXICITY

HMIS: Health, Flammability, Reactivity rating	-	1/1/0; 0/1/0	
Carcinogenic effect	-	not listed by ACGIH, NIOSH, NTP	
TLV, ACGIH	mg m^{-3}	3 (respirable), 10 (total)	
OSHA	mg m^{-3}	5 (respirable), 15 (total)	

PROCESSING

Typical processing methods	-	coating, electrospinning, extrusion, extrusion foaming; film extrusion, injection molding, monofilament extrusion, rotational molding	
Preprocess drying: temperature/ time/residual moisture	°C/h/%	80-130/2-8/0.02-0.04	
Processing temperature	°C	235-270 (injection molding); 230-290 (coating); 230-270 (pipe extrusion); 230-270 (extrusion)	
Processing pressure	bar	30 (extrusion melt pressure); <100 (extrusion backpressure); 20-150 (injection pressure); 0.3-0.7 (injection molding back-pressure)	
Residence time	min	<35; 2.5 (max cycle time)	
Additives used in final products	-	Fillers: antimony trioxide, aramid, barium sulfate, boron nitride, calcinated kaolin, carbon black, carbon fiber, glass fiber, glass spheres, mica, montmorillonite, talc, titanium dioxide, zinc borate; Antistatics: antimony-doped tin oxide, carbon nano-tubes, polyaniline, polyisonaphthalene; Antiblocking: calcium carbonate, diatomaceous earth, silicone fluid, spherical sili-cone resin, synthetic silica; Release: calcium stearate, fluorine compounds, glycerol bistearate, pentaerythritol ester, silane modified silica, zinc stearate; Slip: spherical silica, silicone oil	
Applications	-	automotive (connectors, distributor caps, door knobs, lamp sockets, lightning bezels, mirror housings, windshield wiper arms), brush bristles, composites, automobile lamps, con-soles, contact carriers, covers, electrical and electronics (capacitors, connectors, circuit breakers, capacitor housings, encapsulation of transformers, fibers, motors, and solenoids, printed circuit boards, relays), housewares, housings, lighting, plumbing, power tools, sporting goods, textiles, tire cords	
Outstanding properties	-	dimensional stability, weather resistance, resistance to fuels, impact resistance, electrical insulation properties	

BLENDS

Suitable polymers	-	ASA, epoxy, ICP, PA12, PC, PCL, PEO, PET, PHB, PLA, PU, PVB, SAN	

ANALYSIS

FTIR (wavenumber-assignment)	cm^{-1}/-	C–H – 2925, 2853, 1456, 1018, 872; C=O – 1707; aromatic ring - 1407; CO–O – 1241; O–CH$_2$ – 1096	Deshmukh, G S; Peshwe, D R; Pathak, S U; Ekhe, J D, Thermo-chim. Acta, 581, 41-53, 2014.
x-ray diffraction peaks	degree	15.7 (011), 16.9 (010), 20.3 (102), 23.1 (100), 24.9 (111) 16.15 (011), 17.01 (010), 20.46 (110), 23.06 (100), 24.77 (111)	Deshmukh, G S; Peshwe, D R; Pathak, S U; Ekhe, J D, Ther-mochim. Acta, 581, 41-53, 2014; Soudmand, B H; Shelesh-Nezhad, K, Tribology Int., 151, 106439, 2020.

PC polycarbonate

PARAMETER	UNIT	VALUE	REFERENCES
GENERAL			
Common name	-	polycarbonate	
IUPAC name	-	poly(oxycarbonyloxy-1,4-phenylene(dimethylmethylene)-1,4-phenylene]	
ACS name	-	carbonic acid, polymer with 4,4'-(1-methylethylidene) bis[phenol]; polycarbonates	
Acronym	-	PC	
CAS number	-	25037-45-0; 25766-59-0	
RTECS number	-	TR1580150	
Formula			
HISTORY			
Person to discover	-	Hermann Schnell	
Date	-	1953; 1958	
Details	-	Bayer chemist synthesized polycarbonate in the first attempt; the experiment was based on analysis of previous research and selection of right building blocks; in order to succeed with implementation, he had to overcome scepticism of his peers that such reaction and outcome are possible; polymer was patented immediately but production begun in 1958	
SYNTHESIS			
Monomer(s) structure	-		
Monomer(s) CAS number(s)	-	80-05-7; 75-44-5	
Monomer(s) molecular weight(s)	dalton, g/mol, amu	228.29; 98.92	
Monomer(s) expected purity(ies)	%	98; 99	
Method of synthesis	-	Bisphenol A is treated with NaOH, which is then reacted with phosgene. It can also be manufactured by transesterification of bisphenol A with diphenyl carbonate.	
Temperature of polymerization	°C	30-40 (polycondensation); 160-240 (transesterification)	Couper, J R; Penney, W R; Fair, J R; Walas, S M, Chemical Equipment, 2nd Ed., Elsevier, 2010, pp. 581-640; Kim, J; Kim, Y J; Kim, J-D; Ahmed, T S; Dong, L B; Roberts, G W; Oh, S-G, Polymer, 2520-26, 2010.
Time of polymerization	h	0.25-4 (polycondensation)	
Pressure of polymerization	Pa	atmospheric	
Catalyst	-	benzyltriethylammonium chloride (polycondesation); LiOH (transesterification)	Couper, J R; Penney, W R; Fair, J R; Walas, S M, Chemical Equipment, 2nd Ed., Elsevier, 2010, pp. 581-640; Kim, J; Kim, Y J; Kim, J-D; Ahmed, T S; Dong, L B; Roberts, G W; Oh, S-G, Polymer, 2520-26, 2010.
Number average molecular weight, M_n	dalton, g/mol, amu	17,500-41,300	

PC polycarbonate

PARAMETER	UNIT	VALUE	REFERENCES
Mass average molecular weight, M_w	dalton, g/mol, amu	19,000-56,000; 73,000-103,000 (isosorbide polycarbonates)	Yum, S, Kim, H, Seo, Y, Polymer, 179, 121685, 2019.
Polydispersity, M_w/M_n	-	1.3-3.2	
Molar volume at 298K	cm^3 mol^{-1}	calc.=213.8; exp.=211.9	
Van der Waals volume	cm^3 mol^{-1}	calc.=136.21; exp.=138.36	
Molecular cross-sectional area, calculated	$cm^2 \times 10^{-16}$	19.8	
Radius of gyration	nm	13	Fornes, T D; Baur, J W; Sabba, Y; Thomas, E L, Polymer, 47, 1704-14, 2006.

STRUCTURE			
Crystallinity	%	typically amorphous because of its rigid backbone; 20-42	Farmer, R, A Study of Crystallization in Bisphenol A Polycarbonate. Dissertation, Virginia Polytgechnic Institute, 2001; Kim, J; Kim, Y J; Kim, J-D; Ahmed, T S; Dong, L B; Roberts, G W; Oh, S-G, Polymer, 2520-26, 2010.
Entanglement molecular weight	dalton, g/mol, amu	calc.=1,734-2,495; exp.=2,495-4,800	
Lamellae thickness	nm	3.18-5.39	Farmer, R, A Study of Crystallization in Bisphenol A Polycarbonate. Dissertation, Virginia Polytgechnic Institute, 2001.

COMMERCIAL POLYMERS			
Some manufacturers	-	Bayer; Sabic	
Trade names	-	Makrolon; Lexan	

PHYSICAL PROPERTIES			
Density at 20°C	g cm^{-3}	1.19-1.22; 1.04 (melt); 1.25-1.52 (10-40% glass fiber)	
Bulk density at 20°C	g cm^{-3}	0.66 (pellets)	
Refractive index, 20°C	-	calc.=1.5773-1.587; exp.=1.586-1.587	
Birefringence	-	-0.001 (surface) to 0.001 (bulk)	Lin, T H; Isayev, A I, Antec, 1664-68, 2006.
Transmittance	%	82-91	
Haze	%	<0.8 to 3	
Melting temperature, DSC	°C	255-267	
Storage temperature (max)	°C	93	
Decomposition onset temperature	°C	420	
Thermal expansion coefficient, -40 to 95°C	°C^{-1}	6.0-7.5E-5; 1.6-6.5E-5 (10-40% glass fiber)	
Thermal conductivity, melt	W m^{-1} K^{-1}	0.19-0.24; 0.2-0.22 (10-40% glass fiber)	
Thermal conductivity, panels	W m^{-1} K^{-1}	0.0.0528-0.0.0668, 0.04529-0.06242 (declared value by manufacturers)	Čekon. M, Šikula, O, J. Build. Eng., 32, 101715, 2020.
U value	W m$_{-2}$ K$_{-1}$	1.3-3	Čekon. M, Šikula, O, J. Build. Eng., 32, 101715, 2020.

PC polycarbonate

PARAMETER	UNIT	VALUE	REFERENCES
Glass transition temperature	°C	calc.=134-158; exp.=137-154	
Specific heat capacity	J K^{-1} kg^{-1}	1,200; 1,040-1,210 (10-40% glass fiber)	
Heat of fusion	kJ kg^{-1}	100-115 (crystal); 22.95-23.11 (heat to melt)	DeLassus, P T; Landes, B G; Harris, L M, Antec, 1636-39, 1997.
Maximum service temperature	°C	-40 to 130	
Temperature index (50% tensile strength loss after 20,000 h/5000 h)	°C	120	
Heat deflection temperature at 0.45 MPa	°C	133-166; 141-154 (10-40% glass fiber)	
Heat deflection temperature at 1.8 MPa	°C	119-156; 135-146 (10-40% glass fiber)	
Vicat temperature VST/A/50	°C	138-150; 144-149 (10-35% glass fiber)	
Vicat temperature VST/B/50	°C	144-170; 154 (10-40% glass fiber)	
Hansen solubility parameters, δ_D, δ_P, δ_H	MPa$^{0.5}$	19.6, 8.8, 5.7	
Interaction radius		10.2	
Hildebrand solubility parameter	MPa$^{0.5}$	22.2	
Surface tension	mN m^{-1}	calc.=33.0-37.9; exp.=28.4-42.9	Liao, C-C; Wang, C-C; Shih, K-C; Chen, C-Y, Eur. Polym. J., 47, 911-24, 2011.
Dielectric constant at 100 Hz/1 MHz	-	3.2/2.9	
Relative permittivity at 100 Hz	-	2.9-3.2; 3.1-3.6 (10-40% glass fiber)	
Relative permittivity at 1 MHz	-	2.8-3.1; 3.0-3.6 (10-40% glass fiber)	
Dissipation factor at 100 Hz	E-4	5-310; 8-13 (10-40% glass fiber)	
Dissipation factor at 1 MHz	E-4	90-120; 67-90 (10-40% glass fiber)	
Volume resistivity	ohm-m	1E12 to 1E15	
Surface resistivity	ohm	1E15 to 1E17	
Electric strength K20/P50, d=0.60.8 mm	kV mm^{-1}	15-67; 18-36 (10-40% glass fiber)	
Comparative tracking index, CTI, test liquid A	-	200-250; 175 (10-35% glass fiber)	
Comparative tracking index, CTIM, test liquid B	-	125M	
Coefficient of friction	ASTM D1894	0.21 (chrome steel); 0.41-0.54 (aluminum)	Maldonado, J E, Antec, 3431-35, 1998.
Permeability to nitrogen, 25°C	cm^3 m^{-2} 24 h^{-1} bar^{-1}	130 (100 µm film), 510 (25.4 µm film)	
Permeability to oxygen, 25°C	cm^3 m^{-2} 24 h^{-1} bar^{-1}	700 (100 µm film), 2,760 (25.4 µm film)	
Permeability to water vapor, 25°C	g m^{-2} 24 h^{-1} bar^{-1}	15	
Contact angle of water, 20°C	degree	81.3-84.0	
Surface free energy	mJ m^{-2}	42.3	
Speed of sound	m s^{-1}	38	
Acoustic impedance		2.69-2.77	

PC polycarbonate

PARAMETER	UNIT	VALUE	REFERENCES
Attenuation	dB cm⁻¹, 5 MHz	22.1-24.9	

MECHANICAL & RHEOLOGICAL PROPERTIES

PARAMETER	UNIT	VALUE	REFERENCES
Tensile strength	MPa	55-88; 45-158 (10-40% glass fiber)	Chang, M C O; Garrett, P D, Antec, 2588-93, 1996.
Tensile modulus	MPa	2,200-3,100; 3,800-9,400 (10-35% glass fiber)	
Tensile stress at yield	MPa	57-74	Chang, M C O; Garrett, P D, Antec, 2588-93, 1996.
Tensile creep modulus, 1000 h, elongation 0.5 max	MPa	1,700-1,900; 2,900-8,500 (10-35% glass fiber)	
Elongation	%	66-140; 1.8-15 (10-40% glass fiber)	Chang, M C O; Garrett, P D, Antec, 2588-93, 1996.
Tensile yield strain	%	5.7-7; 8 (10-40% glass fiber)	
Flexural strength	MPa	94-120; 103-186 (10-40% glass fiber)	
Flexural modulus	MPa	2,220-2,600; 3,440-9,600 (10-40% glass fiber)	
Elastic modulus	MPa	1,600	
Compressive strength	MPa	70	
Young's modulus	MPa	2,390-2,600	Chang, M C O; Garrett, P D, Antec, 2588-93, 1996.
Charpy impact strength, unnotched, 23°C	kJ m⁻²	no break; 40-150 (10-35% glass fiber)	
Charpy impact strength, unnotched, -30°C	kJ m⁻²	no break	
Charpy impact strength, notched, 23°C	kJ m⁻²	11-80; 8-12 (10-35% glass fiber)	
Charpy impact strength, notched, -30°C	kJ m⁻²	9-14	
Izod impact strength, unnotched, 23°C	J m⁻¹	12-90 to NB; 1280-2140 (10-40% glass fiber)	
Izod impact strength, notched, 23°C	J m⁻¹	12-736; 8-133 (10-40% glass fiber)	
Izod impact strength, notched, -30°C	J m⁻¹	55-618	
Shear modulus	MPa	805	Ozcelik, B; Sonat, I, Mater. Design, 30, 367-75, 2009.
Abrasion resistance (ASTM D1044)	mg/1000 cycles	11 (10-40% glass fiber)	
Poisson's ratio	-	calc.=0.424; exp.=0.401-0.420	Lin, T H; Isayev, A I, Antec, 1664-68, 2006.
Rockwell hardness	-	L89; M85, R124 (10-40% glass fiber)	
Ball indention hardness at 358 N/30 S (ISO 2039-1)	MPa	116	
Shrinkage	%	0.4-0.9; 0.2-0.55 (10-35% glass fiber)	
Melt volume flow rate (ISO 1133, procedure B), 300°C/1.2 kg	cm³/10 min	1.25-36	
Pressure coefficient of melt viscosity, b	G Pa⁻¹	26.6	Aho, J; Syrjala, S, J. Appl. Polym. Sci., 117, 1076–84, 2010.
Melt index, 300°C/1.2 kg	g/10 min	6-80	
Water absorption, equilibrium in water at 23°C	%	0.12-0.40; 0.23 (10-40% glass fiber)	

PC polycarbonate

PARAMETER	UNIT	VALUE	REFERENCES
Moisture absorption, equilibrium 23°C/50% RH	%	0.09-0.3	

CHEMICAL RESISTANCE

PARAMETER	UNIT	VALUE	REFERENCES
Acid dilute/concentrated	-	good/poor	
Alcohols	-	good	
Alkalis	-	good-poor	
Aliphatic hydrocarbons	-	good	
Aromatic hydrocarbons	-	poor	
Esters	-	poor	
Greases & oils	-	good-poor	
Halogenated hydrocarbons	-	poor	
Ketones	-	poor	
Θ solvent, Θ-temp.=170°C	-	n-butyl benzyl ether	
Good solvent	-	acetophenone (hot), aniline (hot), benzene (hot), chloroform, cresol, 1,2-dichloroethane, methylene chloride	
Non-solvent	-	amyl alcohol, ethylene glycol, heptane, isopropyl alcohol	
Chemicals causing environmental stress cracking	list	acetone, benzyl alcohol, carbon tetrachloride, cyclohexanone, nitrobenzene	Wang, H T; Pan, Q G; Du, Q C; Li, Y Q, Polym. Test., 22, 125-28, 2003.
Effect of EtOH sterilization (tensile strength retention)	%	98-100	Navarrete, L; Hermanson, N, Antec, 2807-18, 1996.

FLAMMABILITY

PARAMETER	UNIT	VALUE	REFERENCES
Ignition temperature	°C	480	
Autoignition temperature	°C	550	
Limiting oxygen index	% O_2	25-30; 30-43 (flame retardant grades)	
Minimum ignition energy	J	0.025	
Heat release	kW m^{-2}	479-548; 124-385 (with fire retardant)	Yu, B; Liu, M; Lu, L; Dong, X; Gao, W; Tang, K, Fire Mater., 34, 251-61, 2010.
NBS smoke chamber	Ds	190	Padey, D; Walling, J; Wood A, Polymers in Defence and Aerospace 2007, Rapra, 2007, paper 15.
Time to ignition	s	48-193 (depending on heat flux)	Ghazzawi, Y M, Osorio, A F, Heitzmann, M T, Constr. Build. Mater., 234, 117889, 2020.
Burning time	s	393-495 (depending on heat flux)	Ghazzawi, Y M, Osorio, A F, Heitzmann, M T, Constr. Build. Mater., 234, 117889, 2020.
Burning rate (Flame spread rate)	mm min^{-1}	passed	
Peak of heat release rate	kW m^{-2}	284-445 (depending on heat flux)	Ghazzawi, Y M, Osorio, A F, Heitzmann, M T, Constr. Build. Mater., 234, 117889, 2020.
Total heat release	MJ m^{-2}	101-113 (depending on heat flux)	Ghazzawi, Y M, Osorio, A F, Heitzmann, M T, Constr. Build. Mater., 234, 117889, 2020.
Char at 500°C	%	21.7	Lyon, R E; Walters, R N, J. Anal. Appl. Pyrolysis, 71, 27-46, 2004.
Heat of combustion	J g^{-1}	31,060-31,530	Walters, R N; Hacket, S M; Lyon, R E, Fire Mater., 24, 5, 245-52, 2000.

PC polycarbonate

PARAMETER	UNIT	VALUE	REFERENCES
Volatile products of combustion	-	carbon monoxide, carbon dioxide, bisphenol A, diphenyl carbonate, phenol and phenol derivatives. Traces of aliphatic and aromatic hydrocarbons, aldehydes and acids.	
UL 94 rating	-	HB to V-2; V-0 (some flame retardant grades); HB to V-0 (10-35% glass fiber)	

WEATHER STABILITY

PARAMETER	UNIT	VALUE	REFERENCES
Spectral sensitivity	nm	<275 (completely absorbed); 260-300 (photo-Fries rearrangement); 280-305; 330-360	
Activation wavelengths	nm	290-320; 310 (chain scission), 330-360	
Activation energy of radiacal formation by UV	kJ mol^{-1}	238.07	
Activation energy of photo-Fries rearrangement producing a hydroxyl group vicinal to the ester functionality	kJ mol^{-1}	-8.3	Motta, A, La Mantia, F P, Ascione, L, Mistretta, M C, J. Mol. Graphics Modelling, 99, 107622, 2020.
Activation energy for yellowing on exposure to UV	kJ mol^{-1}	21-24	Pickett, J E, Kuvshinnikova, O, Sung, L-P, Ermi, B D, Polym. Deg. Stab., 166, 135-44, 2019.
Important initiators and accelerators	-	4-hydroxystilbene; products of thermal degradation; bisphenol A, stilbene-like structures, water, bis(3-hydroxyphenyl)ether structures in main chain, some inorganic pigments	
Products of degradation	-	photo-Fries rearrangement; chain scissions, crosslinks, free radicals, hydroxyl groups, ethers, unsaturations (photolysis); chain scissions, hydroperoxides, free radicals, hydroxyl groups, carbonyl groups (photooxidation)	
Stabilizers	-	UVA: 2-hydroxy-4-octyloxybenzophenone; 2,2',4,4'-tetrahydroxybenzophenone; 2-(2H-benzotriazol-2-yl)-p-cresol; 2-benzotriazol-2-yl-4,6-di-tert-butylphenol; 2-(2H-benzotriazole-2-yl)-4,6-di-tert-pentylphenol; 2-(2H-benzotriazole-2-yl)-4-(1,1,3,3-tetraethylbutyl)phenol; 2-(2H-benzotriazol-2-yl)-4,6-bis(1-methyl-1-phenylethyl)phenol; 2,2'-methylenebis(6-(2H-benzotriazol-2-yl)-4-1,1,3,3-tetramethylbutyl)phenol; 2-(2H-benzotriazol-2-yl)-6-dodecyl-4-methylphenol, branched & linear; 2-(2'-hydroxy-5'-methacryloxyethylphenyl)-2H-benzotriazole; 2-(2H-benzotriazol-2-yl)-4,6-bis(1-methyl-1-phenylethyl)phenol; 2-(3-sec-butyl-5-tert-butyl-2-hydroxyphenyl)benzotriazole; reaction product of methyl 3(3-(2H-benzotriazol-2-yl)-5-t-butyl-4-hydroxyphenyl propionate/PEG 300; 2-(4,6-diphenyl-1,3,5-triazin-2-yl)-5-hexyloxy-phenol; 2-[4,6-bis(2,4-dimethylphenyl)-1,3,5-triazin-2-yl]-5-(octyloxy)phenol; 1,3-bis-[(2'-cyano-3',3'-diphenylacryloyl)oxy]-2,2-bis-{[(2'-cyano-3',3'-diphenylacryloyl)oxy]methyl}-propane; propanedioic acid, [(4-methoxyphenyl)-methylene]-dimethyl ester; 2,2'-(1,4-phenylene)bis[4H-3,1-benzoxazin-4-one]; Screener: zinc oxide; Phenolic antioxidant: 2-(1,1-dimethylethyl)-6-[[3-(1,1-dimethylethyl)-2-hydroxy-5-methylphenyl] methyl-4-methylphenyl acrylate; Phosphite: tris (2,4-di-tert-butylphenyl)phosphite; isodecyl diphenyl phosphite; di(p-butoxyphenyl)cyclohexylphosphine oxide; Optical brightener: 2,2'-(2,5-thiophenediyl)bis(5-tert-butylbenzoxazole); 2,2'-(1,2-ethylenediyl-di-4,1-phenylene)bisbenzoxazole; Plasticizer: dicyclohexyl phthalate (reduces yellowing on exposure to gamma radiation	
Results of exposure	-	yellowness index: 45 after 60 month exposure in Florida without stabilization and 22 after 60 month exposure in Florida with stabilization; 13 and 26 after 120 month exposure in Engerfeld, Germany with and without stabilization, respectively	

PC polycarbonate

PARAMETER	UNIT	VALUE	REFERENCES
Low earth orbit erosion yield	cm^3 atom^{-1} x 10^{-24}	4.29	Waters, D L; Banks, B A; De Groh, K K; Miller, S K R; Thorson, S D, High Performance Polym., 20, 512-22, 2008.

BIODEGRADATION

Colonized products	-	compact disks, medical devices	
Typical biodegradants	-	microorganisms producing esterase are capable of damage; marine environment	Artham, T; Doble, M, J. Polym. Environ., 17, 170-80, 2009.
Stabilizers	-	silver-based biocides	

TOXICITY

NFPA: Health, Flammability, Reactivity rating	-	0/1/0	
Carcinogenic effect	-	not listed by ACGIH, NIOSH, NTP	
OSHA	mg m^{-3}	5 (respirable), 15 (total)	
Oral rat, LD$_{50}$	mg kg^{-1}	>10,000	
Skin rabbit, LD$_{50}$	mg kg^{-1}	>2,000	

ENVIRONMENTAL IMPACT

Aquatic toxicity, *Daphnia magna*, LC$_{50}$, 48 h	mg l^{-1}	5,000	Lither, D; Damberg, J; Dave, G; Larsson, A, Chemosphere, 74, 1198^{-1}200, 2009.
Cradle to grave non-renewable energy use	MJ/kg	111	
Cradle to pellet greenhouse gasses	kg CO$_2$ kg^{-1} resin	7.8	

PROCESSING

Typical processing methods	-	blow molding, calendering, electrospinning, extrusion, gas-assisted injection molding, injection molding, solution casting, thermoforming	
Preprocess drying: temperature/ time/residual moisture	°C/h/%	120-135/2-12/0.02	
Processing temperature	°C	280-355; 310-340 (10-40% glass fiber)	
Processing pressure	MPa	0.3-0.7 (back pressure)	
Additives used in final products	-	Fillers: boric oxide, glass beads, carbon black and graphite fibers for EMI shielding, glass fiber, graphite, molybdenum sulfide, nanosilica, nickel-coated graphite fibers, PTFE, single-walled carbon nanotubes, steel fibers, titanium dioxide, wollastonite; Plasticizers: dibutyl phthalate, dioctyl phthalate, dicyclohexyl phthalate, mineral oil, pentaerythritol tetraborate, trimellitic acid tridecyloctyl ester, tritolyl phosphate, tetra-ethylene glycol dimethyl ether, tri-(2-ethylhexyl) phosphate; Antistatics: carbon black, carbon nanotubes, copper oxide, glycerol mono-iso-stearate, indium tin oxide, nickel-coated carbon fiber, polyetheresteramide, polyoxyethylene fatty acid ester; Antiblocking: amorphous silica, calcium carbonate, dimethylsiloxane grafting, siloxane particles; Release: glycerol monostearate, pentaerythritol tetrastearate, siloxane, zinc stearate; impact modifier; UV stabilizers; release agents	

340　　　**HANDBOOK OF POLYMERS** 3rd Edition, Copyrights 2022; ChemTec Publishing

PC polycarbonate

PARAMETER	UNIT	VALUE	REFERENCES
Applications	-	bearings, blood collector containers, camera components, computer printers, copying machines, corrective eyeglasses, dental applications, data storage (CD, DVD, etc.), dinnerware, drinking cups, disposable syringes, head lamp covers and housings, gears, glazing, goggles, golf tees, guide pins, helmets, instrument panels, laminated walls, lenses, medical tubing, microfibers, needles syringes, optical lenses, pacemaker components, projection screens, rollers, roofing, safety glasses, skylights, speedometer needles, solar modules, tool boxes, toys, water bottles, windows, windscreens	
BLENDS			
Suitable polymers	-	ABS, PBT, PET, PHEMA, PLA, PVDF	
ANALYSIS			
FTIR (wavenumber-assignment)	cm^{-1}/-	C-H – 3047, 2877; C-O-C – 989; C=O – 1778	Abdel-Salam, M H; Nouh, S A; Radwan, Y E; Fouad, S S, Mater. Chem. Phys., 127, 305-9, 2011.
Raman (wavenumber-assignment)	cm^{-1}/-	C=O – 1769, 1597, 712; CH_3 – 1453	Hoeller, T L, Antec, 3124-30, 2007.

PCL poly(ε-caprolactone)

PARAMETER	UNIT	VALUE	REFERENCES
GENERAL			
Common name	-	poly(ε-caprolactone)	
CAS name	-	2-oxepanone, homopolymer; poly[oxy(1-oxo-1,6-hexanediyl)]	
Acronym	-	PCL	
CAS number	-	24980-41-4; 25248-42-4	
Formula		$-\left[O(CH_2)_5\overset{\overset{\textstyle O}{\|\|}}{C}\right]n$	
HISTORY			
Person to discover	-	Hostettler, F	Hostettler, F, US Patent 2,933,477, Union Carbide, Apr. 19, 1960.
Date	-	1960	
Details	-	polycaprolactone synthesis	
SYNTHESIS			
Monomer(s) structure	-		
Monomer(s) CAS number(s)	-	502-44-3	
Monomer(s) molecular weight(s)	dalton, g/mol, amu	114.14	
Method of synthesis	-	ring-opening addition polymerization of ε-caprolactone at 170ºC under nitrogen with dibutyl stanneous oxide as the catalyst; also condensation of 6-hydroxycaproic acid gives PCL; polymerization is frequently conducted in extruder	Kim, B J; White, J L, Antec, 224-8, 2000; Labet, M; Thielemans, Chem. Soc. Rev., 38, 3484-3504, 2009.
Temperature of polymerization	ºC	130	
Catalyst	-	99 (reactive extrusion)	Raquez, J-M; Degee, P; Dubois, P; Balakrishnan, S; Narayan, R, Polym. Eng. Sci., 45, 622-29, 2005.
Number average molecular weight, M_n	dalton, g/mol, amu	530-630,000	
Mass average molecular weight, M_w	dalton, g/mol, amu	10,000-200,000	
Polydispersity, M_w/M_n	-	1.08-1.53	
STRUCTURE	-		
Crystallinity	%	57-76	Labet, M; Thielemans, Chem. Soc. Rev., 38, 3484-3504, 2009; Gumusderelioglu, M; Kaya, F B; Beskardes, I G, J. Colloid Interface Sci., 358, 444-53, 2011.
Cell type (lattice)	-	orthorhombic	
Cell dimensions	nm	a:b:c=0.745:0.498:1.705	
Number of chains per unit cell	-	4	
Chain conformation	-	nearly planar	
Rapid crystallization temperature	ºC	58-150	

PCL poly(ε-caprolactone)

PARAMETER	UNIT	VALUE	REFERENCES
COMMERCIAL POLYMERS			
Some manufacturers	-	BASF; eSun; Ingevity	
Trade names	-	Capromer; eMate-PCL; Capa	
PHYSICAL PROPERTIES			
Density at 20°C	g cm^{-3}	1.07-1.20	
Color	-	white	
Odor	-	odorless	
Melting temperature, DSC	°C	58-63; 57.8	-; Selli, F, Gooneie, A, Erdoğan, U H, Hufenus, R, Perret, E, Data in Brief, 32, 106223, 2020.
Crystallization temperature	°C	18.1	Selli, F, Gooneie, A, Erdoğan, U H, Hufenus, R, Perret, E, Data in Brief, 32, 106223, 2020.
Decomposition temperature	°C	200-220; 385.1 (TGA)	-; Selli, F, Gooneie, A, Erdoğan, U H, Hufenus, R, Perret, E, Data in Brief, 32, 106223, 2020.
Glass transition temperature	°C	-60 to -72	
Heat of fusion	kJ mol^{-1}	8.9	
Melting enthalphy	J g^{-1}	65.2	Selli, F, Gooneie, A, Erdoğan, U H, Hufenus, R, Perret, E, Data in Brief, 32, 106223, 2020.
Contact angle of water, 20°C	degree	77-141 (advancing); 35-54 (receding)	Gumusderelioglu, M; Kaya, F B; Beskardes, I G, J. Colloid Interface Sci., 358, 444-53, 2011.
MECHANICAL & RHEOLOGICAL PROPERTIES			
Tensile strength	MPa	7.6-58	
Tensile modulus	MPa	200-1,380	
Elongation	%	300-600	
Flexural modulus	MPa	200-500	
Elastic modulus	MPa	237-288 (microfiber)	
Young's modulus	MPa	210-440	
Izod impact strength, notched, 23°C	J m^{-1}	120-375	Wei, X; Wong, S-C; Baji, A, Antec, 2737-41, 2009.
Viscosity number	ml g^{-1}	70-130	
Melt viscosity, shear rate=1000 s^{-1}	Pa s	1500	
CHEMICAL RESISTANCE			
Alcohols	-	good	
Aromatic hydrocarbons	-	poor	
Halogenated hydrocarbons	-	poor	
Ketones	-	poor	
Good solvent	-	benzene, chloroform, dimethylacetamide, THF	
Non-solvent	-	methanol	
FLAMMABILITY			
Ignition temperature	°C	275	
Heat release	kJ g^{-1}	24.4	Lyon, R E; Walters, R N, J. Anal. Appl. Pyrolysis, 71, 27-46, 2004.

PCL poly(ε-caprolactone)

PARAMETER	UNIT	VALUE	REFERENCES
Char at 500°C	%	0	Lyon, R E; Walters, R N, J. Anal. Appl. Pyrolysis, 71, 27-46, 2004.
Heat of combustion	J g⁻¹	24,400	
BIODEGRADATION			
Typical biodegradants	-	*Pseudomonas, Alcanivorax,* and *Tenacibaculum;* lipase and cutinase *from Pichia pastoris*	Sekiguchi, T; Saika, A; Nomura, K; Watanabe, T; Watanabe, T; Fujimoto, Y; Enoki, M; Sata, T; Kato, C; Kanehiro, H, Polym. Deg. Stab., 96, 1397-1403, 2011; Liu, M, Zhang, T, Long, L, Zhang, R, Ding, S, Polym. Deg. Stab., 160, 120-5, 2019.
TOXICITY			
Carcinogenic effect	-	not listed by ACGIH, NIOSH, NTP	
OSHA	mg m⁻³	5 (respirable), 15 (total)	
Oral rat, LD$_{50}$	mg kg⁻¹	10,000	
PROCESSING			
Processing method	-	compounding, extrusion, electrospinning, PU prepolymer synthesis	
Applications	-	component of polyurethanes, fibers, medical, solid electrolyte for all-solid-state batteries	Zhang, B, Liu, Y, Liu, J, Sun, L, Pan, X, J. Energy Chem., 52, 318-25, 2021.
Outstanding properties	-	biodegradable	
BLENDS			
Suitable polymers	-	ABS, CB, NC, PBA, PC, PHB, PLA, PPy, PVAl, PVC, chitosan, starch	
ANALYSIS			
FTIR (wavenumber-assignment)	cm⁻¹/-	C=O – 1778; C=C – 1642; C-O – 1164, 1107;	Vogel, C; Siesler, H W, Macromol. Symp., n265, 1483-94, 2008.

PCS polycarbodihydridosilane

PARAMETER	UNIT	VALUE	REFERENCES
GENERAL			
Common name	-	polycarbodihydridosilane, hydridopolycarbosilane	
Acronym	-	PCS, HPCS	
CAS number	-	74056-94-3	
Linear formula		$(CH_4Si)_n$	
PHYSICAL PROPERTIES			
Density at 20°C	g cm^{-3}	0.95	
Refractive index, 20°C	-	1.509	
FLAMMABILITY			
Flash point	°C	26	
TOXICITY			
HMIS: Health, Flammability, Reactivity rating	-	0/3/0	
PROCESSING			
Applications	-	silicon (oxy)carbide precursor, wear coating and sealer coating for ceramic composites and precursor for porous composites for high-temperature filters and catalysts	Yan, X, Sahimi, M, Tsotsis, T T, Microporous Mesoporous Mater., 241, 338-45, 2017.

PCT poly(cyclohexylene terephthalate)

PARAMETER	UNIT	VALUE	REFERENCES
GENERAL			
Common name	-	poly(cyclohexylene terephthalate)	
CAS name	-	poly(oxycarbonyl-1,4-phenylenecarbonyloxymethylene-1,4-cyclohexanediylmethylene); 1,3-benzenedicarboxylic acid, polymer with 1,4-benzenedicarboxylic acid and 1,4-cyclohex-anedimethanol	
Acronym	-	PCT	
CAS number	-	24936-69-4; 26124-27-6	
Formula			
SYNTHESIS			
Monomer(s) structure	-		
Monomer(s) CAS number(s)	-	121-91-5; 105-08-8	
Monomer(s) molecular weight(s)	dalton, g/mol, amu	166.13; 144.24	
COMMERCIAL POLYMERS			
Some manufacturers	-	Celanese	
Trade names	-	Thermx	
PHYSICAL PROPERTIES			
Density at 20°C	g cm^{-3}	1; 1.38-1.55 (20-40% glass fiber)	
Bulk density at 20°C	g cm^{-3}	0.6-0.9 (20-40% glass fiber)	
Refractive index, 20°C	-	1.597-1.605	
Odor		slight, specific	
Melting temperature, DSC	°C	>200; 285 (20-40% glass fiber)	
Decomposition temperature	°C	350	
Thermal expansion coefficient, 23-80°C	10^{-4} °C^{-1}	0.3-0.8	
Thermal conductivity, melt	W m^{-1} K^{-1}	0.2	
Glass transition temperature	°C	87-105	
Specific heat capacity	J K^{-1} kg^{-1}	1,470	
Heat deflection temperature at 1.8 MPa	°C	253-262 (20-40% glass fiber)	
Volume resistivity	ohm-m	1E13	
Surface resistivity	ohm	1E15	
Electric strength K20/P50, d=0.60.8 mm	kV mm^{-1}	41	
MECHANICAL & RHEOLOGICAL PROPERTIES			
Tensile strength	MPa	45; 100-130 (20-40% glass fiber)	
Tensile modulus	MPa	6,400-13,000 (20-40% glass fiber)	

346   **HANDBOOK OF POLYMERS** 3ʳᵈ Edition, Copyrights 2022; ChemTec Publishing

PCT poly(cyclohexylene terephthalate)

PARAMETER	UNIT	VALUE	REFERENCES
Tensile creep modulus, 1 h/1,000 h, elongation 0.5 max	MPa	6,000/4,600 (20-40% glass fiber)	
Elongation	%	1.3-2.3 (20-40% glass fiber)	
Flexural strength	MPa	155-178 (20-40% glass fiber)	
Flexural modulus	MPa	5,800-12,800 (20-40% glass fiber)	
Charpy impact strength, unnotched, 23°C	kJ m^{-2}	35-55 (20-40% glass fiber)	
Charpy impact strength, unnotched, -30°C	kJ m^{-2}	30 (20-40% glass fiber)	
Charpy impact strength, notched, 23°C	kJ m^{-2}	7-8 (20-40% glass fiber)	
Charpy impact strength, notched, -30°C	kJ m^{-2}	7 (20-40% glass fiber)	
Izod impact strength, notched, 23°C	J m^{-1}	6 (20-40% glass fiber)	
Shrinkage	%	0.3-0.8 (20-40% glass fiber)	
Melt volume flow rate (ISO 1133, procedure B), 300°C/2.16 kg	cm^3/10 min	30 (20-40% glass fiber)	
Water absorption, equilibrium in water at 23°C	%	1.1	
Moisture absorption, equilibrium 23°C/50% RH	%	0.1	

CHEMICAL RESISTANCE

PARAMETER	UNIT	VALUE	REFERENCES
Acid dilute/concentrated	-	good	
Alcohols	-	good	
Alkalis	-	poor	
Aliphatic hydrocarbons	-	good	
Aromatic hydrocarbons	-	fair	
Esters	-	good	
Greases & oils	-	good	
Ketones	-	good	

FLAMMABILITY

PARAMETER	UNIT	VALUE	REFERENCES
Limiting oxygen index	% O$_2$	33 (flame resistant grade)	
Volatile products of combustion	-	CO, CO_2	
UL 94 rating	-	HB; V-0 (flame resistant grade)	

TOXICITY

PARAMETER	UNIT	VALUE	REFERENCES
NFPA: Health, Flammability, Reactivity rating	-	1/1/0; 0/1/0 (HMIS)	
Carcinogenic effect	-	not listed by ACGIH, NIOSH, NTP	
TLV, ACGIH	mg m^{-3}	3 (respirable), 10 (total)	
OSHA	mg m^{-3}	5 (respirable), 15 (total)	

PCT poly(cyclohexylene terephthalate)

PARAMETER	UNIT	VALUE	REFERENCES
PROCESSING			
Typical processing methods	-	injection molding	
Preprocess drying: temperature/ time/residual moisture	ºC/h/%	95/4-6/	
Applications	-	automotive ignition system, circuit board connectors, automotive connectors (headers), lamp sockets and relays	
Outstanding properties	-	hydrolysis resistance, high temperature resistance (short-term temperature resistance up to 255ºC)	
BLENDS			
Suitable polymers	-	LCP, PA, PBT, PET	

348   **HANDBOOK OF POLYMERS** 3rd Edition, Copyrights 2022; ChemTec Publishing

PCTFE polychlorotrifluoroethylene

PARAMETER	UNIT	VALUE	REFERENCES
GENERAL			
Common name	-	polychlorotrifluoroethylene	
IUPAC name	-	poly(chlorotrifluoroethylene)	
ACS name	-	ethene, 1-chloro-1,2,2-trifluoro-, homopolymer	
Acronym	-	PCTFE	
CAS number	-	9002-83-9	
RTECS number	-	KM6555000	
Formula		(formula structure)	
HISTORY			
Date	-	1937, 1953 (commercialization)	
Details	-	first polymerization was reported by IG Farben	
SYNTHESIS			
Monomer(s) structure	-	$F_2C{=}CFCl$	
Monomer(s) CAS number(s)	-	79-38-9	
Monomer(s) molecular weight(s)	dalton, g/mol, amu	116.469	
Monomer ratio	-	100%	
Formulation example	-	water, CTFE, potassium persulfate, sodium bisulfate, perfluorooctanoic acid	Ebnesajjad, S, Fluoroplastics. Vol. 1. Non-melt Processible Fluoroplastics, William Andrew, 2000.
Method of synthesis	-	emulsion polymerization is used in industry	Ebnesajjad, S, Fluoroplastics. Vol. 1. Non-melt Processible Fluoroplastics, William Andrew, 2000.
Temperature of polymerization	°C	21-53	
Pressure of polymerization	MPa	0.34-1	
Mass average molecular weight, M_w	dalton, g/mol, amu	250,000	
Molar volume at 298K	cm^3 mol^{-1}	53.2 (crystalline); 60.7 (amorphous)	
Van der Waals volume	cm^3 mol^{-1}	36.9 (crystalline); 36.9 (amorphous)	
STRUCTURE			
Crystallinity	%	29-33 (DSC)	
Cell type (lattice)	-	pseudohexagonal	
Cell dimensions	nm	a:c=0.644:4.15	
Chain conformation	-	helical	
COMMERCIAL POLYMERS			
Some manufacturers	-	Honeywell	
Trade names	-	Aclar	

PCTFE polychlorotrifluoroethylene

PARAMETER	UNIT	VALUE	REFERENCES
PHYSICAL PROPERTIES			
Density at 20°C	g cm^{-3}	2.1-2.14; 2.077 (amorphous); 2.187 (crystalline)	Ebnesajjad, S, Fluoroplastics. Vol. 2. Melt Processible Fluoroplastics, William Andrew, 2003.
Color	-	clear	
Refractive index, 20°C	-	1.435	
Haze	%	<1	
Odor	-	odorless	
Melting temperature, DSC	°C	210-212	McKeen, L W, The Effect of UV Light and Weather on Plastics and Elastomers, 4th Ed., William Andrew, 2019, 361-91,
Decomposition temperature	°C	260	
Thermal expansion coefficient, 23-80°C	°C^{-1}	4.5E-5 to 7E-5	
Thermal conductivity, melt	W m^{-1} K^{-1}	0.26-0.27	Ebnesajjad, S, Fluoroplastics. Vol. 1. Non-melt Processible Fluoroplastics, William Andrew, 2000.
Glass transition temperature	°C	40-75	Ebnesajjad, S, Fluoroplastics. Vol. 1. Non-melt Processible Fluoroplastics, William Andrew, 2000.
Specific heat capacity	J K^{-1} kg^{-1}	900	
Heat of fusion	kJ kg^{-1}	1.2	
Maximum service temperature	°C	-240 and 200	
Long term service temperature	°C	180	
Heat deflection temperature at 0.45 MPa	°C	126-129	
Heat deflection temperature at 1.8 MPa	°C	75	
Hansen solubility parameters, δ_D, δ_P, δ_H	MPa$^{0.5}$	15.6, 2.5, 4.7	
Interaction radius		5.8	
Hildebrand solubility parameter	MPa$^{0.5}$	16.5	
Surface tension	mN m^{-1}	30.9	Wu, S, Adhesion, 5, 39, 1973.
Dielectric constant at 100 Hz/1 MHz	-	2.7/2.2-2.6	
Dissipation factor at 100 Hz		0.01-0.06	
Dissipation factor at 1 MHz		0.023-0.027	
Volume resistivity	ohm-m	1E14	
Surface resistivity	ohm	1E15	
Electric strength K20/P50, d=0.60.8 mm	kV mm^{-1}	157	
Permeability to nitrogen, 25°C	cm^3 cm cm^{-2} s^{-1} Pa^{-1} x 10^{12}	0.000375	
Permeability to water vapor, 25°C	cm^3 cm cm^{-2} s^{-1} Pa^{-1} x 10^{12}	0.0218	

PCTFE polychlorotrifluoroethylene

PARAMETER	UNIT	VALUE	REFERENCES
Permeability to helium 35°C	Barrer	26	Yavari, M, Okamoto, Y, Lin, H, J. Membrane Sci., 548, 380-9, 2018.
Contact angle of water, 20°C	degree	90	Kwok, D Y; Neumann, A W, Colloids Surfaces A, 161, 49-62, 2000.
Surface free energy	mJ m^{-2}	30.7	

MECHANICAL & RHEOLOGICAL PROPERTIES

Tensile strength	MPa	28-78	
Tensile modulus	MPa	1300-1800	
Elongation	%	50-300	
Elastic modulus	MPa	900-1360	
Compressive strength	MPa	9-12	
Izod impact strength, unnotched, 23°C	J m^{-1}	267	
Shore D hardness	-	75-90	
Rockwell hardness	R	75-112	
Shrinkage	%	2-15	
Water absorption, equilibrium in water at 23°C	%	0.00	

CHEMICAL RESISTANCE

Acid dilute/concentrated	-	very good	
Alcohols	-	very good	
Alkalis	-	very good	
Aliphatic hydrocarbons	-	very good	
Aromatic hydrocarbons	-	very good	
Esters	-	very good	
Greases & oils	-	very good	
Halogenated hydrocarbons	-	poor	
Ketones	-	very good	
Good solvent	-	hot carbon tetrachloride, cyclohexane, mesitylene, toluene	
Non-solvent	-	common organic solvents at room temperature	

FLAMMABILITY

Ignition temperature	°C	n/a	
Autoignition temperature	°C	n/a	
Limiting oxygen index	% O$_2$	100	
Heat of combustion	J g^{-1}	6,170	Hsieh, F-U; Stoltzfus, J M; Beeson, H D, Fire Mater., 20, 301-3, 1996.
Volatile products of combustion	-	HF, HCl, ClFC=CF$_2$, CO, CO$_2$	
UL 94 rating	-	V-0	

WEATHER STABILITY

Low earth orbit erosion yield	cm^3 atom^{-1} x 10^{-24}	0.831	Waters, D L; Banks, B A; De Groh, K K; Miller, S K R; Thorson, S D, High Performance Polym., 20, 512-22, 2008.

PCTFE polychlorotrifluoroethylene

PARAMETER	UNIT	VALUE	REFERENCES
TOXICITY			
NFPA: Health, Flammability, Reactivity rating	-	1/0/0	
Carcinogenic effect	-	not listed by ACGIH, NIOSH, NTP	
Oral rat, LD$_{50}$	mg kg^{-1}	>9,200	
PROCESSING			
Applications	-	glucose/oxygen biofuel cell, pharmaceutical blister packaging	Jeerapan, I, Sempionatto, J R, You, J-M, Wang, J, Biosensors Bioelectronics, 122, 284-9, 2018.
BLENDS			
Suitable polymers	-	PMMA	Tripathi, J, Sharma, A, Tripathi, S, Bisen, R, Agrawal, A, Mater. Chem. Phys., 194, 172-81, 2017.
ANALYSIS			
NMR (chemical shifts)	ppm	F NMR: CFClCF$_2$Cl – 63.7; CF$_3$CF$_2$ – 72.3 and more in ref	Hill, D J T; Thurecht, K J; Whittaker, A K, Radiat. Phys. Chem., 67, 729-36, 2003.

PCTG poly(ethylene-co-1,4-cyclohexylenedimethylene terephthalate)

PARAMETER	UNIT	VALUE	REFERENCES
GENERAL			
Common name	-	poly(ethylene-co-1,4-cyclohexylenedimethylene terephthalate)	
ACS name	-	1,4-benzenedicarboxylic acid, polymer with 1,4-cyclohexanedi-methanol and 1,2-ethanediol	
Acronym	-	PCTG	
CAS number	-	25038-91-9	
SYNTHESIS			
Monomer(s) structure	-		
Monomer(s) CAS number(s)	-	100-21-0; 105-08-8; 107-21-1	
Monomer(s) molecular weight(s)	dalton, g/mol, amu	166.0; 144.21; 62.07	
Monomer(s) expected purity(ies)	%	-;99;>99	
Ethylene content	%	67	Matsuda, H; Nagasaka, B; Asakure, T, Polymer, 44, 4681-87, 2003.
Method of synthesis	-	melt polycondensation	
Mass average molecular weight, M_w	dalton, g/mol, amu	30,000-52,600	
STRUCTURE			
Cis content	%	30	Matsuda, H; Nagasaka, B; Asakure, T, Polymer, 44, 4681-87, 2003.
Entanglement molecular weight	dalton, g/mol, amu	4,900	Barany, T; Czigany, T; Karger-Kotsis, J, Prog. Polym. Sci., 35, 1257-87, 2010.
COMMERCIAL POLYMERS			
Some manufacturers	-	Eastman; Celanese	
Trade names	-	Eastar; Thermx	
PHYSICAL PROPERTIES			
Density at 20°C	g cm^{-3}	1.23; 1.39-1.55 (20-40% glass fiber)	
Color	-	colorless	
Transmittance	%	89-91	
Haze	%	<1	
Odor	-	slight	
Melting temperature, DSC	°C	225-285	
Softening point	°C	83-88	
Thermal expansion coefficient, 23-80°C	10^{-4} °C^{-1}	0.70-0.87	
Thermal conductivity, 20°C	W m^{-1} K^{-1}	0.19	
Glass transition temperature	°C	77-91	
Specific heat capacity	J K^{-1} kg^{-1}	1,340	
Heat deflection temperature at 0.45 MPa	°C	74-89; 280 (20-40% glass fiber)	

PCTG poly(ethylene-co-1,4-cyclohexylenedimethylene terephthalate)

PARAMETER	UNIT	VALUE	REFERENCES
Heat deflection temperature at 1.8 MPa	°C	64-78; 256-265 (20-40% glass fiber)	
Vicat temperature VST/A/50	°C	83-88	
Dielectric constant at 1000 Hz/1 MHz	-	2.8/2.6	
Dissipation factor at 1000 Hz	E-4	60	
Dissipation factor at 1 MHz	E-4	190	
Volume resistivity	ohm-m	1E13	
Surface resistivity	ohm	1E16	
Electric strength K20/P50, d=0.60.8 mm	kV mm^{-1}	16.1-19.7 (20-40% glass fiber)	
Comparative tracking index, CTI, test liquid A	-	560 (20-40% glass fiber)	
Permeability to nitrogen, 25°C	cm^3 mm m^{-2} day^{-1} atm^{-1}	3	
Permeability to oxygen, 25°C	cm^3 mm m^{-2} day^{-1} atm^{-1}	10	

MECHANICAL & RHEOLOGICAL PROPERTIES

PARAMETER	UNIT	VALUE	REFERENCES
Tensile strength	MPa	30-53; 100-128 (20-40% glass fiber)	
Tensile modulus	MPa	2,010	
Tensile stress at yield	MPa	45	
Elongation	%	340; 1.9-2.3 (20-40% glass fiber)	
Tensile yield strain	%	5	
Flexural strength	MPa	67; 155-200 (20-40% glass fiber)	
Flexural modulus	MPa	1,800; 5,900-11,000 (20-40% glass fiber)	
Charpy impact strength, unnotched, 23°C	kJ m^{-2}	NB	
Izod impact strength, unnotched, 23°C	J m^{-1}	NB; 520-800 (20-40% glass fiber)	
Izod impact strength, notched, 23°C	J m^{-1}	49; 60-75 (20-40% glass fiber)	
Rockwell hardness	-	R105	
Shrinkage	%	0.2-0.5	
Water absorption, equilibrium in water at 23°C	%	0.13	

CHEMICAL RESISTANCE

PARAMETER	UNIT	VALUE	REFERENCES
Acid dilute/concentrated	-	fair	
Alcohols	-	good-fair	
Alkalis	-	good	
Aliphatic hydrocarbons	-	good	
Aromatic hydrocarbons	-	poor	
Esters	-	poor	
Greases & oils	-	good	
Halogenated hydrocarbons	-	poor	
Ketones	-	poor	

PCTG poly(ethylene-co-1,4-cyclohexylenedimethylene terephthalate)

PARAMETER	UNIT	VALUE	REFERENCES
Good solvent	-	methylene chloride	
FLAMMABILITY			
UL rating	-	HB; HB (20-40% glass fiber)	
WEATHER STABILITY			
Excitation wavelengths	nm	330; 260-390	Allen, N S; Rivalle, G; Edge, M; Roberts, I; Fagerburg, D R, Polym. Deg. Stab., 67, 325-34, 2000.
Emission wavelengths	nm	380, 460; 450, 550	Allen, N S; Rivalle, G; Edge, M; Roberts, I; Fagerburg, D R, Polym. Deg. Stab., 67, 325-34, 2000.
TOXICITY			
HMIS: Health, Flammability, Reactivity rating	-	1/1/0	
Oral rat, LD_{50}	mg kg^{-1}	>3,200	
Skin rabbit, LD_{50}	mg kg^{-1}	>1,000 (highest dose tested)	
PROCESSING			
Typical processing methods	-	injection molding	
Preprocess drying: temperature/ time/residual moisture	ºC/h/%	65-95/4-12/0.05	
Processing temperature	ºC	215-295	
Processing pressure	MPa	0.3-1 (back)	
Process time	min	4 (max residence time)	
Applications	-	medical (anesthesia manifold, wound healing systems)	Sashi, V R, Plastics in Medical Devices, Elsevier, 2010, pp. 121-173.
Outstanding properties	-	heat resistance, resistance to automotive fluids, low dielectric constant, low moisture absorption	
BLENDS			
Suitable polymers	-	PAR, PC; PEI, PET	

PDCPD polydicyclopentadiene

PARAMETER	UNIT	VALUE	REFERENCES
GENERAL			
Common name	-	polydicyclopentadiene	
Acronym	-	PDCDP	
SYNTHESIS			
Method of synthesis	-	ring opening methathesis polymerization of dicyclopentadiene	
Catalyst	-	2nd generation Grubbs catalyst (cure)	Rhode, B J; Robertson, M L; Krishnamoorti, R, Polymer, 69, 204-14, 2015.
COMMERCIAL POLYMERS			
Some manufacturers	-	Materia, Inc; Rimtec Co.	
Trade names	-	Proxima; Telene	
PHYSICAL PROPERTIES			
Density at 20°C	g cm^{-3}	1.03-1.05; 0.97 (liquid)	
Decomposition temperature	°C	450 (TGA)	Gottschalk, D, M. Sc. Thesis, Iowa State UNiversity, 2011.
Thermal expansion coefficient, -40 to 40°C	°C^{-1}	79E-06	
Thermal conductivity, melt	W m^{-1} K^{-1}	0.17	Le Gac, P Y; Choqueuse, D; Paris, M; Recher, M; Zimmer, C; Melot, D, Polym. Deg. Stab., 98, 809-17, 2013.
Glass transition temperature	°C	124-255 (DMA)	Vallons, K A M, Drozdzak, R; Charret, M; Lomov, S V; Verpoest, I, Compos. Part A, 78, 191-200, 2015.
Maximum service temperature	°C	180	
Minimum service temperature	°C	-60	
Heat deflection temperature at 0.45 MPa	°C	105-118	
Hildebrand solubility parameter	MPa$^{0.5}$	17	Huang, J, Derue, I, Le Gac, P Y, Richaud, E, Polym. Deg. Stab., 180, 109219, 2020.
Relative permittivity at 1 MHz	-	2.75-3.02 (depending on catalyst used)	Gottschalk, D, M. Sc. Thesis, Iowa State UNiversity, 2011.
Coefficient of friction	-	0.7	Pan, B; Zhao, J; Zhang, Y; Zhang, Y, IERI Procedia, 1, 19-24, 2012.
MECHANICAL & RHEOLOGICAL PROPERTIES			
Tensile strength	MPa	35-73	Vallons, K A M, Drozdzak, R; Charret, M; Lomov, S V; Verpoest, I, Compos. Part A, 78, 191-200, 2015.
Tensile modulus	MPa	1770-3,100	Knorr, D B; Masser, K A; Elder, R M; Sirk, T W; Hindenlang, M D; Yu, J H; Richardson D A; Boyd, S E; Spurgeon, W A; Lenhart, J L, Compos. Sci. Technol., 114, 17-25, 2015.
Tensile stress at yield	MPa	52.4	
Elongation	%	2.7	Hu, Y; Lang, A W; Li, X; Nutt, S R, Polym. Deg. Stab., 110, 465-72, 2014.
Tensile yield strain	%	4-5	

PDCPD polydicyclopentadiene

PARAMETER	UNIT	VALUE	REFERENCES
Flexural strength	MPa	67-75 (yield)	
Flexural modulus	MPa	1,850-2,000	
Elastic modulus	MPa	1,870-1,980	
Fracture toughness	MPa m$^{1/2}$	3.3	Gottschalk, D, M. Sc. Thesis, Iowa State UNiversity, 2011.
Charpy impact strength, notched, 23°C	kJ m^{-2}	118	
Izod impact strength, notched, 23°C	kJ m^{-1}	22-30	
Shear modulus	MPa	680-710	
Poisson's ratio	-	0.39	
Rockwell hardness	-	R114	
Water absorption, equilibrium in water at 23°C	%	1	

THERMAL STABILITY			
Stabilizers	-	Tinuvin 123 and Chimassorb 2020	Huang, J, Le Gac, P E, Richaud, E, Polym. Deg. Stab., in press, 109267, 2020.

FLAMMABILITY			
Flammability according to UL-standard; thickness 1.6/0.8 mm	class	HB to V-0	

PROCESSING			
Typical processing methods	-	cure, RIM	
Processing temperature	°C	120	
Process time	min	60	
Post-cure	min	60 @ 190°C	
Additives used in final products	-	organic photoredox mediators	Goetz, A E; Boydston, A J, J. Am. Chem. Sco., 137, 24, 7572-5, 2015.
Applications	-	agricultural and construction equipment, trucks and buses (body panels), the chlor-alkali industry (electrolyzer cell covers or butterfly valves), containers (wastewater systems, waste containers, military boxes)	

ANALYSIS			
FTIR (wavenumber-assignment)	cm^{-1}/-	*trans* C=C-H bending – 974; *cis* C=C-H bending – 754, 735	Rhode, B J; Robertson, M L; Krishnamoorti, R, Polymer, 69, 204-14, 2015.
^{13}C NMR (chemical shifts)	ppm	C6 – 35; C4 – 42; C2+C3 – 46; C5 – 55; carbon double bond – 131	Le Gac, P Y; Choqueuse, D; Paris, M; Recher, M; Zimmer, C; Melot, D, Polym. Deg. Stab., 98, 809-17, 2013.

PDL polylysine

PARAMETER	UNIT	VALUE	REFERENCES
GENERAL			
Common name	-	polylysine, ε-polysine, poly-L-lysine; poly-D-lysine	
IUPAC name	-	poly[imino[(2S)-2-amino-1-oxo-1,6-hexanediyl]	
CAS name	-	28211-04-3; 25104-18-1	
SYNTHESIS			
Method of synthesis	-	biosynthesis	Hiraki, J; Suzuki, E, US Patent 5,900,363, Chisso Corp., May 4, 1999.
Mass average molecular weight, M_w	dalton, g/mol, amu	5,000-150,000	
Polydispersity, M_w/M_n	-	1.14	
Typical forms	-	α-polylysine is a synthetic polymer, which can be composed of either L-lysine or D-lysine. "L" and "D" refer to the chirality at lysine's central carbon. This results in poly-L-lysine and poly-D-lysine, respectively.; ε-polylysine is a natural polypeptide composed of 25-35 homogeneous l-lysine, linked through peptide bonds between ε-amino and α-carboxyl	Li, T, Wen, C, Dong, Y, Li, D, Song, S, Food Hydroclloids, 95, 212-8, 2019.
COMMERCIAL POLYMERS			
Some manufacturers	-	Chisso Corp./JNC Corp.	
PHYSICAL PROPERTIES			
Color	-	light yellow	
Melting temperature, DSC	°C	142-172	
pKa	-	9-10	
Maximum service temperature	°C	120	
CHEMICAL RESISTANCE			
Good solvent		water and alcohol	
TOXICITY			
Acceptable daily intake	μgrams kg^{-1} body weight day^{-1}	recognized by FDA as GRAS material	
Oral rat, LD_{50}	mg kg^{-1}	>5,000	
PROCESSING			
Additives used in final products	-	iron acetylacetate (magnetic particles)	Yang, G; Zhang, B; Wang, J; Xie, S; Li, X, J. Magnet. Magnet. Mater., 374, 205-8, 2015.
Applications	-	coatings, fiber, food industry, medicine, pharmaceutical	
Outstanding properties	-	biodegradability, biocompatibility, antimicrobial properties	Vidal, L; Thuault, V; Mangas, A; Covenas, R; Thienpont, A; Geffard, M, J. Amino Acids, 672367, 2014.
BLENDS			
Suitable polymers	-	chitosan, polyacrylamide	Yu, Z, Rao, G, Wei, Y, Yu, J, Fang, Y, Int. J. Biol. Macromol., 141, 545-52, 2019.

PDMS polydimethylsiloxane

PARAMETER	UNIT	VALUE	REFERENCES

GENERAL

Common name	-	polydimethylsiloxane	
CAS name	-	poly[oxy(dimethylsilylene)]; poly[oxy(dimethylsilylene)], α-(trimethylsilyl)-omega-[(trimethylsilyl)oxy]-	
Acronym	-	PDMS	
CAS number	-	9016-00-6; 42557-10-8	
EC number	-	226-171-3	
RTECS number	-	TQ2690000	
Formula			

$$\left[-\underset{\underset{CH_3}{|}}{\overset{\overset{CH_3}{|}}{Si}} - O - \right]_n$$

HISTORY

Person to discover	-	Frederick Stanley Kipping	
Date	-	1901-1930; 1935	
Details	-	Kipping developed foundations of organosilicone chemistry and technology; first practical application of silicones in 1935	

SYNTHESIS

Monomer(s) structure	-		

$$Cl - \underset{\underset{CH_3}{|}}{\overset{\overset{CH_3}{|}}{Si}} - Cl$$

Monomer(s) CAS number(s)	-	1066-35-9	
Monomer(s) molecular weight(s)	dalton, g/mol, amu	94.62	
Monomer ratio	-	100%	
Method of synthesis	-	the direct reaction between silicon metal and methyl chloride in a fluid bed reactor yields a complex mixture of methyl chlorosilanes; the chlorosilanes are distilled or purified, and the primary product - dimethyldichlorosilane, $(CH_3)_2SiCl_2$ – is reacted with water (hydrolysis) to give poly(dimethylsiloxane) oligomers: $[Me_2SiO]_n$.	
Catalyst	-	DOTM	
Cure mechanism		acetoxy, neutral, neutral catalytic (moisture cured systems)	
Number average molecular weight, M_n	dalton, g/mol, amu	300-66,000, 1.49	Tugui, C, Tiron, V, Dascalu, M, Sacarescu, L, Cazacu, M, Eur. Polym. J., 120, 109243, 2019.
Mass average molecular weight, M_w	dalton, g/mol, amu	500-423,000; 1,395,000	
Polydispersity, M_w/M_n	-	1.6-3.9	
Molar volume at 298K	cm^3 mol^{-1}	69.1 (crystalline)	
Van der Waals volume	cm^3 mol^{-1}	47.6 (crystalline)	
Radius of gyration	nm	1-1.2; 3.1 (partially crosslinked); 5.5-6.1 (final crosslink density in membrane)	Jadav, G L; Aswal, V K; Singh, P S, J. Membrane Sci., in press, 2011; Serbescu, A; Saalwaechter, K, Polymer, 50, 5434-42, 2009.

PDMS polydimethylsiloxane

PARAMETER	UNIT	VALUE	REFERENCES
Chain-end groups	-	methyl, vinyl, hydrogen, hydroxyl	Mrozek, R A; Cole, P J; Otim, K J; Shull, K R; Lenhart, J L, Polymer, in press, 2011; Jadav, G L; Aswal, V K; Singh, P S, J. Membrane Sci., in press, 2011.

STRUCTURE

PARAMETER	UNIT	VALUE	REFERENCES
Crystallinity	%	24-95 (depends on cooling rate); 0-14 (with 10-40% fumed silica)	Aranguren, M I, Polymer, 39, 20, 4897-4906, 1998.
Cell type (lattice)	-	monoclinic	
Cell dimensions	nm	a:b:c=1.3:0.775:0.83	
Unit cell angles	degree	α:β:γ=90:60:90	
Entanglement molecular weight	dalton, g/mol, amu	calc.=8,160; exp.=21,000-33,000	
Rapid crystallization temperature	°C	-56 to -65	

COMMERCIAL POLYMERS

PARAMETER	UNIT	VALUE	REFERENCES
Some manufacturers	-	Dow Corning	

PHYSICAL PROPERTIES

PARAMETER	UNIT	VALUE	REFERENCES
Density at 20°C	g cm^{-3}	0.97; 0.98 (amorphous); 1.07 (crystalline)	
Color	-	clear fluid	
Refractive index, 20°C	-	1.414; 1.4-1.5 (LED applications)	Kaneko, Y, Hayashi, H, Ishii, Y, Kada, W, Nishikawa, H, Nucl. Instr. Meth. Phys. Res. Sect. B: Bean Interactions Mater. Atoms, 459, 94-7, 2019.
Gloss, 60°, Gardner (ASTM D523)	%	47	
Odor	-	odorless	
Melting temperature, DSC	°C	-35 to -55	
Decomposition temperature	°C	>343 (silicone oil); 235 (sealant)	
Thermal expansion coefficient, 23-80°C	°C^{-1}	9.0-9.6E-4	
Thermal conductivity, 15-80°C	W m^{-1} K^{-1}	0.16; 0.151-0.167; 0.15-1 (sealants); 1.9-6.8 (specially formulated thermally conductive adhesives and greases)	Wei, J, Liao, M, Ma, A, Chen, Y, Yu, J, Compos. Commun., 17, 141-6, 2020.
Glass transition temperature	°C	-123 to -127; -121 to -122 (with 10-40% fumed silica)	Aranguren, M I, Polymer, 39, 20, 4897-4906, 1998.
Specific heat capacity	J K^{-1} kg^{-1}	1,350-1,700	
Maximum service temperature	°C	350 (coatings, up to 1000 h); -40 to 260 (lead-free solder reflow)	
Long term service temperature	°C	-45 to 150 (sealants); -55 to 200 (coatings)	
Hildebrand solubility parameter	MPa$^{0.5}$	14.9-15.6	
Surface tension	mN m^{-1}	20.3-21.5	
Dielectric constant at 20 Hz/100 Hz/1 MHz	-	6.5/3.8/3.0	Ishaq, S, Kanwal, F, Atiq, S, Moussa, M, Losic, D, Ceramics Int., 45, 7A, 8713-20, 2019.
Dissipation factor at 100 Hz	-	0.00012-0.001	
Volume resistivity	ohm-m	4E13	
Electric strength K20/P50, d=0.60.8 mm	kV mm^{-1}	12-27	

PDMS polydimethylsiloxane

PARAMETER	UNIT	VALUE	REFERENCES
Contact angle of water, 20°C	degree	107-110	
Surface free energy	mJ m^{-2}	20.4	
Speed of sound	m s^{-1}	837-987	

MECHANICAL & RHEOLOGICAL PROPERTIES

PARAMETER	UNIT	VALUE	REFERENCES
Tensile strength	MPa	0.5-9.7	
Tensile modulus	MPa	0.69-3.45	
Elongation	%	220-1,600	
Tear strength	kN m^{-1}	5-40	
Tension adhesion	MPa	1.2	
Peel strength	kg cm^{-1}	2.7-7.1	
Poisson's ratio	-	0.5	
Shore A hardness	-	15-70	

CHEMICAL RESISTANCE

PARAMETER	UNIT	VALUE	REFERENCES
Acid dilute/concentrated	-	good	
Alcohols	-	fair-poor	
Alkalis	-	good	
Aliphatic hydrocarbons	-	good	
Aromatic hydrocarbons	-	good-poor	
Greases & oils	-	good	
Halogenated hydrocarbons	-	poor	
Ketones	-	poor	
Other	-	poor	
Θ solvent, Θ-temp.=-5.2, -38.2, -81.4, 4.8°C	-	benzene, n-butyl acetate, cyclohexane, ethyl acetate	
Good solvent	-	amyl acetate, chlorobenzene, chloroform, cyclohexyl acetate, dichlorobenzene, ethyl bromide, ethyl acetate, hydrocarbons, trichloroethylene	
Non-solvent	-	acetone, acetonitrile, acetophenone, benzyl alcohol, benzyl acetate, gamma-butyrolactone, cyclohexanone, cyclohexanol, dichlorobenzene, dioxane, diphenyl oxide, ethyl formate, ethyl benzoate, methanol, nitrobenzene	

FLAMMABILITY

PARAMETER	UNIT	VALUE	REFERENCES
Ignition temperature	°C	>320 (silicone oil), >100 (sealant)	
Autoignition temperature	°C	418-490	
Limiting oxygen index	% O$_2$	26-42	
Heat of combustion	J g^{-1}	19,530	Walters, R N; Hacket, S M; Lyon, R E, Fire Mater., 24, 5, 245-52, 2000.
Volatile products of combustion	-	CO, CO$_2$, SiO$_2$	

WEATHER STABILITY

PARAMETER	UNIT	VALUE	REFERENCES
Spectral sensitivity	nm	300-350	
Activation wavelengths	nm	318, 325, 330, 360	
Transmittance	%	100 – 76.5; 300 nm – 44.8	Matsuzawa, N N; Oizumi, H; Mori, S; Irie, S; Shirayone, S; Yano, E; Okazaki, S; Ishitani, S; Dixon, D A, Jpn. J. Appl. Phys., 38, 7109-13, 1999.

PDMS polydimethylsiloxane

PARAMETER	UNIT	VALUE	REFERENCES
Important initiators and accelerators	-	benzophenone, nitrous oxide, flame retardants containing halogens, ozone, stress	
Products of degradation	-	hydrogen, water, carbon dioxide, ketone, unsaturations, hydro-peroxides, radicals, chain scissions, crosslinks, quinomethane structures, benzene, acetophenone, benzaldehyde, benzene, formic acid, acetic acid, benzoic acid, conjugated double bonds	
Stabilizers	-	UVA: 2,4-dihydroxybenzophenone; 2-(2H-benzotriazol-2-yl)-p-cresol; 2-(2H-benzotriazole-2-yl)-4,6-di-tert-pentylphenol; 2-(2H-benzotriazole-2-yl)-4-(1,1,3,3-tetraethylbutyl)phe-nol; HAS: 1,3,5-triazine-2,4,6-triamine, N,N'''[1,2-ethane-diyl-bis[[[4,6-bis[butyl(1,2,6,6-pentamethyl-4-piperidinyl) amino]-1,3,5-triazine-2-yl]imino]-3,1-propanediyl]bis[N',N''-dibutyl-N',N''-bis(1,2,2,6,6-pentamethyl-4-piperidinyl)-; bis(2,2,6,6-tetramethyl-4-piperidyl) sebacate; 2,2,6,6-tetra-methyl-4-piperidinyl stearate	

BIODEGRADATION

PARAMETER	UNIT	VALUE	REFERENCES
Colonized products		coatings, dental materials, insulating rubber, medical devices, mortar protective coatings, sealants, stone protective coating	
Typical biodegradants	-	fungal growth decreases hydrophobicity of silicone products	
Stabilizers	-	2,3,5,6-tetrachloro-4-methylsulfonyl-pyridine, 2,3,5,6-tetrachlo-ro-4-methylsulfonylpyridine, N-chloramine, sodium benzoate, triclosan, zosteric acid	

TOXICITY

PARAMETER	UNIT	VALUE	REFERENCES
NFPA: Health, Flammability, Reactivity rating	-	1/1/0	
Carcinogenic effect	-	not listed by ACGIH, NIOSH, NTP	
Reproductive toxicity	-	adverse reproductive effects have occurred in experimental animals	
Oral rat, LD_{50}	mg kg^{-1}	>4990	
Skin rabbit, LD_{50}	mg kg^{-1}	>18,400	

ENVIRONMENTAL IMPACT

PARAMETER	UNIT	VALUE	REFERENCES
Aquatic toxicity, *Rainbow trout*, LC_{50}, 48 h	mg l^{-1}	>10,000	
Mean degradation half-life	months	1-2 (coatings)	

PROCESSING

PARAMETER	UNIT	VALUE	REFERENCES
Typical processing methods	-	calendering, casting, compression molding, extrusion, injec-tion molding, room temperature, moisture or chemical cure of premixed compounds, transfer molding, vulcanization	
Processing temperature	ºC	200-316 (vulcanization); 0-50 (moisture cure); 250-500 (extru-sion vulcanization); 180-200 (continuous vulcanization in steam); 60-90 followed by 130-200 (peroxide cure)	
Processing pressure	MPa	3.45-13.78 (injection)	
Process time	s	5-10 (injection time); 30-90 (molding time for peroxide cured)	

PDMS polydimethylsiloxane

PARAMETER	UNIT	VALUE	REFERENCES
Additives used in final products	-	Fillers: aluminum oxide, aluminum powder, boron nitride, calcium carbonate, carbon black, carbon nanotubes, fumed silica, glass beads, metal powders, mica, montmorillonite, nano-CaCO$_3$, nanosilica, precipitated silica, silver, spherical alumina, zinc oxide; Plasticizers: acetyl triethyl citrate, diethylene glycol dibenzoate, dimethylsiloxane oligomer, epoxidized soybean oil, ethylene or propylene glycols, glycerin, hydroxy-terminated polydimethylsiloxane, phthalates, polyisobutylene, silicone oil, tricresyl phosphate, tripropylene glycol monoethyl ether; Antistatics: acicular conductive titanium oxide, acrylamidoamidosiloxane, chitin, graphite, nickel, silver, sulfonated silicone; Antiblocking: ailica; Release: PDMS compounds; Slip: silicone fluid; Peroxides (benzoyl peroxide or 2,4-dichlorobenzoyl peroxide); Thickener (polyacrylic acid polymers, e.g., Carbopol, hydroxyethyl cellulose); Surfactant (e.g., cocoamidopropyl betaine, amine type, sorbitan monoisostearate); Silicone oil (improves gloss, scratch resistance, etc.); Solvent (e.g., heptane, Exxsol); Wax, (e.g., carnauba, paraffin)	
Applications	-	automotive (shaft sealing rings, spark plug caps, o-rings, gaskets, ignition cables, coolant and heater hose), caulking and sealants, coatings, cookware, conductive rubber, defoaming, door and windows seal, drycleaning, electronics, firestops, general tubing, heat dissipative grease, lithium ion battery, lubricants, medicine, moldmaking, personal care, toys, transfusion and dialysis tubing, waveguide	

BLENDS

Suitable polymers	-	PC, PET, PPS, TPU	

ANALYSIS

FTIR (wavenumber-assignment)	cm^{-1}/-	ester linkage – 1720; CH=CH – 1600; Si-C – 1374, Si–O–Si – 1007 and 1080, Si–C – 785	Rao, H; Zhang, Z; Song, C; Qiao, T, Reactive Functional Polym., 71, 537-43, 2011; Ishaq, S, Kanwal, F, Atiq, S, Moussa, M, Losic, D, Ceramics Int., 45, 7A, 8713-20, 2019.
Raman (wavenumber-assignment)	cm^{-1}/-	Si-O-Si – 437-441; C-H – 1567	Lin, L-H; Liu, H-H; Hwang, J-J; Chen, K-M; Chao, J-C, Mater. Chem. Phys., 127, 248-52, 2011.

PDPD poly(dicyclopentadiene-co-p-cresol)

PARAMETER	UNIT	VALUE	REFERENCES
GENERAL			
Common name	-	poly(dicyclopentadiene-co-p-cresol)	
Acronym	-	DCPD	Hsiue, G-H, Wei, H-F, Shia, S-J, Kuo, W-J, Sha, Y-A, Polym. Deg. Stab., 73, 309-18, 2001.
CAS number	-	68610-51-5	
EC number	-	271-867-2	
PHYSICAL PROPERTIES			
Density at 20°C	g cm^{-3}	1.04-1.1	
Color	-	white	
Melting temperature, DSC	°C	105-118	
Glass transition temperature	°C	92	
FLAMMABILITY			
Flash point	°C	46.8	
TOXICITY			
HMIS: Health, Flammability, Reactivity rating	-	0/0/0	
Oral rat, LD$_{50}$	mg kg^{-1}	5,000	
Skin rabbit, LD$_{50}$	mg kg^{-1}	2,000	
ENVIRONMENTAL IMPACT			
Aquatic toxicity, *Daphnia magna*, LC$_{50}$, 48 h	mg l^{-1}	0.2	
Aquatic toxicity, *Rainbow trout*, LC$_{50}$, 48 h	mg l^{-1}	0.2	
Partition coefficient	log Pow	7.56	
BLENDS			
Suitable polymers	-	epoxy	Hsiue, G-H, Wei, H-F, Shia, S-J, Kuo, W-J, Sha, Y-A, Polym. Deg. Stab., 73, 309-18, 2001.

PDS polydioxanone

PARAMETER	UNIT	VALUE	REFERENCES
GENERAL			
Common name	-	polydioxanone	
CAS name	-	poly[oxy(1-oxo-1,2-ethanediyl)oxy-1,2-ethanediyl]	
Acronym	-	PDS, PDO, PDX	
CAS number	-	31621-87-1	
Formula			
HISTORY			
Person to discover	-	Weipert, E A; Schultz, H S	Weipert, E A, US Patent 3,020,289, Wyandotte, Feb. 6, 1962. Schultz, H S, US Patent 3,063,967, General Aniline & Film Corp., Nov. 13, 1962.
Date	-	1962 (filed 1960); 1962 (filed 1959)	
Details	-	Weipert patented polymerization of dioxanone in the presence of sulfuric acid; Schultz patented dioxanone polymerization in presence of organoaluminum catalyst	
SYNTHESIS			
Monomer(s) structure	-		
Monomer(s) CAS number(s)	-	3041-16-5	
Monomer(s) molecular weight(s)	dalton, g/mol, amu	102.09	
Monomer(s) expected purity(ies)	%	99.5	
Method of synthesis	-	ring-opening polymerization of p-dioxanone in the presence of organometallic catalyst (e.g., zirconium acetylacetone)	Li, Y; Wang, X-L, Yang, K-K; Wang, Y-Z, Polym. Bull., 57, 873-880, 2006.
Temperature of polymerization	°C	room temp.	
Catalyst	-	organoaluminum	
Yield	%	60-78 (reactive extrusion)	Raquez, J-M; Degee, P; Dubois, P; Balakrishnan, S; Narayan, R, Polym. Eng. Sci., 45, 622-29, 2005.
Typical impurities	ppm	500 (solvents); 10 (heavy metals); 100 (catalyst)	
Typical concentration of residual monomer	ppm	<1	
Mass average molecular weight, M_w	dalton, g/mol, amu	58,000-175,000	
Polydispersity, M_w/M_n	-	1.6-1.7	
STRUCTURE			
Crystallinity	%	40-57; 70.4-70.7 (monofilament); 68, 53 (after compression molding)	Liu, Z-P; Ding, S-D; Sui, Y-J; Wang, Y-Z, J. Appl. Polym. Sci., 112, 3079-86, 2009; Liu, Y, Zhu, G, Yang, H, Wang, C, Han, G, J. Mech. Beh. Biomed. Mater., 77, 157-63, 2018; Ahlinder, A, Fuoco, T, Finne-Wistrand, A, Polym. Testing, 72, 214-22, 2018.
Cell type (lattice)	-	orthorhombic	Jaidann, M; Brisson, J, J. Polym. Sci. B, 46, 406-17, 2008.

PDS polydioxanone

PARAMETER	UNIT	VALUE	REFERENCES
Cell dimensions	nm	a:b:c=0.97:0.742:0.682	Jaidann, M; Brisson, J, J. Polym. Sci. B, 46, 406-17, 2008.
Chain conformation	-	2/1 helix	Jaidann, M; Brisson, J, J. Polym. Sci. B, 46, 406-17, 2008.
Lamellae thickness	nm	7	Gesti, S; Lotz, B; Casas, M T; Aleman, C; Puiggalli, Eur. Polym. J., 43, 4662-74, 2007.
Avrami constants, k/n	-	n=3.2-4.0	Zheng, G-C; Ding, S-D; Zeng, J-B; Wang, Y-Z; Li, Y-D, J. Macromol. Sci. B, 49, 269-85, 2010.

COMMERCIAL POLYMERS

Some manufacturers	-	Evonik/ ITV Denkendorf Produktservice GmbH	
Trade names	-	Resomer X 206 S	

PHYSICAL PROPERTIES

Density at 20°C	g cm^{-3}	1.34-1.38	
Color	-	white, off-white	
Odor	-	light	
Melting temperature, DSC	°C	106-115, 119, 109 (after compression molding)	-; Ahlinder, A, Fuoco, T, Finne-Wistrand, A, Polym. Testing, 72, 214-22, 2018.
Decomposition temperature	°C	130-190	
Storage temperature	°C	-20	
Glass transition temperature	°C	-16 to 0; -17, -11 (after compression molding)	-; Ahlinder, A, Fuoco, T, Finne-Wistrand, A, Polym. Testing, 72, 214-22, 2018.
Complete absorption time (medical applications)	days	232	Greenberg, J A, Sugrue, R, Am. J. Obstetrics Gynecology, 219, 6, 631, 2018.

MECHANICAL & RHEOLOGICAL PROPERTIES

Tensile strength	MPa	6.3-48.3	McClure, M J; Sell, S A; Barnes, C P; Bowen, W C; Bowlin, G L, 3, 1, 1-10, 2008; Boland, E D; Coleman, B D; Barnes, C P; Simpson, D G; Wnek, G E; Bowlin, G L, Acta Biomater., 1, 115-123, 2005.
Tensile modulus	MPa	30	McClure, M J; Sell, S A; Barnes, C P; Bowen, W C; Bowlin, G L, 3, 1, 1-10, 2008.
Elongation	%	60-600	McClure, M J; Sell, S A; Barnes, C P; Bowen, W C; Bowlin, G L, 3, 1, 1-10, 2008.
Elastic modulus	MPa	2,100	Boland, E D; Coleman, B D; Barnes, C P; Simpson, D G; Wnek, G E; Bowlin, G L, Acta Biomater., 1, 115-123, 2005.
Young's moduluis	MPa	600-800	
Intrinsic viscosity, 25°C	dl g^{-1}	1.5-2.2	
Water absorption, equilibrium in water at 23°C	%	0.5	

CHEMICAL RESISTANCE

Acid dilute/concentrated	-	poor	
Alcohols	-	good	

PDS polydioxanone

PARAMETER	UNIT	VALUE	REFERENCES
Aliphatic hydrocarbons	-	good	
BIODEGRADATION			
Typical biodegradants	-	lipase	Gesti, S; Lotz, B; Casas, M T; Aleman, C; Puiggalli, Eur. Polym. J., 43, 4662-74, 2007.
TOXICITY			
Carcinogenic effect	-	not listed by ACGIH, NIOSH, NTP	
PROCESSING			
Processing methods	-	extrusion, injection molding, spinning	
Processing temperature	ºC	100-115	
Processing pressure	MPa	3-10 (injection)	
Additives used in final products	-	Dye (C. I. Solvent Violet 13(=C. I. # 60725))	
Applications	-	adhesives, coatings, foams, laminates, medical (adhesion barrier, bone fixation pins, cardiovascular, drug delivery, ligating clips, meshes, orthopedics, plastic surgery, scaffolds, staples, sutures, tissue engineering. vascular closure devices)	
Outstanding properties	-	biodegradable	
BLENDS			
Suitable polymers	-	PCD, PCL	
ANALYSIS			
Raman (wavenumber-assignment)	cm^{-1}/-	C=O – 1730; C-C – 1050; C-O-C – 868	Jaidann, M; Brisson, J, J. Polym. Sci. B, 46, 406-17, 2008.

PE polyethylene

PARAMETER	UNIT	VALUE	REFERENCES
GENERAL			
Common name	-	polyethylene	
IUPAC name	-	polyethylene	
Acronym	-	PE	
CAS number	-	9002-88-4	
Formula		$\left[CH_2\right]_n$	
RTECS number	-	TQ3325000	
HISTORY			
Person to discover	-	Hans von Pechmann (first synthesis); Eric Fawcett, Reginald Gibson, Michael Perrin (industrial synthesis) of ICI; Ziegler (Nobel Prize)	
Date	-	1898; 1933; 1963	
SYNTHESIS			
Monomer(s) structure	-	$CH_2=CH_2$	
Monomer(s) CAS number(s)	-	74-85-1	
Monomer(s) molecular weight(s)	dalton, g/mol, amu	28.05	
Comonomers		butene, hexene, octene	
Monomer ratio	-	100% (some polymers)	
Catalyst	-	Ziegler-Natta, single-site; nickel-based precatalyst (bimodal PE and UHMWPE)	-; Chen, M, Chen, Y, Li, W, Dong, C, Yang, Y, Eur. Polym. J., 135, 109878, 2020.
Heat of polymerization	kJ mol^{-1}	93.6	Kaminsky, W, Adv. Catalysis, 46, 89-159, 2001.
Mass average molecular weight, M_w	dalton, g/mol, amu	28,000-6,300,000	
Polydispersity, M_w/M_n	-	1.9-14.1	
Polymerization degree (number of monomer units)	-	30,121	
Xylene extractables	%	1.15-9.63	Auger, J; Duff, A; Weber, M; Bellehumeur, C, Antec, 456-60, 2004.
Hexene extractables	%	0.87-4.49	Auger, J; Duff, A; Weber, M; Bellehumeur, C, Antec, 456-60, 2004.
Molar volume at 298K	cm^3 mol^{-1}	calc.=28.1-32.8; exp.=33.1	
Van der Waals volume	cm^3 mol^{-1}	calc.=20.46; 20.50 (amorphous)	
Degree of branching	#/1000C	$C_2 - 1$; $C_4 - 5$-36; $C_5 - 2$-8; C_6 or more $- 2$	Mieda, N; Okamoto, K; Yamaguchi, M, Antec, 1-3, 2009.
STRUCTURE			
Crystallinity	%	35-94; 41-35 (dicumyl peroxide crosslinking 1-7%); bulk crystallinity 72.9 (DSC) and linear crystallinity 90.5 (WAXD) of UHMWPE catalyzed by metallocene supported on confined polystyrene	Nilsson, S; Hjertberg, T; Smedberg, A, Eur. Polym. J., 46, 1759-69, 2010; Wu, Y, Yang, H, Li, W, Mattea, C, Ye, X, Catalysis Today, in press, 2019.
Cell type (lattice)	-	orthorhombic (stable form); monoclinic (metastable form), and hexagonal (high pressure form)	
Cell dimensions	nm	a:b:c=0.736-0.742:0.493-0.495:0.253-0.255 (orthorhombic)	Kavesh, S; Schultz, J M, J. Polym. Sci., A-2, 8, 243, 1970.

PE polyethylene

PARAMETER	UNIT	VALUE	REFERENCES
Number of chains per unit cell	-	2	
Crystallite size	nm	5.1-23.9 (thickness); 7,000-1,000 (dicumyl peroxide crosslinking 1-7%)	Nilsson, S; Hjertberg, T; Smedberg, A, Eur. Polym. J., 46, 1759-69, 2010.
Spacing between crystallites	nm	22-42	
Chain conformation	-	helix 1*1/1	Androsch, R; Di Lorenzo, M L; Schick, C; Wunderlich, B, Polymer, 51, 4639-62, 2010.
Entanglement molecular weight	dalton, g/mol, amu	calc.=1422, 1854; 1143 (UHMWPE)	-; Zhang, X, Zhao, S, Xin, Z, Polymer, 202, 122631, 2020.
Entanglement density	-	753.6 (UHMWPE)	Zhang, X, Zhao, S, Xin, Z, Polymer, 202, 122631, 2020.
Degree of branching	CH_3/100C	0.05-6.94	
Lamellae thickness	nm	9.25-35.5 (HDPE); 12.6-20.6 (LDPE); 6.7-9.8 (low crystallization temperature); 26.3 (PE); 27.1 (XPE)	Bistolfi, A; Bellare, A, Acta Biomaterialia, in press, 2011.
Lamellae long spacing	nm	29-48 (SAXS), 31 (AFM) (PE fibers)	Henry, C K, Sandoz-Rosado, E, Roenbeck, M R, Magagnosc, D J, Alvarez, N J, Polymer, 202, 122589, 2020.
Lamellae diameter	nm	13-18 (SAXS), 63-64 (AFM) (PE fibers)	Henry, C K, Sandoz-Rosado, E, Roenbeck, M R, Magagnosc, D J, Alvarez, N J, Polymer, 202, 122589, 2020.
Shish domain length	nm	415-446 (SAXS)	Henry, C K, Sandoz-Rosado, E, Roenbeck, M R, Magagnosc, D J, Alvarez, N J, Polymer, 202, 122589, 2020.

COMMERCIAL POLYMERS

Some manufacturers	-	LyondellBasell; Formosa; Nova/Entec	
Trade names	-	Petrothene; Formolene: Sclair	

PHYSICAL PROPERTIES

Density at 20°C	g cm^{-3}	0.90-0.98	
Color	-	milky transparency	
Refractive index, 20°C	-	calc.=1.4648-1.4939; exp.=1.4728-1.52	
Clarity	%	7-98	
Odor	-	odorless	
Melting temperature, DSC	°C	99-138; 138.4-140.9 (equilibrium melting temperature), 141.4 (infinitely long linear polyethylene chain)	Mohammadi, H, Vincent, M, Marand, H, Polymer, 146, 344-60, 2018.
Decomposition temperature	°C	335	Patel, P; Hull, T R; McCabe, R W; Flath, D; Grasmeder, J; Percy, M, Polym. Deg. Stab., 95, 709-18, 2010.
Thermal expansion coefficient	°C^{-1}	5.13E-4 (175-250K) and 7.64E-4 (250-325K);	Bowman, A L, Mun, S, Nouranian, S, Huddleston, B D, Horstemeyer, M F, Polymer, 170, 85-100, 2019.
Thermal conductivity, melt	W m^{-1} K^{-1}	0.2-0.5 (bulk polymer), 1400 (single PE chain in the axial direction), 237 (single PE crystal lamella), 65-104 (formulated PE with improved thermal conductivity)	Zhang, R-C, Huang, Z, Huang, Z, Zhong, M, Sun, D, Compos. Sci. Technol., 196, 108154, 2020.
Glass transition temperature	°C	calc.=-59 to -133; exp.=-20 to -128	
Vicat temperature VST/A/50	°C	88-132	
Hansen solubility parameters, δ_D, δ_P, δ_H	MPa$^{0.5}$	16.8, 3.8, 3.8; 18.0, 1.2, 1.4 (PE surface), 16.5, 15.3, 8.2 (plasma treated fibers)	-; Kusano, Y, Teodoru, S, Hansen, C M, Surf. Coat. Technol., 205, 8-9, 2793-8, 2011.
Interaction radius		6.6	

PE polyethylene

PARAMETER	UNIT	VALUE	REFERENCES
Hildebrand solubility parameter	MPa$^{0.5}$	calc.=16.2-19.93; exp.=15.76-17.6	
Surface tension	mN m^{-1}	calc.=35.1-37.6; exp.=31.0-35.7	
Dielectric constant at 100 Hz/1 MHz	-	2.4/-	Huang, Y, Wei, X, Liu, L, Yu, H, Yang, J, Mater. Lett., 232, 86-91, 2018.
Electric strength K20/P50, d=0.60.8 mm	kV mm^{-1}	39	
Coefficient of friction	-	0.6	
Contact angle of water, 20°C	degree	94.9-97.2	
Surface free energy	mJ m^{-2}	33.5	
Speed of sound	m s^{-1}	32.5	
Acoustic impedance		1.73	

MECHANICAL & RHEOLOGICAL PROPERTIES

PARAMETER	UNIT	VALUE	REFERENCES
Tensile strength	MPa	30-48.6; 3153 (UHMWPE, Dyneema); 1100 (UHMW-PE fiber with diameter of 150 µm); 830 (oriented UHMW film)	-; Shah, K, Sockalingam, S, Defence Technol., 16, 1, 35-42, 2020; Kakiage, M, Fukagawa, D, Mater. Today, 23, 100864, 2020; Dayyoub, T, Maksimkin, A V, Senatov, F S, Kaloshkin, S D, Kolesnikov, E A, Int. J. Adhesion Adhesives, 98, 102535, 2020.
Tensile modulus	MPa	190-240	
Tensile stress at yield	MPa	10.6-20.7	
Elongation	%	180-1000	
Tensile yield strain	%	19-24	
Young's modulus	MPa	498	Bistolfi, A; Bellare, A, Acta Biomaterialia, in press, 2011.
Tenacity (fiber) (standard atmosphere)	cN tex^{-1} (daN mm-2)	32-70	Fourne, F, Synthetic Fibers. Machines and Equipment Manufacture, Properties. Carl Hanser Verlag, 1999.
Tenacity (wet fiber, as % of dry strength)	%	100	Fourne, F, Synthetic Fibers. Machines and Equipment Manufacture, Properties. Carl Hanser Verlag, 1999.
Fineness of fiber (titer)	dtex	10-25	Fourne, F, Synthetic Fibers. Machines and Equipment Manufacture, Properties. Carl Hanser Verlag, 1999.
Length (elemental fiber)	mm	38-200, filament	Fourne, F, Synthetic Fibers. Machines and Equipment Manufacture, Properties. Carl Hanser Verlag, 1999.
Poisson's ratio	-	exp.=0.47-0.49	
Shrinkage	%	1.7-1.85	
Residual stress	MPa	2-5 (extruded pipes)	
Melt viscosity, shear rate=1000 s^{-1}	Pa s	150-400	
Melt index, 230°C/3.8 kg	g/10 min	3.2-9	

CHEMICAL RESISTANCE

PARAMETER	UNIT	VALUE	REFERENCES
Acid dilute/concentrated	-	very good	
Alcohols	-	good	
Alkalis	-	very good	

PE polyethylene

PARAMETER	UNIT	VALUE	REFERENCES
Aliphatic hydrocarbons	-	poor	
Aromatic hydrocarbons	-	poor	
Esters	-	poor	
Greases & oils	-	good to poor	
Halogenated hydrocarbons	-	poor	
Ketones	-	poor	
Θ solvent, Θ-temp.=192.2, -73, 173.9, 133.3°C	-	amyl alcohol, carbon disulfide, n-heptane, n-hexane	
Good solvent	-	1,2,4-trichlorobenzene, decalin, halogenated hydrocarbons, aliphatic ketones, xylene (all above 60°C)	
Non-solvent	-	most common solvents	

FLAMMABILITY

PARAMETER	UNIT	VALUE	REFERENCES
Ignition temperature	°C	340-343	
Autoignition temperature	°C	440	
Limiting oxygen index	% O_2	17.4-18.5	
Minimum ignition energy	J	0.01	
Heat of combustion	J g^{-1}	47,740	Walters, R N; Hacket, S M; Lyon, R E, Fire Mater., 24, 5, 245-52, 2000.
Volatile products of combustion	-	CO, CO_2, aldehydes, acrolein, oligomers, waxes, oxygenated hydrocarbons	

WEATHER STABILITY

PARAMETER	UNIT	VALUE	REFERENCES
Spectral sensitivity	nm	<300	
Activation wavelengths	nm	300, 330-360	
Important initiators and accelerators	-	unsaturations, aromatic carbonyl compounds (deoxyanisoin, dibenzocycloheptadienone, flavone, 4-methoxybenzophenone, 10-thioxanthone), hydrogen bound to tertiary carbon at branching points, aromatic amines, groups formed on oxidation (hydroperoxides, carbonyl, carboxyl, hydroxyl) substituted benzophenones, complexes with ground-state oxygen, quinones (anthraquinone, 2-chloroanthraquinone, 2-tert-butyl-athraquinone, 1-methoxyanthraquinone, 2-ethylanthraquinone, 2-methylanthraquinone), transition metal compounds (Ni < Zn < Fe < Co), ferrocene derivatives, titanium dioxide (anatase), ferric stearate, polynuclear aromatic compounds (anthracene, phenanthrene, pyrene, naphthalene	
Products of degradation	-	free radicals, hydroperoxides, carbonyl groups, chain scission, crosslinking	

PE polyethylene

PARAMETER	UNIT	VALUE	REFERENCES
Stabilizers	-	UVA: 2-hydroxy-4-octyloxybenzophenone; phenol, 2-(5-chloro-2H-benzotriazole-2-yl)-6-(1,1-dimethylethyl)-4-methyl-; 2,2'-methylenebis(6-(2H-benzotriazol-2-yl)-4-1,1,3,3-tetramethylbutyl)phenol; 2,4-di-tert-butyl-6-(5-chloro-2H-benzotriazol-2-yl)-phenol; reaction product of methyl 3(3-(2H-benzotriazole-2-yl)-5-t-butyl-4-hydroxyphenyl propionate/PEG 300; 2-[4,6-bis(2,4-dimethylphenyl)-1,3,5-triazin-2-yl]-5-(octyloxy) phenol; Screener: titanium dioxide; zinc oxide; carbon black; Acid scavenger: hydrotalcite; Fiber: carbon nanotube; HAS: 1,3,5-triazine-2,4,6-triamine, N,N'''[1,2-ethane-diyl-bis[[[4,6-bis[butyl(1,2,6,6-pentamethyl-4-piperidinyl)amino]-1,3,5-triazine-2-yl]imino]-3,1-propanediyl]bis[N',N''-dibutyl-N',N''-bis(1,2,2,6,6-pentamethyl-4-piperidinyl)-; bis(1,2,2,6,6-pentamethyl-4-piperidyl)sebacate + methyl-1,2,2,6,6-pentamethyl-4-piperidyl sebacate; 2,2,6,6-tetramethyl-4-piperidinyl stearate; reaction products of N,N'-ethane-1,2-diylbis(1,3-propanediamine), cyclohexane, peroxidized 4-butylamino-2,2,6,6-tetramethylpiperidine and trichloro-1,3,5-triazine; poly[[(6-[1,1,3,3-tetramethylbutyl)amino]-1,3,5-triazine-2,4-diyl][2,2,6,6-tetramethyl-4-piperidinyl)imino]-1,6-hexanediyl[2,2,6,6-tetramethyl-4-piperidinyl)imino]]; Phenolic antioxidant: 2,6,-di-tert-butyl-4-(4,6-bis(octylthio)-1,3,5,-triazine-2-ylamino) phenol; 3,3',3',5,5',5'-hexa-tert-butyl-a,a',a'-(mesitylene-2,4,6-triyl)tri-p-cresol; ,4,6(1H,3H,5H)-trione; 3,4-dihydro-2,5,7,8-tetramethyl-2-(4,8,12-trimethyltridecyl)-2H-1-benzopyran-6-ol; 2',3-bis[[3-[3,5-di-tert-butyl-4-hydroxy-phenyl]propionyl]]propionohydrazide; isotridecyl-3-(3,5-di-tert-butyl-4-hydroxyphenyl) propionate; 2,2'-ethylidenebis (4,6-di-tert-butylphenol); ethylene bis[3,3-bis[3-(1,1-dimethylethyl)-4-hydroxyphenyl]butanoate];1,3,5-tris(4-tert-butyl-3-hydroxy-2,6-dimethylbenzyl)-1,3,5-triazine-2,4,6-(1H,3H,5H)-trione; 2,2'-methylenebis(4-methyl-6-tertbutylphenol); 3,5-bis(1,1-dimethyethyl)-4-hydroxy-benzenepropanoic acid, C13-15 alkyl esters; 2,2'-isobutylidenebis(2,4-dimethylphenol); 1,1,3-tris(2'methyl-4'-hydroxy-5'tert-butylphenyl)butane; Phosphite: bis-(2,4-di-t-butylphenol) pentaerythritol diphosphite; tris (2,4-di-tert-butylphenyl)phosphite; trinonylphenol phosphite; distearyl pentaerythritol diphosphite; trilauryl trithiophosphite; Thiosynergist: didodecyl-3,3'-thiodipropionate; dioctadecyl 3,3'-thiodipropionate; 2,2'-thiodiethylene bis[3-(3,5-ditert-butyl-4-hydroxyphenyl)propionate]; 4,4'-thiobis(2-t-butyl-5-methylphenol); 2,2'-thiobis(6-tert-butyl-4-methylphenol); pentaerythritol tetrakis(b-laurylthiopropionate); Quencher: (2,2'-thiobis(4-tert-octyl-phenolato)-N-butylamine-nickel(II); Optical brightener: 2,2'-(2,5-thiophenediyl)bis(5-tert-butylbenzoxazole); Vitamin E in medical applications	Micheli, B R; Wannomae, K K; Lozynski, A J; Christensen, S D; Muratoglu, O K, J. Arthroplasty, in press, 2011; Wypych, G, Handbook of Materials Weathering, 6th Ed., ChemTec Publishing, 2018.
Low earth orbit erosion yield	cm^3 $atom^{-1}$ x 10^{-24}	3.97	Waters, D L; Banks, B A; De Groh, K K; Miller, S K R; Thorson, S D, High Performance Polym., 20, 512-22, 2008.

BIODEGRADATION			
Colonized products		bags for oranges, food films, oxidized materials from LDPE, pipe wrap, polyethylene containing starch, shrink-wrap film, wood flower containg films	
Stabilizers	-	1-(2-(2,4-dichlorophenyl)-2-(2-propenyloxy)ethyl)-1H-imidazole, 2,4,4'-trichloro-2'-hydroxydiphenyl ether, benzoic anhydride, nisin, silver, zinc	

TOXICITY			
NFPA: Health, Flammability, Reactivity rating	-	0/1/0	
Carcinogenic effect	-	not listed by ACGIH, NIOSH, NTP	

PE polyethylene

PARAMETER	UNIT	VALUE	REFERENCES
Reproductive toxicity	-	not expected to occur	
TLV, ACGIH	mg m^{-3}	3 (respirable) 10 (inhalable)	
OSHA	mg m^{-3}	5 (respirable), 15 (total)	
Oral rat, LD$_{50}$	mg kg^{-1}	>2,000; 4,000	
Skin rabbit, LD$_{50}$	mg kg^{-1}	non-irritant	

ENVIRONMENTAL IMPACT

PARAMETER	UNIT	VALUE	REFERENCES
Aquatic toxicity, *Daphnia magna*, LC$_{50}$, 48 h	mg l^{-1}	75,000	Lithner, Ph D Thesis, University of Gothenburg, 2011.
Cradle to grave non-renewable energy use	MJ/kg	73.7-81.8	Harding, K G; Dennis, J S; von Blottnitz, H; Harrison, S T L, J. Biotechnol., 130, 57-66, 2007.

PROCESSING

PARAMETER	UNIT	VALUE	REFERENCES
Typical processing methods	-	blown film extrusion, cast film extrusion, extrusion, injection molding, molding, rotational molding, thin wall injection molding	
Processing pressure	MPa	5-6 (injection)	
Additives used in final products	-	Fillers: aluminum, barium sulfate, calcium carbonate, calcium sulfate whiskers, carbon black, diatomaceous earth, ferromagnetic powder, glass fiber, glass spheres, ground tire rubber, hollow silicates, hydrotalcite, kaolin, lignin, magnesium hydroxide, marble, mica, nickel fibers, red mud, sand, silica, soot, starch, superconductor ($YBa_2Cu_3O_{7-x}$), talc, wollastonite, wood flour, zirconium silicate; Plasticizers: dioctyl phthalate, EPDM, EVA, glycerin, glyceryl tribenzoate, mineral oil, paraffin oil, polyethylene glycol, sunflower oil; Antistatics: carbon black, copper complex of polyacrylic acid, ethoxylated amines, fatty diethanol amines, glycerol monostearate, graphite, ionomer, lauric diethanolamide, polyethylene glycol, quaternary ammonium compound, trineoalkoxy zirconate; Antiblocking: diatomaceous earth, natural silica, siloxane spheres, synthetic silica, talc, zeolite; Release: stearyl erucamide; Slip: erucamide, ethylene bisoleamide, oleamide	
Applications	-	packaging and film are the major applications, many other applications are part of normal industrial production, such as medical (artificial tendons, orthopaedic implants, wound dressing)	

ANALYSIS

PARAMETER	UNIT	VALUE	REFERENCES
FTIR (wavenumber-assignment)	cm^{-1}/-	CH$_2$ – 2916, 2848, 1463, 719; C=C – 1640	Shi, L-S; Wang, L-Y; Wang, Y-N, Eur. Polym. J., 42, 1625-33, 2006.
Raman (wavenumber-assignment)	cm^{-1}/-	CH$_2$ – 1464, 1443, 1420, 1372; C-C – 1138, 1070	Kim, J; Kim, Y; Chung, H, Talanta, 83, 879-84, 2011.
NMR (chemical shifts)	ppm	all-trans – 33.3; amorphous – 30.5, and more	Chaiyut, N; Amornsakchai, T; Kaji, H; Horii, F, Polym., 2470-81, 2006.

PEA poly(ethyl acrylate)

PARAMETER	UNIT	VALUE	REFERENCES
GENERAL			
Common name	-	poly(ethyl acrylate)	
CAS name	-	2-propenoic acid, ethyl ester, homopolymer	
Acronym	-	PEA	
CAS number	-	9003-32-1	
Linear formula		$\left[\begin{array}{c} CH-CH_2 \\ | \\ COOCH_2CH_3 \end{array} \right]_n$	
SYNTHESIS			
Monomer(s) structure	-	$H_2C{=}CHCOCH_2CH_3$ (with O double bonded)	
Monomer(s) CAS number(s)	-	140-88-5	
Monomer(s) molecular weight(s)	dalton, g/mol, amu	100.11	
Monomer(s) expected purity(ies)	%	99	
Method of synthesis	-	radical polymerization in the presence of 0.1 wt% benzoin as a photoinitiator	Mas Estelles, J; Krakovsky, I; Rodriguez Hernandz, J C; Piotrowska, A M; Monleon Pradas, M, J. Mater. Sci., 42, 8629-35, 2007; Rico, P; Gonzalez-Garcia, C; Petrie, T A; Garcia, A J, Colloids Surfaces: Biointerfaces, 78, 310-16, 2010.
Number average molecular weight, M_n	dalton, g/mol, amu	7,700	
Mass average molecular weight, M_w	dalton, g/mol, amu	14,000-1,600,000	
Polydispersity, M_w/M_n	-	1.06-3.44	
Molar volume at 298K	cm³ mol⁻¹	calc.=86.6-87.6; 89.4 (amorphous); exp.=89.4	
Van der Waals volume	cm³ mol⁻¹	calc.=56.1 (amorphous); exp.=56.89	
End-to-end distance of unperturbed polymer chain	nm	95.7-177.8	Srinivasan, K S V; Santapa, M, Polymer, 14, 1, 5-8, 1973.
Degree of branching	%	1.31	Castignolles, P; Graf, R; Parkinson, M; Wilhelm, M; Gaborieau, M, Polymer, 50, 2373-83, 2009.
STRUCTURE			
Cis content	%	33	McNeill, I C; Mohammed, M H, Polym. Deg. Stab., 48, 175-87, 1995.
Entanglement molecular weight	dalton, g/mol, amu	calc.=9,093	
PHYSICAL PROPERTIES			
Density at 20°C	g cm⁻³	1.08; 1.12 (amorphous)	
Refractive index, 20°C	-	calc.=1.468-1.477; exp.=1.469-1.54	
Decomposition temperature	°C	234	Castignolles, P; Graf, R; Parkinson, M; Wilhelm, M; Gaborieau, M, Polymer, 50, 2373-83, 2009.

PEA poly(ethyl acrylate)

PARAMETER	UNIT	VALUE	REFERENCES
Thermal conductivity, melt	W m^{-1} K^{-1}	0.1658	
Glass transition temperature	°C	-24 to -21; calc.=-6 to -22; exp.=-21 to -27; -24 (syndiotactic) -25 (isotactic); -17 (grafted)	Zampano, G; Bertoldo, M; Bronco, S, Carbohydrate, 75, 22-31, 2009.
Hildebrand solubility parameter	MPa$^{0.5}$	calc.=18.27; exp.=19.3	
Surface tension	mN m^{-1}	37.0	
Contact angle of water, 20°C	degree	89	Rico, P; Gonzalez-Garcia, C; Petrie, T A; Garcia, A J, Colloids Surfaces: Biointerfaces, 78, 310-16, 2010.
Surface free energy	mJ m^{-2}	35.1	

MECHANICAL & RHEOLOGICAL PROPERTIES

Poisson's ratio	-	calc.=0.373	
Water absorption, equilibrium in water at 23°C	%	1.7	

CHEMICAL RESISTANCE

Alcohols	-	poor	
Aliphatic hydrocarbons	-	poor	
Aromatic hydrocarbons	-	poor	
Esters	-	poor	
Halogenated hydrocarbons	-	poor	
Ketones	-	poor	
Θ solvent, Θ-temp.=44.9, 37.4, 20.5°C	-	n-butanol, ethanol, methanol	Llopis, J; Albert, A; Usobiaga, Eur. Polym. J., 3, 259-65, 1967.
Good solvent	-	acetone, aromatic hydrocarbons, butanol, chlorinated hydrocarbons, ester, glycol ether, ketones, methanol, THF, p-xylene	
Non-solvent	-	aliphatic hydrocarbons, cyclohexanol, diethyl ether, higher aliphatic alcohols, tetrahydrofurfuryl alcohol	

FLAMMABILITY

Volatile products of combustion	-	CO_2, ethylene, ethanol, ethyl acrylate	McNeill, I C; Mohammed, M H, Polym. Deg. Stab., 48, 175-87, 1995.

TOXICITY

NFPA: Health, Flammability, Reactivity rating	-	1/3/0	
Carcinogenic effect	-	not listed by ACGIH, NIOSH, NTP	
Mutagenic effect	-	none	
TLV, ACGIH	ppm	188	
Oral rat, LD$_{50}$	mg kg^{-1}	636	
Skin rabbit, LD$_{50}$	mg kg^{-1}	12,124	

PROCESSING

Additives used in final products	-	Plasticizers: dipropylene glycol dibenzoate, isodecyl diphenyl phosphate, dibutyl phthalate, 2,2,4-trimethyl-1,3-pentanediol monoisobutyrate and dibutyl, dihexyl, and dioctyl phthalates	
Outstanding properties	-	latex paints, tablet coating	

PEA poly(ethyl acrylate)

PARAMETER	UNIT	VALUE	REFERENCES
BLENDS			
Suitable polymers	-	EEA, epoxy, LDPE, PBA, PFS, PPA, PVC, PVDF-HFA, PVF	
ANALYSIS			
NMR (chemical shifts)	ppm	CH_3 – 1.2; CH_2 – 4.05, 1.84, 1.58; CH – 2.24;	McNeill, I C; Mohammed, M H, Polym. Deg. Stab., 48, 175-87, 1995.

PEC poly(ester carbonate)

PARAMETER	UNIT	VALUE	REFERENCES
GENERAL			
Common name	-	poly(ester carbonate)	Pinna, I; Hellums, M W; Koros, W J, Polymer, 32, 14, 2612-17, 1991.
ACS name	-	1,3-benzenedicarbonyl dichloride, polymer with 1,4-benzenedi-carbonyl dichloride, carbonic dichloride and 4,4'-(1-methylethy-lidene)bis[phenol]	
Acronym	-	PEC	
CAS number	-	71519–80–7	
HISTORY			
Person to discover	-	Cotter, R J; Sulzberg, T	Cotter, R J; Sulzberg, T, US Patent 3,536,781, Union Carbide, Oct. 27, 1970.
Date	-	1970	
SYNTHESIS			
Monomer(s) structure	-		
Monomer(s) CAS number(s)	-	100-20-9; 99-63-8; 80-05-7; 75-44-5	
Monomer(s) molecular weight(s)	dalton, g/mol, amu	203.02; 203.02; 228.29; 98.92	
Formulation example	-	the resins are manufactured using phthaloyl chloride/carbonyl chloride mole ratio of 0.81/1 or greater	
Method of synthesis	-	polyestercarbonate resins are produced by the condensation of 4,4'-isopropyllidenediphenol, carbonyl chloride, terephtha-loy chloride, and isophthaloyl chloride, such that the finished resins are composed of 45 to 85 mole% ester of which up to 55 mole% is the terephthaloyl isomer	
Number average molecular weight, M_n	dalton, g/mol, amu	3,390-4,400; 50,000	-; Socka, M, Sitko, M, Boncel, S, Kost, B, Brzeziński, M, Polym. Deg. Stab., 170, 109000, 2019.
Mass average molecular weight, M_w	dalton, g/mol, amu	3,700-21,800; 144,000	-; Beck-Broichsitter, M, Polym. Deg. Stab., 177, 109186, 2020.
Polydispersity, M_w/M_n	-	3.4	Beck-Broichsitter, M, Polym. Deg. Stab., 177, 109186, 2020.
STRUCTURE			
Crystallinity	%	0 (bisphenol A-based); 36.2-52.7 (1,3-propanediol-based)	Murthy, N S; Aharoni, S M, Polymer, 28, 13, 2171-75, 1987; Chandure, A S; Umare, S S; Pandley, R A, Eur. Polym. J., 44, 2068-86, 2008.
COMMERCIAL POLYMERS			
Some manufacturers	-	Empower Materials	
Trade names	-	QPAC	
PHYSICAL PROPERTIES			
Density at 20°C	g cm^{-3}	1.13-1.2	
Color	-	white	
Refractive index, 20°C	-	1.6	

PEC poly(ester carbonate)

PARAMETER	UNIT	VALUE	REFERENCES
Transmittance	%	85	
Haze	%	1	
Odor	-	slight	
Melting temperature, DSC	°C	41.4-45.5	
Thermal expansion coefficient, 23-80°C	10^{-4} °C^{-1}	0.6-0.92	
Thermal conductivity, melt	W m^{-1} K^{-1}	0.21	
Glass transition temperature	°C	19	Beck-Broichsitter, M, Polym. Deg. Stab., 177, 109186, 2020.
Specific heat capacity	J K^{-1} kg^{-1}	1,250	
Temperature index (50% tensile strength loss after 20,000 h/5000 h)	°C	125-130	
Heat deflection temperature at 1.8 MPa	°C	143-149	
Vicat temperature VST/B/50	°C	160-185	
Relative permittivity at 60 Hz	-	3.15-3.27	
Relative permittivity at 1 MHz	-	3-3.1	
Dissipation factor at 60 Hz	E-4	12-16	
Volume resistivity	ohm-m	2.5E15	
Electric strength K20/P50, d=0.60.8 mm	kV mm^{-1}	20-20.2	
Permeability to nitrogen, 25°C	barrer (cm^3 cm cm^{-2} s^{-1} cmHg^{-1})	9.1	Pinna, I; Hellums, M W; Koros, W J, Polymer, 32, 14, 2612-17, 1991.
Permeability to oxygen, 25°C	barrer (cm^3 cm cm^{-2} s^{-1} cmHg^{-1})	1.85	Pinna, I; Hellums, M W; Koros, W J, Polymer, 32, 14, 2612-17, 1991.
Contact angle of water, 20°C	degree	74	Beck-Broichsitter, M, Polym. Deg. Stab., 177, 109186, 2020.

MECHANICAL & RHEOLOGICAL PROPERTIES

PARAMETER	UNIT	VALUE	REFERENCES
Tensile strength	MPa	71-78	
Tensile modulus	MPa	2,090-2,250	
Tensile stress at yield	MPa	65-66	
Elongation	%	78-122	
Tensile yield strain	%	6-8	
Flexural strength	MPa	95-97	
Flexural modulus	MPa	2,020-2,350	
Charpy impact strength, notched, 23°C	kJ m^{-2}	15	
Izod impact strength, unnotched, 23°C	J m^{-1}	3,200 to NB	
Izod impact strength, notched, 23°C	J m^{-1}	300-640	
Izod impact strength, notched, -30°C	J m^{-1}	84-144	

PEC poly(ester carbonate)

PARAMETER	UNIT	VALUE	REFERENCES
Rockwell hardness	-	M85-92; R122-127	
Shrinkage	%	0.7-1	
Melt index, 300°C/1.2 kg	g/10 min	2-8	
Water absorption, 24h at 23°C	%	0.15-0.19; 0.35 (saturation)	
Moisture absorption, equilibrium 23°C/50% RH	%	0.15-0.35	

CHEMICAL RESISTANCE

Acid dilute/concentrated	-	poor	
Alcohols	-	good	
Alkalis	-	poor	
Aliphatic hydrocarbons	-	good	
Aromatic hydrocarbons	-	poor	
Esters	-	good	
Halogenated hydrocarbons	-	poor	
Ketones	-	poor	
Good solvent	-	concentrated sulfuric acid	
Non-solvent	-	m-cresol; sym-tetrachloroethane; 1-methyl-2-pyrollidinone; N,N-dimethylacetamide; N,N-dimethylformamide; dimethylsulfoxide; and hexamethylphosphoric triamide	

FLAMMABILITY

Autoignition temperature	°C	630	
Volatile products of combustion	-	CO, CO_2, methylene chloride, phenol, diphenylcarbonate	
UL rating	-	HB; V-0 (fire-resistant grade)	Shen, D; van de Grampel, R D; Lambertus, T; Singh, H R K; Lend, J-P, US Patent 20090062439, Sabic, 2009.

BIODEGRADATION

Typical biodegradants	-	hydrolysis by *Rhizopus delemar*	Chandure, A S; Umare, S S; Pandley, R A, Eur. Polym. J., 44, 2068-86, 2008.

TOXICITY

Oral rat, LD_{50}	mg kg^{-1}	5,000	
Skin rabbit, LD_{50}	mg kg^{-1}	very slight irritant	

PROCESSING

Typical processing methods	-	injection molding, spraying	
Preprocess drying: temperature/ time/residual moisture	°C/h/%	120/3-4/0.02	
Processing temperature	°C	320-370	
Processing pressure	MPa	0.3-0.7 (back)	
Applications	-	3D printing, automotive bezels, automotive lightning, cookware, electronic passive components, fire helmets, headlamp reflectors, medical applications (1,3-propanediol-based), MEMS fabrication, mold release, pharmaceutical	

PEC poly(ester carbonate)

PARAMETER	UNIT	VALUE	REFERENCES
Outstanding properties	-	biocompatible, sterilizable	
BLENDS			
Suitable polymers	-	CB, PBS, PC, PMP, PU	
ANALYSIS			
FTIR (wavenumber-assignment)	cm^{-1}/-	C=O – 1736; ester – 1730-1728; C-O-C – 1173	Chandure, A S; Umare, S S; Pandley, R A, Eur. Polym. J., 44, 2068-86, 2008.
NMR (chemical shifts)	ppm	1,3-propanediol residue – 1.87-1.99; CH$_2$ – 2.33, 1.63-1.66	Chandure, A S; Umare, S S; Pandley, R A, Eur. Polym. J., 44, 2068-86, 2008.

PEDOT poly(3,4-ethylenedioxythiophene)

PARAMETER	UNIT	VALUE	REFERENCES
GENERAL			
Common name	-	poly(3,4-ethylenedioxythiophene)	Elschner, A; Kirchmeyer, S; Lovenich, W; Merker, U; Reuter, K, PEDOT: Principles and Applications of an Intrinsically Conductive Polymer, CRC Press, 2011.
IUPAC name	-	poly(thiophene-2,5-diyl)	
ACS name	-	thiophene, homopolymer	
Acronym	-	PEDOT	
CAS number	-	25233-34-5	
Linear formula			
HISTORY			
Person to discover	-	Louvar, J J; Jonas, F; Heywang, G; Schmidtberg, W	Louvar, J J, US Patent 3,574,072, Universal Oil Products, Apr. 6, 1971; Jonas, F; Heywang, G; Schmidtberg, W, DE Patent 38 13 589, Apr. 22, 1988, Bayer AG.
Date	-	1971; 1988	
Details	-	electrolysis of monomer (thiophene) in acetic acid leads to polymerization; Bayer's patent	
SYNTHESIS			
Monomer(s) structure	-		
Monomer(s) CAS number(s)	-	126213-50-1	
Monomer(s) molecular weight(s)	dalton, g/mol, amu	142.18	
Monomer(s) expected purity(ies)	%	97	
Monomer ratio	-	1	
PEDOT/PSS ratio		1:1 to 1:20	Petrosino, M; Rubino, A, Organic Electronics, 12, 1159-65, 2011.
Methods of synthesis	-	*in situ* polymerization, plasma polymerization, electropolymerization, coupling reaction catalyzed by metal, and chemical oxidative polymerization	
Time of polymerization	h	30	
Pressure of polymerization	Pa	67-267	
Number average molecular weight, M_n	dalton, g/mol, amu	6,000-82,300	Perepichka, I F; Perepichka, D F; Meng, H; Wudl, F, Adv. Mater. 17, 2281-2305, 2005.
Polydispersity, M_w/M_n	-	1.27-9.1	Perepichka, I F; Perepichka, D F; Meng, H; Wudl, F, Adv. Mater. 17, 2281-2305, 2005.
STRUCTURE	-		
Crystallinity	%	40; 63 (poly(3-dodecylthiophene))	Saxena, V; Malhotra, B D, Handbook of polymers in Elecronics, Ed. Malhotra, B D, Rapra, 2002; Liu, S L; Chung, T S, Antec, 1999.

PEDOT poly(3,4-ethylenedioxythiophene)

PARAMETER	UNIT	VALUE	REFERENCES
Cell type (lattice)	-	orthorhombic	
Cell dimensions	nm	a:b:c=0.78:0.555:0.803; a:b:c=1.52:0.68:0.77; a:b:c=0.78:1.18:0.69	Niu, L; Kvarnstrom, C; Froberg, K; Ivaska, A, Synthetic Metals, 122, 425-29, 2001; Lenz, A; Kariis, H; Pohl, A; Persson, P; Ojamae, Chem. Phys., in press, 2011.
Number of chains per unit cell	-	4	
Spacing between chains	nm	0.38	Vukmirovic, N; Wang, L-W, J. Phys. Chem. B, 113, 409-15, 2009.

COMMERCIAL POLYMERS

Some manufacturers	-	Hereaus	
Trade names	-	Clevios	

PHYSICAL PROPERTIES

Density at 20°C	g cm^{-3}	1.4-1.6	
Refractive index, 20°C	-	1.5-1.55	
Melting temperature, DSC	°C	>350	
Softening point	°C	70 (poly(3-undecyl bithiophene))	
Decomposition temperature	°C	325 (air); 375 (N_2)	
Glass transition temperature	°C	64.4	
Maximum service temperature	°C	250	
Conductivity	S cm^{-1}	2.79E-2 (PEDOT), 1.32 (with 15 wt% MWCNTs)	Taj, M, Manohara, S R, Synthetic Metals, 269, 116572, 2020.
Dielectric constant at 1 kHz	-	1.2E4, 3.7E4 (with 15 wt% MWCNT)	Taj, M, Manohara, S R, Synthetic Metals, 269, 116572, 2020.
Volume resistivity	ohm-m	2.5-3.3E-3 to 5E6	Saxena, V; Malhotra, B D, Handbook of polymers in Elecronics, Ed. Malhotra, B D, Rapra, 2002; Petrosino, M; Rubino, A, Organic Electronics, 12, 1159-65, 2011.
Optical absorption edge	eV	2	Saxena, V; Malhotra, B D, Handbook of polymers in Elecronics, Ed. Malhotra, B D, Rapra, 2002.

MECHANICAL & RHEOLOGICAL PROPERTIES

Tensile strength	MPa	84-178; 17.2-53.2 (PEDOT/PSS); 94-130 (PEDOT/PSS, microfiber)	Wang, X-S; Feng, X-Q, J. Mater. Sci. Lett., 21, 715-17, 2002; Lang, U; Naujoks, N; Dual, J, Synthetic Metals, 159, 473-79, 2009; Okuzaki, H; Harashina, Y; Yan, H, Eur. Polym. J., 45, 256-61, 2009.
Elongation	%	4.5-12 (PEDOT/PSS, microfiber)	Okuzaki, H; Harashina, Y; Yan, H, Eur. Polym. J., 45, 256-61, 2009.
Young's modulus	MPa	3,200-6,800; 1,100-2,800 (PEDOT/PSS); 2,500-4,000 (PEDOT/PSS, microfiber)	Wang, X-S; Feng, X-Q, J. Mater. Sci. Lett., 21, 715-17, 2002; Lang, U; Naujoks, N; Dual, J, Synthetic Metals, 159, 473-79, 2009; Okuzaki, H; Harashina, Y; Yan, H, Eur. Polym. J., 45, 256-61, 2009.
Compression set, 24h 70°C	%	0.32-0.35 (PEDOT/PSS)	Lang, U; Naujoks, N; Dual, J, Synthetic Metals, 159, 473-79, 2009.
Water absorption, equilibrium in water at 23°C	%	1.15	Elschner, A; Kirchmeyer, S; Lovenich, W; Merker, U; Reuter, K, PEDOT: Principles and Applications of an Intrinsically Conductive Polymer, CRC Press, 2011.

PEDOT poly(3,4-ethylenedioxythiophene)

PARAMETER	UNIT	VALUE	REFERENCES
CHEMICAL RESISTANCE			
Alcohols	-	very good	
Aliphatic hydrocarbons	-	very good	
Aromatic hydrocarbons	-	very good	
Esters	-	very good	
Greases & oils	-	very good	
Halogenated hydrocarbons	-	very good	
Ketones	-	very good	
Good solvent	-	diphenyl ether and nitrobenzene at 220-250°C (poly(3-undecyl bithiophene))	
FLAMMABILITY			
Autoignition temperature	°C	>500	
WEATHER STABILITY			
Activation wavelengths	nm	<320 (PEDOT:PSS)	Elschner, A; Kirchmeyer, S; Lovenich, W; Merker, U; Reuter, K, PEDOT: Principles and Applications of an Intrinsically Conductive Polymer, CRC Press, 2011; Schaefer, M; Holtkamp, J; Gillner, A, Synthetic Metals, 161, 1051-57, 2011.
Excitation wavelengths	nm	303-577	Perepichka, I F; Perepichka, D F; Meng, H; Wudl, F, Adv. Mater. 17, 2281-2305, 2005.
Emission wavelengths	nm	470-783	Perepichka, I F; Perepichka, D F; Meng, H; Wudl, F, Adv. Mater. 17, 2281-2305, 2005.
TOXICITY			
HMIS: Health, Flammability, Reactivity rating	-	1/0/0	
Carcinogenic effect	-	not listed by ACGIH, NIOSH, NTP	
Oral rat, LD$_{50}$	mg kg^{-1}	>2,000	
ENVIRONMENTAL IMPACT			
Aquatic toxicity, *Bluegill sunfish*, LC$_{50}$, 48 h	mg l^{-1}	>10,000	
PROCESSING			
Typical processing methods	-	wet spinning (microfibers), spin coating, inkjet printing	
Applications	-	antistatic coatings, capacitors, displays, electrodes, electrodialysis, heat conductive coatings, OLED displays, printed wiring boards, smart windows, solar cells, switchable mirrors	
Outstanding properties	-	conductivity, transparency, thermal stability	
BLENDS			
Suitable polymers	-	fullerene, P3TMA	

PEDOT poly(3,4-ethylenedioxythiophene)

PARAMETER	UNIT	VALUE	REFERENCES
ANALYSIS			
FTIR (wavenumber-assignment)	cm^{-1}/-	C-C – 1518; thiophene ring – 1050; C-S – 970, 698	Ye, S; Shen, C; Pang, H; Wang, J; Lu, Y, Polymer, 52, 2542-49, 2011.
Raman (wavenumber-assignment)	cm^{-1}/-	band assignments for PEDOT, P3CT, and P3MT is ref.	Chen, F; Shi, G, Zhang, J; Fu, M, Thin Solid Films, 424, 283-90, 2003.

PEEK polyetheretherketone

PARAMETER	UNIT	VALUE	REFERENCES
GENERAL			
Common name	-	polyetheretherketone	
IUPAC name	-	poly(oxy-1,4-phenyleneoxy-1,4-phenylenecarbonyl-1,4-phenylene)	
CAS name	-	poly(oxy-1,4-phenyleneoxy-1,4-phenylenecarbonyl-1,4-phenylene)	
Acronym	-	PEEK	
CAS number	-	29658-26-2; 31694-16-3	
Linear formula			
HISTORY			
Date	-	1962; 1964; 1982	
Details	-	first produced in DuPont laboratories in 1962; ICI chemists synthesized it in 1964; Victrex PEEK commercialized by ICI in 1982	
SYNTHESIS			
Monomer(s) structure	-		
Monomer(s) CAS number(s)	-	345-92-6; 123-31-9	
Monomer(s) molecular weight(s)	dalton, g/mol, amu	218.20; 110.01	
Monomer ratio	-	molar equivalent amounts	
Method of synthesis	-	polycondensation of monomers can be conducted in high boiling solvent (e.g., diphenyl sulfone)	
Temperature of polymerization	°C	280-350	Lu, Q; Yang, Z; Li, X; Jin, S, J. Appl. Polym. Sci., 114, 2060-70, 2009.
Time of polymerization	h	6.5	Lu, Q; Yang, Z; Li, X; Jin, S, J. Appl. Polym. Sci., 114, 2060-70, 2009.
Pressure of polymerization	Pa	atmospheric, under N_2 blanket	
Number average molecular weight, M_n	dalton, g/mol, amu	6,200-15,800	
Mass average molecular weight, M_w	dalton, g/mol, amu	14,300-100,000	
Radius of gyration	nm	15.5-28	Devaux, J; Delimoy, D; Daoust, D; Legras, R; Mercier, Strazielle, C; Nield, E, Polymer, 26, 13, 1994-2000, 1985.
STRUCTURE			
Crystallinity	%	16-47; 28-44 (yarn); 49 (max); 8.6-19 (with 5% silica)	Welsh, W J; Collantes, E; Gahimer, T, Grayson, M, Antec, 2172-75, 1996; Shekar, R I; Kotresh, T M; Rao, P M D; Kumar, K, J. Appl. Polym. Sci., 112, 2497-2510, 2009; Jaekel, D J; MacDonald, D W; Kurtz, S M, J. Mech. Behavior Biomed. Mater., in press, 2011; Kuo, M C; Kuo, J S; Yang, M H; Huang, J C, Mater. Chem. Phys., 123, 471-80, 2010.

PEEK polyetheretherketone

PARAMETER	UNIT	VALUE	REFERENCES
Cell type (lattice)	-	orthorhombic	
Cell dimensions	nm	a:b:c=0.743-0.788:0.584-0.596:0.986-3.037	Doumeng, M, Makhlouf, L, Berthet, F, Marsan, O, Chabert, F, Polym. Testing, in press, 106878, 2020.
Unit cell angles	degree	α:β:γ=90:90:90	
Number of chains per unit cell	-	2	
Crystallite size	nm	3.34-9.50	Karacan, I, Fibers Polym., 6, 3, 206-18, 2005.
Crystallization exotherm	°C	172	Arous, M; Amor, I B; Kallel, A; Fakhafakh, Z, Perrier, J. Phys. Chem. Solids, 1405-14, 2007.
Avrami constants, k/n	-	3.84-6.28	Kuo, M C; Kuo, J S; Yang, M H; Huang, J C, Mater. Chem. Phys., 123, 471-80, 2010.

COMMERCIAL POLYMERS

Some manufacturers	-	Ensinger; Evonik; Solvay; Victrex	
Trade names	-	Tecapeek; Vestakeep; KetaSpire; Victrex PEEK	

PHYSICAL PROPERTIES

Density at 20°C	g cm^{-3}	1.26-1.45; 1.260-1.267 (amorphous); 1.384-1.401 (crystalline); 1.53 (30% glass fiber); 1.41 (30% carbon fiber)	
Color	-	white	
Refractive index, 20°C	-	1.65-1.77	
Birefringence	-	0.00-0.04 (low crystallinity, 12-20%; 0.10-0.14 (high crystallinity, 30-42%); 0.354 (maximum birefringence for fully crystalline perfectly oriented fiber)	Bicakci, S; Cakama, M, Polymer, 43, 9, 2737-46, 2002; Karacan, I, Fibers Polym., 6, 3, 206-18, 2005.
Odor	-	odorless	
Melting temperature, DSC	°C	334-350	
Decomposition onset temperature	°C	575	Patel, P; Hull, T R; McCabe, R W; Flath, D; Grasmeder, J; Percy, M, Polym. Deg. Stab., 95, 709-18, 2010.
Thermal expansion coefficient, 23-80°C	°C^{-1}	0.43-1.6E-4; 1.9E-5 (30% glass fiber); 5.2-6.7E-6 (30% carbon fiber); 6.69E-4 (melt)	
Thermal conductivity, melt	W m^{-1} K^{-1}	0.25	
Glass transition temperature	°C	143-158	Padey, D; Walling, J; Wood A, Polymers in Defence and Aerospace 2007, Rapra, 2007, paper 15; Arous, M; Amor, I B; Kallel, A; Fakhafakh, Z, Perrier, J. Phys. Chem. Solids, 1405-14, 2007.
Specific heat capacity	J K^{-1} kg^{-1}	2160	
Heat of fusion	kJ mol^{-1}	36.8-37.5	
Maximum service temperature	°C	315-400	
Continuous use temperature	°C	260	Patel, P; Hull, T R; McCabe, R W; Flath, D; Grasmeder, J; Percy, M, Polym. Deg. Stab., 95, 709-18, 2010.
Heat deflection temperature at 1.8 MPa	°C	155-162; 315 (30% glass fiber and 30% carbon fiber)	
Enthalpy of crystallization	J g^{-1}	18.59-48.72	Kuo, M C; Kuo, J S; Yang, M H; Huang, J C, Mater. Chem. Phys., 123, 471-80, 2010.

PEEK polyetheretherketone

PARAMETER	UNIT	VALUE	REFERENCES
Hildebrand solubility parameter	$MPa^{0.5}$	22.8	
Surface tension	$mN\ m^{-1}$	21.2-22.6 (calc); 23.7 (medical grade)	Zou, F, Lv, F, Ma, X, Xia, X, Jiang, J, Mater. Design, 188, 108433, 2020.
Volume resistivity	ohm-m	1E14	
Electric strength K20/P50, d=0.60.8 mm	$kV\ mm^{-1}$	19	
Coefficient of friction	-	0.22; 0.08-0.09 (lubricated conditions); 0.25-0.3 (dry conditions); 0.42 (implant grade)	Xiong, D; Xiong, L; Liu, L, J. Biomed. Mater. Res. B, 93, 492-96, 2010; Lv, X, Wang, X, Tang, S, Wang, D, Yang, L, He, A, Tang, T, Wei, J, Colloids Surf. B: Biointerfaces, 189, 110819, 2020.
Contact angle of water, 20°C	degree	90; 80 (medical grade)	Zhang, S; Awaja, F; James, N; McKenzie, D R; Ruys, A J, Colloid Surfaces A: Physicochem. Eng. Aspects, 374, 88-95, 2011; Zou, F, Lv, F, Ma, X, Xia, X, Jiang, J, Mater. Design, 188, 108433, 2020.
Speed of sound	$m\ s^{-1}$	1,860-3,040	Shekar, R I; Kotresh, T M; Rao, P M D; Kumar, K, J. Appl. Polym. Sci., 112, 2497-2510, 2009.

MECHANICAL & RHEOLOGICAL PROPERTIES

Tensile strength	MPa	75-100; 158-162 (30% glass fiber); 201-223 (30% carbon fiber)	
Tensile modulus	MPa	3,500-4,400; 10,500-10,800 (30% glass fiber); 19,700-20,900 (30% carbon fiber)	
Elongation	%	20-50; 2.7-2.8 (30% glass fiber); 1.7-2.0 (30% carbon fiber)	
Flexural strength	MPa	146-170; 260-261 (30% glass fiber); 317-321 (30% carbon fiber)	
Flexural modulus	MPa	3,700-4,300; 10,400-10,500 (30% glass fiber); 17,500-17,900 (30% carbon fiber)	
Elastic modulus	MPa	3,500-4,000	
Compressive strength	MPa	118-169	
Izod impact strength, unnotched, 23°C	$J\ m^{-1}$	no break; 640-850 (30% glass fiber); 640-750 (30% carbon fiber)	
Izod impact strength, notched, 23°C	$J\ m^{-1}$	77-91; 69 (30% glass fiber); 64-69 (30% carbon fiber)	
Shear strength	MPa	53	
Poisson's ratio	-	0.4-0.41	Ramani, K; Zhao, W, Antec, 1160-64, 1997.
Shore D hardness	-	88; 91 (30% glass fiber); 92 (30% carbon fiber)	
Rockwell hardness	-	R120	
Shrinkage	%	1.2-1.8; 0.2-1.5 (30% glass fiber); 0.1-1.6 (30% carbon fiber)	
Brittleness temperature (ASTM D746)	°C	-65	
Intrinsic viscosity, 25°C	$dl\ g^{-1}$	0.45-1.59	
Melt viscosity, shear rate=1000 s^{-1}	Pa s	380-440; 350 (30% glass fiber)	
Melt index, 230°C/3.8 kg	g/10 min	3-36; 0.7-14 (30% glass fiber); 1.1-11 (30% carbon fiber)	
Water absorption, 24h at 23°C	%	0.1-0.5; 0.1 (30% glass fiber)	

PEEK polyetheretherketone

PARAMETER	UNIT	VALUE	REFERENCES
Moisture absorption, equilibrium 23°C/50% RH	%	0.5	

CHEMICAL RESISTANCE

PARAMETER	UNIT	VALUE	REFERENCES
Acid dilute/concentrated	-	good	
Alcohols	-	very good	
Alkalis	-	very good	
Aliphatic hydrocarbons	-	very good	
Aromatic hydrocarbons	-	very good	
Esters	-	very good	
Greases & oils	-	very good	
Halogenated hydrocarbons	-	very good	
Ketones	-	good	

FLAMMABILITY

PARAMETER	UNIT	VALUE	REFERENCES
Ignition temperature	°C	575-595	
Autoignition temperature	°C	595	
Limiting oxygen index	$\% \, O_2$	35-37.3	Patel, P; Hull, T R; Lyon, R E; Stoliarov, S I; Walters, R N; Crowley, S; Safronova, N; Polym. Deg. Stab., 96, 12-22, 2011.
Heat release	$kJ \, g^{-1}$	10.7	Patel, P; Hull, T R; Lyon, R E; Stoliarov, S I; Walters, R N; Crowley, S; Safronova, N; Polym. Deg. Stab., 96, 12-22, 2011.
NBS smoke chamber	Ds	30	
Burning length	mm	30.9	Patel, P; Hull, T R; Lyon, R E; Stoliarov, S I; Walters, R N; Crowley, S; Safronova, N; Polym. Deg. Stab., 96, 12-22, 2011.
Char at 500°C	%	41-52; 67 (carbon fiber); 63 (glass fiber)	Patel, P; Hull, T R; McCabe, R W; Flath, D; Grasmeder, J; Percy, M, Polym. Deg. Stab., 95, 709-18, 2010; Lyon, R E; Walters, R N, J. Anal. Appl. Pyrolysis, 71, 27-46, 2004;
Heat of combustion	$J \, g^{-1}$	22,100-31,480	
Activation energy of decomposition	$kJ \, mol^{-1}$	220	Swallowe, G M; Dawson, P C; Tang, T B; Xu, Q L, J. Mater. Sci., 30, 3853-55, 1995.
Volatile products of combustion	-	CO, CO_2, diphenyl ether, phenol, benzene and more	Walters, R N; Hacket, S M; Lyon, R E, Fire Mater., 24, 5, 245-52, 2000; Patel, P; Hull, T R; McCabe, R W; Flath, D; Grasmeder, J; Percy, M, Polym. Deg. Stab., 95, 709-18, 2010.
UL 94 rating	-	V-0 to V-1	

WEATHER STABILITY

PARAMETER	UNIT	VALUE	REFERENCES
Spectral sensitivity	nm	286, 306, 345	Giancaterina, S; Rossi, A; Rivaton, A; Gardette, J L, Polym. Deg. Stab., 68, 133-44, 2000.
Excitation wavelengths	nm	280, 310	Giancaterina, S; Rossi, A; Rivaton, A; Gardette, J L, Polym. Deg. Stab., 68, 133-44, 2000.

PEEK polyetheretherketone

PARAMETER	UNIT	VALUE	REFERENCES
Emission wavelengths	nm	315, 400	Giancaterina, S; Rossi, A; Rivaton, A; Gardette, J L, Polym. Deg. Stab., 68, 133-44, 2000.
Depth of UV penetration	μm	<250	Nakamura, H; Nakamura, T; Noguchi, T; Imagawa, K, Polym. Deg. Stab., 91, 740-6, 2006.

BIODEGRADATION

Colonized products	-	graphite containing composites	
Stabilizers	-	sodium o-phenylphenate	

TOXICITY

Carcinogenic effect	-	not listed by ACGIH, NIOSH, NTP	
OSHA	mg m^{-3}	5 (respirable), 15 (total)	
Oral rat, LD$_{50}$	mg kg^{-1}	15,000	

PROCESSING

Typical processing methods	-	extrusion blow molding, film extrusion, injection molding, mixing, profile extrusion, thermoforming, wire and cable extrusion	
Preprocess drying: temperature/ time/residual moisture	°C/h/%	150/4/	
Processing temperature	°C	355-380	
Additives used in final products	-	Fillers: carbon fiber, glass fiber, graphite, nano-zirconium oxide, PTFE, titanium dioxide; Other: melt stabilizers (e.g., zinc oxide or zinc sulfide, phosphites, phosphonites); Antistatics: fatty quaternary ammonium compounds, quaternary or tertiary ammonium ions and bis(perfluoroalkanesulfonyl)imide	
Applications	-	aerospace, automotive, bearing cages, belts, bolts and nuts, bone screws, butterfly valve seatings, chemically resistant bearings and cams, cryogenic propellant tank for supersonic aircrafts, dental applications, ducting, electrical (cable ties, cable insulation, rechargeable batteries), film, fracture fixation plates, fuel valves, heat-resistant gears, high performance conveyors, horizontal stabilizers for helicopters, hot melt adhesive, implants, machine tools, medical (compression plates, catheter body, arthroereisis prosthesis, bone substitutes), nuclear power plants, oil/gas, orthopedic implant, piston rings, pump impellers, satellites, seals, semiconductor wafer carriers, soil well data logging tools, sterilization equipment for medical and dental applications, surgical instruments, tennis racket strings,tubing, vacuum pump blades, valve linings, valve seats	
Outstanding properties	-	chemical and thermal resistance, high strength, wear resistance	

BLENDS

Suitable polymers	-	PI, PEI, PTFE, PVP, UHMWPE	

PEEK polyetheretherketone

PARAMETER	UNIT	VALUE	REFERENCES
ANALYSIS			
FTIR (wavenumber-assignment)	cm^{-1}/-	1653 − C=O stretching vibration band, 1597 and 1502 − in-plane vibrational band of R−O−R benzene ring, 1309 − R−CO−R benzene ring in-plane vibrational band, 1225 − R−O−R asymmetric stretching vibration band, 1164 − C−H in-plane bending vibration absorption band of benzene ring in aryl ether or aryl ketone structure, 929 − R−CO−R symmetric stretching vibration band, 841 and 746 − C−H out-of-plane bending vibration absorption bands of benzene ring	Lv, X, Wang, X, Tang, S, Wang, D, Yang, L, He, A, Tang, T, Wei, J, Colloids Surf. B: Biointerfaces, 189, 110819, 2020.
Raman (wavenumber-assignment)	cm^{-1}/-	C=O − 1651 (amorphous), 1644 − (crystalline), 1607, 1595, 1575 − C=C, 1146 − C−CO−C	Doumeng, M, Makhlouf, L, Berthet, F, Marsan, O, Chabert, F, Polym. Testing, in press, 106878, 2020.
x-ray diffraction peaks	degree	18.7, 20.6, 22.9, 28.8; 18.9, 20.9, 23.1, and 28.9 (medical grade) 18.8, 20.7, 22.7, and 28.8 (artificial joint applications)	Diez-Pascual, A M; Naffakh, M; Gonzalez-Dominiguez, J M; Anson, A; Martizez-Rubi, Y; Martinez, M T; Simard, B; Gomez, M A, Carbon, 48, 3485-99, 2010; Zou, F, Lv, F, Ma, X, Xia, X, Jiang, J, Mater. Design, 188, 108433, 2020; Lv, X, Wang, X, Tang, S, Wang, D, Yang, L, He, A, Tang, T, Wei, J, Colloids Surf. B: Biointerfaces, 189, 110819, 2020.

PEF poly(ethylene furanoate)

PARAMETER	UNIT	VALUE	REFERENCES
GENERAL			
Common name	-	poly(ethylene furanoate)	
Acronym	-	PEF	
SYNTHESIS			
Method of synthesis	-	synthesis from ethylene glycol and 2,5-furandicarboxylic acid	Burgess, S K; Karvan, O; Johnson, J R; Kriegel, R M; Koros, W J, Polymer, 55, 4748-56, 2014.
Number average molecular weight, M_n	dalton, g/mol, amu	47,000-66,000	Mao, Y, Bucknall, D G, Kriegel, R M, Polymer, 143, 228-36, 2018.
Mass average molecular weight, M_w	dalton, g/mol, amu	87,000-110,000	Mao, Y, Bucknall, D G, Kriegel, R M, Polymer, 143, 228-36, 2018.
STRUCTURE			
Crystallinity	%	25-33; 48.8	Mao, Y, Bucknall, D G, Kriegel, R M, Polymer, 143, 228-36, 2018; Terzopoulou, Z, Wahbi, M, Kasmi, N,Papageorgiou, G Z, Bikiaris, D N, Thermochim. Acta, 686, 178549, 2020.
Lamellae thickness	nm	4	Mao, Y, Bucknall, D G, Kriegel, R M, Polymer, 143, 228-36, 2018.
COMMERCIAL POLYMERS			
Some manufacturers	-	Avantium	
PHYSICAL PROPERTIES			
Density at 20°C	g cm^{-3}	1.39-1.4299	
Melting temperature, DSC	°C	219-235	Mao, Y, Bucknall, D G, Kriegel, R M, Polymer, 143, 228-36, 2018.
Glass transition temperature	°C	83.6-84.6	Terzopoulou, Z, Wahbi, M, Kasmi, N,Papageorgiou, G Z, Bikiaris, D N, Thermochim. Acta, 686, 178549, 2020.
Permeability to oxygen, 35°C	barrer	0.0107 (PET - 0.114; 11 times more permeable)	Burgess, S K; Karvan, O; Johnson, J R; Kriegel, R M; Koros, W J, Polymer, 55, 4748-56, 2014.
MECHANICAL & RHEOLOGICAL PROPERTIES			
Tensile modulus	MPa	3,600	
Elongation	%	36	
PROCESSING			
Applications	-	potential replacement for PET in drink packaging	Kriegel, R M; Shi, Y; Moffitt, R D, US Patent 20150064383, Coca-Cola, Mar 5, 2015; Kriegel, R M; Moffitt, R D; Schultheis, M W; Shi, Y; You, X, US Patent 20150110983, Coca-Cola, Mar 5, 2015.
Thermal stabilizers	-	phenolic antioxidant (Irganox 1098), and phosphoric acid and triphenyl phosphate	Terzopoulou, Z, Wahbi, M, Kasmi, N,Papageorgiou, G Z, Bikiaris, D N, Thermochim. Acta, 686, 178549, 2020.
Outstanding properties	-	biologically sourced polymer because 2,5-furandicarboxylic acid can be produced from renewable sugars; 100% plant-based, 100% recyclable and degradable plastic	Burgess, S K; Karvan, O; Johnson, J R; Kriegel, R M; Koros, W J, Polymer, 55, 4748-56, 2014.

PEF poly(ethylene furanoate)

PARAMETER	UNIT	VALUE	REFERENCES
BLENDS			
Suitable polymers	-	PET	Niaounakis, M, Eur. Polym. J., 114, 464-75.

PEI poly(ether imide)

PARAMETER	UNIT	VALUE	REFERENCES
GENERAL			
Common name	-	poly(ether imide)	
CAS name	-	1,3-isobenzofurandione, 5,5'-[(1-methylethylidene)bis(4,1-phenyleneoxy)]bis-, polymer with 1,3-benzenediamine	
Acronym	-	PEI	
CAS number	-	61128-46-9	
Linear formula			
HISTORY			
Person to discover	-	Wirth, J G; Heath, D R	Wirth, J G; Heath, D R, US Patent 3,787,364, General Electric, Jan. 22, 1974.
Date	-	1974	
Details	-	polymerization patented	
SYNTHESIS			
Monomer(s) structure	-		
Monomer(s) CAS number(s)	-	38103-06-9; 108-45-2	
Monomer(s) molecular weight(s)	dalton, g/mol, amu	520.49; 108.14	
Method of synthesis	-	reaction of bis(chlorophthalimide) with alkalimetal salt of divalent carbocyclic aromatic radical in the presence of appropriate solvent (can be water)	Chiefari, J; Dao, B; Groth, A M; Hodgikin, J H, High Performance Polym., 15, 269-79, 2003.
Temperature of polymerization	^{o}C	180	
Time of polymerization	h	3-4	
Number average molecular weight, M_n	dalton, g/mol, amu	10,000-42,000	
Mass average molecular weight, M_w	dalton, g/mol, amu	30,000-75,000	
Polydispersity, M_w/M_n	-	1.6	
STRUCTURE	-		
Crystallinity	%	amorphous	Rath, T; Kumar, S; Mahaling, R N; Mukherjee, M; Das, C K; Pandley, K N; Saxena, A K, Polym. Compos., 27, 533-38, 2006; Rath, T; Kumar, S; Mahaling, R N; Khatua, B B; Das, C K; Yadaw, S B, Mater. Sci. Eng., 490A, 198-207, 2008.
Entanglement molecular weight	dalton, g/mol, amu	1,850	Yi, J H; Won, H J; Dong, I Y, Polym. Bull., 39, 2, 257-63, 1997.
COMMERCIAL POLYMERS			
Some manufacturers	-	Sabic	
Trade names	-	Ultem, Extem	

PEI poly(ether imide)

PARAMETER	UNIT	VALUE	REFERENCES
PHYSICAL PROPERTIES			
Density at 20°C	g cm^{-3}	1.27-1.31; 1.52 (30% glass fiber); 1.27 (amorphous)	
Color	-	white	
Refractive index, 20°C	-	1.630-1.687	
Transmittance	%	58	
Haze	%	2	
Odor	-	none	
Melting temperature, DSC	°C	229	
Thermal expansion coefficient, -40 to 150°C	°C^{-1}	1-5.6E-5	
Thermal conductivity, melt	W m^{-1} K^{-1}	0.17-0.26	
Glass transition temperature	°C	209-249	da Conceicao, T F; Scharnagl, N; Dietzel, W; Kainer, K U, Corrosion Sci., 53, 338-46, 2011.
Long term service temperature	°C	170	
Temperature index (50% tensile strength loss after 20,000 h/5000 h)	°C	170	Padey, D; Walling, J; Wood A, Polymers in Defence and Aerospace 2007, Rapra, 2007, paper 15.
Heat deflection temperature at 0.45 MPa	°C	210-250; 257 (30% glass fiber)	
Heat deflection temperature at 1.8 MPa	°C	190-235; 254 (30% glass fiber)	
Vicat temperature VST/B/50	°C	219-260; 267 (30% glass fiber)	
Hansen solubility parameters, δ_D, δ_P, δ_H	MPa$^{0.5}$	19.6, 7.6, 9.0	Hansen, C M; Just, L, Ind. Eng. Chem. Res., 40, 21-25, 2001.
Interaction radius		6.0	Hansen, C M; Just, L, Ind. Eng. Chem. Res., 40, 21-25, 2001.
Hildebrand solubility parameter	MPa$^{0.5}$	19.8	
Work of adhesion	mN m^{-1}	80.8	Saraswathi, M S A, Divya, K, Selvapandian, P, Mohan, D, Nagendran, A, Mater. Chem. Phys., 231, 159-67, 2019.
Relative permittivity at 100 Hz/1 MHz	-	3.19	Zhang, Q, Chen, X, Zhang, T, Zhang, Q M, Nano Energy, 64, 103916, 2019.
Dissipation factor at 100 Hz		0.0015-0.025	
Dissipation factor at 1 MHz		0.0025-0.007	
Volume resistivity	ohm-m	1E13 to 1E15	
Electric strength K20/P50, d=0.60.8 mm	kV mm^{-1}	14-33	
Comparative tracking index, CTI, test liquid A	V	175	
Coefficient of friction	-	0.15-0.7 (against steel; depending on sliding speed, and applied pressure)	Mimaroglu, A; Unal, H; Arda, T, Wear, 262, 1407-13, 2007.
Permeability to nitrogen, 25°C	barrer	0.07	Lopez-Gonzalez, M M; Compan, V; Saiz, E; Riande, E; Guzman, J, J. Membrane Sci., 253, 175-81, 2005.
Permeability to oxygen, 25°C	barrer	0.5	Lopez-Gonzalez, M M; Compan, V; Saiz, E; Riande, E; Guzman, J, J. Membrane Sci., 253, 175-81, 2005.
Diffusion coefficient of nitrogen	cm^2 s^{-1} x10^9	1.9	Lopez-Gonzalez, M M; Compan, V; Saiz, E; Riande, E; Guzman, J, J. Membrane Sci., 253, 175-81, 2005.

PEI poly(ether imide)

PARAMETER	UNIT	VALUE	REFERENCES
Diffusion coefficient of oxygen	cm² s⁻¹ x10⁹	6.1	Lopez-Gonzalez, M M; Compan, V; Saiz, E; Riande, E; Guzman, J. J. Membrane Sci., 253, 175-81, 2005.
Contact angle of water, 20°C	degree	64; 60-81.5	Kratz, K, Heuchel, M, Weigel, T, Lendlein, A, Polymer, 210, 123045, 2020; Selvapandian, P, Mohan, D, Nagendran, A, Mater. Chem. Phys., 231, 159-67, 2019.

MECHANICAL & RHEOLOGICAL PROPERTIES

PARAMETER	UNIT	VALUE	REFERENCES
Tensile strength	MPa	53-124; 156 (30% glass fiber)	
Tensile modulus	MPa	3420-3,700; 10,230 (30% glass fiber)	
Tensile stress at yield	MPa	96-103; 156 (30% glass fiber)	
Elongation	%	14-50; 3 (30% glass fiber)	
Tensile yield strain	%	6-7; (30% glass fiber)	
Flexural strength	MPa	145-168; 206 (30% glass fiber)	
Flexural modulus	MPa	3,040-3,810; 8,960 (30% glass fiber)	
Elastic modulus	MPa	3,000-3,300	
Compressive strength	MPa	118	
Young's modulus	MPa	3,000	
Charpy impact strength, unnotched, 23°C	kJ m⁻²	NB	
Charpy impact strength, unnotched, -30°C	kJ m⁻²	NB	
Charpy impact strength, notched, 23°C	kJ m⁻²	21; 24 (30% glass fiber)	
Izod impact strength, unnotched, 23°C	J m⁻¹	1,340-1870 to NB	
Izod impact strength, notched, 23°C	J m⁻¹	27-69; 89 (30% glass fiber)	
Izod impact strength, notched, -30°C	J m⁻¹	74	
Poisson's ratio	-	0.36	Ramani, K; Zhao, W, Antec, 1160-64, 1997.
Rockwell hardness	-	R127	
Shrinkage	%	0.5-1.2	
Melt viscosity, shear rate=1000 s⁻¹	Pa s	350	Lou, J; Shabazi, A; Harinath, V, Antec, 2193-7, 2003.
Melt volume flow rate (ISO 1133, procedure B), 360°C/5 kg	cm³/10 min	8	
Melt index, 230°C/3.8 kg	g/10 min	14-18; 4 (30% glass fiber)	
Water absorption, equilibrium in water at 23°C	%	0.65 (24 h immersion); 1.75 (30 days immersion)	
Moisture absorption, equilibrium 23°C/50% RH	%	0.15 (24 h); 0.76 (30 days)	

CHEMICAL RESISTANCE

PARAMETER	UNIT	VALUE	REFERENCES
Acid dilute/concentrated	-	very good	
Alcohols	-	good	
Alkalis	-	very good	
Aliphatic hydrocarbons	-	good	

PEI poly(ether imide)

PARAMETER	UNIT	VALUE	REFERENCES
Aromatic hydrocarbons	-	good	
Esters	-	good	
Greases & oils	-	very good	
Halogenated hydrocarbons	-	very good	
Ketones	-	good	
Good solvent	-	N-methylpyrrolidone, dichloromethane	
FLAMMABILITY			
Autoignition temperature	°C	385	
Limiting oxygen index	% O_2	45-47	
NBS smoke chamber	Ds	50	Padey, D; Walling, J; Wood A, Polymers in Defence and Aerospace 2007, Rapra, 2007, paper 15.
Char at 700°C	%	49.0	Selvapandian, P, Mohan, D, Nagendran, A, Mater. Chem. Phys., 231, 159-67, 2019.
Heat of combustion	J g^{-1}	29,060-35,220	Walters, R N; Hacket, S M; Lyon, R E, Fire Mater., 24, 5, 245-52, 2000.
UL 94 rating	-	V-0	
WEATHER STABILITY			
Spectral sensitivity	nm	293	
Excitation wavelengths	nm	350	
Emission wavelengths	nm	470; 550	
Products of degradation	-	acetophenone, phenyl acetic acid, phenols, benzoic acid, phthalic anhydride and phthalic acid end-groups; chain scission, photooxidative degradation of the isopropylidene bridge of BPA units, photooxidation of phthalimide units to phthalic anhydride end groups, hydrolysis of phthalic anhydride end groups	
Stabilizers	-	triphenyl; phosphate	
TOXICITY			
HMIS: Health, Flammability, Reactivity rating	-	0/1/0	
Carcinogenic effect	-	not listed by ACGIH, NIOSH, NTP	
Oral rat, LD$_{50}$	mg kg^{-1}	>5,000	
PROCESSING			
Typical processing methods	-	blow molding, extrusion, injection molding	
Preprocess drying: temperature/ time/residual moisture	°C/h/%	150-175/4-6/0.02	
Processing temperature	°C	260-320 (molding)	
Processing pressure	MPa	0.3-0.7 (back pressure)	
Process time	min	5	
Additives used in final products	-	Fillers: calcium silicate, carbon fibers, glass fibers, graphite flakes, mica, multiwalled carbon nanotubes, titaniun dioxide; Plasticizers: pentaerythritotetrabenzoate ester; Antistatics: fatty quaternary ammonium compounds, potassium titanate whisker; Release: fatty acid amide, p-tallow toluenesulfonamide, pentaerythritol tetrastearate, polyolefin	

396   **HANDBOOK OF POLYMERS** 3rd Edition, Copyrights 2022; ChemTec Publishing

PEI poly(ether imide)

PARAMETER	UNIT	VALUE	REFERENCES
Applications	-	aircraft interiors, automotive engine sensors, bulb sockets, electronic connectors, microwaveable cookware, steam sterilizable surgical components thermally resistant film for copper laminated boards,231 vacuum pump vanes	
BLENDS			
Suitable polymers	-	epoxy, LCP, PA (amorphous), PBT, PC, PET, PPT, PSF, poly(ether-methyl-ether-urea), silicone	
ANALYSIS			
FTIR (wavenumber-assignment)	cm^{-1}/-	C=O – 1725, 1779; Ar-O-Ar – 1200, 1300	da Conceicao, T F; Scharnagl, N; Dietzel, W; Kainer, K U, Corrosion Sci., 53, 338-46, 2011.

PEK polyetherketone

PARAMETER	UNIT	VALUE	REFERENCES
GENERAL			
Common name	-	polyetherketone, poly(4,4'-oxydiphenylene ketone)	
IUPAC name	-	poly(oxy-1,4-pheleneoxy-1,4-phenylene-carbonyl-1,4-phenylene)	
ACS name	-	poly(oxy-1,4-phenylenecarbonyl-1,4-phenylene)	
Acronym	-	PEK	
CAS number	-	27380-27-4	
Formula			
RTECS number	-	DC0875000	
HISTORY			
Person to discover	-	Staniland, P A	Staniland, P A, US Patent 4,056,511, ICI, Nov. 1, 1977
Date	-	1977	
Details	-	patent for polymerization	
SYNTHESIS			
Monomer(s) structure	-		Fink, J K, High Performance Polymers, William Andrew, 2008.
Monomer(s) CAS number(s)	-	345-92-6; 611-99-4	
Monomer(s) molecular weight(s)	dalton, g/mol, amu	218.2; 214.22	
Monomer ratio	-	equimolar quantities	
Method of synthesis	-	nucleophilic route, ketimine route, or electrophilic process; PEK can be obtained by reaction of 4,4'-difuorobenzophenone with 4,4'-dihydroxybenzophenone in the presence of potassium carbonate, using diphenyl sulfone as solvent	Ben-Haida, A; Colquhoun, H M; Hodge, P; Williams, D J, J. Mater. Chem., 10, 2011-16, 2000.
Temperature of polymerization	°C	0-30	
Catalyst	-	Friedel-Crafts catalyst	Fink, J K, High Performance Polymers, William Andrew, 2008.
STRUCTURE	-		
Crystallinity	%	28-44	Hamdan, S; Swallowe, G M, J. Polym. Sci. B, 34, 699-705, 1996.
Cell type (lattice)	-	orthorhombic	
Cell dimensions	nm	a:b:c=0.763-0.776:0.596-0.6:1.0-1.009	
COMMERCIAL POLYMERS			
Some manufacturers	-	Gharda Chemicals, RTP	
PHYSICAL PROPERTIES			
Density at 20°C	g cm^{-3}	1.26-1.30; 1.272 (amorphous); 1.43 (crystalline); 1.41-1.45 (15-40% glass fiber)	Shekar, R I; Kotresh, T M; Rao, P M D; Kumar, K, J. Appl. Polym. Sci., 112, 2497-2510, 2009.
Color	-	opaque	
Odor		odorless	
Melting temperature, DSC	°C	340-373	
Decomposition temperature	°C	>500	

398   **HANDBOOK OF POLYMERS** 3ʳᵈ Edition, Copyrights 2022; ChemTec Publishing

PEK polyetherketone

PARAMETER	UNIT	VALUE	REFERENCES
Thermal conductivity, 23°C	W m^{-1} K^{-1}	0.29	
Glass transition temperature	°C	152-154	
Specific heat capacity	J K^{-1} kg^{-1}	2,200	
Heat deflection temperature at 1.8 MPa	°C	163-167; 316 (10-40% glass fiber)	
Volume resistivity	ohm-m	1E14	
Surface resistivity	ohm	1E16	
Electric strength K20/P50, d=0.60.8 mm	kV mm^{-1}	16-17	
Comparative tracking index, CTI, test liquid A	-	145-150	
Arc resistance	s	175	

MECHANICAL & RHEOLOGICAL PROPERTIES

Tensile strength	MPa	105-110; 138-290 (10-40% glass fiber)	
Tensile modulus	MPa	3,700-4,200; 8,274-37,200 (10-40% glass fiber)	
Tensile stress at yield	MPa	110-115	
Elongation	%	10-20; 1.3-3 (10-40% glass fiber)	
Flexural strength	MPa	185-190; 207-427 (10-40% glass fiber)	
Flexural modulus	MPa	4,100-4,200; 6,206-31,000 (10-40% glass fiber)	
Compressive strength	MPa	140	
Charpy impact strength, notched, 23°C	kJ m^{-2}	3.8	
Izod impact strength, unnotched, 23°C	J m^{-1}	NB; 530-800 (10-40% glass fiber)	
Izod impact strength, notched, 23°C	J m^{-1}	55-90 (10-40% glass fiber)	
Shore D hardness	-	86-87	
Rockwell hardness	-	M103	
Shrinkage	%	1-1.3; 0.05-0.4 (10-40% glass fiber)	
Intrinsic viscosity, 25°C	dl g^{-1}	1.05	
Melt viscosity, shear rate=1000 s^{-1}	Pa s	200	
Water absorption, 24h at 23°C	%	0.07; 0.6 (saturation)	

CHEMICAL RESISTANCE

Acid dilute/concentrated	-	very good	
Alcohols	-	very good	
Alkalis	-	very good	
Aliphatic hydrocarbons	-	very good	
Aromatic hydrocarbons	-	very good	
Esters	-	very good	
Greases & oils	-	very good	
Ketones	-	very good	

PEK polyetherketone

PARAMETER	UNIT	VALUE	REFERENCES
FLAMMABILITY			
Ignition temperature	°C	400	
Char at 500°C	%	52.9-56	Lyon, R E; Walters, R N, J. Anal. Appl. Pyrolysis, 71, 27-46, 2004.
Heat of combustion	J g⁻¹	31.070	Walters, R N; Hacket, S M; Lyon, R E, Fire Mater., 24, 5, 245-52, 2000.
Activation energy of decomposition	kJ mol⁻¹	230	Swallowe, G M; Dawson, P C; Tang, T B; Xu, Q L, J. Mater. Sci., 30, 3853-55, 1995.
UL rating	-	V-0; V-0 (10-40% glass fiber)	
PROCESSING			
Typical processing methods	-	extrusion, injection molding, soft litography, transfer molding	
Preprocess drying: temperature/ time/residual moisture	°C/h/%	150-180/2-6/0.1 (10-40% glass fiber)	
Processing temperature	°C	385-425 (melt) (10-40% glass fiber)	
Processing pressure	MPa	69-138 (10-40% glass fiber)	
Applications	-	automotive, aircraft, coatings, composites, electronics, fiber, food, medical, oil & gas, textiles	
Outstanding properties	-	high tensile, high temperature resistance	
BLENDS			
Suitable polymers	-	epoxy	

PEKK polyetherketoneketone

PARAMETER	UNIT	VALUE	REFERENCES
GENERAL			
Common name	-	poly(4,4'-oxydiphenylene m-phenylene diketone), polyetherketoneketone	
CAS name	-	poly(oxy-1,4-phenylenecarbonyl-1,4-phenylenecarbonyl-1,4-phenylene)	
Acronym	-	PEKK	
CAS number	-	74970-25-5	
Linear formula			
HISTORY			
Person to discover	-	Bonner, W H	Bonner, W H, US Patent 3,065,205, DuPont, 1962.
Date	-	1962	
Details	-	reported synthesis of PEEK in DuPont labs	
SYNTHESIS			
Monomer(s) structure	-		
Monomer(s) CAS number(s)	-	100-20-9; 101-84-8	
Monomer(s) molecular weight(s)	dalton, g/mol, amu	203.02; 170.21	
Method of synthesis	-	by Fiedel-Crafts acylation, Bonner condensed isophthaloyl chloride or terephthaloyl chloride with diphenyl ether using nitrobenzene as solvent and aluminum trichloride as a catalyst	
Catalyst	-	aluminum tetrachloride	
STRUCTURE			
Crystallinity	%	0-35	
Cell type (lattice)	-	orthorhombic	
Cell dimensions	nm	a:b:c=0.766:0.611:1.576	
Rapid crystallization temperature	°C	240-270	de Vries, H, Influence of processing parameters on mechanical properties of PEKK, NRL, 2006.
Crystallization half-time	min	7-9	de Vries, H, Influence of processing parameters on mechanical properties of PEKK, NRL, 2006.
Processing parameters for maximum crystallinity	-	340-350°C, <3°C/min cooling rate	de Vries, H, Influence of processing parameters on mechanical properties of PEKK, NRL, 2006.
COMMERCIAL POLYMERS			
Some manufacturers	-	Arkema; Hexcel; RTP	
Trade names	-	Kepstan; HexAM; PEKK	
PHYSICAL PROPERTIES			
Density at 20°C	g cm^{-3}	1.28-1.31; 1.44-1.6 (20-40% glass fiber)	
Color	-	amber	

PEKK polyetherketoneketone

PARAMETER	UNIT	VALUE	REFERENCES
Melting temperature, DSC	°C	363-386, 305 (PEKK with 60% straight and 40% kinked segments), 360 (PEEK with 80% straight and 20% kinked segments)	Alqurashi, H, Khurshid, Z, Yaqin, S A U, Habib, S R, Zafar, M S, J. Adv. Res., in press, 2020.
Thermal expansion coefficient, 23-80°C	10^{-4} °C^{-1}	0.21-0.38	
Thermal conductivity, melt	W m^{-1} K^{-1}	0.25	
Thermal stability	°C	335.8	Sun, F, Shen, X, Zhou, N, Gao, Y, Wu, G, J. Prosthetic Dentistry, 124, 4, 495-9, 2020.
Glass transition temperature	°C	154-171	Martin-Franch, P; Martin, T; Tunnicliffe, D L; Das-Gupta, D K, Sensors Actuators A: Phys., 99, 3, 236-43, 2002.
Heat of fusion	kJ kg^{-1}	57.5	
Long term service temperature	°C	260	
Heat deflection temperature at 1.8 MPa	°C	141-175; >299 to >316 (20-40% glass fiber)	
Dielectric constant at 1000 Hz/1 MHz	-	3.3-3.6/	
Dissipation factor at 1000 Hz	E-4	40	
Volume resistivity	ohm-m	1E14	
Surface resistivity	ohm	2E16	
Electric strength K20/P50, d=0.60.8 mm	kV mm^{-1}	24	
Coefficient of friction	-	0.17-0.18 (dynamic); 0.26-0.28 (static)	

MECHANICAL & RHEOLOGICAL PROPERTIES			
Tensile strength	MPa	90-115; 134-190 (20-40% glass fiber)	Alqurashi, H, Khurshid, Z, Yaqin, S A U, Habib, S R, Zafar, M S, J. Adv. Res., in press, 2020.
Tensile modulus	MPa	3,450-4,400; 8,960-13,800 (20-40% glass fiber)	
Elongation	%	12; 1.9-2.5 (20-40% glass fiber)	
Elastic modulus	MPa	5100	Alqurashi, H, Khurshid, Z, Yaqin, S A U, Habib, S R, Zafar, M S, J. Adv. Res., in press, 2020.
Flexural strength	MPa	140-200; 214-280 (20-40% glass fiber)	Alqurashi, H, Khurshid, Z, Yaqin, S A U, Habib, S R, Zafar, M S, J. Adv. Res., in press, 2020.
Flexural modulus	MPa	3,380-4,600; 7,580-12,400 (20-40% glass fiber)	
Compressive strength	MPa	103-207; 246	-; Alqurashi, H, Khurshid, Z, Yaqin, S A U, Habib, S R, Zafar, M S, J. Adv. Res., in press, 2020.
Izod impact strength, unnotched, 23°C	J m^{-1}	480-900 (20-40% glass fiber)	
Izod impact strength, notched, 23°C	J m^{-1}	43-69; 50-110 (20-40% glass fiber)	
Shear strength	MPa	138	
Poisson's ratio	-	0.40	
Rockwell hardness	-	M88	
Shrinkage	%	0.01-1.4; 0.2-0.3 (20-40% glass fiber)	
Melt index, 380°C/8.4 kg	g/10 min	25-120	
Water absorption, 24h at 23°C	%	0.2-0.3	

PEKK polyetherketoneketone

PARAMETER	UNIT	VALUE	REFERENCES
CHEMICAL RESISTANCE			
Acid dilute/concentrated	-	very good	
Alcohols	-	very good	
Alkalis	-	very good	
Aliphatic hydrocarbons	-	very good	
Aromatic hydrocarbons	-	very good	
Esters	-	very good	
Greases & oils	-	good	
Ketones	-	very good	
FLAMMABILITY			
Limiting oxygen index	% O_2	40	
NBS smoke chamber density	-	10	
Char at 500°C	%	60.7	Lyon, R E; Walters, R N, J. Anal. Appl. Pyrolysis, 71, 27-46, 2004.
Heat of combustion	J g^{-1}	31,150	Walters, R N; Hacket, S M; Lyon, R E, Fire Mater., 24, 5, 245-52, 2000.
UL 94 rating	-	V-0; V-0 (20-40% glass fiber)	
PROCESSING			
Typical processing methods	-	3D printing, compression molding, injection molding	
Preprocess drying: temperature/ time/residual moisture	°C/h/%	149-232/3/0.1	
Processing temperature	°C	377-382 (20-40% glass fiber)	
Processing pressure	MPa	103-138 (injection)	
Applications	-	bearings, body implants, capillary tubing, composites, tubing	
Outstanding properties	-	biocompatibility, chemical resistance, low smoke toxicity	
BLENDS			
Suitable polymers	-	PI	

PEM poly(ethylene-co-methacrylic acid)

PARAMETER	UNIT	VALUE	REFERENCES
GENERAL			
Common name	-	poly(ethylene-co-methacrylic acid)	
CAS name	-	2-propenoic acid, 2-methyl-, polymer with ethene	
Acronym	-	PEM, EMAA	
CAS number	-	25053-53-6	
HISTORY			
Person to discover	-	DuPont	
Date	-	1960s	
SYNTHESIS			
Monomer(s) structure	-	$H_2C{=}CH_2 \quad H_2C{=}\overset{O}{\overset{\|}{C}}COH$ $\underset{CH_3}{\|}$	
Monomer(s) CAS number(s)	-	74-85-1; 79-41-4	
Monomer(s) molecular weight(s)	dalton, g/mol, amu	28.05; 86.06	
Methacrylic acid content	wt%	4-20 (Nucrel, Surlyn)	
Sodium cations content	% of COOH groups	30-60	
Number average molecular weight, M_n	dalton, g/mol, amu	13,900-16,600	
Mass average molecular weight, M_w	dalton, g/mol, amu	69,000-103,700	
Polydispersity, M_w/M_n	-	5.0-6.4	
STRUCTURE			
Crystallinity	%	32.7-36.5	
Avrami constants, k/n	-	n=3.54-3.9	Huang, J-W; Wen, Y-L; Kang, C-C; Yeh, M-Y; Wen, S-B, Thermochim. Acta, 465, 48-58, 2007.
COMMERCIAL POLYMERS			
Some manufacturers	-	Dow (Entec)	
Trade names	-	Conpol, Nucrel, Surlyn (ionomer)	
PHYSICAL PROPERTIES			
Density at 20°C	g cm^{-3}	0.93-0.95	
Color	-	white	
Odor	-	mild methacrylic acid	
Melting temperature, DSC	°C	76-105	
Softening point	°C	62	
Decomposition temperature	°C	>325	
Glass transition temperature	°C	45.8	Cai, X; Riedl, B; Ait-Kadi, A; Composites: Part A, 34, 1075-84, 2003.
Maximum processing temperature	°C	235-310	
Vicat temperature VST/A/50	°C	60-95	

PEM poly(ethylene-co-methacrylic acid)

PARAMETER	UNIT	VALUE	REFERENCES
MECHANICAL & RHEOLOGICAL PROPERTIES			
Tensile strength	MPa	1.3-530.8	Kraemer, R H; Raza, M A; Gedde, U W, Polym. Deg. Stab., 92, 1795-1802, 2007.
Tensile stress at yield	MPa	17.9-21.2	
Elongation	%	630-1,180	
Flexural modulus	MPa	400	
Young's modulus	MPa	276	Cai, X; Riedl, B; Ait-Kadi, A; Composites: Part A, 34, 1075-84, 2003.
Izod impact strength, notched, 23°C	J m^{-1}	370 to NB	
Shore D hardness	-	46	
Melt index, 190°C/2.16 kg	g/10 min	1.5-60	Dolog, R, Weiss, R A, Polymer, 128, 128-34, 2017.
FLAMMABILITY			
Ignition temperature	°C	335-365	
Volatile products of combustion	-	CO, organic acids, aldehydes, alcohols, other hydrocarbon oxidation products	
BIODEGRADATION			
Typical biodegradants	-	*Aspergillus, Penicillium*	Weng, Y-M; Chen, M-J; Chen, W, Lebensm.-Wiss. u-Technol., 32, 191-5, 1999.
Stabilizers	-	benzoic and sorbic acids	
TOXICITY			
Carcinogenic effect	-	not listed by ACGIH, NIOSH, NTP	
OSHA	mg m^{-3}	5 (respirable dust), 15 (total dust)	
PROCESSING			
Typical processing methods	-	extrusion, injection molding	
Processing temperature	°C	135-235	
Additives used in final products	-	slip agent, antifog agent, antiblocking agent	
Applications	-	adhesives, footwear, glass coating, metal coating, modification of surface properties of films and coatings; refrigerated food packaging, wire and cable; ionomer (Surlyn): golf ball covers, hockey helmets, ski boots	
Outstanding properties	-	inherently flexible, self-healing following ballistic puncture	
BLENDS			
Compatible polymers	-	ionomer (Surlyn), ethylene-methacrylic acid-acrylate terpolymer (Nucrel), PBT, PE, PEO, PET	
Compatibilizers	-	compatibilizer of PET/LLDPE blends	
ANALYSIS			
FTIR (wavenumber-assignment)	cm^{-1}/-	C=O – 1756, 1750, 1743, 1727; calcium carbonate – 1580-1300, CH$_2$ – 1435, CH$_3$ – 1365; C-O – 1255	Kraemer, R H; Raza, M A; Gedde, U W, Polym. Deg. Stab., 92, 1795-1802, 2007.
WAXD	2θ	21.5	Huang, J-W; Wen, Y-L; Kang, C-C; Yeh, M-Y; Wen, S-B, Thermochim. Acta, 465, 48-58, 2007.

PEN poly(ethylene 2,6-naphthalate)

PARAMETER	UNIT	VALUE	REFERENCES
GENERAL			
Common name	-	poly(ethylene 2,6-naphthalate)	
CAS name	-	poly(oxy-1,2-ethanediyloxycarbonyl-2,6-naphthalenediyl-carbonyl) (24968-11-4); 2,6-naphthalenedicarboxylic acid, polymer with 1,2-ethanediol (25230-87-9)	
Acronym	-	PEN	
CAS number	-	24968-11-4; 25230-87-9	
Formula			
HISTORY			
Person to discover	-	Cook, J G; Hugggill, H P W; Lowe, A R	
Date	-	1948	
Details	-	ICI obtained first patent in 1948	
SYNTHESIS			
Monomer(s) structure	-		
Monomer(s) CAS number(s)	-	1141-38-4; 107-21-1	
Monomer(s) molecular weight(s)	dalton, g/mol, amu	216.193; 62.07	
Monomer(s) expected purity(ies)	%	99.6	
Monomer ratio	-	3.48:1	
Method of synthesis	-	polymerization occurs in two stages: in the first stage, low molecular ester is produced followed by polycondensation, which produces high molecular weight polymer	Fink, J K, High Performance Polymers, William Andrew, 2008.
Temperature of polymerization	℃	280-300	
Pressure of polymerization	Pa	133	
Catalyst	-	antimony trioxide	
STRUCTURE	-		
Crystallinity	%	1-4.17 (amorphous); 6.76-49.6 (annealed at 123-170℃ for various times); 31-38 (draw ratio 3.6-6.1); 50.6 (biaxially oriented); 50.8-56.5 (fibers containing 5% TLCP); 62.5-87.2 (high pressure crystallized)	Scott, A; Hakme, C; Stevenson, I; Voice, A, Macromol. Symp., 230, 78-86, 2005; Zegnini, B; Boudou, L; Martinez-Vega, J, J. Appl. Sci., 8, 7, 1206-13, 2008; Khemici, M W; Gourari, A; Doulache, N, Int. J. Polym. Anal. Charact., 14, 322-35, 2009; Li, L; Wang, C; Huang, R; Zhang, L; Hong, S, Polymer, 8867-72, 2001.
Cell type (lattice)	-	triclinic	
Cell dimensions	nm	a:b:c=0.651:0.575:1.32 (α-form); a:b:c=0.926:1.559:1.273 (β-form)	Zhang, Y; Mukoyama, S; Hu, Y; Yan, C; Ozaki, Y; Takahashi, I, Macromolecules, 40, 4009-15, 2007.
Unit cell angles	degree	α:β:γ=81:114:100 (α-form); α:β:γ=121.6:95.57:122.52 (β-form)	Zhang, Y; Mukoyama, S; Hu, Y; Yan, C; Ozaki, Y; Takahashi, I, Macromolecules, 40, 4009-15, 2007.

PEN poly(ethylene 2,6-naphthalate)

PARAMETER	UNIT	VALUE	REFERENCES
Number of chains per unit cell	-	1	
Polymorphs	-	α, β	Ulcer, Y; Cakmak, M, Polymer, 38, 2907-10, 1997.
Chain conformation	-	nearly planar	
Entanglement molecular weight	dalton, g/mol, amu	1,810; 25,600	Yi, J H; Won, H J; Dong, I Y, Polym. Bull., 39, 2, 257-63, 1997; Barany, T; Cziganý, T; Karger-Kotsis, J, Prog. Polym. Sci., 35, 1257-87, 2010.
Rapid crystallization temperature	^oC	198-203	
Avrami constants, k/n	-	2.8-3.0	Wu, T-M; Liu, C-Y, Polymer, 5621-29, 2005.

COMMERCIAL POLYMERS

	UNIT	VALUE	REFERENCES
Some manufacturers	-	Teijin	
Trade names	-	Teonex	

PHYSICAL PROPERTIES

	UNIT	VALUE	REFERENCES
Density at 20°C	g cm^{-3}	1.33-1.36; 1.325 (amorphous); 1.407 (crystalline)	
Refractive index, 20°C	-	1.50-1.65	
Birefringence	-	0.001-0.0018; 0.48	Suzuki, A; Tojyo, M, Eur. Polym. J., 2922-27, 2007; Yin, J, Luo, F, Wang, D. Sun, X, Eur. Polym. J., 114, 66-71, 2019.
Transmittance	%	84	
Melting temperature, DSC	^oC	261-290; 268	-; Yin, J, Luo, F, Wang, D. Sun, X, Eur. Polym. J., 114, 66-71, 2019.
Thermal expansion coefficient, 23-80°C	$^oC^{-1}$	4.4E-5, 1.3E-5	-; Yin, J, Luo, F, Wang, D. Sun, X, Eur. Polym. J., 114, 66-71, 2019.
Glass transition temperature	^oC	120	Yin, J, Luo, F, Wang, D. Sun, X, Eur. Polym. J., 114, 66-71, 2019.
Heat of fusion	kJ mol^{-1}	9.2	
Long term service temperature	^oC	155	
Enthalpy of melting	J g^{-1}	54.3-66.4 (high pressure crystallization)	Li, L; Wang, C; Huang, R; Zhang, L; Hong, S, Polymer, 8867-72, 2001.
Dielectric constant at 100 Hz/1 MHz	-	3.2	
Dissipation factor at 1 MHz	E-4	48	
Volume resistivity	ohm-m	1E13	
Surface resistivity	ohm	1E14	
Electric strength K20/P50, d=0.60.8 mm	kV mm^{-1}	160	
Coefficient of friction	-	0.27	
Permeability to carbon dioxide, 25°C	cm^3 cm cm^{-2} s^{-1} Pa^{-1} x 10^{12}	0.01	
Permeability to oxygen, 25°C	cm^3 cm cm^{-2} s^{-1} Pa^{-1} x 10^{12}	0.006	

PEN poly(ethylene 2,6-naphthalate)

PARAMETER	UNIT	VALUE	REFERENCES
Permeability to water vapor, 25°C	cm^3 cm cm^{-2} s^{-1} Pa^{-1} x 10^{12}	40	

MECHANICAL & RHEOLOGICAL PROPERTIES

PARAMETER	UNIT	VALUE	REFERENCES
Tensile strength	MPa	60-68	
Tensile modulus	MPa	2,000	
Elongation	%	250-340	
Flexural strength	MPa	99.5	
Flexural modulus	MPa	2,500	
Young's modulus	MPa	2,850; 6100	Caligiuri, L; Stagnaro, P; Valenti, B; Canalini, G, Eur. Polym. J., 45, 217-25, 2009; Yin, J, Luo, F, Wang, D. Sun, X, Eur. Polym. J., 114, 66-71, 2019.
Izod impact strength, unnotched, 23°C	J m^{-1}	420	
Shrinkage	%	0.8	
Intrinsic viscosity, 25°C	dl g^{-1}	0.59	
Melt viscosity, shear rate=1000 s^{-1}	Pa s	950	Kim, S Y; Kim, S H; Lee, S H; Youn, J R, Composites: Part A, 40, 607-12, 2009.
Melt index, 300°C/3.8 kg	g/10 min	2.5	Kim, S Y; Kim, S H; Lee, S H; Youn, J R, Composites: Part A, 40, 607-12, 2009.

CHEMICAL RESISTANCE

PARAMETER	UNIT	VALUE	REFERENCES
Acid dilute/concentrated	-	good/poor	
Alcohols	-	good	
Alkalis	-	good	
Aromatic hydrocarbons	-	good	
Greases & oils	-	good	
Halogenated hydrocarbons	-	poor	
Ketones	-	good	
Good solvent	-	phenol/o-chlorobenzene	

FLAMMABILITY

PARAMETER	UNIT	VALUE	REFERENCES
Ignition temperature	°C	>400	
Autoignition temperature	°C	587	
Char at 500°C	%	18.2	Lyon, R E; Walters, R N, J. Anal. Appl. Pyrolysis, 71, 27-46, 2004.
Heat of combustion	J g^{-1}	25,920	Walters, R N; Hacket, S M; Lyon, R E, Fire Mater., 24, 5, 245-52, 2000.
UL rating	-	V-2	

WEATHER STABILITY

PARAMETER	UNIT	VALUE	REFERENCES
Spectral sensitivity	nm	290-390	
Activation wavelengths	nm	370-380	
Excitation wavelengths	nm	330; 375	

PEN poly(ethylene 2,6-naphthalate)

PARAMETER	UNIT	VALUE	REFERENCES
Emission wavelengths	nm	435; 435; 580	
Important initiators and accelerators	-	ferrocene, cobalt octoates and naphthenates, compounds containing aromatic keto-ester groups	
Products of degradation	-	radicals, crosslinks, hydroperoxides, hydroxyl and carbonyl groups, CO, CO_2	
Stabilizers	-	UVA: 2-hydroxy-4-octyloxybenzophenone; 2-hydroxy-4-methoxybenzophenone; 2,4-dihydroxybenzophenone; 2,2',4,4'-tetrahydroxybenzophenone; 2,2'-dihydroxy-4-methoxybenzophenone; 2-(2H-benzotriazol-2-yl)-p-cresol; 2-benzotriazol-2-yl-4,6-di-tert-butylphenol; 2-(2H-benzotriazole-2-yl)-4,6-di-tert-pentylphenol; 2-(2H-benzotriazole-2-yl)-4-(1,1,3,3-tetraethylbutyl)phenol; 2-(2H-benzotriazol-2-yl)-4,6-bis(1-methyl-1-phenylethyl) phenol; 2,2'-methylenebis(6-(2H-benzotriazol-2-yl)-4-1,1,3,3-tetramethylbutyl)phenol; 2-(2H-benzotriazol-2-yl)-6-dodecyl-4-methylphenol, branched & linear; 2-(2H-benzotriazol-2-yl)-4,6-bis(1-methyl-1-phenylethyl) phenol; 2,4-di-tert-butyl-6-(5-chloro-2H-benzotriazole-2-yl)-phenol; 2-(3-sec-butyl-5-tert-butyl-2-hydroxyphenyl) benzotriazole; 2-[4-[(2-hydroxy-3-(2'-ethyl)hexyl)oxy]-2-hydroxyphenyl]-4,6-bis(2,4-dimethylphenyl)-1,3,5-triazine; 2-(4,6-diphenyl-1,3,5-triazin-2-yl)-5-hexyloxy-phenol; 2-[4,6-bis(2,4-dimethylphenyl)-1,3,5-triazin-2-yl]-5-(octyloxy) phenol; (2-ethylhexyl)-2-cyano-3,3-diphenylacrylate; 1,3-bis-[(2'-cyano-3',3'-diphenylacryloyl)oxy]-2,2-bis-{[(2'-cyano-3',3'-diphenylacryloyl)oxy]methyl}-propane; propane-dioic acid, [(4-methoxyphenyl)-methylene]-dimethyl ester; 2,2'-(1,4-phenylene)bis[4H-3,1-benzoxazin-4-one]; Screener: carbon black, zinc oxide; Acid scavenger: hydrotalcite; HAS: 1,3,5-triazine-2,4,6-triamine, N,N'''[1,2-ethane-diyl-bis[[[4,6-bis[butyl(1,2,6,6-pentamethyl-4-piperidinyl)amino]-1,3,5-triazine-2-yl]imino]-3,1-propanediyl]bis[N',N''-dibutyl-N',N''-bis(1,2,2,6,6-pentamethyl-4-piperidinyl)-; decanedioic acid, bis(2,2,6,6-tetramethyl-1-(octyloxy)-4-piperidinyl)ester, reaction products with 1,1-dimethylethylhydroperoxide and octane; bis(1,2,2,6,6-pentamethyl-4-piperidyl)sebacate + methyl-1,2,2,6,6-pentamethyl-4-piperidyl sebacate; alkenes, C20-24-. alpha.-, polymers with maleic anhydride, reaction products with 2,2,6,6-tetramethyl-4-piperidinamine; 1, 6-hexanediamine, N, N'-bis(2,2,6,6-tetramethyl-4-piperidinyl)-, polymers with 2,4-dichloro-6-(4-morpholinyl)-1,3,5-triazine; 1,6-hexanedi-amine, N,N'-bis(2,2,6,6-tetramethyl-4-piperidinyl)-, polymers with morpholine-2,4,6-trichloro-1,3,5-triazine reaction products, methylated; Phenolic antioxidants: ethylene-bis(oxyethylene)-bis(3-(5-tert-butyl-4-hydroxy-m-tolyl)-propionate); N,N'-hexane-1,6-diylbis(3-(3,5-di-tert-butyl-4-hydroxy-phenylpropionamide)); 3,3',3',5, 5',5'-hexa-tert-butyl-a,a',a'-(mesitylene-2,4,6-triyl)tri-p-cresol; 1,3,5-tris(3,5-di-tert-butyl-4-hydroxybenzyl)-1,3,5-triazine-2,4,6(1H,3H,5H)- trione; Phosphite: bis-(2,4-di-t-butylphenol) pentaerythritol diphosphite; tris (2,4-di-tert-butylphenyl)phosphite; phosphoric acid, (2,4-di-butyl-6-methylphenyl)ethylester; distearyl pentaerythritol diphosphite; Optical brightener: 2,2'-(2,5-thio-phenediyl)bis(5-tert-butylbenzoxazole)	

TOXICITY

PARAMETER	UNIT	VALUE	REFERENCES
Carcinogenic effect	-	not listed by ACGIH, NIOSH, NTP	
Mutagenic effect	-	none	
Teratogenic effect	-	none	

PEN poly(ethylene 2,6-naphthalate)

PARAMETER	UNIT	VALUE	REFERENCES
PROCESSING			
Typical processing methods	-	extrusion, injection molding, spinning	
Additives used in final products	-	Antiblocking: calcium carbonate, diatomaceous earth, silicone fluid, spherical silicone resin, synthetic silica; Release: calcium stearate, fluorine compounds, glycerol bistearate, pentaerythritol ester, silane modified silica, zinc stearate; Slip: spherical silica, silicone oil	
Applications	-	fibers, magnetic recording media, microelectrodes, OLED, solar cells	
Outstanding properties	-	thermal and mechanical properties	
BLENDS			
Suitable polymers	-	LCP, PET, PBT, PA6,6, HDPE, PTT, PC	
Compatibilizers	-	ethylene/methyl acrylate copolymer compatibilizes blend with HDPE	
ANALYSIS			
FTIR (wavenumber-assignment)	cm^{-1}/-	C=O – 1729; CH_2 *trans* - 1332; C–O *gauche* – 1092; crystalline – 1004, 838, 814	Zhang, Y; Mukoyama, S; Hu, Y; Yan, C; Ozaki, Y; Takahashi, I, Macromolecules, 40, 4009-15, 2007.
Raman (wavenumber-assignment)	cm^{-1}/-	*gauche* – 1107; *trans* – 1098	Schoukens, G; De Clerck, K, Polymer, 46, 845-57, 2005.
NMR (chemical shifts)	ppm	H NMR: ethylene protons – 4.86; methylene protons – 4.45	Woo, E M; Hou, S-S; Huang, D-H; Lee, L-T, Polymer, 46, 7425-35, 2005.
x-ray diffraction peaks	degree	23.3-23.4; 26.8-27.0; 15.6 peaks of (010), (100), (110) at 16.4, 22.7, and 25.8, respectively	Zhang, Y; Mukoyama, S; Hu, Y; Yan, C; Ozaki, Y; Takahashi, I, Macromolecules, 40, 4009-15, 2007; Yin, J, Luo, F, Wang, D. Sun, X, Eur. Polym. J., 114, 66-71, 2019.

PEO poly(ethylene oxide)

PARAMETER	UNIT	VALUE	REFERENCES
GENERAL			
Common name	-	poly(ethylene oxide), poly(ethylene glycol)	
IUPAC name	-	poly(ethylene oxide)	
ACS name	-	poly(oxy-1,2-ethanediyl), alpha-hydro-omega-hydroxy-	
Acronym	-	PEO; PEG	
CAS number	-	25322-68-3	
RTECS number	-	TQ3500000; TQ3520000; TQ3560000; TQ3580000; TQ3600000; TQ3610000; TQ3620000; TQ3630000; TQ3650000; TQ3675000; TQ3700000; TQ3800000; TQ3850000; TQ4025000; TQ4026000; TQ4027000; TQ4028000; TQ4030000; TQ4040000; TQ4041000; TQ4050000; TQ4070000; TQ4100000; TQ4105000; TQ4110000; TQ4950000; TQ5090000	
Linear formula		$H-[OCH_2CHO]_h-H$	
HISTORY			
Person to discover	-	Schoeller, C; Wittwer, M	Schoeller, C; Wittwer, M, US Patent n1,970,578, IG Farber, Aug. 21, 1934.
Date	-	1934 (application in Germany in 1930)	
SYNTHESIS			
Monomer(s) structure	-		
Monomer(s) CAS number(s)	-	75-21-8	
Monomer(s) molecular weight(s)	dalton, g/mol, amu	44.05	
Monomer ratio	-	100%	
Method of synthesis	-	ethylene glycols are used in synthesis because they form polymers of low polydispersity. Anionic polymerization is used more frequently as it produces polymers of low polydispersity.	
Number average molecular weight, M_n	dalton, g/mol, amu	120-136,000	
Mass average molecular weight, M_w	dalton, g/mol, amu	200-8,000,000	Wang, H; Rren, J; Yan, M, J. Colloid Interface Sci., 354, 160-7, 2011.
Polydispersity, M_w/M_n	-	1.1-1.3	
Molar volume at 298K	$cm^3 mol^{-1}$	calc.=39.9; 34.5 (crystalline); 38.9 (amorphous); exp.=39.2	
Van der Waals volume	$cm^3 mol^{-1}$	24.2 (crystalline); 24.2 (amorphous); exp.=25.11	
Radius of gyration	nm	38-187	Sung, J H; Lee, D C; Park, H J, Polymer, 48, 4205-12, 2007.
End-to-end distance of unperturbed polymer chain	nm	6-15	Li, T; Park, K, Computational Theor. Polym. Sci., 11, 133-42, 2001.
Chain-end groups	-	OH	

PEO poly(ethylene oxide)

PARAMETER	UNIT	VALUE	REFERENCES
STRUCTURE			
Crystalline structure			
Crystallinity	%	70-82	Narh, K A; Khanolkar, M; Umbrajkar, S M; Dreizin, E, Antec, 1776-80, 2007.
Cell type (lattice)	-	monoclinic	
Cell dimensions	nm	a:b:c=0.795-0.805:1.299-1.34:1.925-1.95	
Unit cell angles	degree	β=124.6-126.9	
Number of chains per unit cell	-	4	
Crystallite size	nm	6.8	Slusarczyk, C, Radiation Phys. Chem., in press, 2011.
Chain conformation	-	helix 7/2, planar zigzag 2/1	
Entanglement molecular weight	dalton, g/mol, amu	calc.=1,718; exp.=2,200	
Lamellae thickness	nm	6.8	Slusarczyk, C, Radiation Phys. Chem., in press, 2011.
Avrami constants, k/n	-	0.0442/1.8	Qiu, Z; Ikehara, T; Nishi, T, Polymer, 44, 3101-6, 2003.
PHYSICAL PROPERTIES			
Density at 20°C	g cm^{-3}	1.1-1.7	
Color	-	clear or white	
Refractive index, 20°C	-	calc.=1.4418-1.4468; exp.=1.4563-1.510	
Odor	-	mild	
Melting temperature, DSC	°C	-23 to 12; PEG 400=4-8; PEG 600=20-25; PEG 1500=44-48; PEG 4000=54-58; PEG 6000=56-63	
Boiling point	°C	>200	
Thermal conductivity, melt	W m^{-1} K^{-1}	calc.=0.9085; exp.=08583	
Glass transition temperature	°C	calc.=-60; -60 to -70	
Heat of fusion	kJ mol^{-1}	8.0-9.4	
Hildebrand solubility parameter	MPa$^{0.5}$	20.2	
Surface tension	mN m^{-1}	calc.=28.2-36.4; exp.=42.9	
Contact angle of water, 20°C	degree	35-63.0	
Surface free energy	mJ m^{-2}	43.1	
Speed of sound	m s^{-1}	37.5	
Acoustic impedance		2.72	
MECHANICAL & RHEOLOGICAL PROPERTIES			
Tensile strength	MPa	11-60; 100-200 (high molecular weight fibers)	
Tensile modulus	MPa	300; 500-1,000 (high molecular weight fibers); 10,000 (theoretical ultimate modulus)	
Elongation	%	30-70 (high molecular weight fibers)	
Poisson's ratio	-	calc.=0.439	
CHEMICAL RESISTANCE			
Alcohols	-	poor	
Aliphatic hydrocarbons	-	good	
Aromatic hydrocarbons	-	poor	

PEO poly(ethylene oxide)

PARAMETER	UNIT	VALUE	REFERENCES
Esters	-	poor	
Θ solvent, Θ-temp.=71°C	-	benzene/isooctane=100/48	
Good solvent	-	alcohols, benzene, cyclohexanone, esters, water (cold)	
Non-solvent	-	aliphatic hydrocarbons, ethers, hot water	

FLAMMABILITY			
Ignition temperature	°C	182-287	
Limiting oxygen index	% O_2	18.5	

BIODEGRADATION			
Typical biodegradants	-	PEO-dehydrogenase, PEO-aldehyde-dehydrogenase and PEO-carboxylate-dehydrogenase act sequentially to produce terminal carbonyl and carboxyl groups from the terminal units of poly(ether) chains, followed by the release of C_2 units as glioxylic acid	

TOXICITY			
NFPA: Health, Flammability, Reactivity rating	-	0/1/0	
Carcinogenic effect	-	not listed by ACGIH, NIOSH, NTP	
TLV, ACGIH	ppm	10	
Oral rat, LD_{50}	mg kg^{-1}	28,000 (MW 200); 38,100 (MW 600); 44,200 (MW 1,000); 50,000 (MW 4,000)	

ENVIRONMENTAL IMPACT			
Aquatic toxicity, *Fathead minnow*, LC_{50}, 48 h	mg l^{-1}	>20,000	
Aquatic toxicity, *Rainbow trout*, LC_{50}, 48 h	mg l^{-1}	>20,000	

PROCESSING			
Typical processing methods	-	coextrusion, compounding, electrospinning (Forcespinning), reacting with other monomers	Sarkar, K; Gomez, C; Zambrano, S; Ramirez, M; de Hoyos, E; Vasquez, H; Lozano, K, Mater. Today, 13, 11, 12-14, 2010.
Additives used in final products	-	Fillers: carbon nanotubes, fumed silica, graphite, Fillers: molybdenum disulfide, montmorillonite, nanosilica, titanium dioxide, vanadium oxide; Plasticizers: dioctyl phthalate, ethylene carbonate, polyoxyethylene-sorbitane monolaureate, propylene carbonate, polyethylene and polypropylene glycols, tetraethylene glycol, tetraglyme; Antistatics: polyoxyethylene sorbitan monolaurate, polyoxyethylene glycol, polyoxyethylene octylphenyl ether	
Applications	-	controlled release drugs, haircare, nanocomposites, oil exploration, pharmaceutical applications, polyester fibers, polyester resins, polymer electrolytes, polyols, solar cells, surfactants, switching elements, unsaturated lithium batteries	

BLENDS			
Suitable polymers	-	chitosan, epoxy, PAA, PCL, PHB, PLA, PMMA, PSU, PVC, PVDF-CTFE, PVF-HFP, PVOH, protein, starch	

PEO poly(ethylene oxide)

PARAMETER	UNIT	VALUE	REFERENCES
ANALYSIS			
FTIR (wavenumber-assignment)	cm^{-1}/-	CH$_2$ – 2640-3080; C–H – 1467, 1359, 1343, 1241, 962; C–O–C – 1061, 1113, 1147	Kaczmarek, H; Bajer, K; Galka, P; Kotnowska, B, Polym. Deg. Stab., 92, 2058-69, 2007.
x-ray diffraction peaks	degree	19.46; 23.52	Kaczmarek, H; Bajer, K; Galka, P; Kotnowska, B, Polym. Deg. Stab., 92, 2058-69, 2007.

PES poly(ether sulfone)

PARAMETER	UNIT	VALUE	REFERENCES
GENERAL			
Common name	-	poly(ether sulfone)	
CAS name	-	poly(oxy-1,4-phenylenesulfonyl-1,4-phenylene)	
Acronym	-	PES, PESU (ISO), PPSU	
CAS number	-	25667-42-9	
Formula			
HISTORY			
Person to discover	-	Khattab, G	Khattab, G, US Patent 3,723,389, Allied Chemical, Mar. 27, 1973.
Date	-	1973	
Details	-	polymeric condensation products of p,p'-dihalodiphenyl sulfone with alkali metal bisphenates	
SYNTHESIS			
Monomer(s) structure	-		
Monomer(s) CAS number(s)	-	80-05-7; 80-07-9	
Monomer(s) molecular weight(s)	dalton, g/ mol, amu	228.29; 287.16	
Method of synthesis	-	reaction between Bisphenol A and bis(4-chlorophenyl)sulfone in the presence of Na_2CO_3 with elimination of NaCl, H_2O, and CO_2	
Number average molecular weight, M_n	dalton, g/ mol, amu	8,600-28,700; 125292	-; Han, X, Jin, S, Zhang, J, Yue, C, Zhang, H, Pang, J, Jiang, Z, Polymer, in press, 123092, 2020.
Mass average molecular weight, M_w	dalton, g/ mol, amu	10,100-38,800	
Polydispersity, M_w/M_n	-	1.17-1.55	
Molar volume at 298K	cm^3 mol^{-1}	157	
Chain-end groups	-	NH_2, OH	
STRUCTURE			
Cell type (lattice)	-	orthorhombic	
Cell dimensions	nm	a:b:c=0.85:0.495:0.67	
Unit cell angles	degree	α:β:γ=90:90:90	
Number of chains per unit cell	-	2	
Chain conformation	-	glide 2/0	
COMMERCIAL POLYMERS			
Some manufacturers	-	BASF; Mitsui Chemical; Solvay	
Trade names	-	Ultrason E; PES; Veradel	
PHYSICAL PROPERTIES			
Density at 20°C	g cm^{-3}	1.37; 1.43; 1.50-1.61 (20-30% glass fiber); 1.23 (melt); 1.35-1.41 (20-30% glass fiber, melt)	

PES poly(ether sulfone)

PARAMETER	UNIT	VALUE	REFERENCES
Bulk density at 20°C	g cm^{-3}	0.7-0.8	
Color	-	transparent, amber	
Refractive index, 20°C	-	1.545-1.65	
Melting temperature, DSC	°C	220-238	
Decomposition temperature	°C	400-584	Duan, Y; Cong, P; Liu, X; Li, T, J. Macromol. Sci. B, 48, 604-16, 2009.
Thermal expansion coefficient, 23-200°C	°C^{-1}	4.3-5.5E-5; 2.0-3.1E-5 (20-30% glass fiber)	
Thermal conductivity, melt	W m^{-1} K^{-1}	0.18 (melt); 0.24; 0.30 (30% glass fiber)	
Glass transition temperature	°C	220-246; 220 (30% glass fiber)	
Specific heat capacity	J K^{-1} kg^{-1}	1,950 (melt); 1,200 (23°C); 1,740 (20-30% glass fiber)	
Maximum service temperature	°C	218; 191 (NSF standard 51)	
Long term service temperature	°C	200 (at 180°C heat resistance is 20 years)	
Temperature index (50% tensile strength loss after 20,000 h/5000 h)	°C	190	Padey, D; Walling, J; Wood A, Polymers in Defence and Aerospace 2007, Rapra, 2007, paper 15.
Heat deflection temperature at 0.45 MPa	°C	214-218; 218-224 (20-30% glass fiber)	
Heat deflection temperature at 1.8 MPa	°C	204-207; 214-216 (20-30% glass fiber)	
Vicat temperature VST/A/50	°C	214-215; 217-218 (20-30% glass fiber)	
Hansen solubility parameters, δ_D, δ_P, δ_H	MPa$^{0.5}$	18.3, 8.2, 6.4; 19.6, 10.8, 9.2	Yune, P S; Kilduff, J E; Belfort, G, J. Membrane Sci., in press, 2011.
Interaction radius		6.3; 6.2	Yune, P S; Kilduff, J E; Belfort, G, J. Membrane Sci., in press, 2011.
Hildebrand solubility parameter	MPa$^{0.5}$	calc.=23.12-24.4; exp.=21.0-22.9	
Dielectric constant at 60 Hz/1 MHz	-	3.51/3.54; 3.7-4.11/3.7-4.17 (20-30% glass fiber)	
Relative permittivity at 100 Hz	-	3.9 (conditioned); 4.2-4.3 (20-30% glass fiber, conditioned)	
Relative permittivity at 1 MHz	-	3.8 (conditioned); 4.2-4.3 (20-30% glass fiber, conditioned)	
Dissipation factor at 100 Hz	E-4	17; 15-20 (20-30% glass fiber)	
Dissipation factor at 1 MHz	E-4	56; 81-100 (20-30% glass fiber)	
Volume resistivity	ohm-m	>1.7E13; 1E14 (20-30% glass fiber)	
Surface resistivity	ohm	>1E15	
Electric strength K20/P50, d=0.60.8 mm	kV mm^{-1}	15; 17 (20-30% glass fiber)	
Comparative tracking index	-	125 (conditioned); 125 (20-30% glass fiber, conditioned)	
Coefficient of friction	-	0.15-0.45 (air); 0.3 (water)	Duan, Y; Cong, P; Liu, X; Li, T, J. Macromol. Sci. B, 48, 604-16, 2009.
Contact angle of water, 20°C	degree	68.5-69.0	Kim, Y; Rana, D; Matsuura, T; Chung, W-J, J. Membrane Sci., 338, 84-91, 2009.
Surface free energy	mJ m^{-2}	47.0	
Speed of sound	m s^{-1}	2260	
Permeability to nitrogen, 35°C	barrer	0.10	Han, X, Jin, S, Zhang, J, Yue, C, Zhang, H, Pang, J, Jiang, Z, Polymer, in press, 123092, 2020.

PES poly(ether sulfone)

PARAMETER	UNIT	VALUE	REFERENCES
Permeability to oxygen, 35°C	barrer	0.52	Han, X, Jin, S, Zhang, J, Yue, C, Zhang, H, Pang, J, Jiang, Z, Polymer, in press, 123092, 2020.
Permeability to CO_2, 35°C	barrer	2.51	Han, X, Jin, S, Zhang, J, Yue, C, Zhang, H, Pang, J, Jiang, Z, Polymer, in press, 123092, 2020.
Diffusion coefficient of nitrogen	cm^2 s^{-1} x10^8	0.28	Han, X, Jin, S, Zhang, J, Yue, C, Zhang, H, Pang, J, Jiang, Z, Polymer, in press, 123092, 2020.
Diffusion coefficient of oxygen	cm^2 s^{-1} x10^8	1.16	Han, X, Jin, S, Zhang, J, Yue, C, Zhang, H, Pang, J, Jiang, Z, Polymer, in press, 123092, 2020.
Diffusion coefficient of CO_2	cm^2 s^{-1} x10^8	0.45	Han, X, Jin, S, Zhang, J, Yue, C, Zhang, H, Pang, J, Jiang, Z, Polymer, in press, 123092, 2020.

MECHANICAL & RHEOLOGICAL PROPERTIES

PARAMETER	UNIT	VALUE	REFERENCES
Tensile strength	MPa	80-95, 105-137 (20-30% glass fiber)	
Tensile modulus	MPa	2,650; 5,700-9,800 (20-30% glass fiber); 20,900 (30% carbon fiber)	
Tensile stress at yield	MPa	90 (conditioned)	
Tensile creep modulus, 1000 h, elongation 0.5 max	MPa	2,300-2,700; 5,600-8,300 (20-30% glass fiber)	
Elongation	%	13-75; 1.9-3.2 (20-30% glass fiber); 1.7 (30% carbon fiber)	
Tensile yield strain	%	5.2; 6.7 (conditioned)	
Flexural strength	MPa	111; 145-190 (20-30% glass fiber)	
Flexural modulus	MPa	2,900; 5,170-8,800 (20-30% glass fiber)	
Compressive strength	Mpa	100; 151-177 (20-30% glass fiber)	
Young's modulus	MPa	4,950	Grunzinger, S J; Watanabe, M; Fukagawa, K; Kikuchi, R; Tominaga, Y; Hayakawa, T; Kakimoto, M, J. Power Sources, 175, 120-26, 2008.
Charpy impact strength, unnotched, 23°C	kJ m^{-2}	no break (conditioned); 42-47 (20-30% glass fiber, conditioned)	
Charpy impact strength, unnotched, -30°C	kJ m^{-2}	no break (conditioned); 45 (20-30% glass fiber, conditioned)	
Charpy impact strength, notched, 23°C	kJ m^{-2}	6.5 (conditioned); 6.5-8 (20-30% glass fiber, conditioned)	
Charpy impact strength, notched, -30°C	kJ m^{-2}	7 (conditioned); 7.5-8 (20-30% glass fiber, conditioned)	
Izod impact strength, unnotched, 23°C	J m^{-1}	no break; 530-640 (20-30% glass fiber)	
Izod impact strength, notched, 23°C	J m^{-1}	85, 59-75 (20-30% glass fiber)	
Shear strength	MPa	55; 61-65 (20-30% glass fiber)	
Poisson's ratio	-	0.41; 0.42 (20-30% glass fiber)	
Shore D hardness	-	88	
Rockwell hardness	R	127; 121 (30% glass fiber)	
Shrinkage	%	0.8-1.4; 0.28-0.58 (20-30% glass fiber)	
Viscosity number	ml g^{-1}	48-52; 56 (20-30% glass fiber)	
Melt viscosity, shear rate=1000 s^{-1}	Pa s	240-800; 350-700 (20-30% glass fiber)	
Melt volume flow rate (ISO 1133, procedure B), 360°C/10 kg	cm^3/10 min	35-150; 25-29 (20-30% glass fiber)	

PES poly(ether sulfone)

PARAMETER	UNIT	VALUE	REFERENCES
Melt index, 380°C/2.16 kg	g/10 min	12-30; 10-18 (20-30% glass fiber)	
Water absorption, 24h at 23°C	%	2.2; 1.6 (20-30% glass fiber)	
Moisture absorption, equilibrium 23°C/50% RH	%	0.8; 0.6 (20-30% glass fiber)	
CHEMICAL RESISTANCE			
Acid dilute/concentrated	-	good/poor	
Alcohols	-	good	
Alkalis	-	good	
Aliphatic hydrocarbons	-	good	
Aromatic hydrocarbons	-	good to fair	
Esters	-	poor	
Greases & oils	-	good	
Halogenated hydrocarbons	-	poor	
Ketones	-	poor	
Θ solvent, Θ-temp.=25°C	-	DMF/methanol=83/17	
FLAMMABILITY			
Flammability according to UL-standard; thickness 1.6/0.8 mm	class	V-0 to V-1; V-0 (20-30% glass fiber)	
Autoignition temperature	°C	580-600	
Limiting oxygen index	% O_2	39; 40-44.5 (20-30% glass fiber)	
NBS smoke chamber, Ds, 4 min	-	1.0	
Char at 500°C	%	12.4-29.3 (air); 48 (nitrogen)	Lyon, R E; Walters, R N, J. Anal. Appl. Pyrolysis, 71, 27-46, 2004; Chen, H; Zhang, K; Xu, J, Polym. Deg. Stab., 96, 197-203, 2011.
Heat of combustion	J g^{-1}	25,420	Walters, R N; Hacket, S M; Lyon, R E, Fire Mater., 24, 5, 245-52, 2000.
Activation energy of decomposition	kJ mol^{-1}	280	Swallowe, G M; Dawson, P C; Tang, T B; Xu, Q L, J. Mater. Sci., 30, 3853-55, 1995.
Volatile products of combustion	-	CO	
TOXICITY			
HMIS: Health, Flammability, Reactivity rating	-	1/1/0	
Carcinogenic effect	-	not listed by ACGIH, NIOSH, NTP	
PROCESSING			
Typical processing methods	-	blow molding, casting, compression molding, film extrusion, injection molding, machining, profile extrusion, spinning, thermoforming, wire and cable extrusion	
Preprocess drying: temperature/time/residual moisture	°C/h/%	140-150/4/0.02-0.05	
Processing temperature	°C	340-390 (injection molding)	
Additives used in final products	-	carbon fiber, glass fiber, graphite	

PES poly(ether sulfone)

PARAMETER	UNIT	VALUE	REFERENCES
Applications	-	aircraft interiors, automotive fuses, coatings, coil bobbins, dip switches, fiber optics connectors, hollow fiber, integrated circuits sockets, medical applications (due to the resistance to different methods of sterilization), membranes, microwave cookware, multipin connectors, printed circuit boards, transformer wire coatings, sight glasses	
BLENDS			
Suitable polymers	-	epoxy, PBI, PC, PEEK, PEO	
ANALYSIS			
NMR (chemical shifts)	ppm	3.83 ppm and 3.86 – protons of $-OCH_3$	Han, X, Jin, S, Zhang, J, Yue, C, Zhang, H, Pang, J, Jiang, Z, Polymer, in press, 123092, 2020.

PET poly(ethylene terephthalate)

PARAMETER	UNIT	VALUE	REFERENCES
GENERAL			
Common name	-	poly(ethylene terephthalate)	
IUPAC name	-	poly(ethylene terephthalate)	
ACS name	-	poly(oxy-1,2-ethanediyloxycarbonyl-1,4-phenylenecarbonyl)	
Acronym	-	PET	
CAS number	-	25038-59-9	
Linear formula			
HISTORY			
Person to discover	-	Whinfield, R J and Dickson, T J	Whinfield, R J and Dickson, T J, US Patent 2,465,319, DuPont, Mar. 22, 1949.
Date	-	1949	
Details	-	first patented by DuPont	
SYNTHESIS			
Monomer(s) structure	-		
Monomer(s) CAS number(s)	-	100-21-0; 107-21-1	
Monomer(s) molecular weight(s)	dalton, g/mol, amu	166.13; 62.07	
Monomer ratio	-	2.676 (72.8:27.2)	
Method of synthesis	-	several processes can be used, including: transesterification, direct esterification, and polycondensation	
Mass average molecular weight, M_w	dalton, g/mol, amu	19,000-66,000; 29,800-45,800 (recycled)	Romao, W; Freanco, M FF; Bueno, M I M S; De Paoli, M-A, Polym. Test., 29, 879-85, 2010.
Molar volume at 298K	cm^3 mol^{-1}	calc.=146.1; 130 (crystalline); 144.5 (amorphous)	
Van der Waals volume	cm^3 mol^{-1}	calc.=96.36; 94.2 (crystalline); 94.2 (amorphous)	
Molecular cross-sectional area, calculated	cm^2 x 10^{-16}	14.0	
STRUCTURE	-		
Crystallinity	%	20-50, 52.5-54.6 (fiber); 85 (oriented film)	Romaao, W; Franco, M F; Corilo, Y E; Ebrlin, M N; Spinace, M A S; De Paoli, M-A, Polym. Deg. Stab., 94, 1849-59, 2009; Blundell, D J; Mahendrasingam, A; Martin, C; Fuller, W; MacKerron, D H; Harvie, J L; Oldman, R J; Riekel, C, Polymer, 41, 7793-7802, 2000.
Cell type (lattice)	-	triclinic	
Cell dimensions	nm	a:b:c=0.448-0.456:0.58-0.594:1.071-1.086	
Unit cell angles	degree	α:β:γ=98-107:118-119:111-112	Kinoshita, Y; Nakamura, R; Kitano, Y; Ashida, T, Polym. Prep., 20, 454, 1979.

PET poly(ethylene terephthalate)

PARAMETER	UNIT	VALUE	REFERENCES
Number of chains per unit cell	-	1	
Crystallite size	nm	27.8-40.8 (fiber)	
Chain conformation	-	nearly planar	
Entanglement molecular weight	dalton, g/mol, amu	1,936	Barany, T; Czigany, T; Karger-Kotsis, J, Prog. Polym. Sci., 35, 1257-87, 2010.
Crystallization temperature	°C	164	Romaao, W; Franco, M F; Corilo, Y E; Ebrlin, M N; Spinace, M A S; De Paoli, M-A, Polym. Deg. Stab., 94, 1849-59, 2009.
Crystallization peak temperature	°C	180	Dong, S, Jia, Y, Xu, X, Luo, J, Han, J, Sun, X, J. Colloid Interface Sci., 539, 54-64, 2019.
Half crystallization time	min	18.97	Dong, S, Jia, Y, Xu, X, Luo, J, Han, J, Sun, X, J. Colloid Interface Sci., 539, 54-64, 2019.
Avrami constants, k/n	-	0.003/1.88	Dong, S, Jia, Y, Xu, X, Luo, J, Han, J, Sun, X, J. Colloid Interface Sci., 539, 54-64, 2019.

COMMERCIAL POLYMERS

Some manufacturers	-	DuPont; Sabic	
Trade names	-	Rynite; PET	

PHYSICAL PROPERTIES

Density at 20°C	g cm^{-3}	1.3-1.4; 1.455 (crystalline); 1.333 (amorphous); 1.47-1.81 (20-55% glass fiber)	
Bulk density at 20°C	g cm^{-3}	0.795-0.88	
Color	-	white to gray	
Refractive index, 20°C	-	calc.=1.5392-1.5557; exp=1.5750	
Birefringence	-	0.0469 (fiber)	Romaao, W; Franco, M F; Corilo, Y E; Ebrlin, M N; Spinace, M A S; De Paoli, M-A, Polym. Deg. Stab., 94, 1849-59, 2009.
Haze	%	0.6	
Odor	-	none	
Melting temperature, DSC	°C	245-265; 254 (20-55% glass fiber)	
Decomposition temperature	°C	285-329	
Storage temperature	°C	<50	
Thermal expansion coefficient, -40 to 160°C	°C^{-1}	1.7E-5 (film); 6.55E-4 (melt); 1.1-9.5E-5 (20-55% glass fiber)	Mark, H F; Gaylord, N G, Encyclopedia of Polymer Science, Vol. 11, Interscience, New York, 1969.
Thermal conductivity, melt	W m^{-1} K^{-1}	calc.=0.1888; exp.=0.147 (melt); 0.29-0.33 (20-55% glass fiber)	
Infrared emissivity	-	0.825	Larciprete, M C, Paoloni, S, Orazi, N, Mercuri, F, Sibilia, C, Int. J. Thermal Sci., 146, 106109, 2019.
Glass transition temperature	°C	calc.=76-88; exp.=60-85; 60-76 (amorphous)	
Specific heat capacity	J K^{-1} kg^{-1}	1,170	
Heat of fusion	kJ mol^{-1}	24.1	
Maximum service temperature	°C	-60 to 105	
Temperature index (50% tensile strength loss after 20,000 h/5000 h)	°C	140 (20-55% glass fiber)	

PET poly(ethylene terephthalate)

PARAMETER	UNIT	VALUE	REFERENCES
Heat deflection temperature at 0.45 MPa	°C	63.9; 240-248 (20-55% glass fiber)	
Heat deflection temperature at 1.8 MPa	°C	61.1; 210-229 (20-55% glass fiber)	
Vicat temperature VST/B/50	°C	77	
Enthalpy of melting	J g^{-1}	27	Romaao, W; Franco, M F; Corilo, Y E; Ebrlin, M N; Spinace, M A S; De Paoli, M-A, Polym. Deg. Stab., 94, 1849-59, 2009.
Hansen solubility parameters, δ_D, δ_P, δ_H	MPa$^{0.5}$	18.7, 6.3, 6.7	
Interaction radius		6.5	
Hildebrand solubility parameter	MPa$^{0.5}$	calc.=14.65-17.15; exp.=17.1-20.8	
Surface tension	mN m^{-1}	calc.=38.7-47.3; exp.=39.5-42.1	
Dielectric constant at 100 Hz/1 MHz	-	3.0-3.3/3.2-3.3; 3.2-4.0/3.0-3.9 (20-55% glass fiber)	
Dissipation factor at 100 Hz	E-4	20-25; 50-100 (20-55% glass fiber)	
Dissipation factor at 1 MHz	E-4	30; 110-150 (20-55% glass fiber)	
Electric conductivity	S cm^{-1}	1E-15	Larciprete, M C; Paoloni, S, Orazi, N, Mercuri, F, Sibilia, C, Int. J. Thermal Sci., 146, 106109, 2019.
Volume resistivity	ohm-m	1E13; 1E13 (20-55% glass fiber)	
Surface resistivity	ohm	1E14 (20-55% glass fiber)	
Electric strength K20/P50, d=0.60.8 mm	kV mm^{-1}	22-26; 24.5-25.5 (20-55% glass fiber)	
Comparative tracking index, CTI, test liquid A	-	175-400	
Arc resistance	s	120-360	
Power factor	-	0.019	
Coefficient of friction	-	0.45 (film to film); 0.17-0.21 (20-55% glass fiber, self); 0.14 (film to steel); 0.17-0.20 (20-55% glass fiber, steel)	
Permeability to oxygen, 25°C	m^3 m m^{-2} s^{-1} Pa^{-1} x 10^{18}	0.4	Colomiones, G; Ducruet, V; Courgneau, C; Guinault, A; Domenek, S, Polym. Int., 59, 818-26, 2010.
Permeability to water vapor, 25°C	cm^3 cm cm^{-2} s^{-1} Pa^{-1} x 10^{12}	11.3	
Contact angle of water, 20°C	degree	72.0-75.0	Donelli, I; Freddi, G; Niefstrasz, V A; Taddei, Polym. Deg. Stab., 95, 1542-50, 2010.
Surface free energy	mJ m^{-2}	44.0	

MECHANICAL & RHEOLOGICAL PROPERTIES			
Tensile strength	MPa	24-41.4; 590 (fiber); 114-189 (20-55% glass fiber)	
Tensile modulus	MPa	2,300; 7,240-17,900 (20-55% glass fiber)	
Tensile stress at yield	MPa	36.5-62.7; 327 (fiber)	
Elongation	%	100-250; 1.6-2.3 (20-55% glass fiber)	
Tensile yield strain	%	4	
Flexural strength	MPa	68.9-78; 172-290 (20-55% glass fiber)	
Flexural modulus	MPa	1,560-2,650; 6,480-17,900 (20-55% glass fiber)	

PET poly(ethylene terephthalate)

PARAMETER	UNIT	VALUE	REFERENCES
Compressive strength	MPa	80; 172-241 (20-55% glass fiber)	
Young's modulus	MPa	2,000-2,700	
Izod impact strength, notched, 23°C	J m^{-1}	640	
Izod impact strength, notched, -30°C	J m^{-1}	190	
Shear strength	MPa	79.0-86.5 (20-55% glass fiber)	
Tenacity (fiber) (standard atmosphere)	cN tex^{-1} (daN mm^{-2})	25-95 (35-130)	Fourne, F, Synthetic Fibers. Machines and Equipment Manufacture, Properties. Carl Hanser Verlag, 1999.
Tenacity (wet fiber, as % of dry strength)	%	95-100	Fourne, F, Synthetic Fibers. Machines and Equipment Manufacture, Properties. Carl Hanser Verlag, 1999.
Fineness of fiber (titer)	dtex	0.6-44	Fourne, F, Synthetic Fibers. Machines and Equipment Manufacture, Properties. Carl Hanser Verlag, 1999.
Length (elemental fiber)	mm	38-200, filament	Fourne, F, Synthetic Fibers. Machines and Equipment Manufacture, Properties. Carl Hanser Verlag, 1999.
Abrasion resistance (ASTM D1044)	mg/1000 cycles	30-44	
Poisson's ratio	-	0.421-0.430; 0.39-0.41 (20-55% glass fiber)	
Rockwell hardness	-	R110; M95-100 (20-55% glass fiber); R120 (20-55% glass fiber)	
Shrinkage	%	0.7-1.2; 0.18-0.35 (20-55% glass fiber, flow); 0.7-0.9 (20-55% glass fiber, transverse)	
Intrinsic viscosity, 25°C	dl g^{-1}	0.75-0.84	
Melt index, 230°C/3.8 kg	g/10 min	7	
Water absorption, 24h at 23°C	%	0.04-0.14 (24 h); 0.6 (eq)	
Moisture absorption, equilibrium 23°C/50% RH	%	0.35	

CHEMICAL RESISTANCE

Acid dilute/concentrated	-	good/fair	
Alcohols	-	good/fair	
Alkalis	-	poor	
Aliphatic hydrocarbons	-	good	
Aromatic hydrocarbons	-	good	
Esters	-	fair-poor	
Greases & oils	-	good	
Halogenated hydrocarbons	-	poor	
Ketones	-	fair	
Good solvent	-	DMSO (hot), halogenated aliphatic carboxylic acids, nitrobenzene, phenol	
Non-solvent	-	aliphatic alcohols, carboxylic esters, chlorinated hydrocarbons, ether, hydrocarbons, ketones	

PET poly(ethylene terephthalate)

PARAMETER	UNIT	VALUE	REFERENCES
FLAMMABILITY			
Ignition temperature	°C	390	
Autoignition temperature	°C	508	
Limiting oxygen index	% O_2	20-23; 20 (20-55% glass fiber)	
Char at 500°C	%	5.1	Lyon, R E; Walters, R N, J. Anal. Appl. Pyrolysis, 71, 27-46, 2004.
Heat of combustion	J g^{-1}	24,130	Walters, R N; Hacket, S M; Lyon, R E, Fire Mater., 24, 5, 245-52, 2000.
Volatile products of combustion	-	CO, CO_2, acrolein, formaldehyde, ethanol, methanol, acetic acid, acetone	
UL rating	-	HB (20-55% glass fiber); V-0 (20-55% glass fiber and flame retardant)	
WEATHER STABILITY			
Spectral sensitivity	nm	290-325; 270-325	Yang, J; Xia, Z; Kong, F; Ma, X, Polym. Deg. Stab., 95, 53-8, 2010.
Activation wavelengths	nm	305, 325	
Excitation wavelengths	nm	280, 320, 344, 357; 300 and 330	Yang, J; Xia, Z; Kong, F; Ma, X, Polym. Deg. Stab., 95, 53-8, 2010.
Emission wavelengths	nm	370, 389, 405, 425, 460; 328 and 387+460	Yang, J; Xia, Z; Kong, F; Ma, X, Polym. Deg. Stab., 95, 53-8, 2010.
Important initiators and accelerators	-	ferrocene, cobalt octoates and naphthenates, compounds containing aromatic keto-ester groups, fluorophores	Gordon, D A, Zhan, Z, Bruckman, L S, Polym. Deg. Stab., 161, 85-94, 2019.
Products of degradation	-	radicals, crosslinks, hydroperoxides, hydroxyl and carbonyl groups, CO, CO_2	
Stabilizers	-	UVA: 2-hydroxy-4-octyloxybenzophenone; 2-hydroxy-4-methoxybenzophenone; 2,4-dihydroxybenzophenone; 2,2',4,4'-tetrahydroxybenzophenone; 2,2'-dihydroxy-4-methoxybenzophenone; 2-(2H-benzotriazol-2-yl)-p-cresol; 2-benzotriazol-2-yl-4,6-di-tert-butylphenol; 2-(2H-benzotriazole-2-yl)-4,6-di-tert-pentylphenol; 2-(2H-benzotriazole-2-yl)-4-(1,1,3,3-tetraethylbutyl)phenol; 2-(2H-benzotriazol-2-yl)-4,6-bis(1-methyl-1-phenylethyl)phenol; 2,2'-methylenebis(6-(2H-benzotriazol-2-yl)-4-1,1,3,3-tetramethylbutyl)phenol; 2-(2H-benzotriazol-2-yl)-6-dodecyl-4-methylphenol, branched & linear; 2-(2H-benzotriazol-2-yl)-4,6-bis(1-methyl-1-phenylethyl)phenol; 2,4-di-tert-butyl-6-(5-chloro-2H-benzotriazole-2-yl)-phenol; 2-(3-sec-butyl-5-tert-butyl-2-hydroxyphenyl)benzotriazole; 2-[4-[(2-hydroxy-3-(2'-ethyl)hexyl)oxy]-2-hydroxyphenyl]-4,6-bis(2,4-dimethylphenyl)-1,3,5-triazine; 2-(4,6-diphenyl-1,3,5-triazin-2-yl)-5-hexyloxy-phenol; 2-[4,6-bis(2,4-dimethylphenyl)-1,3,5-triazin-2-yl]-5-(octyloxy)phenol; (2-ethylhexyl)-2-cyano-3,3-diphenylacrylate; 1,3-bis-[(2'-cyano-3',3'-diphenylacryloyl)oxy]-2,2-bis-{[(2'-cyano-3',3'-diphenylacryloyl)oxy]methyl}-propane; propanedioic acid, [(4-methoxyphenyl)-methylene]-dimethyl ester; 2,2'-(1,4-phenylene)bis[4H-3,1-benzoxazin-4-one]; Screener: carbon black, zinc oxide;	Wypych, G, Handbook of Materials Weathering, 6th Ed., ChemTec. Publishing, 2018.

PET poly(ethylene terephthalate)

PARAMETER	UNIT	VALUE	REFERENCES
Stabilizers (continuation)	-	Acid scavenger: hydrotalcite; HAS: 1,3,5-triazine-2,4,6-triamine, N,N'''[1,2-ethane-diyl-bis[[[4,6-bis[butyl(1,2,6,6-pentamethyl-4-piperidinyl)amino]-1,3,5-triazine-2-yl]imino]-3,1-propanediyl]bis[N',N''-dibutyl-N',N''-bis(1,2,2,6,6-pentamethyl-4-piperidinyl)-; decanedioic acid, bis(2,2,6,6-tetramethyl-1-(octyloxy)-4-piperidinyl)ester, reaction products with 1,1-dimethylethylhydroperoxide and octane; bis(1,2,2,6,6-pentamethyl-4-piperidyl)sebacate + methyl-1,2,2,6,6-pentamethyl-4-piperidyl sebacate; alkenes, C20-24-.alpha.-, polymers with maleic anhydride, reaction products with 2,2,6,6-tetramethyl-4-piperidinamine; 1, 6-hexanediamine, N, N'-bis(2,2,6,6-tetramethyl-4-piperidinyl)-, polymers with 2,4-dichloro-6-(4-morpholinyl)-1,3,5-triazine; 1,6-hexanedi-amine, N,N'-bis(2,2,6,6-tetramethyl-4-piperidinyl)-, polymers with morpholine-2,4,6-trichloro-1,3,5-triazine reaction products, methylated; Phenolic antioxidants: ethylene-bis(oxyethylene)-bis(3-(5-tert-butyl-4-hydroxy-m-tolyl)-propionate); N,N'-hexane-1,6-diylbis(3-(3,5-di-tert-butyl-4-hydroxy-phenylpropionamide)); 3,3',3',5, 5',5'-hexa-tert-butyl-a,a',a'-(mesitylene-2,4,6-triyl)tri-p-cresol; 1,3,5-tris(3,5-di-tert-butyl-4-hydroxybenzyl)-1,3,5-triazine-2,4,6(1H,3H,5H)- trione; Phosphite: bis-(2,4-di-t-butylphenol) pentaerythritol diphosphite; tris (2,4-di-tert-butylphenyl)phosphite; phosphoric acid, (2,4-di-butyl-6-methylphenyl)ethylester; distearyl pentaerythritol diphosphite; Optical brightener: 2,2'-(2,5-thiophenediyl)bis(5-tert-butylbenzoxazole)	Wypych, G, Handbook of Materials Weathering, 6th Ed., ChemTec. Publishing, 2018.

BIODEGRADATION

PARAMETER	UNIT	VALUE	REFERENCES
Typical biodegradants	-	abiotic hydrolysis is the most important reaction for initiating the environmental degradation of poly(ethylene terephthalate). Also cutinases and carboxylesterases have both shown the potential to hydrolyze polyester bonds similarly to lipases; *Alcanivorax, Hyphomonas*, and *Cycloclasticus* species were found to degrade marine polluting plastics; *Yarrowia lipolytica* enzymatically hydrolyzes poly(ethylene terephthalate) using lipase	Donelli, I; Freddi, G; Niefstrasz, V A; Taddei, Polym. Deg. Stab., 95, 1542-50, 2010; Denaro, R, Aulenta, F, Crisafi, F, Di Pippo, F, Rossetti, S, Sci Total Environ., 749, 141608, 2020; da Costa, A M, de Oliveira Lopes, V R, Vidal, L, Nicaud, J-M, Zarur Coelho, M A, Process Biochem., 95, 81-90, 2020.
Stabilizers	-	2,4,4'-trichloro-2'-hydroxydiphenyl ether, N-hexyl-N'-(4-vinylbenzyl)-4,4'-bipyridinium bromide chloride	

TOXICITY

PARAMETER	UNIT	VALUE	REFERENCES
HMIS: Health, Flammability, Reactivity rating	-	1/1/0	
Carcinogenic effect		not listed by ACGIH, NIOSH, NTP	
Oral rat, LD_{50}	mg kg^{-1}	>10,000	

ENVIRONMENTAL IMPACT

PARAMETER	UNIT	VALUE	REFERENCES
Cradle to grave non-renewable energy use	MJ/kg	69.7-85.5	
Cradle to pellet greenhouse gasses	kg CO_2 kg^{-1} resin	2.5-3.5	

PROCESSING

PARAMETER	UNIT	VALUE	REFERENCES
Typical processing methods	-	blow molding, extrusion, injection blow molding, injection molding, monofilament extrusion	
Preprocess drying: temperature/time/residual moisture	°C/h/%	110-175/2-16/0.003-0.02	
Processing temperature	°C	260-300	

HANDBOOK OF POLYMERS 3rd Edition, Copyrights 2022; ChemTec Publishing

PET poly(ethylene terephthalate)

PARAMETER	UNIT	VALUE	REFERENCES
Additives used in final products	-	Fillers: antimony doped tin oxide, aramid, carbon black, carbon fiber, clays, fly ash, glass fiber, glass spheres, mica, montmorillonite, multiwalled carbon nanotubes, silica, talc, titanium dioxide, wollastonite; Antistatics: antimony-doped tin oxide, carbon nanotubes, polyaniline, polyisonaphthalene; Antiblocking: calcium carbonate, diatomaceous earth, silicone fluid, spherical silicone resin, synthetic silica; Release: calcium stearate, fluorine compounds, glycerol bistearate, pentaerythritol ester, silane modified silica, zinc stearate; Slip: spherical silica, silicone oil	
Applications	-	automotive, bottles, brush bristles, composites, electrical, fiber, film, housings, housewares, lighting, medical (orthopaedic bandages, artificial kidneys, sutures, artificial tendons, cardiovascular implants), motor parts, packaging, plumbing, power tools, sporting goods, support brackets, textiles, water pipes	
Outstanding properties	-	high strength, stiffness, excellent, dimensional stability, outstanding chemical, excellent flow characteristics, and heat resistance, and good electrical properties	

BLENDS

PARAMETER	UNIT	VALUE	REFERENCES
Suitable polymers	-	PA-6, PP, poly(m-xylylene adipamide)	

ANALYSIS

PARAMETER	UNIT	VALUE	REFERENCES
FTIR (wavenumber-assignment)	cm^{-1}/-	detailed assignment for amorphous and crystalline in ref.	Donelli, I; Freddi, G; Niefstrasz, V A; Taddei, Polym. Deg. Stab., 95, 1542-50, 2010.
NMR (chemical shifts)	ppm	H NMR: terephthalic acid – 4.80; ethylene glycol – 8.13; isophthalic acid – 8.71, 8.29, 7.60; diethylene glycol – 4.64, 4.13	Romaao, W; Franco, M F; Corilo, Y E; Ebrlin, M N; Spinace, M A S; De Paoli, M-A, Polym. Deg. Stab., 94, 1849-59, 2009.
x-ray diffraction peaks	degree	21.3	Sun, B; Lu, Y; Ni, H; Wang, C, Polymer, 39, 1, 159-63, 1998.

PEX silane-crosslinkable polyethylene

PARAMETER	UNIT	VALUE	REFERENCES
GENERAL			
Common name	-	silane-crosslinkable polyethylene	
Acronym	-	PEX	
HISTORY			
Person to discover	-	Ishino, I; Ohno, A; Isaka, T in 1987; Giacobbi, E; Miglioli, C in 2007	Ishino, I; Ohno, A; Isaka, T, US Patent 4,689,369, Mitsubishi Petrochemical, Aug. 25, 1987. Giacobbi, E; Miglioli, C, US Patent 2007/0117933 A1, Solvay, May 24, 2007.
Date	-	1987; 2007	
Details	-	patent for copolymerization; polyethylene grafting	
SYNTHESIS			
Monomer(s) structure	-	CH_2CHSiH_3	
Monomer(s) CAS number(s)	-	7291-09-0	
Monomer(s) molecular weight(s)	dalton, g/mol, amu	55.1307	
Method of synthesis	-	these copolymers can either be produced in a reactor by polymerizing ethylene with vinyl-silane, or by extruder grafting of polyethylene with the vinyl-silane; these methods replace previously used peroxide or irradiation methods both leading to crosslinking. Two methods are used: Monosil, which is one step process with grafting taking place during fabrication of the product (e.g., pipe), Sioplas, which is two step process (first step is that of grafting silane, the second step is that of its moisture cure to obtain crosslinking)	Wu, T-S, Plastics Additives Compounding, 9, 6, 40-3, 2007.
Gel content	%	65-78	
STRUCTURE			
Crystallinity	%	22-41; decreases with crosslinking and gel content increase	Rahman, W A W A; Hoong, C C; Fareed, A, J. Teknologi, 46A, 73-86, 2007.
COMMERCIAL POLYMERS			
Some manufacturers	-	Solvay	
Trade names	-	Polidan	
PHYSICAL PROPERTIES			
Density at 20°C	g cm^{-3}	0.900-1.01; 0.938	-; Holder, S L, Hedenqvist, M S, Nilsson, F, Water Res., 157, 301-9, 2019.
Bulk density at 20°C	g cm^{-3}	0.59	
Color	-	white	
Odor		very faint	
Melting temperature, DSC	°C	110	
Decomposition temperature	°C	255-285 (PE 245)	
Maximum service temperature	°C	250 (a few seconds)	
Long term service temperature	°C	130	
Surface tension	mN m^{-1}	31	

PEX silane-crosslinkable polyethylene

PARAMETER	UNIT	VALUE	REFERENCES
Dielectric constant at 100 Hz/1 MHz	-	2-2.31	
Dielectric loss factor at 1 kHz	-	3E-4	
Volume resistivity	ohm-m	1E14	
Electric strength K20/P50, d=0.60.8 mm	kV mm^{-1}	21-38	
Power factor	-	0.0003-0.0017	

MECHANICAL & RHEOLOGICAL PROPERTIES			
Tensile strength	MPa	9-26	
Elongation	%	350-600	
Shrinkage	%	<2	
Brittleness temperature (ASTM D746)	°C	-76	
Melt index, 190°C/2.16 kg	g/10 min	0.35-8	

CHEMICAL RESISTANCE			
Acid dilute/concentrated	-	resistant	
Alcohols	-	resistant	
Alkalis	-	resistant	
Aliphatic hydrocarbons	-	resistant/non-resistant	
Aromatic hydrocarbons	-	resistant	
Esters	-	non-resistant	
Greases & oils	-	resistant	
Ketones	-	non-resistant	

FLAMMABILITY			
Autoignition temperature	°C	260-320	
Limiting oxygen index	% O_2	17.5; 29 (flame retarded)	Wang, Z; Hu, Y; Gui, Z; Zong, R, Polym. Test., 22, 533-38, 2003.
Heat release	kW m^{-2}	930; 151-319 (80-150 phr of magnesium hydroxide)	Wang, Z; Hu, Y; Gui, Z; Zong, R, Polym. Test., 22, 533-38, 2003.
Char at 500°C	%	2.4; 33.4-48.5 (flame retarded)	Wang, Z; Hu, Y; Gui, Z; Zong, R, Polym. Test., 22, 533-38, 2003.
Volatile products of combustion	-	CO, CO_2, NO_x, aldehydes	
UL 94 rating	-	VW-1	

PROCESSING			
Typical processing methods	-	coextrusion, extrusion, film, injection molding	
Preprocess drying: temperature/ time/residual moisture	°C/h/%	60/8 (color masterbatches)	
Processing temperature	°C	160-180 (Syncure, two step process)	
Additives used in final products	-	catalyst masterbatch for moisture curing (Sioplast method); magnesium hydroxide as flame retardant; distearyl thiodipropionate (Irganox PS802)	Xu, A, Roland, S, Colin, X, Polym. Deg. Stab., 181, 109276, 2020.

PEX silane-crosslinkable polyethylene

PARAMETER	UNIT	VALUE	REFERENCES
Applications	-	engineering systems, gas distribution, geothermal and district heating, industrial, offshore and onshore, plumbing and heating, pressure pipe, signal and power cables, wire & cable	
Outstanding properties	-	high and low working temperatures, chemical resistance, abrasion resistance, memory effect, thermal and aging stability	Wu, T-S, Plastics Additives Compounding, 9, 6, 40-3, 2007.
ANALYSIS			
FTIR (wavenumber-assignment)	cm^{-1}/-	peak area from 1200-1000 correlates with gel content	Giacobbi, E; Miglioli, C, US Patent Application 20070117933, Solvay, 2007.

PF phenol-formaldehyde resin

PARAMETER	UNIT	VALUE	REFERENCES
GENERAL			
Common name	-	phenol-formaldehyde resin	
CAS name	-	phenol, polymer with formaldehyde	
Acronym	-	PF	
CAS number	-	9003-35-4; 25104-55-6	
HISTORY			
Person to discover	-	Leo Baekeland	
Date	-	1907	
Details	-	invention of thermosetting phenol formaldehyde resin called Bakelite	
SYNTHESIS			
Monomer(s) structure	-	$\bigcirc\!\!-OH \qquad H_2C{=}O$	
Monomer(s) CAS number(s)	-	108-95-2; 50-00-0	
Monomer(s) molecular weight(s)	dalton, g/mol, amu	94.11; 30.03	
Monomer(s) expected purity(ies)	%	90-99; 98	
Monomer ratio	-	1.5 (resols) to <1 (novolacs)	
Method of synthesis	-	step-growth polymerization	Gogotov, A F; Varfolomeev, A A; Sinegibskaya, A D; Kanitskaya, L V; Rokhin, A V, Russ, J. Appl. Chem., 82, 6, 1002-5, 2009.
Temperature of polymerization	$^{\circ}C$	90	
Catalyst	-	NaOH; acid in novolacs	
Mass average molecular weight, M_w	dalton, g/mol, amu	620-6,600	
Polydispersity, M_w/M_n	-	1.41-1.72	
Molecular cross-sectional area, calculated	$cm^2 \times 10^{-16}$	19.2	
PHYSICAL PROPERTIES			
Density at 20°C	g cm^{-3}	1.24-1.32	
Bulk density at 20°C	g cm^{-3}	0.58-0.88	
Refractive index, 20°C	-	1.7000	
Melting temperature, DSC	$^{\circ}C$	90-107	
Decomposition temperature	$^{\circ}C$	420 (hollow microspheres)	Liu, X; Li, H; Ma, T; Li, K, Polym. Int., 58, 465-68, 2009.
Thermal expansion coefficient, 23-80°C	$10^{-4}\ ^{\circ}C^{-1}$	0.17-0.22	
Glass transition temperature	$^{\circ}C$	200 (hollow microspheres)	Liu, X; Li, H; Ma, T; Li, K, Polym. Int., 58, 465-68, 2009.
Heat deflection temperature at 1.8 MPa	$^{\circ}C$	204	
Dielectric constant at 100 Hz/1 MHz	-	3.5-5.0	
Relative permittivity at 100 Hz	-	12	
Volume resistivity	ohm-m	1E9-1E10	

430 **HANDBOOK OF POLYMERS** 3rd Edition, Copyrights 2022; ChemTec Publishing

PF phenol-formaldehyde resin

PARAMETER	UNIT	VALUE	REFERENCES
Surface resistivity	ohm	1E10-1E11	
Electric strength K20/P50, d=0.60.8 mm	kV mm^{-1}	16	
Speed of sound	m s^{-1}	26.5	
Acoustic impedance	-	3.63	

MECHANICAL & RHEOLOGICAL PROPERTIES

PARAMETER	UNIT	VALUE	REFERENCES
Tensile strength	MPa	34.4-62	
Tensile modulus	MPa	7,580	
Tensile stress at yield	MPa	110	
Elongation	%	1.5-2.2	
Tensile yield strain	%	1.1	
Flexural strength	MPa	75-120	
Flexural modulus	MPa	6,500-9,100	
Compressive strength	MPa	69-200	
Charpy impact strength, unnotched, 23°C	kJ m^{-2}	6-7	
Charpy impact strength, notched, 23°C	kJ m^{-2}	1.3-1.5	
Izod impact strength, notched, 23°C	J m^{-1}	870	
Poisson's ratio	-	0.402	
Rockwell hardness	-	M93-128	
Shrinkage	%	0.35-0.9	
Water absorption, 24h at 23°C	%	0.1-0.4	

CHEMICAL RESISTANCE

PARAMETER	UNIT	VALUE	REFERENCES
Acid dilute/concentrated	-	good	
Alcohols	-	very good	
Alkalis	-	good	
Aliphatic hydrocarbons	-	very good	
Aromatic hydrocarbons	-	good	
Esters	-	good	
Greases & oils	-	good	
Halogenated hydrocarbons	-	very good	
Ketones	-	good	

FLAMMABILITY

PARAMETER	UNIT	VALUE	REFERENCES
Ignition temperature	^{o}C	520	
Autoignition temperature	^{o}C	570	
Limiting oxygen index	% O$_2$	29-66	
Volatile products of combustion	-	phenol, formaldehyde, CO, CO$_2$, and more	Wang, J; Jiang, H; Jiang, N, Thermochim. Acta, 496, 136-42, 2009.

BIODEGRADATION

PARAMETER	UNIT	VALUE	REFERENCES
Colonized products		ceiling tiles, composites, fireboard	

PF phenol-formaldehyde resin

PARAMETER	UNIT	VALUE	REFERENCES
Typical biodegradants	-	fungal growth occurs when materials are wet; *Pseudomonas cepacia* and *Bacillus brevis* are capable of utilizing phenols	Arutchelvan, V; Kanakasabai, V; Nagarajan, S; Muralikrishnan, V, J. Hazardous Mater., B127, 238-443, 2005.
Stabilizers	-	4,5-dichloro-2-n-octyl-3(2H)-isothiazolone	

TOXICITY			
NFPA: Health, Flammability, Reactivity rating	-	0/0/0	
Carcinogenic effect	-	not listed by ACGIH, NIOSH, NTP (it may contain small amount of formaldehyde which is carcinogenic)	
Oral rat, LD$_{50}$	mg kg^{-1}	8,394	
Free formaldehyde content	%	0.09-0.14	Wei, Q, Wang, W-H, Int. J. Adhesion Adhesives, 84, 166-72, 2018.

PROCESSING			
Typical processing methods	-	casting, compounding, molding, reactive processing	
Curing temperature	°C	128-130	Wei, Q, Wang, W-H, Int. J. Adhesion Adhesives, 84, 166-72, 2018.
Additives used in final products	-	silane coupling agent KH550 (γ-aminopropyltriethoxysilane)	Wei, Q, Wang, W-H, Int. J. Adhesion Adhesives, 84, 166-72, 2018.
Applications	-	abrasive products, adhesives, binder, electrodes, membranes, microspheres, particle boards, thermal insulation	

BLENDS			
Suitable polymers	-	PA-6	

ANALYSIS			
FTIR (wavenumber-assignment)	cm^{-1}/-	OH – 3389-3400, 1370; C=C aromatic ring – 1633, 1552, 1513; C-C-OH – 1235; C-O – 1154, 1058	Poljansek, I; Krajnc, M, Acta Chim. Slov., 52, 238-44, 2005.

PFA perfluoroalkoxy resin

PARAMETER	UNIT	VALUE	REFERENCES

GENERAL

Common name	-	perfluoroalkoxy resin; poly[tetrafluoroethylene-co-perfluoro(alkyl vinyl ether)]	
CAS name	-	propane, 1,1,1,2,2,3,3-heptafluoro-3-[(1,2,2-trifluoroethenyl)oxy]-, polymer with 1,1,2,2-tetrafluoroethene	
Acronym	-	PFA	
CAS number	-	26655-00-5	
Formula		$\left[\begin{array}{c}CF_2CF_2CFCF_2 \\ \quad\quad\ \ | \\ \quad\quad OCF_2CF_2CF_3\end{array}\right]_n$	

HISTORY

Person to discover	-	Carlson, D P	
Date	-	1970; 1973 (commercialization)	
Details	-	copolymerization in halogenated solvents	Ebnesajjad, S, Fluoroplastics. Vol. 2. Melt Processible Fluoroplastics, William Andrew, 2003.

SYNTHESIS

Monomer(s) structure	-	$F_3CCF_2CF_2OCF{=}CF_2 \qquad F_2C{=}CF_2$	
Monomer(s) CAS number(s)	-	1623-05-8; 116-14-3	
Monomer(s) molecular weight(s)	dalton, g/mol, amu	266.04; 100.02	
Monomer(s) expected purity(ies)	%	98	
Formulation example	-	monomers, water, water-soluble initiator, surfactant	
Method of synthesis	-	aqueous polymerization	
Temperature of polymerization	^{o}C	70-95	
Time of polymerization	h	3-6	
Pressure of polymerization	MPa	1.7-2.4	
Drying temperature	^{o}C	125-150	

STRUCTURE

Crystallinity	%	45-70	
Chain conformation	-	13/6 helix	

COMMERCIAL POLYMERS

Some manufacturers	-	3M; Daikin; Solvay	
Trade names	-	Dyneon; Neoflon; Hyflon	

PHYSICAL PROPERTIES

Density at 20°C	g cm^{-3}	2.12-2.18	
Bulk density at 20°C	g cm^{-3}	0.25-0.6	
Color	-	translucent to white	
Refractive index, 20°C	-	1.34-1.35	
Transmittance	%	71-91 (UV); 91-96 (vis); 96-98 (IR)	Ebnesajjad, S, Fluoroplastics. Vol. 2. Melt Processible Fluoroplastics, William Andrew, 2003.

PFA perfluoroalkoxy resin

PARAMETER	UNIT	VALUE	REFERENCES
Haze	%	4	
Odor	-	odorless	
Melting temperature, DSC	°C	285-315	
Decomposition temperature	°C	270; 400	
Thermal expansion coefficient, 23-80°C	°C^{-1}	1.2-2.1E-4	
Thermal conductivity, melt	W m^{-1} K^{-1}	0.19-0.195	
Glass transition temperature	°C	2-15	
Specific heat capacity	J K^{-1} kg^{-1}	1,172	
Heat of fusion	J g^{-1}	17.1	Zhong, X; Yu, L; Zhao, W; Zhang, Y; Sun, J, Polym. Deg. Stab., 40, 115-16, 1993.
Maximum service temperature	°C	-240 to 260	
Long term service temperature	°C	250-260	
Heat deflection temperature at 0.45 MPa	°C	72	
Heat deflection temperature at 1.8 MPa	°C	49; 85 (20% glass fiber)	
Surface tension	mN m^{-1}	22.0	Becker, K, Int. Biodet. Biodeg., 41, 93-100, 1998.
Dielectric constant at 100 Hz/1 MHz	-	2.04/2.04	
Dissipation factor at 100 Hz	E-4	2	
Dissipation factor at 1 MHz	E-4	7	
Volume resistivity	ohm-m	1E15 to 1E16	
Surface resistivity	ohm	1E15	
Electric strength K20/P50, d=0.60.8 mm	kV mm^{-1}	260	
Coefficient of friction	-	0.1-0.3 (kinetic, PFA/steel); 0.13-0.16 (20% glass fiber)	
Permeability to nitrogen, 25°C	cm^3 m^{-2} 24 h^{-1} atm^{-1}	2,000	
Permeability to oxygen, 25°C	cm^3 m^{-2} 24 h^{-1} atm^{-1}	6,700	
Permeability to water vapor, 25°C	g m^{-2} day^{-1}	2	
Diffusion coefficient of water vapor	cm^2 s^{-1} x10^7	4.05 (20°C); 8.5 (90°C)	Hansen, C M, Prog. Org. Coat., 42, 167-78, 2001.
Contact angle of water, 20°C	degree	122	
MECHANICAL & RHEOLOGICAL PROPERTIES			
Tensile strength	MPa	20-31.7; 34 (20% glass fiber)	
Tensile modulus	MPa	276	
Tensile stress at yield	MPa	12-15	
Elongation	%	150-360; 6 (20% glass fiber)	
Flexural modulus	MPa	550-740; 1,145 (20% glass fiber)	
Elastic modulus	MPa	480	

PFA perfluoroalkoxy resin

PARAMETER	UNIT	VALUE	REFERENCES
Tear strength	MPa	12	Drobny, J G, Applications of Fluoropolymer Films, William Andrew, 2020, pp.153-5.
Shore D hardness	-	55-65	
Shrinkage	%	1 (150°C)	
Water absorption, equilibrium in water at 23°C	%	<0.02 to <0.03	

CHEMICAL RESISTANCE

Acid dilute/concentrated	-	resistant	
Alcohols	-	resistant	
Alkalis	-	resistant	
Aliphatic hydrocarbons	-	resistant	
Aromatic hydrocarbons	-	resistant	
Esters	-	resistant	
Greases & oils	-	resistant	
Halogenated hydrocarbons	-	resistant	
Ketones	-	resistant	
Other	-	reacts with fluorine and molten alkalis	

FLAMMABILITY

Ignition temperature	°C	530-550	
Autoignition temperature	°C	530-560	
Limiting oxygen index	% O_2	>95	
Heat of combustion	J g^{-1}	6,110	
Volatile products of combustion	-	CO, CO_2, HF, smoke	
UL rating	-	V-0	

TOXICITY

HMIS: Health, Flammability, Reactivity rating	-	2/1/0	
Carcinogenic effect	-	not listed by ACGIH, NIOSH, NTP	
OSHA	mg m^{-3}	5 (respirable), 15 (total)	
Oral rat, LD_{50}	mg kg^{-1}	>11,000	

ENVIRONMENTAL IMPACT

Aquatic toxicity, *Daphnia magna*, LC_{50}, 48 h	mg l^{-1}	>10,000	
Aquatic toxicity, *Fathead minnow*, LC_{50}, 48 h	mg l^{-1}	>10,000	

PROCESSING

Typical processing methods	-	coating, extrusion, injection molding	
Additives used in final products	-	Fillers: calcium hydroxide, carbon black, graphite, magnesium oxide, metal particles, molybdenum disulfide, PTFE	
Applications	-	automotive weather seals for doors and windows, coating for hostile environments, column packing, filtration, fittings, marine coatings, pipes, pump, silicon wafer carriers, tubing, wear resistant products	

HANDBOOK OF POLYMERS 3rd Edition, Copyrights 2022; ChemTec Publishing

PFA perfluoroalkoxy resin

PARAMETER	UNIT	VALUE	REFERENCES
Outstanding properties	-	weather resistance, thermal resistance, chemical resistance	
ANALYSIS			
FTIR (wavenumber-assignment)	cm^{-1}/-	COF – 1884; C=O – 1813, 1775; CONH$_2$ – 1768, 1587	Pianca, M; Barchiesi, E; Esposto, G; Radice, S, J. Fluorine Chem., 95, 71-84, 1999.

PFI perfluorinated ionomer

PARAMETER	UNIT	VALUE	REFERENCES
GENERAL			
Common name	-	perfluorinated ionomer	
CAS name	-	2-[1-difluoro[(trifluoroethenyl)oxy]methy]-1,2,2,-tetrafluoroethoxy]-1,1,2,2-tetrafluoro-ethanesulfonic acid, polymer with tetrafluoroethylene	
Acronym	-	PFI	
CAS number	-	31175-20-9; 66796-30-3	
HISTORY			
Person to discover	-	Walther Grot	
Date	-	1968	
Details	-	PTFE modification	
SYNTHESIS			
Monomer(s) structure	-	$CF_2=CF_2$; perfluorosulfonic acid	
Mass average molecular weight, M_w	dalton, g/mol, amu	100,000-1,000,000	Heitner-Wirguin, C, J. Membrane Sci., 120, 1-33, 1996.
STRUCTURE			
Crystallinity	%	14-23 (nonionic and carboxylated forms); 3-12 (sulfonated form); with increase of equivalent weight from 1100 to 1500, the degree of crystallinity increases from 12 to 22%	Mauritz, K A; Moore, R B, Chem. Rev., 104, 4535-85, 2004.
Crystallite size	nm	3.9-4.4	Mauritz, K A; Moore, R B, Chem. Rev., 104, 4535-85, 2004.
Spacing between crystallites	nm	1-3.3	Mauritz, K A; Moore, R B, Chem. Rev., 104, 4535-85, 2004.
Chain conformation	-	zigzag (34 carbon atoms)	Mauritz, K A; Moore, R B, Chem. Rev., 104, 4535-85, 2004.
COMMERCIAL POLYMERS			
Some manufacturers	-	Chemours; Solvay	
Trade names	-	Nafion; Hyflon	
PHYSICAL PROPERTIES			
Density at 20°C	g cm^{-3}	1.97-1.98	
Color	-	white	
Odor	-	odorless	
Decomposition temperature	°C	280	
Glass transition temperature	°C	110-165	Ghielmi, A; Vaccarono, P; Troglia, C; Arcella, V, J. Power Sources, 145, 1008-15, 2005.
Available acid capacity	meq/g	0.9-0.92	
Maximum service temperature	°C	240	
Long term service temperature	°C	175	
Conductivity	S cm^{-1}	0.13	Liu, X, Luo, X, Chen, X, Zou, S, Dong, D, J. Electroanal. Chem., 871, 114283, 2020.
Diffusion coefficient of water vapor	cm^2 s^{-1} x10^7	1-7.5	Hallinan, D T; De Angelis, M G; Baschetti, M C; Sarti, G S; Elabd, Y A, Macromolecules, 43, 4667-78, 2010.

PFI perfluorinated ionomer

PARAMETER	UNIT	VALUE	REFERENCES
MECHANICAL & RHEOLOGICAL PROPERTIES			
Tensile strength	MPa	23-43	
Tensile modulus	MPa	64-249 (depends on moisture content and temperature)	
Tensile stress at yield	MPa	14-18	
Elongation	%	250-350	
Young's modulus	MPa	250-340	
Tear strength	g mm^{-1}	3,000-6,000	
Water absorption, equilibrium in water at 23°C	%	38	
CHEMICAL RESISTANCE			
Acid dilute/concentrated	-	very good	
Alcohols	-	poor	
Alkalis	-	very good	
Aliphatic hydrocarbons	-	very good	
Aromatic hydrocarbons	-	very good	
Esters	-	very good	
Greases & oils	-	very good	
Halogenated hydrocarbons	-	very good	
Ketones	-	very good	
Other	-	only sodium attacks PFI	
FLAMMABILITY			
Limiting oxygen index	% O$_2$	95	
Heat of combustion	J g^{-1}	5,800	
Volatile products of combustion	-	CO, CO$_2$, HF, SO$_2$, COF$_2$, COS	
UL 94 rating	-	V-0	
WEATHER STABILITY			
Products of degradation	-	oxygen radicals in side chain and perfluorinated radicals obtained by photo-Fenton reactions	Bosnjakovic, A; Kadirov, M K; Schlick, S, Res. Chem. Intermed., 33, 8-9, 677-87, 2007.
TOXICITY			
HMIS: Health, Flammability, Reactivity rating	-	1/1/1	
Carcinogenic effect	-	not listed by ACGIH, NIOSH, NTP	
Oral rat, LD$_{50}$	mg kg^{-1}	20,000	
PROCESSING			
Applications	-	fuel cells, ion-exchange membranes, metal-free-catalyst cathode for anion exchange membrane fuel cells, moisture regulator, proton-exchange membranes, super-acid catalyst	
Outstanding properties	-	chemical resistance, ion conductive, thermal stability	

PFI perfluorinated ionomer

PARAMETER	UNIT	VALUE	REFERENCES
ANALYSIS			
FTIR (wavenumber-assignment)	cm^{-1}/-	1647 – C–N bond in the –SO$_2$NH–CH$_2$–, ~1198 and ~1143 - CF$_2$ stretching, SO$_3$ – 1056; C–O–C – 969 (main chain), 982 (side chain), and more	Danilczuk, M; Lin, L; Schlick, S; Hamrock, S J; Schaberg, M S, J. Power Sources, in press, 2011; Liu, X, Luo, X, Chen, X, Zou, S, Dong, D, J. Electroanal. Chem., 871, 114283, 2020.
NMR (chemical shifts)	ppm	H NMR: hydrogen ions – 5.5-9; water physically sorbed – 3.5	Nosaka, A Y; Nosaka, Y, J. Power Sources, 180, 733-37, 2008.

PFPE perfluoropolyether

PARAMETER	UNIT	VALUE	REFERENCES
GENERAL			
Common name	-	perfluoropolyether; poly(perfluoropropylene oxide-co-perfluoro-methylene oxide)	
IUPAC name	-	trifluoromethyl-poly[oxy-2-trifluoromethyl-1,1,2-trifluo-roethylene]-poly[oxy-difluoromethylene]-trifluoromethyl ether	
ACS name	-	1-propene, 1,1,2,3,3,3-hexafluoro-, oxidized and polymerized	
Acronym	-	PFPE	
CAS number	-	69991-67-9; 60164-51-4	
HISTORY			
Person to discover	-	Arbogast, F L	Arbogast, F L, US Patent 3,412,148, DuPont, Nov. 19, 1968.
Date	-	1968	
SYNTHESIS			
Monomer(s) structure	-	hexafluoropropylene oxide, C_3F_6O	
Monomer(s) CAS number(s)	-	428-59-1	
Monomer(s) molecular weight(s)	dalton, g/mol, amu	166.02	
Number average molecular weight, M_n	dalton, g/mol, amu	870-2400 (lubricants)	Fan, A, Waltman, R J, Chemical Data Collections, 23, 100269, 2019.
Mass average molecular weight, M_w	dalton, g/mol, amu	870-1210 (vapor phase soldering); 1,500-6,250 (cosmetics); 1200-7,500 (lubricants); 17,500-374,000 (Fomblin polymers)	Sanguineti, A; Guarda, P A; Marchionni, G; Ajroldi, G, Polymer, 36, 19, 3697-3703, 1995.
Polydispersity, M_w/M_n	-	1.11-1.32 (Fomblin polymers)	Sanguineti, A; Guarda, P A; Marchionni, G; Ajroldi, G, Polymer, 36, 19, 3697-3703, 1995.
Polymerization degree (number of monomer units)	-	10-60 (oil)	
Chain-end groups	-	OH (lubricants)	
COMMERCIAL POLYMERS			
Some manufacturers	-	Chemours; Solvay	
Trade names	-	Krytox; Fluorolink, Fomblin, Galden	
PHYSICAL PROPERTIES			
Density at 20°C	g cm^{-3}	1.79-1.92; 2.13-2.20 (Algoflon)	
Color	-	colorless	
Refractive index, 20°C	-	1.293-1.302	
Odor	-	odorless	
Pour point	°C	-25 to -62	
Boiling point	°C	200-260 (MW: 870-1210)	
Decomposition temperature	°C	350 (air); 470 (nitrogen)	
Long term service temperature	°C	175	Baik, J-H, Kim, D-G, Lee, J-H, Kim, S, Hong, D G, Lee, J-C, J. Ind. Eng. Chem., 64, 453-60, 2018.
Thermal expansion coefficient, 25-99°C	10^{-4} °C^{-1}	9.5-10.9	
Thermal conductivity	W m^{-1} K^{-1}	0.0831-0.0934 (38°C); 0.0692-0.0883 (260°C)	

PFPE perfluoropolyether

PARAMETER	UNIT	VALUE	REFERENCES
Specific heat capacity	J K^{-1} kg^{-1}	960-1,000 (23°C); 1,210-1,260 (204°C)	
Heat of vaporization	kJ kg^{-1}	63	
Maximum service temperature	°C	-75 to 350; 250 (fuel cells); 270 (heat transfer fluids)	
Long term service temperature	°C	288 (in the presence of certain metal oxides)	
Surface tension	mN m^{-1}	16-25	Tao, Z; Bhushan, B, Wear, 259, 1352-61, 2005.
Volume resistivity	ohm-m	1E13	
Electric strength K20/P50, d=0.60.8 mm	kV mm^{-1}	15.7	
Coefficient of friction	-	0.2-0.23	Tao, Z; Bhushan, B, Wear, 259, 1352-61, 2005.
Contact angle of water, 20°C	degree	144-147; 85-99 (lubricants)	Gallo Stampino, P; Molina, D; Omati, L; Turri, S; Levi, M; Cristiani, C; Dotelli, G, J. Power Sources, in press, 2011; Tao, Z; Bhushan, B, Wear, 259, 1352-61, 2005.

CHEMICAL RESISTANCE

Alcohols	-	insoluble	
Aromatic hydrocarbons	-	insoluble	
Greases & oils	-	insoluble	
Halogenated hydrocarbons	-	insoluble	
Ketones	-	insoluble	
Good solvent	-	fluorinated solvents	

FLAMMABILITY

Ignition temperature	°C	400 (in gaseous oxygen at pressure of 13 MPa)	

TOXICITY

NFPA: Health, Flammability, Reactivity rating	-	1/0/0	
Carcinogenic effect	-	not listed by ACGIH, NIOSH, NTP	
Mutagenic effect	-	did not cause genetic damage in cultured bacterial cells	
Oral rat, LD$_{50}$	mg kg^{-1}	15,000; >37,400	
Skin rabbit, LD$_{50}$	mg kg^{-1}	non-irritating; >17,000	

ENVIRONMENTAL IMPACT

Aquatic toxicity, *Daphnia magna*, LC$_{50}$, 48 h	mg l^{-1}	no effect; >1,000	
Aquatic toxicity, *Rainbow trout*, LC$_{50}$, 48 h	mg l^{-1}	no effect; >1,000	

PROCESSING

Additives used in final products	-	pigments, mineral powders, emulsifiers, thickeners, surfactants	

PFPE perfluoropolyether

PARAMETER	UNIT	VALUE	REFERENCES
Applications	-	cosmetics (emollients, hair conditioners, hand and body care, lotions, shaving products, skin protectants and feel improvers, water and oil repellents); ferrofluids, fuel cells; lubricants (anti-lock brakes, bearings working at high temperatures, gasoline tank floats,lithium batteries, missile catapult system, oxygen and chlorine valves, space rockets), oxygen selective membranes, proton exchange membrane fuel cells, solar industry; vapor phase soldering)	
Outstanding properties	-	biological inertness, chemical resistance, high solubility of respiratory gases, thermal and electrical resistance	
BLENDS			
Compatible polymers	-	compatible with most common elastomers and plastics	

PGA poly(glycolic acid)

PARAMETER	UNIT	VALUE	REFERENCES
GENERAL			
Common name	-	poly(glycolic acid)	
IUPAC name	-	poly[ox(1-oxo-1,2-ethanediyl)]	
CAS name	-	acetic acid, 2-hydroxy-, homopolymer	
Acronym	-	PGA	
CAS number	-	26124-68-5; 26009-03-0	
Linear formula			

$$\left[OCCH_2O \right]_n$$ with O double-bonded above the first carbon

HISTORY			
Person to discover	-	Norton Higgins, DuPont; Edward Schmitt and Rocco Polistina, American Cyanamid	
Date	-	1954; 1960	
Details	-	Higgins patented production process and in 1960 it was first used for production chirurgical gut (sutures), known as Dextron	

SYNTHESIS			
Monomer(s) structure	-		

(glycolide ring structure; $HOCCH_2OH$ with O double-bonded)

Monomer(s) CAS number(s)	-	502-97-6; 79-14--1	
Monomer(s) molecular weight(s)	dalton, g/mol, amu	116.07	
Monomer ratio	-	100%	
Method of synthesis	-	several methods can be used, including polycondensation of glycolic acid, ring-opening polymerization of glycolide, or reaction of formaldehyde with carbon monoxide in the presence of acid	Takahashi, K; Taniguchi, I; Miyamoto, M; Kimura, Y, Polymer, 41, 8725-28, 2000.
Catalyst	-	$SnCl_2$	Saigusa, K, Saijo, H, Yamazaki, M, Takarada, W, Kikutani, T, Polym. Deg. Stab., 172, 109054, 2020.
Number average molecular weight, M_n	dalton, g/mol, amu	60,000-79,000	Saigusa, K, Saijo, H, Yamazaki, M, Takarada, W, Kikutani, T, Polym. Deg. Stab., 172, 109054, 2020.
Mass average molecular weight, M_w	dalton, g/mol, amu	100,000-200,000	
Polydispersity, M_w/M_n	-	1.7-2.3	Takahashi, K; Taniguchi, I; Miyamoto, M; Kimura, Y, Polymer, 41, 8725-28, 2000.

STRUCTURE			
Crystallinity	%	33-55	Sekine, S; Yamauchi, K; Aoki, A; Asakura, T, Polymer, 50, 6083-90, 2009.
Cell type (lattice)	-	orthorhombic	Chatani, Y; Suehiro, K; Okita, Y; Tadokoro, H; Chujo, K, Makromol. Chem., 113, 215, 1968.
Cell dimensions	nm	a:b:c=0.522:0.619:0.702	Sekine, S; Yamauchi, K; Aoki, A; Asakura, T, Polymer, 50, 6083-90, 2009.

PGA poly(glycolic acid)

PARAMETER	UNIT	VALUE	REFERENCES
Number of chains per unit cell	-	2	
Crystallite size	nm	27.5, 6.8, 2.1 (in three crystalline directions)	Marega, C; Marigo, A; Zannetti, R; Paganetto, G, Eur. Polym. J, 28, 12, 1485-86, 1992.
Chain conformation	-	planar zig-zag	

COMMERCIAL POLYMERS

Some manufacturers	-	Kureha	
Trade names	-	Kuredux	

PHYSICAL PROPERTIES

Density at 20°C	g cm^{-3}	1.46-1.74; 1.70 (crystalline); 1.5 (amorphous)	Nakafuku, C; Yoshimura, H, Polymer, 45, 3583-85, 2004.
Refractive index, 20°C	-	1.45-1.51	
Birefringence	-	1.556, 1.466	
Haze	%	<1	
Melting temperature, DSC	°C	200-231	
Storage temperature	°C	2-8	
Thermal expansion coefficient, 23-80°C	10^{-4} °C^{-1}	0.54	
Thermal conductivity, melt	W m^{-1} K^{-1}	0.35	
Glass transition temperature	°C	35-53	
Specific heat capacity	J K^{-1} kg^{-1}	1,120	
Heat of fusion	J g^{-1}	183.2	Nakafuku, C; Yoshimura, H, Polymer, 45, 3583-85, 2004.
Enthalpy of melting	J g^{-1}	50.77	Shum, A W T; Mak, A F T, Polym. Deg. Stab., 81, 141-9, 2003.
Hansen solubility parameters, dD, dP, dH	(J cm-3)^{0,5}	17.70, 6.21, 12.50	Agrawal, A; Saran, A D; Rath, S S; Khanna, A, Polymer, 45, 8603-12, 2004.
Molar volume	(J cm-3)^{0,5}	1.92	Agrawal, A; Saran, A D; Rath, S S; Khanna, A, Polymer, 45, 8603-12, 2004.

MECHANICAL & RHEOLOGICAL PROPERTIES

Tensile strength	MPa	61-72; 1,100 (highly oriented fibers)	de Oca, H M; Farrar, D F; Ward, I M, Acta Biomater., 7, 1535-41, 2011.
Elongation	%	5-20	
Flexural strength	MPa	178	
Young's modulus	MPa	6,080-7,180	
Melt viscosity, shear rate=1000 s^{-1}	Pa s	360-950	
Melt index, 250°C/2.16 kg	g/10 min	6-22	
Water absorption, equilibrium in water at 23°C	%	28	

CHEMICAL RESISTANCE

Acid dilute/concentrated	-	poor	Saigusa, K, Saijo, H, Yamazaki, M, Takarada, W, Kikutani, T, Polym. Deg. Stab., 172, 109054, 2020.

PGA poly(glycolic acid)

PARAMETER	UNIT	VALUE	REFERENCES
Alkalis	-	poor	
Esters	-	good	
Halogenated hydrocarbons	-	good	
Ketones	-	good	
Good solvent	-	hexafluoroisopropanol	

FLAMMABILITY

Heat of combustion	J g^{-1}	12,000	

BIODEGRADATION

Typical biodegradants	-	decomposes in 6 month at 37°C at pH=9.0	

TOXICITY

NFPA: Health, Flammability, Reactivity rating	-	1/1/0	
Carcinogenic effect	-	not listed by ACGIH, NIOSH, NTP	

PROCESSING

Typical processing methods	-	electrospinning, extrusion	
Processing temperature	$^{\circ}$C	240	
Applications	-	monofilaments, orthopaedics, packaging, suture, wound dressing	
Outstanding properties	-	biodegradable, resorbable	

BLENDS

Suitable polymers	-	PCL, PLA	

ANALYSIS

FTIR (wavenumber-assignment)	cm^{-1}/-	C=O (ester) – 1744; C=O (acetate end group) – 1630; C-O – 1229; C-OH – 1096	Shum, A W T; Mak, A F T, Polym. Deg. Stab., 81, 141-9, 2003; Kister, G; Cassanas, G; Vert, M, Spectrochim. Acta, 53A, 1399-1403, 1999.
Raman (wavenumber-assignment)	cm^{-1}/-	CH$_2$ – 2988; C=O – 1776, 1759; C-O-C – 1165, 1087, 1032; and more for crystalline and amorphous	Kister, G; Cassanas, G; Vert, M, Spectrochim. Acta, 53A, 1399-1403, 1999.
NMR (chemical shifts)	ppm	C NMR: amorphous peak – 61; all *trans* conformation – 62.5; CH$_2$ – 63.5; C=O – 168	Sekine, S; Yamauchi, K; Aoki, A; Asakura, T, Polymer, 50, 6083-90, 2009.
x-ray diffraction peaks	degree	22.5, 29	Marega, C; Marigo, A; Zannetti, R; Paganetto, G, Eur. Polym. J, 28, 12, 1485-86, 1992.

PHB poly(3-hydroxybutyrate)

PARAMETER	UNIT	VALUE	REFERENCES
GENERAL			
Common name	-	poly(3-hydroxybutyrate)	
CAS name	-	butanoic acid, 3-hydroxy-, homopolymer (26063-00-3); poly[oxy(1-methyl-3-oxo-1,3-propanediyl)] (26744-04-7); poly[oxy[(1R)-1-methyl-3-oxo-1,3-propanediyl]] (31759-58-7)	
Acronym	-	PHB	
CAS number	-	26063-00-3; 26744-04-7; 31759-58-7	
Linear formula		$\left[\begin{array}{c} CH_3 \\ | \\ OCHCH_2C \end{array}\right]_n$	
HISTORY			
Person to discover	-	Maurice Lemoigne	
Date	-	1925	
Details	-	isolated and characterized	
SYNTHESIS			
Monomer(s) structure	-		
Monomer(s) CAS number(s)	-	3068-88-0	
Monomer(s) molecular weight(s)	dalton, g/mol, amu	86.09	
Monomer(s) expected purity(ies)	%	98 (glycerol)	Posada, J A; Naranjo, J M; Lopez, J A; Higuita, J C; Cardona, C A, Process Biochem., 46, 310-17, 2011.
Method of synthesis	-	produced by biosynthesis by bacteria and plants in response to physiological stress; also *Haloarcula sp.* can be used to produce PHB from petrochemical wastewater; economical high scale production from glycerol is feasible	Taran, M, J. Hazardous Mater., 188, 26-28, 2011; Posada, J A; Naranjo, J M; Lopez, J A; Higuita, J C; Cardona, C A, Process Biochem., 46, 310-17, 2011.
Temperature of polymerization	°C	139 (sterilization); 35 (fermentation)	Posada, J A; Naranjo, J M; Lopez, J A; Higuita, J C; Cardona, C A, Process Biochem., 46, 310-17, 2011.
Time of polymerization	h	21-22.5	
Pressure of polymerization	Pa	atmospheric	
Number average molecular weight, M_n	dalton, g/mol, amu	22,000-768,000	
Mass average molecular weight, M_w	dalton, g/mol, amu	12,000 (*Eubacteria, Archaebacteria,* and *Eukaryotes*); 200,000-3,000,000 (microbial cell cytoplasm); ultra high molecular weight >3,000,000 (*Escherichia coli*)	Bastioli, C, Handbook of Biodegradable Polymers, Rapra, 2005.
Polydispersity, M_w/M_n	-	1.95-3.7, 1.17 (Hydal)	Oliveira, F C; Dias, M L; Castilho, L R; Freire, D M G, Biosource Technol., 98, 633-38, 2007; Kovalcik, A, Sangroniz, L, Kalina, M. Skopalova, K, Müller, A J, Int. J. Niol. Macromol., 161, 364-76, 2020.
Polymerization degree (number of monomer units)	-	120-200 (low molecular weight)	Bastioli, C, Handbook of Biodegradable Polymers, Rapra, 2005.

PHB poly(3-hydroxybutyrate)

PARAMETER	UNIT	VALUE	REFERENCES
STRUCTURE			
Crystallinity	%	30-80	
Cell type (lattice)	-	orthorhombic	Cornibert, J; Marchessault, R H, J. Mol. Biol. 71, 735, 1972.
Cell dimensions	nm	a:b:c=0.576:1.320:0.596	
Crystallization temperature	°C	41.1	Kovalcik, A, Sangroniz, L, Kalina, M. Skopalova, K, Müller, A J, Int. J. Niol. Macromol., 161, 364-76, 2020.
Unit cell angles	degree	90	
Rapid crystallization temperature	°C	90-110	
Lamellar diameter	nm	13.1-16.4	Perret, E, Reifler, F A, Gooneie, A, Chen, K, Hufenus, R, Data in Brief, 31, 105675, 2020.
COMMERCIAL POLYMERS			
Some manufacturers	-	Mirel Bioplastics; PHB Industrial	
Trade names	-	Mirel; Biocycle	
PHYSICAL PROPERTIES			
Density at 20°C	g cm^{-3}	1.17-1.25; 1.177 (amorphous); 1.262 (crystalline); 1.2 (fiber)	
Color	-	white to yellow	
Odor	-	mild	
Melting temperature, DSC	°C	166-185; 163.7-169.3	Bastioli, C, Handbook of Biodegradable Polymers, Rapra, 2005; Kovalcik, A, Sangroniz, L, Kalina, M. Skopalova, K, Müller, A J, Int. J. Niol. Macromol., 161, 364-76, 2020.
Storage temperature temperature	°C	25	
Glass transition temperature	°C	-4 to +2.4; -10.2	-; Kovalcik, A, Sangroniz, L, Kalina, M. Skopalova, K, Müller, A J, Int. J. Niol. Macromol., 161, 364-76, 2020.
Maximum service temperature	°C	130	
Vicat temperature VST/B/50	°C	53-96	
Enthalpy of fusion	J g^{-1}	93.56; 66.9 (enthalpy of melting)	Suttiwijitpukdee, N; Sato, H; Zhang, J; Hashimoto, T; Ozaki, Y, Polymer, 52, 461-71, 2011.
Hildebrand solubility parameter	MPa$^{0.5}$	19.2	
Dielectric constant at 100 Hz/1 MHz	-	-/3	
Volume resistivity	ohm-m	1E14	
Contact angle of water, 20°C	degree	104	Brunetti, L, Degli Esposti, M, Morselli, D, Boccaccini, A R, Liverani, L, Mater. Lett., 278, 128389, 2020.
Avrami parameters	n/k	2.2-41/0.0001/0.0544	Kovalcik, A, Sangroniz, L, Kalina, M. Skopalova, K, Müller, A J, Int. J. Niol. Macromol., 161, 364-76, 2020.

PHB poly(3-hydroxybutyrate)

PARAMETER	UNIT	VALUE	REFERENCES
MECHANICAL & RHEOLOGICAL PROPERTIES			
Tensile strength	MPa	40-62; 105-182 (fiber)	Bastioli, C, Handbook of Biodegradable Polymers, Rapra, 2005; Perret, E, Reifler, F A, Gooneie, A, Chen, K, Hufenus, R, Data in Brief, 31, 105675, 2020.
Tensile creep modulus, 1000 h, elongation 0.5 max	MPa	10-27	
Elongation	%	5-58; 22-91	Perret, E, Reifler, F A, Gooneie, A, Chen, K, Hufenus, R, Data in Brief, 31, 105675, 2020.
Flexural modulus	MPa	1,000-2,000	
Young's modulus	MPa	3,500	
Thoughness	MPa	27-57	Perret, E, Reifler, F A, Gooneie, A, Chen, K, Hufenus, R, Data in Brief, 31, 105675, 2020.
Melt index, 230ºC/3.8 kg	g/10 min	5-100	
CHEMICAL RESISTANCE			
Acid dilute/concentrated	-	non-resistant	
Alcohols	-	resistant	
Esters	-	non-resistant	
Halogenated hydrocarbons	-	non-resistant	
Good solvent	-	chloroform, dichloroacetic acid, 1,2-dichloroethane, DMF, ethylacetoacetate, glacial acetic acid, trichloroethyelene	
Non-solvent	-	acetone, n-butanol, carbon tetrachloride, ethanol, ethyl acetate, methanol	
WEATHER STABILITY			
Depth of UV penetration	μm	>100	Sadi, R K; Fechine, G J M; Demarquette, N R, Polym. Deg. Stab., 95, 2318-27, 2010.
BIODEGRADATION			
Typical biodegradants	-	enzymes, animal tissue active components, basic environment; biodegradation is hindered by previous UV degradation	Artsis, M I; Bonartsev, A P; Iordanskii, A L; Bonartseva, G A; Zaikov, G E, Mol. Cryst. Liq. Crys, 523, 21-49, 2010; Sadi, R K; Fechine, G J M; Demarquette, N R, Polym. Deg. Stab., 95, 2318-27, 2010.
TOXICITY			
NFPA: Health, Flammability, Reactivity rating	-	0/1/0	
Carcinogenic effect	-	not listed by ACGIH, NIOSH, NTP	
ENVIRONMENTAL IMPACT			
Cradle to grave non-renewable energy use	MJ/kg	67.0-113.7	Harding, K G; Dennis, J S; von Blottnitz, H; Harrison, S T L, J. Biotechnol., 130, 57-66, 2007.
PROCESSING			
Typical processing methods	-	film, injection molding	
Processing temperature	ºC	160 (injection)	

PHB poly(3-hydroxybutyrate)

PARAMETER	UNIT	VALUE	REFERENCES
Applications	-	biodegradable plastics, medical (surgical film manufactured by Tepha, surgical mesh, sutures), packaging, pharmaceutical (controlled drug release)	
Outstanding properties	-	biodegradable, sustainable	

BLENDS

Suitable polymers	-	C, CA, cellulose ester, PCL, PEG, PEO, PHA, PHBV, PLA, PVAc, PVF	

ANALYSIS

FTIR (wavenumber-assignment)	cm^{-1}/-	C–O stretching – 1053, 1130, 1181; C–C–O stretching – 1276; CH_3 symmetric deformation – 1378; C=O (free, amorphous) – 1747; C=O (intra, crystal) – 1723	Suttiwijitpukdee, N; Sato, H; Zhang, J; Hashimoto, T; Ozaki, Y, Polymer, 52, 461-71, 2011.
Raman (wavenumber-assignment)	cm^{-1}/-	C=O – 1740, 1725; helical structure bands – 3009, 2998, 1725, 1402, 1220; disordered domain bands – 2990, 2938, 2881, 1740, 1453; and much more in ref.	Izumi, C M S; Temperini, M L A, Vibrational Spectroscopy, 54, 127-32, 2010.
NMR (chemical shifts)	ppm	H NMR: methyl – 1.268; methylene – 2.430-2.620; methylidene – 5.232	Zhang, X; Wei, C; He, Q; Ren, Y, J. Env. Sci., 22, 8, 1267-72, 2010.
x-ray diffraction peaks	degree	13.4, 16.1, 16.7, 19.9, 21.7, 22.5, 30.3	Suttiwijitpukdee, N; Sato, H; Zhang, J; Hashimoto, T; Ozaki, Y, Polymer, 52, 461-71, 2011.

PHBV poly(3-hydroxybutyrate-co-3-hydroxyvalerate)

PARAMETER	UNIT	VALUE	REFERENCES
GENERAL			
Common name	-	poly(3-hydroxybutyrate-co-3-hydroxyvalerate)	
CAS number	-	80181-31-3	
Acronym	-	PHBV	
HISTORY			
Date	-	1983	
Details	-	Imperial Chemical Industries (first manufacturer)	
SYNTHESIS			
3-Hydroxyvalerate content	%	2-80; 15.23	Scheithauer, E C; Li, W; Ding, Y; Harhaus, L; Roether, J A; Boccaccini, A R, Mater. Lett., 158, 66-9, 2015; Berezina, N; Yada, B, New Biol., in press, 2015; Rivera-Briso, A L, Aachmann, F L, Moreno-Manzano, V, Serrano-Aroca, A, Int. J. Biol. Macromol., 143, 1000-8, 2020.
Method of synthesis	-	biosynthesis; atom transfer radical polymerization, RAFT	Tebaldi, M L, Chaves Maia, A L, Poletto, F, de Andrade, F V, Ferreira Soares, D C, J. Drug Delivery Sci. Technol., 51, 115-26, 2019.
Mass average molecular weight, M_w	dalton, g/mol, amu	127,000-590,000	Jost, V; Langowski, H-C, Eur. Polym. J., 68, 302-12, 2015; Arcos-Hernandez, M V; Laycock, B; Donose, B C; Pratt, S; Halley, P; Al-Luaibi, S; Werker, A; Lant, P A, Eur. Polym. J., 49, 904-13, 2013.
Polydispersity	-	1.99	de Almeida Neto, G R, Valinhos Barcelos, M, Araújo Ribeiro, M A, Manhaes Folly, M, Sánchez Rodríguez, R J, Mater. Sci. Eng. C, 104, 110004, 2019.
STRUCTURE			
Crystallinity	%	59; 57.1-64	Jost, V; Langowski, H-C, Eur. Polym. J., 68, 302-12, 2015; de Almeida Neto, G R, Valinhos Barcelos, M, Araújo Ribeiro, M A, Manhaes Folly, M, Sánchez Rodríguez, R J, Mater. Sci. Eng. C, 104, 110004, 2019.
Crystallite size	nm	3.7	de Almeida Neto, G R, Valinhos Barcelos, M, Araújo Ribeiro, M A, Manhaes Folly, M, Sánchez Rodríguez, R J, Mater. Sci. Eng. C, 104, 110004, 2019.
COMMERCIAL POLYMERS			
Some manufacturers	-	TianAn; Metabilix	
Trade names	-	Enmat; Biomer L	
PHYSICAL PROPERTIES			
Melting temperature, DSC	°C	165-176; 169 (DSC)	Jost, V; Langowski, H-C, Eur. Polym. J., 68, 302-12, 2015. Daitx, T S; Carli, L N; Crespo, J S; Mauler, R S, Appl. Clay Sci., 115, 157-64, 2015.

450   **HANDBOOK OF POLYMERS** 3rd Edition, Copyrights 2022; ChemTec Publishing

PHBV poly(3-hydroxybutyrate-co-3-hydroxyvalerate)

PARAMETER	UNIT	VALUE	REFERENCES
Glass transition temperature	°C	-20.4 to -12.4	Arcos-Hernandez, M V; Laycock, B; Donose, B C; Pratt, S; Halley, P; Al-Luaibi, S; Werker, A; Lant, P A, Eur. Polym. J., 49, 904-13, 2013.
Permeability to nitrogen, 25°C	barrer	0.033	Cretois, R; Follain, N; Dargent, E; Soulestin, J; Bourbigot, S; Marais, S; Lebrun, L, J. Membrane Sci., 467, 56-66, 20147.
Permeability to oxygen, 25°C	barrer	0.031	Cretois, R; Follain, N; Dargent, E; Soulestin, J; Bourbigot, S; Marais, S; Lebrun, L, J. Membrane Sci., 467, 56-66, 20147.
Diffusion coefficient of water vapor	$cm^2\ s^{-1}$ $\times 10^6$	0.00124	Cretois, R; Follain, N; Dargent, E; Soulestin, J; Bourbigot, S; Marais, S; Lebrun, L, J. Membrane Sci., 467, 56-66, 20147.

MECHANICAL & RHEOLOGICAL PROPERTIES

PARAMETER	UNIT	VALUE	REFERENCES
Tensile strength	MPa	31.3-37.8	Jost, V; Langowski, H-C, Eur. Polym. J., 68, 302-12, 2015. Daitx, T S; Carli, L N; Crespo, J S; Mauler, R S, Appl. Clay Sci., 115, 157-64, 2015.
Elongation	%	0.8-1.7	
Young's modulus	MPa	3,761-4,583	Jost, V; Langowski, H-C, Eur. Polym. J., 68, 302-12, 2015. Daitx, T S; Carli, L N; Crespo, J S; Mauler, R S, Appl. Clay Sci., 115, 157-64, 2015.
Melt index, 190°C/10 kg	g/10 min	16.3	Masood, F; Yasin, T; Hameed, A, Int. Biodet. Biodeg., 87, 1-8, 2014.

BIODEGRADATION

PARAMETER	UNIT	VALUE	REFERENCES
Typical biodegradants	-	composting conditions	Weng, Y-X; Wang, Y; Wang, X-L; Wang, Y-Z, Polym. Testing, 29, 579-87, 2010.

TOXICITY

PARAMETER	UNIT	VALUE	REFERENCES
HMIS: Health, Flammability, Reactivity rating	-	0/0/0	

PROCESSING

PARAMETER	UNIT	VALUE	REFERENCES
Additives used in final products	-	halloysite	
Applications	-	food packaging, microspheres for drug release, medical implants	
Outstanding properties	-	biodegradability, biocompatibility, thermoplastic behavior	

BLENDS

PARAMETER	UNIT	VALUE	REFERENCES
Suitable polymers	-	EVA, PBAT, PCL, PEO, PLA	Mofokeng, J P; Luyt, A S, Thermochim Acta, 613, 41-53, 2015.

PHEMA poly(2-hydroxyethylmethacrylate)

PARAMETER	UNIT	VALUE	REFERENCES
GENERAL			
Common name	-	poly(2-hydroxyethylmethacrylate)	
Acronym	-	PHEMA	
CAS number	-	25249-16-5	
Formula			
HISTORY			
Person to discover	-	Lim, D	
Date	-	1955	
Details	-	Lim, working for inventor of contact lenses (Otto Wichterle) synthesized PHEMA which was used for contact lenses	
SYNTHESIS			
Monomer(s) structure	-		
Monomer(s) CAS number(s)	-	868-77-9	
Monomer(s) molecular weight(s)	dalton, g/mol, amu	130.1	
Monomer(s) expected purity(ies)	%	99	
Monomer ratio	-	100%	
Method of synthesis	-	solution polymerization with AIBN or AMBN as initiator COBF as transfer agent; RAFT	Jackson, A T; Thalassinos, K; John, R O; McGuire, N; Freeman, D; Scrivens, J H, Polymer, 51, 1418-24, 2010; Dagys, L, Klimkevičius, V, L,, V, Aidas, K, Balevicius, X, Solid State NMR, 105, 101641, 2020.
Mass average molecular weight, M_w	dalton, g/mol, amu	20,000-1,000,000	
Molar volume at 298K	cm^3 mol^{-1}	calc.=106.7	
Van der Waals volume	cm^3 mol^{-1}	71.04	
STRUCTURE			
Crystallinity	%	amorphous	
Entanglement molecular weight	dalton, g/mol, amu	calc.=13,918	
PHYSICAL PROPERTIES			
Density at 20°C	g cm^{-3}	1.15-1.16	
Color	-	white	
Refractive index, 20°C	-	calc.=1.4898-1.4973; exp.=1.437-1.472	Rossos, A K, Banti, C N, Kalampounias, A G, Papachristodoulou, C, Hadjikakou, S K, Mater. Sci., Eng. C, 111, 110770, 2020.

PHEMA poly(2-hydroxyethyl methacrylate)

PARAMETER	UNIT	VALUE	REFERENCES
Decomposition temperature	°C	299; 330 (thermal analysis)	
Thermal expansion coefficient, 23-80°C	°C^{-1}	3.7E-4	
Thermal conductivity, melt	W m^{-1} K^{-1}	0.1666	
Glass transition temperature	°C	85-100 (atactic); 31 (isotactic); 109 (syndiotactic)	
Hildebrand solubility parameter	MPa$^{0.5}$	calc.=18.45	
Surface tension	mN m^{-1}	calc.=38.4-42.8; exp.=28.8	
Diffusion coefficient of water vapor	cm^2 s^{-1} x10^6	0.59-5.37	
Contact angle of water, 20°C	degree	68.2	Lai, S; Hudiono, Y; Lee, L J; Dauneri, S; Madou, M J, Antec, 2703-7, 2002.
Surface free energy	mJ m^{-2}	35.4-36.0	

MECHANICAL & RHEOLOGICAL PROPERTIES

Tensile strength	MPa	0.6	Pereira, A T, Henriques, P C, Costa, P C, Martins, M C L, Gonçalves, I C, Compos. Sci. Technol., 184, 107819, 2019.
Elongation	%	20	Pereira, A T, Henriques, P C, Costa, P C, Martins, M C L, Gonçalves, I C, Compos. Sci. Technol., 184, 107819, 2019.
Young's modulus	MPa	2	Pereira, A T, Henriques, P C, Costa, P C, Martins, M C L, Gonçalves, I C, Compos. Sci. Technol., 184, 107819, 2019.
Poisson's ratio	-	0.357	

CHEMICAL RESISTANCE

Alcohols	-	poor	
Alkalis	-	poor	
Esters	-	poor	
Θ solvent, Θ-temp.=15.8, 32.1, 15.3°C	-	ethanol, n-propanol, water	
Good solvent	-	DMF, methanol, methyl cellosolve	

TOXICITY

HMIS: Health, Flammability, Reactivity rating	-	2/1/0	
Carcinogenic effect	-	not listed by ACGIH, NIOSH, NTP	

PROCESSING

Typical processing methods	-	initiated chemical vapor deposition	Bose, R K; Lau, K K S, Thin Solid Films, in press, 2011.
Applications	-	contact lenses, hydrogel nerve guide, pharmaceutical capsules, polymer electrolytes, tissue engineering	Rajan, A; Crugnola, A, Antec, 3640-43, 2002.

BLENDS

Suitable polymers	-	PANI, PDMS, PHA, PHB, PCL, PLA, PU, PVP	

PHEMA poly(2-hydroxyethylmethacrylate)

PARAMETER	UNIT	VALUE	REFERENCES
ANALYSIS			
FTIR (wavenumber-assignment)	cm⁻¹/-	C=O (free) – 1730; C=O (H-bonded) – 1703, 3536	Tang, Q; Yu, J-R; Chen, L; Zhu, J; Hu, Z-M, Current Appl. Phys., 11, 945-50, 2011; Morita, S; Kitagawa, K; Ozaki, Y, Vibrational Spectroscopy, 51, 28-33, 2009.

PHSQ polyhydridosilsesquioxane

PARAMETER	UNIT	VALUE	REFERENCES
GENERAL			
Common name	-	polyhydridosilsesquioxane, polyhydrosilsesquioxane	
Acronym	-	PHSQ	
SYNTHESIS			
Monomer(s) structure	-	$SiH(OCH_2CH_3)_3$	
Monomer(s) CAS number(s)	-	998-30-1	
Monomer(s) molecular weight(s)	dalton, g/mol, amu	164.28	
Formulation example	-	160 g of triethoxysilane was mixed with 800 g acetone. 11.7 g of water and 14.6 g of 0.02N nitric acid were added to the triethoxysilane/water solution. The final solution was stored at 22°C for 10 days.	Leung, R Y-k; Case, S, US Patent 6,413,882, Allied Signal, 2002.
Temperature of polymerization	°C	22	
Time of polymerization	days	10	
Pressure of polymerization	Pa	atmospheric	
Catalyst	-	nitric acid	
Mass average molecular weight, M_w	dalton, g/mol, amu	2,000-10,000	
PHYSICAL PROPERTIES			
Refractive index, 20°C	-	1.383	
Decomposition temperature	°C	300 (change from cage to ladder structure with releasing hydrogen); PSQ is completely transformed to SiO_2 at 600°C	
Dielectric constant at 100 Hz/1 MHz	-	3.1	
WEATHER STABILITY			
Absorbance at 157 nm	μm^{-1}	0.01-0.02	Ando, S; Fujigaya, T; Ueda, M, J. Photopolym. Sci. Technol., 15, 4, 231-36, 2002.
PROCESSING			
Typical processing methods	-	spin-coating	
Applications	-	integrated circuits; 157 nm lithography	
Outstanding properties	-	low dielectric properties	
BLENDS			
Suitable polymers	-	PPO	

HANDBOOK OF POLYMERS 3rd Edition, Copyrights 2022; ChemTec Publishing

PHT polyhexahydrotriazine

PARAMETER	UNIT	VALUE	REFERENCES
GENERAL			
Common name	-	polyhexahydrotriazine	
HISTORY			
Person to discover	-	Jeanette Garcia, IBM	Garcia, J M; Jones, G O, Virvani, K; McCkey, B D; Boday, D J; ter Huurne, G H; Horn, H W; Coady, D J; Bintaleb, A M; Alabdulrahman, A M S; Alwailem, F; Almegren, H A A, Hedrick, J L, Science, 344, 6185, 732-5, 2014.
Date	-	May, 2014	
SYNTHESIS			
Method of synthesis	-	a reaction of paraformaldehyde and 4,4-oxydianiline	
Temperature of polymerization	°C	50-185	
Time of polymerization	min	10-30	
PHYSICAL PROPERTIES			
Color	-	yellow	
Decomposition temperature	°C	300	
Glass transition temperature	°C	190	
MECHANICAL & RHEOLOGICAL PROPERTIES			
Young's modulus	MPa	10,000	
CHEMICAL RESISTANCE			
Other	-	stable to solvents at pH>3	
PROCESSING			
Outstanding properties	-	strongest thermosetting plastics, self-healing (forming spontaneous bonds between ruptured and damaged polymeric links, deep razor cut heals itself in 2 hours), recyclable (thermoset!!), dipped in sulfuric acid reverts to viscous state from which material can be remolded	

PI polyimide

PARAMETER	UNIT	VALUE	REFERENCES
GENERAL			
Common name	-	polyimide; poly(pyromellitimide-1,4-diphenyl ether)	
CAS name	-	poly[(5,7-dihydro-1,3,5,7-tetraoxobenzo[1,2-c:4,5-c']dipyrrole-2,6(1H,3H)-diyl)-1,4-phenyleneoxy-1,4-phenylene] (25036-53-7); 1H,3H-benzo[1,2-c:4,5-c']difuran-1,3,5,7-tetrone, polymer with 4,4'-oxybis[benzenamine] (25038-81-7)	
Acronym	-	PI	
CAS number	-	25036-53-7; 25038-81-7	
Formula			
HISTORY			
Person to discover	-	Paul John Flory; Edwards, W M and Maxwell, R I	
Date	-	1951; 1955	
Details	-	Flory reported condensation of sebacyl chloride and potassium phthalimide (first polyimide) and Edwards and Maxwell patented for DuPont PI made from pyromellitic acid	Fink, J K, High Performance Polymers, William Andrew, 2008.
SYNTHESIS			
Monomer(s) structure	-		
Monomer(s) CAS number(s)	-	89-32-7; 101-80-4	
Monomer(s) molecular weight(s)	dalton, g/mol, amu	218.12; 200.24	
Monomer(s) expected purity(ies)	%	97-98; 98-99	
Monomer ratio	-	1.11:1	
Formulation example	-	monomers, solvent (e.g., N-methyl-2-pyrrolidone), xylene (for azeotropic distallation to remove water)	
Method of synthesis	-	several routes can be used to obtain polyimides, including reaction between polyamic acid and diamine, isocyanate route, aqueous route, transimidization, and chemical vapor deposition	Fink, J K, High Performance Polymers, William Andrew, 2008.
Temperature of polymerization	°C	100	
Yield	%	94-100	
Number average molecular weight, M_n	dalton, g/mol, amu	10,000-100,000	
Mass average molecular weight, M_w	dalton, g/mol, amu	10,000-210,000	
Polydispersity, M_w/M_n	-	1.2-2.6	
Polymerization degree (number of monomer units)	-	25-275	
Molar volume at 298K	cm^3 mol^{-1}	calc.=275.5; 247 (crystalline)	
Van der Waals volume	cm^3 mol^{-1}	188.02; 184.1 (crystalline)	

HANDBOOK OF POLYMERS 3rd Edition, Copyrights 2022; ChemTec Publishing

PI polyimide

PARAMETER	UNIT	VALUE	REFERENCES
STRUCTURE			
Crystallinity	%	44-60	
Cell type (lattice)	-	orthorhombic	
Cell dimensions	nm	a:b:c=0.635:0.405:3.26	
Unit cell angles	degree	$\alpha:\beta:\gamma$=90:90:90	
Chain conformation	-	planar zig-zag	Chang, C-J; Chou, R-L; Lin, Y-C; Liang, B-J; Chen, J-J, Thin Solid Films, 519, 5013-16, 2011.
Entanglement molecular weight	dalton, g/mol, amu	1,894	
Lamellae thickness	nm	5-15	Verker, R; Grossman, E; Gouzman, I; Eliaz, N, Composites Sci. Technol., 69, 2178-84, 2009.
Crystallization temperature	°C	282; 220 (peak)	
Avrami constants, k/n	-	n=2.6	Chung, T S; Liu, S L; Oikawa, H; Yamaguchi, A, Antec, 1494-8, 1998.
COMMERCIAL POLYMERS			
Some manufacturers	-	DuPont; Ensinger; Sabic	
Trade names	-	Cirlex, Kapton, Vespel; Tecasint; Extem	
PHYSICAL PROPERTIES			
Density at 20°C	g cm^{-3}	1.31-1.43	
Color	-	light amber	
Refractive index, 20°C	-	calc.=1.5932-.16429; exp.=1.61-1.68	
Birefringence	-	0.011	Wang, Y-W; Chen, W-C, Composites Sci. Technol., 70, 769-75, 2010.
Transmittance	%	30-60	Yi, C, Li, W, Shi, S, He, K, Yang, C, Solar Energy, 195, 340-54, 2020.
Odor	-	none	
Melting temperature, DSC	°C	340-408	
Decomposition temperature	°C	495-520	Lian, R, Lei, X, Xue, S, Chen, Y, Zhang, Q, Appl. Surf. Sci., 535, 147654, 2021.
Thermal expansion coefficient, 23-80°C	°C^{-1}	2-5.4E-5	
Thermal conductivity, 23°C	W m^{-1} K^{-1}	0.12-0.35	
Glass transition temperature	°C	190-385 (some like Vespel do not have T_g); 482	-; Ke, H, Zhao, L, Zhang, X, Qiao, Y, Wang, Z, Polym. Testing, 90, 106746, 2020.
Specific heat capacity	J K^{-1} kg^{-1}	1,090-1,130	
Heat of fusion	J g^{-1}	139	Huo, P P; Cebe, P, Polymer, 34, 4, 696-704, 1993.
Maximum service temperature	°C	-269 to 400-500	Cousins, K, Polymers in Electronics. Market Report, Rapra, 2006.

PI polyimide

PARAMETER	UNIT	VALUE	REFERENCES
Long term service temperature	°C	300	
Heat deflection temperature at 0.45 MPa	°C	260-263	
Heat deflection temperature at 1.8 MPa	°C	221-360	
Vicat temperature VST/A/50	°C	257	
Vicat temperature VST/B/50	°C	262-263	
Surface tension	mN m^{-1}	calc.=37.7-41.0	
Dielectric constant at 1000 Hz/1 MHz	-	2.74-3.6/3.55	Jacobs, J D; Arlen, M J; Wang, D H; Ounaies, Z; Berry, R; Tan, L-S; Garrett, P H; Vaia, R A, Polymer, 51, 3139-46, 2010.
Dielectric strength	V μm^{-1}	280	Yi, C, Li, W, Shi, S, He, K, Yang, C, Solar Energy, 195, 340-54, 2020.
Dielectric loss factor at 1 kHz	-	0.0033	
Dissipation factor at 1000 Hz		0.0014-0.003	
Dissipation factor at 1 MHz		0.0034	
Volume resistivity	ohm-m	1E13-1E16	
Surface resistivity	ohm	1E15-1E16	
Electric strength K20/P50, d=0.60.8 mm	kV mm^{-1}	22-506	
Arc resistance	s	165	
Coefficient of friction	-	0.29-0.48 (kinetic); 0.35-0.63 (static)	
Permeability to oxygen, 25°C	barrer	160	Cui, L; Qiu, W; Paul, D R, Koros, W J, Polymer, in press, 2011.
Diffusion coefficient of water	cm^2 s^{-1} x10^9	5.6-8.1	Musto, P; Ragosta, G; Mensitieri, G; Lavorgna, M, Macromolecules, 40, 9614-27, 2007.
Contact angle of water, 20°C	degree	71.5-79.9	
Surface free energy	mJ m^{-2}	43.8	

MECHANICAL & RHEOLOGICAL PROPERTIES

Tensile strength	MPa	81-241; 89 (Janus PI); 60.7 (thermosetting), 102 (thermoplastic)	-; Lian, R, Lei, X, Xue, S, Chen, Y, Zhang, Q, Appl. Surf. Sci., 535, 147654, 2021; Ke, H, Zhao, L, Zhang, X, Qiao, Y, Wang, Z, Polym. Testing, 90, 106746, 2020.
Tensile modulus	MPa	1,200-3,800; 3130 (thermosetting), 2330 (thermoplastic)	-; Ke, H, Zhao, L, Zhang, X, Qiao, Y, Wang, Z, Polym. Testing, 90, 106746, 2020.
Tensile stress at yield	MPa	112-120	
Elongation	%	7-95	
Tensile yield strain	%	9	
Flexural strength	MPa	110-155; 104 (thermosetting), 141 (thermoplastic)	Ke, H, Zhao, L, Zhang, X, Qiao, Y, Wang, Z, Polym. Testing, 90, 106746, 2020.
Flexural modulus	MPa	2,900-3,520; 3220 (thermosetting), 2740 (thermoplastic)	-; Ke, H, Zhao, L, Zhang, X, Qiao, Y, Wang, Z, Polym. Testing, 90, 106746, 2020.
Compressive strength	MPa	150-234	

PI polyimide

PARAMETER	UNIT	VALUE	REFERENCES
Toughness	MPa	7.7-10	Lian, R, Lei, X, Xue, S, Chen, Y, Zhang, Q, Appl. Surf. Sci., 535, 147654, 2021.
Young's modulus	MPa	2,500	
Charpy impact strength, notched, 23°C	kJ m^{-2}	20-22	
Izod impact strength, unnotched, 23°C	J m^{-1}	NB to 750	
Izod impact strength, notched, 23°C	J m^{-1}	43-110	
Shear strength	MPa	55-90	
Poisson's ratio	-	0.15-0.42	
Rockwell hardness	-	M112	
Shrinkage	%	0.004-1.3 (molding); 0.03-0.17 (30 min/150°C); 1.25 (120 min/400°C)	
Intrinsic viscosity, 25°C	dl g^{-1}	0.75-2.18	
Melt index, 400°C/6.6 kg	g/10 min	10	
Water absorption, equilibrium in water at 23°C	%	0.39-2.9 (24 h)	
Moisture absorption, equilibrium 23°C/50% RH	%	1-1.8	

CHEMICAL RESISTANCE			
Acid dilute/concentrated	-	non-resistant	
Alcohols	-	resistant	
Alkalis	-	non-resistant	
Aliphatic hydrocarbons	-	resistant	
Aromatic hydrocarbons	-	resistant	
Esters	-	resistant	
Greases & oils	-	resistant	
Halogenated hydrocarbons	-	resistant	
Ketones	-	resistant	
Good solvent	-	hot p-chlorophenol and m-cresol	Wu, Z; Yoon, Y; Harris, F W; Cheng, Z D; Chuang, K C, Antec, 3038-42, 1996.

FLAMMABILITY			
Ignition temperature	°C	>540; chars but does not burn in air	
Autoignition temperature	°C	>540	
Limiting oxygen index	% O$_2$	37-53	
Heat release	kW m^{-2}	21	
NBS smoke chamber	DM	<1	
Burning rate (Flame spread rate)	mm min^{-1}		
Char at 800°C	%	62.5	Ke, H, Zhao, L, Zhang, X, Qiao, Y, Wang, Z, Polym. Testing, 90, 106746, 2020.
Heat of combustion	J g^{-1}	26,030	Walters, R N; Hacket, S M; Lyon, R E, Fire Mater., 24, 5, 245-52, 2000.

PI polyimide

PARAMETER	UNIT	VALUE	REFERENCES
Volatile products of combustion	-	CO, CO_2, H_2O	Pramoda, K P; Chung, T S; Liu, S L; Oikawa, H; Yamaguchi, A, Polym. Deg. Stab., 67, 2, 365-74, 2000.
UL 94rating	-	V-0	

WEATHER STABILITY

PARAMETER	UNIT	VALUE	REFERENCES
Spectral sensitivity	nm	<500; vacuum ultraviolet (e.g., 172; it also has synergistic action with atomic oxygen)	Yokota, K; Ohmae, N; Tagawa, M, High Performance Polym., 16, 221-34, 2004.
Excitation wavelengths	nm	380, 450	
Emission wavelengths	nm	505, 508, 566	
Depth of UV penetration	μm	0.5; limited to surface because of strong intrinsic absorption	
Products of degradation	-	only surface erosion	
Stabilizers	-	resistance to γ-radiation, atomic oxygen, and Lyman emission	
Results of exposure	Florida	1300 h to reduce elongation by 50%	
Low earth orbit erosion yield	cm^3 $atom^{-1}$ x 10^{-24}	2.81-3.0	Waters, D L; Banks, B A; De Groh, K K; Miller, S K R; Thorson, S D, High Performance Polym., 20, 512-22, 2008.

BIODEGRADATION

PARAMETER	UNIT	VALUE	REFERENCES
Typical biodegradants	-	two steps are involved in degradation: an initial decline of resistance related to the partial ingress of water and ionic species into the polymer matrix. This is followed by further deterioration of the polymer by activity of the fungi, resulting in a large decrease in resistivity	

TOXICITY

PARAMETER	UNIT	VALUE	REFERENCES
Carcinogenic effect	-	not listed by ACGIH, NIOSH, NTP	
TLV, ACGIH	$mg\ m^{-3}$	3 (respirable); 10 (total)	
OSHA	$mg\ m^{-3}$	5 (respirable); 15 (total)	
Oral rat, LD_{50}	$mg\ kg^{-1}$	15,600	

ENVIRONMENTAL IMPACT

PARAMETER	UNIT	VALUE	REFERENCES
Aquatic toxicity, *Daphnia magna*, LC_{50}, 48 h	$mg\ l^{-1}$	16	
Aquatic toxicity, *Bluegill sunfish*, LC_{50}, 48 h	$mg\ l^{-1}$	380	
Aquatic toxicity, *Rainbow trout*, LC_{50}, 48 h	$mg\ l^{-1}$	340	

PROCESSING

PARAMETER	UNIT	VALUE	REFERENCES
Typical processing methods	-	casting, compression molding, drawing of oriented films, extrusion, injection molding, sintering, spin coating, spinning, vapor phase deposition	
Preprocess drying: temperature/ time/residual moisture	ºC/h/%	175/4-6/0.02	
Processing temperature	ºC	380-430	
Processing pressure	MPa	0.3-0.7 (back)	

PI polyimide

PARAMETER	UNIT	VALUE	REFERENCES
Additives used in final products	-	Fillers: aluminum nitride, barium titanate, aluminum nitride, antimony trioxide, aramide fiber, attapulgite, carbon fiber, carbon nanofiber, carbon nanotubes, clay, glass fiber, graphite, molybdenum sulfide, montmorillonite, PTFE, silica, smectite, titanium oxide whisker; Plasticizers: diethylene glycol dibenzoate, dimethyl phthalate, triallyl phthalate, diethynyldiphenyl methane, phenylethynyldiphenyl methane, 4-hydroxybenzophenone; Antistatics: antimony-containing tin oxide, carbon black, carbon, nanotubes, indium oxide microspheres, polythiophene; Release: polyethylene wax, PTFE, silicone oil, zirconium chelate	
Applications	-	aerospace, composites, electronics (mostly films and coatings), foam composites, hollow fiber membranes, electronics, fibers, mechanical parts (bearings, piston rings, valve seats, washers), microprocessor chip carriers, non-lubricated applications, nuclear power plants, photosensitive materials for positive imaging, photovoltaic film, solar cells, space shuttle, structural adhesives, ultrafiltration membranes	
Outstanding properties	-	broad range of temperature resistance, low moisture uptake, excellent electric properties	

BLENDS

Suitable polymers	-	PEI, PBI, PEEK, PES, PTFE, TPU	

ANALYSIS

PARAMETER	UNIT	VALUE	REFERENCES
FTIR (wavenumber-assignment)	cm^{-1}/-	imide absorption bands: C=O – 1780 and 725, C–N – 1380; carboxylic acid band of polyamic acid - 1700;	
Raman (wavenumber-assignment)	cm^{-1}/-	C–CO–C – 1788, 1728; C–N–C – 1394, 1124; aromatic dianhydride – 1614, 753	Samyn, P; De Baets, P; Van Craenenbroeck, J; Verpoort, F; Schoukens, Antec, 121-5; 2005.
NMR (chemical shifts)	ppm	C=O – 166.6-168.5	Powell, C E; Duthie, X J; Kentish, S E; Qiao, G G; Stevens, G W, J. Membrane Sci., 291, 199-209, 2007.
x-ray diffraction peaks	degree	5, 18	Goodwin, A A; Whittaker, A K; Jack, K S; Hay, J N; Forsythe, J, Polymer, 41, 7263-71, 2000.

PIB polyisobutylene

PARAMETER	UNIT	VALUE	REFERENCES

GENERAL

Common name	-	polyisobutylene, polyisobutene	
IUPAC name	-	poly(2-methylpropene), polyisobutylene	
CAS name	-	1-propene, 2-methyl-, homopolymer	
Acronym	-	PIB	
CAS number	-	9003-27-4; 9003-29-6 (oligomers)	
RTECS number	-	UD1010000	
Linear formula	-	$\left[\begin{array}{c} CH_3 \\ | \\ CCH_2 \\ | \\ CH_3 \end{array} \right]_n$	

HISTORY

Person to discover	-	IG Farben	
Date	-	1931	

SYNTHESIS

Monomer(s) structure	-	$\begin{array}{c} CH_2 \\ \| \\ H_3CCCH_3 \end{array}$	
Monomer(s) CAS number(s)	-	115-11-7	
Monomer(s) molecular weight(s)	dalton, g/mol, amu	56.11	
Monomer ratio	-	100%	
Number average molecular weight, M_n	dalton, g/mol, amu	180-6,000 (oligomers)	
Mass average molecular weight, M_w	dalton, g/mol, amu	900-1,100,000	
Polydispersity, M_w/M_n	-	1.06-2.10	
Functionality	-	epoxide (telechelic prepolymer)	Yang, B, Store, R F, Reactive Functional Polym., 150, 104563, 2020.
Molar volume at 298K	cm³ mol⁻¹	calc.=59.2 (crystalline)	
Van der Waals volume	cm³ mol⁻¹	calc.=40.9 (crystalline)	
Radius of gyration	nm	0.57 (M_w=390), 19.6 (4,040), 83.1 (73,200)	Frick, B; Dosseh, G; Cailliaux, A; Alba-Simionesco, C, Chem. Phys., 292, 311-23, 2003.
Chain-end groups	-	H, CH_2-CH=CH_2; Cl, COOH (derivatives)	Nagy, L; Palfi, V; Narmandakh, M; Kuki, A; Nyiri, A; Ivan, B; Zsuga, M; Keki, J. Am. Soc. Mass Spectrom., 20, 2342-51, 2009.

STRUCTURE

Cell type (lattice)	-	orthorhombic	
Cell dimensions	nm	a:b:c=0.688:1.191:1.86	
Unit cell angles	degree	α:β:γ=90:90:90	
Number of chains per unit cell	-	4	
Chain conformation	-	helix	

PIB polyisobutylene

PARAMETER	UNIT	VALUE	REFERENCES
Entanglement molecular weight	dalton, g/mol, amu	calc.=8,818	

COMMERCIAL POLYMERS

Some manufacturers	-	BASF; INEOS	
Trade names	-	Oppanol; Indopol (oligomers)	

PHYSICAL PROPERTIES

Density at 20°C	g cm^{-3}	0.788-0.921 (oligomers); 0.972 (crystalline)	
Color	-	clear to transparent	
Refractive index, 20°C	-	1.445-1.508 (oligomers); 1.5050-1.5100	
Odor	-	odorless	
Decomposition temperature	°C	>120	
Thermal conductivity, melt	W m^{-1} K^{-1}	0.19-0.26	
Glass transition temperature	°C	calc.=-71; exp.=-72; -62 to -70	
Maximum service temperature	°C	-40 to 90	
Hansen solubility parameters, δ_D, δ_P, δ_H	MPa$^{0.5}$	16.4, 1.7, 4.7; 16.9, 2.5, 4.0	
Interaction radius		7.9; 7.2	
Hildebrand solubility parameter	MPa$^{0.5}$	17.1	
Surface tension	mN m^{-1}	33.6	Roe, J R, J. Phys. Chem., 72, 2013, 1968.
Electric strength K20/P50, d=0.60.8 mm	kV mm^{-1}	42	
Contact angle of water, 20°C	degree	112.1	
Surface free energy	mJ m^{-2}	33.2	

MECHANICAL & RHEOLOGICAL PROPERTIES

Tensile strength	MPa	1.7-2.5	
Elongation	%	50-700	
Compression set, 24h 70°C	%	15	
Melt index, 230°C/3.8 kg	g/10 min	200-300	

CHEMICAL RESISTANCE

Acid dilute/concentrated	-	good	
Alcohols	-	good	
Alkalis	-	good	
Aliphatic hydrocarbons	-	poor	
Aromatic hydrocarbons	-	poor	
Esters	-	poor	
Greases & oils	-	poor	
Halogenated hydrocarbons	-	poor	
Ketones	-	poor	
Good solvent	-	benzene, carbon bisulfide, carbon tetrachloride, cyclohexanone, paraffin wax, toluene, xylene	

PIB polyisobutylene

PARAMETER	UNIT	VALUE	REFERENCES
FLAMMABILITY			
Ignition temperature	°C	>40 to >280 (oligomers)	
Autoignition temperature	°C	>200	
Char at 500°C	%	0	Lyon, R E; Walters, R N, J. Anal. Appl. Pyrolysis, 71, 27-46, 2004.
Volatile products of combustion	-	CO, CO_2, monomer and 52 more products	Lyon, R E; Walters, R N, J. Anal. Appl. Pyrolysis, 71, 27-46, 2004; Lehre, R S; Pattenden, Polym. Deg. Stab., 63, 321-40, 1999.
WEATHER STABILITY			
Important initiators and accelerators	-	ozone	
Products of degradation	-	hydroperoxides, radicals, ketones, carboxyl groups, hydroxyls, peresters, esters, lactones, chain scission, crosslinking, hydroxyls, double bonds	
TOXICITY			
NFPA: Health, Flammability, Reactivity rating	-	0-1/1/0	
Carcinogenic effect	-	not listed by ACGIH, NIOSH, NTP	
Mutagenic effect	-	none	
Teratogenic effect	-	none	
Reproductive toxicity	-	none	
TLV, ACGIH	mg m^{-3}	2 (inhalable)	
Oral rat, LD$_{50}$	mg kg^{-1}	>2,000; >34,600	
Skin rabbit, LD$_{50}$	mg kg^{-1}	non-irritant	
ENVIRONMENTAL IMPACT			
Aquatic toxicity, *Daphnia magna*, LC$_{50}$, 48 h	mg l^{-1}	>1,000	
Aquatic toxicity, *Fathead minnow*, LC$_{50}$, 48 h	mg l^{-1}	>100	
Aquatic toxicity, *Rainbow trout*, LC$_{50}$, 48 h	mg l^{-1}	>1,000	
PROCESSING			
Typical processing methods	-	compounding, vulcanization, coating, sheeting	
Additives used in final products	-	Fillers: aluminum hydroxide, calcium carbonate, carbon black, cellulose, clay, kaolin, magnesium hydroxide, zinc oxide; Plasticizers: mineral oil, silicone oil, octyl palmitate	
Applications	-	cling film, glazing spacers, lubricants, pressure-sensitive adhesives, roofing membranes, sealants, transdermal administration of hypermic active substances, vascular graft	
BLENDS			
Suitable polymers	-	PE, PS	

PIB polyisobutylene

PARAMETER	UNIT	VALUE	REFERENCES
ANALYSIS			
FTIR (wavenumber-assignment)	cm^{-1}/-	C-H – 1365, 1385; C=O – 1832, 1730, 1702	Small, C M; McNally, G M; Marks, A; Murphy, W R, Antec, 2882-86, 2002; Gonon, L; Troquet, M; Fanton, E; Gardette, J-L, Polym. Deg. Stab., 62, 541-49, 1998.

PIB, *cis* cis-polyisoprene

PARAMETER	UNIT	VALUE	REFERENCES
GENERAL			
Common name	-	*cis*-polyisoprene, natural rubber	
CAS name	-	natural rubber; 1,3-butadiene, 2-methyl-, homopolymer	
Acronym	-	*cis*-PIP	
CAS number	-	9003-31-0; 9006-04-6 (natural rubber); 104389-31-3	
Linear formula			
HISTORY			
Person to discover	-	Horne, S E	Horne, S E, US Patent 3,114,743, Goodrich-Gulf Chemicals, Dec. 17, 1963.
Date	-	1963 (filling 1954)	
SYNTHESIS			
Monomer(s) structure	-	$H_2C=CHC=CH_2$ $\quad\quad\quad CH_3$	
Monomer(s) CAS number(s)	-	78-79-5	
Monomer(s) molecular weight(s)	dalton, g/mol, amu	68.12	
Monomer(s) expected purity(ies)	%	99	
Rubber content	%	94 (natural), >99 synthetic	
Method of synthesis	-	solvent, monomer and catalyst (all high purity) are added to reactor, polymerization is stopped by addition of catalyst deactivator, and rubber protected by addition of a non-staining antioxidant	
Catalyst	-	organoaluminum	
Formulation example (compounding)	phr	rubber – 100, carbon black – 20, zinc oxide – 3, stearic acid – 2, antioxidant (6PPD) –1, sulfur – 1.5, vulcanization accelerator (CBS) – 1.5	Xiang, F, Schneider, K, Heinrich, G, Int. J. Fatigue, 135, 105508, 2020.
Mass average molecular weight, M_w	dalton, g/mol, amu	40,000-1,240,000 (natural rubber); 1,500,000-2,500,000 (synthetic)	
Polydispersity, M_w/M_n	-	1.02-1.04	
Molar volume at 298K	cm^3 mol^{-1}	68.1 (crystalline)	
Van der Waals volume	cm^3 mol^{-1}	47.5 (crystalline)	
STRUCTURE			
Crystallinity	%	30	
Cell type (lattice)	-	orthorhombic	
Cell dimensions	nm	a:b:c=1.241:0.881:0.843	
Unit cell angles	degree	β=94.6	
Number of chains per unit cell	-	4	
Crystallite size	nm	6-25 (filaments)	

PIB, *cis* cis-polyisoprene

PARAMETER	UNIT	VALUE	REFERENCES
Tacticity	%	69.5-98	Hyun, K; Hoefl, S; Kahle, S; Wilhelm, M, J. Non-Newtonian Fluid Mech., 160, 93-103, 2009; Bussiere, P-O; Gardette, J-L; Lacoste, J; Baba, M, Polym. Deg. Stab., 88, 182-88, 2005.
Cis content	%	100 (natural rubber); 90-98 (synthetic)	
Chain conformation	-	$P2_{1/a}-C_{2h}$	
Rapid crystallization temperature	°C	-25	
PHYSICAL PROPERTIES			
Density at 20°C	g cm^{-3}	0.906-0.93; 0.95 (vulcanized)	
Color	-	nearly colorless	
Refractive index, 20°C	-	1.5191-1.52	
Odor	-	odorless	
Melting temperature, DSC	°C	370-384	Ginting, E M, Bukit, N, Frida, E, Bukit, B F, Case Studies thermal Eng., 17, 100575, 2020.
Thermal expansion coefficient, 23-80°C	°C^{-1}	6.7E-4	
Thermal conductivity, melt	W m^{-1} K^{-1}	0.134	
Glass transition temperature	°C	-75 (natural rubber), -70 to -72 (polyisoprene)	
Specific heat capacity	J K^{-1} kg^{-1}	1905	
Maximum service temperature	°C	-50 to 50; (unvulcanized); -55 to 80 (vulcanized)	
Hansen solubility parameters, δ_D, δ_P, δ_H	MPa$^{0.5}$	18.1, 2.4, 2.3	
Interaction radius		10.3	
Hildebrand solubility parameter	MPa$^{0.5}$	17-18.4	
Surface tension	mN m^{-1}	32	Lee, L H, J. Polym. Sci. A-2, 5, 1103, 1967.
Dielectric constant at 1 Hz/1 MHz	-	2.37-2.45/2.6; 2.9 (vulcanized)	
Dielectric loss factor at 1 kHz	-	0.001-0.003	
Dissipation factor at 1000 Hz	E-4	20	
Volume resistivity	ohm-m	1E13	
Electric strength K20/P50, d=0.60.8 mm	kV mm^{-1}	17; 50 (vulcanized)	
Permeability to oxygen, 25°C	cm^3 cm cm^{-2} s^{-1} Pa^{-1} x 10^{12}	1.76	
Diffusion coefficient of oxygen	cm^2 s^{-1} x10^6	1.73	
MECHANICAL & RHEOLOGICAL PROPERTIES			
Tensile strength	MPa	15-28 (vulcanized)	
Tensile stress at yield	MPa	21.6	
Elongation	%	100-800	
Rebound resilience at 23°C	%	25.9-39.7	Salim, M A, Saad, A M, Rosszainily, I R A, Encyclopedia of Renewable and Sustainable Materials, Vol. 3, Elsevier, 2020, pp. 71-8.

PIB, *cis* cis-polyisoprene

PARAMETER	UNIT	VALUE	REFERENCES
Compression set	%	16.5-20.5	Salim, M A, Saad, A M, Rosszainily, I R A, Encyclopedia of Renewable and Sustainable Materials, Vol. 3, Elsevier, 2020, pp. 71-8.
Tear strength	N mm^{-1}	57-94	Salim, M A, Saad, A M, Rosszainily, I R A, Encyclopedia of Renewable and Sustainable Materials, Vol. 3, Elsevier, 2020, pp. 71-8.
Shore A hardness	-	30-90 (unvulcanized); 30-100 (vulcanized)	
Shore D hardness	-	30-45 (vulcanized)	

CHEMICAL RESISTANCE			
Acid dilute/concentrated	-	poor	
Alkalis	-	fair	
Aliphatic hydrocarbons	-	poor	
Aromatic hydrocarbons	-	poor	
Esters	-	poor	
Greases & oils	-	poor	
Halogenated hydrocarbons	-	poor	
Ketones	-	fair	
Θ solvent, Θ-temp.=25, 32.1, 14.5°C	-	butanone, dioxane, 2-pentanone	
Good solvent	-	chlorinated hydrocarbons, cyclohexane, hydrocarbons, MIBK, toluene	
Non-solvent	-	acetone, alcohols, carboxylic acids	

FLAMMABILITY			
Ignition temperature	$^{\circ}$C	>113	
Heat of combustion	J g^{-1}	45,200	

WEATHER STABILITY			
Important initiators and accelerators	-	ozone, singlet oxygen, mechanical stress, FeCl$_3$ (photo-Fenton process)	
Products of degradation	-	radicals, hydroperoxides, epoxy groups, ketone groups, crosslinking, chain scission	
Stabilizers	-	cabon black, antioxidants, antiozonants, 3-mercapto-1,2,4-triazin-5-one derivatives, encapsulated butylated hydroxy toluene	

TOXICITY			
Carcinogenic effect	-	not listed by ACGIH, NIOSH, NTP	

PROCESSING			
Typical processing methods	-	Banbury mixer, calendering, coating, Gordon plasticator, sheeting, skim coating, tubing, vulcanization	

PIB, *cis* *cis*-polyisoprene

PARAMETER	UNIT	VALUE	REFERENCES
Additives used in final products	-	Fillers: barium and strontium ferrites, boron carbide, calcinated clays, calcium carbonate, carbon black, carbon-silica dual phase filler, clays, dolomite, fumed silica, iron oxide, magnesium aluminum silicate, magnesium carbonate, mica, montmorillonite, nickel zinc ferrite, nylon fibers, pulverized polyurethane foam, quartz, silica carbide, soapstone, talc, zinc oxide; Processing aids: naphthenic oil (liquid guayule natural rubber (LGNR) was produced by thermal degradation of guayule natural rubber and tested as a renewable alternative to naphthenic oil), polybutene, aromatic oil, esters of dicarboxylic acid; Plasticizers: adipates, aromatic mineral oil, paraffin oil, phosphates, phthalates, polyethylene glycol, processing oil, sebacates; Antistatics: dihydrogen phosphate of ε-aminocaproic acid, iodine doping, carbon black, quaternary ammonium salt, zinc oxide whisker; Antiblocking: diatomaceous earth; Release: propylene wax; Slip: erucamide+stearamide	Ren, X, Barrera, C S, Tardiff, J L, Cornish, K, J. Cleaner Prod., 276, 122933, 2020.
Applications	-	boots, conveyor belts, electrician's gloves, gloves, heels and soles, hoses, instrument panels, latex foams, machined components, pipes, plugs, pumps, shock absorbers, sockets, storage-battery cases, switchboard panels, telephone receivers, tire cord impregnation, tires, toys, tubes, valves, vibration dampers, waterproof clothing and bathing apparel, wire and cables	

BLENDS			
Suitable polymers	-	PANI, PBAT, PLA, PU	

ANALYSIS			
NMR (chemical shifts)	ppm	H NMR: CH_2 – 4.7; CH – 5.18; C NMR: *cis* – 26.5 and 31.1, *trans* – 32.1	Pilichowski, J-F; Morel, M; Tamboura, F; Chmela, S; Baba, M; Lacoste, J, Polym. Deg. Stab., 95, 1575-80, 2010.

PIP, *trans* trans-polyisoprene

PARAMETER	UNIT	VALUE	REFERENCES
GENERAL			
Common name	-	*trans*-polyisoprene, guttapercha	
CAS name	-	1,3-butadiene, 2-methyl-, homopolymer	
Acronym	-	*trans*-PIP	
CAS number	-	104389-32-4	
Linear formula			
HISTORY			
Person to discover	-	Saltman, W M	Saltman, W M, US Patent 3,008,945, Goodyear Tire & Rubber Company, Nov. 14, 1961.
Date	-	1961	
Details	-	patented polymerization permits synthesis of product containing 97% *trans*-form	
SYNTHESIS			
Monomer(s) structure	-	$H_2C=CHC=CH_2$ CH_3	
Monomer(s) CAS number(s)	-	78-79-5	
Monomer(s) molecular weight(s)	dalton, g/mol, amu	68.12	
Method of synthesis	-	bulk precipitation polymerization of isoprene catalyzed by supported titanium catalyst $TiCl_4/MgCl_2$	
Biosynthesis		few species of higher plants have been shown to produce polyisoprene in the all *trans*-l,4 configuration; these include Gutta Percha from *Palaqium gutta* and Balata from *Mimusops balata* which are typical high molecular weight trans-polyisoprenes occurring as latex; *Achras sapota* produces low molecular-weight *trans*-polyisoprene as a mixture with high molecular weight *cis*-polisoprene in latex form. In contrast to the rubber tree (*Hevea brasiliensis*), which produces cis-polyisoprene, *Eucommia ulmoides* has evolved to synthesize long-chain trans-polyisoprene via farnesyl diphosphate synthases	Wuyun, T-n, Wang, L, Liu, H, Du, H, Molecular Plant, 11, 3, 429-42, 2018.
Temperature of polymerization	°C	20-40	
Time of polymerization	h	40	
Catalyst	-	$TiCl_4/MgCl_2$	Huang, B; Zhao, Z; Yao, W; Du, A; Zhao, Y, US Patent 7,718,742, Qingdao Qust Fangtai Material Engineering, 2010.
Yield	%	95	
Number average molecular weight, M_n	dalton, g/mol, amu	2,560-1,199,400	
Mass average molecular weight, M_w	dalton, g/mol, amu	5,000-500,000	
Polydispersity, M_w/M_n	-	1.04-1.11	
Molar volume at 298K	cm^3 mol^{-1}	64.7 (crystalline)	

HANDBOOK OF POLYMERS 3rd Edition, Copyrights 2022; ChemTec Publishing

PIP, *trans* trans-polyisoprene

PARAMETER	UNIT	VALUE	REFERENCES
Van der Waals volume	cm^3 mol^{-1}	47.5 (crystalline)	

STRUCTURE

PARAMETER	UNIT	VALUE	REFERENCES
Crystallinity	%	34	
Cell type (lattice)	-	monoclinic (α), orthorhombic (β)	
Trans content	%	97 (guttapercha); 92-99+ (synthetic)	

COMMERCIAL POLYMERS

PARAMETER	UNIT	VALUE	REFERENCES
Some manufacturers	-	Kuraray	

PHYSICAL PROPERTIES

PARAMETER	UNIT	VALUE	REFERENCES
Density at 25°C	g cm^{-3}	0.90-0.95	
Color	-	white	
Odor		odorless	
Melting temperature, DSC	°C	58-67	
Glass transition temperature	°C	-63 to -68	
Surface tension	mN m^{-1}	31	Lee, L H, J. Polym. Sci. A-2, 5, 1103, 1967.
Permeability to nitrogen, 25°C	cm^3 cm cm^{-2} s^{-1} Pa^{-1} x 10^{12}	0.711	
Permeability to water vapor, 25°C	cm^3 cm cm^{-2} s^{-1} Pa^{-1} x 10^{12}	172	
Diffusion coefficient of nitrogen	cm^2 s^{-1} x10^6	1.17	

MECHANICAL & RHEOLOGICAL PROPERTIES

PARAMETER	UNIT	VALUE	REFERENCES
Tensile strength	MPa	19.6-36.7 (vulcanized)	
Elongation	%	250-400 (vulcanized)	
Shore A hardness	-	90-95 (vulcanized)	
Mooney viscosity	-	20-90	

CHEMICAL RESISTANCE

PARAMETER	UNIT	VALUE	REFERENCES
Acid dilute/concentrated	-	poor	
Alkalis	-	fair	
Aliphatic hydrocarbons	-	poor	
Aromatic hydrocarbons	-	poor	
Esters	-	poor	
Greases & oils	-	poor	
Halogenated hydrocarbons	-	poor	
Ketones	-	fair	
Θ solvent, Θ-temp.=47.7, 60.0°C	-	dioxane, n-propyl acetate	
Good solvent	-	chlorinated hydrocarbons, cyclohexane, hydrocarbons, MIBK, toluene	

PIP, *trans* trans-polyisoprene

PARAMETER	UNIT	VALUE	REFERENCES
Non-solvent	-	acetone, alcohols, carboxylic acids	
TOXICITY			
HMIS: Health, Flammability, Reactivity rating	-	0/1/0	
Carcinogenic effect	-	not listed by ACGIH, NIOSH, NTP	
OSHA	mg m^{-3}	5	
PROCESSING			
Typical processing methods	-	calendering, extrusion, mixing, molding, vulcanization	
Additives used in final products	-	Fillers: calcium carbonate, carbon black, carbon fiber, graphite, kaolin, montmorillonite, silica, silicates, titanium dioxide, zinc oxide	
Applications	-	use in more than 40,000 products; ablatives, aircraft tires, dental (root filling material), electrochemical cell components, pressure-sensitive adhesives, and many more; sealing root apexes in dental procedures	Bisag, A, Manzini, M, Simoncelli, E, Colombo, V, Clinical Plasma Med., 19-20, 100100, 2020.
BLENDS			
Suitable polymers	-	PP	
ANALYSIS			
FTIR (wavenumber-assignment)	cm^{-1}/-	C=O: 1690 (carboxyl), 1720 (aldehyde), 1745 (ketone); 855 – *trans*-isoprenyl unit	
Raman (wavenumber-assignment)	cm^{-1}/-	CH$_3$ – 2970, 2960; CH$_2$ – 2932, 2906; C=C – 1662; C–C – 1098, 1010, 985	Arjunan, Subramanian, S; Mohan, S; Spectrochim. Acta, 57A, 2547-54, 2001.

PK polyketone

PARAMETER	UNIT	VALUE	REFERENCES
GENERAL			
Common name	-	polyketone	
CAS name	-	1-propene, polymer with carbon monoxide and ethene	
Acronym	-	PK	
CAS number	-	88995-51-1	
Linear formula		$-[CH_2CH_2C(=O)]_n-$	
HISTORY			
Person to discover	-	Brubaker, M M 1950. Van Broekhoven, J A M; Drent, E; Klei, E; Nozaki, K	Brubaker, M M, US Patent 2,495,286, DuPont, Jan. 24, 1950. Van Broekhoven, J A M; Drent, E; Klei, E; Nozaki, K, US Patent 4,880,903, Shell, Nov. 14, 1989.
Date	-	1950; 1989	
Details	-	monomers are polymerized in the presence of benzoyl peroxide; currently used technology has been patented by Shell in 1989	
SYNTHESIS			
Monomer(s) structure	-	$CH_2=CH_2$; CO	
Monomer(s) CAS number(s)	-	74-85-1; 630-08-0	
Monomer(s) molecular weight(s)	dalton, g/mol, amu	28.05; 28.01	
Monomer ratio	-	1; some in addition have 6 mol% propylene (e.g, Carilon P1000)	Zuiderduin, W C J; Huetink, J; Gaymans, R J, J. Appl. Polym. Sci., 91, 2558-75, 2004.
Method of synthesis	-	polyketone can be made with a palladium(II) catalyst from ethylene and carbon monoxide (e.g., Carilon) cross-metathesis polymerization is an efficient and feasible method for the synthesis of aliphatic polyketones	-; Zeng, F-R, Xu, J, Xiong, Q, Qin, K-X, Li, Z-C, Polymer, 185, 121936, 2019.
Temperature of polymerization	°C	85	
Time of polymerization	h	5	
Pressure of polymerization	MPa	5.5	
Catalyst	-	palladium(ii) acetate+	
Yield	%	29.7-70 (oligomer)	Mul, W P; Dirkzwager, H; Broekhuis, A A; Heeres, H J; van der Linden, A J; Orpen, A G, Inorganica Chim. Acta, 327, 147-59, 2002.
Number average molecular weight, M_n	dalton, g/mol, amu	40,000-250,000; 1,500-5,000 (Carilite oligomer); 12200-14,000 (cross-metathesis polymerization)	Mul, W P; Dirkzwager, H; Broekhuis, A A; Heeres, H J; van der Linden, A J; Orpen, A G, Inorganica Chim. Acta, 327, 147-59, 2002.
Mass average molecular weight, M_w	dalton, g/mol, amu	100,000-296,000	Zuiderduin, W C J; Homminga, D S; Huetink, H J; Gaymans, R J, Polymer, 44, 6361-70, 2003.
Polydispersity	-	1.91-2.49 (cross-metathesis polymerization)	Zeng, F-R, Xu, J, Xiong, Q, Qin, K-X, Li, Z-C, Polymer, 185, 121936, 2019.

PK polyketone

PARAMETER	UNIT	VALUE	REFERENCES
STRUCTURE			
Crystallinity	%	30-58; 30-40 (bulk-crystallized); 24-29.7 (aliphatic polyketone terpolymer)	Lin, H, Pearson, A, Kazemi, Y,, H E, Polym. Deg. Stab., 179, 109260, 2020.
Cell type (lattice)	-	orthorhombic	
Cell dimensions	nm	a:b:c= 0.691:0.512:0.76 (α); a:b:c=0.797:0.476:0.757 (β)	Chatani, Y; Takizawa, T; Murahashi, S; Sakata, Y; Nishimura, Y, J. Polym. Sci., 55, 162, 811-19, 1961; Waddon, A J; Karttunen, N R; Lesser, A J, Macromolecules, 32, 423-28, 1999.
Unit cell angles	degree	26 (for angle between molecular plane and bc plane in α-form); 40 (in β-form)	Ohsawa, O; Lee, K-H; Kim, B-S; Lee, S; Kim, I-S, Polymer, 2007-12, 2010.
Number of chains per unit cell	-	4	
Polymorphs	-	α, β	
Chain conformation	-	planar zigzag	
Entanglement molecular weight	dalton, g/mol, amu	1,700	Zuiderduin, W C J; Homminga, D S; Huetink, H J; Gaymans, R J, Polymer, 44, 6361-70, 2003.
Lamellae thickness	nm	3.48-5.7	Blundell, D J; Liggat, J J; Flory, A, Polymer, 33, 12, 2475-82, 1992.
Avrami constants, k/n	-	n=2.02-2.85	Holt, G A; Spruiell, J E, Antec, 1780-88, 1996.
COMMERCIAL POLYMERS			
Some manufacturers	-	Shell	
Trade names	-	Carilon	
PHYSICAL PROPERTIES			
Density at 20°C	g cm^{-3}	1.24; 1.383 (crystalline)	Ohsawa, O; Lee, K-H; Kim, B-S; Lee, S; Kim, I-S, Polymer, 2007-12, 2010.
Melting temperature, DSC	°C	220-260 (replacement of some ethylene by propylene lowers MP)	
Thermal expansion coefficient, 23-80°C	10^{-4} °C^{-1}	1.1	
Thermal conductivity, melt	W m^{-1} K^{-1}	0.27	Lee, S, Kim, H M, Seong, D G, Lee, D, Carbon, 143, 650-9, 2019.
Glass transition temperature	°C	13-15, -1.5 to -57.7 (cross-metathesis polymerization)	-; Zeng, F-R, Xu, J, Xiong, Q, Qin, K-X, Li, Z-C, Polymer, 185, 121936, 2019.
Specific heat capacity	J K^{-1} kg^{-1}	1730-1,800	
Heat of fusion	J g^{-1}	62.6-73 (melt quenched material); 90.9-97.1 (solution crystallized)	Waddon, A J; Karttunen, N R; Lesser, A J, Macromolecules, 32, 423-28, 1999.
Temperature index (50% tensile strength loss after 20,000 h/5000 h)	°C	90	
Heat deflection temperature at 0.45 MPa	°C	210	
Heat deflection temperature at 1.8 MPa	°C	105	
Vicat temperature VST/A/50	°C	215	
Vicat temperature VST/B/50	°C	205	

PK polyketone

PARAMETER	UNIT	VALUE	REFERENCES
Dielectric constant at 1000 Hz/1 MHz	-	5.7/5.2	
Dissipation factor at 1000 Hz	E-4	120	
Dissipation factor at 1 MHz	E-4	400	
Volume resistivity	ohm-m	1E11	
Surface resistivity	ohm	1E14	
Arc resistance	s	60-120	
Coefficient of friction	-	0.07 (static); 0.49 (dynamic)	Kelley, J W, Antec, 3028-34, 1998.
Permeability to nitrogen, 25°C	$cm^3\ cm\ cm^{-3}\ min^{-1}\ atm^{-1}\ x\ 10^9$	1.23	De Nobile, M A; Mensitieri, G; Sommazzi, A, Polymer, 36, 26, 4943-50, 1995.
Permeability to oxygen, 25°C	$cm^3\ cm\ cm^{-3}\ min^{-1}\ atm^{-1}\ x\ 10^9$	6.43	De Nobile, M A; Mensitieri, G; Sommazzi, A, Polymer, 36, 26, 4943-50, 1995.
Diffusion coefficient of nitrogen	$cm^2\ s^{-1}\ x10^9$	2.28	De Nobile, M A; Mensitieri, G; Sommazzi, A, Polymer, 36, 26, 4943-50, 1995.
Diffusion coefficient of oxygen	$cm^2\ s^{-1}\ x10^9$	5.6-7.73	Backman, A; Lange, J; Hedenqvist, M S, J. Polym. Sci. B, 42, 947-55, 2004; De Nobile, M A; Mensitieri, G; Sommazzi, A, Polymer, 36, 26, 4943-50, 1995.
Diffusion coefficient of water vapor	$cm^2\ s^{-1}\ x10^9$	29.3	De Nobile, M A; Mensitieri, G; Sommazzi, A, Polymer, 36, 26, 4943-50, 1995.

MECHANICAL & RHEOLOGICAL PROPERTIES

PARAMETER	UNIT	VALUE	REFERENCES
Tensile strength	MPa	55-63	
Tensile modulus	MPa	1,600-2,300	
Tensile stress at yield	MPa	55-70	
Elongation	%	300-350	
Tensile yield strain	%	17-25	
Flexural strength	MPa	55	
Flexural modulus	MPa	1,600	
Young's modulus	MPa	1,500	
Charpy impact strength, unnotched, 23°C	$kJ\ m^{-2}$	NB	
Charpy impact strength, unnotched, -30°C	$kJ\ m^{-2}$	NB	
Charpy impact strength, notched, 23°C	$kJ\ m^{-2}$	1.8-20	
Charpy impact strength, notched, -30°C	$kJ\ m^{-2}$	0.4	
Izod impact strength, unnotched, 23°C	$J\ m^{-1}$	NB	
Izod impact strength, notched, 23°C	$J\ m^{-1}$	2.4-2.7	
Izod impact strength, notched, -40°C	$J\ m^{-1}$	0.5	

PK polyketone

PARAMETER	UNIT	VALUE	REFERENCES
Abrasion resistance (ASTM D1044)	mg/1000 cycles	12	
Shore D hardness	-	75	
Rockwell hardness	-	R105	
Shrinkage	%	2-2.1	
Intrinsic viscosity, 25°C	dl g^{-1}	1.21-2.84	Zuiderduin, W C J; Homminga, D S; Huetink, H J; Gaymans, R J, Polymer, 44, 6361-70, 2003.
Melt index, 240°C/2.16 kg	g/10 min	2.7-80	
Water absorption, equilibrium in water at 23°C	%	2.1	
Moisture absorption, equilibrium 23°C/50% RH	%	0.5	

CHEMICAL RESISTANCE

Acid dilute/concentrated	-	good	
Alcohols	-	very good	
Alkalis	-	good	
Aliphatic hydrocarbons	-	very good	
Aromatic hydrocarbons	-	very good	
Esters	-	very good	
Greases & oils	-	very good	
Halogenated hydrocarbons	-	very good	
Ketones	-	very good	

FLAMMABILITY

Limiting oxygen index	% O$_2$	21; 25	-; Lin, H, Pearson, A, Kazemi, Y,, H E, Polym. Deg. Stab., 179, 109260, 2020.
Char at 900°C	%	18.06	Lin, H, Pearson, A, Kazemi, Y,, H E, Polym. Deg. Stab., 179, 109260, 2020.
Total heat release	MJ m^{-2}	41	Lin, H, Pearson, A, Kazemi, Y,, H E, Polym. Deg. Stab., 179, 109260, 2020.
Peak of heat release rate	kW m^{-2}	464.4	Lin, H, Pearson, A, Kazemi, Y,, H E, Polym. Deg. Stab., 179, 109260, 2020.
UL rating	-	HB	

PROCESSING

Processing methods	-	blow molding, electrospinning, extrusion, injection molding, melt spinning	
Processing temperature	°C	240-260; 220-260 (injection)	
Processing pressure	MPa	55 (holding); 40-80 (injection)	
Process time	s	23 (cycle time)	
Applications	-	appliance, automotive, bottles, electrical, fibers, fuel cells, membranes, nanofibers, powder coatings, support for enzyme immobilization	
Outstanding properties	-	chemical resistance, fast crystallization, stiffness, toughness	

PK polyketone

PARAMETER	UNIT	VALUE	REFERENCES
BLENDS			
Suitable polymers	-	CSR (core-shell rubber), PF, PP, PVC	
ANALYSIS			
FTIR (wavenumber-assignment)	cm^{-1}/-	CH$_2$ – 1408, 1331, 1261	Ohsawa, O; Lee, K-H; Kim, B-S; Lee, S; Kim, I-S, Polymer, 2007-12, 2010.
Raman (wavenumber-assignment)	cm^{-1}/-	CH$_2$ – 1260, 1350-1500; C=O – 1710	Lagaron, J M; Powell, A K; Davidson, N S, Macromolecules, 33, 1030-35, 2000.
NMR (chemical shifts)	ppm	C NMR: –CO– – 208.1; –Ph – 139.9, 133.1, 128.5, 16.6, –CHCH$_2$– – 54.2, 45.4 (product of copolymerization of CO and styrene)	Guo, J T; Ye, Y Q; Gao, S; Feng, Y K, J. Molecular Catalysis, 307A, 121-27, 2009.
x-ray diffraction peaks	degree	17.38, 21.75, 24.65, 25.97, 31.47, 37.87, 39.38, 41.84	Ohsawa, O; Lee, K-H; Kim, B-S; Lee, S; Kim, I-S, Polymer, 2007-12, 2010.

PLA poly(lactic acid)

PARAMETER	UNIT	VALUE	REFERENCES
GENERAL			
Common name	-	poly(lactic acid)	
CAS name	-	poly[oxy(1-methyl-2-oxo-1,2-ethanediyl)]; 1,4-dioxane-2,5-dione, 3,6-dimethyl-, homopolymer	
Acronym	-	PLA	
CAS number	-	51063-13-9, 26680-10-4, 34346-01-5	
Formula	-		
HISTORY			
Person to discover	-	Carother, W H; Dourough, G L; Van Natta F J. Filachione, E M	Carother, W H; Dourough, G L; Van Natta F j, J. Am. Chem. Soc., 54, 761-72, 1932; Filachione, E M, US Patent 2,396,994, USA, Mar. 19, 1946.
Date	-	1932, 1946	
Details	-	first synthesis by DuPont scientists; lactic acid was polymerized in the presence of p-toluenesulfonic acid	
SYNTHESIS			
Monomer(s) structure	-		
Monomer(s) CAS number(s)	-	4511-42-6	
Monomer(s) molecular weight(s)	dalton, g/mol, amu	144.13	
Monomer ratio	-	100% or less (in blends)	
Concentration of L-lactide	%	94-98	
Formulation example	-	lactic acid and tin catalyst	
Method of synthesis	-	lactic acid is heated at 150°C to obtain oligomeric PLA (polymerization degree: 1-8). Oligomers are heated at 180°C under vacuum for 5 hours to give PLA having molecular weight of 100,000	Bastioli, C, Handbook of Biodegradable Polymers, Rapra, 2005.
Temperature of polymerization	°C	150-180	
Time of polymerization	h	5	
Pressure of polymerization	Pa	vacuum	
Catalyst	-	tin	
Number average molecular weight, M_n	dalton, g/mol, amu	74,000-660,000	
Mass average molecular weight, M_w	dalton, g/mol, amu	80,000-380,000; 4000-6000 (DL); 100000 (L)	
Polydispersity, M_w/M_n	-	1.5-3.79	Bastioli, C, Handbook of Biodegradable Polymers, Rapra, 2005.

HANDBOOK OF POLYMERS 3rd Edition, Copyrights 2022; ChemTec Publishing

PLA poly(lactic acid)

PARAMETER	UNIT	VALUE	REFERENCES
STRUCTURE			
Crystallinity	%	20-47; 25-70 (L-PLA); 10-20 (film); 65 (fiber), 20-36 (DSC); 20-44 (WAXD)	Bastioli, C, Handbook of Biodegradable Polymers, Rapra, 2005; Rudnik, E; Briassoulis, D, Ind. Crops Prod., 33, 648-58, 2011; Tsai, C-C; Wu, R-J; Cheng, H-Y; Li, S-C; Siao, Y-Y; Kong, D-C; Jang, G-W, Polym. Deg. Stab., 95, 1292-98, 2010.
Cell type (lattice)	-	orthorhombic (α), hexagonal (α'), trigonal (β), monoclinic (γ)	
Cell dimensions	nm	a:b:c=1.06:0.61:2.88 (orthorhombic, α); a=b:c=1.052:0.88 (trigonal, β); a:b:c=0.995:0.625:0.88 (monoclinic, γ)	Johnson, C M; Sugiharto, A B; Roke, S, Chem. Phys. Lett., 449, 191-95, 2007; Lin, T T; Liu, X Y; He, C, Polymer, 51, 2779-85, 2010.
Polymorphs	-	α, α', β, γ	Kalish, J P; Zeng, X; Yang, X; Hsu, S L, Polymer, in press, 2011.
Lamellae thickness	nm	2.03-28.6	Tsai, C-C; Wu, R-J; Cheng, H-Y; Li, S-C; Siao, Y-Y; Kong, D-C; Jang, G-W, Polym. Deg. Stab., 95, 1292-98, 2010.
Avrami constants, k/n	-	n=1.8-2.3	Tsai, C-C; Wu, R-J; Cheng, H-Y; Li, S-C; Siao, Y-Y; Kong, D-C; Jang, G-W, Polym. Deg. Stab., 95, 1292-98, 2010.
COMMERCIAL POLYMERS			
Some manufacturers	-	Cargill Dow; Durect	
Trade names	-	PLA; Lactel	
PHYSICAL PROPERTIES			
Density at 20°C	g cm^{-3}	1.21-1.29	
Refractive index, 20°C	-	1.35-1.45	
Transmittance	%	2.2	
Melting temperature, DSC	°C	164-178; 180-184 (L-PLA)	Bastioli, C, Handbook of Biodegradable Polymers, Rapra, 2005.
Decomposition temperature	°C	>200	
Glass transition temperature	°C	55-75	Bastioli, C, Handbook of Biodegradable Polymers, Rapra, 2005.
Specific heat capacity	J K^{-1} kg^{-1}	540-600	
Heat of fusion	kJ mol^{-1}	146	
Vicat temperature VST/A/50	°C	55-60	
Enthalpy of fusion	J g^{-1}	21.9-43.8	
Hansen solubility parameters, dD, dP, dH	(J cm^{-3})$^{0.5}$	18.50, 9.70, 6.0	Agrawal, A; Saran, A D; Rath, S S; Khanna, A, Polymer, 45, 8603-12, 2004.
Radius of interaction	(J cm^{-3})$^{0.5}$	13.53	Agrawal, A; Saran, A D; Rath, S S; Khanna, A, Polymer, 45, 8603-12, 2004.
Hildebrand solubility parameter	MPa$^{0.5}$	calc.=19.2-20.3; exp.=19.0-21.0	Auras, R; Harte, B; Selke, S, Antec, 2862-6, 2003.
Surface resistivity	ohm	1.9E11	Khoddami, A; Avinc, O; Ghahremanzadeh, F; Prog. Org. Coat., in press, 2011.
Permeability to nitrogen, 25°C	cm^3 cm^{-3} cmHg^{-1} x 10^4	2.2	Bao, L; Dorgan, J R; Knauss, D; Hait, S; Oliveira, N S; Maruccho, I M, J. Membrane Sci., 166-172, 2006.

480   **HANDBOOK OF POLYMERS** 3rd Edition, Copyrights 2022; ChemTec Publishing

PLA poly(lactic acid)

PARAMETER	UNIT	VALUE	REFERENCES
Permeability to oxygen, 25°C	$cm^3 cm^{-3}$ $cmHg^{-1}$ x 10^4	2.2-4.9	Bao, L; Dorgan, J R; Knauss, D; Hait, S; Oliveira, N S; Maruccho, I M, J. Membrane Sci., 166-172, 2006.
Permeability to water vapor, 25°C	$cm^3 m^{-2}$ $24h^{-1}$	110	Zenkiewicz, M; Richert, J; Rytlewski, P; Moraczewski, K; Stepczynska, M; Karasiewicz, T, Polym. Test., 28, 412-18, 2009.
Diffusion coefficient of nitrogen	$cm^2 s^{-1}$ $x10^8$	2.4	Bao, L; Dorgan, J R; Knauss, D; Hait, S; Oliveira, N S; Maruccho, I M, J. Membrane Sci., 166-172, 2006.
Diffusion coefficient of oxygen	$cm^2 s^{-1}$ $x10^8$	5.6-7.6	Bao, L; Dorgan, J R; Knauss, D; Hait, S; Oliveira, N S; Maruccho, I M, J. Membrane Sci., 166-172, 2006.

MECHANICAL & RHEOLOGICAL PROPERTIES

PARAMETER	UNIT	VALUE	REFERENCES
Tensile strength	MPa	52-72; 27-41 (DL); 55-82 (L)	
Tensile modulus	MPa	2700-16000	
Tensile stress at yield	MPa	65.6-77	Carrasco, F; Pages, P; Gamez-Perez, J; Santana, O O; Maspoch, M L, Polym. Deg. Stab., 95, 116-25, 2010.
Tensile creep modulus, 1000 h, elongation 0.5 max	MPa	48-70	
Elongation	%	4-6; 3-10 (DL); 5-10 (L)	
Tensile yield strain	%	2.4-10	
Flexural strength	MPa	83	
Flexural modulus	MPa	1,000-3,800	
Elastic modulus	MPa	3400	Perić, M, Putz, R, Paulik, C, Eur. Polym. J., 114, 426-33, 2019.
Young's modulus	MPa	3,700-4,100	
Izod impact strength, notched, 23°C	$J m^{-1}$	13-24.6	
Tenacity (fiber)	$cN tex^{-1}$	32-36	
Intrinsic viscosity, 25°C	$dl g^{-1}$	0.15-1.2	
Melt index, 210°C/2.16 kg	g/10 min	7	Li, J, Peng, W-J, Fu, Z-J, Wang, M, Compos. Pat B: Eng. 171, 204-13, 2019.
Water absorption, equilibrium in water at 23°C	%	0.5	

CHEMICAL RESISTANCE

PARAMETER	UNIT	VALUE	REFERENCES
Alcohols	-	poor	
Aromatic hydrocarbons	-	poor	
Esters	-	poor	
Greases & oils	-	good	
Halogenated hydrocarbons	-	poor	
Ketones	-	poor	
Good solvent	-	acetone, benzene, chloroform, m-cresol, dichloromethane, dioxane, DMF, ethyl acetate, isoamyl alcohol, toluene, xylene	

PLA poly(lactic acid)

PARAMETER	UNIT	VALUE	REFERENCES
FLAMMABILITY			
Limiting oxygen index	% O_2	19; 23-26 (with flame retardant)	
Heat release	kW m^{-2}	427	Wei, L-L; Wang, D-Y; Chen, H-B; Chen, L; Wang, X-L; Wang, Y-Z, Polym. Deg. Stab., in press, 2011.
NBS smoke chamber	m2 kg^{-1}	63	
Char at 500°C	%	0-1.4; 1.4-3.2 (with flame retardant)	Wei, L-L; Wang, D-Y; Chen, H-B; Chen, L; Wang, X-L; Wang, Y-Z, Polym. Deg. Stab., in press, 2011.
Heat of combustion	J g^{-1}	19,000	Perepelkin, K E, Fibre Chem., 34, 2, 2002.
UL 94 rating	-	V-0 (FR)	
WEATHER STABILITY			
Depth of UV penetration	μm	bulk erosion	
Important initiators and accelerators	-	nano-titanium dioxide	
Products of degradation	-	double bonds, chain cleavage	
Stabilizers	-	Phenolic antioxidant: pentaerythritol tetrakis(3-(3,5-di-tert-butyl-4-hydroxyphenyl)propionate); HAS: decanedioic acid, bis(2,2,6,6-tetramethyl-1-(octyloxy)-4-piperidinyl) ester, reaction products with 1,1-dimethylethylhydroperoxide and octane; Phosphite: bis(2-ethylhexyl)phosphite	
BIODEGRADATION			
Typical biodegradants	-	composting: complete fragmentation in 15 days; degradation complete in 4.8 years at 25°C; lipases from *Cryptococcus sp.* and proteases from *Bacillus* strains	Hartmann, M; Whiteman, N, Antec, 4-8, 2001; Kawai, F; Nakadai, KK; Nishioka, E; Nakajima, H; Ohara, H; Masaki, Iefuji, H, Polym. Deg. Stab., 96, 1343-48, 2011.
TOXICITY			
NFPA: Health, Flammability, Reactivity rating	-	1/1/0	
Carcinogenic effect	-	not listed by ACGIH, NIOSH, NTP	
Oral rat, LD$_{50}$	mg kg^{-1}	>5,000	
Skin rabbit, LD$_{50}$	mg kg^{-1}	>2,000	
ENVIRONMENTAL IMPACT			
Aquatic toxicity, *Daphnia magna*, LC$_{50}$, 48 h	mg l^{-1}	1,000	
Power consumption for production	MJ kg^{-1}	92 (fiber)	Perepelkin, K E, Fibre Chem., 34, 2, 2002.
CO2 liberation	kg kg^{-1}	4.1-6.5	Perepelkin, K E, Fibre Chem., 34, 2, 2002.
PROCESSING			
Typical processing methods	-	extrusion, extrusion coating, injection molding, microcellular foaming, reactive extrusion, spinning	Liao, J, Brosse, N, Hoppe, S, Pizzi, A, Mater. Design, 191, 108603, 2020.
Preprocess drying: temperature/time/residual moisture	°C/h/%	80 (vac)/8/	
Processing temperature	°C	220-255 (extrusion); 280-300 (fibers)	

PLA poly(lactic acid)

PARAMETER	UNIT	VALUE	REFERENCES
Processing pressure	MPa	82 (injection)	
Additives used in final products	-	Plasticizers: polyethylene glycol, polypropylene glycol, partial fatty ester, glucose monoester, citrate, adipate and azelate esters, epoxidized soybean oil, acetylated coconut oil, linseed oil, acetyl tributyl citrate, glycerol triacetate, glycerol tripropionate; Antistatics: ethoxylated fatty amines, polyethylene glycol ester, quaternary ammonium salt; Antiblocking: diatomaceous earth, talc; Slip: erucamide	
Applications	-	clip, dermal fillers, drug delivery, envelope with window, fabrics, fibers, film, sheet, shopping bags, synthetic paper, tissue engineering, trash bags	Liu, S, Qin, S, He, M, Wang, H, Compos. Part B: Eng, 199, 108238, 2020.
Outstanding properties	-	sustainable, biodegradable	

BLENDS

Suitable polymers	-	chitosan, PC, PCL, PEG, PET, PHB, PR, PVP, starch	

ANALYSIS

FTIR (wavenumber-assignment)	cm^{-1}/-	C=O − 1748; reference − 1451	Rudnik, E; Briassoulis, D, Ind. Crops Prod., 33, 648-58, 2011.
Raman (wavenumber-assignment)	cm^{-1}/-	C–O − 1128; C–C − 1044	Yang, X; Kang, S; Yang, Y; Aou, K; Hsu, S L, Polymer, 45, 4241-48, 2004.
NMR (chemical shifts)	ppm	C NMR: C=O − 170.8; −CH − 70.5; −CH$_3$ − 18.1	Zhang, X; Espiritu, M; Bilyk, A; Kurniawan, L, Polym. Deg. Stab., 93, 1964-70, 2008.

PLGA poly(DL-lactide-co-glycolide)

PARAMETER	UNIT	VALUE	REFERENCES
GENERAL			
Common name	-	poly(DL-lactide-co-glycolide)	
CAS name	-	26780-50-7	
Acronym	-	PLGA	
SYNTHESIS			
Monomer ratio	-	lactide:glycolide=50:50 to 85:15; 75/25	Pereira, M C; Hill, L E; Zambiazi, R C; Metens-Talcott, S; Talcott, S; Gomes, C L, LWT - Food Sci. Technol., 63, 100-7, 2015; Monge, M, Fornaguera, C, Quero, C, Solans, S, Eur. J. Pharm. Biopharm., 156, 155-64, 2020.
Method of synthesis	-	biosynthesis	
Mass average molecular weight, M_w	dalton, g/mol, amu	3,000-15,000; 10,000; 32,595-81,919	Pereira, M C; Hill, L E; Zambiazi, R C; Metens-Talcott, S; Talcott, S; Gomes, C L, LWT - Food Sci. Technol., 63, 100-7, 2015; Monge, M, Fornaguera, C, Quero, C, Solans, S, Eur. J. Pharm. Biopharm., 156, 155-64, 2020; Hadar, J, Skidmore, S, Garner, J, Jiang, X, J. Controlled Release, 304, 75-89, 2019.
Number average molecular weight, M_n	dalton, g/mol, amu	22,779-50589	Hadar, J, Skidmore, S, Garner, J, Jiang, X, J. Controlled Release, 304, 75-89, 2019.
COMMERCIAL POLYMERS			
Some manufacturers	-	Evonik; Mitsui; Akina	
Trade names	-	Resomer 752H; Akina	Monge, M, Fornaguera, C, Quero, C, Solans, S, Eur. J. Pharm. Biopharm., 156, 155-64, 2020.
PHYSICAL PROPERTIES			
Color	-	white to light tan	
Odor	-	odorless	
Glass transition temperature	°C	45-55	
CHEMICAL RESISTANCE			
Good solvent		CH_2Cl_2, $CHCl_3$, DMF, DMSO, THF	
PROCESSING			
Additives used in final products	-	wollastonite	
Applications	-	bone substitute, drug delivery, nanoencapsulation (encapsulation efficiency of bioflavonoid - 70%)	Maity, S, Chakraborti, A S, Eur. Polym. J., 134, 109818, 2020.
Outstanding properties	-	biosynthesis, biodegradation	
BLENDS			
Suitable polymers	-	chitosan, PEG	

PLS poly(L-serine)

PARAMETER	UNIT	VALUE	REFERENCES
GENERAL			
Common name	-	poly(L-serine)	
Acronym	-	PSL	
CAS number	-	25821-52-7	
SYNTHESIS			
Method of synthesis	-	electropolymerization; ring-opening polymerization	Chitravathi, S; Swamy, B E K; Mamatha, G P; Sherigara, J. Mol. Liq., 160, 193-9, 2011.
Mass average molecular weight, M_w	dalton, g/mol, amu	3,000-10,000	
Polymerization degree (number of monomer units)	-	20-100	
STRUCTURE			
Crystallinity	%	15	Gupta, A; Tandon, P; Gupta, V D; Rastogi, S, Polymer, 38, 10, 2389-97, 1997.
PHYSICAL PROPERTIES			
Specific heat capacity	$J\ K^{-1}\ kg^{-1}$	959	

PLT poly(L-tyrosine)

PARAMETER	UNIT	VALUE	REFERENCES
GENERAL			
Common name	-	poly(L-tyrosine)	
CAS name	-	25619-78-7	
Acronym	-	PLT	
SYNTHESIS			
Mass average molecular weight, M_w	dalton, g/mol, amu	10,000-100,000	Higashi, T; Nakajima, Y; Kojima, M; Ishii, K; Inoue, A; Maekawa, T, Hanajiri, T, Chem. Phys. Lett., 501, 451-4, 2011.
PHYSICAL PROPERTIES			
Specific heat capacity	$J\,K^{-1}\,kg^{-1}$	183.6	
TOXICITY			
Carcinogenic effect	-	not listed by IARC and ACGIH	
ANALYSIS			
FTIR (wavenumber-assignment)	cm^{-1}/-	amide I – 1695, 1632	Loksztejn, A; Dzwolak, W; Krysinski, Bioelectochemistry, 72, 1, 34-40, 2008.

PMA poly(methyl acrylate)

PARAMETER	UNIT	VALUE	REFERENCES
GENERAL			
Common name	-	poly(methyl acrylate)	
IUPAC name	-	poly(methyl propenoate)	
CAS name	-	2-propenoic acid, methyl ester, homopolymer	
Acronym	-	PMA	
CAS number	-	9003-21-8	
Formula		$\left[\begin{array}{c} CH_2CH \\ | \\ O=COCH_3 \end{array} \right]_n$	
HISTORY			
Person to discover	-	Neher, H T	Neher, H T, US Patent 2,032,663, Roehm and Haas, Mar. 3, 1936.
Date	-	1936	
Details	-	laminated glass obtained with *in situ* polymerization of PMA	
SYNTHESIS			
Monomer(s) structure	-	$H_2C=CHCOCH_3$ (with O double bond)	
Monomer(s) CAS number(s)	-	96-33-3	
Monomer(s) molecular weight(s)	dalton, g/mol, amu	86.04	
Monomer ratio	-	100%	
Method of synthesis	-	because of high heat of polymerization, the best conditions of temperature control are given in emulsion polymerization; catalyst of polymerization is dissolved in water	
Time of polymerization	min	20-25	
Pressure of polymerization	Pa	atmospheric	
Catalyst	-	ammonium peroxydisulfate or potassium peroxydisulfate	
Mass average molecular weight, M_w	dalton, g/mol, amu	38,000-555,000	
Polydispersity, M_w/M_n	-	1.15-1.26	
Molar volume at 298K	cm^3 mol^{-1}	calc.=70.6 (amorphous)	
Van der Waals volume	cm^3 mol^{-1}	calc.=45.9 (amorphous)	
Degree of branching	%	1.92	Castignolles, P; Graf, R; Parkinson, M; Wilhelm, M; Gaborieau, M, Polymer, 50, 2373-83, 2009.
STRUCTURE			
Entanglement molecular weight	dalton, g/mol, amu	calc.=9070	
PHYSICAL PROPERTIES			
Density at 20°C	g cm^{-3}	1.19-1.22	
Refractive index, 20°C	-	1.4793	

PMA poly(methyl acrylate)

PARAMETER	UNIT	VALUE	REFERENCES
Decomposition temperature	°C	227	Castignolles, P; Graf, R; Parkinson, M; Wilhelm, M; Gaborieau, M, Polymer, 50, 2373-83, 2009.
Glass transition temperature	°C	calc.=6-9; exp.=3.5-21; 17.8-19.9 (absorbed on treated silica); 23.2-24.3 and 40.4-42.3 and 60.7 (absorbed on untreated silica); 14.3	Castignolles, P; Graf, R; Parkinson, M; Wilhelm, M; Gaborieau, M, Polymer, 50, 2373-83, 2009; Metin, B; Blum, F D, Langmuir, 26, 7, 5226-31, 2010; Azmar, A, Saaid, F, Winie, T, Mater. Today: Proc. 4, 4C, 5100-7, 2017.
Hildebrand solubility parameter	$MPa^{0.5}$	calc.=18.21; exp.=20.7	
Surface tension	$mN\ m^{-1}$	41.0-42.7	
Surface free energy	$mJ\ m^{-2}$	39.8	

CHEMICAL RESISTANCE			
Alcohols	-	good	
Aliphatic hydrocarbons	-	good	
Aromatic hydrocarbons	-	poor	
Esters	-	poor	
Halogenated hydrocarbons	-	poor	
Ketones	-	poor	
Good solvent	-	aromatic hydrocarbons, chlorinated hydrocarbons, esters, glycolic ester ethers, THF	
Non-solvent	-	alcohols, aliphatic hydrocarbons, carbon tetrachloride, diethyl ether	

FLAMMABILITY			
Ignition temperature	°C	>250	
Autoignition temperature	°C	304	
Volatile products of combustion	-	CO, CO_2, methyl formate, formaldehyde, methanol	

WEATHER STABILITY			
Products of degradation	-	formaldehyde, methanol, and methyl formate	

TOXICITY			
Carcinogenic effect	-	not listed by ACGIH, NIOSH, NTP	

PROCESSING			
Applications	-	rheological additive for biodiesel	Monirul, I M, Kalam, M A, Masjuki, H H, Ruhul, A M, Rewable Energy, 101, 702-12, 2017.

BLENDS			
Suitable polymers	-	PEO, PLA, PMMA, PS, PVAc, PVF	

ANALYSIS			
NMR (chemical shifts)	ppm	H NMR: CH_3 – 2.1-14 triplet; methine – 2.3; C NMR: C=O – 174.6-175	Brar, A S; Goyal, A K; Hooda, S, J. Molecular Structure, 885, 15-17, 2008.

PMAA poly(methacrylic acid)

PARAMETER	UNIT	VALUE	REFERENCES

GENERAL

Common name	-	poly(methacrylic acid)	
CAS name	-	2-propenoic acid, 2-methyl-, homopolymer	
Acronym	-	PMAA	
CAS number	-	25087-26-7	
Formula			

HISTORY

Person to discover	-	Strain, D E	Strain, D E, US Patent 2,133,257, DuPont, Oct. 11, 1938.
Date	-	1938	
Details	-	polymerization conducted in the presence of emulsifying agent and initiator	

SYNTHESIS

Monomer(s) structure	-		
Monomer(s) CAS number(s)	-	79-41-4	
Monomer(s) molecular weight(s)	dalton, g/mol, amu	86.06	
Monomer ratio	-	100%	
Method of synthesis	-	free radical bulk polymerization in the presence of benzoyl peroxide	Vinu, R; Madrs, G, Polym. Deg. Stab., 93, 1440-49, 2008.
Temperature of polymerization	°C	60	
Catalyst	-	anatase titania	
Yield	%	67-77	Bai, F; Huang, B; Yang, X; Huang, W, Eur. Polym. J., 43, 3923-32, 2007.
Mass average molecular weight, M_w	dalton, g/mol, amu	25,000-350,000; 30,000	-; Kagkoura, A, Sentoukas, T, Nakanishi, Y, Tagmatarchis, N, Chem. Phys. Lett., 716, 1-5, 2019.
Polydispersity, M_w/M_n	-	1.03-2.9; 2.4	-; Vshivkov, T. S. Soliman, E. S. Kluzhin, A. A. Kapitanov, A A, J. Mol. Liquids, 294, 111551, 2019.
Radius of gyration	nm	3.6-4	Muroga, Y; Yoshida, T; Kawaguchi, S, Biophys. Chem., 81, 45-57, 1999.
Hydrodynamic radius	nm	10-13	Sitar, S, Aseyev, V, Žagar, E, Kogej, K, Polymer, 174, 1-10, 2019.
Mean-square radius of chain's cross-section,	nm	0.3-0.42	Muroga, Y; Yoshida, T; Kawaguchi, S, Biophys. Chem., 81, 45-57, 1999.
Mean distance between chain ends	nm	19	Vshivkov, T. S. Soliman, E. S. Kluzhin, A. A. Kapitanov, A A, J. Mol. Liquids, 294, 111551, 2019.

STRUCTURE

Stereoregularity	%	iso – 13, hetero – 52, syndio – 35	Muroga, Y; Yoshida, T; Kawaguchi, S, Biophys. Chem., 81, 45-57, 1999.

PMAA poly(methacrylic acid)

PARAMETER	UNIT	VALUE	REFERENCES
Tacticity	%	93 of isotactic, 3 of syndiotactic, 4 of atactic triads	

PHYSICAL PROPERTIES

PARAMETER	UNIT	VALUE	REFERENCES
Density at 20°C	g cm^{-3}	1.285	
Melting temperature, DSC	°C	205	
Decomposition temperature	°C	220	
Glass transition temperature	°C	228-230	
Speed of sound	m s^{-1}	3,350	

MECHANICAL & RHEOLOGICAL PROPERTIES

PARAMETER	UNIT	VALUE	REFERENCES
Tensile strength	MPa	2.53	
Intrinsic viscosity, 25°C	dl g^{-1}	1.0	

CHEMICAL RESISTANCE

PARAMETER	UNIT	VALUE	REFERENCES
Acid dilute/concentrated	-	non resistant	
Alcohols	-	non resistant	
Alkalis	-	non resistant	
Aliphatic hydrocarbons	-	resistant	
Aromatic hydrocarbons	-	resistant	
Esters	-	resistant	
Greases & oils	-	resistant	
Ketones	-	resistant	
Θ solvent, Θ-temp.=27.1°C	-	DMF/dioxane=5/7	Sivadasa, K; Gundiah, S, Polymer, 28, 8, 1426-28, 1987.
Good solvent	-	alcohols, dioxane, DMF, ethanol, methanol, water,	
Non-solvent	-	acetone, aliphatic hydrocarbons, benzene, diethyl ether, esters, ketones	

FLAMMABILITY

PARAMETER	UNIT	VALUE	REFERENCES
Autoignition temperature	°C	500	
Char at 500°C	%	0.5	

WEATHER STABILITY

PARAMETER	UNIT	VALUE	REFERENCES
Excitation wavelengths	nm	290	Ruiz-Perez, L; Pryke, A; Sommer, M; Battaglia, G; Soutar, I; Swanson, L; Geoghegan, M, Macromolecules, 41, 2203-11, 2008.
Transmittance	%	100 nm – 58.4; 300 nm – 19.9	Matsuzawa, N N; Oizumi, H; Mori, S; Irie, S; Shirayone, S; Yano, E; Okazaki, S; Ishitani, S; Dixon, D A, Jpn. J. Appl. Phys., 38, 7109-13, 1999.

TOXICITY

PARAMETER	UNIT	VALUE	REFERENCES
NFPA: Health, Flammability, Reactivity rating	-	1/1/0; 0/1/0 (HMIS)	
Carcinogenic effect	-	not listed by ACGIH, NIOSH, NTP	

PMAA poly(methacrylic acid)

PARAMETER	UNIT	VALUE	REFERENCES
PROCESSING			
Applications	-	cosmetics (thickening and viscosity enhancement), flocculants, super absorbent	
BLENDS			
Suitable polymers	-	casein, chitosan, PS, starch	
ANALYSIS			
FTIR (wavenumber-assignment)	cm^{-1}/-	C=O – 1736, 1719, 1695, 1679; C-C-O – 1262, 1242, C-O – 1185, 1154	Tajiri, T; Morita, S; Ozaki, Y, Polymer, 50, 5765-70, 2009.

PMAN polymethacrylonitrile

PARAMETER	UNIT	VALUE	REFERENCES
GENERAL			
Common name	-	polymethacrylonitrile	
CAS name	-	2-propenenitrile, 2-methyl-, homopolymer	
Acronym	-	PMAN	
CAS number	-	25067-61-2	
Formula			

$$\left[CH_2\underset{\underset{CH_3}{|}}{\overset{\overset{CN}{|}}{C}} \right]_n$$

PARAMETER	UNIT	VALUE	REFERENCES
HISTORY			
Person to discover	-	Howk, B W	Howk, B W, US Patent 2,232,785, Du Pont, Feb. 25, 1941.
Date	-	1941	
Details	-	monomer was polymerized in the presence of hydroquinone with 50% yield	
SYNTHESIS			
Monomer(s) structure	-		

$$H_2C{=}\underset{\underset{CH_3}{|}}{C}CN$$

PARAMETER	UNIT	VALUE	REFERENCES
Monomer(s) CAS number(s)	-	126-98-7	
Monomer(s) molecular weight(s)	dalton, g/mol, amu	67.09	
Monomer ratio	-	100%	
Method of synthesis	-	methacrylonitrile is polymerized in the presence of free radical initiator (AIBN or BP)	Saunier, J; Chaix, N; Alloin, F; Bilieres, J-P; Sanchez, J-Y, Electro-chim. Acta, 47, 1321-26, 2002.
Temperature of polymerization	°C	60-80	
Time of polymerization	h	48	
Catalyst	-	diethylmagnesium	
Yield	%	60	
Number average molecular weight, M_n	dalton, g/mol, amu	16,000-59,000	
Mass average molecular weight, M_w	dalton, g/mol, amu	140,000	
Polydispersity, M_w/M_n	-	1.71-1.82	
STRUCTURE			
Cell type (lattice)	-	monoclinic	
Cell dimensions	nm	a:b:c=1.35:0.771:0.762	
Unit cell angles	degree	β=97.49	
Tacticity	%	19 – isotactic, 49 – heterotactic, 32 – syndiotactic	Bashir, Z; Packer, E J; Herbert, I R; Price, D M, Polymer, 33, 2, 373-78, 1992.
PHYSICAL PROPERTIES			
Density at 20°C	g cm^{-3}	1.13	
Color	-	white	

PMAN polymethacrylonitrile

PARAMETER	UNIT	VALUE	REFERENCES
Refractive index, 20°C	-	1.52; 1.5932	
Odor	-	odorless	
Melting temperature, DSC	°C	220-250	
Decomposition temperature	°C	250 (depolymerization)	
Glass transition temperature	°C	120	
Hansen solubility parameters, δ_D, δ_P, δ_H	MPa$^{0.5}$	17.2, 14.4, 7.6	
Interaction radius		3.8	
Hildebrand solubility parameter	MPa$^{0.5}$	21.9-23.7	
Dielectric constant at 60 Hz/1 MHz	-	4.14/3.3	
Dissipation factor at 60 Hz		0.046	
Dissipation factor at 1 MHz		0.025	
Volume resistivity	ohm-m	1.14E14	
Surface free energy	mJ m^{-2}	42.3	

MECHANICAL & RHEOLOGICAL PROPERTIES			
Water absorption, equilibrium in water at 23°C	%	0.24	

CHEMICAL RESISTANCE			
Acid dilute/concentrated	-	good	
Alcohols	-	good	
Aliphatic hydrocarbons	-	good	
Aromatic hydrocarbons	-	good	
Esters	-	good	
Halogenated hydrocarbons	-	poor	
Ketones	-	good	
Good solvent	-	acetonitrile, acrylonitrile, aniline, m-cresol, cyclohexanone, DMF, DMSO, formic acid, propylene carbonate, pyridine, trifluoroacetic acid	
Non-solvent	-	acetic acid, 1-butanone, n-butyl acetate, chlorobenzene, cyclohexane, diethyl ether, diisobutyl ketone, isoamyl alcohol, isopropyl alcohol, tetralin, trichloroethane, toluene	

FLAMMABILITY			
Volatile products of combustion	-	CO, monomer, HCN, isobutene	

TOXICITY			
NFPA: Health, Flammability, Reactivity rating	-	1/1/0	

PROCESSING			
Applications	-	coatings, containers, fibers, laminates, lithium batteries, vesicular systems	

PMAN polymethacrylonitrile

PARAMETER	UNIT	VALUE	REFERENCES
BLENDS			
Suitable polymers	-	ETFE (grafting PMAN improves gas barrier properties of membrane)	Youcef, H B, Henkensmeier, D, Balog, S, Gubler, L, Int. Hydrogen Energy, 45, 11, 7059-68, 2020.

PMFS polymethyltrifluoropropylsiloxane

PARAMETER	UNIT	VALUE	REFERENCES
GENERAL			
Common name	-	polymethyltrifluoropropylsiloxane	
CAS name	-	polysiloxanes, Me 3,3,3-trifluoropropyl	
Acronym	-	PMFS	
CAS number	-	63148-56-1	
Formula			

$$\left[\begin{array}{c} CH_3 \\ | \\ Si-O \\ | \\ CH_2CH_2CF_3 \end{array}\right]_n$$

PARAMETER	UNIT	VALUE	REFERENCES
HISTORY			
Date of discovery	-	1950	
SYNTHESIS			
Monomer(s) structure	-		

$$\begin{array}{c} CH_3 \\ | \\ Cl-Si-Cl \\ | \\ CH_2CH_2CF_3 \end{array}$$

PARAMETER	UNIT	VALUE	REFERENCES
Mass average molecular weight, M$_w$	dalton, g/mol, amu	900-340,000	
Producer	-	Dow Corning	
PHYSICAL PROPERTIES			
Density at 20°C	g cm^{-3}	1.24-1.30	
Color	-	clear	
Refractive index, 20°C	-	1.381-1.383	
Odor		slight	
Melting temperature, DSC	°C	-47	
Decomposition temperature	°C	300	
Glass transition temperature	°C	-66 to -75	
Maximum service temperature	°C	-40 to 285	
Hildebrand solubility parameter	MPa$^{0.5}$	18.0	
Surface tension	mN m^{-1}	25.7-28.7	
Dielectric constant at 100 Hz/1 MHz	-	6.95-7.35	
Volume resistivity	ohm-m	1E11	
MECHANICAL & RHEOLOGICAL PROPERTIES			
Tensile strength	MPa	9	
Elongation	%	240	
CHEMICAL RESISTANCE			
Alcohols	-	good	
Aliphatic hydrocarbons	-	good	
Aromatic hydrocarbons	-	poor	
Esters	-	poor	
Halogenated hydrocarbons	-	poor	

PMFS polymethyltrifluoropropylsiloxane

PARAMETER	UNIT	VALUE	REFERENCES
Ketones	-	poor	
Θ solvent, Θ-temp.=25.7°C	-	cyclohexyl acetate, methyl hexanoate	

FLAMMABILITY			
Ignition temperature	°C	325 (open cup); 101.1 (closed cup)	
Volatile products of combustion	-	CO, CO_2, formaldehyde, fluorine compounds	

TOXICITY			
NFPA: Health, Flammability, Reactivity rating	-	2/1/0	
Oral rat, LD_{50}	mg kg^{-1}	500-5,000	
Skin rabbit, LD_{50}	mg kg^{-1}	1,000-2,000	

PROCESSING			
Applications	-	automotive, dentures, electrical contacts, greases, flotation medium for inertial guidance systems, lubricants in aerospace, precision timing devices, sonar lenses	

PMMA polymethylmethacrylate

PARAMETER	UNIT	VALUE	REFERENCES
GENERAL			
Common name	-	polymethylmethacrylate	
IUPAC name	-	poly(methyl methacrylate)	
CAS name	-	2-propenoic acid, 2-methyl-, methyl ester, homopolymer	
Acronym	-	PMMA	
CAS number	-	9011-14-7	
RTECS number	-	TR0400000	
Linear formula			
HISTORY			
Person to discover	-	Wilhelm Rudolph Fitting; Otto Roehm	
Date	-	1877; 1933; 1936	
Details	-	Fitting found in 1877 that polymerization of MMA gives PMMA; Roehm patented polymer and registered Plexiglas as a brand-name in 1933 and started its production in 1936	
SYNTHESIS			
Monomer(s) structure	-		
Monomer(s) CAS number(s)	-	80-62-6	
Monomer(s) molecular weight(s)	dalton, g/mol, amu	100.1	
Monomer ratio	-	100% (less in copolymers)	
Method of synthesis	-	produced by emulsion polymerization, solution polymerization and bulk polymerization; radical initiation is used (including living polymerization methods), but anionic polymerization of MMA can also be performed	
Heat of polymerization	$J\ g^{-1}$	-577	
Mass average molecular weight, M_w	dalton, g/mol, amu	13,000-2,200,000	
Mass average molecular weight, M_w	dalton, g/mol, amu	43,000 (nanocellular product)	Bernardo, V, Martin-de Leon, J, Rodriguez-Perez, M A,Mater. Lett., 255, 126587, 2019.
Polydispersity, M_w/M_n	-	1.8	Hirschberg, V, Lacroix, F, Wilhelm, M, Rodrigue, D, Mech. Mater., 137, 103100, 2019.
Molar volume at 298K	$cm^3\ mol^{-1}$	calc.=85.9; 81.8 (crystalline); 102.0 (amorphous); exp.=85.6	
Van der Waals volume	$cm^3\ mol^{-1}$	calc.=56.14; 56.1 (crystalline); 66.3 (amorphous)	
Molecular cross-sectional area, calculated	$cm2\ x\ 10-16$	37.2	
STRUCTURE			
Crystallinity	%	amorphous, 48 (isotactic)	
Cell type (lattice)	-	orthorhombic	

PMMA polymethylmethacrylate

PARAMETER	UNIT	VALUE	REFERENCES
Cell dimensions	nm	a:b:c=2.098:1.206:1.040; a:b:c=4.196:2.434:10.050 (isotactic)	Tadokoro, H, Structures of Crystalline Polymers. Wiley, New York, 1979.
Unit cell angles	degree	$\alpha{:}\beta{:}\gamma{=}90{:}90{:}90$	
Number of chains per unit cell	-	4	
Tacticity	%	6-10 (isotactic), 30-38 (heterotactic), 50-70 (syndiotactic)	Witttmann, J C; Kovacs, A J, J. Polym. Sci., 16, 4443, 1969.
Chain conformation	-	double helix 10/1	
Entanglement molecular weight	dalton, g/mol, amu	calc.=8,782, 9,200; exp.=9,200	

COMMERCIAL POLYMERS			
Some manufacturers	-	Arkema; Röhm GmbH	
Trade names	-	Plexiglas; Acrylite	

PHYSICAL PROPERTIES			
Density at 20°C	g cm^{-3}	1.17-1.20; 1.26 (crystalline); 1.21 (isotactic)	
Bulk density at 20°C	g cm^{-3}	0.66	
Color	-	white	
Refractive index, 20°C	-	calc.=1.4846-1.4922; exp.=1.448-1.50	
Birefringence	-	0.00002	ise, R J; Thomas, R, Antec, 1283-7, 2000.
Transmittance	%	92	
Haze	%	<1	
Odor	-	odorless	
Melting temperature, DSC	°C	105-160	
Decomposition temperature	°C	170	Patel, P; Hull, T R; McCabe, R W; Flath, D; Grasmeder, J; Percy, M, Polym. Deg. Stab., 95, 709-18, 2010.
Thermal expansion coefficient, 23-80°C	°C^{-1}	7E-5 to 6E-4; 2.5E-4 (below T_g), 5.74E-4 (above T_g)	
Thermal conductivity, melt	W m^{-1} K^{-1}	calc.=0.6862; exp.=0.19	
Glass transition temperature	°C	calc.=82-105; exp.=104-105; atactic=105-122; isotactic=51-107; syndiotactic=105-120	
Heat deflection temperature at 1.8 MPa	°C	96-100	
Vicat temperature VST/A/50	°C	105-117	
Hansen solubility parameters, δ_D, δ_P, δ_H	MPa$^{0.5}$	17.9, 10.1, 5.4; 18.64, 10.52, 7.51	
Interaction radius		11.0; -	
Hildebrand solubility parameter	MPa$^{0.5}$	21.3	
Surface tension	mN m^{-1}	22.69	
Dielectric constant at 100 Hz/1 MHz	-	3.6/2.2-2.6	
Dissipation factor at 1 MHz		0.014	
Volume resistivity	ohm-m	1E10	

PMMA polymethylmethacrylate

PARAMETER	UNIT	VALUE	REFERENCES
Permeability to nitrogen, 25°C	$cm^3 \, cm \, cm^{-2} \, s^{-1} \, Pa^{-1} \, x \, 10^{12}$	0.615	
Permeability to oxygen, 25°C	$cm^3 \, cm \, cm^{-2} \, s^{-1} \, Pa^{-1} \, x \, 10^{12}$	0.0115	
Permeability to water vapor, 25°C	$cm^3 \, cm \, cm^{-2} \, s^{-1} \, Pa^{-1} \, x \, 10^{12}$	48	
Diffusion coefficient of oxygen	$cm^2 \, s^{-1} \, x10^6$	0.005	
Contact angle of water, 20°C	degree	69.1-74.7	
Surface free energy	$mJ \, m^{-2}$	41.8	
Roughness	nm	0.259-0.328	Grzywacz, H, Milczarek, M, Jenczyk, P, Jarząbek, D M, Measurement, 168, 108267, 2021.

MECHANICAL & RHEOLOGICAL PROPERTIES

Tensile strength	MPa	63-78	
Tensile modulus	MPa	3,200-3,400	
Elongation	%	2-6	
Tensile yield strain	%	4-6	
Flexural strength	MPa	107-117	
Flexural modulus	MPa	3,400-3,500	
Young's modulus	MPa	1,300; 5,000	Jimenez, G A; Jana, S C, Composites, 38A, 983-93, 2007; Grzywacz, H, Milczarek, M, Jenczyk, P, Jarząbek, D M, Measurement, 168, 108267, 2021.
Izod impact strength, notched, 23°C	$J \, m^{-1}$	19	
Poisson's ratio	-	0.350-0.400; 0.38	-; Grzywacz, H, Milczarek, M, Jenczyk, P, Jarząbek, D M, Measurement, 168, 108267, 2021.
Rockwell hardness	M	89-95	
Shrinkage	%	0.3-0.7	
Melt index, 230°C/3.8 kg	g/10 min	2.2-24	
Water absorption, 24h at 23°C	%	0.1-0.3	

CHEMICAL RESISTANCE

Acid dilute/concentrated	-	non-resistant	
Alcohols	-	non-resistant	
Aliphatic hydrocarbons	-	resistant	
Aromatic hydrocarbons	-	non-resistant	
Esters	-	non-resistant	
Greases & oils	-	non-resistant	
Halogenated hydrocarbons	-	non-resistant	
Ketones	-	non-resistant	

PMMA polymethylmethacrylate

PARAMETER	UNIT	VALUE	REFERENCES
Θ solvent, Θ-temp.=-126, 41, -20, 27, 74°C	-	acetone, n-amyl acetate, n-butyl acetate, carbon tetrachloride, toluene	
Good solvent	-	ethanol/water, ethanol/carbon tetrachloride, formic acid, MEK, nitroethane	
Non-solvent	-	butylene glycol, carbon tetrachloride, diethyl ether, m-cresol, ethanol (absolute), turpentine	
Chemicals causing environmental stress cracking	list	benzene, benzyl alcohol, carbon tetrachloride, cyclohexanone, ethanol, nitrobenzene	Wang, H T; Pan, Q G; Du, Q C; Li, Y Q, Polym. Test., 22, 125-28, 2003.

FLAMMABILITY			
Ignition temperature	°C	>250	
Autoignition temperature	°C	304-460	
Minimum ignition energy	J	0.015	
Char at 500°C	%	0	Lyon, R E; Walters, R N, J. Anal. Appl. Pyrolysis, 71, 27-46, 2004.
Heat of combustion	J g^{-1}	26,750-26,860	Walters, R N; Hacket, S M; Lyon, R E, Fire Mater., 24, 5, 245-52, 2000.
Volatile products of combustion	-	CO, CO_2, CH_4, C_2H_8, methylmethacrylate (main product), MeOH, EtOH	Czech, Z; Pelech, Z, J. Therm. Anal. Calorim., 101, 309-13, 2010.

WEATHER STABILITY			
Spectral sensitivity	nm	290-320	
Activation wavelengths	nm	260, 280, 300	
Transmittance	%	100 nm – 61.9; 300 nm – 23.7	Matsuzawa, N N; Oizumi, H; Mori, S; Irie, S; Shirayone, S; Yano, E; Okazaki, S; Ishitani, S; Dixon, D A, Jpn. J. Appl. Phys., 38, 7109-13, 1999.
Products of degradation	-	lower molecular weight, chain scission of side groups	
Stabilizers	-	UVA: 2,4-dihydroxybenzophenone; 2-(2H-benzotriazol-2-yl)-p-cresol; phenol, 2-(5-chloro-2H-benzotriazol-2-yl)-6-(1,1-dimethylethyl)-4-methyl-; 2-(2H-benzotriazole-2-yl)-4,6-di-tert-pentylphenol; 2-(2H-benzotriazole-2-yl)-4-(1,1,3,3-tetraethylbutyl)phenol; 2-(2H-benzotriazol-2-yl)-4,6-bis(1-methyl-1-phenylethyl)phenol; 2-(2H-benzotriazol-2-yl)-6-dodecyl-4-methylphenol, branched & linear; 2,4-di-tert-butyl-6-(5-chloro-2H-benzotriazole-2-yl)-phenol; reaction product of methyl 3(3-(2H-benzotriazole-2-yl)-5-t-butyl-4-hydroxyphenyl propionate/PEG 300; 2-[4,6-bis(2,4-dimethylphenyl)-1,3,5-triazin-2-yl]-5-(octyloxy)phenol; (2-ethylhexyl)-2-cyano-3,3-diphenylacrylate; HAS: 1,3,5-triazine-2,4,6-triamine, N,N'''[1,2-ethane-diyl-bis[[4,6-bis[butyl(1,2,6,6-pentamethyl-4-piperidinyl)amino]-1,3,5-triazine-2-yl]imino]-3,1-propanediyl]bis[N',N''-dibutyl-N',N''-bis(1,2,2,6,6-pentamethyl-4-piperidinyl)-; 1, 6-hexanediamine, N, N'-bis(2,2,6,6-tetramethyl-4-piperidinyl)-, polymers with 2,4-dichloro-6-(4-morpholinyl)-1,3,5-triazine; 1,6-hexanediamine, N,N'-bis(2,2,6,6-tetramethyl-4-piperidinyl)-, polymers with morpholine-2,4,6-trichloro-1,3,5-triazine reaction products, methylated; Screener: ZnO (nano; 5 nm); Phenolic antioxidant: 2-(1,1-dimethylethyl)-6-[[3-(1,1-dimethylethyl)-2-hydroxy-5-methylphenyl] methyl-4-methylphenyl acrylate; Optical brightener: 2,2'-(1,2-ethylenediyldi-4,1-phenylene)bisbenzoxazole	

500 **HANDBOOK OF POLYMERS** 3rd Edition, Copyrights 2022; ChemTec Publishing

PMMA polymethylmethacrylate

PARAMETER	UNIT	VALUE	REFERENCES
Low earth orbit erosion yield	cm^3 $atom^{-1}$ x 10^{-24}	>5.60	Waters, D L; Banks, B A; De Groh, K K; Miller, S K R; Thorson, S D, High Performance Polym., 20, 512-22, 2008.
Effect of neutron and gamma radiation		optical transmission of optical fibers was initially improved followed by radiation damage	Toh, K; Sakasai, K; Nakamura, T; Soyama, K; Shikama, T, J. Nuclear Mater., in press, 2011.

BIODEGRADATION			
Colonized products	-	self-polishing paints	
Typical biodegradants	-	PMMA paint undergoes a controlled hydrolysis and is completely released into sea water	
Stabilizers	-	chitosan, copper thiocyanate, cuprous oxide, quaternary ammonium compound, tolylfluanid; antibiotics in detal prostheses: gentamicin and vancomycin; silver nanoparticles	Bertazzoni Minelli, E; Della Bora, T; Benini, A, Anaerobe, in press, 1-4, 2011; Singh, N; Khanna, P K, Mater. Chem. Phys., 104, 2-3, 367-72, 2007.

TOXICITY			
NFPA: Health, Flammability, Reactivity rating	-	1/1/0	
Carcinogenic effect	-	not listed by ACGIH, NIOSH, NTP	
TLV, ACGIH	ppm	50	
OSHA	ppm	100	
Oral rat, LD_{50}	mg kg^{-1}	8,000	

ENVIRONMENTAL IMPACT			
Cradle to grave non-renewable energy use	MJ/kg	115	
Cradle to pellet greenhouse gasses	kg CO_2 kg^{-1} resin	7	
Energy required for depolymerization	kWh kg^{-1}	0.5-2.2	Breyer, K; Michaeli, W, Antec, 2942-45, 1998.
Aquatic toxicity	-	marine plankton (*Tetraselmis chuii, Nannochloropsis gaditana, Isochrysis galbana, Thalassiosira weissflogii,* and *Brachionus plicatilis*)	Venâncio, C, Ferreira, I, Martins, M A, Oliveira, M, Ecotox. Environ. Safety, 184, 109632, 2019.

PROCESSING			
Typical processing methods	-	casting, compression molding, electrophoretic deposition, extrusion, injection molding, inkjet printing	D'Elia, A, Deering, J, Clifford, A, Zhitomirsky, I, Colloids Surf. B: Biointerfaces, 188, 110763, 2020.
Preprocess drying: temperature/ time/residual moisture	°C/h/%	80/3-4/	
Processing temperature	°C	210-260	
Additives used in final products	-	Fillers: aluminum, barium sulfate, aluminum hydroxide, glass fiber, mica, montmorillonite, Ni-BaTiO$_3$, nickel, silica, titanium dioxide, titanium fiber; Plasticizers: di-(2-ethylhexyl) phthalate, 2-hydroxyethyl methacrylate, 4-cyanophenyl 4-heptylbenzoate; Antistatics: copper dimethacrylate, glycerol monolaureate, indium tin oxide, lauramide diethanolamide, polyaniline, polypyrrole; Antiblocking: crosslinked siloxane particles; Release: magnesium stearate, methylpolysilsiquioxane, silicon nitride, stearic acid; Slip: erucamide; Nucleating agents: sepiolite and a polymethylmethacrylate-polybutylacrylate-polymethylmethacrylate	

PMMA polymethylmethacrylate

PARAMETER	UNIT	VALUE	REFERENCES
Applications	-	artificial stones (filled products) for injection molded bath sinks and kitchen worktops, automotive cluster lenses, bone cement, car rear lights, composites, dials, Fresnel lenses, household items, light guides, medical applications (e.g., bone cement, dental prosthetics, electronic instrument lenses, hip spacer. implants), ophthalmology, optical components, optical fibers, point-of-purchase displays, rods, solar collector lenses	
Outstanding properties	-	optical clarity, weather resistance	
BLENDS			
Suitable polymers	-	PEO, PP, PPy, PVAc, PVC, PVDF	
ANALYSIS			
FTIR (wavenumber-assignment)	cm^{-1}/-	C=O – 1724; C-O – 1267, 1239, 725, 595, 361, 291; more in ref.	Haris, M R H M; Kathiresan, S; Mohan, S, Der Pharma Chem., 2, 4, 316-23, 2010.
Raman (wavenumber-assignment)	cm^{-1}/-	C=O – 810; C-O – 733, 600, 367, 300; more in ref.	Haris, M R H M; Kathiresan, S; Mohan, S, Der Pharma Chem., 2, 4, 316-23, 2010.

PMP polymethylpentene

PARAMETER	UNIT	VALUE	REFERENCES
GENERAL			
Common name	-	polymethylpentene, Nalgene	
CAS name	-	pentene, methyl-, homopolymer	
Acronym	-	PMP	
CAS number	-	9016-80-2	
Formula		$\left[CH_2CH\right]n$ CH_2 H_3C-CH CH_3	
HISTORY			
Date	-	1965	
Details	-	introduced by ICI	
SYNTHESIS			
Monomer(s) structure	-	$H_2C=CCH_2CH_2CH_3$ CH_3	
Monomer(s) CAS number(s)	-	763-29-1	
Monomer(s) molecular weight(s)	dalton, g/mol, amu	84.16	
Method of synthesis	-	dimerization of propylene	
Catalyst	-	Ziegler-Natta	
Mass average molecular weight, M_w	dalton, g/mol, amu	200,000-700,000	
Polydispersity, M_w/M_n	-	2.8-4.1	
Molar volume at 298K	cm^3 mol^{-1}	92.0 (crystalline); 100.2 (amorphous)	
Van der Waals volume	cm^3 mol^{-1}	61.4 (crystalline); 61.4 (amorphous)	
STRUCTURE			
Crystallinity	%	55-85	
Cell type (lattice)	-	tetragonal	
Cell dimensions	nm	a:b:c= 1.86:1.86, 1.38	
Unit cell angles	degree	$\alpha=\beta=\gamma=90$	
Tacticity	%	60-90 (isotactic)	
Chain conformation	-	7/2 (isotactic); 24/7 (syndiotactic)	
COMMERCIAL POLYMERS			
Some manufacturers	-	Mitsui, RTP	
Trade names	-	TPX, Polymethylpentene	
PHYSICAL PROPERTIES			
Density at 20°C	g cm^{-3}	0.83-0.84	
Refractive index, 20°C	-	1.463	
Transmittance	%	80-93	

HANDBOOK OF POLYMERS 3rd Edition, Copyrights 2022; ChemTec Publishing

PMP polymethylpentene

PARAMETER	UNIT	VALUE	REFERENCES
Haze	%	1-2	
Melting temperature, DSC	°C	235-240; 245 (isotactic)	
Softening point	°C	47-52	
Thermal expansion coefficient, 23-80°C	°C^{-1}	1.2E-4	
Thermal conductivity, melt	W m^{-1} K^{-1}	0.167-0.17	
Glass transition temperature	°C	23-50	
Specific heat capacity	J K^{-1} kg^{-1}	1,970-2,180	
Heat of fusion	kJ mol^{-1}	5.3	
Maximum service temperature	°C	170	Hainberger, R; Bruck, R; Kataeva, N; Heer, R; Koeck, A; Czepl, P; Kaiblinger, K; Pipelka, F; Bilenberg, B, Microelectronic Eng., 87, 821-23, 2010.
Long term service temperature	°C	75	
Heat deflection temperature at 0.45 MPa	°C	80-90	
Heat deflection temperature at 1.8 MPa	°C	48-50	
Vicat temperature VST/A/50	°C	145-178	
Enthalpy of melting	J g^{-1}	26.9	Danch, A; Osoba, W; Wawryszczuk, J, Radiation Phys. Chem., 76, 150-2, 2007.
Hildebrand solubility parameter	MPa$^{0.5}$	15.1-16.4	
Surface tension	mN m^{-1}	25	Lee, L H, J. Polym. Sci. A-2, 5, 1103, 1967.
Dielectric constant at 100 Hz/1 MHz	-	2.12	
Dielectric loss factor at 1 kHz	-	0.0003	
Volume resistivity	ohm-m	1E14	
Electric strength K20/P50, d=0.60.8 mm	kV mm^{-1}	42-65	
Permeability to oxygen, 25°C	cm^3 mm m^{-2} day^{-1} MPa^{-1}	12,000	
Speed of sound,	m s^{-1}	1080-2180	

MECHANICAL & RHEOLOGICAL PROPERTIES			
Tensile strength	MPa	16-33; 67 (30% glass fiber)	
Tensile modulus	MPa	1,520-2,000	
Tensile stress at yield	MPa	15-30	
Elongation	%	50-120	
Flexural strength	MPa	19.6-46; 97 (30% glass fiber)	
Flexural modulus	MPa	1170; 5,900 (30% glass fiber)	
Young's modulus	MPa	1,900; 1090	-; Golubenko, D V, Yaroslavtsev, A B, Mendeleev Commun., 27, 6, 572-3, 2017.
Izod impact strength, unnotched, 23°C	J m^{-1}	534	
Izod impact strength, notched, 23°C	J m^{-1}	267	

PMP polymethylpentene

PARAMETER	UNIT	VALUE	REFERENCES
Poisson's ratio	-	0.34-0.43	
Rockwell hardness	R	35-85	
Shrinkage	%	1.6-3	
Melt index, 230°C/3.8 kg	g/10 min	22-26	
Water absorption, 24h at 23°C	%	<0.01	

CHEMICAL RESISTANCE

Acid dilute/concentrated	-	good	
Alcohols	-	very good	
Alkalis	-	very good	
Aliphatic hydrocarbons	-	good	
Aromatic hydrocarbons	-	good	
Esters	-	poor	
Greases & oils	-	good	
Halogenated hydrocarbons	-	poor	
Ketones	-	good	
Θ solvent, Θ-temp.=194°C	-	diphenyl	
Good solvent	-	cyclohexane, decalin, tetralin, xylene (above 100°C)	
Non-solvent	-	all organic solvent	

FLAMMABILITY

Burning rate (Flame spread rate)	mm min^{-1}	25.4	
UL 94 rating	-	HB	

WEATHER STABILITY

Spectral sensitivity	nm	>350 (some absorption)	

TOXICITY

HMIS: Health, Flammability, Reactivity rating	-	0/1/0	
Carcinogenic effect	-	not listed by ACGIH, NIOSH, NTP	

PROCESSING

Typical processing methods	-	blow molding, extrusion, injection molding	
Preprocess drying: temperature/ time/residual moisture	°C/h/%	79.4/2	
Processing temperature	°C	266-304	
Applications	-	animal cages, chemical tubes, cosmetic caps and tubes, fibers, heat resistant nonwoven, LED molds, laboratory wares, mandrels and sheaths for rubber hose production, medical equipment, nanopatterned photonic materials, release film, release paper for synthetic leather, syringes, tubing	
Outstanding properties	-	chemical resistance, clarity, gas permeability, heat resistance, transparency	

PMP polymethylpentene

PARAMETER	UNIT	VALUE	REFERENCES
BLENDS			
Suitable polymers	-	PP	

PMPS polymethylphenylsilylene

PARAMETER	UNIT	VALUE	REFERENCES
GENERAL			
Common name	-	polymethylphenylsilylene	
CAS name	-	poly(methylphenylsilylene)	
Acronym	-	PMPS	
CAS number	-	76188-55-1	
Formula			
HISTORY			
Person to discover	-	Yajima, S; Okamura, K; Hasegawa, Y	Yajima, S; Okamura, K; Hasegawa, Y, US Patent 4,220,600, The Research Institute for Special Inorganic Materials, Sept. 2, 1980.
Date	-	1980	
SYNTHESIS			
Monomer(s) structure	-		
Monomer(s) CAS number(s)	-	149-74-6	
Monomer(s) molecular weight(s)	dalton, g/mol, amu	191.13	
Method of synthesis	-	thermal reductive coupling of the corresponding dichlorosilane with a dispersion of sodium in toluene	Demoustier-Champagne, S; Cordier, S; Devaux, J, Polymer, 36, 5, 1003-7, 1995.
Number average molecular weight, M_n	dalton, g/mol, amu	4,250-283,000	Demoustier-Champagne, S; Cordier, S; Devaux, J, Polymer, 36, 5, 1003-7, 1995.
Mass average molecular weight, M_w	dalton, g/mol, amu	28,500-653,000	Demoustier-Champagne, S; Cordier, S; Devaux, J, Polymer, 36, 5, 1003-7, 1995.
Polydispersity, M_w/M_n	-	2.3-23.1	Demoustier-Champagne, S; Cordier, S; Devaux, J, Polymer, 36, 5, 1003-7, 1995.
PHYSICAL PROPERTIES			
Density at 20°C	g cm^{-3}	1.08-1.12	
Color	APHA	40	
Refractive index, 20°C	-	1.69	Sato, T; Nagayama, N; Yokoyama, M, J. Photopolym. Sci. Technol., 16, 5, 679-84, 2003.
Melting temperature, DSC	°C	35	
Decomposition temperature	°C	200	
Thermal conductivity, melt	W m^{-1} K^{-1}	0.147	
Glass transition temperature	°C	-37 to -21	
Specific heat capacity	J K^{-1} kg^{-1}	1.52	

PMPS polymethylphenylsilylene

PARAMETER	UNIT	VALUE	REFERENCES
Maximum service temperature	ºC	-70 to 260	
Surface tension	mN m^{-1}	26.1-28.5	
Dielectric constant at 100 Hz/1 MHz	-	2.98/2.98	
Dissipation factor at 100 Hz		13E-4	
Dissipation factor at 1 MHz		10E-4	
Volume resistivity	ohm-m	1E11	
Speed of sound	m s^{-1}	1372	

CHEMICAL RESISTANCE			
Alcohols	-	good	
Aromatic hydrocarbons	-	poor	
Esters	-	poor	
Halogenated hydrocarbons	-	poor	
Ketones	-	poor	
Θ solvent	-	diisobutylamine	
Good solvent	-	acetone (hot), chloroform, diethyl ether, ethyl acetate, toluene	
Non-solvent	-	ethanol, ethylene glycol, methanol, n-propanol	

FLAMMABILITY			
Ignition temperature	ºC	302	
Autoignition temperature	ºC	487	

WEATHER STABILITY			
Spectral sensitivity	nm	194, 259, 264, 266, 270, 280, 332, 334, 355	Schauer, F; Kuritka, I; Saha, P; Nespurek, S, J. Pys.: Condens. Matter, 19, 076101, 1-11; 2007.
Activation wavelengths	nm	266, 355	
Excitation wavelengths	nm	313	Skryshevskii, Y A, J. appl. Spectroscopy, 71, 5, 671-675, 2004.
Emission wavelengths	nm	355, 415	

BLENDS			
Suitable polymers	-	PS	

ANALYSIS			
FTIR (wavenumber-assignment)	cm^{-1}/-	Si-H – 2100-2150 and 880-890 and 640; Si-C – 1420-1430 and 1090-1120; C-H – 1240-1260 and 1020-1040; Si-O – 1000-1080 and 795-840	Kuritka, I; Horvath, P; Schauer, F; Zemek, J, Polym. Deg. Stab., 91, 2901-10, 2006.

PMS poly(p-methylstyrene)

PARAMETER	UNIT	VALUE	REFERENCES
GENERAL			
Common name	-	poly(p-methylstyrene)	
ACS name	-	benzene, 1-ethenyl-4-methyl homopolymer	
Acronym	-	PMS	
CAS number	-	24936-41-2	
Formula			
HISTORY			
Person to discover	-	Soday, F J	Soday, F J, US Patent 2,394,407, United Gas Improvement Company, Feb. 5, 1946.
Date	-	1946	
Details	-	polymerization of heat polymerizable aromatic olefins	
SYNTHESIS			
Monomer(s) structure	-		
Monomer(s) CAS number(s)	-	622-97-9	
Monomer(s) molecular weight(s)	dalton, g/mol, amu	118.2	
Monomer ratio	-	100%	
Method of synthesis	-	several methods can be used including: radical, anionic, and photoinitiated cationic polymerization	
Catalyst	-	titanium or zirconium compounds and methylalumoxane	
Heat of polymerization	$J\ g^{-1}$	283-330	Worsfold, D J; Bywater, S, J. Polym., Sci., 26, 299, 1957.
Number average molecular weight, M_n	dalton, g/mol, amu	28,000-141,000	
Mass average molecular weight, M_w	dalton, g/mol, amu	25,000-293,000	
Polydispersity, M_w/M_n	-	1.0-4.3	
Molar volume at 298K	$cm^3\ mol^{-1}$	calc.=113.2; 102.0 (crystalline); 111.0 (amorphous); exp.=115.0	
Van der Waals volume	$cm^3\ mol^{-1}$	74.66; 78.0 (crystalline); 73.0 (amorphous)	
Molecular cross-sectional area, calculated	$cm^2\ x\ 10^{-16}$	48.4	
STRUCTURE			
Crystallinity	%	20-30 (syndiotactic)	
Cell type (lattice)	-	orthorhombic	

HANDBOOK OF POLYMERS 3rd Edition, Copyrights 2022; ChemTec Publishing

PMS poly(p-methylstyrene)

PARAMETER	UNIT	VALUE	REFERENCES
Cell dimensions	nm	a:b:c=1.336:2.321:0.512 (form III)	De Rosa, C; Petraccone, V; Guerra, G, Polymer, 37, 23, 5247-53, 1996.
Unit cell angles	degree	$\alpha:\beta:\gamma$=90:90:90	
Number of chains per unit cell	-	6	
Crystallite size	nm	3 (length of form III)	de Ballesteros, O R; Auriemma, F; De Rosa, C; Floridi, C; Petraccone, V, Polymer, 39, 15, 3523+28, 1998.
Polymorphs	-	I, II, III, IV, V	Rizzo, P; de Ballesteros, O R; De Rosa, C; Auriemma, F; La Camera, D; Petraccone, V; Lotz, B, Polymer, 41, 3745, 49, 2000.
Tacticity	%	95 (syndiotactic)	Esposito, G; Tarallo, O; Petraccone, V; Eur. Polym. J., 43, 1278-87, 2007.
Chain conformation	-	helical s(2/1)2 (I, II) (minimum energy conformation); *trans*-plannar (III, IV,V)	Esposito, G; Tarallo, O; Petraccone, V; Eur. Polym. J., 43, 1278-87, 2007.
Space group		Pnam (form III)	de Ballesteros, O R; Auriemma, F; De Rosa, C; Floridi, C; Petraccone, V, Polymer, 39, 15, 3523+28, 1998.
Entanglement molecular weight	dalton, g/mol, amu	calc.=13,477	

PHYSICAL PROPERTIES

Density at 20°C	g cm^{-3}	1.01-1.04	
Color	-	white	
Refractive index, 20°C	-	calc.=1.5921-1.595; exp.=1.5874-1.610	
Odor	-	odorless	
Melting temperature, DSC	°C	225	
Thermal expansion coefficient, 23-80°C	°C^{-1}	6.6-7.1E-4	
Thermal conductivity, melt	W m^{-1} K^{-1}	0.1323	
Glass transition temperature	°C	calc.=81-126; exp.=93-110; 113 (atactic)	Camelio, P; Lazzeri, V; Waegell, B; Cypcar, C; Mathias, L J, Macromolecules, 31, 2305-11, 1998.
Heat deflection temperature at 1.8 MPa	°C	92	
Hildebrand solubility parameter	MPa$^{0.5}$	calc.=19.33	
Surface tension	mN m^{-1}	calc.=38.8-48.6, exp.=38.7	
Surface free energy	mJ m^{-2}	38.7	

MECHANICAL & RHEOLOGICAL PROPERTIES

Poisson's ratio	-	0.341-0.345	
Melt index, 230°C/3.8 kg	g/10 min	5	

CHEMICAL RESISTANCE

Acid dilute/concentrated	-	good	
Alcohols	-	very good	
Aliphatic hydrocarbons	-	good	
Aromatic hydrocarbons	-	poor	
Esters	-	poor	

PMS poly(p-methylstyrene)

PARAMETER	UNIT	VALUE	REFERENCES
Halogenated hydrocarbons	-	poor	
Ketones	-	good	
Θ solvent, Θ-temp.=34.3, 10, 85°C	-	cyclohexane, *trans*-decalin, n-hexyl acetate	
Good solvent	-	benzene, butyl acetate, carbon disulfide, chlorinated aliphatic hydrocarbons, chloroform, cyclohexanone, dioxane, ethyl acetate, ethylbenzene, MEK, NMP, THF	
Non-solvent	-	acetic acid, acetone, alcohols, ethyl ether, saturated hydrocarbons	

FLAMMABILITY

Char at 500°C	%	0	Lyon, R E; Walters, R N, J. Anal. Appl. Pyrolysis, 71, 27-46, 2004.
Volatile products of combustion	-	CO, CO_2, and more in ref.	Zuev, V V; Bertini, F; Audisio, G, Polym. Deg. Stab., 71, 213-21, 2001.

WEATHER STABILITY

Activation wavelengths	nm	265	
Excitation wavelengths	nm	294, 336, 425	Al Ani, K E; Ramadhan, A E, Polym. Deg. Stab., 93, 1590-96, 2008.
Products of degradation	-	chain scission	

TOXICITY

HMIS: Health, Flammability, Reactivity rating	-	0/1/0	
Carcinogenic effect	-	not listed by ACGIH, NIOSH, NTP	

BLENDS

Suitable polymers	-	PC, PMMA, PS	

ANALYSIS

NMR (chemical shifts)	ppm	C NMR: quaternary carbon – 142.3; phenyl carbons – 126.8, 127.7, 133.8; CH – 39.9; CH_2 – 43.2; CH_3 – 20.2	Zhang, X; Yan, W; Li, H; Shen, X, Polymer, 46, 11958-61, 2005.

PMSQ polymethylsilsesquioxane

PARAMETER	UNIT	VALUE	REFERENCES
GENERAL			
Common name	-	polymethylsilsesquioxane	
ACS name	-	silsesquioxanes, Me	
Acronym	-	PMSQ	
CAS number	-	68554-70-1	
Linear formula		$\left[CH_3Si_{1\cdot 5} \right]_n$	Haussmann, M; Reznik, B; Bockhorn, H; Denev, J A, J. Anal. Appl. Pyrolysis, 91, 224-31, 2011.
HISTORY			
Person to discover	-	Gordon, D J; Wessel, J K	Gordon, D J; Wessel, J K, US Patent 4,290,896, Dow Corning, Sept. 22, 1981.
Date	-	1981	
Details	-	dewatering fine coal slurries using organopolysiloxanes	
SYNTHESIS			
Monomer(s) structure	-	H₃CO–Si(OCH₃)(CH₃)–OCH₃	
Method of synthesis	-	ladder polymethylsilsesquioxane nanoparticles with average diameter size of 15 and 20 nm were synthesized by hydrolysis and condensation of methyltrimethoxysilane	Baatti, A, Erchiqui, F, Bébin, P, Bussières, D, Adv. Powder Technol., 28, 3, 1038-46, 2017.
Monomer(s) CAS number(s)	-	1185-55-3	
Monomer(s) molecular weight(s)	dalton, g/mol, amu	136.22	
Catalyst	-	zinc acetate	
Mass average molecular weight, M_w	dalton, g/mol, amu	1,800-31,000 (prepolymer); 756-6985 (dendrimers)	-; Boldyrev, K, Tatarinova, E, Meshkov, I, Muzafarov, A, Polymer, 174, 159-69, 2019.
Radius of gyration	nm	0.15-0.78 (dendrimers)	Boldyrev, K, Tatarinova, E, Meshkov, I, Muzafarov, A, Polymer, 174, 159-69, 2019.
COMMERCIAL POLYMERS			
Some manufacturers	-	Momentive; Wacker	
Trade names	-	Tospearl; Belsil	
PHYSICAL PROPERTIES			
Density at 20°C	g cm⁻³	1.32-1.43	
Bulk density at 20°C	g cm⁻³	0.17-0.46	
Color	-	white	
Refractive index, 20°C	-	1.41-1.42	
Odor	-	characteristic	
Melting temperature, DSC	°C	>1000	
Thermal conductivity	W m⁻¹ K⁻¹	0.0321-0.0384 (aerogel)	Guo, T, Yun, S, Li, Y, Gao, Y, Vacuum, in press, 109825, 2021.
Maximum service temperature	°C	400	Xiang, H; Zhang, L; Wang, Z; Yu, X; Long, Y; Zhang, X; Zhao, N; Xu, J, J. Colloid Interface Sci., 359, 296-303, 2011.

PMSQ polymethylsilsesquioxane

PARAMETER	UNIT	VALUE	REFERENCES
Dielectric constant at 100 Hz/1 MHz	-	2.6--2.8	Kim, B R; Kim, Y D; Moon, M S; Choi, B K; Ko, M J, Microelectronic Eng., 85, 74-80, 2008.
Contact angle of water, 20°C	degree	146	Yao, B, Zhang, X, Yang, F, Li, C, Mu, Z, Powder Technol., 329, 137-48, 2018.
Specific surface area	$m^2\,g^{-1}$	15-45; 397 (ladder PMSQ)	-; Baatti, A, Erchiqui, F, Bébin, P, Bussières, D, Adv. Powder Technol., 28, 3, 1038-46, 2017.

MECHANICAL & RHEOLOGICAL PROPERTIES

PARAMETER	UNIT	VALUE	REFERENCES
Elastic modulus	MPa	8,500-10,000	Kim, B R; Kim, Y D; Moon, M S; Choi, B K; Ko, M J, Microelectronic Eng., 85, 74-80, 2008.
Water absorption, equilibrium in water at 23°C	%	2	

CHEMICAL RESISTANCE

PARAMETER	UNIT	VALUE	REFERENCES
Alcohols	-	good	
Aliphatic hydrocarbons	-	good	
Esters	-	poor	
Greases & oils	-	good	
Ketones	-	good	
Good solvent	-	DMF	

FLAMMABILITY

PARAMETER	UNIT	VALUE	REFERENCES
Volatile products of combustion	-	CO, CO_2, dense smoke, H_2O, CH_4	Haussmann, M; Reznik, B; Bockhorn, H; Denev, J A, J. Anal. Appl. Pyrolysis, 91, 224-31, 2011.

TOXICITY

PARAMETER	UNIT	VALUE	REFERENCES
NFPA: Health, Flammability, Reactivity rating	-	1/0/1	
Carcinogenic effect	-	not listed by ACGIH, NIOSH, NTP	
Oral rat, LD_{50}	$mg\,kg^{-1}$	>6,000	

PROCESSING

PARAMETER	UNIT	VALUE	REFERENCES
Typical processing methods	-	electrospinning, impregnation	
Applications	-	aerogels, antiblocking agent in plastic films, copying machines and laser printers (fluid control and prevention of static electricity), cosmetics (lipsticks, skin lotions, skin creams), paints and inks (moisture resistance, viscosity control)	
Outstanding properties	-	water repellency (superhydrophobic), insoluble in organic solvents, heat resistance	

ANALYSIS

PARAMETER	UNIT	VALUE	REFERENCES
FTIR (wavenumber-assignment)	cm^{-1}/-	Si–C – 1273, 781; Si–O–Si – 1000-1130	Shirgholami, M A; Khalil-Abad, M S; Khajavi, R; Yazdanshenas, M E, J. Colloid Interface Sci., 359, 530-35, 2011.

PN polynorbornene

PARAMETER	UNIT	VALUE	REFERENCES
GENERAL			
Common name	-	polynorbornene, poly(1,3-cyclopentylenevinylene)	
CAS name	-	bicyclo[2.2.1]hept-2-ene, homopolymer	
Acronym	-	PN	
CAS number	-	25038-76-0	
HISTORY			
Person to discover	-	Rinehart, R E	Rinehart, R E, US Patent 3,367,924, Uniroyal, Feb. 6, 1968.
Date	-	1968	
Details	-	emulsion polymerization of norbornenes in the presence of ruthenium or iridium catalysts	
SYNTHESIS			
Monomer(s) structure	-		
Monomer(s) CAS number(s)	-	498-66-8	
Monomer(s) molecular weight(s)	dalton, g/mol, amu	94.15	
Monomer ratio	-	100%	
Formulation example	g	DI water – 95; norbornene – 5; acetone – 0.5; Na dodecyl sulfate – 0.25; catalyst – 0.0032; activator – 0.0056; THF – 1	Crosbie, D; Stubbs, J; Sundberg, D, Macromolecules, 40, 5743-49, 2007.
Method of synthesis	-	ring opening polymerization of norbornene	
Temperature of polymerization	°C	60	
Catalyst	-	ß -diketonate titanium, methylaluminoxane, ruthenium chloride, palladium compound	Casares, J A; Espinet, P; Salas, G, Organometallics, 27, 3761-69, 2008.
Heat of polymerization	J g^{-1}	652-690	Lebedev, B V; Smirnova, N; Kiparisova, Y, Makromol. Chem., 193, 1399, 1992.
Mass average molecular weight, M$_w$	dalton, g/mol, amu	2,000,000-3,000,000	
Van der Waals volume	cm^3 mol^{-1}	calc.=108; exp.=150	
STRUCTURE			
Crystallinity	%	amorphous; only crystallizes when *cis* is predominant	
Cell type (lattice)	-	monoclinic	
Cell dimensions	nm	a:b:c=4.64-5.13:4.22-4.78:9.84:11.56	
Unit cell angles	degree	γ=68.1-73.5	
Tacticity	%	75-81 (*trans*)	
Chain conformation	-	helix	Karafilidis, C; angermund, K; Gabor, B; Rufinska, A; Mynott, R J; Breitenbruch, G; Thiel, W; Fink, G, Angew. Chem. Int. Ed., 46, 3745-49, 2007.
Entanglement molecular weight	dalton, g/mol, amu	41,000	

PN polynorbornene

PARAMETER	UNIT	VALUE	REFERENCES
COMMERCIAL POLYMERS			
Some manufacturers	-	Astrotech	
Trade names	-	Norsorex	
PHYSICAL PROPERTIES			
Density at 20°C	g cm^{-3}	0.94-0.96	
Color	-	white	
Refractive index, 20°C	-	1.534	
Odor	-	characteristic	
Decomposition temperature	°C	456 (*trans*), 466 (*cis*)	
Thermal expansion coefficient, 23-80°C	10^{-4} °C^{-1}	0.6	
Glass transition temperature	°C	35-45; 37 (trans)	
Long term service temperature	°C	-40 to 80	
Dielectric constant at 100 Hz/1 MHz	-	2.6	
Dielectric loss factor at 1 kHz	-	0.0007	
MECHANICAL & RHEOLOGICAL PROPERTIES			
Tensile strength	MPa	50-60	
Tensile modulus	MPa	1,400	
Elongation	%	10-20	
Shore A hardness	-	18-80	
Intrinsic viscosity, 30°C	dl g^{-1}	3.4-5.0	
Water absorption, equilibrium in water at 23°C	%	0.1	
CHEMICAL RESISTANCE			
Acid dilute/concentrated	-	poor	
Alcohols	-	poor	
Alkalis	-	poor	
Aliphatic hydrocarbons	-	poor	
Aromatic hydrocarbons	-	poor	
Esters	-	poor	
Greases & oils	-	poor	
FLAMMABILITY			
Heat release	kW m^{-2}	3,300	Mizuno, K; Ueno, T; Hirata, A; Ishikawa, T; Takeda, K; Polym. Deg. Stab., 92, 2257-63, 2007.
Char at 500°C	%	6	
Heat of combustion	J g^{-1}	35,400	Mizuno, K; Ueno, T; Hirata, A; Ishikawa, T; Takeda, K; Polym. Deg. Stab., 92, 2257-63, 2007.
Volatile products of combustion	-	CO, CO$_2$, more in ref.	Mizuno, K; Ueno, T; Hirata, A; Ishikawa, T; Takeda, K; Polym. Deg. Stab., 92, 2257-63, 2007.

PN polynorbornene

PARAMETER	UNIT	VALUE	REFERENCES
WEATHER STABILITY			
Important initiators and accelerators	-	singlet oxygen	Wu, S K; Lucki, J; Rabek, J F; Ranby, B, Polym. Photochem., 2, 125-32, 1982.
Products of degradation	-	alkoxy and hydroxy radicals, hydrogen abstraction, formation of carbonyls, and hydroxyl groups	
TOXICITY			
Carcinogenic effect	-	not listed by ACGIH, NIOSH, NTP	
PROCESSING			
Additives used in final products	-	Activators (zinc oxide, stearic acid), Crosslinkers (sulfur); Plasticizers (DOP, DIDP, DIDA, DOZ, DOA, DOS, DTDA); Process aids (stearic acid)	
Applications	-	bumpers, door sealing, electronic equipment, grip improvement, oil cleaning, rail, shoe parts, ski parts, tires, transmission belts, transport rolls	
Outstanding properties	-	high friction, high glass transition temperature, optical clarity, vibration dumping	
BLENDS			
Suitable polymers	-	NBR, PO, PVC	
ANALYSIS			
FTIR (wavenumber-assignment)	cm^{-1}/-	OOH – 3450	Wu, S K; Lucki, J; Rabek, J F; Ranby, B, Polym. Photochem., 2, 125-32, 1982.

PNR phthalonitrile resin

PARAMETER	UNIT	VALUE	REFERENCES
GENERAL			
Common name	-	phthalonitrile resin	
Acronym	-	PNR	
HISTORY			
Person to discover	-	Keller, T M; US Navy Laboratory	Keller, T M, US Patent 4,410,676, US Navy Laboratory, Oct 18, 1983
Date	-	1983	
SYNTHESIS			
Method of synthesis	-	phthalonitrile resin is obtained by heating a phenol with a diphthalonitrile monomer at a elevated temperature	Keller, T M, US Patent 4,410,676, US Navy Laboratory, Oct 18, 1983.
Temperature of polymerization	°C	200 (cure)	Laskowski, M; Shear, M B; Neal, A; Dominguez, D D; Ricks-Laskoski, H L; Hervey, J; Keller, T M, Polymer, 67, 185-91, 2015.
COMMERCIAL POLYMERS			
Some manufacturers	-	Akron Polymer Systems; Maverick Corp.	
PHYSICAL PROPERTIES			
Glass transition temperature	°C	300	Derradji, M; Ramdani, N; Zhang, T; Wang, J; Lin, Z-w; Yang, M; Xu, X-d; Liu, W-b, Mater. Lett., 149, 81-4, 2015.
Maximum service temperature	°C	290 (still strong at 500); 375 (thermooxidative stability)	
Long term service temperature	°C	>250	
MECHANICAL & RHEOLOGICAL PROPERTIES			
Flexural strength	MPa	1,400	
Shear strength	MPa	90	
FLAMMABILITY			
Peak heat release	kW m^{-2}	118	
PROCESSING			
Typical processing methods	-	resin transfer molding, resin infusion molding, filament winding, prepreg consolidation	

POE very highly branched polyethylene

PARAMETER	UNIT	VALUE	REFERENCES
GENERAL			
Common name	-	very highly branched polyethylene, ultralow density ethylene copolymer	
CAS name	-	1-octene, polymer with ethene	
Acronym	-	POE	
CAS number	-	26221-73-8	
HISTORY			
Person to discover	-	Finlayson, M F; Garrison, C C; Guerra, R E; Guest, M J; Kolthammer, B W S; Parikh, D R; Ueligger, S M	Finlayson, M F; Garrison, C C; Guerra, R E; Guest, M J; Kolthammer, B W S; Parikh, D R; Ueligger, S M, US Patent 6,723,810, Dow Chemical, Apr. 20, 2004.
Date	-	2004	
Details	-	process of making ultralow density PE	
SYNTHESIS			
Monomer(s) structure	-	$H_2C=CH_2 \qquad H_2C=CH(CH_2)_5CH_3$	
Monomer(s) CAS number(s)	-	74-85-1; 111-66-0	
Monomer(s) molecular weight(s)	dalton, g/mol, amu	28.05; 112.21	
Monomer(s) expected purity(ies)	%	99.99; 97-99	
Ethylene content	%	75-98	
C3-C12 alpha-olefin content	%	1-25	
C4-C20 diene content	%	0-2	
Octene content	%	5.9-25	Yang, K; Yu, W; Zhou, C, J. Appl. Polym. Sci., 105, 846-52, 2007.
Catalyst	-	metallocene (Insite technology)	
Number average molecular weight, M_n	dalton, g/mol, amu	36,000-77,000	Shan, H; White, J L; deGroot, A W, Int. J. Polym. Anal. Charact., 12, 231-49, 2007.
Mass average molecular weight, M_w	dalton, g/mol, amu	76,000-192,000	Shan, H; White, J L; deGroot, A W, Int. J. Polym. Anal. Charact., 12, 231-49, 2007.
Polydispersity, M_w/M_n	-	1.5-2.5	
Degree of branching	per 1000 carbons	13.7-32	Parkinson, M; Klimke, K; Spiess, H W; Wilhelm, M, Macromol. Chem. Phys., 208, 2128-33, 2007; Yang, K; Yu, W; Zhou, C, J. Appl. Polym. Sci., 105, 846-52, 2007; Crosby, B J; Mangnus, M; de Groot, W; Daniels, R; McLeish, T C B; J. Rheol., 46, 2, 401-26, 2002.
Type of branching	-	octene	
STRUCTURE			
Crystallinity	%	12-40	
Entanglement molecular weight	dalton, g/mol, amu	calc.=2,200	
Lamellae thickness	nm	3.2-5.3; 40-100 (lamellar length)	
Heat of crystallization	kJ kg^{-1}	33-82	
Rapid crystallization temperature	°C	40-54	Shan, H; White, J L; deGroot, A W, Int. J. Polym. Anal. Charact., 12, 231-49, 2007.

POE very highly branched polyethylene

PARAMETER	UNIT	VALUE	REFERENCES
COMMERCIAL POLYMERS			
Some manufacturers	-	DuPont	
Trade names	-	Engage	
PHYSICAL PROPERTIES			
Density at 20°C	g cm^{-3}	0.863-0.885	
Color	-	white	
Melting temperature, DSC	°C	49-76	
Decomposition temperature	°C	300	
Glass transition temperature	°C	-29 to -52	
Heat of fusion	kJ mol^{-1}	0.35-1.1	
MECHANICAL & RHEOLOGICAL PROPERTIES			
Mooney viscosity	-	5-35	
Melt index, 190°C/2.16 kg	g/10 min	0.5-30	
CHEMICAL RESISTANCE			
Acid dilute/concentrated	-	very good	
Alcohols	-	good	
Alkalis	-	very good	
Aliphatic hydrocarbons	-	poor	
Aromatic hydrocarbons	-	poor	
Esters	-	poor	
Greases & oils	-	good to poor	
Halogenated hydrocarbons	-	poor	
Ketones	-	poor	
Effect of EtOH sterilization (tensile strength retention)	%	86-131	Navarrete, L; Hermanson, N, Antec, 2807-18, 1996.
FLAMMABILITY			
Autoignition temperature	°C	330-410	
Heat of combustion	J g^{-1}	47,740	
WEATHER STABILITY			
Spectral sensitivity	nm	<300	
Activation wavelengths	nm	300, 330-360	
Important initiators and accelerators	-	unsaturations, aromatic carbonyl compounds (deoxyanisoin, dibenzocycloheptadienone, flavone, 4-methoxybenzophenone, 10-thioxanthone), hydrogen bound to tertiary carbon at branching points, aromatic amines, groups formed on oxidation (hydroperoxides, carbonyl, carboxyl, hydroxyl) substituted benzophenones, complexes with ground-state oxygen, quinones (anthraquinone, 2-chloroanthraquinone, 2-tert-butyl-athraquinone, 1-methoxyanthraquinone, 2-ethylanthraquinone, 2-methylanthraquinone), transition metal compounds (Ni < Zn < Fe < Co), ferrocene derivatives, titanium dioxide (anatase), ferric stearate, polynuclear aromatic compounds (anthracene, phenanthrene, pyrene, naphthalene	
Products of degradation	-	free radicals, hydroperoxides, carbonyl groups, chain scission, crosslinking	

POE very highly branched polyethylene

PARAMETER	UNIT	VALUE	REFERENCES
Stabilizers	-	UVA: 2-hydroxy-4-octyloxybenzophenone; phenol, 2-(5-chloro-2H-benzotriazole-2-yl)-6-(1,1-dimethylethyl)-4-methyl-; 2,2'-methylenebis(6-(2H-benzotriazol-2-yl)-4-1,1,3,3-tetramethylbutyl)phenol; 2,4-di-tert-butyl-6-(5-chloro-2H-benzotriazol-2-yl)-phenol; reaction product of methyl 3(3-(2H-benzotriazol-2-yl)-5-t-butyl-4-hydroxyphenyl propionate/PEG 300; 2-[4,6-bis(2,4-dimethylphenyl)-1,3,5-triazin-2-yl]-5-(octyloxy) phenol; Screener: titanium dioxide; zinc oxide; carbon black; Acid scavenger: hydrotalcite; Fiber: carbon nanotube; HAS: 1,3,5-triazine-2,4,6-triamine, N,N'''[1,2-ethane-diyl-bis[[[4,6-bis[butyl(1,2,6,6-pentamethyl-4-piperidinyl)amino]-1,3,5-triazine-2-yl]imino]-3,1-propanediyl] bis[N',N''-dibutyl-N',N''-bis(1,2,2,6,6-pentamethyl-4-piperidinyl)-; bis(1,2,2,6,6-pentamethyl-4-piperidyl)sebacate + methyl-1,2,2,6,6-pentamethyl-4-piperidyl sebacate; 2,2,6,6-tetramethyl-4-piperidinyl stearate; reaction products of N,N'-ethane-1,2-diylbis(1,3-propanediamine), cyclohexane, peroxidized 4-butylamino-2,2,6,6-tetramethylpiperidine and trichloro-1,3,5-triazine; poly[[(6-[1,1,3,3-tetramethylbutyl)amino]-1,3,5-triazine-2,4-diyl][2,2,6,6-tetramethyl-4-piperidinyl)imino]-1,6-hexanediyl[2,2,6,6-tetramethyl-4-piperidinyl)imino]]; 1,6-hexanediamine- N,N'-bis(2,2,6,6-tetramethyl-4-piperidinyl)-polymer with 2,4,6-trichloro-1,3,5-triazine, reaction products with N-butyl-1-butanamine an N-butyl-2,2,6,6-tetramethyl-4-piperidinamine; butanedioic acid, dimethylester, polymer with 4-hydroxy-2,2,6,6-tetramethyl-1-piperidine ethanol; alkenes, C20-24-.alpha.-, polymers with maleic anhydride, reaction products with 2,2,6,6-tetramethyl-4-piperidinamine; 1,6-hexanediamine, N,N'-bis(2,2,6,6-tetramethyl-4-piperidinyl)-, polymers with morpholine-2,4,6-trichloro-1,3,5-triazine reaction products, methylated; Phenolic antioxidant: 2,6,-di-tert-butyl-4-(4,6-bis(octylthio)-1,3,5,-triazine-2-ylamino) phenol; pentaerythritol tetrakis(3-(3,5-di-tert-butyl-4-hydroxyphenyl) propionate); octadecyl-3-(3,5-di-tert-butyl-4-hydroxyphenyl)-propionate; 3,3',3',5,5',5'-hexa-tert-butyl-a,a',a'-(mesitylene-2,4,6-triyl)tri-p-cresol; 2-(1,1-dimethylethyl)-6-[[3-(1,1-dimethylethyl)-2-hydroxy-5-methylphenyl] methyl-4-methylphenyl acrylate; 1,3,5-tris(3,5-di-tert-butyl-4-hydroxybenzyl)-1,3,5-triazine-2,4,6(1H,3H,5H)-trione; 3,4-dihydro-2,5,7,8-tetramethyl-2-(4,8,12-trimethyltridecyl)-2H-1-benzopyran-6-ol; 2',3-bis[[3-[3,5-di-tert-butyl-4-hydroxyphenyl]propionyl]]propionohydrazide; isotridecyl-3-(3,5-di-tert-butyl-4-hydroxyphenyl) propionate; 2,2'-ethylidenebis(4,6-di-tert-butylphenol); ethylene bis[3,3-bis[3-(1,1-dimethylethyl)-4-hydroxyphenyl]butanoate];1,3,5-tris(4-tert-butyl-3-hydroxy-2,6-dimethylbenzyl)-1,3,5-triazine-2,4,6-(1H,3H,5H)-trione; 2,2'-methylenebis(4-methyl-6-tertbutylphenol); 3,5-bis(1,1-dimethyethyl)-4-hydroxy-benzenepropanoic acid, C13-15 alkyl esters; 2,2'-isobutylidenebis(2,4-dimethylphenol); 1,1,3-tris(2'methyl-4'-hydroxy-5'tert-butylphenyl)butane; Phosphite: bis-(2,4-di-t-butylphenol) pentaerythritol diphosphite; tris(2,4-di-tert-butylphenyl)phosphite; trinonylphenol phosphite; distearyl pentaerythritol diphosphite; trilauryl trithiophosphite; Thiosynergist: didodecyl-3,3'-thiodipropionate; dioctadecyl 3,3'-thiodipropionate; 2,2'-thiodiethylene bis[3-(3,5-ditert-butyl-4-hydroxyphenyl)propionate]; 4,4'-thiobis(2-t-butyl-5-methylphenol); 2,2'-thiobis(6-tert-butyl-4-methylphenol); pentaerythritol tetrakis(b-laurylthiopropionate); Quencher: (2,2'-thiobis(4-tert-octyl-phenolato))-N-butylamine-nickel(II); Optical brightener: 2,2'-(2,5-thiophenediyl)bis(5-tert-butylbenzoxazole)	

POE very highly branched polyethylene

PARAMETER	UNIT	VALUE	REFERENCES
TOXICITY			
NFPA: Health, Flammability, Reactivity rating	-	1/1/0	
Carcinogenic effect	-	not listed by ACGIH, NIOSH, NTP	
OSHA	mg m^{-3}	5 (respirable), 15 (total)	
PROCESSING			
Applications	-	adhesives, caulks, fibers, inks, oil modifiers, processing aids, sealants, viscosity modifiers, wax substitutes	
BLENDS			
Suitable polymers	-	HDPE, PP, PS	Li, X; Wu, H; Wang, Y; Bai, H; Liu, L; Huang, T, Mater. Sci. Eng., 527A, 3, 531-38, 2010.
ANALYSIS			
NMR (chemical shifts)	ppm	peak assignments for ^{13}C NMR spectra in ref.	Qiu, X H; Redwine, D; Gobbi, G; Naumthanom, A; Rinaldi, P L, Macromolecules, 40, 6879-84, 2007.

POM polyoxymethylene

PARAMETER	UNIT	VALUE	REFERENCES
GENERAL			
Common name	-	polyoxymethylene, polyacetal	
IUPAC name	-	poly(oxymethylene)	
CAS name	-	poly(oxymethylene) (9002-81-7); poly(oxymethylene), α-acetyl-ω-(acetyloxy)- (25231-38-3)	
Acronym	-	POM	
CAS number	-	9002-81-7, 25231-38-3	
Formula		$\left[CH_2O \right]_n$	
HISTORY			
Person to discover	-	Hermann Staudinger	
Date	-	1920, 1956 (commercial application by DuPont)	
SYNTHESIS			
Monomer(s) structure	-		
Monomer(s) CAS number(s)	-	110-88-3	
Monomer(s) molecular weight(s)	dalton, g/mol, amu	90.08	
Monomer ratio	%	at least 95 monomer units	
Comonomer	-	ethylene oxide (thermal stabilization by Celanese method); formaldehyde+cyclic ethers (decelerator of thermal degradation)	Yang, F; Li, H; Cai, L; Lan, F; Xiang, M, Polym.-Plast. Technol. Eng., 48, 530-34, 2009; Ramirez, N V; Sanchez-Soto, M; Illescas, S; Gordillo, A, Polym.-Plast. Techn. Eng., 48, 470-77, 2009.
Method of synthesis	-	anhydrous formaldehyde is polymerized in the presence of anionic catalysis and the resulting polymer stabilized by reaction with acetic anhydride	
Temperature of polymerization	°C	80	
Catalyst	-	$BF_3 \cdot OEt_2$	
Heat of polymerization	J g^{-1}	-706	
Number average molecular weight, M_n	dalton, g/mol, amu	20,000-110,000	
Mass average molecular weight, M_w	dalton, g/mol, amu	21,000-1,000,000	
Polydispersity, M_w/M_n	-	1.84-14.7	Sukhanova, T; Bershtein, V; Keating, M; Matveeva, G; Vylegzhanina, M; Egorov, V; Peschanskaya, N; Yakushev, P; Flexman, E; Greulich, S; Sauer, B; Schodr, K, Maacromol. Symp., 214, 135-145, 2004.
Molar volume at 298K	cm^3 mol^{-1}	19.3 (crystalline)	
Van der Waals volume	cm^3 mol^{-1}	13.9 (crystalline)	
Radius of gyration	nm	2.8-3.3	Sukhanova, T; Bershtein, V; Keating, M; Matveeva, G; Vylegzhanina, M; Egorov, V; Peschanskaya, N; Yakushev, P; Flexman, E; Greulich, S; Sauer, B; Schodr, K, Maacromol. Symp., 214, 135-145, 2004.

POM polyoxymethylene

PARAMETER	UNIT	VALUE	REFERENCES
Chain-end groups	-	acetyl, acetyloxy (improve thermal stability by DuPont method)	Yang, F; Li, H; Cai, L; Lan, F; Xiang, M, Polym.-Plast. Technol. Eng., 48, 530-34, 2009.

STRUCTURE

PARAMETER	UNIT	VALUE	REFERENCES
Crystallinity	%	48-85; 72-92 (highly oriented)	Zhao, X; Ye, L; Mater. Sci. Eng., A528, 4585-91, 2011.
Orientation factor	-	0.84-0.99 (draw ratio up to 1000%)	Zhao, X; Ye, L; Mater. Sci. Eng., A528, 4585-91, 2011.
Cell type (lattice)	-	hexagonal	Zhao, X; Ye, L; Mater. Sci. Eng., A528, 4585-91, 2011.
Cell dimensions	nm	a:b:c=0.443-0.447:0.443-0.447:1.725-1.739	Zhao, X; Ye, L; Mater. Sci. Eng., A528, 4585-91, 2011.
Unit cell angles	degree	1	
Crystallite size	nm	10-15.6	
Chain conformation	-	helix 9/5	
Entanglement molecular weight	dalton, g/mol, amu	calc.=2540	
Lamellae thickness	nm	200-5,400	Sukhanova, T; Bershtein, V; Keating, M; Matveeva, G; Vylegzhanina, M; Egorov, V; Peschanskaya, N; Yakushev, P; Flexman, E; Greulich, S; Sauer, B; Schodr, K, Maacromol. Symp., 214, 135-145, 2004.
Avrami constant, k/n	-	n=1.5-3.9; n=1.3-1.5 (non-isothermal crystallization)	Shu, Y; Ye, L; Zhao, X, Polym.-Plast. Technol. Eng., 45, 963-70, 2006; Zhao, X; Ye, L, Composites, B42, 926, 33, 2011; Li, Y; Zhou, T; Chen, Z; Hui, J; Li, L; Zhang, A, Polymer, 52, 2059-69, 2011.

COMMERCIAL POLYMERS

PARAMETER	UNIT	VALUE	REFERENCES
Some manufacturers	-	BASF; DuPont; Mitsubishi Chemical	
Trade names	-	Ultraform; Delrin; Lupital	

PHYSICAL PROPERTIES

PARAMETER	UNIT	VALUE	REFERENCES
Density at 20°C	g cm^{-3}	1.35-1.44; 1.49-1.53 (crystalline); 1,49-1.60 (10-25% glass fiber); 1.14-1.16 (melt); 1.33 (melt, 20% glass fiber)	
Bulk density at 20°C	g cm^{-3}	0.85	
Refractive index, 20°C	-	1.545-1.553; 1.47	
Odor	-	odorless	
Melting temperature, DSC	°C	166-185; 168-178 (10-25% glass fiber)	
Decomposition temperature	°C	230-250 (air); 277-326 (nitrogen)	Yang, F; Li, H; Cai, L; Lan, F; Xiang, M, Polym.-Plast. Technol. Eng., 48, 530-34, 2009; Archodoulaki, V-M; Lueftl, S; Seidler, S, Polym. Deg. Stab., 86, 75-83, 2004.
Freezing temperature	°C	140	
Thermal expansion coefficient, -40 to 100°C	°C^{-1}	1-1.2E-4; 0.14-1.0E-4 (10-25% glass fiber)	
Thermal conductivity, melt	W m^{-1} K^{-1}	0.294-0.312; 0.13-0.15 (melt); 0.2 (melt, 20% glass fiber); 0.78 (11.6 vol% carbon nanotubes)	Zhao, X; Ye, L, Composites, B42, 926, 33, 2011.
Glass transition temperature	°C	calc.=-50; exp.=-60 to -90	
Specific heat capacity	J K^{-1} kg^{-1}	2,200-3,100 (melt)	
Heat of fusion	kJ kg^{-1}	80-140	

POM polyoxymethylene

PARAMETER	UNIT	VALUE	REFERENCES
Maximum service temperature	°C	140	
Long term service temperature	°C	90	
Heat deflection temperature at 0.45 MPa	°C	150-160; 174-176 (10-25% glass fiber)	
Heat deflection temperature at 1.8 MPa	°C	85-116; 161-172 (10-25% glass fiber)	
Vicat temperature VST/A/50	°C	150-160; 160-163 (20-25% glass fiber)	
Vicat temperature VST/B/50	°C	170-190	
Surface tension	mN m^{-1}	calc.=44.6	
Dielectric constant at 100 Hz/1 MHz	-	3.0-3.7	
Dielectric loss factor at 1 kHz	-	0.001	
Relative permittivity at 100 Hz	-	3.7-3.9; 3.7-4.0 (10-25% glass fiber)	
Relative permittivity at 1 MHz	-	3.7-3.9; 3.9-4.1 (10-25% glass fiber)	
Dissipation factor at 100 Hz	E-4	10-200; 40 (10-25% glass fiber)	
Dissipation factor at 1 MHz	E-4	40-70; 70 (10-25% glass fiber)	
Volume resistivity	ohm-m	1E12-1E13; 1E11 to 1E12 (10-25% glass fiber)	
Surface resistivity	ohm	1E12-1E16; 1E14 (25% glass fiber)	
Electric strength K20/P50, d=0.60.8 mm	kV mm^{-1}	21-40; 28-29 (10-25% glass fiber)	
Comparative tracking index	-	600; 600 (10-25% glass fiber)	
Arc resistance	s	120-200	
Coefficient of friction	-	0.10-0.38 (against itself); 0.18-0.41 (against steel); 0.11 (dynamic)	De Baets, P; Ost, W; Samyn, P; Schoukens, G; Van Parys, F, J. Synthetic Lubrication, 19, 2, 109-18,2002; Hu, K H; Wang, J; Schraube, Y F; Xu, Y F; Hu, X G; Stengler, R, Wear, 266, 1198-1207, 2009.
Permeability to water vapor, 25°C	cm^3 cm cm^{-2} s^{-1} Pa^{-1} x 10^{12}	68.3	
Diffusion coefficient of water vapor	cm^2 s^{-1} x10^6	0.027	
Contact angle of water, 20°C	degree	74.5-79.0	
Surface free energy	mJ m^{-2}	38.6	

MECHANICAL & RHEOLOGICAL PROPERTIES

PARAMETER	UNIT	VALUE	REFERENCES
Tensile strength	MPa	52-60; 95-145 (10-25% glass fiber); 900-2,000 (highly oriented)	Zhao, X; Ye, L; Mater. Sci. Eng., A528, 4585-91, 2011.
Tensile modulus	MPa	2,400-3,200; 5,500-9,400 (10-25% glass fiber); 12,000-25,000 (highly oriented)	Zhao, X; Ye, L; Mater. Sci. Eng., A528, 4585-91, 2011.
Tensile stress at yield	MPa	43-74	
Tensile creep modulus, 1000 h, elongation 0.5 max	MPa	1,300-1,700; 4,500-5,800 (20-25% glass fiber)	
Elongation	%	20-120; 3 (20-25% glass fiber)	
Tensile yield strain	%	7.5-30; 4.3 (10-25% glass fiber)	
Flexural strength	MPa	88-98	
Flexural modulus	MPa	2,500-3,000; 4,800-8,500 (10-25% glass fiber)	
Elastic modulus	MPa	3,000	

POM polyoxymethylene

PARAMETER	UNIT	VALUE	REFERENCES
Compressive strength	MPa	85; 100 (30% glass fiber)	
Charpy impact strength, unnotched, 23°C	kJ m^{-2}	70-280; 50-55 (10-25% glass fiber)	
Charpy impact strength, unnotched, -30°C	kJ m^{-2}	22-230; 50-60 (10-25% glass fiber)	
Charpy impact strength, notched, 23°C	kJ m^{-2}	5.3-8; 5-9 (10-25% glass fiber)	
Charpy impact strength, notched, -30°C	kJ m^{-2}	2.5-5.5; 5.0-8.5 (10-25% glass fiber)	
Shear modulus	MPa	3,100	
Poisson's ratio	-	0.27	Benabdallah, H S; Wei, J J, J. Tribology, 127, 766-75, 2005.
Rockwell hardness	-	M79-92; R117-120	
Shrinkage	%	1.9-2.1; 0.4-1.6 (10-25% glass fiber)	
Intrinsic viscosity, 25°C	ml g^{-1}	63.2-96.8	Oner, M; White, D H, Polym. Deg. Stab., 40, 297-303, 1993.
Melt viscosity, shear rate=1000 s^{-1}	Pa s	100-500	
Melt volume flow rate (ISO 1133, procedure B), 190°C/2.16 kg	cm^3/10 min	2.2-45; 4-4.3 (20% glass fiber)	
Melt index, 190°C/3.8 kg	g/10 min	1.9-52	
Water absorption, equilibrium in water at 23°C	%	0.8-1.65; 0.9-1.1 (10-25% glass fiber)	
Moisture absorption, equilibrium 23°C/50% RH	%	0.2-0.4; 0.15-0.17 (10-25% glass fiber)	

CHEMICAL RESISTANCE

Acid dilute/concentrated	-	good-poor	
Alcohols	-	good	
Alkalis	-	good-poor	
Aliphatic hydrocarbons	-	very good	
Aromatic hydrocarbons	-	good	
Esters	-	good	
Greases & oils	-	good	
Halogenated hydrocarbons	-	good	
Ketones	-	good	
Good solvent	-	aniline, benzyl alcohol, bromobenzene, γ-butyrolactone, chlorophenols, diphenyl ether, DMF, formamide, phenol (all at elevated temperatures)	
Non-solvent	-	lower alcohols, diethyl ether, lower esters, hydrocarbons	

FLAMMABILITY

Flammability according to UL-94 standard; thickness 1.6/0.8 mm	class	HB	
Ignition temperature	°C	320-340	
Autoignition temperature	°C	375	
Limiting oxygen index	% O$_2$	15-16; 21 (10-25% glass fiber); 31-46 (with flame retardant)	Wang, Z-Y; Liu, Y; Wang, Q, Polym. Deg. Stab., 95, 945-54, 2010.

POM polyoxymethylene

PARAMETER	UNIT	VALUE	REFERENCES
Heat release	kW m^{-2}	268.8	Wang, Z-Y; Liu, Y; Wang, Q, Polym. Deg. Stab., 95, 945-54, 2010.
Burning rate (Flame spread rate)	mm min^{-1}	50	
Char at 500°C	%	0	Lyon, R E; Walters, R N, J. Anal. Appl. Pyrolysis, 71, 27-46, 2004.
Heat of combustion	J g^{-1}	17,390	Walters, R N; Hacket, S M; Lyon, R E, Fire Mater., 24, 5, 245-52, 2000.
Volatile products of combustion	-	formaldehyde, CO, CO_2, and more	Pan, G; Li, H; Cao, Y, J. Appl. Polym. Sci., 93, 577-83, 2004.
UL 94 rating	-	HB, V-1 (with flame retardant)	Wang, Z-Y; Liu, Y; Wang, Q, Polym. Deg. Stab., 95, 945-54, 2010.

WEATHER STABILITY			
Excitation wavelengths	nm	290	
Emission wavelengths	nm	312, 435, 450	
Important initiators and accelerators	-	$CuCl_2$, water, acids	
Products of degradation	-	hydroperoxides, carbonyls, chain scission, hydrogen, carbon oxides, methane, ethane, formaldehyde	
Stabilizers	-	UVA: 2-(2H-benzotriazol-2-yl)-p-cresol; 2-(2H-benzotriazole-2-yl)-4,6-di-tert-pentylphenol; 2-(2H-benzotriazole-2-yl)-4-(1,1,3,3-tetraethylbutyl)phenol; 2,2'-methylenebis(6-(2H-benzotriazol-2-yl)-4-1,1,3,3-tetramethylbutyl)phenol; 2-(2H-benzotriazol-2-yl)-4,6-bis(1-methyl-1-phenylethyl)phenol; 2-(3-sec-butyl-5-tert-butyl-2-hydroxyphenyl)benzotriazole; HAS: 1,3,5-triazine-2,4,6-triamine, N,N'''[1,2-ethanediyl-bis[[[4,6-bis[butyl (1,2,6,6-pentamethyl-4-piperidinyl)amino]-1,3,5-triazine-2-yl]imino]-3,1-propanediyl] bis[N',N''-dibutyl-N',N''-bis(1,2,2,6,6-pentamethyl-4-piperidinyl)-; bis(2,2,6,6-tetramethyl-4-piperidyl) sebacate; poly[[(6-[1,1,3,3-tetramethylbutyl)amino]-1,3,5-triazine-2,4-diyl][2,2,6,6-tetramethyl-4-piperidinyl)imino]-1,6-hexanediyl[2,2,6,6-tetramethyl-4-piperidinyl)imino]]; butanedioic acid, dimethylester, polymer with 4-hydroxy-2,2,6,6-tetramethyl-1-piperidine ethanol; 1, 6-hexanediamine, N, N'-bis(2,2,6,6-tetramethyl-4-piperidinyl)-, polymers with 2,4-dichloro-6-(4-morpholinyl)-1,3,5-triazine; 1,6-hexanediamine, N,N'-bis(2,2,6,6-tetramethyl-4-piperidinyl)-, polymers with morpholine-2,4,6-trichloro-1,3,5-triazine reaction products, methylated; Phenolic antioxidant: ethylene-bis(oxyethylene)-bis(3-(5-tert-butyl-4-hydroxy-m-tolyl)-propionate); pentaerythritol tetrakis(3-(3,5-di-tert-butyl-4-hydroxyphenyl)propionate); N,N'-hexane-1,6-diylbis(3-(3,5-di-tert-butyl-4-hydroxyphenylpropionamide)); 1,3,5-tris(4-tert-butyl-3-hydroxy-2,6-dimethyl benzyl)-1,3,5-triazine-2,4,6-(1H,3H,5H)-trione; 2,2'-methylenebis(4-methyl-6-tertbutylphenol); Optical brightener: Fluorescent Brightener 378 (Clariant)	
Results of exposure, tensile strength retention	SAE J-1885, 1000 h; %	about 50; 98 (UV stabilized)	
Low earth orbit erosion yield	cm^3 atom^{-1} x 10^{-24}	9.14	Waters, D L; Banks, B A; De Groh, K K; Miller, S K R; Thorson, S D, High Performance Polym., 20, 512-22, 2008.

POM polyoxymethylene

PARAMETER	UNIT	VALUE	REFERENCES
γ-radiation	Mrad	5 (dose sufficient to make compacts brittle)	Kassem, M E; Bassiouni, M E; El-Muraikhi, J. Mater. Sci., Mater. Electronics, 13, 717-19, 2002.

BIODEGRADATION

Typical biodegradants	-	not biodegradable	
Stabilizers	-	POM releases its own product of degradation which is formaldehyde acting as biocide	

TOXICITY

NFPA: Health, Flammability, Reactivity rating	-	0-1/1/0	
Carcinogenic effect	-	not listed by ACGIH, NIOSH, NTP (decomposition produces formaldehyde, which is Group 1 carcinogen according to IARC)	

PROCESSING

Typical processing methods	-	blow molding, compression molding, extrusion, foam molding, injection molding, machining, rotational molding, stamping, transfer molding	
Preprocess drying: temperature/ time/residual moisture	°C/h/%	80-100/2-4/0.1-0.2	
Processing temperature	°C	180-230, 205-215 (recommended) (injection molding); 175-180 (film and profile extrusion)	
Processing pressure	MPa	80-100 (hold pressure); 50-100 (injection)	
Process time	s	42 (cycle), 4 (injection)	Ramirez, N V; Sanchez-Soto, M; Illescas, S; Gordillo, A, Polym.-Plast. Techn. Eng., 48, 470-77, 2009.
Additives used in final products	-	Fillers: aramid fiber, calcium carbonate, carbon black, carbon fiber, carbon nanotubes, glass beads, glass fiber, iron powder, metal flakes, nano-$CaCO_3$ (nucleating agent), PTFE fiber, talc, zinc whisker; Antistatics: polyetheresteramide, quaternary ammonium compound, superconductive carbon black; Release: fluoropolymer, N,N'-ethylene bisstearamide, paintable silicone; Slip: PTFE; Thermal stabilizer: triethanolamine	
Special grades		toughened, UV resistant, low-friction and low wear	
Applications	-	appliances, automotive parts (door handles, window winders, tank filler necks and caps, carburetor, screw caps for cooling system expansion tanks, fuel pumps), bearings, cams, clips, containers, home electronics and hardware, parts of textile machines, phones (dialing units and slider guideways), pneumatic components, pump impellers, rollers, shower parts, springs, and many other applications	
Outstanding properties	-	fatigue endurance, high resistance to repeated impacts, dimensional stability, electrical insulating capabilities	

BLENDS

Suitable polymers	-	HDPE, MBS, PA, PAH, PE, PTFE (nucleation), PUR	

ANALYSIS

FTIR (wavenumber-assignment)	cm⁻¹/-	C=O – 1733; CH_2 – 1470, 1430	Ramirez, N V; Sanchez-Soto, M; Illescas, S; Gordillo, A, Polym.-Plast. Techn. Eng., 48, 470-77, 2009.

POM polyoxymethylene

PARAMETER	UNIT	VALUE	REFERENCES
Raman (wavenumber-assignment)	cm^{-1}/-	C–O–C – 936, 545; CH$_2$ – 1491, 1324;	Zhao, X; Ye, L; Mater. Sci. Eng., A528, 4585-91, 2011.
NMR (chemical shifts)	ppm	CH$_2$–O – 5.20, 4.63, 4.5; CH$_2$–CH$_2$–O – 3.73	Pan, G; Li, H; Cao, Y, J. Appl. Polym. Sci., 93, 577-83, 2004.
x-ray diffraction peaks	degree	22.9, 34.6, 48.4	Zhao, X; Ye, L; Mater. Sci. Eng., A528, 4585-91, 2011.

PP polypropylene

PARAMETER	UNIT	VALUE	REFERENCES
GENERAL			
Common name	-	polypropylene	
IUPAC name	-	poly(propene)	
CAS name	-	1-propene, homopolymer	
Acronym	-	PP	
CAS number	-	9003-07-0	
RTECS number	-	UD1842000	
Formula		$\left[CH_2CH \atop\quad\ CH_3 \right]_n$	
HISTORY			
Person to discover	-	Paul Hogan and Robert Banks	
Date	-	1951	
Details	-	Paul Hogan and Robert Banks obtained in laboratories of Phillips Petroleum "crystalline polypropylene"	
SYNTHESIS			
Monomer(s) structure	-	$H_2C = CHCH_3$	
Monomer(s) CAS number(s)	-	115-07-1	
Monomer(s) molecular weight(s)	dalton, g/mol, amu	42.08	
Monomer(s) expected purity(ies)	%	99.2	
Monomer ratio	-	100% and less	
Formulation example	-	monomer(s), hydrogen (molecular mass control), catalyst	Maier, R-D; Bidell, Encyclopedia of Materials: Science & Technology, 7694-97, Elsevier, 2008.
Method of synthesis	-	gaseous propylene is polymerized under strict control of heat and pressure in the presence of catalyst; major polypropylene process technologies include slurry, bulk loop, stirred gas, fluid gas, and stirred bulk	Maier, C; Calafut, T, Polypropylene. The Definitive User's Guide and Databook, William Andrew, 1998.
Temperature of polymerization	ºC	70 (bulk); 60-100 (gas-phase)	Maier, R-D; Bidell, Encyclopedia of Materials: Science & Technology, 7694-97, Elsevier, 2008.
Pressure of polymerization	MPa	3.-3.5 (bulk); 1-4.6 (gas-phase)	Maier, R-D; Bidell, Encyclopedia of Materials: Science & Technology, 7694-97, Elsevier, 2008.
Catalyst	-	morphology controlled Ziegler-Natta catalyst (Lynx)	
Mass average molecular weight, M_w	dalton, g/mol, amu	1,000-5,400,000	Maier, C; Calafut, T, Polypropylene. The Definitive User's Guide and Databook, William Andrew, 1998.
Polydispersity, M_w/M_n	-	1.1-5.5	
Polymerization degree (number of monomer units)	-	24-240,000	
Molar volume at 298K	cm³ mol⁻¹	calc.=49.5 (amorphous)	
Van der Waals volume	cm³ mol⁻¹	calc.=30.7 (amorphous)	
Radius of gyration	nm	11.3-41.3 (rapidly quenched); 14.8-47.7 (isothermally crystallized)	

HANDBOOK OF POLYMERS 3rd Edition, Copyrights 2022; ChemTec Publishing

PP polypropylene

PARAMETER	UNIT	VALUE	REFERENCES
Chain-end groups	-	unsaturated: vinylidene, i-butenyl, 2-butenyl, 4-butenyl; saturated: n-prropyl, i-propyl, n-butyl, ethyl	Kawahara, N; Kojoh, S-I; Matsuo, S; Kaneko, H; Matsugi, T; Toda, Y; Mizuno, A; Kashiwa, N, Polymer, 45, 2883-88, 2004.

STRUCTURE

PARAMETER	UNIT	VALUE	REFERENCES
Crystallinity	%	3.2-67	
Cell type (lattice)	-	orthorhombic (syndiotactic); monoclinic (isotactic)	
Cell dimensions	nm	orthorhombic (syndiotactic): a:b:c=1.45:1.12:0.74; monoclinic (isotactic): a:b:c=0.666:2.078:0.6495	
Unit cell angles	degree	99.62 (monoclinic)	
Number of chains per unit cell	-	2 (monoclinic); 2 (orthorhombic)	
Polymorphs	-	α (monoclinic), β (hexagonal)	Shieh, Y-T; Lee, M-S; Chen, S-A, Polymer, 42, 4439-48, 2001.
Tacticity	%	26 (rr, atactic); 32-67 (rr, semi-syndiotactic)	Sevegney, M S; Kannan, R M; Siedle, A R; Percha, P A, J. Polym. Sci. B, 43, 439-61, 2005.
Chain conformation	-	helix 3/1 (monoclinic, hexagonal)	Shieh, Y-T; Lee, M-S; Chen, S-A, Polymer, 42, 4439-48, 2001.
Entanglement molecular weight	dalton, g/mol, amu	7,050 (metallocene)	
Lamellae thickness	nm	7-13	Huang, W; Alamo, R G, Antec, 3546-50, 2000.
Heat of crystallization	kJ kg^{-1}		
Crystallization temperature	°C	116-140	Shieh, Y-T; Lee, M-S; Chen, S-A, Polymer, 42, 4439-48, 2001; Naguib, H E; Xu, J X; Park, C B, Antec, paper 438, 2001.
Avrami constants, K/n	-	n=2.5-2.8	Shieh, Y-T; Lee, M-S; Chen, S-A, Polymer, 42, 4439-48, 2001.

COMMERCIAL POLYMERS

PARAMETER	UNIT	VALUE	REFERENCES
Some manufacturers	-	DOW; ExxonMobil; Total	
Trade names	-	Polypropylene; Achieve; Polypropylene	

PHYSICAL PROPERTIES

PARAMETER	UNIT	VALUE	REFERENCES
Density at 20°C	g cm^{-3}	0.84-0.91; 0.97-1.33 (10-50% glass fiber); 0.98-1.25 (10-40% talc)	
Color	-	translucent to white to off-white	
Refractive index, 20°C	-	1.49-1.51	
Molar polarizability	cm^3 x 10^{-25}	6.1213	
Haze	%	14	
Gloss, 60°, Gardner (ASTM D523)	%	34-52	
Odor	-	may have acrid odor	
Melting temperature	°C	120-176; 147-158 (metallocene); 160-176 (monoclinic); 140-153 (hexagonal)	Shieh, Y-T; Lee, M-S; Chen, S-A, Polymer, 42, 4439-48, 2001; Cheng, C Y, Antec, 2019-2026, 1996.
Softening point	°C	155-161	
Decomposition onset temperature	°C	328	Patel, P; Hull, T R; McCabe, R W; Flath, D; Grasmeder, J; Percy, M, Polym. Deg. Stab., 95, 709-18, 2010.

PP polypropylene

PARAMETER	UNIT	VALUE	REFERENCES
Thermal expansion coefficient, 23-80°C	°C^{-1}	1.05E-4	
Thermal conductivity, melt	W m^{-1} K^{-1}	0.17-0.22	
Glass transition temperature	°C	calc.=-15; exp.=-8; -3.2 (isotactic); -9 to -51 (elastomeric)	
Heat of fusion	J g^{-1}	209 (perfectly crystalline PP)	Fan, Y; Zhang, C; Xue, Y; Zhang, X; Ji, X; Bo, S, Polymer, 52, 557-63, 2011.
Maximum service temperature	°C	100	
Heat deflection temperature at 0.45 MPa	°C	85-107; 143-154 (10-50% glass fiber); 149 (10-50% glass fiber, chemically coupled)	
Heat deflection temperature at 1.8 MPa	°C	42-54; 93-121 (10-50% glass fiber); 113 (10-50% glass fiber, chemically coupled); 60-82 (10-40% talc)	
Vicat temperature VST/A/50	°C	138-155	
Vicat temperature VST/B/50	°C	82-96	
Hansen solubility parameters, δ_D, δ_P, δ_H	MPa$^{0.5}$	17.7, 2.9, 1.2	
Interaction radius		6.2	
Hildebrand solubility parameter	MPa$^{0.5}$	18.0-19.2	
Surface tension	mN m^{-1}	20.4	
Dielectric constant at 100 Hz/1 MHz	-	2.2-2.6	
Dissipation factor at 100 Hz		0.0005	
Dissipation factor at 1 MHz		0.0005	
Volume resistivity	ohm-m	1E-12 to 1E-15; 9.6E1 (with 0.6 vol faction of Ni coated mica)	Kandasubramanian, B; Gilbert, M, Macromol. Symp., 211, 185-95, 2005.
Shielding effectiveness	dB	20-28 (with 0.6 vol fraction of Ni-coated mica)	Kandasubramanian, B; Gilbert, M, Macromol. Symp., 211, 185-95, 2005.
Coefficient of friction	ASTM D1894	0.27-0.29 (chrome steel); 0.35-0.36 (aluminum)	Maldonado, J E, Antec, 3431-35, 1998.
Permeability to nitrogen, 25°C	cm^3 cm cm^{-2} s^{-1} Pa^{-1} x 10^{12}	0.033	
Permeability to oxygen, 25°C	cm^3 cm cm^{-2} s^{-1} Pa^{-1} x 10^{12}	0.17	
Permeability to water vapor, 25°C	cm^3 cm cm^{-2} s^{-1} Pa^{-1} x 10^{12}	1.58	
Contact angle of water, 20°C	degree	94.9-107.3	
Surface free energy	mJ m^{-2}	30.2	
Speed of sound	m s^{-1}	44.3-45.7	
Attenuation	dB cm^{-1}, 5 MHz	5.1-18.2	

PP polypropylene

PARAMETER	UNIT	VALUE	REFERENCES
MECHANICAL & RHEOLOGICAL PROPERTIES			
Tensile strength	MPa	26-32; 39-63 (10-50% glass fiber); 46-97 (10-50% glass fiber, chemically coupled); 30-33 (10-40% talc)	
Tensile modulus	MPa	1,700; 2,900-11,700 (10-50% glass fiber); 3,100-11,700 (10-50% glass fiber, chemically coupled); 2,550-5,200 (10-40% talc)	
Tensile stress at yield	MPa	31-35.2	
Elongation	%	10-140; 1-8.5 (10-50% glass fiber); 2.5-8 (10-50% glass fiber, chemically coupled); 10 (10-40% talc)	
Tensile yield strain	%	7-12	
Flexural strength	MPa	41; 56-98 (10-50% glass fiber); 66-150 (10-50% glass fiber, chemically coupled); 49-54 (10-40% talc)	
Flexural modulus	MPa	1,240-1,600; 2,100-8,900 (10-50% glass fiber); 2,400-8,900 (10-50% glass fiber, chemically coupled); 2.,100-3,900 (10-40% talc)	
Compressive strength	MPa	40	
Young's modulus	MPa	1,200-2,000; 27,000 (fiber from ultrahigh molecular weight PP)	Chen, J; Si, X; Hu, S; Wang, Y; Wang, Y, J. Macromol. Sci. Eng., Part B, 47, 1, 192-200, 2008.
Izod impact strength, unnotched, 23°C	$J\ m^{-1}$	1600; 190-480 (10-50% glass fiber); 530-640 (10-50% glass fiber, chemically coupled); 370-1,175 (10-40% talc)	
Izod impact strength, notched, 23°C	$J\ m^{-1}$	18-69; 37-53 (10-50% glass fiber); 80-110 (10-50% glass fiber, chemically coupled); 43-59 (10-40% talc)	
Tenacity (fiber) (standard atmosphere)	$cN\ tex^{-1}$ $(daN\ mm^{-2})$	15-60	Fourne, F, Synthetic Fibers. Machines and Equipment Manufacture, Properties. Carl Hanser Verlag, 1999.
Tenacity (wet fiber, as % of dry strength)	%	100	Fourne, F, Synthetic Fibers. Machines and Equipment Manufacture, Properties. Carl Hanser Verlag, 1999.
Fineness of fiber (titer)	dtex	1.5-40	Fourne, F, Synthetic Fibers. Machines and Equipment Manufacture, Properties. Carl Hanser Verlag, 1999.
Length (elemental fiber)	mm	38-200	Fourne, F, Synthetic Fibers. Machines and Equipment Manufacture, Properties. Carl Hanser Verlag, 1999.
Shore A hardness	-	64-90 (elastomeric)	Myers, C; Allen, C; Ernst, A; Naim, H, Antec, 2050-55, 1999.
Rockwell hardness	-	R102-103	
Shrinkage	%	0.72-2; 0.1-0.8 (10-50% glass fiber); 0.7-1.6 (10-40% talc)	Chang, T C; Faison, E, Polym. Eng. Sci., 41, 5, 703-10, 2001.
Intrinsic viscosity, 25°C	$dl\ g^{-1}$	1.12-1.87	
Melt viscosity, shear rate=100 s^{-1}	Pa s	100	
Melt volume flow rate (ISO 1133, procedure B), 230°C/2.16 kg	$cm^3/10$ min	4-26	
Pressure coefficient of melt viscosity, b	$G\ Pa^{-1}$	20.5	Aho, J; Syrjala, S, J. Appl. Polym. Sci., 117, 1076-84, 2010.
Melt index, 230°C/2.16 kg	g/10 min	0.3-40	
Water absorption, equilibrium in water at 23°C	%	0.02-0.04	

PP polypropylene

PARAMETER	UNIT	VALUE	REFERENCES
CHEMICAL RESISTANCE			
Acid dilute/concentrated	-	very good	
Alcohols	-	very good	
Alkalis	-	very good	
Aliphatic hydrocarbons	-	fair to poor	
Aromatic hydrocarbons	-	poor	
Esters	-	fair	
Greases & oils	-	good to fair	
Halogenated hydrocarbons	-	poor	
Ketones	-	good	
Θ solvents	-	i-amyl acetate, i-butyl acetate, cyclohexanone, diphenyl ether	
Good solvent	-	chlorinated hydrocarbons, cyclohexane, diethyl ether, toluene	
Non-solvent	-	many polar solvents	
Effect of EtOH sterilization (tensile strength retention)	%	100 to 106	Navarrete, L; Hermanson, N, Antec, 2807-18, 1996.
FLAMMABILITY			
Ignition temperature	°C	>200; 93.3 (fibers & yarns)	
Autoignition temperature	°C	570	
Limiting oxygen index	% O_2	17-19	
Minimum ignition energy	J	0.03	
Heat release	kW m^{-2}	101-727 (with flame retardants)	Yu, B; Liu, M; Lu, L; Dong, X; Gao, W; Tang, K, Fire Mater., 34, 251-61, 2010.
Char at 500°C	%	0	Lyon, R E; Walters, R N, J. Anal. Appl. Pyrolysis, 71, 27-46, 2004.
Heat of combustion	J g^{-1}	45,800	Walters, R N; Hacket, S M; Lyon, R E, Fire Mater., 24, 5, 245-52, 2000.
Volatile products of combustion	-	CO, CO_2, soot	
CO yield	%	5-16 (with flame retardants)	Yu, B; Liu, M; Lu, L; Dong, X; Gao, W; Tang, K, Fire Mater., 34, 251-61, 2010.
UL 94 rating	-	HB; V-0 (flame retarded grades)	
WEATHER STABILITY			
Spectral sensitivity	nm	320-360; 300-350	
Activation wavelengths	nm	310, 300-350	
Excitation wavelengths	nm	230, 283, 287, 290, 295, 323, 330 (thermally degraded film); 230, 270, 285, 290, 330	
Emission wavelengths	nm	295, 320-,330, 332, 340, 342, 400, 430, 470, 480, 520 (thermally degraded film); 309, 320, 420, 445, 480, 510	
Depth of UV penetration	μm	100	

PP polypropylene

PARAMETER	UNIT	VALUE	REFERENCES
Important initiators and accelerators	-	unsaturations, aromatic carbonyl compounds (deoxyanisoin, dibenzocycloheptadienone, flavone, 4-methoxybenzophenone, 10-thioxanthone), hydrogen bound to tertiary carbon at branching points, aromatic amines, groups formed on oxidation (hydroperoxides, carbonyl, carboxyl, hydroxyl) substituted benzophenones, complexes with ground-state oxygen, quinones (anthraquinone, 2-chloroanthraquinone, 2-tert-butyl-athraquinone, 1-methoxyanthraquinone, 2-ethylanthraquinone, 2-methylanthraquinone), transition metal compounds (Ni < Zn < Fe < Co), ferrocene derivatives, titanium dioxide (anatase), ferric stearate, polynuclear aromatic compounds (anthracene, phenanthrene, pyrene, naphthalene, titanium polymerization catalyst	
Products of degradation	-	free radicals, hydroperoxides, carbonyl groups, chain scissions, crosslinks	
Stabilizers	-	UVA: phenol, 2-(5-chloro-2H-benzotriazole-2-yl)-6-(1,1-dimethylethyl)-4-methyl-; 2-(2H-benzotri-azole-2-yl)-4,6-di-tert-pentylphenol; 2-(2H-benz-otriazole-2-yl)-4-(1,1,3,3-tetraethylbutyl)phenol; 2-(2H-benzotriazol-2-yl)-4,6-bis(1-methyl-1-phenylethyl)phenol; 2,2'-methylene-bis(6-(2H-benzotriazol-2-yl)-4-1,1,3,3-tetramethylbutyl)phenol; 2,4-di-tert-butyl-6-(5-chloro-2H-benzotriazole-2-yl)-phenol; 2-[4,6-bis(2,4-dimethylphenyl)-1,3,5-triazin-2-yl]-5-(octyloxy) phenol; Screener: titanium dioxide, zinc oxide, carbon black; Acid neutralizer: hydrotalcite; Fiber: carbon nanotubes; HAS: 1,3,5-triazine-2,4,6-triamine, N,N'''[1,2-ethane-diyl-bis[[[4,6-bis[butyl(1,2,6,6-pentam-ethyl-4-piperidinyl)amino]-1,3,5-triazine-2-yl]imino]-3,1-propanediyl]bis[N',N''-dibutyl-N',N''-bis(1,2,2,6,6-pentamethyl-4-piperidinyl)-; bis(2,2,6,6-tetramethyl-4-piperidyl) sebacate; 2,2,6,6-tetramethyl-4-piperidinyl stearate; N,N'-bisformyl-N,N'-bis-(2,2,6,6-tetramethyl-4-piperidinyl)-hexamethylendiamine; 1,6-hexanediamine- N,N'-bis(2,2,6,6-tetramethyl-4-piperidinyl)-polymer with 2,4,6-trichloro-1,3,5-triazine, reaction products with N-butyl-1-butanamine an N-butyl-2,2,6,6-tetramethyl-4-piperidinamine; butanedioic acid, dimethylester, polymer with 4-hydroxy-2,2,6,6-tetramethyl-1-piperidine ethanol; 1,6-hexanediamine, N, N'-bis(2,2,6,6-tetramethyl-4-piperidinyl)-, polymers with 2,4-dichloro-6-(4-morpholinyl)-1,3,5-triazine; 1,6-hexanediamine, N,N'-bis(2,2,6,6-tetramethyl-4-piperidi-nyl)-, polymers with morpholine-2,4,6-trichloro-1,3,5-triazine reaction products, methylated; Phenolic antioxidant: 2,6,-di-tert-butyl-4-(4,6-bis(octylthio)-1,3,5,-triazine-2-ylamino) phenol; pentaerythritol tetrakis(3-(3,5-di-tert-butyl-4-hydroxyphenyl) propionate); octadecyl-3-(3,5-di-tert-butyl-4-hydroxyphenyl)-propionate; 3,3',3',5,5',5'-hexa-tert-butyl-a,a',a'-(mesitylene-2,4,6-triyl)tri-p-cresol; 2-(1,1-dimethylethyl)-6-[[3-(1,1-dimethylethyl)-2-hydroxy-5-methylphenyl]methyl-4-methylphenyl acrylate; 1,3,5-tris(3,5-di-tert-butyl-4-hydroxybenzyl)-1,3,5-triazine-2,4,6(1H,3H,5H)-trione; 2',3-bis[[3-[3,5-di-tert-butyl-4-hydroxyphenyl]propionyl]] propionohydrazide; ethylene bis[3,3-bis[3-(1,1-dimethylethyl)-4-hydroxyphenyl]butanoate]; 1,3,5-tris(4-tert-butyl-3-hy-droxy-2,6-dimethyl benzyl)-1,3,5-triazine-2,4,6-(1H,3H,5H)-trione; 2,2'-methylenebis(4-methyl-6-tertbutylphenol); 1,1,3-tris(2'methyl-4'-hydroxy-5'tert-butylphenyl)butane; Phosphites: bis-(2,4-di-t-butylphenol) pentaerythritol diphosphite; tris (2,4-di-tert-butylphenyl)phosphite; distearyl pentaerythritol diphosphite; trilauryl trithiophosphite; Quencher: (2,2'-thiobis(4-tert-octyl-phenolato))-N-butylamine-nickel(II); Optical brightener: 2,2'-(2,5-thiophenediyl)bis(5-tert-butylbenzoxazole); 2,2'-(1,2-ethylenediyldi-4,1-phenylene)bisbenzoxazole	
Effect of exposure		cracks are fomed after 228 h in Xenotest	

PP polypropylene

PARAMETER	UNIT	VALUE	REFERENCES
BIODEGRADATION			
Colonized products	-	construction materials, membranes, thin films	
Typical biodegradants	-	formation of hydroperoxides which destabilize the polymeric carbon chain to form a carbonyl group	
Stabilizers	-	1, 2-benzisothiazolin-3-one, N-halamine precursor, silver nanoparticles, silver powder, TiO_2-anatase; surface functionalization	Yao, F; Fu, G-D; Zhao, J; Kang, E-T; Neoh, K G, J. Membrane Sci., 319, 1-2, 149-57, 2008.
TOXICITY			
NFPA: Health, Flammability, Reactivity rating	-	1/1/0	
Carcinogenic effect	-	not listed by ACGIH, NIOSH, NTP	
Mutagenic effect	-	not known	
Teratogenic effect	-	not known	
Reproductive toxicity	-	not known	
TLV, ACGIH	$mg\ m^{-3}$	3 (respiratory), 10 (total)	
OSHA	$mg\ m^{-3}$	5 (respiratory), 15 (total)	
Oral rat, LD_{50}	$mg\ kg^{-1}$	>5,000	
Skin rabbit, LD_{50}	$mg\ kg^{-1}$	>2,000	
ENVIRONMENTAL IMPACT			
Aquatic toxicity, *Daphnia magna*, LC_{50}, 48 h	$mg\ l^{-1}$	3,000-75,000	Lithner, Ph D Thesis, Univrsity of Gothenburg, 2011.
Cradle to grave non-renewable energy use	MJ/kg	65-72	
Cradle to pellet greenhouse gasses	$kg\ CO_2$ kg^{-1} resin	1.5-2.0	
Life cycle value analysis	mPt	85 (the same part from aluminum - 96)	Ibeh, C C; Bhattarai, D, Antec, 2858-61, 2003.
PROCESSING			
Typical processing methods	-	blow molding, extrusion, injection molding, injection-stretch blow molding, thermoforming	
Preprocess drying: temperature/ time/residual moisture	°C/h/%	79/2	
Processing temperature	°C	191-250	
Processing pressure	MPa	69-103 (injection); 8 (back); 49 (holding)	

PP polypropylene

PARAMETER	UNIT	VALUE	REFERENCES
Additives used in final products	-	Fillers: aluminum flakes, antimony trioxide, barium sulfate, bismuth carbonate, calcium carbonate, calcium sulfate, carbon black, carbon nanotube, clay, fly ash, glass beads, glass fiber, glass flakes, hydromagnesite-huntite, hydrotalcite, magnesium hydroxide, metal powders (aluminum, iron, nickel), mica, montmorillonite, nano-calcium carbonate, phenolic microspheres, poly(alkylene terephthalate) fiber, potassium-magnesium aluminosilicate, red phosphorus, sepiolite, silica flour, silicium carbide, silver powder, stainless steel fiber, talc, wollastonite, wood fiber and flour, zinc borate; Plasticizers: dioctyl sebacate, glycerin, paraffinic oil, isooctyl tallate, paraffinic, naphthenic, and aromatic processing oils, polybutenes; Antistatics: alkyl-bis(2-hydroxyethyl)amine, carbon nanotubes, glycerol monostearate, lauric diethanol amide, N,N-bis(2-hydroxyethyl) alkoxypropylbetaine, polypyrrole, stearyldiethanolamine; Antiblocking: calcium carbonate, crosslinked silicone spheres, diatomaceous earth, natural silica, synthetic silica; Release: calcium stearate, glyceryl monostearate; Slip: behenamide, erucamide, N,N'-bisethylene oleamide, oleamide, silicone oil	
Applications	-	automotive, electrical components, fibers, furniture, packaging, tapes, many other applications, such as for example, mechanical lungs, orthopaedic bandages, sutures	
Outstanding properties	-	sterilizable (autoclave and ethylene oxide), low extractables	
BLENDS			
Suitable polymers	-	EOC, EPDM, PA6, PANI, PE, PCL, PHB, PPy, PS, SEBS	
ANALYSIS			
FTIR (wavenumber-assignment)	cm^{-1}/-	degradation products: hydroxyl – 3600-3200; carbonyl – 1800-1700; ketone – 1725-15; carboxylic acid – 1712-1705, vinyl – 909	Rajakumar, K; Sarasvathy, V; Thamarai Chelvan, A; Chitra, R; Vijayakumar, C T, J. Polym. Environ., 17, 191-202, 2009.
NMR (chemical shifts)	ppm	pentad structure determination	Harding, G W; van Reenen, Eur. Polym. J., 47, 1, 70-77, 2011.
x-ray diffraction peaks	degree	effect of UV exposure on crystallinity retention	Wanasekara, N; Chalivendra, V; Calvert, P, Polym. Deg. Stab., 96, 4, 432-37, 2011.

PP, iso isotactic-polypropylene

PARAMETER	UNIT	VALUE	REFERENCES
GENERAL			
Common name	-	isotactic polypropylene	
CAS name	-	1-propene, homopolymer, isotactic	
Acronym	-	iso-PP	
CAS number	-	25085-53-4	
RTECS number	-	UD1842000	
Formula		$-CH_2CHCH_2CHCH_2CH-$ $CH_3 \quad CH_3 \quad CH_3$	
HISTORY			
Person to discover	-	Natta, G; Pino, P; Mazzanti, G	Natta, G; Pino, P; Mazzanti, G, US Patent 3,112,300, Montecatini, Nov. 26, 1963.
Date	-	1963	
Details	-	isotactic polypropylene	
SYNTHESIS			
Monomer(s) structure	-	$H_2C=CHCH_3$	
Monomer(s) CAS number(s)	-	115-07-1	
Monomer(s) molecular weight(s)	dalton, g/mol, amu	42.08	
Monomer(s) expected purity(ies)	%	99	
Monomer ratio	-	100% and less	
Formulation example	-	hydrogen is used to control molecular weight	Harding, G W; van Reenen, A J, Eur. Polym. J., 47, 70-77, 2011.
Method of synthesis	-	polymerization carried in the liquid propylene or in a gas-phase reactors	
Temperature of polymerization	^{o}C	40	
Catalyst	-	titanium halide/aluminum alkyl or metallocene	
Number average molecular weight, M_n	dalton, g/mol, amu	5,000-166,000	Fayolle, B; Tchakhtchi, A; Verdu, J, Polym. Testing, 23, 939-47, 2004.
Mass average molecular weight, M_w	dalton, g/mol, amu	158,000-580,000	Fayolle, B; Tchakhtchi, A; Verdu, J, Polym. Testing, 23, 939-47, 2004.
Polydispersity, M_w/M_n	-	1.9-9.7; 3.0-3.9 (metallocene)	Hanyu, A; Wheat, R, J. Plast. Film Sheeting, 15, 2, 109-19, 1999.
Molar volume at 298K	cm^3 mol^{-1}	44.4 (crystalline)	
Van der Waals volume	cm^3 mol^{-1}	30.7 (crystalline)	
Radius of gyration	nm	29.7-30.5	Logotheti, G E; Theodorou, D N, Macromolecules, 40, 2235-45, 2007.
End-to-end distance of unperturbed polymer chain	nm	187-189	Logotheti, G E; Theodorou, D N, Macromolecules, 40, 2235-45, 2007.
STRUCTURE			
Crystalline structure	-	mesomorphic	Van der Burgt, F, Crystallization of isotactic polypropylene, Technical University of Eindhoven, 2002.

PP, iso isotactic-polypropylene

PARAMETER	UNIT	VALUE	REFERENCES
Crystallinity	%	29-75; 31-50 (non-spherulitic); 44-67 (non-spherulitic, quenched) 40-57 (spherulitic); 67-69 (uniaxially stretched)	Fayolle, B; Tchakhtchi, A; Verdu, J, Polym. Testing, 23, 939-47, 2004; Mileva, D; Androsch, R; Radusch, H-J, Polym. Bull., 62, 561-71, 2009; Hedesiu, C; Demco, D E; Remerie, K; Bluemich, B; Litvinov, V M, Macromol. Chem. Phys., 209, 734-45, 2008.
Cell type (lattice)	-	monoclinic (α), hexagonal (β), orthorhombic or triclinic (γ; the most thermodynamically stable)	Chen, J-H; Tsai, F-C; Nien, Y-H; Yeh, P-H, Polymer, 46, 5680-88, 2005.
Cell dimensions	nm	a:b:c=0.639:2.044:0.647 (monoclinic); a:b:c=0.854:0.993:4.241 (orthorhombic); a:b:c=0.655:2.157:0.655 (triclinic)	
Unit cell angles	degree	$\alpha:\beta:\gamma$=90:99.2:90 (monoclinic); $\alpha:\beta:\gamma$=90:90:90 (orthorhombic); $\alpha:\beta:\gamma$=97.4:98.8:97.4	
Crystallite size	nm	3.97-6.36 (α); 4-8 (oriented)	Romanos, N A; Theodorou, D N, Macromolecules, 43, 5455-69, 2010; Kang, Y-A; Kim, K-H; Ikehata, S; Ohhoshi, Y; Gotoh, Y; Nagura, M; Urakawa, H, Polymer, 52, 2044-50, 2011.
Polymorphs	-	α, β (metastable), γ	Zhao, S; Xin, Z, J. Polym. Sci. B, 48, 653-65, 2010.
Beta-crystallinity, K value vs. cast roll temperature	-/°C	0.35/60, 0.78/90, 0.85/104	Kim, S; Townsend, E B, Antec, 2002.
Tacticity	%	90.5-99.5 (isotactic)	Capt, L; Kamal, M R; Rettenberger, S; Muenstedt, H, Antec, 997-1001, 2003; Harding, G W; van Reenen, A J, Eur. Polym. J., 47, 70-77, 2011.
Chain conformation	-	helix, 3/1	
Entanglement molecular weight	dalton, g/mol, amu	6,900 (metallocene)	
Lamellae thickness	nm	15.1-18.5	White, H M; Bassett, D C; Jaaskelainen, P, Polymer, 50, 5559-64, 2009.
Heat of crystallization	kJ kg^{-1}	83.7	Chen, J-H; Tsai, F-C; Nien, Y-H; Yeh, P-H, Polymer, 46, 5680-88, 2005.
Rapid crystallization temperature	°C	138-144	Pantani, R; Coccorullo, I; Volpe, V; Titomanlio, G, Macromolecules, 43, 9030-38, 2010.
Avrami constants, k/n	-	n=2-3 for monoclinic and 0.45-0.55 for mesomorphic	La Carrubba, V; Piccarolo, S; Brucato, V, J. Appl. Polym. Sci., 104, 1358-67, 2007.
Crystallization activation energy	J mol^{-1}	211.1-316.6	Zhao, S; Xin, Z, J. Polym. Sci. B, 48, 653-65, 2010.
Crystal growth rate	μm s^{-1}	0.1-0.8	Pantani, R; Coccorullo, I; Volpe, V; Titomanlio, G, Macromolecules, 43, 9030-38, 2010.

COMMERCIAL POLYMERS

Some manufacturers	-	Atofina; Daelin; LyondellBasell; Sunoco	
Trade names	-	Fiinacene; Polypropylene; Moplen; Polypropylene	

PHYSICAL PROPERTIES

Density at 20°C	g cm^{-3}	0.90-0.91	
Color	-	white	
Refractive index, 20°C	-	1.4900-1.503	

538 **HANDBOOK OF POLYMERS** 3rd Edition, Copyrights 2022; ChemTec Publishing

PP, iso isotactic-polypropylene

PARAMETER	UNIT	VALUE	REFERENCES
Transmittance	%	60-90 (quenched); 50-65 (slowly cooled)	Mileva, D; Androsch, R; Radusch, H-J, Polym. Bull., 62, 561-71, 2009.
Haze	%	0.3; 0.2-0.9 (biaxially oriented, metallocene)	Hanyu, A; Wheat, R, J. Plast. Film Sheeting, 15, 2, 109-19, 1999.
Gloss, 60°, Gardner (ASTM D523)	%	99; 94-98 (biaxially oriented, metallocene)	
Odor	-	odorless	
Melting temperature, DSC	°C	157-171; 151-166 (commercial); 148-151 (biaxially oriented, metallocene)	Hanyu, A; Wheat, R, J. Plast. Film Sheeting, 15, 2, 109-19, 1999.
Softening point	°C	155-156	
Decomposition temperature	°C	240	He, P; Xiao, Y; Zhang, P; Xing, C; Zhu, N; Zhu, X; Yan, D, Polym. Deg. Stab., 88, 473-79, 2005.
Activation energy of thermal degradation	kJ mol^{-1}	265 (TGA); 254 (IR)	He, P; Xiao, Y; Zhang, P; Xing, C; Zhu, N; Zhu, X; Yan, D, Polym. Deg. Stab., 88, 473-79, 2005.
Thermal expansion coefficient, 23-80°C	°C^{-1}	1.1-1.4E-4; 6.6E-4 (melt)	
Thermal conductivity, melt	W m^{-1} K^{-1}	0.12-0.22	
Glass transition temperature	°C	-10	
Specific heat capacity	J K^{-1} kg^{-1}	2,500-3,400 (depending on annealing temperature)	Zia, Q; Radusch, H-J; Androsch, R, Polymer, 48, 3504-11, 2007.
Heat of fusion	kJ kg^{-1}	177	Masirek, R; Piorkowska, E, Eur. Polym. J., 46, 1436-45, 2010.
Heat deflection temperature at 0.45 MPa	°C	88-107	
Heat deflection temperature at 1.8 MPa	°C	55	
Vicat temperature VST/A/50	°C	150-155	
Hildebrand solubility parameter	MPa$^{0.5}$	17.2-18.8	
Surface tension	mN m^{-1}	20.2-22.5	
Dielectric constant at 100 Hz/1 MHz	-	2.2-2.3	
Dissipation factor at 100 Hz		0.0003-0.001	
Dissipation factor at 1 MHz		0.0001-0.0003	
Volume resistivity	ohm-m	1E14 to 1E15	
Permeability to oxygen, 25°C	cm^3 m^{-2} day^{-1}	2,600; 2,300-2,900 (biaxially oriented, metallocene)	
Permeability to water vapor, 25°C	cm^3 m^{-2} day^{-1}	3.4; 2.6-3.2 (biaxially oriented, metallocene)	
Contact angle of water, 20°C	degree	116	
Speed of sound	m s^{-1}	2100-125000	

MECHANICAL & RHEOLOGICAL PROPERTIES

PARAMETER	UNIT	VALUE	REFERENCES
Tensile strength	MPa	30 (commercial); 130 (MD, biaxially stretched); 300 (TD, biaxially stretched); 185-219 (equibiaxially stretched); 140-160 (MD; metallocene); 250-300 (TD; metallocene)	Capt, L; Kamal, M R; Rettenberger, S; Muenstedt, H, Antec, 997-1001, 2003 Hanyu, A; Wheat, R, J. Plast. Film Sheeting, 15, 2, 109-19, 1999.
Tensile modulus	MPa	825 (commercial); 910 (alpha-form); 820 (beta-form)	Mezghani, K S; Gasem, Z; Faheem, M, antec, 2884-91, 2004.
Tensile stress at yield	MPa	33-36	

PP, iso isotactic-polypropylene

PARAMETER	UNIT	VALUE	REFERENCES
Elongation	%	90-500; 100-119 (equibiaxially stretched); 120-170 (MD; biaxially oriented, metallocene); 40-60 (TD; biaxially oriented, metallocene)	Capt, L; Kamal, M R; Rettenberger, S; Muenstedt, H, Antec, 997-1001, 2003; Hanyu, A; Wheat, R, J. Plast. Film Sheeting, 15, 2, 109-19, 1999.
Tensile yield strain	%	10-12	
Flexural strength	MPa	38.9	
Flexural modulus	MPa	1,150-1,570	
Elastic modulus	MPa	2,357-3,450 (equibiaxially stretched)	Capt, L; Kamal, M R; Rettenberger, S; Muenstedt, H, Antec, 997-1001, 2003; Hanyu, A; Wheat, R, J. Plast. Film Sheeting, 15, 2, 109-19, 1999.
Izod impact strength, unnotched, 23°C	$J\ m^{-1}$	33.8; 30.9-74.0 (nucleated)	Zhao, S; Xin, Z, J. Polym. Sci. B, 48, 653-65, 2010.
Izod impact strength, notched, 23°C	$J\ m^{-1}$	25-39	
Poisson's ratio	-	0.38	
Rockwell hardness	-	R95-105	
Shrinkage	%	7; (MD); 10 (TD); 4-11 (MD; biaxially oriented, metallocene); 8-21 (TD; biaxially oriented, metallocene)	
Melt viscosity, shear rate=0 s^{-1}	kPa s	2.9-9.9	
Melt index, 230°C/2.16 kg	g/10 min	1.9-31	

CHEMICAL RESISTANCE

PARAMETER	UNIT	VALUE	REFERENCES
Acid dilute/concentrated	-	very good	
Alcohols	-	very good	
Alkalis	-	very good	
Aliphatic hydrocarbons	-	fair to poor	
Aromatic hydrocarbons	-	poor	
Esters	-	fair	
Greases & oils	-	good to fair	
Halogenated hydrocarbons	-	poor	
Ketones	-	good	
Θ solvent, Θ-temp.=122, 206, 142.8, 184°C	-	n-butyl alcohol, p-cresol, diphenyl ether, p-ethyl phenol	
Good solvent	-	1,2,4-trichlorobenzene, decalin, halogenated hydrocarbons, aliphatic ketones, xylene (all above 80°C)	
Non-solvent	-	most common solvents	

FLAMMABILITY

PARAMETER	UNIT	VALUE	REFERENCES
Ignition temperature	°C	>200; 93.3 (fibers & yarns)	
Autoignition temperature	°C	570	
Limiting oxygen index	% O_2	17	
Heat release	$kW\ m^{-2}$	101-727 (with flame retardants)	
Char at 500°C	%	0	Lyon, R E; Walters, R N, J. Anal. Appl. Pyrolysis, 71, 27-46, 2004.
Heat of combustion	$J\ g^{-1}$	45,800	
Volatile products of combustion	-	CO, CO_2, soot	
UL rating	-	HB; V-0 (flame retarded grades)	

PP, iso isotactic-polypropylene

PARAMETER	UNIT	VALUE	REFERENCES
WEATHER STABILITY			
Effect of tacticity		sPP is substantially more stable than iPP	Kato, M; Tsuruta, A; Kuroda, S; Osawa, Z, Polym. Deg. Stab., 67, 1-5, 2000.
TOXICITY			
NFPA: Health, Flammability, Reactivity rating	-	1/1/0	
Carcinogenic effect	-	not listed by ACGIH, NIOSH, NTP	
Mutagenic effect	-	not known	
Teratogenic effect	-	not known	
Reproductive toxicity	-	not known	
TLV, ACGIH	mg m^{-3}	3 (respiratory), 10 (total)	
OSHA	mg m^{-3}	5 (respiratory), 15 (total)	
Oral rat, LD$_{50}$	mg kg^{-1}	>5,000	
Skin rabbit, LD$_{50}$	mg kg^{-1}	>2,000	
PROCESSING			
Typical processing methods	-	blown film, extrusion, injection molding	
Additives used in final products	-	antiblocking; slip; antioxidant; nucleating agent	
Applications	-	bags, fibers, film, food packaging	
Outstanding properties	-	clarity, stiffness	
BLENDS			
Suitable polymers	-	EPR, HDPE, PA66, PB, PET, s-PP, PS	
ANALYSIS			
FTIR (wavenumber-assignment)	cm^{-1}/-	isotactic sequences of different length – 808, 841, 900, 973, 998	Li, L; Liu, T; Zhao, L; Yuan, W-k, J. Supercritical Fluids, in press, 2011.
NMR (chemical shifts)	ppm	pentad sequence content determined by C NMR	Harding, G W; van Reenen, A J, Eur. Polym. J., 47, 70-77, 2011.
x-ray diffraction peaks	degree	α-form: 14.08, 16.95, 18.5, 21.2, 21.85 (other forms see reference)	Chen, J-H; Tsai, F-C; Nien, Y-H; Yeh, P-H, Polymer, 46, 5680-88, 2005.

PP, syndio syndiotactic-polypropylene

PARAMETER	UNIT	VALUE	REFERENCES
GENERAL			
Common name	-	polypropylene, syndiotactic	
CAS name	-	1-propene, homopolymer, syndiotactic	
Acronym	-	s-PP	
CAS number	-	26063-22-9	
RTECS number	-	UD1842000	
Formula		$-CH_2CHCH_2CHCH_2CH-$ with CH_3 substituents	
HISTORY			
Person to discover	-	Natta, G; Corradini, P; Pasquon, I; Pegoraro, M; Peraldo, M	Natta, G; Corradini, P; Pasquon, I; Pegoraro, M; Peraldo, M, US Patent 3,258,455, Montecatini, June 28, 1966.
Date	-	1966	
Details	-	syndiotactic polypropylene	
SYNTHESIS			
Monomer(s) structure	-	$H_2C=CHCH_3$	
Monomer(s) CAS number(s)	-	115-07-1	
Monomer(s) molecular weight(s)	dalton, g/mol, amu	42.08	
Monomer ratio	-	100% and less	
Temperature of polymerization	°C	-78	
Catalyst	-	special class of metallocene, vanadium-based (Natta)	De Rosa, C; Auriemma, F, Prog. Polym. Sci., 31, 145-237, 2006.
Yield	%	100	
Number average molecular weight, M_n	dalton, g/mol, amu	35,000-119,000	
Mass average molecular weight, M_w	dalton, g/mol, amu	87,000-1,190,000	
Polydispersity, M_w/M_n	-	1.29-6.2	
STRUCTURE			
Crystallinity	%	25-63	
Cell type (lattice)	-	orthorhombic	
Cell dimensions	nm	a:b:c=1.45:1.12:0.74 (form I); a:b:c=1.45:0.6:0.74 (form II); a:b:c=0.522:1.117:0.506 (form III)	Yamashita, K; Fujiwara, N; Fujikawa, Y; Nakaoki, T; Chiu, W-Y; Stroeve, P, Polym. Eng. Sci., 49, 740-46, 2009.
Crystallite size	nm	2.7	Arranz-Andres, J; Guevara, J L; Velilla, T; Quijada, R; Benavente, R; Perez, E; Cerrada, M L, Polymer, 46, 12287-97, 2005.
Space group		Ibca (form I); C222/1 (form II)	Razavi, Encyclopedia of Materials: Science and Technology, 7708-11, Elsevier, 2008.
Polymorphs	-	I, II, III, IV	De Rosa, C; Auriemma, F, Prog. Polym. Sci., 31, 145-237, 2006.

542   **HANDBOOK OF POLYMERS** 3rd Edition, Copyrights 2022; ChemTec Publishing

PP, syndio syndiotactic-polypropylene

PARAMETER	UNIT	VALUE	REFERENCES
Tacticity	%	90.0-96.8 (syndiotactic)	
Chain conformation	-	helical (t2g2)n (I and II); transplanar (t6g2t2g2)n (III and IV); zigzag (tttt) (on stretching)	Bonnet, M; Yan, S; Petermann, J; Zhang, B; Yang, D, J. Mater. Sci., 36, 3, 635-41, 2001; Tian, N; Lv, R; Na, B; Xu, W; Li, Z, J. Phys. Chem. B, 113, 14920-24, 2009.
Entanglement molecular weight	dalton, g/ mol, amu	2,700 (metallocene)	

COMMERCIAL POLYMERS

Some manufacturers	-	Total	
Trade names	-	Polypropylene	

PHYSICAL PROPERTIES

Density at 20°C	g cm^{-3}	0.88; 0.856 (amorphous); 0.93 (crystalline)	
Color	-	white	
Transmittance	%	91	
Haze	%	2	
Odor	-	odorless	
Melting temperature, DSC	°C	117-156	
Decomposition temperature	°C	260	He, P; Xiao, Y; Zhang, P; Xing, C; Zhu, N; Zhu, X; Yan, D, Polym. Deg. Stab., 88, 473-79, 2005.
Activation energy of thermal degradation	kJ mol^{-1}	268 (TGA); 269 (IR)	He, P; Xiao, Y; Zhang, P; Xing, C; Zhu, N; Zhu, X; Yan, D, Polym. Deg. Stab., 88, 473-79, 2005.
Glass transition temperature	°C	-15 to 3	
Heat of fusion	kJ mol^{-1}	4.4-8.2	
Vicat temperature VST/A/50	°C	111	

MECHANICAL & RHEOLOGICAL PROPERTIES

Tensile strength	MPa	15.2-25.2	
Tensile modulus	MPa	483	
Elongation	%	250-300	
Tensile yield strain	%	10-11	
Flexural modulus	MPa	345	
Izod impact strength, unnotched, 23°C	J m^{-1}	640	
Shrinkage	%	33-38 (fibers)	Guadagno, L; D'Aniello, C; Naddeo, C; Vittoria, V, Macromolecules, 34, 2512-21, 2001.
Melt index, 230°C/2.16 kg	g/10 min	2-20	

CHEMICAL RESISTANCE

Acid dilute/concentrated	-	very good	
Alcohols	-	very good	
Alkalis	-	very good	
Aliphatic hydrocarbons	-	fair to poor	
Aromatic hydrocarbons	-	poor	
Esters	-	fair	
Greases & oils	-	good to fair	

PP, syndio syndiotactic-polypropylene

PARAMETER	UNIT	VALUE	REFERENCES
Halogenated hydrocarbons	-	poor	
Ketones	-	good	
Θ solvent, Θ-temp.=45, 36°C	-	i-amyl acetate, cyclohexane	
Good solvent	-	chlorinated hydrocarbons, cyclohexane, diethyl ether, toluene	
Non-solvent	-	many polar solvents	

FLAMMABILITY			
Ignition temperature	°C	>200; 93.3 (fibers & yarns)	
Autoignition temperature	°C	570	
Limiting oxygen index	% O_2	17	
Heat release	kW m^{-2}	101-727 (with flame retardants)	
Char at 500°C	%	0	Lyon, R E; Walters, R N, J. Anal. Appl. Pyrolysis, 71, 27-46, 2004.
Heat of combustion	J g^{-1}	45,800	
Volatile products of combustion	-	CO, CO_2, soot	
UL 94 rating	-	HB; V-0 (flame retarded grades)	

WEATHER STABILITY			
Effect of tacticity	-	sPP is substantially more stable than iPP	Kato, M; Tsuruta, A; Kuroda, S; Osawa, Z, Polym. Deg. Stab., 67, 1-5, 2000.
Effect of exposure	-	300-500 h in Xenotest decrease tensile strength of sPP by 65-90%	Barany, T; Foldes, E; Czigany, T; Karger-Kocsis, J, J. Appl. Polym. Sci., 3462-3469, 2004.

TOXICITY			
NFPA: Health, Flammability, Reactivity rating	-	1/1/0	
Carcinogenic effect	-	not listed by ACGIH, NIOSH, NTP	
Mutagenic effect	-	not known	
Teratogenic effect	-	not known	
Reproductive toxicity	-	not known	
TLV, ACGIH	mg m^{-3}	3 (respiratory), 10 (total)	
OSHA	mg m^{-3}	5 (respiratory), 15 (total)	
Oral rat, LD$_{50}$	mg kg^{-1}	>5,000	
Skin rabbit, LD$_{50}$	mg kg^{-1}	>2,000	

PROCESSING			
Typical processing methods	-	extrusion, injection molding, injection blow molding	
Applications	-	fiber, film, impact modifier, medicine, sheet	
Outstanding properties	-	narrow MW, melt strength	

BLENDS			
Suitable polymers	-	EPR, PE, i-PP	

PP, syndio syndiotactic-polypropylene

PARAMETER	UNIT	VALUE	REFERENCES
ANALYSIS			
FTIR (wavenumber-assignment)	cm^{-1}/-	1715 – carbonyl; 3420 – OH; more in refs.	He, P; Xiao, Y; Zhang, P; Xing, C; Zhu, N; Zhu, X; Yan, D, Polym. Deg. Stab., 88, 473-79, 2005; Sevegney, M S; Kannan, R M; Siedle, A R; Percha, P A, J. Polym. Sci. B, 43, 439-61, 2005.
NMR (chemical shifts)	ppm	syndiotacticity index – 303-313; planar zigzag – 375; helical – 537, 550, 776; amorphous – 845, 970, 996 and more	Sevegney, M S; Kannan, R M; Siedle, A R; Naik, R; Naik, V M, Vibrational Spectroscopy, 40, 246-56, 2006.
x-ray diffraction peaks	degree	12.2, 15.8, 18.8, 20.6	Razavi, Encyclopedia of Materials: Science and Technology, 7708-11, Elsevier, 2008.

PPA polyphthalamide

PARAMETER	UNIT	VALUE	REFERENCES
GENERAL			
Common name	-	polyphthalamide	
CAS name	-	1,3-benzenedicarboxylic acid, polymer with 1,4-benzenedicarboxylic acid and 1,6-hexanediamine (25750-23-6); 1,3-benzenedicarboxylic acid, polymer with 1,4-benzenedicarboxylic acid, 1,6-hexanediamine and hexanedioic acid (27135-32-6)	
Acronym	-	PPA	
CAS number	-	25750-23-6; 27135-32-6	
HISTORY			
Date	-	1991	
Details	-	commercialization	
SYNTHESIS			
Monomer(s) structure	-		
Monomer(s) CAS number(s)	-	100-21-0+121-91-5; 124-09-4	
Monomer(s) molecular weight(s)	dalton, g/mol, amu	166.13; 116.21	
Mass average molecular weight, M_w	dalton, g/mol, amu	11,000-13,700	Singletary, N; Bates, R B; Jacobsen, N; Lee, A K; Lin, G; Somogyi, A; Streeter, M J; Hall, H K, Macromolecules, 42, 2336-43, 2009.
STRUCTURE			
Crystallinity	%	33-45	Moyak, D M, Antec, 3505-10, 1996.
COMMERCIAL POLYMERS			
Some manufacturers	-	Desco; EMS; Solvay	
Trade names	-	Destron; Grivory; Amodel	
PHYSICAL PROPERTIES			
Density at 20°C	g cm⁻³	1.18-1.19; 1.48-1.59 (33-45% glass fiber)	
Refractive index, 20°C	-	1.57-1.59	
Odor		nearly odorless	
Melting temperature, DSC	°C	294-335	
Thermal expansion coefficient, 23-80°C	°C⁻¹	8E-5; 1.8-2.4E-5 (33-45% glass fiber)	
Thermal conductivity, melt	W m⁻¹ K⁻¹	0.289-0.372 15-45% glass fiber)	
Glass transition temperature	°C	121-138	Pini, N; Zaniboni, C; Busato, S; Ermanni, P, J. Thermoplast. Composite Mater., 19, 207-16, 2006.
Specific heat capacity	J K⁻¹ kg⁻¹	1,500-2,400 (23°C); 4,200-6,000 (melt)	
Long term service temperature	°C	260	
Temperature index (50% tensile strength loss after 20,000 h/5000 h)	°C	160	Padey, D; Walling, J; Wood A, Polymers in Defence and Aerospace 2007, Rapra, 2007, paper 15.

PPA polyphthalamide

PARAMETER	UNIT	VALUE	REFERENCES
Heat deflection temperature at 1.8 MPa	°C	120; 285-300 (33-45% glass fiber)	
Vicat temperature VST/A/50	°C	301-314 (33-45% glass fiber)	
Enthalpy of melting	J g⁻¹	54.1 (*in situ* polymerized); 40.7 (melt-crystallized)	Pini, N; Zaniboni, C; Busato, S; Ermanni, P, J. Thermoplast. Composite Mater., 19, 207-16, 2006.
Dielectric constant at 100 Hz/1 MHz	-	4.6-5.1/3.6-4.2 (33-45% glass fiber)	
Dissipation factor at 60 Hz	E-4	40-50	
Dissipation factor at 1 MHz	E-4	12-17	
Volume resistivity	ohm-m	1E14 (33-45% glass fiber)	
Surface resistivity	ohm	1E15 (33-45% glass fiber)	
Electric strength K20/P50, d=3.2 mm	kV mm⁻¹	21-23 (33-45% glass fiber)	
Comparative tracking index, CTI, test liquid A	-	550 (33-45% glass fiber)	

MECHANICAL & RHEOLOGICAL PROPERTIES

PARAMETER	UNIT	VALUE	REFERENCES
Tensile strength	MPa	90; 200-259 (33-45% glass fiber)	
Tensile modulus	MPa	13,100-17,200 (33-45% glass fiber)	
Elongation	%	6; 1.9-2.6 (33-45% glass fiber)	
Flexural strength	MPa	290-363 (33-45% glass fiber)	
Flexural modulus	MPa	11.0-13.8 (33-45% glass fiber)	
Compressive strength	MPa	148-194 (33-45% glass fiber)	
Young's modulus	MPa	2,500-3,500	
Charpy impact strength, unnotched, 23°C	kJ m⁻²	60-93 (33-45% glass fiber)	
Charpy impact strength, notched, 23°C	kJ m⁻²	9.2-10.7 (33-45% glass fiber)	
Izod impact strength, unnotched, 23°C	J m⁻¹	770-1105 (33-45% glass fiber)	
Izod impact strength, notched, -30°C	J m⁻¹	80-110 (33-45% glass fiber)	
Shear strength	MPa	88-108 (33-45% glass fiber)	
Poisson's ratio	-	0.39-0.41 (33-45% glass fiber)	
Shrinkage	%	0.18-1.0; 0.2-1.0 (33-45% glass fiber)	
Intrinsic viscosity, 25°C	dl g⁻¹	0.85-1.06	
Moisture absorption, 24h 23°C/50% RH	%	0.1-0.3	

CHEMICAL RESISTANCE

PARAMETER	UNIT	VALUE	REFERENCES
Acid dilute/concentrated	-	good to very good	
Alcohols	-	good	
Alkalis	-	good to very good	
Aliphatic hydrocarbons	-	good	
Greases & oils	-	good to very good	

FLAMMABILITY

PARAMETER	UNIT	VALUE	REFERENCES
NBS smoke chamber, Ds, 4 min	-	3-12	

PPA polyphthalamide

PARAMETER	UNIT	VALUE	REFERENCES
UL 94 rating	-	HB (33-45% glass fiber)	
TOXICITY			
NFPA: Health, Flammability, Reactivity rating	-	1/1/0	
PROCESSING			
Typical processing methods	-	electroplating, injection molding	
Processing temperature	°C	330-350	
Processing pressure	MPa	4-5 (hold)	
Applications	-	metal replacement	
Outstanding properties	-	dimensional stability, heat resistance, chemical and moisture resistance	

PPG polypropylene glycol

PARAMETER	UNIT	VALUE	REFERENCES
GENERAL			
Common name	-	polypropylene glycol, polypropylene oxide	
ACS name	-	poly[oxy(methyl-1,2-ethanediyl)], α-hydro-ω-hydroxy-	
Acronym	-	PPG	
CAS number	-	25322-69-4	
EC number	-	233-239-6; 500-039-8	
RTECS number	-	TR5250000 TR5300000 TR5425000 TR5600000 TR5775000 TR5785000 TR5800000 TR5950000 TR6125000 TR6129000 TR6130000 TR6200000 TR6210000 TR6215000 TR6220000	
Formula		$\left[\begin{array}{c}CH_2CHO \\ | \\ CH_3\end{array}\right]_n$	
HISTORY			
Person to discover	-	Morris, R C; Snider, A V	Morris, R C; Snider, A V, US Patent 2,520,733, Shell, Aug. 29, 1950.
Date	-	1950	
Details	-	polymers of trimethylene glycol	
SYNTHESIS			
Monomer(s) structure	-		
Monomer(s) CAS number(s)	-	75-56-9	
Monomer(s) molecular weight(s)	dalton, g/mol, amu	58.08	
Monomer ratio	-	100%	
Mass average molecular weight, M_w	dalton, g/mol, amu	76-18,200	Gainaru, C; Hiller, W; Boehmer, R, Macromolecules, 43. 1907-14, 2010.
Polydispersity, M_w/M_n	-	1.0-1.07	
Polymerization degree (number of monomer units)	-	3-180	
Molar volume at 298K	$cm^3 mol^{-1}$	calc.=50.5 (crystalline); 58.1 (amorphous)	
Van der Waals volume	$cm^3 mol^{-1}$	calc.-34.4 (crystalline); 34.4 (amorphous)	
Free volume fraction at T_g		0.010-0.029	Consolati, G; Levi, M; Messa, L; Tieghi, G, Europhys. Lett., 53, 4, 497-503, 2001.
Free volume at T_g	$cm^3 g^{-1}$	0.895-0.898	Consolati, G; Levi, M; Messa, L; Tieghi, G, Europhys. Lett., 53, 4, 497-503, 2001.
Chain-end groups	-	OH	
STRUCTURE			
Cell type (lattice)	-	orthorhombic	
Cell dimensions	nm	a:b:c=1.046-1.052:0.464-0.469:0.692-0.716	
Unit cell angles	degree	$\alpha{:}\beta{:}\gamma$=90:90:90	
Number of chains per unit cell	-	4	
Chain conformation	-	planar zigzag	

PPG polypropylene glycol

PARAMETER	UNIT	VALUE	REFERENCES
COMMERCIAL POLYMERS			
Some manufacturers	-	DOW	
Trade names	-	PPGs	
PHYSICAL PROPERTIES			
Density at 20°C	g cm^{-3}	1.002-1.09; 1.126 (crystalline)	
Color	-	clear	
Refractive index, 20°C	-	1.447-1.459	
Odor		sweet	
Melting temperature, DSC	°C	-40 to 73	
Pour point	°C	-18 to -45	
Boiling temperature	°C	188 to >300	
Storage temperature	°C	24	
Thermal expansion coefficient, 23-80°C	10^{-4} °C^{-1}	7.2-8.3 (liquid); 2.35-4.2 (glassy)	Consolati, G; Levi, M; Messa, L; Tieghi, G, Europhys. Lett., 53, 4, 497-503, 2001.
Glass transition temperature	°C	-77 to -74	Consolati, G; Levi, M; Messa, L; Tieghi, G, Europhys. Lett., 53, 4, 497-503, 2001.
Hildebrand solubility parameter	MPa$^{0.5}$	exp.=16.1-16.3	
Surface tension	mN m^{-1}	31.2-51.3	
Surface free energy	mJ m^{-2}	30.6	
CHEMICAL RESISTANCE			
Alcohols	-	soluble	
Aliphatic hydrocarbons	-	miscible	
Aromatic hydrocarbons	-	soluble	
Esters	-	miscible	
Greases & oils	-	insoluble	
Halogenated hydrocarbons	-	soluble	
Ketones	-	soluble	
Θ solvent, Θ-temp.=50.5°C	-	isooctane	
Good solvent	-	acetone, benzene, chloroform, dioxane, ethanol, methanol (hot), THF	
Non-solvent	-	diethyl ether, N,N-dimethylacetamide	
FLAMMABILITY			
Ignition temperature	°C	185-246	
Autoignition temperature	°C	>350	
Heat of combustion	J g^{-1}	10,510-11,410	
Volatile products of combustion	-	CO, CO_2	
WEATHER STABILITY			
Important initiators and accelerators	-	Fe acetylacetone	Semsarzadeh, M; Salehi, H, Eur. Polym. J., 36, 5, 1001-10, 2000.

PPG polypropylene glycol

PARAMETER	UNIT	VALUE	REFERENCES
BIODEGRADATION			
Typical biodegradants	-	*Sphingopyxis terrae, S. macrogoltabida, Sphingomonas sp., Sphingobium* species, *Pseudomonas* species, and *S. maltophilia*	Hu, X; Fukutani, A; Liu, X; Kimbara, K; Kawai, F, Appl. Microbiol. Biotechnol., 73, 1407-13, 2007.
TOXICITY			
NFPA: Health, Flammability, Reactivity rating	-	1/1/0	
Carcinogenic effect	-	not listed by ACGIH, NIOSH, NTP	
Oral rat, LD$_{50}$	mg kg^{-1}	2,150-21,000	
Skin rabbit, LD$_{50}$	mg kg^{-1}	>10,000 to >30,000	
ENVIRONMENTAL IMPACT			
Aquatic toxicity, *Bluegill sunfish*, LC$_{50}$, 48 h	mg l^{-1}	1,700	
Aquatic toxicity, *Rainbow trout*, LC$_{50}$, 48 h	mg l^{-1}	10,000	
PROCESSING			
Applications	-	fiber and textile processing, food, metalworking, paper processing, personal care, plastics, polyurethane synthesis, water and waste water treatment	
ANALYSIS			
FTIR (wavenumber-assignment)	cm^{-1}/-	C=O – 1728; CH$_2$ – 2975	Semsarzadeh, M; Salehi, H, Eur. Polym. J., 36, 5, 1001-10, 2000.

PPMA polypropylene, maleic anhydride modified

PARAMETER	UNIT	VALUE	REFERENCES
GENERAL			
Common name	-	polypropylene, maleic anhydride modified	
Acronym	-	PPMA	
CAS number	-	25722-45-6	
HISTORY			
Person to discover	-	Nogues, P	Nogues, P, US Patent 4,735,992, Atochem, Apr. 5, 1988.
Date	-	1988	
SYNTHESIS			
Maleic anhydride content	%	1-10	
Number average molecular weight, M_n	dalton, g/mol, amu	3,900	
Mass average molecular weight, M_w	dalton, g/mol, amu	9,100	
COMMERCIAL POLYMERS			
Some manufacturers	-	Clariant; DuPont	
Trade names	-	Licocene; Fusabond P	
PHYSICAL PROPERTIES			
Density at 20°C	g cm^{-3}	0.90-0.95	
Color	-	yellowish	
Odor		mild hydrocarbon	
Melting temperature, DSC	°C	152-157	Wong, S-C; Lee, H; Qu, S; Mall, S; Chen, L, Polymer, 47, 7477-84, 2006.
Softening point	°C	135-162	
Decomposition temperature	°C	>300	
Vicat temperature VST/A/50	°C	112	
MECHANICAL & RHEOLOGICAL PROPERTIES			
Elastic modulus	MPa	2,000	
Melt viscosity, shear rate=1000 s^{-1}	mPa s	50-1,100 (170°C)	
Melt index, 190°C/2.16 kg	g/10 min	11-400	
CHEMICAL RESISTANCE			
Alcohols	-	very good	
Aliphatic hydrocarbons	-	fair to poor	
Aromatic hydrocarbons	-	poor	
Esters	-	fair	
Greases & oils	-	good to fair	
Halogenated hydrocarbons	-	poor	
Ketones	-	good	

PPMA polypropylene, maleic anhydride modified

PARAMETER	UNIT	VALUE	REFERENCES
FLAMMABILITY			
Volatile products of combustion	-	CO,CO_2, acids, aldehydes, alcohols, acrolein	
TOXICITY			
HMIS: Health, Flammability, Reactivity rating	-	2/1/0	
Carcinogenic effect	-	not listed by ACGIH, NIOSH, NTP	
OSHA	mg m^{-3}	5 (respirable), 15 (total)	
Oral rat, LD$_{50}$	mg kg^{-1}	>2,000	
Skin rabbit, LD$_{50}$	mg kg^{-1}	>2,000	
PROCESSING			
Typical processing methods	-	extrusion, pultrusion, glass mat process	
Applications	-	compatibilizer, coupling agent for composite building panels, coupling agent for wire and cable; coupling agent for short and long glass fiber filled PP	
BLENDS			
Suitable polymers	-	NR; PP	

PPO poly(phenylene oxide)

PARAMETER	UNIT	VALUE	REFERENCES
GENERAL			
Common name	-	poly(phenylene oxide); poly(2,6-dimethyl-1,4-phenyle oxide)	
IUPAC name	-	poly[oxy(2,6-dimethyl-1,4-phenylene)]	
CAS name	-	poly(oxyphenylene)	
Acronym	-	PPO, PPE	
CAS number	-	9041-80-9	
Linear formula			
HISTORY			
Person to discover	-	A S Hay	
Date	-	1956, 1960 (commercialization)	
Details	-	Hay discovered polymer and GE commercialized it	
SYNTHESIS			
Monomer(s) structure	-		
Monomer(s) CAS number(s)	-	526-26-1	
Monomer(s) molecular weight(s)	dalton, g/mol, amu	122.17	
Formulation example	-	monomer, solvent, catalyst	
Method of synthesis	-	three methods are used for synthesis, including oxidative coupling, radical polymerization, and Ullmann reaction	Fink, J K, High Peformance Polymers, William Andrew, 2008.
Temperature of polymerization	°C	35-55	
Catalyst	-	Mn, Cu, or Co derivatives	
Number average molecular weight, M_n	dalton, g/mol, amu	15,000-164,000	
Mass average molecular weight, M_w	dalton, g/mol, amu	35,000-320,000; 350,000 (Sabic)	-; Rizzo, P, Gallo, C, Vitale, V, Guerra, G, Polymer, 167, 193-201, 2019.
Polydispersity, M_w/M_n	-	1.3-2.4	
Molar volume at 298K	cm^3 mol^{-1}	calc.=75.0; 92.0 (crystalline)	
Van der Waals volume	cm^3 mol^{-1}	49.42; 69.3 (crystalline)	
Molecular cross-sectional area, calculated	$cm^2 \times 10^{-16}$	27.6	
STRUCTURE			
Crystallinity	%	40-58	
Cell type (lattice)	-	orthorhombic	
Cell dimensions	nm	a:b:c=0.807:0.554:1.026	
Unit cell angles	degree	$\alpha{:}\beta{:}\gamma$=90:90:90	

554   **HANDBOOK OF POLYMERS** 3rd Edition, Copyrights 2022; ChemTec Publishing

PPO poly(phenylene oxide)

PARAMETER	UNIT	VALUE	REFERENCES
Number of chains per unit cell	-	4	
Chain conformation	-	2/1 helix	
Entanglement molecular weight, M_e	dalton, g/mol, amu	calc.=1461, 3620	

COMMERCIAL POLYMERS

Some manufacturers	-	Evonik; Sabic	
Trade names	-	Vestoran; Noryl	

PHYSICAL PROPERTIES

Density at 25°C	g cm^{-3}	1.04-1.06; 0.96 (melt); 1.16 (crystalline)	
Refractive index, 20°C	-	calc.=1.608-1.6209; exp.=1.6400	
Melting temperature, DSC	°C	240-267	
Decomposition temperature	°C	300 (under vacuum and N_2)	
Thermal expansion coefficient, -30 to 30°C	°C^{-1}	2.5-5.2E-5	
Thermal conductivity, melt	W m^{-1} K^{-1}	calc.=0.2060	
Glass transition temperature	°C	calc.=85-115; exp.=205-215	
Heat of fusion	kJ mol^{-1}	7.8	
Heat deflection temperature at 0.45 MPa	°C	106	
Hansen solubility parameters, δ_D, δ_P, δ_H	MPa$^{0.5}$	16.9, 8.9, 2.7	
Interaction radius		11.7	
Hildebrand solubility parameter	MPa$^{0.5}$	19.3	
Surface tension	mN m^{-1}	calc.=44.5; exp.=42.8	Pozniak, G; Gancarz, I; Tylus, W, Desalination, 198, 215-224, 2006.
Dielectric constant at 100 Hz/1 MHz	-	4.6-4.7/4.5-4.8	
Dielectric loss factor at 1 kHz	-	0.0027	
Permeability to nitrogen, 25°C	cm^3 cm cm^{-2} s^{-1} Pa^{-1} x 10^{12}	0.286	
Permeability to oxygen, 25°C	cm^3 cm cm^{-2} s^{-1} Pa^{-1} x 10^{12}	0.119	
Permeability to water vapor, 25°C	cm^3 cm cm^{-2} s^{-1} Pa^{-1} x 10^{12}	304.5	
Surface free energy	mJ m^{-2}	91.3-93.9	Khayet, M; Villaluega, J P G; Godino, M P; Mengual J I; Seoane, B; Khulbe, K C; Matsuura, T, J. Colloid Interface Sci., 278, 410-422, 2004.

PPO poly(phenylene oxide)

PARAMETER	UNIT	VALUE	REFERENCES
MECHANICAL & RHEOLOGICAL PROPERTIES			
Tensile modulus	MPa	2,700	
Tensile stress at yield	MPa	98	
Elongation	%	20-40	
Tensile yield strain	%	7	
Flexural strength	MPa	114-137	
Flexural modulus	MPa	5,880-10,000	
Izod impact strength, notched, 23°C	J m^{-1}	69	
Poisson's ratio	-	0.410-0.492	
Shrinkage	%	0.25-0.35	
Intrinsic viscosity, 25°C	dl g^{-1}	0.7-1.57	Khayet, M; Villaluega, J P G; Godino, M P; Mengual J I; Seoane, B; Khulbe, K C; Matsuura, T, J. Colloid Interface Sci., 278, 410-422, 2004.
CHEMICAL RESISTANCE			
Acid dilute/concentrated	-	very good	
Alcohols	-	good	
Alkalis	-	very good	
Aliphatic hydrocarbons	-	poor	
Aromatic hydrocarbons	-	poor	
Esters	-	poor	
Greases & oils	-	good	
Halogenated hydrocarbons	-	poor	
Ketones	-	poor	
Θ solvent, Θ-temp.=69°C	-	methylene chloride	
Good solvent	-	benzene, halogenated hydrocarbons, toluene	
Non-solvent	-	acetone, alcohols, THF	
FLAMMABILITY			
Char at 500°C	%	25.5	Lyon, R E; Walters, R N, J. Anal. Appl. Pyrolysis, 71, 27-46, 2004.
Heat of combustion	J g^{-1}	34,210	Walters, R N; Hacket, S M; Lyon, R E, Fire Mater., 24, 5, 245-52, 2000.
WEATHER STABILITY			
Spectral sensitivity	nm	320	
Depth of UV penetration	µm	20	
Important initiators and accelerators	-	products of thermal degradation, hydroperoxides, phenyl radicals, phenoxy radicals, benzyl radicals, hydroxyl groups	
Products of degradation	-	chain scission (oxygen atmosphere), crosslinking (under nitrogen)	

556   **HANDBOOK OF POLYMERS** 3rd Edition, Copyrights 2022; ChemTec Publishing

PPO poly(phenylene oxide)

PARAMETER	UNIT	VALUE	REFERENCES
Stabilizers	-	UVA: 2,2'-methylenebis(6-(2H-benzotriazol-2-yl)-4-1,1,3,3-tetramethylbutyl)phenol; 2-(4,6-diphenyl-1,3,5-triazin-2-yl)-5-hexyloxy-phenol; HAS: 1,3,5-triazine-2,4,6-triamine, N,N'''[1,2-ethane-diyl-bis[[[4,6-bis[butyl(1,2,6,6-pentamethyl-4-piperidinyl)amino]-1,3,5-triazine-2-yl]imino]-3,1-propanediyl]bis[N',N''-dibutyl-N',N''-bis(1,2,2,6,6-pentamethyl-4-piperidinyl)-; Electron transfer quencher: 1,2,4-trimethoxybenzene	

PROCESSING

PARAMETER	UNIT	VALUE	REFERENCES
Typical processing methods	-	blow molding, casting, extrusion, injection molding, thermoforming	
Additives used in final products	-	Fillers: aluminum flake, calcium carbonate, carbon fiber, cellulose fiber, glass fiber, graphite fiber, nickel coated graphite fiber, PTFE, zinc borate, wood flour; Other: blowing agents (e.g., azodicarbonamide), flame retardants (e.g., antimony trioxide, brominated PS), impact modifiers (e.g., HIPS); Plasticizers: aromatic phosphates, diphenyl phthalate, pentaerythritol tetrabenzoate, polybutene, triphenyl trimellitate; Antistatics: carbon black (including superconductive), carbon fibers, lithium chloride, polyether ester amide, potassium titanate, sodium alkanesulfonate, stainless steel fiber; Release: fluororesin, stearic acid salt	Fink, J K, High Performance Polymers, William Andrew, 2008.
Applications	-	adhesives, air conditioner housings, automotive (instrument panels, interior and exterior trim, glove compartments, wheel covers, electric connectors, fuse boxes), electronics (computer and television housings, keyboard frames, interface boxes), hospital and office furniture, membranes, production of blends; UV dosimetry	

BLENDS

PARAMETER	UNIT	VALUE	REFERENCES
Suitable polymers	-	epoxy, HIPS, PA6, PA66, PAE, PBI, PE, PP, PPS, PS, PSF, PVME, SI	Fink, J K, High Performance Polymers, William Andrew, 2008.
Compatibilizers	-	maleic anhydride, fumaric acid, methacrylic anhydride, epichlorohydrin, benzoyl chloride	Fink, J K, High Performance Polymers, William Andrew, 2008.

ANALYSIS

PARAMETER	UNIT	VALUE	REFERENCES
x-ray diffraction peaks	degree	7.7, 13.0, 16.0, 21.7	Khayet, M; Villaluega, J P G; Godino, M P; Mengual J I; Seoane, B; Khulbe, K C; Matsuura, T, J. Colloid Interface Sci., 278, 410-422, 2004.

PPP poly(1,4-phenylene)

PARAMETER	UNIT	VALUE	REFERENCES
GENERAL			
Common name	-	poly(1,4-phenylene)	
IUPAC name	-	poly(1,4-phenylene)	
CAS name	-	poly(1,4-phenylene)	
Acronym	-	PPP	
CAS number	-	25190-62-9	
Formula			
SYNTHESIS			
Monomer(s) structure	-		
Monomer(s) CAS number(s)	-	71-43-2	
Monomer(s) molecular weight(s)	dalton, g/mol, amu	78.11	
Monomer ratio	-	100%	
Mass average molecular weight, M_w	dalton, g/mol, amu	10,000	
Molar volume at 298K	cm^3 mol^{-1}	calc.=66.7	
Van der Waals volume	cm^3 mol^{-1}	44.5	
STRUCTURE			
Crystallinity	%	0	
Cell type (lattice)	-	orthorhombic	
Cell dimensions	nm	a:b:c=0.779-0.806:0.553-0.562:0.42-0.43	Yamamoto, T; Kanbara, T; Mori, C, Synthetic Metals, 38, 399-402, 1990.
Entanglement molecular weight	dalton, g/mol, amu	calc.=2,222	
COMMERCIAL POLYMERS			
Some manufacturers	-	Solvay	
Trade names	-	PrimoSpire	
PHYSICAL PROPERTIES			
Density at 20°C	g cm^{-3}	1.19-1.24	
Refractive index, 20°C	-	calc.=1.6401-1.651	
Melting temperature, DSC	°C	>300	
Decomposition temperature	°C	370	
Thermal expansion coefficient, 23-80°C	°C^{-1}	3.1E-5	
Thermal conductivity, melt	W m^{-1} K^{-1}	0.1892	
Glass transition temperature	°C	150-180	

PPP poly(1,4-phenylene)

PARAMETER	UNIT	VALUE	REFERENCES
Heat deflection temperature at 1.8 MPa	°C	151-171	
Surface tension	mN m^{-1}	calc.=34.5-58.3	
Dielectric constant at 100 Hz/1 MHz	-	3.12/3.01	
Dissipation factor at 100 Hz		0.007	
Dissipation factor at 1 MHz		0.007	
Volume resistivity	ohm-m	>7E13	
Electric strength K20/P50, d=0.60.8 mm	kV mm^{-1}	20	
Optical absorption edge	eV	3	Saxena, V; Malhotra, B D, Handbook of polymers in Elecronics, Ed. Malhotra, B D, Rapra, 2002.

MECHANICAL & RHEOLOGICAL PROPERTIES

PARAMETER	UNIT	VALUE	REFERENCES
Tensile strength	MPa	115-152	
Tensile modulus	MPa	3,900-5,520	
Tensile stress at yield	MPa	148	
Elongation	%	10-15	
Flexural strength	MPa	164-234	
Flexural modulus	MPa	4,000-6,000	
Compressive strength	MPa	620	Friedrich, K; Burkhart, T; Almajid, A A; Haupert, F, Int. J. Polym. Mater., 59, 680-92, 2010.
Young's modulus	MPa	8,300	Friedrich, K; Burkhart, T; Almajid, A A; Haupert, F, Int. J. Polym. Mater., 59, 680-92, 2010.
Izod impact strength, unnotched, 23°C	J m^{-1}	1,600	
Izod impact strength, notched, 23°C	J m^{-1}	59-69	
Poisson's ratio	-	0.425	
Rockwell hardness	B	32	
Melt index, 380°C/5 kg	g/10 min	8-15	
Water absorption, equilibrium in water at 23°C	%	0.1	

CHEMICAL RESISTANCE

PARAMETER	UNIT	VALUE	REFERENCES
Aliphatic hydrocarbons	-	very good	
Aromatic hydrocarbons	-	very good	
Greases & oils	-	good	
Halogenated hydrocarbons	-	poor	
Good solvent	-	methyl chloride	

FLAMMABILITY

PARAMETER	UNIT	VALUE	REFERENCES
Limiting oxygen index	% O$_2$	55	

PPP poly(1,4-phenylene)

PARAMETER	UNIT	VALUE	REFERENCES
WEATHER STABILITY			
Spectral sensitivity	nm	318, 341	Mulazzi, E; Ripamonti, A; Athouel, L; Wery, J; Lefrant, S, Phys Rev. B, 65, 08520,1-9, 2002.
Maximum absorption	nm	380-390	Aboulkassim, A; Chevrot, C, Polymer, 34, 2, 401-5, 1993.
PROCESSING			
Typical processing methods	-	injection molding, machining	
Preprocess drying: temperature/ time/residual moisture	ºC/h/%	150/3/	
Processing temperature	ºC	320-370	
Process time	min	15 (residence time)	
Applications	-	aerospace components, bearings, bushings, gears, high-strength tubing, medical devices, military articles, semiconductor components, surgical instruments, water processing and test components	
Outstanding properties	-	exceptional strength and stiffness without reinforcements; inherent flame resistance	
BLENDS			
Suitable polymers	-	PPS, PVK	

PPS poly(p-phenylene sulfide)

PARAMETER	UNIT	VALUE	REFERENCES
GENERAL			
Common name	-	poly(p-phenylene sulfide)	
CAS name	-	poly(thio-1,4-phenylene)	
Acronym	-	PPS	
CAS number	-	25212-74-2; 26125-40-6	
Linear formula			
HISTORY			
Person to discover	-	Charles Friedel and James Mason Crafts; Wayne Hill and James Edmonds	
Date	-	1888; 1967, 1972	
Details	-	PPS was discovered by Fridel and Crafts 1888, and method of production was developed by Hill and Edmonds in 1967, and commercialization of PPS by Phillips Petroleum Company in 1972	Fink, J K, High Peformance Polymers, William Andrew, 2008.
SYNTHESIS			
Monomer(s) structure	-		
Monomer(s) CAS number(s)	-	106-46-7; 1313-82-2	
Monomer(s) molecular weight(s)	dalton, g/mol, amu	147. 004; 78.04	
Monomer ratio	-	1.88	
Formulation example	-	reagents, solvent, catalyst, molecular weight modifier	Fink, J K, High Peformance Polymers, William Andrew, 2008.
Method of synthesis	-	PPS is manufactured based on reaction between sodium sulfide and p-dichlorobenzene	Fink, J K, High Peformance Polymers, William Andrew, 2008.
Temperature of polymerization	$^{\circ}$C	160-260	
Time of polymerization	h	88	
Catalyst	-	lithium salts, sodium acetate, cyclic amine compounds	Fink, J K, High Peformance Polymers, William Andrew, 2008.
Mass average molecular weight, M_w	dalton, g/mol, amu	12,000-1,400,000	
Polydispersity, M_w/M_n	-	1.4-2.0	
Molar volume at 298K	cm^3 mol^{-1}	75.3 (crystalline)	
Van der Waals volume	cm^3 mol^{-1}	54.1 (crystalline)	
STRUCTURE	-		
Crystallinity	%	40-83	Lu, J; Huang, R; Oh, I-K, Macromol. Chem. Phys., 208, 405-14, 2007.
Cell type (lattice)	-	orthorhombic	Tabor, B J; Magre, E P; Boon, J, Eur. Polym. J., 7, 1127, 1971.
Cell dimensions	nm	a:b:c=0.867:0.561:1.026	Tabor, B J; Magre, E P; Boon, J, Eur. Polym. J., 7, 1127, 1971.
Unit cell angles	degree	α:β:γ=90:90:90	

PPS poly(p-phenylene sulfide)

PARAMETER	UNIT	VALUE	REFERENCES
Space group	-	Pbcn	Napolitano, R; Pirozzi, B; Iannelli, P, Macromol. Theory Simul., 10, 9, 827-32, 2001.
Tacticity	%	*trans* (100)	
Chain conformation	-	helix 2/1	
Entanglement molecular weight	dalton, g/mol, amu	20,000	
Avrami constant, n	-	1.4-3	Nohara, L B; Nohara, E L; Moura, A; Goncalves, J M R P; Costa, M L; Rezende, M C, Polimeros: Ciencia Tecnologia, 16, 2, 104-110, 2006; D'Ilario, L; Martinelli, A, Eur. Phys. J. E, 19, 37-45, 2006.
Activation energy of molecular motion	kJ mol^{-1}	70 (amorphous); 43 (crystalline)	Jurga, J; Wozniak-Braszak, Fojud, Z; Jurga, K, Solid State Magnetic Resonance, 25, 47-52, 2004.

COMMERCIAL POLYMERS			
Some manufacturers	-	Solvay; Celanese; Toyobo	Fortron, Ticona, May 2007.
Trade names	-	Ryton, Primef; Fortron; Procon (fiber)	

PHYSICAL PROPERTIES			
Density at 20°C	g cm^{-3}	1.34-1.36; 1.425-1.44 (crystalline); 1.32 (amorphous)	
Color	-	white to pale yellow	
Refractive index, 20°C	-	1.83	
Birefringence	-	0.27; 0..3 (theoretical maximum)	
Odor	-	mild	
Melting temperature, DSC	°C	285-295	
Decomposition temperature	°C	450-480; 532	Duan, Y; Cong, P; Liu, X; Li, T, J. Macromol. Sci. B, 48, 604-16, 2009.
Explosion temperature	°C	500	
Thermal expansion coefficient, 23-80°C	°C^{-1}	4.9-5.2E5	
Thermal conductivity, melt	W m^{-1} K^{-1}	0.29; 0.20 (40% glass fiber)	
Glass transition temperature	°C	74-92	
Heat of fusion	kJ mol^{-1}	4.5-5.5	
Maximum service temperature	°C	218-232	
Long term service temperature	°C	<240	
Temperature index (50% tensile strength loss after 20,000 h/5000 h)	°C	230	Padey, D; Walling, J; Wood A, Polymers in Defence and Aerospace 2007, Rapra, 2007, paper 15.
Heat deflection temperature at 1.8 MPa	°C	150-263 (depends on mold temperature; 40% glass fiber)	Greer, M R; Reaume, A; Kowalski, G, Antec, 504-8, 2010.
Enthalpy of melting	J g^{-1}	43.1	Nohara, L B; Nohara, E L; Moura, A; Goncalves, J M R P; Costa, M L; Rezende, M C, Polimeros: Ciencia Tecnologia, 16, 2, 104-110, 2006.
Hansen solubility parameters, δ_D, δ_P, δ_H	MPa$^{0.5}$	18.7, 5.3, 3.7	
Interaction radius		6.7	
Hildebrand solubility parameter	MPa$^{0.5}$	19.8	

PPS poly(p-phenylene sulfide)

PARAMETER	UNIT	VALUE	REFERENCES
Dielectric constant at 100 Hz/1 MHz	-	3.8-5.2/3.8-4.9	
Relative permittivity at 10 kHz	-	2.7-3.2; 4 (40% glass fiber)	
Dissipation factor at 100 Hz	E-4	40-300	
Dissipation factor at 1 MHz	E-4	11; 62 (40% glass fiber)	
Volume resistivity	ohm-m	1E9; >10E13 (40% glass fiber)	
Surface resistivity	ohm	>10E15 (40% glass fiber)	
Electric strength K20/P50, d=0.60.8 mm	kV mm^{-1}	18; 28 (40% glass fiber)	
Comparative tracking index, CTI, test liquid A	-	125	
Arc resistance	s	125-185	
Coefficient of friction	-	0.4 (air); 0.22 (water); 0.6 (40% glass fiber)	Duan, Y; Cong, P; Liu, X; Li, T, J. Macromol. Sci. B, 48, 604-16, 2009.
Contact angle of water, 20°C	degree	80.3	
Surface free energy	mJ m^{-2}	46.8	

MECHANICAL & RHEOLOGICAL PROPERTIES

PARAMETER	UNIT	VALUE	REFERENCES
Tensile strength	MPa	90; 131-195 (40-65% glass fiber)	
Tensile modulus	MPa	3,800; 14,500-19,100 (40% glass fiber)	
Elongation	%	3-8; 0.9-1.9 (40-65% glass fiber)	
Flexural strength	MPa	125-145; 200-285 (40-65% glass fiber)	
Flexural modulus	MPa	3,750-4,200; 14,500-19,400 (40-65% glass fiber)	
Compressive strength	MPa	112; 260-296 (40-65% glass fiber)	
Charpy impact strength, unnotched, 23°C	kJ m^{-2}	53 (40% glass fiber)	
Charpy impact strength, notched, 23°C	kJ m^{-2}	10 (40% glass fiber)	
Izod impact strength, unnotched, 23°C	J m^{-1}	35-82; 34 (40% glass fiber)	
Izod impact strength, notched, 23°C	J m^{-1}	2.6-3.5; 10 (40% glass fiber)	
Rockwell hardness	-	M90-95; M100 (40% glass fiber)	
Shrinkage	%	1.2-1.8; 0.13-0.7 (40% glass fiber)	Greer, M R; Reaume, A; Kowalski, G, Antec, 504-8, 2010.
Water absorption, equilibrium in water at 23°C	%	0.01-0.03	

CHEMICAL RESISTANCE

PARAMETER	UNIT	VALUE	REFERENCES
Acid dilute/concentrated	-	very good to fair (nitric acid, see ref.)	Tanthapanichakoon, W; Hata, M; Nitta, K-h; Faruuchi, M; Otani, Y, Polym. Deg. Stab., 91, 2614-21, 2006.
Alcohols	-	very good	
Alkalis	-	very good	
Aliphatic hydrocarbons	-	very good	
Aromatic hydrocarbons	-	very good	
Esters	-	very good	
Greases & oils	-	very good	
Halogenated hydrocarbons	-	good	

PPS poly(p-phenylene sulfide)

PARAMETER	UNIT	VALUE	REFERENCES
Ketones	-	very good	
Good solvent	-	not soluble below 200ºC; above 200ºC soluble in 1-chloro-naphthalene and biphenyl	
Non-solvent	-	all solvents below 200ºC	

FLAMMABILITY			
Ignition temperature	ºC	480-500	
Autoignition temperature	ºC	540	
Limiting oxygen index	% O_2	40; 47 (40% glass fiber)	
Char at 500ºC	%	41.6	Lyon, R E; Walters, R N, J. Anal. Appl. Pyrolysis, 71, 27-46, 2004.
Heat of combustion	J g^{-1}	28,390-29,620	Walters, R N; Hacket, S M; Lyon, R E, Fire Mater., 24, 5, 245-52, 2000.
Volatile products of combustion	-	CO, CO_2, SO_2, CS	
UL 94 rating	-	V-0	

WEATHER STABILITY			
Spectral sensitivity	nm	290-370	
Activation wavelengths	nm	330	
Excitation wavelengths	nm	300-320; 308	
Emission wavelengths	nm	396; 396	
Products of degradation	-	yellowing, conjugated double bonds, crosslinking, chains scission, carbonyls	
Stabilizers	-	UVA: benzotriazole (derivative of Tinuvin 327 in which chlorine atom is replaced by phenylthio or phenylsulfonyl groups); Screener: carbon black	Das, P K; DesLauriers, P J; Fahey, D R; Wood, F K; Cornforth, F J, Polym. Deg. Stab., 48, 1-10, 1995 and ibid. 11-23.
Effect of exposure	WOM	2,000 h with little change in tensile and impact strength	

TOXICITY			
NFPA: Health, Flammability, Reactivity rating	-	1/0/0	
TLV, ACGIH	mg m^{-3}	3 (respirable), 10 (total)	
OSHA	mg m^{-3}	5 (respirable), 15 (total)	

PROCESSING			
Typical processing methods	-	blending, coating, compression molding, extrusion, injection molding, lamination, thermoforming	
Preprocess drying: temperature/ time/residual moisture	ºC/h/%	120/1-2 (unfilled); 140/4 (reinforced)	
Processing temperature	ºC	300-340 (injection molding); 285-310 (extrusion)	
Processing pressure	MPa	50-110 (injection); 30-70 (holding)	
Additives used in final products	-	Fillers: aramid fiber, calcium carbonate, carbon fiber, ferrosoferric oxide, glass fiber, glass flake, mica, talc, PTFE, zinc oxide; Plasticizers: diphenyl phthalate, hydrogenated terphenyl; Antistatics: carbon nanofiber, expandable graphite, octadecyltriethoxysilane; Other: decolorants; Release, high density polyethylene, silicone	

PPS poly(p-phenylene sulfide)

PARAMETER	UNIT	VALUE	REFERENCES
Applications	-	aerospace, automotive lighting, carburetor parts, chip carriers, coil bobbins, electrical and electronic parts, food choppers, fuel components, halogen lamp sockets, IC card connectors, ignition and braking systems, impeller diffusers, lamp sockets, magnets, microwave oven components, motor fans, optical drive, phone jacks, oil well valves, plastic housing for a high speed motor, pump housings, relay components, sockets, steam hair drier parts, tape recorder head mounts, technical parts (pumps, automotive, printer components, liquid crystal-line display projectors), tissue engineering, thermally-conductive materials, transistor encapsulation	
Outstanding properties	-	temperature resistance, chemical resistance, inherently flame retardant, high rigidity	

BLENDS			
Suitable polymers	-	chitosan, EVA, PA66, PAR, PE, PET, PP, PS	

ANALYSIS			
FTIR (wavenumber-assignment)	cm^{-1}/-	benzene ring – 1574, 1073; SO – 1159, 1320; phenolic group – 3400, more in ref.	Zimmerman, D A; Koenig, J L; Ishida, H, Spectrochim. Acta A51, 2397-2409, 1995.
Raman (wavenumber-assignment)	cm^{-1}/-	C–C – 1574; C–H – 1180, 1074, 840; C–S – 743	Zimmerman, D A; Koenig, J L; Ishida, H, Spectrochim. Acta A51, 2397-2409, 1995.
x-ray diffraction peaks	degree	4.3, 6.2	Langer, L; Billaud, D; Issi, J-P, Solid State Commun., 126, 353-57, 2003.

PPSE poly(trimethylsilyl phosphate)

PARAMETER	UNIT	VALUE	REFERENCES
GENERAL			
Common name	-	poly(trimethylsilyl phosphate)	
CAS name	-	40623-46-9	
Acronym	-	PPSE	
Linear formula	-		
PHYSICAL PROPERTIES			
Density at 20°C	g cm^{-3}	1.18	
Refractive index, 20°C	-	1.434	
Freezing point	°C	4.4	

PPSQ polyphenylsilsesquioxane

PARAMETER	UNIT	VALUE	REFERENCES
GENERAL			
Common name	-	polyphenylsilsesquioxane	
CAS name	-	poly[(1,3-diphenyl-1,3:1,3-disiloxanediylidene)-1,3-bis(oxy)]	
Acronym	-	PPSQ	
CAS number	-	51350-55-1	
HISTORY			
Person to discover	-	Brown, J F	
Date	-	1960	
Details	-	proposed structure	
SYNTHESIS			
Monomer(s) structure	-		
Monomer(s) CAS number(s)	-	98-13-5	
Monomer(s) molecular weight(s)	dalton, g/mol, amu	211.55	
Method of synthesis	-	a synthesis of a non-functional acyclic form of polyphenylsilsesquioxane by Piers-Rubinsztajn reaction of hyperbranched polyphenylethoxysiloxane with dimethylphenylsilane. polycyclic form of PPSQ was obtained by hydrolytic polycondensation of PPEOS in acetic acid followed by the addition of dimethylphenylethoxysilane as the blocking agent one-step condensation of all-*cis*-1,3,5,7-tetraphenyl-1,3,5,7 tetrahydroxycyclotetrasiloxane in liquid ammonia produces ladder-like polyphenylsilsesquioxane	Temnikov, M N, Vasil'ev, V G, Buzin, M I, Muzafarov, A M, Eur. Polym. J., 130, 109676, 2020; Anisimov, A A, Polshchikova, N V, Vysochinskaya, Y S, Muzafarov, A M, Medeleev Commun., 29, 4, 421-3, 2019.
Temperature of polymerization	ºC	0-10	
Time of polymerization	h	0.25	
Pressure of polymerization	Pa	atmospheric	
Mass average molecular weight, M$_w$	dalton, g/mol, amu	6,800-77,100, 86,000-164,000	-; Anisimov, A A, Polshchikova, N V, Vysochinskaya, Y S, Muzafarov, A M, Medeleev Commun., 29, 4, 421-3, 2019.
STRUCTURE	**-**		
Crystallinity	%	amorphous; ladder-like PPSQ forms single crystals from solutions	Li, G Z; Yamamoto, T; Nozaki, K; Hikosaka, M, Polymer, 42, 8435-41, 2001; Li, G Z; Yamamoto, T; Nozaki, K; Hikosaka, M, Polymer, 42, 2827-30, 2000.
Cis content	%	prevailing	
Interchain spacing	nm	1.25; 0.29 (ladder-like)	Li, G Z; Yamamoto, T; Nozaki, K; Hikosaka, M, Polymer, 41, 2827-30, 2000; Liu, C; Liu, Z; Shen, Z; Xie, P; Zhang, R; Yang, J; Bai, F, Macromol. Chem. Phys., 202, 1581-85, 2001.
COMMERCIAL POLYMERS			
Some manufacturers	-	Wacker-Belsil	
Trade names	-	SPR	

PPSQ polyphenylsilsesquioxane

PARAMETER	UNIT	VALUE	REFERENCES
PHYSICAL PROPERTIES			
Density at 20°C	g cm^{-3}	1.34-1.35	
Bulk density at 20°C	g cm^{-3}	0.65	
Color	-	clear to white	
Refractive index, 20°C	-	1.55-1.57	Yasuda, N; Yamamoto, S; Hasegawa, Y; Nobutoki, H; Yanagida, S, Chem. Leet. (Jap), 244-5, 2002.
Melting temperature, DSC	°C	>50	
Glass transition temperature	°C	-34 (acyclic), 29-148 (polycyclic)	Temnikov, M N, Vasil'ev, V G, Buzin, M I, Muzafarov, A M, Eur. Polym. J., 130, 109676, 2020.
Softening point	°C	50-70	
Storage temperature	°C	<30	
Decomposition temperature	°C	>250; 530 (N$_2$)	
Heat resistance	°C	405-480 (air), 500 (inert)	Temnikov, M N, Vasil'ev, V G, Buzin, M I, Muzafarov, A M, Eur. Polym. J., 130, 109676, 2020.
CHEMICAL RESISTANCE			
Alcohols	-	poor	
Aliphatic hydrocarbons	-	good	
Aromatic hydrocarbons	-	poor	
Esters	-	poor	
Good solvent	-	butyl lactate, diisobutyl adipate, ethanol, ethyl lactate, isopropyl myristate, isostearyl alcohol, oleyl alcohol	
FLAMMABILITY			
Ignition temperature	°C	>100	
Autoignition temperature	°C	>400	
Volatile products of combustion	-	CO, CO$_2$, formaldehyde	
PROCESSING			
Typical processing methods	-	spin-coating	
Applications	-	aerogels, anti-reflective coating, hair care, microspheres, lipstick, production of polymetalorganosiloxanes, skin care, sunscreens	Libanov, V, Kapustina, A, Shapkin, N, Puzyrkov, Z, Polymer, 194, 122367, 2020.
BLENDS			
Suitable polymers	-	EPDM, i-PS	
ANALYSIS			
FTIR (wavenumber-assignment)	cm^{-1}/-	C–H – 3075, 3061, 1423, 1191, 743, 730, 500; C–C – 1596; Si–O–Si – 1200-1000	Prado, L A S; Radovanovic, E; Pastore, H O; Yoshida, I V P; Torriani, I L, J. Polym. Sci. A, 38, 1580-89, 2000.

568   **HANDBOOK OF POLYMERS** 3rd Edition, Copyrights 2022 ChemTec Publishing

PPSU poly(phenylene sulfone)

PARAMETER	UNIT	VALUE	REFERENCES
GENERAL			
Common name	-	poly(phenylene sulfone)	
CAS name	-	poly(sulfonyl-1,4-phenylene)	
Acronym	-	PPSU	
CAS number	-	31833-61-1; 877322-41-3	
Linear formula			
HISTORY			
Person to discover	-	Umezawa, M; Tsubota, T, Imai, S	Umezawa, M; Tsubota, T, Imai, S, US Patent 4,942,091, Toray Industries, Jul. 17, 1990.
Date	-	1990	
Details	-	PPSU fibers obtained by oxidation of PPS	
SYNTHESIS			
Monomer(s) structure	-		
Monomer(s) CAS number(s)	-	92-88-6; 80-07-9	
Monomer(s) molecular weight(s)	dalton, g/mol, amu	186.21; 287.16	
Method of synthesis	-	PPSU can be prepared from PPS by oxidation with hydrogen peroxide	Tago, T; Kuwashiro, N; Nishide, H, Bull. Chem. Soc. Jpn., 80, 7, 1429-34, 2007.
Mass average molecular weight, M_w	dalton, g/mol, amu	61,000; 21,400-23,700	-; Gronwald, O, Frost, I, Ulbricht, M, Weber, M, Separation Purification Technol., 250, 117107, 2020.
STRUCTURE	-		
Crystallinity	%	85	Robello, D R; Ulman, A; Urankar, E J, Macromolecules, 26, 6718-21, 1993.
Crystallite size	nm	1.0-2.7	Umezawa, M; Tsubota, T; Imai, S, US Patent 5,244,467, Toray industries, sep. 14, 1993.
COMMERCIAL POLYMERS			
Some manufacturers	-	BASF; Solvay	
Trade names	-	Ultrason P; Radel	
PHYSICAL PROPERTIES			
Density at 20°C	g cm^{-3}	1.29	
Color	-	light yellow to brownish	
Odor		odorless	
Decomposition temperature	°C	>400	
Thermal expansion coefficient, 23-80°C	°C^{-1}	0.55-0.56E-4; 0.18E-4 (30% glass fiber)	
Thermal conductivity, melt	W m^{-1} K^{-1}	0.30	
Glass transition temperature	°C	220	

PPSU poly(phenylene sulfone)

PARAMETER	UNIT	VALUE	REFERENCES
Heat deflection temperature at 0.45 MPa	°C	212-214	
Heat deflection temperature at 1.8 MPa	°C	196-207; 210 (30% glass fiber)	
Dielectric constant at 100 Hz/1 MHz	-	3.44/3.45	
Relative permittivity at 100 Hz	-	3.8	
Relative permittivity at 1 MHz	-	3.8	
Dissipation factor at 100 Hz	E-4	6-17	
Dissipation factor at 1 MHz	E-4	76-90	
Volume resistivity	ohm-m	>1-9E13	
Surface resistivity	ohm	>1E15	
Electric strength K20/P50, d=0.60.8 mm	kV mm^{-1}	15-44	
Comparative tracking index, CTI, test liquid A	-	150	
Contact angle of water, 20°C	degree	86.7	Gronwald, O, Frost, I, Ulbricht, M, Weber, M, Separation Purification Technol., 250, 117107, 2020.

MECHANICAL & RHEOLOGICAL PROPERTIES

PARAMETER	UNIT	VALUE	REFERENCES
Tensile strength	MPa	70; 120 (30% glass fiber)	
Tensile modulus	MPa	2,270-2,340; 9,170 (30% glass fiber)	
Tensile stress at yield	MPa	74	
Tensile creep modulus, 1000 h, elongation 0.5 max	MPa	1,930	
Elongation	%	60-120; 2.4 (30% glass fiber)	
Tensile yield strain	%	7.2-7.8	
Flexural strength	MPa	105; 173 (30% glass fiber)	
Flexural modulus	MPa	2,410; 8,070 (30% glass fiber)	
Compressive strength	MPa	99	
Charpy impact strength, unnotched, 23°C	kJ m^{-2}	no break	
Charpy impact strength, unnotched, -30°C	kJ m^{-2}	no break	
Charpy impact strength, notched, 23°C	kJ m^{-2}	65	
Charpy impact strength, notched, -30°C	kJ m^{-2}	24	
Izod impact strength, unnotched, 23°C	J m^{-1}	no break; 640 (30% glass fiber)	
Izod impact strength, notched, 23°C	J m^{-1}	694-750; 75 (30% glass fiber)	
Shear strength	MPa	61	
Poisson's ratio	-	0.43	
Rockwell hardness	R	122	
Shrinkage	%	0.9-1	
Viscosity number	ml g^{-1}	71	
Intrinsic viscosity, 25°C	dl g^{-1}	0.73	

PPSU poly(phenylene sulfone)

PARAMETER	UNIT	VALUE	REFERENCES
Melt volume flow rate (ISO 1133, procedure B), 360°C/10 kg	cm³/10 min	20	
Water absorption, equilibrium in water at 23°C	%	0.7-1	
Moisture absorption, equilibrium 23°C/50% RH	%	0.6	

CHEMICAL RESISTANCE

Acid dilute/concentrated	-	good	
Alkalis	-	good	
Aliphatic hydrocarbons	-	good	
Aromatic hydrocarbons	-	good	
Halogenated hydrocarbons	-	poor	
Good solvent	-	dichloromethane, ethylene dichloride	

FLAMMABILITY

Autoignition temperature	°C	570	
Limiting oxygen index	% O_2	38-44	
NBS smoke chamber, Ds, 4 min	-	0.4	
Char at 500°C	%	38.4	Lyon, R E; Walters, R N, J. Anal. Appl. Pyrolysis, 71, 27-46, 2004.
Volatile products of combustion	-	CO, CO_2, SO_2; SO_3	
UL rating	-	V-0	

WEATHER STABILITY

Spectral sensitivity	nm	<320, 365	
Excitation wavelengths	nm	245-255, 270, 320,	
Emission wavelengths	nm	310, 360, 450	
Depth of UV penetration	µm	50	
Important initiators and accelerators	-	residual monomer, copper stearate	
Products of degradation	-	products of photooxidation: chain scissions, free radicals, carbonyl groups, acetic acid, sulfoacetic acid, benzoic acid, crosslinks, unsaturations, hydroxyl groups, sulfonic acid, SO_2	

PROCESSING

Typical processing methods	-	injection molding	
Preprocess drying: temperature/ time/residual moisture	°C/h/%	140/4/0.02	
Processing temperature	°C	350-390 (injection molding)	
Applications	-	aircraft interiors, cases and trays for healthcare, dental instruments, food service equipment, fuel cells, medical device components, pipe fittings and manifolds, ultrafiltration membranes	
Outstanding properties	-	unlimited steam sterilizability, excellent resistance to hot chlorinated water	

BLENDS

Suitable polymers	-	PSU	

PPT poly(propylene terephthalate)

PARAMETER	UNIT	VALUE	REFERENCES
GENERAL			
Common name	-	poly(propylene terephthalate)	
CAS name	-	poly[oxy(methyl-1,2-ethanediyl)oxycarbonyl-1,4-phenylenecarbonyl]	
Acronym	-	PPT	
CAS number	-	9022-20-2	
Formula			
HISTORY			
Person to discover	-	Winfield, J R; Dickson, J T	
Date	-	1941, 1946	
Details	-	first synthesis; British patent by ICI	
SYNTHESIS			
Monomer(s) CAS number(s)	-		
Monomer(s) molecular weight(s)	dalton, g/mol, amu	504-63-2; 166.13	
Method of synthesis	-	esterification or transesterification (polycondensation) in the presence of catalyst	Berti, C; Bonora, V; Colonna, M; Lotti, N; Sisti, L, Eur. Polym. J., 39, 1595-1601, 2003.
Temperature of polymerization	°C	225-250	
Time of polymerization	h	2-3	
Pressure of polymerization	Pa	vacuum	
Catalyst	-	tetrabutoxytitanium; many other catalysts are discussed in ref.	Mitra, K; Majumdar, S, Mater. Manufac. Proces., 22, 532-40, 2007; Karayannidis, G P; Roupakias, C P; Bikiaris, D N; Achilias, D S, Polymer, 44, 931-42, 2003.
Number average molecular weight, M_n	dalton, g/mol, amu	28,000-34,800	
Mass average molecular weight, M_w	dalton, g/mol, amu	32,000-80,500	
Polydispersity, M_w/M_n	-	2.3-2.4	
Chain-end groups	-	OH, COOH	
STRUCTURE			
Crystallinity	%	60	Motori, A; Saccani, A; Sisti, L, J. Appl. Polym. Sci., 85, 2271-75, 2002.
Cell type (lattice)	-	triclinic	
Cell dimensions	nm	a:b:c=0.453:0.615:1.861	
Unit cell angles	degree	α:β:γ=97:92:111	
Number of chains per unit cell	-	1	
Avrami constants, k/n	-	n=2.53-2.81	Achilias, D S; Bikiaris, D N; Papastergiadis, E; Giliopoulos, D; Papageorgiu, G Z, Macromol. Chem. Phys., 211, 66-79, 2010.

PPT poly(propylene terephthalate)

PARAMETER	UNIT	VALUE	REFERENCES
COMMERCIAL POLYMERS			
Some manufacturers	-	Shell; DuPont	
Trade names	-	Corterra; Sorona	
PHYSICAL PROPERTIES			
Density at 20°C	g cm^{-3}	1.448 (crystalline)	Achilias, D S; Papageorgiou, G Z; Karaayannidis, G P, J. Polym. Sci. B, 42, 3775-96, 2004.
Melting temperature, DSC	°C	223-239; 230	-; Totaro, G, Sisti, L, Cionci, N B, Celli, A, Polym. Testing, 90, 106719, 2020.
Glass transition temperature	°C	54	Totaro, G, Sisti, L, Cionci, N B, Celli, A, Polym. Testing, 90, 106719, 2020.
Heat of fusion	J g^{-1}	67-69	Berti, C; Bonora, V; Colonna, M; Lotti, N; Sisti, L, Eur. Polym. J., 39, 1595-1601, 2003.
Dielectric constant at 100 Hz/1 MHz	-	2.27	
MECHANICAL & RHEOLOGICAL PROPERTIES			
Intrinsic viscosity, 25°C	dl g^{-1}	0.72	
PROCESSING			
Typical processing methods	-	injection molding, spinning	
Applications	-	fibers	
Outstanding properties	-	recovery rate, stain resistance, UV stability	

PPTA poly(p-phenylene terephthalamide)

PARAMETER	UNIT	VALUE	REFERENCES
GENERAL			
Common name	-	poly(p-phenylene terephthalamide), aramid, Kevlar	
IUPAC name	-	poly(imino-1,4-phenyleneiminocarbonyl-1,4-phenylenecarbonyl)	
CAS name	-	poly(imino-1,4-phenyleneiminocarbonyl-1,4-phenylenecarbonyl)	
Acronym	-	PPTA	
CAS number	-	24938-64-5; 26125-61-1	
Formula			
RTECS number	-	TQ9875000	
HISTORY			
Person to discover	-	Stephanie Kwolek	
Date	-	1965; 1971	
Details	-	discovery in 1965 in DuPont laboratories, production of aramid fiber in 1971	
SYNTHESIS			
Monomer(s) structure	-		
Monomer(s) CAS number(s)	-	100-20-9; 106-50-3	
Monomer(s) molecular weight(s)	dalton, g/mol, amu	203.02; 108.1	
Monomer ratio	-	1.87:1	
Method of synthesis	-	condensation reaction yielding hydrochloric acid as a by-product	
Number average molecular weight, M_n	dalton, g/mol, amu	10,300-19,800	
Mass average molecular weight, M_w	dalton, g/mol, amu	31,400-48,350; 40,600	-; Uchida, T, Hara, Y, Takaki, T, Polymer, 202, 122672, 2020.
Polydispersity, M_w/M_n	-	2.4-5.3; 1.98	-; Chen, Z, Shen, Z, Liu, Y, Zhang, Y, Dong, L, Energy Storage Mater., 33, 1-10, 2020.
Molar volume at 298K	cm^3 mol^{-1}	160 (crystalline)	
Molecular length	nm	220	Uchida, T, Hara, Y, Takaki, T, Polymer, 202, 122672, 2020.
Van der Waals volume	cm^3 mol^{-1}	112.6 (crystalline)	
Chain-end groups	mequiv kg	0.42 (amine)	Horta, A; Coca, J; Diez, F V, Adv. Polym. Technol., 22, 1, 15-21, 2003.
STRUCTURE			
Crystallinity	%	72.2-91.0	Mooney, D A; Don McElroy, J M, Chem. Eng. Sci., 59, 2159-70, 2004.

PPTA poly(p-phenylene terephthalamide)

PARAMETER	UNIT	VALUE	REFERENCES
Cell type (lattice)	-	monoclinic; orthorhombic single crystals had a (110) growth plane, where the thickness of each single crystal was approximately equal to the molecular chain length of the PPTA	Chatzi, E G; Koenig. J L, Polym. Plast. Technol. Eng., 26, 229, 1987; Uchida, T, Hara, Y, Takaki, T, Polymer, 202, 122672, 2020.
Cell dimensions	nm	a:b:c=0.78:0.519:1.29; unit cell dimensional changes at low temp. in ref.	Chatzi, E G; Koenig. J L, Polym. Plast. Technol. Eng., 26, 229, 1987; Iyer, R V; Sooryanarayana, K; Guru Row, T N; Vijayan, K, J. Mater. Sci., 38, 133-39, 2003.
Unit cell angles	degree	$\alpha:\beta:\gamma$=90:90:90	Chatzi, E G; Koenig. J L, Polym. Plast. Technol. Eng., 26, 229, 1987.
Number of chains per unit cell	-	2	Chatzi, E G; Koenig. J L, Polym. Plast. Technol. Eng., 26, 229, 1987.
Crystallite size	nm	5 x 5 x 20; 52-55; 60	Pauw, B R; Vigild, M E; Mortensen, K; Andreasen, J W; Klop, E A, J. Appl. Cryst. 43, 837-49, 2010; Knijnenberg, A; Bos, J; Dingemans, T J, Polymer, 1887-97, 2010; Mooney, D A; Don McElroy, J M, Chem. Eng. Sci., 59, 2159-70, 2004.
Chain conformation	-	TTTT	
Radius of gyration	nm	479 (all *trans* PPTA)	Tonelli, A E, Edwards, J F, Polymer, 193, 122342, 2020.

COMMERCIAL POLYMERS

Some manufacturers	-	DuPont	
Trade names	-	Kevlar	

PHYSICAL PROPERTIES

Density at 20°C	g cm^{-3}	1.44-1.47, 1.48 (crystalline)	
Bulk density at 20°C	g cm^{-3}	0.56	
Refractive index, 20°C	-	2.12 and 1.61	
Birefringence	-	2.0499, 1.5886	
Melting temperature, DSC	°C	551-554	
Decomposition temperature	°C	427-482 (air)	
Thermal expansion coefficient, 23-80°C	°C^{-1}	-2-5.7E-6	Jain, A; Polym. Eng. Sci., 48, 211-15, 2008.
Thermal conductivity, melt	W m^{-1} K^{-1}	5.03 -5.07e(-0.00487) @ 7-290K	Ventura, G; Martelli, V, Cryogenics, 49, 735-37, 2009.
Glass transition temperature	°C	425	
Specific heat capacity	J K^{-1} kg^{-1}	1,420 (25°C); 2,515 (180°C)	
Maximum service temperature	°C	300-350	
Long term service temperature	°C	149-177	De Groh, K K; Banks, B A; McCarthy, C E; Rucker, R N; Roberts, L M; Berger, L A, High Performance Polym., 20, 388-409, 2008.
Contact angle of water, 20°C	degree	63.7-64.6	
Surface free energy	mJ m^{-2}	45.7	

PPTA poly(p-phenylene terephthalamide)

PARAMETER	UNIT	VALUE	REFERENCES
MECHANICAL & RHEOLOGICAL PROPERTIES			
Tensile strength	MPa	2,920-3,600 (yarns); 3,600 (strands impregnated with resin)	
Tensile modulus	MPa	70,500-138,000 (yarns); 83,000-124,000 (strands impregnated with resin); 200,000 (fully crystalline); 192,000 (fully unordered), 289,000 (fully crystalline) both calculated	Deng, L; Young, R J; van der Zwaag, S; Picken, S, Polymer, 51, 2033-39, 2010; Yilmaz, D E, van Duin, A C T, Polymer, 154, 172-81, 2018.
Elongation	%	2.4-3.6	
Compressive strength	MPa	650	Knijnenberg, A; Bos, J; Dingemans, T J, Polymer, 1887-97, 2010.
Young's modulus	MPa	40,000-78,000; 252,000 (theoretical)	Rao, Y; Waddon, A J; Farris, R J, Polymer, 5925-35, 2001; Avanzini, L, Brambilla, L, Marano, C, Milani, A, Polymer, 116, 133-42, 2017.
Tenacity (fiber) (standard atmosphere)	cN tex^{-1} (daN mm^{-2})	44-270 (60-400)	Fourne, F, Synthetic Fibers. Machines and Equipment Manufacture, Properties. Carl Hanser Verlag, 1999.
Tenacity (wet fiber, as % of dry strength)	%	75-100	Fourne, F, Synthetic Fibers. Machines and Equipment Manufacture, Properties. Carl Hanser Verlag, 1999.
Fineness of fiber (titer)	dtex	1.1-12	Fourne, F, Synthetic Fibers. Machines and Equipment Manufacture, Properties. Carl Hanser Verlag, 1999.
Length (elemental fiber)	mm	38-120	Fourne, F, Synthetic Fibers. Machines and Equipment Manufacture, Properties. Carl Hanser Verlag, 1999.
Poisson's ratio	-	0.63	Nakamae, K; Nishino, T; Airu, X, Polymer, 33, 23, 4898-4900, 1992.
Shrinkage	%	<0.1	
Intrinsic viscosity, 25°C	dl g^{-1}	3-7	Allen, S R; Roche, E J; Bennett, B; Molaison, R, Polymer, 33, 9, 1849-54, 1992.
Water absorption, equilibrium in water at 23°C	%	3.5-7.0 (as shipped); 3.5-4.5 (regained from dried)	
Moisture absorption, equilibrium 23°C/50% RH	%	0.96-3.9	Mooney, D A; Don McElroy, J M, Chem. Eng. Sci., 59, 2159-70, 2004.
CHEMICAL RESISTANCE			
Acid dilute/concentrated	-	good-poor	
Alcohols	-	resistant	
Alkalis	-	good-fair	
Aliphatic hydrocarbons	-	resistant	
Aromatic hydrocarbons	-	resistant	
Esters	-	resistant	
Good solvent	-	H_2SO_4, polar aprotic solvents	
Non-solvent	-	aliphatic and aromatic hydrocarbons, alcohols, ethers, esters	
FLAMMABILITY			
Ignition temperature	°C	>550	
Limiting oxygen index	% O_2	28-29	
Char at 500°C	%	36.1	Lyon, R E; Walters, R N, J. Anal. Appl. Pyrolysis, 71, 27-46, 2004.

PPTA poly(p-phenylene terephthalamide)

PARAMETER	UNIT	VALUE	REFERENCES
Heat of combustion	$J\ g^{-1}$	26,450; 35,000	Walters, R N; Hacket, S M; Lyon, R E, Fire Mater., 24, 5, 245-52, 2000.
Volatile products of combustion	-	H_2, CO, CO_2, HCN, H_2O, NH_3, benzene, benzonitrile, toluene	

WEATHER STABILITY

Spectral sensitivity	nm	300-450	
Activation wavelengths	nm	310	
Effect of exposure	Fadeo-meter	50% tensile strength lost in 900 h	

TOXICITY

NFPA: Health, Flammability, Reactivity rating	-	0/1/0	
Carcinogenic effect	-	not listed by ACGIH, NIOSH, NTP	
Mutagenic effect	-	no known effect	
Teratogenic effect	-	animal testing showed effects on embryo-faetal development at levels below those causing maternal toxicity	
Oral rat, LD_{50}	$mg\ kg^{-1}$	7,500	
Skin rabbit, LD_{50}	$mg\ kg^{-1}$	no skin irritation	

PROCESSING

Additives used in final products	-	UV stabilizer: nano-titanate	Xing, Y; Ding, X, J. Appl. Polym. Sci., 103, 3113-19, 2007.
Applications	-	fibers, papers, yarns; some examples: body armor, protection vests, military helmets, heat resistant gloves, vehicle armor, automotive components, fiber optics, adhesives, sealants, composites, oil, gas, lithium ion batteries, and many more	
Outstanding properties	-	structural rigidity, flame resistance, chemical resistance	

BLENDS

Suitable polymers	-	PPy, PVDF	Wang, Q, Chen, Z, Deng, S, Dong, L, Chem. Eng. J., 328, 343-52, 2017.

ANALYSIS

x-ray diffraction peaks	degree	20.5, 23.4	Nakamae, K; Nishino, T; Airu, X, Polymer, 33, 23, 4898-4900, 1992.

PPTI poly(m-phenylene isophthalamide)

PARAMETER	UNIT	VALUE	REFERENCES
GENERAL			
Common name	-	poly(m-phenylene isophthalamide)	
Acronym	-	PPTI, PMIA	
CAS number	-	25765-47-3	
Linear formula			
HISTORY			
Person to discover	-	Wilfred Sweeney	
Date	-	1953; 1963, 1967	
Details	-	invention in DuPont Labs; name coined; commercialized	
SYNTHESIS			
Monomer(s) structure	-		
Monomer(s) CAS number(s)	-	108-45-2; 99-63-8	
Monomer(s) molecular weight(s)	dalton, g/mol, amu	108.1; 203.022	
Monomer ratio	-	1:1.88	
Method of synthesis	-	it is prepared from m-phenylenediamine and isophthaloyl chloride in an amide solvent	
Number average molecular weight, M_n	dalton, g/mol, amu	36,800	
STRUCTURE			
Cell type (lattice)	-	triclinic	
Cell dimensions	nm	a:b:c=0.527:0.525:1.13	
Unit cell angles	degree	α:β:γ=111.5:111.4:88	
Number of chains per unit cell	-	1	
COMMERCIAL POLYMERS			
Some manufacturers	-	DuPont	
Trade names	-	Nomex	
PHYSICAL PROPERTIES			
Density at 20°C	g cm^{-3}	1.38 (yarn); 1.47 (crystalline); 1.32 (amorphous)	
Bulk density at 20°C	g cm^{-3}	0.31-1.12 (paper)	
Color	-	creamy white to gold	
Melting temperature, DSC	°C	435-473	
Decomposition temperature	°C	300	
Thermal expansion coefficient, 23-80°C	°C^{-1}	6.2-18E-6	
Thermal conductivity, melt	W m^{-1} K^{-1}	0.25 (yarn); 0.13-0.14 (paper)	

PPTI poly(m-phenylene isophthalamide)

PARAMETER	UNIT	VALUE	REFERENCES
Glass transition temperature	°C	230-280	
Specific heat capacity	J K^{-1} kg^{-1}	300	
Maximum service temperature	°C	260-300 (yarn); 220 (paper)	
Dielectric constant at 60 Hz/10,000 Hz	-	1.2-2.7/2.6 (paper)	
Dielectric loss factor at 1 kHz	-	0.006	
Dissipation factor at 60 Hz	E-4	30-60 (paper)	
Dissipation factor at 10,000 Hz	E-4	140 (paper)	
Volume resistivity	ohm-m	1E13 to 2E14 (paper)	
Surface resistivity	ohm	2E16 (paper)	
Electric strength K20/P50, d=0.60.8 mm	kV mm^{-1}	9-32 (paper)	
Contact angle of water, 20°C	degree	42, 80	Ren, X; Zhao, C; Du, S; Wang, T; Luan, Z; Wang, J; Hou, D, J. Environ. Sci., 22, 9, 1335-41, 2010; Pramila, J, Melbiah, J S B, Rana, D, Mohan, D, Polym. Testing, 67, 218-27, 2018.

MECHANICAL & RHEOLOGICAL PROPERTIES

Tensile strength	MPa	595-610 (yarn); 41-110 (paper, MD); 29-59 ((paper, TD)	
Elongation	%	19-31 (yarn); 3.4-19 (paper)	
Elmendorf tear strength	g μm^{-1}	0.75-4.25	
Tenacity (fiber)	N tex^{-1}	0.43-0.44 (yarn)	
Shrinkage	%	1.1-4 (yarn); 0.1-0.9	
Intrinsic viscosity, 30°C	dl g^{-1}	1.86-2.11	
Water absorption, equilibrium in water at 23°C	%	4-8.3 (yarn, as shipped); 4.5 (yarn, equilibrium)	

CHEMICAL RESISTANCE

Acid dilute/concentrated	-	poor	
Alcohols	-	very good	
Alkalis	-	poor	
Aliphatic hydrocarbons	-	very good	
Aromatic hydrocarbons	-	very good	
Esters	-	very good	
Greases & oils	-	very good	
Halogenated hydrocarbons	-	good to poor	
Ketones	-	very good	
Good solvent	-	dimethylacetamide, dimethylsulfoxide, DMF, methylpyrrolidone, sulfuric acid	
Non-solvent	-	m-cresol, formic acid, hexamethylphosphoramide	

FLAMMABILITY

Ignition temperature	°C	>500	
Limiting oxygen index	% O$_2$	27-32	
Char at 500°C	%	48.4	Lyon, R E; Walters, R N, J. Anal. Appl. Pyrolysis, 71, 27-46, 2004.

PPTI poly(m-phenylene isophthalamide)

PARAMETER	UNIT	VALUE	REFERENCES
Heat of combustion	J g^{-1}	28,100	Walters, R N; Hacket, S M; Lyon, R E, Fire Mater., 24, 5, 245-52, 2000.
Volatile products of combustion	-	CO, CO_2, HCN	

TOXICITY

NFPA: Health, Flammability, Reactivity rating	-	0/1/0	
Carcinogenic effect	-	not listed by ACGIH, NIOSH, NTP	
Mutagenic effect	-	not known	
Reproductive toxicity	-	not known	
Oral rat, LD$_{50}$	mg kg^{-1}	7,500 (para); 11,000 (meta)	
Skin rabbit, LD$_{50}$	mg kg^{-1}	no irritation	

ENVIRONMENTAL IMPACT

Aquatic toxicity, *Daphnia magna*, LC$_{50}$, 48 h	mg l^{-1}	>500	
Aquatic toxicity, *Fathead minnow*, LC$_{50}$, 48 h	mg l^{-1}	1,500	

PROCESSING

Typical processing methods	-	spinning	
Preprocess drying: temperature/time/residual moisture	°C/h/%	100/5	
Applications	-	aircraft parts, automotive (heat shields, spark plug leads), boats, electric heaters, electrical applications, firefighter clothing, hot gas filtration, insulation paper, lithium-ion battery separators, safety clothing, skin bioconstruct, ultrafiltration membranes	
Outstanding properties	-	flame resistance	

ANALYSIS

FTIR (wavenumber-assignment)	cm^{-1}/-	3309-3306 – unsubstituted amides, 2940 – C–H stretching vibration, 1666 – amide C=O stretching, 1537 – N–H in plane bending and C–N stretching coupled modes, 784 & 710 – disubstituted benzene in a meta substitution	Pramila, J, Melbiah, J S B, Rana, D, Mohan, D, Polym. Testing, 67, 218-27, 2018.

PPV poly(1,4-phenylene vinylene)

PARAMETER	UNIT	VALUE	REFERENCES
GENERAL			
Common name	-	poly(1,4-phenylene vinylene)	
IUPAC name	-	poly(1,4-phenyleneethene-1,2-diyl)	
CAS name	-	poly(1,4-phenylene-1,2-ethenediyl)	
Acronym	-	PPV	
CAS number	-	26009-24-5	
HISTORY			
Person to discover	-	Burroughes, J H; Bradley, D D C; Brown, A R; Marks, R N; Mackay, K; Friend, R H; Burns, P L; Holmes, A B	
Date	-	1990; 1991	
Details	-	demonstration, commercialization	
SYNTHESIS			
Monomer(s) structure	-	BrC_6H_4Br; $H_2C=CH_2$ (for each technique different set of monomers is used)	
Monomer(s) CAS number(s)	-	106-37-6; 74-85-1	
Monomer(s) molecular weight(s)	dalton, g/mol, amu	235,92; 28.05	
Monomer ratio	-	8.4:1	
Method of synthesis	-	polymer in final form cannot be processed, therefore precursor polymer is synthesized first and then converted into film or the final forms. Precursor polymer can be obtained by one of the following methods: Wessling route, ring opening polymerization, chemical vapor deposition, electropolymerization, condensation, phase transfer catalysis, or anionic polymerization	Fink, J K, High Performance Polymers, William Andrew, 2008.
Temperature of polymerization	°C	0-5	
Mass average molecular weight, M_w	dalton, g/mol, amu	100,000	
Polydispersity, M_w/M_n	-	2.0	
STRUCTURE			
Crystallinity	%	45-80	Saxena, V; Malhotra, B D, Handbook of polymers in Elecronics, Ed. Malhotra, B D, Rapra, 2002.
Cell type (lattice)	-	monoclinic	Chen, D; Winokur, M J; Masse, M A; Karasz, F E, Polymer 33, 3116, 1992.
Cell dimensions	nm	a:b:c=0.80:0.60:0.66	
Unit cell angles	degree	$\alpha=123$	
Crystallite size	nm	7.5	Okuzaki, H; Takahashi, T; Miyajima, N; Suzuki, Y; Kuwabara, T, Macromolecules, 39, 4276-78, 2006.
COMMERCIAL POLYMERS			
Some manufacturers	-	DuPont	
Trade names	-	Luxprint (polymer compositions)	
PHYSICAL PROPERTIES			
Refractive index, 20°C	-	2.085, 1.610	
Decomposition temperature	°C	>500	

PPV poly(1,4-phenylene vinylene)

PARAMETER	UNIT	VALUE	REFERENCES
Glass transition temperature	°C	500	
Dielectric constant at 100 Hz/1 MHz	-	3.2	
Volume resistivity	ohm-m	1E-4	Saxena, V; Malhotra, B D, Handbook of polymers in Elecronics, Ed. Malhotra, B D, Rapra, 2002.
Contact angle of water, 20°C	degree	97.8 (advancing); 89.9 (receding)	Ma, K-X; Ho, C H; Chung, T-S, Antec, 2212-15, 1999.
Optical absorption edge	eV	2.4	Saxena, V; Malhotra, B D, Handbook of polymers in Elecronics, Ed. Malhotra, B D, Rapra, 2002.
CHEMICAL RESISTANCE			
Acid dilute/concentrated	-	very good	
Alcohols	-	very good	
Alkalis	-	very good	
Aliphatic hydrocarbons	-	very good	
Aromatic hydrocarbons	-	very good	
Esters	-	very good	
Greases & oils	-	very good	
Halogenated hydrocarbons	-	very good	
Ketones	-	very good	
Good solvent	-	insoluble	
WEATHER STABILITY			
Spectral sensitivity	nm	290-390	Sun, Z; Yang, X; Huang, Y; Ding, L; Qin, L; Wang, Z, Optics Commun., 160, 289-91, 1999.
Absorption maximum	nm	428	Sun, Z; Yang, X; Huang, Y; Ding, L; Qin, L; Wang, Z, Optics Commun., 160, 289-91, 1999.
Excitation wavelengths	nm	326, 402, 457, 473	Holzer, W; Penzkofer, A; Schrader, S; Grim, B, Optical Quantum Electronics, 37, 475-94, 2005.
PROCESSING			
Typical processing methods	-	chemical vapor deposition polymerization; electrospinning; spin coating	
Additives used in final products	-	dopants (e.g., arsenic pentafluoride, iodine, ferric chloride, and others)	Fink, J K, High Performance Polymers, William Andrew, 2008.
Applications	-	electroluminescent devices, photovoltaic devices, polymer light-emitting diodes, nanofibers and tubes, and sensors	
Outstanding properties	-	electroluminescent conjugation; emits light when electric current is passed through them	Fink, J K, High Performance Polymers, William Andrew, 2008.
ANALYSIS			
x-ray diffraction peaks	degree	20.5, 22.0, 28.2	Okuzaki, H; Takahashi, T; Miyajima, N; Suzuki, Y; Kuwabara, T, Macromolecules, 39, 4276-78, 2006.

582   **HANDBOOK OF POLYMERS** 3rd Edition, Copyrights 2022; ChemTec Publishing

PPX poly(p-xylylene)

PARAMETER	UNIT	VALUE	REFERENCES
GENERAL			
Common name	-	poly(p-xylylene), parylene	
IUPAC name	-	poly(1,4-phenyleneethylene)	
ACS name	-	poly(1,4-phenylene-1,2-ethanediyl)	
Acronym	-	PPX	
CAS number	-	25722-33-2	
Linear formula			

PARAMETER	UNIT	VALUE	REFERENCES
HISTORY			
Person to discover	-	Michael Szwarc, William Gorham	
Date	-	1947, 1965	
Details	-	Michael Szwarc was able to identify PPX in products of decomposition of p-xylene. William Gorham developed its synthesis from di-p-xylene, and Union Carbide commercialized it in 1965.	
SYNTHESIS			
Monomer(s) structure	-		

PARAMETER	UNIT	VALUE	REFERENCES
Monomer(s) CAS number(s)	-	1633-22-3	
Monomer(s) molecular weight(s)	dalton, g/mol, amu	208.3	
Method of synthesis	-	chemical vapor polymerization: paracyclophane is evaporated at 150-180ºC in vacuo. Pyrolysis at 680-700ºC is the next stage in which diradicals are formed. The reactive vapor polymerizes on a cold surface kept at ambient temperature. A similar method is used for production of copolymers. It is generally referred to as chemical vapor deposition	Fink, J K, High Performance Polymers, William Andrew, 2008; Smalara, K; Gieldon, A; Bobrowski, M; Rybiscki, J; Czaplewski, C, J. Phys. Chem., 114, 4296-4303, 2010; Pu, H; Jiang, F; Wang, Y; Yan, B, Colloids SurfacesA361, 62-65, 2010.
Temperature of polymerization	ºC	680-700	
Pressure of polymerization	Pa	13.3	
Yield	%	24-26	
Mass average molecular weight, M_w	dalton, g/mol, amu	190,000-500,000; 500,000 (Parylene N)	
Polymerization degree (number of monomer units)	-	2,000-4,000	
Molar volume at 298K	cm^3 mol^{-1}	87.5 (crystalline)	
Van der Waals volume	cm^3 mol^{-1}	63.8 (crystalline)	
STRUCTURE			
Crystallinity	%	35-66	
Cell type (lattice)	-	monoclinic (α form); trigonal (β form)	
Cell dimensions	nm	a:b:c=0.592:1.064:0.655; a:b:c=2.052:2.052:0.655	

HANDBOOK OF POLYMERS 3rd Edition, Copyrights 2022; ChemTec Publishing

PPX poly(p-xylylene)

PARAMETER	UNIT	VALUE	REFERENCES
Unit cell angles	degree	β=134.7; γ=120	
Number of chains per unit cell	-	2; 16	
Lamellae thickness	nm	10-25	

COMMERCIAL POLYMERS			
Some manufacturers	-	Specialty Coating Systems	
Trade names	-	Parylene N	

PHYSICAL PROPERTIES			
Density at 20°C	g cm^{-3}	1.11	
Color	-	transparent	
Refractive index, 20°C	-	1.59-1.6690	
Birefringence	-	0.000069-0.000235	Senkevich, J J; Desu, S B; Simkovic, V, Polymer, 41, 2379-90, 2000.
Melting temperature, DSC	°C	400-427	
Decomposition temperature	°C	<425	
Thermal conductivity, melt	W m^{-1} K^{-1}	0.13	
Glass transition temperature	°C	230-240; 13 (amorphous)	
Specific heat capacity	J K^{-1} kg^{-1}	837	
Long term service temperature	°C	expected to survive exposure to 100°C for 10 years	
Dielectric constant at 100 Hz/1 MHz	-	2.6-2.8/2.8	
Dielectric loss factor at 1 kHz	-	0.002	
Dissipation factor at 100 Hz	E-4	2	
Dissipation factor at 1 MHz	E-4	6	
Volume resistivity	ohm-m	1E13 to 1.4E15	
Surface resistivity	ohm	1E13	
Electric strength K20/P50, d=0.60.8 mm	kV mm^{-1}	276	
Coefficient of friction	-	0.25 (static and dynamic)	
Surface free energy	mJ m^{-2}	46.3	
Contact angle of water, 20°C	degree	73.3-81.6	Chang, C-W, Guan, Z-Y, Kan, M-Y, Lee, L-W, Chen, H-Y, Kang, D-Y, J. Membrane Sci., 539, 101-7, 2017.

MECHANICAL & RHEOLOGICAL PROPERTIES			
Tensile strength	MPa	45-62; 3,000 (high strength fiber); 19,000-23,000 (theoretically calculated values)	
Tensile stress at yield	MPa	42.1	
Tensile creep modulus, 1000 h, elongation 0.5 max	MPa	43	
Elongation	%	40*-140	
Young's modulus	MPa	2,100-14,000; 102,000 (high strength fibers); 280,000 (theoretically calculated value)	Lee, C, Solid State Technol., 28-33, Nov. 2008.
Rockwell hardness	R	85	

PPX poly(p-xylylene)

PARAMETER	UNIT	VALUE	REFERENCES
Water absorption, equilibrium in water at 23°C	%	0.1, 0.01 (24 h)	

CHEMICAL RESISTANCE

PARAMETER	UNIT	VALUE	REFERENCES
Acid dilute/concentrated	-	good	
Alcohols	-	very good	
Alkalis	-	good	
Aliphatic hydrocarbons	-	very good	
Aromatic hydrocarbons	-	very good	
Esters	-	very good	
Greases & oils	-	very good	
Halogenated hydrocarbons	-	good	
Ketones	-	very good	
Good solvent	-	chlorinated biphenyl, methylene chloride, chloroform, toluene	

WEATHER STABILITY

PARAMETER	UNIT	VALUE	REFERENCES
Spectral sensitivity	nm	266 (laser ablation); <340	Bera, M; Rivaton, A; Gandon, C; Gardette, Eur. Polym. J., 36, 1765-77, 2000; Jeong, Y S; Ratier, B; Moliton, A; Guyard, L, Synthetic Metals, 127, 1-3, 189-93, 2002.
Products of degradation	-	methylene group oxidation, chain scission	Bera, M; Rivaton, A; Gandon, C; Gardette, Eur. Polym. J., 36, 1753-64, 2000.

PROCESSING

PARAMETER	UNIT	VALUE	REFERENCES
Typical processing methods	-	coating, vapor deposition	
Applications	-	bobbins, electronics (capacitors, circuit boards, cores, fiber optic components, magnets, power supplies, relays, semiconductors), heat exchangers, medical (implants, membranes for gas separation, needles, pacemakers, stents, surgical instruments), metal primer (derivative), probes	
Outstanding properties	-	barrier properties, easy processing, insulation properties	

ANALYSIS

PARAMETER	UNIT	VALUE	REFERENCES
FTIR (wavenumber-assignment)	cm^{-1}/-	water – 1633; =C–O–C – 1017; CH – 960	Wu, X; Shi, G; Qu, L; Zhang, J; Chen, F, J. Polym. Sci. A, 41, 449-55, 2003.

PPy polypyrrole

PARAMETER	UNIT	VALUE	REFERENCES
GENERAL			
Common name	-	polypyrrole	
IUPAC name	-	poly(pyrrole-2,5-diyl)	
CAS name	-	1H-oyrrole, homopolymer	
Acronym	-	PPy	
CAS number	-	30604-81-0	
Formula			
HISTORY			
Person to discover	-	Bolto, B A; Weiss, D E	
Date	-	1963	
Details	-	published paper on high conductivity of polypyrrole	
SYNTHESIS			
Monomer(s) structure	-		
Monomer(s) CAS number(s)	-	109-97-7	
Monomer(s) molecular weight(s)	dalton, g/mol, amu	67.09	
Method of synthesis	-	dry pyrrole was polymerized in the presence of oxidant and dopant in water medium	
Temperature of polymerization	ºC	50 to -50	Minisy, I M, Bober, P, Šeděnková, I, Stejskal, J, Polymer, 207, 122854, 2020.
Time of polymerization	h	40	
Catalyst	-	ammonium persulfate (oxidant)	
Yield	%	42	
Heat of polymerization	$J\ g^{-1}$	894-2161	
Number average molecular weight, M_n	dalton, g/mol, amu	240,000	
Mass average molecular weight, M_w	dalton, g/mol, amu	22,000-356,000	Bose, S; Kuila, T; Uddin, M E; Kim, N H; Lau, A K T; Lee, J H, Polymer, 51, 5921-28, 2010.
Polydispersity, M_w/M_n	-	1.48	
Polymerization degree (number of monomer units)	-	303	
Radius of gyration	nm	42.4	Huang, K; Wan, M; Long, Y; Chen, Z; Wei, Y, Synthetic Metals, 155, 495-500, 2005.
STRUCTURE			
Crystallinity	%	50	Saxena, V; Malhotra, B D, Handbook of polymers in Elecronics, Ed. Malhotra, B D, Rapra, 2002.
Cell type (lattice)	-	monoclinic	

PPy polypyrrole

PARAMETER	UNIT	VALUE	REFERENCES
Cell dimensions	nm	a:b:c=0.82:0.735:0.682	
Unit cell angles	degree	$\alpha:\beta:\gamma$=90:90:117	
Chain conformation	-	anti-gauche	Fonner, J M; Schmidt, C E; Ren, P, Polymer, 51, 4985-93, 2010.

PHYSICAL PROPERTIES

Density at 20°C	g cm^{-3}	1.3-1.48	
Melting temperature, DSC	°C	>300	
Decomposition temperature	°C	290	
Storage temperature	°C	25	
Glass transition temperature	°C	80	
Volume resistivity	ohm-m	50.7	Bose, S; Kuila, T; Uddin, M E; Kim, N H; Lau, A K T; Lee, J H, Polymer, 51, 5921-28, 2010.
Surface resistivity	ohm	4E2 to 2.8E4	Lee, J Y; Kim, K T; Kim, S Y; Kim, C Y, Antec, 1422-26, 1996.
Conductivity	S cm^{-1}	1.5-4.4	Minisy, I M, Bober, P, Šeděnková, I, Stejskal, J, Polymer, 207, 122854, 2020.
Optical absorption edge	eV	2.5	Saxena, V; Malhotra, B D, Handbook of polymers in Elecronics, Ed. Malhotra, B D, Rapra, 2002.

MECHANICAL & RHEOLOGICAL PROPERTIES

Tensile strength	MPa	30-68; 25 (fibers); 127 (electrochemical film)	
Elongation	%	2 (fibers); 26 (electrochemical film)	
Elastic modulus	MPa	1,500 (fibers); 1,950 (film)	

CHEMICAL RESISTANCE

Alcohols	-	very good	
Aliphatic hydrocarbons	-	very good	
Aromatic hydrocarbons	-	very good	
Esters	-	very good	
Greases & oils	-	very good	
Halogenated hydrocarbons	-	very good	
Ketones	-	very good	

FLAMMABILITY

Volatile products of combustion	-	H_2O, CO, CO_2, HCN, SO_2	

WEATHER STABILITY

Spectral sensitivity	nm	230, 338, 410, 430, 435, 500, 900	Li, X-G; Li, A; Huang, M-R; Liao, Y; Lu, G-Y, J. Phys. Chem C, 144, 19244-55, 2010.

BIODEGRADATION

Properties	-	antibacterial	Gniadek, M, Wichowska, A, Antos-Bielska, M, Donten, M, Synthetic Metals, 266, 116430, 2020.

PPy polypyrrole

PARAMETER	UNIT	VALUE	REFERENCES
TOXICITY			
NFPA: Health, Flammability, Reactivity rating	-	0/1/0; 2/0/0 (HMIS)	
Carcinogenic effect	-	not listed by ACGIH, NIOSH, NTP	
PROCESSING			
Typical processing methods	-	chemical oxidation of pyrrole in carbon black suspension, electrochemical anodic polymerization, Langmuir-Blodgen technique of monolayer production, solution polymerization over the substrate	
Additives used in final products	-	Fillers: carbon black, silica, tin oxide	
Applications	-	antiseptic wound dressing, battery electrodes, capacitors, controlled release agents for other components, electronic displays, EMI shielding, food packaging, nonmetallic conductors, optoelectronic systems, sensors	
BLENDS			
Suitable polymers	-	chitosan, PA6, PAN, PANI, PCL, PLA, PP, PU, PVAc,PVC, PVDF, PVP, SI	
ANALYSIS			
FTIR (wavenumber-assignment)	cm^{-1}/-	H–O – 3558; N–H – 3377, 3388, 1035, 759, 558; CN – 2219, 2871; C=O – 1698; C–O – 1115 and more in ref.	Zhang, N; van Ooij, W J; Luo, S, Antec, 2495-2501, 1999; Bose, S; Kuila, T; Uddin, M E; Kim, N H; Lau, A K T; Lee, J H, Polymer, 51, 5921-28, 2010.
Raman (wavenumber-assignment)	cm^{-1}/-	polaronic form – 962, 1053; bipolaronic form – 930, 1086; C=C – 1589; skeletal band – 1475	Foroughi, J; Spinks, G M; Wallace, G G, Sensors Actuators, B155, 278-84, 2011.
NMR (chemical shifts)	ppm	C NMR: β carbons – 113.6; α, β carbons – 129.3; α carbons – 123.7, 142.9; OH substituted α carbons – 149.2	Rizzi, M; Trueba, M; Trasatti, S P, Synthetic Metals, 161, 23-31, 2011.
x-ray diffraction peaks	degree	23.5-25.1 (amorphous PPy)	Li, X-G; Li, A; Huang, M-R; Liao, Y; Lu, G-Y, J. Phys. Chem C, 144, 19244-55, 2010; Bose, S; Kuila, T; Uddin, M E; Kim, N H; Lau, A K T; Lee, J H, Polymer, 51, 5921-28, 2010.

PR proteins

PARAMETER	UNIT	VALUE	REFERENCES
GENERAL			
Common name	-	proteins	
Acronym	-	PR	
CAS number	-	9010-10-0, 70084-87-6	
EC number	-	232-720-8	
HISTORY			
Date	-	1923	
Details	-	soy-based adhesives developed	
SYNTHESIS			
Method of production	-	soybean processing: first the oil and husk are removed, the remaining flakes are subjected to protein extraction; pH control permits isolation of required range of protein molecules; next step involves chemical modification, which imparts required properties; grades obtained in this technology include: unhydrolyzed grades, hydrolyzed grades, carboxylated soy protein, and proteinates	
Mass average molecular weight, M_w	dalton, g/mol, amu	50,000-300,000; 30,000-1,000,000 (molecular weight of raw soy protein); 19,000-25,200 (casein); 10,000-15,000 (albumins); 150,000-450,000 (globulins)	
Hydrodynamic radius	nm	3.12-3.26 (ovalbumin, main protein in egg white)	Hulse, W L; Forbes, R T, Int. J. Pharmaceutics, 411, 64-68, 2011.
Radius of gyration	nm	3.7-12.8	Vorup-Jensen, T; Boesen, T, adv. Drug Delivery Rev., in press, 2011.
STRUCTURE			
Crystalline structure	-	Proteins exhibit a variety of dense phases ranging from gels, aggregates, and precipitates to crystalline phases and dense liquids.	Greene, D G, Modla,S, Wagner, N J, Lenhoff, A M, Biophys. J., 109, 8, 1716-23, 2015
Cell dimensions	nm	2.2-4.5x3.0-4.4x3.0-5.7	Schwenke, K D, Studies in Interface Science, Vol. 7, pp 1-50, Elsevier, 1998.
Chain conformation	-	α-helix (most), β-sheet, unordered	Sinha, S; Li, Y; Williams, T D; Topp, E M, Biophys. J., 95, 12, 5951-61, 2008.
COMMERCIAL POLYMERS			
Some manufacturers	-	DuPont	
Trade names	-	Pro-Cote, SoBind	
PHYSICAL PROPERTIES			
Density at 20°C	g cm^{-3}	1.36	
Bulk density at 20°C	g cm^{-3}	0.28-0.6	
Color	-	off-white to light brown	
Odor		odorless	
Denaturation temperature	°C	118-124 (lentil protein)	Joshi, M; Adhikari, B; Aldred, P; Panozzo, J F; Kasapis, S, Food Chem., in press, 2011.
Glass transition temperature	°C	181 (wheat glutenin); 192 (collagen); 217 (gelatin); 252 (elastin)	Matveev, Y I; Grinberg, V Y; Sochava, I V; Tolstoguzov, V B, Food Hydrocolloids, 11, 2, 125-33, 1997.

PR proteins

PARAMETER	UNIT	VALUE	REFERENCES
MECHANICAL & RHEOLOGICAL PROPERTIES			
Tensile strength	MPa	40-50 (soy protein); 99 (soy protein drawn, 2.5 draw ratio); 12.7 (zein); 5-6 (zein processed by casting)	Kurose, T; Urman, K; Otaigbe, J U; Lochhead, R Y; Thames, S F, Antec, 1489-93, 2006.
Elongation	%	4.6 (soy protein); 61-122 (zein)	
Young's modulus	MPa	104-1,200	
CHEMICAL RESISTANCE			
Acid dilute/concentrated	-	poor	
Alcohols	-	good/poor	
Alkalis	-	poor	
Aliphatic hydrocarbons	-	good	
Aromatic hydrocarbons	-	good	
Esters	-	good	
WEATHER STABILITY			
Absorption	nm	280 – tyrosine and tryptophan, 260 – phenylalanine	Davies, M J; Truscott, R J W, Comprehensive Series in Photosciences, Vol. 3, pp 251-275, Elsevier, 2001.
Spectral sensitivity	nm	250-300 (disulfide bond)	Davies, M J; Truscott, R J W, Comprehensive Series in Photosciences, Vol. 3, pp 251-275, Elsevier, 2001.
Emission wavelengths	nm	280 – phenylalanine, 300 – tyrosine, 350 – tryptophan	Davies, M J; Truscott, R J W, Comprehensive Series in Photosciences, Vol. 3, pp 251-275, Elsevier, 2001.
BIODEGRADATION			
Typical biodegradants	-	composting according to ASTM D5338 (fast biodegradation of PR/PVOH film)	Su, J-F; Yuan, X-Y; Hung, Z; Xia, W-L, Polym. Deg. Stab., 95, 1226-37, 2010.
TOXICITY			
Carcinogenic effect	-	not listed by ACGIH, NIOSH, NTP	
Mutagenic effect	-	none	
Teratogenic effect	-	none	
Reproductive toxicity	-	none	
ENVIRONMENTAL IMPACT			
Aquatic toxicity, *Daphnia magna*, LC_{50}, 48 h	mg l^{-1}	>1,000	
Aquatic toxicity, *Rainbow trout*, LC_{50}, 48 h	mg l^{-1}	>1,000	
PROCESSING			
Typical processing methods	-	casting, compounding, compression molding, extrusion, injection molding, mixing, solution processing	
Processing temperature	°C	70-105 (extrusion); 130 (molding); 135-165 (compression)	

PR proteins

PARAMETER	UNIT	VALUE	REFERENCES
Additives used in final products	-	Plasticizers: ethylene glycol, glycerin, propylene glycol, sorbitol, triacetin, triethylene glycol; Antistatics: cationic polysoap, N-acyl derivative of a protein hydrolysate; Release: calcium salt, magnesium stearate, stearic acid	
Applications	-	adhesives, animal pharmaceuticals, ceiling tiles, fibers, horticultural pots, leather finishing, mushroom fertilizer, paper and paperboard coatings, rheology modifiers	

BLENDS

Suitable polymers	-	poly(hydroxy ester ether), PLA, PVOH	

ANALYSIS

FTIR (wavenumber-assignment)	cm^{-1}/-	hydrogen bonding – 2900-3100; NH – 1636-1680, 1533-1559; C-C – 750, 900, 920; C-O – 1110, more in ref.	Su, J-F; Yuan, X-Y; Hung, Z; Xia, W-L, Polym. Deg. Stab., 95, 1226-37, 2010.
Raman (wavenumber-assignment)	cm^{-1}/-	C=C – 1634, 1546; C-H – 1168	Chen, L; Han, X; Yang, J; Zhou, J; Song, W; Zhao, B; Xu, W; Ozaki, Y, J. Colloid Interface Sci., 360, 482-87, 2011.
x-ray diffraction peaks	degree	10, 24 (lentil protein); 22 (soy protein)	Joshi, M; Adhikari, B; Aldred, P; Panozzo, J F; Kasapis, S, Food Chem., in press, 2011; Su, J-F; Yuan, X-Y; Hung, Z; Xia, W-L, Polym. Deg. Stab., 95, 1226-37, 2010.

PS polystyrene

PARAMETER	UNIT	VALUE	REFERENCES
GENERAL			
Common name	-	polystyrene	
IUPAC name	-	poly(1-phenylethane-1,2-diyl)	
CAS name	-	benzene, ethenyl-, homopolymer	
Acronym	-	PS	
CAS number	-	9003-53-6	
EC number	-	203-066-0	
RTECS number	-	DA0520000 WL6475000	
Formula			
HISTORY			
Person to discover	-	Eduard Simon; Herman F. Mark; Munters, C G and Tandberg J G	Mark, H; Wulff, C, German Patent, 550 055, I G Farben, 1929; Munters, C G and Tandberg J G, US Patent 2,023,204, 1935.
Date	-	1839; 1929; 1935	
Details	-	Eduard Simon distilled the resin of the Turkish sweetgum tree (*Liquidambar orientalis*) and obtained an oily substance, a monomer, which he named styrol. Several days later styrol had thickened, presumably from oxidation, into a jelly; The first commercial production of polystyrene was by BASF in 1931; patent for manufacture was obtained by Mark and Wulff in 1929; Munters and Tandberg obtained patent for food polystyrene	
SYNTHESIS			
Monomer(s) structure	-		
Monomer(s) CAS number(s)	-	100-42-5	
Monomer(s) molecular weight(s)	dalton, g/mol, amu	104.2	
Monomer ratio	-	100%	
Method of synthesis	-	free radical polymerization in the presence of initiator (peroxide) and solvent (e.g., ethyl benzene), and, frequently, chain transfer agents (mercaptans)	
Temperature of polymerization	^{o}C	100-180	
Heat of polymerization	$J\ g^{-1}$	657-700	
Number average molecular weight, M_n	dalton, g/mol, amu	103,000-1,998,000	
Mass average molecular weight, M_w	dalton, g/mol, amu	258,000-2,038,000, 3,700-979,200	-; Lee, B W, Jeong, M-S, Choi, J S, Park, J, Ko, J-H, Current Appl. Phys., 17, 11, 1396-1400, 2017.
Polydispersity, M_w/M_n	-	1.02-3.5; 1.78-6.0 (syndiotactic)	-; Chen, C-M, Hsieh, T-E, Ju, M-Y, J. Alloys Compounds, 480, 2, 658-61, 2009.
Molar volume at 298K	$cm^3\ mol^{-1}$	calc.=97.0; 92.0 (crystalline); 99.0 (amorphous); exp.=99.1	
Van der Waals volume	$cm^3\ mol^{-1}$	64.5; 64.3 (crystalline); 62.9 (amorphous)	

PS polystyrene

PARAMETER	UNIT	VALUE	REFERENCES
Molecular cross-sectional area, calculated	cm^2 x 10^{-16}	41.1	
Radius of gyration	nm	5.3-8.6	Lee, H; Ahn, H; Naidu, S; Seong, B S; Ryu, D Y; Trombly, D M; Ganesan, V, Macromolecules, 43, 23, 9892-8, 2010.
End-to-end distance of unperturbed polymer chain	nm	10.0-14.2 (good solvent); 8-10.9 (Θ solvent)	Davankov, V; Tsyurupa, M, Comprehensive Analytical Chemistry, 56, 2-62, Elsevier, 2011.

STRUCTURE

PARAMETER	UNIT	VALUE	REFERENCES
Crystallinity	%	atactic: non-crystalline	
Cell type (lattice)	-	orthorhombic (β), hexagonal and trigonal (α)	Woo, E M, Sun, Y S, Yang, C-P, Prog. Polym. Sci., 26, 6, 945-83, 2001.
Polymorphs	-	four crystal forms (α, β, γ, and δ) exist in thermal- or solvent-treated sPS; β form ~50% in isotactic PS containing β-nucleating agent	Woo, E M, Sun, Y S, Yang, C-P, Prog. Polym. Sci., 26, 6, 945-83, 2001; Wang, C, Lee, J L, Landingin, J, Ko, C-H, polymer, 176, 236-43, 2019.
Cell dimensions	nm	a:b:c=0.881:2.882:0.551 (orthorhombic)	Woo, E M, Sun, Y S, Yang, C-P, Prog. Polym. Sci., 26, 6, 945-83, 2001.
Unit cell angles	degree	α:β:γ=90:90:120	
Number of chains per unit cell	-	6	
Chain conformation	-	helix 3/1	
Tacticity	%	~100 atactic	Kong, F, Chang, C, Ma, Y, Zhang, C, Ren, C, Shao, T, Appl. Surf. Sci., 459, 300-8, 2018.
Entanglement molecular weight	dalton, g/mol, amu	calc.=10,544, 12,300 (18 entanglements per chain)	-:Wang, Y, Wang, W, Hong, K, Do, C, Chen, W-R, Polymer, 204, 122698, 2020.

COMMERCIAL POLYMERS

PARAMETER	UNIT	VALUE	REFERENCES
Some manufacturers	-	BASF; Sabic	
Trade names	-	Polystyrol; EPS and PS	

PHYSICAL PROPERTIES

PARAMETER	UNIT	VALUE	REFERENCES
Density at 20°C	g cm^{-3}	1.04-1.06; 1.02-1.03 (impact modified); 0.936 (melt)	
Bulk density at 20°C	g cm^{-3}	0.6	
Color	-	white	
Refractive index, 20°C	-	calc.=1.603-1.6037; exp.=1.5894-1.600	
Transmittance	%	89-90	
Haze	%	1-1.2	
Gloss, 60°, Gardner (ASTM D523)	%	80-95	
Odor		odorless	
Melting temperature, DSC	$^{\circ}$C	275	
Softening point	$^{\circ}$C	74	
Decomposition onset temperature	$^{\circ}$C	285	Patel, P; Hull, T R; McCabe, R W; Flath, D; Grasmeder, J; Percy, M, Polym. Deg. Stab., 95, 709-18, 2010.

PS polystyrene

PARAMETER	UNIT	VALUE	REFERENCES
Activation energy of thermal degradation	kJ mol^{-1}	477 (aPS); 200 (inert atmosphere), 125 (presence of air), 105-124 (waste PS)	Chen, K; Harris, K; Vyazovkin, S, Macromol. Chem. Phys., 208, 2525-32, 2007; Nisar, J, Ali, G, Shah, A, Akhter, M S, Waste Mng., 88, 236-47, 2019.
Thermal expansion coefficient, 23-80°C	°C^{-1}	6-8E-5; 1E-4 (impact modified)	
Thermal conductivity, melt	W m^{-1} K^{-1}	0.105-0.128; 0.155 (melt); 0.165 (melt, impact modified)	
Glass transition temperature	°C	calc.=99-108; exp.=85-102; 61 (bimodal PS with low molecular weight component)	Lee, H; Ahn, H; Naidu, S; Seong, B S; Ryu, D Y; Trombly, D M; Ganesan, V, Macromolecules, in print, 2011; Pradipkanti, L, Satapathy, D K, Thin Solid Films, 651, 18-23, 2018.
Specific heat capacity	J K^{-1} kg^{-1}	1300 (room); 2,300 (melt); 2,290 (melt, impact modified)	Koh, Y P; McKenna, G B; Simon, S L, 1506-9, 2006.
Heat deflection temperature at 0.45 MPa	°C	82-100	
Heat deflection temperature at 1.8 MPa	°C	72-86	
Vicat temperature	°C	84-104; 82-88.5 (impact modified)	
Hansen solubility parameters, δ_D, δ_P, δ_H	MPa$^{0.5}$	22.3, 5.8, 4.3; 21.3, 5.7, 4.3; 17.6, 6.1, 4.1	-: -; Bussi, Golan, Dosoretz, G, Eisen, M S, Desalination, 431, 35-46, 2018.
Interaction radius		4.3; 12.7	
Hildebrand solubility parameter	MPa$^{0.5}$	calc.=18.66-21.1; exp.=17.45-23.9	
Surface tension	mN m^{-1}	calc.=43.3; exp.=39.3-40.7; 26-35 (200°C)	Moreira, J C; Demarquette, N, Antec, 2000.
Dielectric constant at 100 Hz/1 MHz	-	2.4-2.7	
Relative permittivity at 100 Hz	-	2.5; 2.7 (EPS)	Calles-Arriaga, C A, López-Hernández, J, Hernández-Ordoñez, M, Ovando-Medina, V M, Ingeniería, Investigación y Tecnología, 17, 1, 15-21, 2016.
Relative permittivity at 1 MHz	-	2.5	
Dissipation factor at 100 Hz	E-4	0.9; 1.5-4 (impact modified)	
Dissipation factor at 1 MHz	E-4	0.5-0.7; 4 (impact modified)	
Volume resistivity	ohm-m	1E18 to 1E20	
Comparative tracking index	-	375-475; 500 (impact modified)	
Coefficient of friction	ASTM D1894	0.26-0.28 (chrome steel); 0.50-0.56 (aluminum)	Maldonado, J E, Antec, 3431-35, 1998.
Permeability to oxygen, 25°C	cm^3 cm cm^{-2} s^{-1} Pa^{-1} x 10^{12}	0.19	
Permeability to water vapor, 25°C	cm^3 cm cm^{-2} s^{-1} Pa^{-1} x 10^{12}	135	
Contact angle of water, 20°C	degree	75	Awasthi, A K, Shivashankar, M, Chandrasekaran, N, Mater. today: Proc. in press, 2020.
Surface free energy	mJ m^{-2}	38.3	
Speed of sound	m s^{-1}	38.7-40.8	

PS polystyrene

PARAMETER	UNIT	VALUE	REFERENCES
Acoustic impedance		2.42-2.52	
Attenuation	dB cm^{-1}, 5 MHz	1.8-3.6	

MECHANICAL & RHEOLOGICAL PROPERTIES

PARAMETER	UNIT	VALUE	REFERENCES
Tensile strength	MPa	40-66; 22 (impact modified)	
Tensile modulus	MPa	2,250-3,300; 1,550-2,200 (impact modified)	
Tensile stress at yield	MPa	23-27 (impact modified)	
Tensile creep modulus, 1000 h, elongation 0.5 max	MPa	2,200-2,600	
Elongation	%	2-3; 25-50 (impact modified)	
Tensile yield strain	%	1.6-1.8 (impact modified)	
Flexural strength	MPa	66-95	
Flexural modulus	MPa	3,530-3,630	
Compressive strength	MPa	20.8-85.9; 0.095-0.4 (expanded foam)	Ali, Y A Y, Fahmy, E H A, AbouZeid, M N, Mooty, M N A, Const. Build. Mater., 242, 118109, 2020; Li, M-E, Yan, Y-W, Zhao, H-B, Jian, R-K, Wang, Y-Z, Compos. Part B: Eng., 185, 107797, 2020.
Young's modulus	MPa	3,000-3,500	
Charpy impact strength, unnotched, 23°C	kJ m^{-2}	150 (impact modified)	
Charpy impact strength, unnotched, -30°C	kJ m^{-2}	120-160 (impact modified)	
Charpy impact strength, notched, 23°C	kJ m^{-2}	3-4; 10-17 (impact modified)	
Charpy impact strength, notched, -30°C	kJ m^{-2}	7 (impact modified)	
Izod impact strength, notched, 23°C	J m^{-1}	12-20	
Poisson's ratio	-	calc.=0.380; exp.=0.33-0.354	
Rockwell hardness	-	L90-94, M58-64	
Shrinkage	%	0.50 (across the flow), 0.73 (along the flow)	Chang, T; Faison, E, Polym. Eng. Sci., 41, 5, 703-10, 2001.
Viscosity number (ISO 307)	ml g^{-1}	96-119	
Melt viscosity, shear rate=1000 s^{-1}	Pa s	70-100 (230°C); 200 (200°C)	
Melt volume flow rate (ISO 1133, procedure B), 200°C/5 kg	cm^3/10 min	1.5-15	
Pressure coefficient of melt viscosity, b	G Pa^{-1}	35.5	Aho, J; Syrjala, S, J. Appl. Polym. Sci., 117, 1076–84, 2010.
Melt index, 230°C/3.8 kg	g/10 min	3.3-14	
Water absorption, equilibrium in water at 23°C	%	0.03-0.1	

CHEMICAL RESISTANCE

PARAMETER	UNIT	VALUE	REFERENCES
Acid dilute/concentrated	-	very good/poor	
Alcohols	-	good	
Alkalis	-	good	

PS polystyrene

PARAMETER	UNIT	VALUE	REFERENCES
Aliphatic hydrocarbons	-	poor	
Aromatic hydrocarbons	-	poor	
Esters	-	poor	
Greases & oils	-	poor	
Halogenated hydrocarbons	-	poor	
Ketones	-	poor	
Common solvents	-	toluene, perchloroethylene, carbon tetrachloride, carbon disufide, MEK, dioxane, cyclohexanone, ethyl acetate	
Θ solvent, Θ-temp.=249.8; -46, 34.5, 43°C	-	benzene, i-butyl acetate, cyclohexane, methyl acetate	
Good solvent	-	benzene, carbon disulfide, chlorinated aliphatic hydrocarbons, chloroform, cyclohexanone, dioxane, ethyl acetate, ethylbenzene, MEK, NMP, THF	
Non-solvent	-	acetic acid, acetone, alcohols, ethyl ether, saturated hydrocarbons,	
Chemicals causing environmental stress cracking	list	anionic, cationic and nonionic surfactants, solutions of sugar, salt, Na_2SO_4, $NaHCO_3$, and fatty acids	Kawaguchi, T; Nishimura, H; Kasahara, K; Kuriyama, T; Narisawa, I, Polym. Eng. Sci., 43, 2, 419-30, 2003.
Effect of EtOH sterilization (tensile strength retention)	%	84-105; 96-100 (HIPS)	Navarrete, L; Hermanson, N, Antec, 2807-18, 1996.

FLAMMABILITY			
Flammability according to UL-94 standard; thickness 1.6/0.8 mm	class	HB	
Ignition temperature	°C	296	
Autoignition temperature	°C	490	
Limiting oxygen index	% O_2	17.8-18.1, 35.2 (with ammonium phosphate starch carbamate)	Ji, W, Wang, D, Guo, J, Fei, B, Gu, X, Zhang, S, Carbohydrate Polym., 233, 115841, 2020.
Minimum ignition energy	J	0.04	
Heat release	kW m^{-2}	734 (HIPS without flame retardant); 283-378 (HIPS with flame retardants)	Yu, B; Liu, M; Lu, L; Dong, X; Gao, W; Tang, K, Fire Mater., 34, 251-61, 2010.
NBS smoke chamber	Ds	470	Padey, D; Walling, J; Wood A, Polymers in Defence and Aerospace 2007, Rapra, 2007, paper 15.
Char at 500°C	%	0	Lyon, R E; Walters, R N, J. Anal. Appl. Pyrolysis, 71, 27-46, 2004.
Heat of combustion	J g^{-1}	43,650	Walters, R N; Hacket, S M; Lyon, R E, Fire Mater., 24, 5, 245-52, 2000.
Peak of heat release rate	kW m^{-2}	666, 316 (with ammonium phosphate starch carbamate)	Ji, W, Wang, D, Guo, J, Fei, B, Gu, X, Zhang, S, Carbohydrate Polym., 233, 115841, 2020.
Volatile products of combustion	-	styrene, oligomers, organic acids, alcohols, aldehydes, ketones, CO, CO_2	
CO yield	%	8-13 (HIPS with flame retardants)	Yu, B; Liu, M; Lu, L; Dong, X; Gao, W; Tang, K, Fire Mater., 34, 251-61, 2010.

WEATHER STABILITY			
Spectral sensitivity	nm	290, <340	Wypych, G. Handbook of Materials Weathering, 6th Ed., ChemTec Publishing, 2018,
Activation wavelengths	nm	318; 310-345	

PS polystyrene

PARAMETER	UNIT	VALUE	REFERENCES
Excitation wavelengths	nm	254, 290, 300, 315-335	
Emission wavelengths	nm	336, 338, 354-355, 368, 372, 390, 395, 398, 422, 425, 430, 455-456, 485, 492, 510	
Transmittance	%	100 nm – 75.7; 300 nm – 43.3	Matsuzawa, N N; Oizumi, H; Mori, S; Irie, S; Shirayone, S; Yano, E; Okazaki, S; Ishitani, S; Dixon, D A, Jpn. J. Appl. Phys., 38, 7109-13, 1999.
Depth of UV penetration	μm	250-300	
Important initiators and accelerators	-	acetophenone, benzophenone, phenylacetylaldehyde, enones, diketones, succinimides, benzoyl peroxide, in-chain peroxide linkage, hydroperoxides, polycyclic aromatic hydrocarbons, iron (III) derivatives, cobalt salts of fatty acids, aluminum chloride, silica-alumina catalyst, rubene, diphenylanthracene, triphenyldiamine, carotene	
Products of degradation	-	water, carbon dioxide, ketone, unsaturations, hydroperoxides, radicals, chain scissions, crosslinks, conjugated double bonds, radicals, methane, ethylene, quinomethane structures, benzene, acetophenone, benzaldehyde, formic acid, acetic acid, benzoic acid	
Stabilizers	-	UVA: 2-hydroxy-4-methoxybenzophenone; 2,4-dihydroxybenzophenone; 2-(2H-benzotriazol-2-yl)-p-cresol; 2-(2H-benzotriazole-2-yl)-4,6-di-tert-pentylphenol; 2-(2H-benzotriazole-2-yl)-4-(1,1,3,3-tetraethylbutyl)phenol; 2-(2H-benzotriazol-2-yl)-4,6-bis(1-methyl-1-phenylethyl) phenol; 2,2'-methylene-bis(6-(2H-benzotriazol-2-yl)-4-1,1,3,3-tetramethylbutyl) phenol; 2-(2H-benzotriazol-2-yl)-6-dodecyl-4-methylphenol, branched & linear; 2,4-di-tert-butyl-6-(5-chloro-2H-benzotriazol-2-yl)-phenol; 2-(3-sec-butyl-5-tert-butyl-2-hydroxyphenyl)benzotriazole; reaction product of methyl 3(3-(2H-benzotriazole)-5-t-butyl-4-hydroxyphenyl propionate/PEG 300; ethyl-2-cyano-3,3-diphenylacrylate; propanedioic acid, [(4-methoxyphenyl)-methylene]-dimethyl ester; Screeners: carbon black, titanium dioxide, zinc oxide; HAS: 1,3,5-triazine-2,4,6-triamine, N,N'''[1,2-ethane-diyl-bis[[[4,6-bis[butyl(1,2,6,6-pentamethyl-4-piperidinyl)amino]-1,3,5-triazine-2-yl]imino]-3,1-propanediyl]bis[N',N''-dibutyl-N',N''-bis(1,2,2,6,6-pentamethyl-4-piperidinyl)-; bis(1,2,2,6,6-pentamethyl-4-piperidyl) sebacate and methyl 1,2,2,6,6-pentamethyl-4-piperidyl sebacate; bis(1,2,2,6,6-pentamethyl-4-piperidyl)sebacate + methyl-1,2,2,6,6-pentamethyl-4-piperidyl sebacate; bis(2,2,6,6-tetramethyl-4-piperidyl) sebacate; 2,2,6,6-tetramethyl-4-piperidinyl stearate; N,N'-bisformyl-N,N'-bis-(2,2,6,6-tetramethyl-4-piperidinyl)-hexamethylendiamine; poly[[(6-[1,1,3,3-tetramethylbutyl)amino]-1,3,5-triazine-2,4-diyl][2,2,6,6-tetramethyl-4-piperidinyl)imino]-1,6-hexanediyl[2,2,6,6-tetramethyl-4-piperidinyl) imino]]; 1, 6-hexanediamine, N, N'-bis(2,2,6,6-tetramethyl-4-piperidinyl)-, polymers with 2,4-dichloro-6-(4-morpholinyl)-1,3,5-triazine; 1,6-hexanediamine, N,N'-bis(2,2,6,6-tetramethyl-4-piperidinyl)-, polymers with morpholine-2,4,6-trichloro-1,3,5-triazine reaction products, methylated; Phenolic antioxidant: ethylene-bis(oxyethylene)-bis(3-(5-tert-butyl-4-hydroxy-m-tolyl)-propionate); 2,6,-di-tert-butyl-4-(4,6-bis(octylthio)-1,3,5,-triazine-2-ylamino) phenol; pentaerythritol tetrakis(3-(3,5-di-tert-butyl-4-hydroxyphenyl) propionate); octadecyl-3-(3,5-di-tert-butyl-4-hydroxyphenyl)-propionate; 3,3',3',5,5',5'-hexa-tert-butyl-a,a',a'-(mesitylene-2,4,6-triyl)tri-p-cresol; 2-(1,1-dimethylethyl)-6-[[3-(1,1-dimethylethyl)-2-hydroxy-5-methylphenyl]methyl-4-methylphenyl acrylate;	

PS polystyrene

PARAMETER	UNIT	VALUE	REFERENCES
Stabilizers (continuation)	-	1,3,5-tris(3,5-di-tert-butyl-4-hydroxybenzyl)-1,3,5-triazine-2,4,6(1H,3H,5H)-trione; 2',3-bis[[3-[3,5-di-tert-butyl-4-hydroxyphenyl]propionyl]]propionohydrazide; 2,2'-ethylidenebis(4,6-di-tert-butylphenol); 2,2'-methylenebis(4-ethyl-6-tert-butylphenol); 1,3,5-tris(4-tert-butyl-3-hydroxy-2,6-dimethyl benzyl)-1,3,5-triazine-2,4,6-(1H,3H,5H)-trione; Thiosynergist: didodecyl-3,3'-thiodipropionate; dioctadecyl 3,3'-thiodipropionate; 2,2'-thiodiethylene bis[3-(3,5-ditert-butyl-4-hydroxyphenyl)propionate]; Optical brightener: 2,2'-(2,5-thiophenediyl)bis(5-tert-butylbenzoxazole); 2,2'-(1,2-ethylenediyldi-4,1-phenylene)bisbenzoxazole	
Low earth orbit erosion yield	cm^3 $atom^{-1}$ x 10^{-24}	3.74	Waters, D L; Banks, B A; De Groh, K K; Miller, S K R; Thorson, S D, High Performance Polym., 20, 512-22, 2008.

BIODEGRADATION

PARAMETER	UNIT	VALUE	REFERENCES
Typical biodegradants	-	UV exposure increases degradation but biotic degradation is slow	Ojeda, T; Freitas, A; Dalmolin, E; Dal Pizzol, M; Vignol, L; Melnik, J; Jacques, R; Bento, F; Camargo, F, Polym. Deg. Stab., 94, 2128-33, 2009.
Stabilizers	-	1,2-benzisothiazolin-3-one, medetomidine	

TOXICITY

PARAMETER	UNIT	VALUE	REFERENCES
HMIS: Health, Flammability, Reactivity rating	-	1/1/0; 1/3/0 (expandable)	
Carcinogenic effect	-	not listed by ACGIH, NIOSH, NTP	

ENVIRONMENTAL IMPACT

PARAMETER	UNIT	VALUE	REFERENCES
Cradle to grave non-renewable energy use	MJ/kg	85-90	
Cradle to pellet greenhouse gases	kg CO_2 kg^{-1} resin	3.0-3.6	

PROCESSING

PARAMETER	UNIT	VALUE	REFERENCES
Typical processing methods	-	blow molding, electrospinning, extrusion, foaming, injection molding, thermoforming	
Processing temperature	°C	160-280 (injection molding); 180-210 (blown film); 200-240 (flat film)	
Processing pressure	MPa	30 (holding)	
Additives used in final products	-	Fillers: barium sulfate, barium titanate, calcium carbonate, carbon nanotubes, copper, glass beads, glass fibers, kaolin, magnesium hydroxide, mica, montmorillonite, nano-CaCO3, PTFE, red phosphorus, silica, talc, titanium dioxide, zeolites; zinc borate; Plasticizers: adipates and glutarates, benzyl-butyl, dimethyl, diethyl, dipropyl, dibutyl, diheptyl, dioctyl, and diisodecyl phthalates, di- and tri-isopropylbiphenyls, dioctyl sebacate, liquid paraffins, mineral oil, polybutenes, tricresyl phosphate; Antistatics: cocobis(2-hydroxyethyl)amine, conductive carbons blacks, glycerol monostearate, N,N-bis(2-hydroxyethyl)-N-(3-dodecyloxy-2-hydroxypropyl)methylammonium methosulfate, polyaniline, polyetheresteramide, quaternary ammonium compounds, zinc oxide whisker; Release: aluminum distearate, butylene glycol montanate, calcium montanate, PDMS, zinc stearate; Slip: ethylene bisstearamide; Blowing agents	

PS polystyrene

PARAMETER	UNIT	VALUE	REFERENCES
Applications	-	blending with other polymers, electrotechnical components, household items, insulating film, insulating foam, packaging, toys, and numerous other applications	
BLENDS			
Suitable polymers	-	LDPE, PA6, PANI, PDMS, PE, PEDOT, PMMA, POSS, PP, PVC, PVME	
Compatibilizers	-	MAStVA	

PS, iso isotactic-polystyrene

PARAMETER	UNIT	VALUE	REFERENCES
GENERAL			
Common name	-	isotactic polystyrene	
CAS name	-	benzene, ethenyl-, homopolymer, isotactic	
Acronym	-	i-PS	
CAS number	-	25086-18-4	
Formula			
SYNTHESIS			
Monomer(s) structure	-		
Monomer(s) CAS number(s)	-	100-42-5	
Monomer(s) molecular weight(s)	dalton, g/ mol, amu	104.15	
Monomer ratio	-	100%	
Mass average molecular weight, M_w	dalton, g/ mol, amu	18,000-1,570,000	
Polydispersity, M_w/M_n	-	2-6.4, 2.52 (fibers)	-; Wang, C, Lee, J L, Landingin, J, Ko, C-H, Polymer, 176, 236-43, 2019.
Radius of gyration	nm	22	
STRUCTURE			
Crystallinity	%	30-68	
Cell type (lattice)	-	hexagonal	Guenet, J-M, Polymer-Solvent Molecular Compounds, pp. 180-209, Elsevier, 2008.
Cell dimensions	nm	a=b=2.19, c=0.665	Guenet, J-M, Polymer-Solvent Molecular Compounds, pp. 180-209, Elsevier, 2008.
Space group		R3c	Guenet, J-M, Polymer-Solvent Molecular Compounds, pp. 180-209, Elsevier, 2008.
Tacticity	%	90-99% isotactic triads	Xue, G; Zhang, J; Chen, J; Li, Y; Ma, J; Wang, G; Sun, P, Macromolecules, 33, 2299-2301, 2000; Tosaka, M; Yamaguchi, K; Tsuji, M, Polymer, 51, 2, 547-553, 2010.
Chain conformation	-	3/1 helix (alternating *trans* and *gauche*)	Xue, G; Zhang, J; Chen, J; Li, Y; Ma, J; Wang, G; Sun, P, Macromolecules, 33, 2299-2301, 2000; Matsuba, G; Kaji, K; Nishida, K; Kanaya, T; Imai, M, Polym. J., 31, 9, 722-27, 1999.
Lamellae thickness	nm	7.8-13	Taguchi, K; Toda, A; Miyamoto, Y, J. Macromol. Sci. B, 45, 1141-47, 2006.
Spherullite size	μm	300	Kajioka, H; Yoshimoto, S; Taguchi, K; Toda, A, Macromolecules, 43, 3837-43, 2010.
Rapid crystallization temperature	°C	180-200	Bu, H; Gu, F; Bao, L; Chen, M, Macromolecules, 31, 7108-10, 1998.

PS, iso isotactic-polystyrene

PARAMETER	UNIT	VALUE	REFERENCES
PHYSICAL PROPERTIES			
Density at 20°C	g cm^{-3}	1.04-1.060	
Melting temperature, DSC	°C	205-240	
Activation energy of thermal degradation	kJ mol^{-1}	504	Chen, K; Harris, K; Vyazovkin, S, Macromol. Chem. Phys., 208, 2525-32, 2007.
Glass transition temperature	°C	80-100	
Volume resistivity	ohm-m	1E15	
MECHANICAL & RHEOLOGICAL PROPERTIES			
Flexural strength	MPa	51	
Flexural modulus	MPa	870	
Young's modulus	MPa	1,048	Thomas, S P; Thomas, S; Bandyopadhyay, S, Composites A40, 36-44, 2009.
Poisson's ratio	-	0.33	Thomas, S P; Thomas, S; Bandyopadhyay, S, Composites A40, 36-44, 2009.
CHEMICAL RESISTANCE			
Acid dilute/concentrated	-	very good/poor	
Alcohols	-	good	
Alkalis	-	good	
Aliphatic hydrocarbons	-	poor	
Aromatic hydrocarbons	-	poor	
Esters	-	poor	
Greases & oils	-	poor	
Halogenated hydrocarbons	-	poor	
Ketones	-	poor	
Other	-	toluene, perchloroethylene, carbon tetrachloride, carbon disufide, MEK, dioxane, cyclohexanone, ethyl acetate	
Θ solvents	-	benzene, i-butyl acetate, cyclohexane, methyl acetate	
Good solvent	-	n-tetradecane/decahydronaphthalene=2/1, trichlorbenzene	
Non-solvent	-	methyl ethyl ketone and methanol at boiling temperatures	
FLAMMABILITY			
Flammability according to UL-94 standard; thickness 1.6/0.8 mm	class	HB	
Ignition temperature	°C	296	
Autoignition temperature	°C	490	
Limiting oxygen index	% O$_2$	17.8-18.1	
Minimum ignition energy	J	0.04	
Heat release	kW m^{-2}	734 (HIPS without flame retardant); 283-378 (HIPS with flame retardants)	
NBS smoke chamber	-	470	
Char at 500°C	%	0	Lyon, R E; Walters, R N, J. Anal. Appl. Pyrolysis, 71, 27-46, 2004.
Heat of combustion	J g^{-1}	43,650	

PS, iso isotactic-polystyrene

PARAMETER	UNIT	VALUE	REFERENCES
Volatile products of combustion	-	styrene, oligomers, organic acids, alcohols, aldehydes, ketones, CO, CO_2	
CO yield	%	8-13 (HIPS with flame retardants)	

ANALYSIS			
FTIR (wavenumber-assignment)	cm^{-1}/-	3/1 helix double band – 1082-1052 and 922-904; GTTG – 548, TTGG and GGTG – 562, GTGT – 567 and 586	Xue, G; Zhang, J; Chen, J; Li, Y; Ma, J; Wang, G; Sun, P, Macromolecules, 33, 2299-2301, 2000; Matsuba, G; Kaji, K; Nishida, K; Kanaya, T; Imai, M, Polym. J., 31, 9, 722-27, 1999.
x-ray diffraction peaks	degree	14.3, 16.5, 18.4, 21.7, 24.5, 27.3	Xue, G; Zhang, J; Chen, J; Li, Y; Ma, J; Wang, G; Sun, P, Macromolecules, 33, 2299-2301, 2000.

PS, syndio syndiotactic-polystyrene

PARAMETER	UNIT	VALUE	REFERENCES
GENERAL			
Common name	-	syndiotactic polystyrene	
IUPAC name	-	poly(1-phenylethane-1,2-diyl)	
CAS name	-	benzene, ethenyl-, homopolymer, syndiotactic	
Acronym	-	s-PS	
CAS number	-	28325-75-9	
Formula			
HISTORY			
Person to discover	-	Ishihara, N; Kuramoto, M; Uoi, M (Idemitsu)	JP Patent, 62 187 708, 1985.
Date	-	1985	
Details	-	polymerization in the presence of titanium complex/methylaluminum catalyst	
SYNTHESIS			
Monomer(s) structure	-		
Monomer(s) CAS number(s)	-	100-42-5	
Monomer(s) molecular weight(s)	dalton, g/mol, amu	104.15	
Monomer ratio	-	100%	
Method of synthesis	-	reactor is charged with toluene, triisobutylaluminum, methyl-aluminum, and styrene and then catalyst	Huang, C-L; Wang, C; Hsiao, T-J; Tsai, J-C, Polym. Mat. Sci. Eng., 101, 1618-20, 2009; Schellenberg, J; Leder, H-J, Adv. Polym. Technol., 25, 3, 141-51, 2006.
Temperature of polymerization	°C	25	
Time of polymerization	h	24	
Catalyst	-	$CpTiCl_3$, methylalumoxane	Perrin, L; Kirillov, E; Carpentier, J-F; Maron, L, Macromolecules, 43, 6330-36, 2010.
Number average molecular weight, M_n	dalton, g/mol, amu	69,000-139,000	
Mass average molecular weight, M_w	dalton, g/mol, amu	13,000-560,000; 152,000	-; Sano, T, Ebihara, H, Sano, S, Okabe, T, Itagaki, H, Eur. Polym. J., 138, 109975, 2020.
Polydispersity, M_w/M_n	-	1.8-6.2; 1.9	-; Sano, T, Ebihara, H, Sano, S, Okabe, T, Itagaki, H, Eur. Polym. J., 138, 109975, 2020.
Radius of gyration	nm	40.6	Koutsoumpis, S, Klonos, P, Raftopoulos, K N, Pissis, P, Polymer, 153, 548-57, 2018.
STRUCTURE			
Crystallinity	%	40-55; 20-35 (isotactic)	
Cell type (lattice)	-	monoclinic (δ), hexagonal (α), orthorhombic (β, ϵ)	

PS, syndio syndiotactic-polystyrene

PARAMETER	UNIT	VALUE	REFERENCES
Cell dimensions	nm	a:b:c=1.75:1.18:0.78 (monoclinic); 0.263:0.263:0.51 (hexagonal); a:b:c=1.62:2.20:0.79 (othorhombic); a=b=2.626, c=0.626 (α crystal has a trigonal unit cell); a:b:c=0.881:2.882:0.51 (β crystal packed in an orthorhombic unit cell)	Figueroa-Gerstenmaier, S; Daniel, C; Milano, G; Vitillo, J G; Zavorotynska, O; Spoto, G; Guerra, Macromolecules, 43, 8594-8601, 2010: Endo, F, Hotta, A, Polymer, 135, 103-10, 2018.
Unit cell angles	degree	γ=97 (monoclinic)	
Number of chains per unit cell	-	4 (orthorhombic)	
Polymorphs	-	$\alpha, \beta, \gamma, \delta, \varepsilon$	Tarallo, O; Petraccone, V; Albunia, A R; Daniel, C; Guerra, G, Macromolecules, 43, 8549-58, 2010.
Tacticity	%	98-99 (syndiotactic; metallocene)	Saga, S; Matsumoto, H; Saito, K; Minagawa, M; Tanioka, A, J. Power Source, 176, 16-22, 2008.
Chain conformation	-	*trans*-planar (α and β); helical s(2/1)2 ($\gamma, \delta, \varepsilon$); helical (TTGG) in solution and all *trans* (TTTT) in melt; 2_1-helical (TTGG)	Tarallo, O; Petraccone, V; Albunia, A R; Daniel, C; Guerra, G, Macromolecules, 43, 8549-58, 2010; Sano, T, Ebihara, H, Sano, S, Okabe, T, Itagaki, H, Eur. Polym. J., 138, 109975, 2020.
Rapid crystallization temperature	°C	210-260	Sorrentino, A; Pantani, R; Titomanlio, G, J. Polym. Sci. B, 48, 1757-66, 2010.
Avrami constants, k/n	-	n=3.03-3.49; 1.3-1.7 (commercial sPS)	Benson, S D; Moore, R B, Polymer, 51, 5462-72, 2010; Chen, C-M; Hsieh, T-E; Ju, M-Y, J. Alloys compounds, 480, 658-61, 2009.

COMMERCIAL POLYMERS

PARAMETER	UNIT	VALUE	REFERENCES
Some manufacturers	-	Dow; Idemitsu	
Trade names	-	Questra; Xarec	

PHYSICAL PROPERTIES

PARAMETER	UNIT	VALUE	REFERENCES
Density at 20°C	g cm^{-3}	1.01-1.05; 1.03 (trans-planar conformation), 1.05 (amorphous region)	Endo, F, Hotta, A, Polymer, 135, 103-10, 2018.
Melting temperature, DSC	°C	265-284	
Thermal expansion coefficient, 23-80°C	°C^{-1}	9.2E-5	
Glass transition temperature	°C	97-130	Endo, F, Hotta, A, Polymer, 135, 103-10, 2018; Koutsoumpis, S, Klonos, P, Raftopoulos, K N, Pissis, P, Polymer, 153, 548-57, 2018.
Heat of fusion	kJ mol^{-1}	5.8	
Long term service temperature	°C	127	
Heat deflection temperature at 0.45 MPa	°C	110	
Heat deflection temperature at 1.8 MPa	°C	95	
Enthalpy of crystallization	J g^{-1}	29	Sorrentino, A; Pantani, R; Titomanlio, G, J. Polym. Sci. B, 48, 1757-66, 2010.
Dielectric constant at 100 Hz/1 MHz	-	2.5	Koutsoumpis, S, Klonos, P, Raftopoulos, K N, Pissis, P, Polymer, 153, 548-57, 2018.
Dielectric loss factor at 1 kHz	-	0.001	
Dissipation factor at 1 MHz	E-4	10	

PS, syndio syndiotactic-polystyrene

PARAMETER	UNIT	VALUE	REFERENCES
Volume resistivity	ohm-m	>E14	
Electric strength K20/P50, d=0.60.8 mm	kV mm^{-1}	66	
Comparative tracking index, CTI, test liquid A	V	600	
Arc resistance	s	91	

MECHANICAL & RHEOLOGICAL PROPERTIES

Tensile strength	MPa	68	
Tensile modulus	MPa	2,400	D'Aniello, C; Rizzo, P; Guerra, G, Polymer, 46, 11435-41, 2005.
Tensile stress at yield	MPa	35	
Tensile yield strain	%	20	
Flexural strength	MPa	65	
Flexural modulus	MPa	2,500	
Rockwell hardness	-	L60	
Shrinkage	%	1.7	
Melt index, 230°C/3.8 kg	g/10 min	2.5-11.1	
Water absorption, 24h at 23°C	%	0.04	

CHEMICAL RESISTANCE

Acid dilute/concentrated	-	very good/poor
Alcohols	-	good
Alkalis	-	good
Aliphatic hydrocarbons	-	poor
Aromatic hydrocarbons	-	poor
Esters	-	poor
Greases & oils	-	poor
Halogenated hydrocarbons	-	poor
Ketones	-	poor
Other	-	toluene, perchloroethylene, carbon tetrachloride, carbon disufide, MEK, dioxane, cyclohexanone, ethyl acetate
Θ solvents	-	benzene, i-butyl acetate, cyclohexane, methyl acetate
Good solvent	-	n-tetradecane/decahydronaphthalene=2/1, trichlorbenzene
Non-solvent	-	methyl ethyl ketone and methanol at boiling temperatures

FLAMMABILITY

Flammability according to UL-94 standard; thickness 1.6/0.8 mm	class	HB
Ignition temperature	°C	296
Autoignition temperature	°C	490
Limiting oxygen index	% O_2	17.8-18.1
Minimum ignition energy	J	0.04
Heat release	kW m^{-2}	734 (HIPS without flame retardant); 283-378 (HIPS with flame retardants)
NBS smoke chamber	-	470

PS, syndio syndiotactic-polystyrene

PARAMETER	UNIT	VALUE	REFERENCES
Char at 500°C	%	0	Lyon, R E; Walters, R N, J. Anal. Appl. Pyrolysis, 71, 27-46, 2004.
Heat of combustion	J g^{-1}	43,650	
Volatile products of combustion	-	styrene, oligomers, organic acids, Alcohols, aldehydes, ketones, CO, CO$_2$	
CO yield	%	8-13 (HIPS with flame retardants)	
Activation energy of pyrolysis	kJ mol^{-1}	164-249	Huang, J, Li, X, Meng, H, Tong, H, Cai, X, Liu, J, Chem. Phys. Lett., 747, 137334, 2020.

WEATHER STABILITY

PARAMETER	UNIT	VALUE	REFERENCES
Spectral sensitivity	nm	290	

PROCESSING

PARAMETER	UNIT	VALUE	REFERENCES
Typical processing methods	-	extrusion, injection molding, reaction injection molding, thermoforming	
Preprocess drying: temperature/ time/residual moisture	°C/h/%	80/2-5/	
Processing temperature	°C	315-335 (extrusion, melt temperature)	
Outstanding properties	-	high melt temperature, high crystallinity, rapid crystallization rate	

BLENDS

PARAMETER	UNIT	VALUE	REFERENCES
Suitable polymers	-	epoxy, EPR, HDPE, NBR, PA6, PC, PMMA, POM, PPO, PPS, aPS, iPS, PVC, SEBS	

ANALYSIS

PARAMETER	UNIT	VALUE	REFERENCES
FTIR (wavenumber-assignment)	cm^{-1}/-	amorphous phase – 841; α crystal – 851; β crystal – 858	Wu, S-C; Chang, F-C, Polymer, 45, 733-38, 2004.
Raman (wavenumber-assignment)	cm^{-1}/-	*trans*-planar chains – 2917, 2845; TTGG conformation – 2850	Zheng, K; Liu, R; Kang, H; Gao, X; Shen, D; Huang, Y, Polymer, in press, 2011.
x-ray diffraction peaks	degree	6.7, 11.7, 12.3, 21.0; reflection peaks at 2θ = 6.9° (1 1 0) and 8.2° (0 2 0) characteristic of ε crystalline phase	Wang, C; Lin, C-C; Chu, C-P, Polymer, 12595-606, 2005; Sano, T, Ebihara, H, Sano, S, Okabe, T, Itagaki, H, Eur. Polym. J., 138, 109975, 2020.

PSM polysilylenemethylene

PARAMETER	UNIT	VALUE	REFERENCES
GENERAL			
Common name	-	polysilylenemethylene	
Acronym	-	PSM	
HISTORY			
Person to discover	-	Sommer L H; Mitch, F A; Goldberg, G M; Goodwin, J T	Interrante, L V; Liu, Q; Rushkin, I; Shen, Q, J. Organomettalic Chem., 521, 1-10, 1996.
Date	-	1949	
Details	-	Sommer *et al.* first reported; Goodwin patented	
SYNTHESIS			
Monomer(s) structure	-		
Monomer(s) CAS number(s)	-	2146-97-6	
Monomer(s) molecular weight(s)	dalton, g/mol, amu	226.04	
Number average molecular weight, M_n	dalton, g/mol, amu	24,000	
Mass average molecular weight, M_w	dalton, g/mol, amu	11,000-460,000	
Polydispersity, M_w/M_n	-	2.8	
STRUCTURE			
Crystallinity	%	70	
Cell type (lattice)	-	monoclinic	Shen, Q; Interrante, L V, Macromolecules 29, 5788, 1996.
Cell dimensions	nm	a:b:c=0.57:0.875:0.325	Shen, Q; Interrante, L V, Macromolecules 29, 5788, 1996.
Unit cell angles	degree	γ=97.5	Shen, Q; Interrante, L V, Macromolecules 29, 5788, 1996.
PHYSICAL PROPERTIES			
Glass transition temperature	°C	-135 to -140	
PROCESSING			
Additives used in final products	-	Fillers: nanoparticles of Au, Pd, Cu, Ag	
Applications	-	optical material, semiconductor	
ANALYSIS			
NMR (chemical shifts)	ppm	Si NMR: *trans* – 14.4, *cis* – 14.7; H NMR – methylene carbon – 124.8	Kienard, M; Wiegand, C; Apple, T; Interrante, L V, J. Organometallic Chem., 686, 272-80, 2003.

PSMS poly(styrene-co-α-methylstyrene)

PARAMETER	UNIT	VALUE	REFERENCES
GENERAL			
Common name	-	poly(styrene-co-α-methylstyrene), hydrocarbon polymer	
Acronym	-	PSMS	
CAS number	-	9011-11-4	
SYNTHESIS			
Monomer(s) structure	-		
Monomer(s) CAS number(s)	-	100-42-5; 98-83-9	
Monomer(s) molecular weight(s)	dalton, g/mol, amu	1004.15; 118.18	
Number average molecular weight, M_n	dalton, g/mol, amu	600-1,750	
Mass average molecular weight, M_w	dalton, g/mol, amu	950-4,900	
Polydispersity, M_w/M_n	-	1.5-3.5	
COMMERCIAL POLYMERS			
Some manufacturers	-	Eastman	
Trade names	-	Kristalex	
PHYSICAL PROPERTIES			
Density at 20°C	$g\ cm^{-3}$	1.05-1.1	
Color	-	colorless	
Odor	-	slight, hydrocarbon	
Melting temperature, DSC	°C	119	
Softening point	°C	70-140	
Decomposition temperature	°C	>250	
Glass transition temperature	°C	32-63	
MECHANICAL & RHEOLOGICAL PROPERTIES			
Melt viscosity, shear rate=1000 s^{-1}	Pa s	10 (220°C), 100 (180°C), 1000 (160°C) (Kristalex 5140)	
CHEMICAL RESISTANCE			
Alcohols	-	good	
Aliphatic hydrocarbons	-	poor	
Aromatic hydrocarbons	-	poor	
Esters	-	poor	
Halogenated hydrocarbons	-	poor	
Ketones	-	poor	
Good solvent	-	t-butyl acetate, perchlorobenzenetetrafluoride	
FLAMMABILITY			
Ignition temperature	°C	125.6-283	
Autoignition temperature	°C	180	

PSMS poly(styrene-co-α-methylstyrene)

PARAMETER	UNIT	VALUE	REFERENCES
Volatile products of combustion	-	CO, CO_2	

TOXICITY

HMIS: Health, Flammability, Reactivity rating	-	0-1/1/0	
Carcinogenic effect	-	not listed by ACGIH, NIOSH, NTP	

ENVIRONMENTAL IMPACT

Aquatic toxicity, *Bluegill sunfish*, LC_{50}, 48 h	mg l^{-1}	>10	

PROCESSING

Typical processing methods	-	compounding, mixing	
Applications	-	modification of rubbers and resins in automotive, building and construction, caulks and sealants, hot melt adhesives, laminating, nonwovens, pressure sensitive adhesives, tapes	
Outstanding properties	-	water clear, non-polar, compatible with many rubbers and resins, tackifying characteristics, thermal stability	

BLENDS

Suitable polymers	-	PPO, SBS, SEBS, SEPS, SIS	

PSR polysulfide

PARAMETER	UNIT	VALUE	REFERENCES
GENERAL			
Common name	-	polysulfide	
IUPAC name	-	poly(oxyethylenedisulfanediylethylene)	
ACS name	-	polysulfide	
Acronym	-	PSR	
CAS number	-	9080-49-3	
Formula		$\left[CH_2CH_2S_x\right]_n$	
HISTORY			
Person to discover	-	Joseph Cecil Patrick	
Date	-	1926	
Details	-	invented as side product of development of antifreeze	
SYNTHESIS			
Monomer(s) structure	-	$ClCH_2CH_2Cl \qquad Na\!-\!\underset{x}{S}\!-\!Na$	
Monomer(s) CAS number(s)	-	107-06-2; 1344-08-7	
Monomer(s) molecular weight(s)	dalton, g/mol, amu	98.96	
Method of synthesis	-	condensation polymerization reaction between organic dihalides and alkali metal salts of polysulfide anions	
COMMERCIAL POLYMERS			
Some manufacturers	-	Morton, Toray	
Trade names	-	Thiokol	
PHYSICAL PROPERTIES			
Density at 20°C	g cm^{-3}	1.28-1.29	
Refractive index, 20°C	-	1.6423	
Decomposition temperature	°C	200	
Glass transition temperature	°C	-20 to -55; -55	-; Li, X, Nie, W, Xu, Y, Zhou, Y, Zhang, C, Compos. Part B: Eng., 198, 108234, 2020.
Maximum service temperature	°C	150	
Long term service temperature	°C	90	
Thermal conductivity	W m^{-1} K^{-1}	0.235	Li, X, Nie, W, Xu, Y, Zhou, Y, Zhang, C, Compos. Part B: Eng., 198, 108234, 2020.
MECHANICAL & RHEOLOGICAL PROPERTIES			
Tensile strength	MPa	1-17	
Elongation	%	30-500, 125-135	-; Li, X, Nie, W, Xu, Y, Zhou, Y, Zhang, C, Compos. Part B: Eng., 198, 108234, 2020.
Flexural strength	MPa	30	
Compressive strength	MPa	124	
Abrasion resistance (ASTM D1044)	mg/1000 cycles	70	

PSR polysulfide

PARAMETER	UNIT	VALUE	REFERENCES
Shore A hardness	-	25-50	
Shore D hardness	-	65-75	
Melt viscosity, shear rate=1000 s^{-1}	Pa s	1-110	

CHEMICAL RESISTANCE

Acid dilute/concentrated	-	poor	
Alcohols	-	good	
Alkalis	-	good	
Aliphatic hydrocarbons	-	good	
Aromatic hydrocarbons	-	poor	
Esters	-	poor	
Greases & oils	-	good	
Halogenated hydrocarbons	-	fair	
Ketones	-	poor	

FLAMMABILITY

Ignition temperature	°C	93.3	

BIODEGRADATION

Typical biodegradants	-	reductases from *Wolinella succinogenes* and *Clostridium*; sulfate reducing bacteria (*Desulfuromonas*) and some spirilloid bacteria (*Thermotoga*)	Takahashi, Y; Suto, K; Inoue, C, J. Biosci. Bioeng., 109, 4, 372-80, 2010.

TOXICITY

NFPA: Health, Flammability, Reactivity rating	-	0/1/0	
Carcinogenic effect	-	not listed by ACGIH, NIOSH, NTP	
Oral rat, LD$_{50}$	mg kg^{-1}	3,000	
Skin rabbit, LD$_{50}$	mg kg^{-1}	>2,000	

ENVIRONMENTAL IMPACT

Aquatic toxicity, *Daphnia magna*, LC$_{50}$, 48 h	mg l^{-1}	>1,000	
Aquatic toxicity, *Bluegill sunfish*, LC$_{50}$, 48 h	mg l^{-1}	>10,000	
Aquatic toxicity, *Rainbow trout*, LC$_{50}$, 48 h	mg l^{-1}	>10,000	

PROCESSING

Typical processing methods	-	compounding, moisture or chemical curing of premixed compounds, vulcanization	
Additives used in final products	-	Fillers: calcium carbonate, calcium hydroxide, calcium oxide, carbon black, graphene oxide, polymeric beads, polystyrene particles, silver particles, zinc oxide; Plasticizers: 1-isobutyrate benzyl phthalate, 2,2,4-trimethyl-1,3-pentanediol, alkyl sulfonic acid esters of phenol and/or cresol, benzyl butyl phthalate, chlorinated paraffins, hydrogenated perphenyl, isooctyl benzyl phthalate; Curatives: metal peroxides, oxy salts (e.g., dioxides of lead, manganese, calcium, etc.)	

HANDBOOK OF POLYMERS 3rd Edition, Copyrights 2022; ChemTec Publishing

PSR polysulfide

PARAMETER	UNIT	VALUE	REFERENCES
Applications	-	additives to epoxy, coatings, Cr removal, electrical potting compounds, fuel-contact sealants, fuel hoses and tubing, insulating glass, linings, rocket propellant binders, sealants	
Outstanding properties	-	chemical resistance, adhesion	
BLENDS			
Suitable polymers	-	PPy	
Incompatible polymers	-	vinylidene-hexafluoropropene copolymer (Viton A) participates in degradation of syntactic PSR	Vance, A L; Alviso, C T; Harvey, C A, Polym. Deg. Stab., 91, 1960-63, 2006.
ANALYSIS			
FTIR (wavenumber-assignment)	cm^{-1}/-	C=O – 1723, S–H – 2560	Mahon, A; Kemp, T J; Coates, R J, Polym. Deg. Stab., 62, 15-24, 1998.
NMR (chemical shifts)	ppm	H NMR and C NMR: peak assignments in ref.; C=O – 7.95 and 8.02 (H NMR) and 160-49 and 162.02 (C NMR)	Mahon, A; Kemp, T J; Coates, R J, Polym. Deg. Stab., 62, 15-24, 1998.

PSU polysulfone

PARAMETER	UNIT	VALUE	REFERENCES
GENERAL			
Common name	-	polysulfone	
IUPAC name	-	poly[oxy-1,4-phenylenesulfonyl-1,4-phenyleneoxy-1,4-phenylene(dimethylmethylene)-1,4-phenylene]	
CAS name	-	poly[oxy-1,4-phenylenesulfonyl-1,4-phenyleneoxy-1,4-phenylene(1-methylethylidene)-1,4-phenylene]	
Acronym	-	PSU, PSF	
CAS number	-	25135-51-7	
Formula			
Relevant literature			Processing Guide for Injection Molding and Extrusion; Udel Polysulfone, Design Guide, Solvay Advanced Polymers.
HISTORY			
Person to discover	-	Shechter, L	Shechter, L, US Patent 3,282,893, Union Carbide, Nov. 1, 1966.
Date	-	1965, 1966 (filled in 1961)	
Details	-	introduced by Union Carbide in 1965 (patented in 1966)	
SYNTHESIS			
Monomer(s) structure	-		
Monomer(s) CAS number(s)	-	80-07-9; 80-05-7	
Monomer(s) molecular weight(s)	dalton, g/mol, amu	287.16; 228.29	
Monomer ratio	-	1.26:1	
Method of synthesis	-	polysulfone is produced by the reaction of a bisphenol A and bis(4-chlorophenyl)sulfone	
Number average molecular weight, M_n	dalton, g/mol, amu	39,000-41,000	
Mass average molecular weight, M_w	dalton, g/mol, amu	20,000-96,000	
Polydispersity, M_w/M_n	-	1.6	
Molecular cross-sectional area, calculated	cm^2 x 10^{-16}	20.1	
Radius of gyration	nm	7.2	Koriyama, H; Oyama, H T; Ougizawa, T; Inoue, T; Weber, M; Koch, E, Polymer, 40, 6381-93, 1999.
End-to-end distance of unperturbed polymer chain	nm	12.4	Koriyama, H; Oyama, H T; Ougizawa, T; Inoue, T; Weber, M; Koch, E, Polymer, 40, 6381-93, 1999.
Chain-end groups	-	OH; modifications: methacrylate functionality; COOH functionality	Dizman, C; Ates, S; Torun, L; Yagci, Y, Bielstein J. Org. Chem., 6, 56, 1-7, 2010; Hoffmann, T; Pospiech, D; Kretzschmar, B; Reuter, U; Haussler, L; Eckert, F; Perez-Graterol, R; Sandler, J K W; Altstadt, V, High Performance Polym., 19, 48-61, 2007.

HANDBOOK OF POLYMERS 3rd Edition, Copyrights 2022; ChemTec Publishing

PSU polysulfone

PARAMETER	UNIT	VALUE	REFERENCES
STRUCTURE			
Crystallinity	%	amorphous	
Entanglement molecular weight	dalton, g/mol, amu	calc.=2,250	
COMMERCIAL POLYMERS			
Some manufacturers	-	BASF; Solvay	
Trade names	-	Ultrason S; Udel	
PHYSICAL PROPERTIES			
Density at 20°C	g cm^{-3}	1.23-1.24; 1.33-1.49 (10-30% glass fiber)	
Color	-	amber to beige	
Refractive index, 20°C	-	1.6330	
Transmittance	%	84-86	
Haze	%	1.5-2.5	
Odor	-	odorless	
Melting temperature, DSC	°C	185	
Decomposition temperature	°C	550	
Thermal expansion coefficient, 23-80°C	°C^{-1}	5.3-5.7E-5; 1.9-4.9E-5	
Thermal conductivity, melt	W m^{-1} K^{-1}	0.26; 0.19-0.22 (10-30% glass fiber)	
Glass transition temperature	°C	185-190; 187 (20-30% glass fiber)	
Specific heat capacity	J K^{-1} kg^{-1}	2,300 (400°C)	
Maximum service temperature	°C	140-160	
Long term service temperature	°C	150 (glass fiber reinforced)	
Heat deflection temperature at 0.45 MPa	°C	183; 187-188 (20-30% glass fiber)	
Heat deflection temperature at 1.8 MPa	°C	174-175; 179-183 (10-30% glass fiber)	
Vicat temperature VST/A/50	°C	183-188; 187-192 (10-30% glass fiber)	
Hansen solubility parameters, δ_D, δ_P, δ_H	MPa$^{0.5}$	18.5, 8.5, 7.0; 19.03, 0, 6.96	
Interaction radius		9.4	
Hildebrand solubility parameter	MPa$^{0.5}$	20.26; 21.5	
Surface tension	mN m^{-1}	46.11	Ioan, S; Ffilimon, A; Avram, E; Ioanid, G, e-Polymers, 031, 1-13, 2007.
Dielectric constant at 60 Hz/1 MHz	-	3.03/3.02; 3.18-3.48/3.15-3.47 (10-30% glass fiber)	
Dielectric loss factor at 1 kHz	-	3.02; 3.47 (30% glass fiber)	
Relative permittivity at 100 Hz	-	3.1-3.5; 3.5-3.7 (20-30% glass fiber)	
Relative permittivity at 1 MHz	-	3.1-3.5; 3.5-3.7 (20-30% glass fiber)	
Dissipation factor at 60 Hz		0.0007-0.0011; 0.0007-0.001 (10-30% glass fiber)	
Dissipation factor at 1 MHz		0.006-0.0071; 0.005-0.006 (10-30% glass fiber)	
Volume resistivity	ohm-m	3E14; 1-3E14 (10-30% glass fiber)	
Surface resistivity	ohm	4E15; 1-6E15 (10-30% glass fiber)	

PSU polysulfone

PARAMETER	UNIT	VALUE	REFERENCES
Electric strength K20/P50, d=0.60.8 mm	kV mm^{-1}	17-37; 19-46 (10-30% glass fiber)	
Comparative tracking index, CTI	-	125-135; 165 (10-30% glass fiber)	
Coefficient of friction	-	0.48 (air); 0.4 (water)	Duan, Y; Cong, P; Liu, X; Li, T, J. Macromol. Sci. B, 48, 604-16, 2009.
Permeability to nitrogen, 25°C	mm^3 m m^{-2} MPa^{-1} day^{-1}	155	
Permeability to oxygen, 25°C	mm^3 m m^{-2} MPa^{-1} day^{-1}	894	
Permeability to water vapor, 25°C	g m^{-1} s^{-1} Pa^{-1} x 10^9	0.146	Vidotti, S E; Pessan L A, J. Appl. Polym. Sci., 101, 2, 825-32, 2006.
Contact angle of water, 20°C	degree	66-79	
Surface free energy	mJ m^{-2}	44.9	
Speed of sound	m s^{-1}	37.33	
Acoustic impedance		2.78	
Attenuation	dB cm^{-1}, 5 MHz	4.25	

MECHANICAL & RHEOLOGICAL PROPERTIES

Tensile strength	MPa	70-77; 77.9-120 (10-30% glass fiber)
Tensile modulus	MPa	2,480-2,600; 3,720-9,400 (10-30% glass fiber)
Tensile stress at yield	MPa	75
Tensile creep modulus, 1000 h, elongation 0.5 max	MPa	2,500; 6,000-8,300 (20-30% glass fiber)
Elongation	%	50-100, 1.7-4 (10-30% glass fiber)
Tensile yield strain	%	5.7; 2.2 (20-30% glass fiber)
Flexural strength	MPa	106; 128-154 (10-30% glass fiber)
Flexural modulus	MPa	2690; 3,790-7580 (10-30% glass fiber)
Compressive modulus	MPa	2,580; 4,070-8,000 (10-30% glass fiber)
Charpy impact strength, unnotched, 23°C	kJ m^{-2}	no break; 5.9-45 (20-30% glass fiber)
Charpy impact strength, unnotched, -30°C	kJ m^{-2}	no break; 45 (20-30% glass fiber)
Charpy impact strength, notched, 23°C	kJ m^{-2}	5.5-6; 7 (20-30% glass fiber)
Charpy impact strength, notched, -30°C	kJ m^{-2}	6; 7 (20-30% glass fiber)
Izod impact strength, unnotched, 23°C	J m^{-1}	no break; 477 (20% glass fiber)
Izod impact strength, notched, 23°C	J m^{-1}	69; 48-69 (10-30% glass fiber)
Izod impact strength, notched, -30°C	J m^{-1}	41-62; 59 (30% glass fiber)

PSU polysulfone

PARAMETER	UNIT	VALUE	REFERENCES
Poisson's ratio	-	0.37; 0.41-0.43 (10-30% glass fiber)	
Rockwell hardness	M	69; 80-86 (10-30% glass fiber)	
Shrinkage	%	0.68-0.77; 0.2-0.52 (10-30% glass fiber)	
Viscosity number	ml g^{-1}	72-81; 63 (20-30% glass fiber)	
Intrinsic viscosity, 25°C	dl g^{-1}	0.36-0.60	
Melt viscosity, shear rate=1000 s^{-1}	Pa s	400-600; 530-550 (20-30% glass fiber)	
Melt volume flow rate (ISO 1133, procedure B), 360°C/10 kg	cm^3/10 min	30-90; 30-40 (20-30% glass fiber)	
Melt index, 343°C/3.8 kg	g/10 min	3.4-17.5; 6.5 (10-30% glass fiber)	
Water absorption, 24h at 23°C	%	0.22-0.3; 0.22-0.29 (10-30% glass fiber)	
Moisture absorption, equilibrium 23°C/50% RH	%	0.3; 0.2 (20-30% glass fiber)	

CHEMICAL RESISTANCE			
Acid dilute/concentrated	-	excellent	
Alcohols	-	excellent	
Alkalis	-	good to excellent	
Aliphatic hydrocarbons	-	excellent	
Aromatic hydrocarbons	-	good to poor	
Esters	-	poor	
Halogenated hydrocarbons	-	poor	
Ketones	-	poor	

FLAMMABILITY			
Ignition temperature	^{o}C	490; 875 (20-30% glass fiber)	
Autoignition temperature	^{o}C	550-590	
Limiting oxygen index	% O$_2$	26-32 ; 31-40 (10-30% glass fiber)	
NBS smoke chamber (max optical density)	4 min.	16-65	
Char at 500°C	%	28.1-29	Perng, L H, J. Polym. Sci. A, 38, 583-93, 2000; Lyon, R E; Walters, R N, J. Anal. Appl. Pyrolysis, 71, 27-46, 2004.
Heat of combustion	J g^{-1}	30,280-30,630	Walters, R N; Hacket, S M; Lyon, R E, Fire Mater., 24, 5, 245-52, 2000.
Volatile products of combustion	-	CO, CO$_2$, oxides of sulfur; and more	Perng, L H, J. Polym. Sci. A, 38, 583-93, 2000.
UL 94 rating	-	HB to V-0; HB to V-1 to V-0 (10-30% glass fiber)	

WEATHER STABILITY			
Spectral sensitivity	nm	<320, 365; 193 (photolithography)	Chen, L; Goh, Y-K; Lawrie, K; Chuang, Y; Piscani, E; Zimmerman, P; Blakey, I; Whittaker, A K, Radiation Phys. Chem., 80, 242-47, 2011.
Excitation wavelengths	nm	245-255, 270, 320,	
Emission wavelengths	nm	310, 360, 450	
Retention of tensile strength and impact after exposure to 50-100 kGy of gamma radiation	%	93-100	

PSU polysulfone

PARAMETER	UNIT	VALUE	REFERENCES
Depth of UV penetration	μm	50	
Important initiators and accelerators	-	residual monomer, copper stearate	
Products of degradation	-	products of photooxidation: chain scissions, free radicals, carbonyl groups, acetic acid, sulfoacetic acid, benzoic acid, crosslinks, unsaturations, hydroxyl groups, sulfonic acid, SO_2	

BIODEGRADATION

PARAMETER	UNIT	VALUE	REFERENCES
Typical biodegradants	-	Gram-positive and Gram-negative bacteria	Filimon, A; Avram, E; Dunca, S; toica, I; Ioan, S, J. Appl. Polym. Sci., 112, 18088-16, 2009.
Stabilizers	-	quaternization	Filimon, A; Avram, E; Dunca, S; toica, I; Ioan, S, J. Appl. Polym. Sci., 112, 18088-16, 2009.

TOXICITY

PARAMETER	UNIT	VALUE	REFERENCES
Carcinogenic effect	-	not listed by ACGIH, NIOSH, NTP	

PROCESSING

PARAMETER	UNIT	VALUE	REFERENCES
Typical processing methods	-	compression molding, electrospinning, extrusion, extrusion blow molding, injection molding, photolithography, thermoforming	
Preprocess drying: temperature/ time/residual moisture	°C/h/%	163/3 or 140-150/4 or 135/5; residual moisture for injection molding is 0.05% and for extrusion 0.01%	
Processing temperature	°C	350-390; 360-390 (10-30% glass fiber)	
Processing pressure	MPa	0.7-2.1 (back pressure)	
Process time	min	10-20 (residence time)	
Regrind	%	25	
Additives used in final products	-	Fillers: activated carbon, glass fiber, graphene oxide, carbon fiber, aramid fiber, montmorillonite, PTFE, silica, titanium dioxide; Plasticizers: benzyl butyl phthalate, diethyl phthalate, methyl phthalyl ethyl glycolate, tricresyl phosphate; Release: silicone oil, zinc stearate	
Applications	-	battery separator, faucet components, fibers, hot water fittings, medical applications which require resistance to hot water and sterilization, membranes (hemodialysis, water treatment, bioprocessing, food and beverage, and gas separation), membranes, microwave cookware, plumbing manifolds, printed circuit boards, tubing, solar hot water applications, utrafiltration membrane	
Outstanding properties	-	high heat deflection temperature, high strength	

BLENDS

PARAMETER	UNIT	VALUE	REFERENCES
Suitable polymers	-	cellulose derivatives, epoxy, PA6, PC, PDMS, PEG, PEI, PEO, PI, PPS, PPSU, PPy, PTFE, PVC, PVOH, PVDF	

ANALYSIS

PARAMETER	UNIT	VALUE	REFERENCES
FTIR (wavenumber-assignment)	cm^{-1}/-	sulfone – 1302, 1143; Ar–SO_2–Ar – 1151; Ar–O–Ar – 1242	Chen, L; Goh, Y-K; Lawrie, K; Chuang, Y; Piscani, E; Zimmerman, P; Blakey, I; Whittaker, A K, Radiation Phys. Chem., 80, 242-47, 2011; Dahe, G J; Teotia, R S; Kadam, S S; Bellare, J R, Biomaterials, 32, 352-65, 2011.

PSU polysulfone

PARAMETER	UNIT	VALUE	REFERENCES
NMR (chemical shifts)	ppm	phenyl ring – 4.52; sulfonyl group – 7.85; Ha, Hb, Hc, Hd and He of ^{1}H NMR spectrum are raised at δ = 7.24, δ = 7.86, δ = 6.94, δ = 7.01 and δ = 1.70 ppm, respectively.	Yilmaz, G; Toiserkani, H; Demirkol, D O; Sakarya, S; Timur, S; Torun, L; Yagci, Y, Mater. Sci., Eng., C31, 1091-97, 2011; Li, Y, Li, M, Zhou, S, Xue, A, Yang, D, Appl. Clay Sci., 195, 105702, 2020.

PTFE polytetrafluoroethylene

PARAMETER	UNIT	VALUE	REFERENCES
GENERAL			
Common name	-	polytetrafluoroethylene	
CAS name	-	ethene, 1,1,2,2-tetrafluoro-, homopolymer	
Acronym	-	PTFE	
CAS number	-	9002-84-0	
EC number	-	204-126-9	
RTECS number	-	KX4025000	
Formula		$\left[CF_2CF_2 \right]_n$	
HISTORY			
Person to discover	-	Roy Plunkett	Jones, R F; Antec, 2763-68, 1998.
Date	-	1938; 1945, 1946 (industrial production)	
Details	-	Plunkett accidentally discovered polymerization in experiment of production of new refrigerant (iron of container acted as a catalyst of polymerization); DuPont coined Teflon's name in 1945 and initiated industrial production in 1946; the first teflon-coated frying-pan was produced in 1961 by Marion Trozzolo	
SYNTHESIS			
Monomer(s) structure	-	$F_2C{=}CF_2$	
Monomer(s) CAS number(s)	-	116-14-3	
Monomer(s) molecular weight(s)	dalton, g/mol, amu	100.02	
Monomer ratio	-	100%	
Formulation example	-	monomer, water persulfate initiator, dispersant	
Method of synthesis	-	granular resin, water dispersions, and powdered resins are produced by free radical polymerization in aqueous medium; TFE polymerizes linearly without branching; micropowders are produced by irradiation of PTFE by high energy electron beam or polymerization controlled to produce lower molecular weight	Drobny, J G, Fluoroplastics, Rapra, 2006.
Temperature of polymerization	^oC	40-90	Ebnesajjad, S, Fluoroplastics. Vol. 1. Non-melt Processible Fluoro-plastics, William Andrew, 2000.
Pressure of polymerization	MPa	0.03-3.5	Ebnesajjad, S, Fluoroplastics. Vol. 1. Non-melt Processible Fluoro-plastics, William Andrew, 2000.
Heat of polymerization	kJ mol^{-1}	172	
Number average molecular weight, M_n	dalton, g/mol, amu	400,000-10,000,000	Ebnesajjad, S, Fluoroplastics. Vol. 1. Non-melt Processible Fluoro-plastics, William Andrew, 2000.
Molar volume at 298K	cm^3 mol^{-1}	41.2 (crystalline); 50 (amorphous)	
Van der Waals volume	cm^3 mol^{-1}	32.0 (crystalline); 32.0 (amorphous)	
STRUCTURE			
Crystallinity	%	58-98	Ebnesajjad, S, Fluoroplastics. Vol. 1. Non-melt Processible Fluoro-plastics, William Andrew, 2000.
Cell type (lattice)	-	triclinic (below 19^oC), hexagonal (above 19^oC)	Ebnesajjad, S, Fluoroplastics. Vol. 2. Melt Processible Fluoroplastics, William Andrew, 2003.

PTFE polytetrafluoroethylene

PARAMETER	UNIT	VALUE	REFERENCES
Cell dimensions	nm	a:b:c=0.559:0.559:1.688	
Unit cell angles	degree	$\alpha{:}\beta{:}\gamma$=90:90:119.3	
Number of chains per unit cell	-	1	
Chain conformation	-	helix 13/6	
Entanglement molecular weight	dalton, g/mol, amu	calc.=5,580, exp.=3,700	

COMMERCIAL POLYMERS

Some manufacturers	-	DuPont; Solvay	
Trade names	-	Teflon; Algoflon	

PHYSICAL PROPERTIES

Density at 20°C	g cm^{-3}	2.16-2.20; 2.077 (amorphous); 2.187 (crystalline); 2.344 (triclinic)	
Bulk density at 20°C	g cm^{-3}	0.3-0.5	
Color	-	white powder	
Refractive index, 20°C	-	1.35-1.37	
Odor	-	odorless	
Melting temperature, DSC	°C	317-345 (irreversible)	
Decomposition temperature	°C	350 (weight loss 0.001%); 508 (decomposition onset temperature)	Patel, P; Hull, T R; McCabe, R W; Flath, D; Grasmeder, J; Percy, M, Polym. Deg. Stab., 95, 709-18, 2010.
Degradation temperature	°C	440	
Continuous service temperature	°C	260	Drobny, J G, Applications of Fluoropolymer Films, William Andrew, 2020, pp. 147-50.
Thermal expansion coefficient, 23-80°C	°C^{-1}	1.1-2.2E-4	
Thermal conductivity, melt	W m^{-1} K^{-1}	0.234-0.25	Boudenne, A; Ibos, L; Gehin, E; Candau, Y, J. Phys. D: Appl. Phys., 37, 132-39, 2004.
Specific heat capacity	J K^{-1} kg^{-1}	1.2-1.5	
Heat of fusion	kJ kg^{-1}	82	Masirek, R; Piorkowska, E, Eur. Polym. J., 46, 1436-45, 2010.
Maximum service temperature	°C	-260 to 260	
Long term service temperature	°C	260	
Heat deflection temperature at 0.45 MPa	°C	122-132	
Heat deflection temperature at 1.8 MPa	°C	45-60	
Hansen solubility parameters, δ_D, δ_P, δ_H	MPa$^{0.5}$	16.2, 1.8, 3.4	
Interaction radius		3.9	
Hildebrand solubility parameter	MPa$^{0.5}$	12.7; 16.7	
Surface tension	mN m^{-1}	18.5-25.6	Wu, S, J. Macromol. Sci., C10, 1, 1974.
Dielectric constant at 100 Hz/1 MHz	-	2.1/2.1	
Dissipation factor at 100 Hz		0.0003	
Dissipation factor at 1 MHz		0.0003	

PTFE polytetrafluoroethylene

PARAMETER	UNIT	VALUE	REFERENCES
Volume resistivity	ohm-m	1E16	
Surface resistivity	ohm	1E16	
Electric strength K20/P50, d=0.60.8 mm	kV mm^{-1}	19.7-60	
Surface arc resistance	s	>300	
Coefficient of friction	-	0.08 (static); 0.06 (dynamic); 0.24-0.31 (glass fiber); 0.20-0.24 (carbon fiber)	
Permeability to nitrogen, 25°C	cm^3 cm cm^{-2} s^{-1} Pa^{-1} x 10^{12}	0.1	
Permeability to oxygen, 25°C	cm^3 cm cm^{-2} s^{-1} Pa^{-1} x 10^{12}	0.32	
Permeability to water vapor, 25°C	cm^3 cm cm^{-2} s^{-1} Pa^{-1} x 10^{12}	0.6	
Diffusion coefficient of oxygen	cm^2 s^{-1} x10^6	0.0152	
Diffusion coefficient of water vapor	cm^2 s^{-1} x10^7	1.47 (20°C); 5.73 (90°C)	Hansen, C M, Prog. Org. Coat., 42, 167-78, 2001.
Contact angle of water, 20°C	degree	108.9-120; 122 (adv) and 94(rec); 150	Lee, S; Park, J-S; Lee, T R, Langmuir, 24, 4817-26, 2008; Zhu, Y, Shen, C, Li, J, Shi, J, Mater. Chem. Phys., 257, 123828, 2021.

MECHANICAL & RHEOLOGICAL PROPERTIES

PARAMETER	UNIT	VALUE	REFERENCES
Tensile strength	MPa	20-35; 19 (20% glass fiber); 13 (25% carbon fiber); 24.1-29.6 (film)	-; -; Drobny, J G, Applications of Fluoropolymer Films, William Andrew, 2020, pp. 147-50.
Tensile modulus	MPa	400-550; 250 (20% glass fiber); 120 (25% carbon fiber)	
Elongation	%	300-550	
Flexural modulus	MPa	340-620; 1,200 (20% glass fiber); 1,100 (25% carbon fiber)	
Elastic modulus	MPa	482	
Young's modulus	MPa	425	Parthasarathi, N L, Borah, U, Davinci, M A, Albert, S K, Mater. Today: Proc., in press, 2020.
Compressive strength	MPa	47.7	Parthasarathi, N L, Borah, U, Davinci, M A, Albert, S K, Mater. Today: Proc., in press, 2020.
Izod impact strength, notched, 23°C	J m^{-1}	188	
Shear modulus	MPa	186	
Tenacity (fiber) (standard atmosphere)	cN tex^{-1} (daN mm^{-2})	5-14 (10-28)	Fourne, F, Synthetic Fibers. Machines and Equipment Manufacture, Properties. Carl Hanser Verlag, 1999.
Tenacity (wet fiber, as % of dry strength)	%	100	Fourne, F, Synthetic Fibers. Machines and Equipment Manufacture, Properties. Carl Hanser Verlag, 1999.

PTFE polytetrafluoroethylene

PARAMETER	UNIT	VALUE	REFERENCES
Fineness of fiber (titer)	dtex	5-25	Fourne, F, Synthetic Fibers. Machines and Equipment Manufacture, Properties. Carl Hanser Verlag, 1999.
Length (elemental fiber)	mm	filament, staple	Fourne, F, Synthetic Fibers. Machines and Equipment Manufacture, Properties. Carl Hanser Verlag, 1999.
Poisson's ratio	-	0.46-0.5	
Shrinkage	%	2-10	
Melt viscosity, shear rate=1000 s^{-1}	Pa s	1E10	
Water absorption, equilibrium in water at 23°C	%	0	

CHEMICAL RESISTANCE			
Acid dilute/concentrated	-	resistant (including fuming nitric acid, and *aqua regia*)	
Alcohols	-	resistant	
Alkalis	-	resistant (attacked by molten alkali metals)	
Aliphatic hydrocarbons	-	resistant	
Aromatic hydrocarbons	-	resistant	
Esters	-	resistant	
Greases & oils	-	resistant	
Halogenated hydrocarbons	-	resistant	
Ketones	-	resistant	
Good solvent	-	perfluorokerosene at 350°C	
Non-solvent	-	all other solvents	

FLAMMABILITY			
Ignition temperature	°C	494	
Autoignition temperature	°C	>530	
Limiting oxygen index	$\% \ O_2$	>99.5	
Burning rate (Flame spread rate)	mm min^{-1}	120	Padey, D; Walling, J; Wood A, Polymers in Defence and Aerospace 2007, Rapra, 2007, paper 15.
Char at 500°C	%	0	Lyon, R E; Walters, R N, J. Anal. Appl. Pyrolysis, 71, 27-46, 2004.
Heat of combustion	$J \ g^{-1}$	6,3806,680	Walters, R N; Hacket, S M; Lyon, R E, Fire Mater., 24, 5, 245-52, 2000.
Volatile products of combustion	-	CF_4, C_2F_4, C_3F_6	
UL 94 rating	-	V-0	

WEATHER STABILITY			
Spectral sensitivity	nm	well below 290	
Stabilizers	-	not known	
Low earth orbit erosion yield	cm^3 $atom^{-1}$ x 10^{-24}	0.142	Waters, D L; Banks, B A; De Groh, K K; Miller, S K R; Thorson, S D, High Performance Polym., 20, 512-22, 2008.

PTFE polytetrafluoroethylene

PARAMETER	UNIT	VALUE	REFERENCES
Threshold value of absorbed dose for vacuum degradation	kGy	50-60	Parthasarathi, N L, Borah, U, Davinci, M A, Albert, S K, Mater. Today: Proc., in press, 2020.

TOXICITY

PARAMETER	UNIT	VALUE	REFERENCES
NFPA: Health, Flammability, Reactivity rating	-	0/0/0	
Carcinogenic effect	-	not listed by ACGIH, NIOSH, NTP	
Mutagenic effect	-	not known	
Teratogenic effect	-	not known	
Reproductive toxicity	-	not known	
Oral rat, LD_{50}	mg kg^{-1}	1,250	

ENVIRONMENTAL IMPACT

PARAMETER	UNIT	VALUE	REFERENCES
Aquatic toxicity, *Daphnia magna*, LC_{50}, 48 h	mg l^{-1}	>1000	

PROCESSING

PARAMETER	UNIT	VALUE	REFERENCES
Typical processing methods	-	3D micro-printing, casting, coating, compression molding, dip coating, film coating, fiber spinning, flow coating, isostatic molding, paste extrusion (mixed with lubricants and forced through cold die followed by lubricant evaporation and sintering), ram extrusion, sintering, solid phase forming, spray coating, spraying	Ebnesajjad, S, Fluoroplastics. Vol. 1. Non-melt Processible Fluoroplastics, William Andrew, 2000; Zhang, Y, Yin, M-J, Ouyang, X, Zhang, A P, Tam, H-Y, Appl. Mater. Today, 19, 100580, 2020.
Processing temperature	°C	400	
Additives used in final products	-	alumina, attapulgite, boron nitride, bronze powder, carbon black, carbon fiber, carbon nanofiber, copper powder, diamond, glass fiber, graphite, molybdenum sulfide, Ni-Zn ferrite, silica, titanium dioxide	
Applications	-	aesthetic correction, aircraft insulated wires, arteriovenous grafts, bearings, dry lubricants, electric insulation applications, film, filtration membranes, friction reduction, fuel cells, gaskets, Gore-Tex™ membranes, laboratory equipment, Li-ion batteries, non-stick coatings, oils and greases, paints and coatings, pipes, piston rings, printing inks, seals, sutures, tank lining, tapes, tubes, valve and pump parts, vascular grafts, wear reduction	Ebnesajjad, S, Fluoroplastics. Vol. 1. Non-melt Processible Fluoroplastics, William Andrew, 2000.
Outstanding properties	-	chemical inertness, heat resistance, low coefficient of friction, insulating properties	

BLENDS

PARAMETER	UNIT	VALUE	REFERENCES
Suitable polymers	-	FEP, PA6, PA66, PEEK, PI, PLA, POM, PPS, PVDF	

ANALYSIS

PARAMETER	UNIT	VALUE	REFERENCES
NMR (chemical shifts)	ppm	CF_2 – 122 (commercial only peak); branching – 130-138	Ikeda, S; Tabata, Y; Suzuki, H; Miyoshi, T; Kudo, H; Katsumura, Y, Radiation Phys. Chem., 77, 1050-56, 2008.
x-ray diffraction peaks	degree	18.6, 31.3	Gao, Y; Zhang, J; Xu, j; Yu, J, Appl. Surf. Sci., 254, 3408-11, 2008.

PTFE-AF poly(tetrafluoroethylene-co-2,2-bis(trifluoromethyl)-4,5-difluoro-1,3-dioxole)

PARAMETER	UNIT	VALUE	REFERENCES
GENERAL			
Common name	-	poly(tetrafluoroethylene-co-2,2-bis(trifluoromethyl)-4,5-difluoro-1,3-dioxole), Teflon AF	
ACS name	-	poly[4,5-difluoro-2,2-bis(trifluoromethyl)-1,3-dioxole-co-tetrafluoroethylene]	
Acronym	-	PTFE-AF	
CAS number	-	37626-13-4; 187475-17-8	
SYNTHESIS			
Monomer(s) structure	-		
Monomer(s) CAS number(s)	-	116-14-3; 37697-64-6	
Monomer(s) molecular weight(s)	dalton, g/mol, amu	100.02; 244.04	
Initiator		bis(perfluoro-2-N-propoxypropionyl) peroxide was used as initiator	Michel, U; Resnik, P; Kipp, B; DeSimone, J M, Macromolecules, 36, 19, 7107-13, 2003.
Yield	%	74	Michel, U; Resnik, P; Kipp, B; DeSimone, J M, Macromolecules, 36, 19, 7107-13, 2003.
STRUCTURE			
Crystallinity	%	amorphous	
COMMERCIAL POLYMERS			
Some manufacturers	-	DuPont	
Trade names	-	Teflon AF	
PHYSICAL PROPERTIES			
Density at 20°C	g cm^{-3}	1.67-1.78	
Refractive index, 20°C	-	1.29-1.31	
Transmittance	%	>95	
Decomposition temperature	°C	360	
Thermal expansion coefficient, 23-80°C	°C^{-1}	2.6-3E-4	
Glass transition temperature	°C	160-240; 334 (for PDD homopolymer)	
Heat deflection temperature at 0.45 MPa	°C	156-200	
Heat deflection temperature at 1.8 MPa	°C	154-174	
Dielectric constant at 100 Hz/1 MHz	-	1.89-1.93 (lowest of any plastic material)	
Dissipation factor at 100 Hz	E-4	1-3	
Dissipation factor at 1 MHz	E-4	1-3	
Electric strength K20/P50, d=0.60.8 mm	kV mm^{-1}	19-21	
Permeability to nitrogen, 25°C	barrer	130-490	
Permeability to oxygen, 25°C	barrer	340-990	

PTFE-AF poly(tetrafluoroethylene-co-2,2-bis(trifluoromethyl)-4,5-difluoro-1,3-dioxole)

PARAMETER	UNIT	VALUE	REFERENCES
Permeability to water vapor, 25°C	barrer	1142-4026	
Contact angle of water, 20°C	degree	104-105	

MECHANICAL & RHEOLOGICAL PROPERTIES

Tensile strength	MPa	26.4-26.9	
Tensile modulus	MPa	1,500-1,600	
Tensile stress at yield	MPa	26.4-27.4	
Elongation	%	7.9-17.1	
Flexural modulus	MPa	1,600-1,800	
Shore D hardness	-	75-77	
Rockwell hardness	-	97.5-103	
Melt viscosity, shear rate=100 s^{-1}	Pa s	2,657 (250°C); 540 (350°C)	
Water absorption, equilibrium in water at 23°C	%	<0.01	

CHEMICAL RESISTANCE

Alcohols	-	very good	
Aliphatic hydrocarbons	-	very good	
Aromatic hydrocarbons	-	very good	
Esters	-	very good	
Greases & oils	-	very good	
Halogenated hydrocarbons	-	very good	
Ketones	-	very good	
Good solvent	-	perfluoromethylcyclohexane, perfluorobenzene, and perfluorodecalin	

FLAMMABILITY

Volatile products of combustion	-	HF, COF_2, CO, HFA	

WEATHER STABILITY

Spectral sensitivity	nm	transparent to solar UV	
Activation wavelengths	nm	157	
Products of degradation	-	hexafluoroacetone, main chain and chain-end radicals, char formation	Blakey, I; George, G A; Hill, D J T; Liu, H; Rasoul, F; Rintoul, L; Zimmerman, P; Whittaker, A K, Macromolecules, 40, 25, 8954-61, 2007.

PROCESSING

Typical processing methods	-	compression molding, extrusion, injection molding	
Processing temperature	°C	240-275 (1600 range) and 340-360 (2400 range)	
Applications	-	microelectronics, optics, clear coatings, semiconductors, dielectric materials, release materials, fiber optics, implantable medical devices, photolithography	
Outstanding properties	-	optical clarity, lowest dielectric constant	

PTMG poly(tetramethylene glycol)

PARAMETER	UNIT	VALUE	REFERENCES
GENERAL			
Common name	-	poly(tetramethylene glycol)	
CAS name	-	poly(oxy-1,4-butanediyl), alpha-hydro-gamma-hydroxy-; glycols, polytetramethylene	
Acronym	-	PTMG	
CAS number	-	25190-06-1	
RTECS number	-	MD0916000	
Formula	-	$\left[OCH_2CH_2CH_2CH_2\right]_n$	
SYNTHESIS			
Monomer(s) structure	-		
Monomer(s) CAS number(s)	-	109-99-9	
Monomer(s) molecular weight(s)	dalton, g/mol, amu	72.11	
Monomer ratio	-	100%	
Formulation example	-	reaction mixture comprising tetrahydrofuran, catalyst, accelerator, and ionic liquid	Harmer, M A; Junk, C P; Manzer, L E, US Patent 7,402,711, DuPont, 2008.
Temperature of polymerization	ºC	0-70	
Catalyst	-	acid	
Number average molecular weight, M_n	dalton, g/mol, amu	248-4,462	
Mass average molecular weight, M_w	dalton, g/mol, amu	400-20,000	
Polydispersity, M_w/M_n	-	1.8-4	
Polymerization degree (number of monomer units)	-	3-62	
Molar volume at 298K	cm^3 mol^{-1}	calc.=61.0 (crystalline); 73.5 (amorphous)	
Van der Waals volume	cm^3 mol^{-1}	calc.=44.6 (crystalline); 44.6 (amorphous)	
COMMERCIAL POLYMERS			
Some manufacturers	-	LyondellBasell; Invista	
Trade names	-	Polymeg, Terathane	
PHYSICAL PROPERTIES			
Density at 20ºC	$g\ cm^{-3}$	0.972-0.987	
Bulk density at 20ºC	$g\ cm^{-3}$	0.973-0.982	
Color	-	colorless	
Odor		odorless	
Melting temperature, DSC	ºC	19-55	
Decomposition temperature	ºC	210-240	
Storage temperature	ºC	20 to <95; stable for 1 year at 65ºC under nitrogen	
Glass transition temperature	ºC	-74 to -85	

PTMG poly(tetramethylene glycol)

PARAMETER	UNIT	VALUE	REFERENCES
Surface free energy	mJ m^{-2}	31.9	

MECHANICAL & RHEOLOGICAL PROPERTIES

Melt viscosity, shear rate=1000 s^{-1}	Pa s	0.4-1.27	
Water absorption, equilibrium in water at 23°C	%	2	

CHEMICAL RESISTANCE

Alcohols	-	soluble	
Aliphatic hydrocarbons	-	insoluble	
Aromatic hydrocarbons	-	soluble	
Esters	-	soluble	
Halogenated hydrocarbons	-	soluble	
Ketones	-	soluble	

FLAMMABILITY

Ignition temperature	°C	193-246	
Autoignition temperature	°C	>245	
Activation energy of thermal degradation	kJ mol^{-1}	60-70	Tago, K; Tsuchiya, M; Gondo, Y; Ishimaru, K; Kojima, T, J. Appl. Polym. Sci., 77, 1538-44, 2000.
Volatile products of combustion	-	CO, CO_2, tetrahydrofuran	

BIODEGRADATION

Typical biodegradants	-	fungus resistance	

TOXICITY

NFPA: Health, Flammability, Reactivity rating	-	1/1/0	
Carcinogenic effect	-	not listed by ACGIH, NIOSH, NTP	
Oral rat, LD$_{50}$	mg kg^{-1}	>5,000 to 18,830	
Skin rabbit, LD$_{50}$	mg kg^{-1}	8,375-10,250	

ENVIRONMENTAL IMPACT

Aquatic toxicity, *Fathead minnow*, LC$_{50}$, 48 h	mg l^{-1}	>1,000 to >2,000	
Biological oxygen demand, BOD$_{32}$	% ThOD	10-50	

PROCESSING

Applications	-	coatings, plasticizer for zein films, polyurethanes, spandex	Kang, K S, Jee, C, Bae, J-H, Jung, H J, Huh, P, Prog. Org. Coat., 123, 238-41, 2018.
Outstanding properties	-	products using PTMG have: low temperature flexibility, tear strength, abrasion resistance, and good hydrolytic stability	

BLENDS

Suitable polymers	-	PVC	

PTT poly(trimethylene terephthalate)

PARAMETER	UNIT	VALUE	REFERENCES
GENERAL			
Common name	-	poly(trimethylene terephthalate)	
CAS name	-	poly(oxy-1,3-propanediyloxycarbonyl-1,4-phenylenecarbonyl)	
Acronym	-	PTT	
CAS number	-	26546-03-2; 26590-75-0; 36619-23-5; 9022-20-2	
Formula			
HISTORY			
Person to discover	-	John Rex Whinfield and James Tennant Dickson	
Date	-	1941; 1998	
Details	-	patent issued for Calico Printing Ink was never used because of lack of low cost monomers; in 1991 Shell developed production of PDO; polymer was commercialized in 1998	
SYNTHESIS			
Monomer(s) structure	-		
Monomer(s) CAS number(s)	-	504-63-2; 100-21-0	
Monomer(s) molecular weight(s)	dalton, g/mol, amu	76.09; 166.13	
Monomer ratio	-	TPA/PDO=1/1.2	
Method of synthesis	-	esterification in the presence of catalyst	
Temperature of polymerization	°C	260	
Time of polymerization	h	3-4	
Pressure of polymerization	Pa	5	
Catalyst	-	$Ti(OC_4H_9)_4$	
Yield	%	80	
Number average molecular weight, M_n	dalton, g/mol, amu	17,300-28,000	
Mass average molecular weight, M_w	dalton, g/mol, amu	10,000-56,000	
STRUCTURE			
Crystallinity	%	15-45	
Cell dimensions	nm	triclinic	
Unit cell angles	degree	a:b:c=0.4637:0.6226:1.864; a:b:c=0.46:0.62:1.83	Vasanthan, N; Yaman, M, J. Polym. Sci. B, 45, 1675-82, 2007; Yun, J H; Kuboyama, K; Chiba, T; Ougizawa, T, Polymer, 47, 4831-38, 2006.
Number of chains per unit cell	-	α:β:γ=98.4:93:111.5	
Chain conformation	-	zigzag (methylene groups in *gauche* conformations)	
Entanglement molecular weight	dalton, g/mol, amu	1175-2100	Chuah, H H, J. Polym. Sci. B, 40, 1513-20, 2002.
Lamellae thickness	nm	4-5	Chuang, W-T; Hong, P-D; Chuah, H H, Polymer, 45, 2413-25, 2004.

PTT poly(trimethylene terephthalate)

PARAMETER	UNIT	VALUE	REFERENCES
Crystallization temperature	°C	152	Zhang, J. J. Appl. Polym. Sci., 91, 1657-66, 2004.
Avrami constants, k/n	-	n=2.41-3.45	Xu, Y; Jia, H-b; Ye, S-r; Huang, J, e-Polymers, 006, 1-7, 2008; Wang, Y; Liu, W; Zhang, H, Polym. Test., 28, 402-11, 2009.

COMMERCIAL POLYMERS

Some manufacturers	-	DuPont; Shell; RTP	
Trade names	-	Sorona; Corterra; PTT	

PHYSICAL PROPERTIES

Density at 20°C	g cm⁻³	1.33-1.35; 1.432 (crystalline); 1.295 (amorphous)	Yun, J H; Kuboyama, K; Chiba, T; Ougizawa, T, Polymer, 47, 4831-38, 2006.
Color	-	white	
Refractive index, 20°C	-	1.60-1.62; 1.636 (uniaxial orientation)	
Birefringence	-	0.029- 0.06	Chuah, H H, J. Polym. Sci. B, 40, 1513-20, 2002; Yun, J, H; Kuboyama, K; Ougizawa, T, Polymer, 47, 1715-21, 2006.
Melting temperature, DSC	°C	226-233	Xue, M-L; Yu, Y-L; Rhee, J M; Kim, N H; Lee, J H, Eur. Polym. J., 43, 9, 3826-37, 2007.
Decomposition temperature	°C	265; 374 (by TGA)	Liu, J; Bian, S G; Xiao, M; Wang, S J; Meng, Y Z, Chin. Chem. Lett., 20, 487-91, 2009.
Thermal expansion coefficient, -40 to 160°C	°C⁻¹	0.25-1.32E-4	
Glass transition temperature	°C	40-75; 40 (DSC), 55 (DMA)	Xue, M-L; Yu, Y-L; Rhee, J M; Kim, N H; Lee, J H, Eur. Polym. J., 43, 9, 3826-37, 2007.
Heat of fusion	J g⁻¹	60.3-145.63	
Enthalpy of melting of hard segment	J g⁻¹	43.2	Yao, C; Yang, G, Polymer, 51, 1516-23, 2010.
Electrical conductivity	S m⁻¹	3.01E-9	Braga, N F, LaChance, A M, Liu, B, Passador, F R, Adv. Ind. Eng. Polym. Res., 2, 3, 121-5, 2019.

MECHANICAL & RHEOLOGICAL PROPERTIES

Tensile strength	MPa	50-65; 110-165 (15-30% glass fiber); 72.5	-; Braga, N F, LaChance, A M, Liu, B, Passador, F R, Adv. Ind. Eng. Polym. Res., 2, 3, 121-5, 2019.
Tensile modulus	MPa	2,400-2,550; 6,200-11,000 (15-30% glass fiber)	
Tensile stress at yield	MPa	60	
Elongation	%	10-15; 2.0-3 (15-30% glass fiber); 36-42 (fiber)	
Tensile yield strain	%	5.5-6	
Flexural strength	MPa	84-103; 170-245 (15-30% glass fiber)	
Flexural modulus	MPa	2,400-2,800; 5,700-9,600 (15-30% glass fiber)	
Young's modulus	MPa	1,560	Braga, N F, LaChance, A M, Liu, B, Passador, F R, Adv. Ind. Eng. Polym. Res., 2, 3, 121-5, 2019.
Toughness	kJ m⁻²	1.9	Braga, N F, LaChance, A M, Liu, B, Passador, F R, Adv. Ind. Eng. Polym. Res., 2, 3, 121-5, 2019.
Charpy impact strength, unnotched, 23°C	kJ m⁻²	25-50 (15-30% glass fiber)	

PTT poly(trimethylene terephthalate)

PARAMETER	UNIT	VALUE	REFERENCES
Charpy impact strength, unnotched, -30°C	kJ m^{-2}	30-45 (15-30% glass fiber)	
Charpy impact strength, notched, 23°C	kJ m^{-2}	4-5; 5-9 (15-30% glass fiber)	
Charpy impact strength, notched, -30°C	kJ m^{-2}	6-9 (15-30% glass fiber)	
Izod impact strength, unnotched, 23°C	J m^{-1}	214	
Izod impact strength, notched, 23°C	J m^{-1}	27-57	
Tenacity (fiber)	cN/dtex	3.4-3.7	
Poisson's ratio	-	0.35 (15-30% glass fiber)	
Shrinkage	%	1-1.3; 0.3-0.8 (15-30% glass fiber)	
Intrinsic viscosity, 25°C	dl g^{-1}	0.56-0.94	
Melt viscosity, shear rate=1000 s^{-1}	Pa s	85	Xue, M-L; Yu, Y-L; Rhee, J M; Kim, N H; Lee, J H, Eur. Polym. J., 43, 9, 3826-37, 2007.
Melt index, 250°C/2.16 kg	g/10 min	35	
Water absorption, equilibrium in water at 23°C	%	0.2-0.4	

CHEMICAL RESISTANCE			
Alcohols	-	very good	
Aliphatic hydrocarbons	-	very good	
Aromatic hydrocarbons	-	very good	
Esters	-	good	
Greases & oils	-	very good	
Halogenated hydrocarbons	-	poor	
Good solvent	-	tetrachloroethane/phenol, hexafluoroisopropanol, trifluoroacetic acid/methylene chloride, phenol/tetrachloroethane	Chuah, H H; Lin-Vien, D; Soni, U, Polymer, 42, 7137-39, 2001.

FLAMMABILITY			
Autoignition temperature	°C	>300	
Activation energy of thermal decomposition	kJ mol^{-1}	192-201	
Volatile products of combustion	-	acrolein, allyl alcohol, CO, CO_2, ethanol, methanol, acetic acid, acetone	

TOXICITY			
NFPA: Health, Flammability, Reactivity rating	-	0/1/0	
Carcinogenic effect	-	not listed by ACGIH, NIOSH, NTP	
OSHA	mg m^{-3}	5 (respirable), 15 (total)	
Oral rat, LD$_{50}$	mg kg^{-1}	>5,000	

PROCESSING			
Typical processing methods	-	compounding, electrospinning, injection molding	

PTT poly(trimethylene terephthalate)

PARAMETER	UNIT	VALUE	REFERENCES
Preprocess drying: temperature/ time/residual moisture	°C/h/%	88-120/2-6/0.01-0.02	
Processing temperature	°C	250-270	
Processing pressure	MPa	<0.3 (back pressure); 69-103 (injection)	
Applications	-	fibers, film, engineering thermoplastics	
Outstanding properties	-	35% renewably sourced ingredients; high elasticity, excellent recovery rate, stain resistance, UV stability, partially from renewable resources	

BLENDS

Second polymer	-	ABS, LLDPE, PC, PEI, PEN, PEO, PET, PLA, PP	

ANALYSIS

FTIR (wavenumber-assignment)	cm^{-1}/-	C=O – 1720; O–H – 1268; C-O-C – 1103	Xue, C H; Wang, D; Xiang, B; Chiou, b-S; Sun, G, Mater. Chem. Phys., 124, 48-51, 2010.
NMR (chemical shifts)	ppm	allyl end-group – 5.40-5.48; butenyl end-group – 5.08-5.25	Kelsey, D R; Kiibler, K S; Tutunjian, P N, Polymer, 46, 8937-46, 2005.

PU polyurethane

PARAMETER	UNIT	VALUE	REFERENCES
GENERAL			
Common name	-	polyurethane	
IUPAC name	-	e.g., polyurea: poly[ureylene(2-methyl-1,3-phenylene)ureylenehexane-1,6-diyl]; polyurethane: poly{oxycarbonylimino(4-methyl-1,3-phenylene)iminocarbonyl[poly(oxyethylene)]}	
Acronym	-	PU; PUR (IUPAC)	
CAS number	-	9009-54-5, 75701-44-9	
HISTORY			
Person to discover	-	Otto Bayer; Scholleberger, C S,	
Date	-	1937, 1954; 1971, 1972	
Details	-	Otto Bayer discovered polyurethane reaction in IG Farben; commercial production begun in 1954 in Bayer AG Thermoplastic polyurethane published and patented	
SYNTHESIS			
Monomer(s) structure	-	diols, triol, tetrafunctional polyols; 1-3 functional isocyanates	
Curatives		for polyols, isocyanates or isocyanurates play a role of curatives; prepolymers are cured either by polyols (frequently multifunctional to obtain tridimensional networks) or by amines; considering that amines are very reactive with isocyanate groups, amines are frequently used in a blocked form of ketimines and aldomines which require moisture to hydrolyze them to free reactive amines and this slows curing process up to the extent that frequently catalytic systems have to be used to bring reaction to a required rate	
Monomer(s) CAS number(s)	-	too many used to make their listing practical	
Monomer(s) molecular weight(s)	dalton, g/mol, amu	400-8,400 (polyols)	
Monomer ratio	-	+-5% stoichiometric	
Formulation example	-	polyol, isocyanate, catalyst	
Hydroxyl number	mg/g KOH	14-865	
NCO content	%	23-48	
Method of synthesis	-	components of formulation are mixed and formed to shape as soon as possible; catalyst selection is part of design to achieve required rate of polycondensation; in many cases prepolymers are produced first (prepolymers are low molecular products containing most frequently one molecule of polyol and two or more molecules of isocyanate)	
Temperature of polymerization	°C	room	
Time of polymerization	s	60-1,500	
Pressure of polymerization	Pa	usually atmospheric	
Catalyst	-	tin catalysts are the most popular in the case of prepolymer synthesis (polyureas), and amines in the case of polycondensation of polyol and isocyanate (polyurethane)	Abdel Hakim, A A; Nassar, M; Emam, A; Sultan, M, Mater. Chem. Phys., 129, 301-7, 2011.
Yield	%	close to 100%	
Typical concentration of residual monomer	ppm	full conversion; if isocyanate would not react with polyol component or curative, it will react with water	Spirkova, M; Pavlicevic, J; Strachota, A; Poremba, R; Bera, O; Kapralkova, L; Baldrian, J; Slouf, M; Lazic, N; Budinski-Simendic, J, Eur. Polym. J., 47, 959-72, 2011.

632   **HANDBOOK OF POLYMERS** 3rd Edition, Copyrights 2022; ChemTec Publishing

PU polyurethane

PARAMETER	UNIT	VALUE	REFERENCES
Number average molecular weight, M_n	dalton, g/mol, amu	3,317-20,377 (thermochromic blocked waterborne PU); 190,000-322,000 (switchable polyurethanes with reversible oxime-carbamate bonds)	Ji, X, Liu, J, Zhang, W, Liu, W, Wang, C, Prog. Org. Coat., 145, 105164, 2020; Qiao, Z, Yang, Z, Liu, W, Xu, J, Chem. Eng. J., 384, 123287, 2020.
Mass average molecular weight, M_w	dalton, g/mol, amu	4,428-33,788 (thermochromic blocked waterborne PU); 47,900-366,800 (thermoplastic polyurethane elastomers)	Ji, X, Liu, J, Zhang, W, Liu, W, Wang, C, Prog. Org. Coat., 145, 105164, 2020; Xiao, S, Sue, H-J, Polymer, 169, 124-30, 2019.
Polydispersity, M_w/M_n	-	1.34-1.66 (thermochromic blocked waterborne PU); 1.32-1.7 (switchable polyurethanes with reversible oxime-carbamate bonds); 1.77-1.99 (thermoplastic polyurethane elastomers)	Ji, X, Liu, J, Zhang, W, Liu, W, Wang, C, Prog. Org. Coat., 145, 105164, 2020; Xiao, S, Sue, H-J, Polymer, 169, 124-30, 2019.
Crosslink density	g mol cm^{-3} x 10^4	1.5-4; 0.016-1.178	Amrollahi, M; Sadeghi, M M; Kashcooli, Y, Mater. Design, 32, 3933-41, 2011; Trzebiatowska, P J, Echart, A S, Correas, T C, Datta, J, Prog. Org. Coat., 115, 41-8, 2018.
Molecular weight between crosslinks	g mol^{-1}	3,200-6,800; 8,486-603,542	Amrollahi, M; Sadeghi, M M; Kashcooli, Y, Mater. Design, 32, 3933-41, 2011; Trzebiatowska, P J, Echart, A S, Correas, T C, Datta, J, Prog. Org. Coat., 115, 41-8, 2018.
Radius of gyration	nm	1.2; 6.89-7.0 (waterborne biodegradable polyurethane)	Yang, H; Li, Z-s; Lu, Z-y; Sun, C-c, Polymer, 45, 6753-59, 2004 Wen, C-H Hsu, S-C, Hsu, S-h, Chang, S-W, Computational Struct. Biotech. J., 17, 110-17, 2019.

STRUCTURE			
Crystallinity	%	0-13; 11 (waterborne PU PCL-based); 7.1-8.3 (thermoplastic polyurethane elastomers)	Spirkova, M; Pavlicevic, J; Strachota, A; Poremba, R; Bera, O; Kapralkova, L; Baldrian, J; Slouf, M; Lazic, N; Budinski-Simendic, J, Eur. Polym. J., 47, 959-72, 2011; Valčić, M D; Cakić, S M; Ristić, I S, János, C J, Int. J. Adhesion Adhesives, 104, 102738, 2021; Xiao, S, Sue, H-J, Polymer, 169, 124-30, 2019.
Cell type (lattice)	-	triclinic	Petrovic, Z S; Ferguson, J, Prog. Polym. Sci., 16, 695-836, 1991.
Cell dimensions	nm	a:b:c=0.492:0.566:3.835 (MDI/BD) (III); a:b:c=0.96:0.56:3.68 (12-PUR)	Petrovic, Z S; Ferguson, J, Prog. Polym. Sci., 16, 695-836, 1991; Fernandez, C E; Bermudez, M; Versteegen, R M; Meijer, E W; Vancso, G J; Munoz-Guerra, S, Eur. Polym. J., 46, 2089-98, 2010.
Unit cell angles	degree	$\alpha{:}\beta{:}\gamma$=124:104.5:86 (MDI/BD) (III); $\alpha{:}\beta{:}\gamma$= 47.3:90:70.9	Petrovic, Z S; Ferguson, J, Prog. Polym. Sci., 16, 695-836, 1991; Fernandez, C E; Bermudez, M; Versteegen, R M; Meijer, E W; Vancso, G J; Munoz-Guerra, S, Eur. Polym. J., 46, 2089-98, 2010.
Number of chains per unit cell	-	2	
Polymorphs	-	I (quiescent crystallization), II (quiescent crystallization), III (orientation)	
Lamellae thickness	nm	21.9; 8-9	Yang, H; Li, Z-s; Lu, Z-y; Sun, C-c, Polymer, 45, 6753-59, 2004; Fernandez, C E; Bermudez, M; Versteegen, R M; Meijer, E W; Vancso, G J; Munoz-Guerra, S, Eur. Polym. J., 46, 2089-98, 2010.

PU polyurethane

PARAMETER	UNIT	VALUE	REFERENCES
Avrami constants, k/n	-	n=3.4-3.9	Fernandez, C E; Bermudez, M; Versteegen, R M; Meijer, E W; Vancso, G J; Munoz-Guerra, S, Eur. Polym. J., 46, 2089-98, 2010.

COMMERCIAL POLYMERS			
Some manufacturers	-	Covestro, DOW	
Trade names of polyols	-	Acclaim, Arcol, Baycoll, Desmophen, Hyperlite, Ultracel, Voranol, Vorapel, Voraguard	
Trade names of isocyanates	-	Baydur, Baymidur, Desmodur, Papi	
Trade names of prepolymers	-	Desmocap, Desmoseal, Desmotherm, Echelon	

PHYSICAL PROPERTIES			
Density at 20°C	g cm^{-3}	1.10-1.25; 1.322 (fully crystalline)	
Color	-	off white to yellow	
Odor	-	none	
Melting temperature, DSC	°C	141-157	Fernandez, C E; Bermudez, M; Versteegen, R M; Meijer, E W; Vancso, G J; Munoz-Guerra, S, Eur. Polym. J., 46, 2089-98, 2010.
Decomposition temperature	°C	120-126	
Thermal expansion coefficient, 23-80°C	10^{-4} °C^{-1}	1.8	
Thermal conductivity, melt	W m^{-1} K^{-1}	0.13	
Glass transition temperature	°C	-60.3 to -19	Sadeghi, M; Semsarzadeh, M A; Barikani, M; Chenar, M P, J. Membrane, Sci., 376, 188-95, 2011.
Maximum service temperature	°C	80-82	
Heat deflection temperature at 0.45 MPa	°C	136-252	
Heat deflection temperature at 1.8 MPa	°C	123-232	
Enthalpy of crystallization	J g^{-1}	32.9-72.5	
Hansen solubility parameters, δ_D, δ_P, δ_H	MPa$^{0.5}$	18.8, 10.0, 8.2; 17.5-18.1, 2.7-4.9, 7.7-11.3 (segmented thermoplastic polyurethanes)	-; Gallu, R, Méchin, F, Dalmas, F, Loup, F, Polymer, 207, 122882, 2020.
Sphere radius	MPa$^{0.5}$	3.7-7.3 (segmented thermoplastic polyurethanes)	Gallu, R, Méchin, F, Dalmas, F, Loup, F, Polymer, 207, 122882, 2020.
Hildebrand solubility parameter	MPa$^{0.5}$	22.8; 33 (containing 25 wt% ITO)	
Surface tension	mN m^{-1}	calc.=36.3-39.0; 39.8, 33-37 (fluorinated)	-; Wen, J, Sun, Z, Xiang, J, Fan, H, Yan, J, Appl. Surf. Sci., 494, 610-8, 2019.
Dielectric constant at 100 Hz/1 MHz	-	2.68 (waterborne thermoplastic polyurethane); 4.8-5.5 (UV curable)	Zhao, H, Zhao, S-Q, Li, Q, Khan, M R, Jiang, T, Polymer, 209, 122992, 2020; Mendes-Felipe, C, Barbosa, J C, Gonçalves, S, Lanceros-Mendez, S, Compos. Sci. Technol., 199, 108363, 2020.
Dissipation factor at 100 Hz	E-4	59-95	
Volume resistivity	ohm-m	4.8E9	

PU polyurethane

PARAMETER	UNIT	VALUE	REFERENCES
Permeability to nitrogen, 25°C	barrer	8.7	Sadeghi, M; Semsarzadeh, M A; Barikani, M; Chenar, M P, J. Membrane, Sci., 376, 188-95, 2011.
Permeability to oxygen, 25°C	barrer	20	Sadeghi, M; Semsarzadeh, M A; Barikani, M; Chenar, M P, J. Membrane, Sci., 376, 188-95, 2011.
Contact angle of water, 20°C	degree	77.5-83.1; 82, 116 (PU-POSS)	Vakili, H, Mohseni, M, Makki, H, Irusta, L, Prog. Org. Coat., 150, 105965, 2021.
Surface free energy	mJ m^{-2}	37.5	

MECHANICAL & RHEOLOGICAL PROPERTIES

PARAMETER	UNIT	VALUE	REFERENCES
Tensile strength	MPa	7.6-66	
Tensile stress at yield	MPa	31-57.2	
Elongation	%	350-1,200	
Flexural strength	MPa	20-120	
Flexural modulus	MPa	540-3,000	
Tear strength	kN m^{-1}	24-119	
Compression set, 24h 70°C	%	27-40	
Shore A hardness	-	60-95	
Shore D hardness	-	36-91	
Shrinkage	%	1.2-2.5	
Brittleness temperature (ASTM D746)	°C	-70	

CHEMICAL RESISTANCE

PARAMETER	UNIT	VALUE	REFERENCES
Acid dilute/concentrated	-	good	
Alcohols	-	good	
Alkalis	-	good	
Aliphatic hydrocarbons	-	very good	
Aromatic hydrocarbons	-	very good	
Esters	-	good	
Greases & oils	-	good	
Ketones	-	good	
Effect of EtOH sterilization (tensile strength retention)	%	103-109	Navarrete, L; Hermanson, N, Antec, 2807-18, 1996.

FLAMMABILITY

PARAMETER	UNIT	VALUE	REFERENCES
Autoignition temperature	°C	340-560	
Limiting oxygen index	% O$_2$	21.5; 29.9 (with flame retardant)	Wang, Y, Zhang, Y, Liu, B, Zhao, Q, Qi, Y, Xie, H, Compos. Commun., 21, 100382, 2020.
Heat release	kW m^{-2}	330-700	Kraemer, R H; Zammarano, M; Linteris, G T; Gedde, U W; Gilman, J W, Polym. Deg. Stab., 95, 1115-22, 2010.
Heat of combustion	J g^{-1}	24,000-28,000	Kraemer, R H; Zammarano, M; Linteris, G T; Gedde, U W; Gilman, J W, Polym. Deg. Stab., 95, 1115-22, 2010.
Volatile products of combustion	-	CO, CO$_2$, HCN, NO$_x$	

PU polyurethane

PARAMETER	UNIT	VALUE	REFERENCES
UL 94 rating	-	V-2	
WEATHER STABILITY			
Activation wavelengths	nm	313, 334, 365, 405, 435	
Excitation wavelengths	nm	320, 372	
Emission wavelengths	nm	420, 423, 455, 489	
Important initiators and accelerators	-	catalysts used in prepolymer synthesis, catalysts used in the curing process, heavy metals, peroxides in polyol, products of reaction of amine catalysts and polyols, nitrous oxide, acids and bases (hydrolysis), traces of solvents of types capable of producing hydroperoxides, products of thermooxidative degradation	
Products of degradation	-	photo-Fries rearrangement, yellowing, chains scission, hydroperoxides, carbonyls	
Stabilizers	-	UVA: 2,2'-dihydroxy-4-methoxybenzophenone; 2-(2H-benzotriazol-2-yl)-p-cresol; 2-benzotriazol-2-yl-4,6-di-tert-butylphenol; phenol, 2-(5-chloro-2H-benzotriazole-2-yl)-6-(1,1-dimethylethyl)-4-methyl-; 2-(2H-benzotriazole-2-yl)-4,6-di-tert-pentylphenol; 2-(2H-benzotriazole-2-yl)-4-(1,1,3,3-tetraethylbutyl)phenol; 2-(2H-benzotriazol-2-yl)-4,6-bis(1-methyl-1-phenylethyl)phenol; 2-(2H-benzotriazol-2-yl)-6-dodecyl-4-methylphenol, branched & linear; 2,4-di-tert-butyl-6-(5-chloro-2H-benzotriazole-2-yl)-phenol; 2-(3-sec-butyl-5-tert-butyl-2-hydroxyphenyl) benzotriazole; reaction product of methyl 3(3-(2H-benzotriazole-2-yl)-5-t-butyl-4-hydroxyphenyl propionate/PEG 300; ethyl-2-cyano-3,3-diphenylacrylate; (2-ethylhexyl)-2-cyano-3,3-diphenylacrylate; N-(2-ethoxyphenyl)-N'-(4-isododecylphenyl)oxamide; N-(2-ethoxyphenyl)-N'-(2-ethylphenyl)oxamide; benzoic acid, 4-[[(methylphenylamino) methylene]amino]-, ethyl ester; Screeners: carbon black; HAS: 1,3,5-triazine-2,4,6-triamine, N,N'''[1,2-ethane-diyl-bis[[[4,6-bis[butyl(1,2,6,6-pentamethyl-4-piperidinyl) amino]-1,3,5-triazine-2-yl]imino]-3,1-propanediyl]bis[N',N''-dibutyl-N',N''-bis(1,2,2,6,6-pentamethyl-4-piperidinyl)-; 2,4-bis[N-butyl-N-(1-cyclohexyloxy-2,2,6,6-tetramethylpiperidin-4-yl)amino]-6-(2-hydroxyethylamine)-1,3,5-triazine; bis(1,2,2,6,6-pentamethyl-4-piperidyl) sebacate and methyl 1,2,2,6,6-pentamethyl-4-piperidyl sebacate; bis(1,2,2,6,6-pentamethyl-4-piperidyl)sebacate + methyl-1,2,2,6,6-pentamethyl-4-piperidyl sebacate; bis(2,2,6,6-tetramethyl-4-piperidyl) sebacate; 2,2,6,6-tetramethyl-4-piperidinyl stearate; 2-dodecyl-N-(2,2,6,6-tetramethyl-4-piperidinyl)succinimide; poly[[(6-[1,1,3,3-tetramethylbutyl)amino]-1,3,5-triazine-2,4-diyl][2,2,6,6-tetramethyl-4-piperidinyl)imino]-1,6-hexanediyl[2,2,6,6-tetramethyl-4-piperidinyl)imino]]; butanedioic acid, dimethylester, polymer with 4-hydroxy-2,2,6,6-tetramethyl-1-piperidine ethanol; alkenes, C20-24-.alpha.-, polymers with maleic anhydride, reaction products with 2,2,6,6-tetramethyl-4-piperidinamine; polymer of 2,2,4,4-tetramethyl-7-oxa-3,20-diaza-dispiro [5.1.11.2]-heneicosan-21-on and epichlorohydrin; 1, 6-hexanediamine, N, N'-bis(2,2,6,6-tetramethyl-4-piperidinyl)-, polymers with 2,4-dichloro-6-(4-morpholinyl)-1,3,5-triazine; 1,6-hexanediamine, N,N'-bis(2,2,6,6-tetramethyl-4-piperidinyl)-, polymers with morpholine-2,4,6-trichloro-1,3,5-triazine reaction products, methylated;	Wypych, G, Handbook of Materials Weathering, 6th Edition, ChemTec Publishing, 2018.

PU polyurethane

PARAMETER	UNIT	VALUE	REFERENCES
Stabilizers (continuation)	-	Phenolic antioxidants: ethylene-bis(oxyethylene)-bis(3-(5-tert-butyl-4-hydroxy-m-tolyl)-propionate); pentaerythritol tetrakis(3-(3,5-di-tert-butyl-4-hydroxyphenyl)propionate); octadecyl-3-(3,5-di-tert-butyl-4-hydroxyphenyl)-propionate; N,N'-hexane-1,6-diylbis(3-(3-(3,5-di-tert-butyl-4-hydroxy-phenylpropionamide); benzopropanoic acid, 3,5-bis(1,1-dimethyl-ethyl)-4-hydroxy-C7-C9 branched alkyl esters; 3,3',3',5,5',5'-hexa-tert-butyl-a,a',a'-(mesitylene-2,4,6-triyl) tri-p-cresol; 1,3,5-tris(3,5-di-tert-butyl-4-hydroxybenzyl)-1,3,5-triazine-2,4,6(1H,3H,5H)-trione; 3,4-dihydro-2,5,7,8-tetramethyl-2-(4,8,12-trimethyltridecyl)-2H-1-benzopyran-6-ol; isotridecyl-3-(3,5-di-tert-butyl-4-hydroxyphenyl) propionate; 2,2'-ethylidenebis (4,6-di-tert-butylphenol); 3,5-tris(4-tert-butyl-3-hydroxy-2,6-dimethyl benzyl)-1,3,5-triazine-2,4,6-(1H,3H,5H)-trione; 3,5-bis(1,1-dimethyethyl)-4-hydroxy-ben-zenepropanoic acid, C13-15 alkyl esters; Phosphite: isodecyl diphenyl phosphite; Thiosynergist: 4,6-bis(dodecylthiomethyl)-o-cresol; 4,4'-thiobis(2-t-butyl-5-methylphenol); 2,2'-thiobis(6-tert-butyl-4-methylphenol); Amine: benzenamine, N-phenyl-, reaction products with 2,4,4-trimethylpentene; Optical brightener: 2,2'-(2,5-thiophenediyl)bis(5-tert-butylbenzoxazole)	Wypych, G, Handbook of Materials Weathering, 6th Edition, ChemTec Publishing, 2018.

BIODEGRADATION

PARAMETER	UNIT	VALUE	REFERENCES
Colonized products		catheters, coatings, fibrous membrane, microporous membranes, military components, paint, tissue repair products	
Typical biodegradants	-	microbial degradation of polyester polyurethane is hypothesized to be mainly due to the hydrolysis of ester bonds by esterase enzymes	
Stabilizers	-	2,4-di-tert-butylphenol, 4,5,-dichloro-2-n-octyl-4-isothiazolone-3-one biocide, 2-n-octyl-4-isothiazolin-3-one, chitosan, gold nanoparticles, grafted 2,2,5,5-tetramethyl-imidozalidin-4-one, nonylphenol disulfide, silver nanoparticles	

TOXICITY

PARAMETER	UNIT	VALUE	REFERENCES
Oral rat, LD$_{50}$	mg kg^{-1}	>5,000	

ENVIRONMENTAL IMPACT

PARAMETER	UNIT	VALUE	REFERENCES
Aquatic toxicity, *Daphnia magna*, LC$_{50}$, 48 h	mg l^{-1}	3,500-4,900; 31,000-38,000 (EC$_{50}$)	Lithner, D; Damberg, J; Dave, G; Larsson, A, Chemosphere, 74, 1195-1200, 2009; Lithner, Ph D Thesis, University of Gothenburg, 2011.
Biodegradation in compost and soil after 12 weeks	%	loss of 30% and 71% mass, and 41% and 71.5% compression force (PU foams)	Gunawan, N R, Tessman, M, Schreiman, A C, Mayfield, S P, Bioresource Technol. Rep., 11, 100513, 2020.

PROCESSING

PARAMETER	UNIT	VALUE	REFERENCES
Typical processing methods	-	PU: chemical or moisture cure, coatings, compounding, electrospinning, mixing; TPU: blow molding, calendering coating, extrusion, injection molding, solution coating of fabrics	
Preprocess drying: temperature/time/residual moisture	°C/h/%	85/3/0.01	
Processing temperature	°C	150-200	

PU polyurethane

PARAMETER	UNIT	VALUE	REFERENCES
Additives used in final products	-	Fillers: aluminum hydroxide, aluminum nitride, bentonites, calcium carbonate, calcium sulfate, fluoromica, graphite, kaolin, mica, montmorillonite, nanosilica, organic fibers, rubber particles, sand, sepiolite, silica; Plasticizers: adipates, azelates, benzoates, citrates, epoxidized soybean oil, phosphates, phthalates, sebacates; Antistatics: carbon black, carbon nanotubes, glycerol monostearate, graphite, polyaniline, polyethylene glycol, polypyrrole, silver-coated basalt particles, sulfonated polyol, tetraorganoboron, vanadium pentoxide; Antiblocking: diatomaceous earth, glass or ceramic spheres, natural silica, starch; Release: crosslinked silicone, fluorocarbon, isononylphenyl isocyanate, lecithin, silicone fluid, sodium myristate, sodium oleate, zinc stearate; Slip: ethylene bisstearamide, metal soap, montan ester wax, silicone oil	
Applications	-	PU: adhesives, coatings, foams, mortars, primers, sealants, tissue engineering scaffolds, and numerous other products; TPU: automotive, coatings, film and sheet, footwear, hose and tubes, self-healing materials, wire and cable, and many other products	

BLENDS

PARAMETER	UNIT	VALUE	REFERENCES
Suitable polymers	-	CA, CHI, PCL, PDMS, PEG, SBS	

ANALYSIS

PARAMETER	UNIT	VALUE	REFERENCES
FTIR (wavenumber-assignment)	cm^{-1}/-	N–H – 3315-3400; carbonyl – 1715-1729; hydrogen-bonded C=O stretching vibration – 1681-1688; C–N stretching and – NH bending vibrations – 1500-1590	Amrollahi, M; Sadeghi, M M; Kashcooli, Y, Mater. Design, 32, 3933-41, 2011; Thiyagu, C, Manjubala, I, Narendrakumar, U, Mater. Today: Proc., in press, 2020.
^{1}H NMR (chemical shifts)	ppm	NH (–NH–CO–O–) groups – 7.25; CH_2 groups attached to oxygen atoms of urethane (–NH–CO–O–CH_2–) and ester (–CO–O–CH_2–) bonds – 3.84-3.86 and 4.01-4.1 (confirming urethane formation); presence of urethane groups comes from the signal at 3.20-3.25, which can be attributed to methylene groups connected to urethane nitrogen atoms (–CO–O–NH–CH_2–); protons of amino methylene (–CH_2–NH–) – 2.66; CH_2 group directly connected to the carbonyl carbon of ester groups (–OCO–CH_2–) – 2.27	Valčić, M D, Cakić, S M, Ristić, I S, János, C J, Int. J. Adhesion Adhesives, 104, 102738, 2021.
x-ray diffraction peaks	degree	23; 13.64, 21.02	Spirkova, M; Pavlicevic, J; Strachota, A; Poremba, R; Bera, O; Kapralkova, L; Baldrian, J; Slouf, M; Lazic, N; Budinski-Simendic, J, Eur. Polym. J., 47, 959-72, 2011; Kumar, H; Siddaramaiah; Somashekar, R; Mahesh, S S, Mater. Sci. Eng, A447, 58-64, 2007.

PVAc poly(vinyl acetate)

PARAMETER	UNIT	VALUE	REFERENCES
GENERAL			
Common name	-	poly(vinyl acetate)	
IUPAC name	-	poly(ethenyl ethanoate)	
CAS name	-	acetic acid ethenyl ester, homopolymer	
Acronym	-	PVAc	
CAS number	-	9003-20-7	
EC number	-	203-545-4	
RTECS number	-	AK0920000	
Formula	-		
HISTORY			
Person to discover	-	Fritz Klatte	Klatte, F; Rollett, A, US Patent 1,241,738, 1917.
Date	-	1912	
SYNTHESIS			
Monomer(s) structure	-		
Monomer(s) CAS number(s)	-	108-05-4	
Monomer(s) molecular weight(s)	dalton, g/mol, amu	86.09	
Monomer ratio	-	100% or less (copolymers)	
Method of synthesis	-	oxidative addition of acetic acid to ethylene	
Catalyst	-	Pd; vanadium complex	Shaver, M P; Hanhan, M E; Jones, M R, Chem. Commun., 46, 2127-29, 2010.
Heat of polymerization	$J\ g^{-1}$	875-1,045	Joshi, R M, J. Polym. Sci., 56, 313, 1962.
Number average molecular weight, M_n	dalton, g/mol, amu	31,400	Alig, I, Böhm, F, Lellinger, D, Thermochim. Acta, 677, 4-11, 2019.
Mass average molecular weight, M_w	dalton, g/mol, amu	13,000-500,000; 71,900	
Polydispersity, M_w/M_n	-	2.0	
Polymerization degree (number of monomer units)	-	100-5000	
Molar volume at 298K	cm^3 mol^{-1}	calc.=74.25; 64.5 (crystalline); 72.4 (amorphous)	
Van der Waals volume	cm^3 mol^{-1}	calc.=45.9 (crystalline); 45.9 (amorphous)	
Radius of gyration	nm	66	Ahmed, I; Pritchard, J G; Blakely, C F, Polymer, 25, 4, 543-50, 1984.
End-to-end distance of unperturbed polymer chain	nm	170-380	Ahmed, I; Pritchard, J G; Blakely, C F, Polymer, 25, 4, 543-50, 1984.

HANDBOOK OF POLYMERS 3rd Edition, Copyrights 2022; ChemTec Publishing

PVAc poly(vinyl acetate)

PARAMETER	UNIT	VALUE	REFERENCES
STRUCTURE			
Crystallinity	%	amorphous	
Entanglement molecular weight	dalton, g/mol, amu	calc.=8667	
COMMERCIAL POLYMERS			
Some manufacturers	-	Vinavil	
Trade names	-	Vinavil	
PHYSICAL PROPERTIES			
Density at 20°C	g cm^{-3}	1.18-1.20	
Bulk density at 20°C	g cm^{-3}	0.7-0.85	
Color	-	colorless	
Refractive index, 20°C	-	1.467-1.469	
Odor		odorless	
Melting temperature, DSC	°C	152-180	
Softening point	°C	>190	
Decomposition temperature	°C	>250	
Thermal expansion coefficient, 23-80°C	°C^{-1}	2.8E-4	
Thermal conductivity, melt	W m^{-1} K^{-1}	0.159	
Glass transition temperature	°C	calc.=31; 28-32; 24-31 (atactic); 26 (isotactic); 28-40 (commercial); 40 (laboratory standard)	Alig, I, Böhm, F, Lellinger, D, Thermochim. Acta, 677, 4-11, 2019.
Hansen solubility parameters, δ_D, δ_P, δ_H	MPa$^{0.5}$	20.9, 11.3, 9.7	
Interaction radius		13.7	
Hildebrand solubility parameter	MPa$^{0.5}$	calc.=19.2-20.93; exp.=18.0-25.7	
Surface tension	mN m^{-1}	36.5	
Dielectric constant at 100 Hz/1 MHz	-	-/3.5	
Dissipation factor at 1 MHz		150	
Permeability to oxygen, 25°C	cm^3 cm cm^{-2} s^{-1} Pa^{-1} x 10^{12}	0.0367	
Diffusion coefficient of oxygen	cm^2 s^{-1} x10^6	0.0562	
Contact angle of water, 20°C	degree	60.6	
Surface free energy	mJ m^{-2}	38.5	
MECHANICAL & RHEOLOGICAL PROPERTIES			
Tensile strength	MPa	6.2-12	
Water absorption, equilibrium in water at 23°C	%	3-6	

PVAc poly(vinyl acetate)

PARAMETER	UNIT	VALUE	REFERENCES
CHEMICAL RESISTANCE			
Acid dilute/concentrated	-	poor	
Alcohols	-	poor	
Alkalis	-	good	
Aliphatic hydrocarbons	-	fair	
Aromatic hydrocarbons	-	poor	
Esters	-	poor	
Halogenated hydrocarbons	-	poor	
Ketones	-	poor	
Θ solvent, Θ-temp.=29, 123, 19, 6°C	-	n-butyl ethyl ketone, cetyl alcohol, ethanol, methanol	
Good solvent	-	acetic acid, acetone, acetonitrile, allyl alcohol, chlorobenzene, chloroform, DMF, DMSO, metanol, THF, toluene	
Non-solvent	-	acids, diluted alkalis, carbon disulfide, cyclohexanol, ethylene glycol, mesitylene	
FLAMMABILITY			
NFPA rating	-	0/0/0	
Autoignition temperature	°C	427	
Char at 500°C	%	1.2	Lyon, R E; Walters, R N, J. Anal. Appl. Pyrolysis, 71, 27-46, 2004.
WEATHER STABILITY			
Spectral sensitivity	nm	313	Ferreira, J L; Melo, M J; Ramos, A, Polym. Deg. Stab., 95, 453-61, 2010.
Products of degradation	-	chain scission, acetic acid	
BIODEGRADATION			
Typical biodegradants	-	plasticizers are attacked by fungi; hydrolysis by lipase	Domenech-Carbo, M T; Bitossi, G; de la Cruz-Canizares, J; Bolivar-Galiano, F; del Mar Lopez-Miras, M; Romero-Noguera, J; Martin-Sanchez, I, J. Anal. Appl. Pyrolysis, 85, 480-86, 2009; Chattopadhyay, S; Sivalingam, G; Madras, G, Polym. Deg. Stab., 80, 477-83, 2003.
TOXICITY			
Carcinogenic effect	-	not listed by ACGIH, NIOSH, NTP	
Mutagenic effect	-	non-mutagenic	
Oral rat, LD$_{50}$	mg kg^{-1}	>25,000; 3,080 mg kg^{-1} day^{-1} (NOAEL)	
Skin rabbit, LD$_{50}$	mg kg^{-1}	non-irritant	
PROCESSING			
Typical processing methods	-	mixing/compounding	

PVAc poly(vinyl acetate)

PARAMETER	UNIT	VALUE	REFERENCES
Additives used in final products	-	Fillers: aluminosilicate, calcium carbonate, clay, mica, talc; Plasticizers: acetyl triethyl citrate, benzyl butyl phthalate, dibutyl phthalate, castor oil, dipropylene glycol dibenzoate, epoxidized soybean oil, ethylene and propylene glycols, glycerin, triethanolamine, triacetin, tributyl citrate, triethyl citrate; Antistatic: quaternary ammonium salt; Release: zinc stearate	
Applications	-	adhesives, paints, paper, production of poly(vinyl alcohol)	
BLENDS			
Suitable polymers	-	NR, P34HB , PCL, PE, PEO, PHB, PLA, PMA, PMMA, PPy, PVC, PVDF, PVF, PVOH, polyacrylate	
ANALYSIS			
FTIR (wavenumber-assignment)	cm^{-1}/-	C=O – 1737; C–H – 1375; O–C – 1020	Asensio, R C; San Andres Moya, M; de la Roja, J M; Gomez, M, Anal. Bioanal. Chem., 395, 2081-96, 2009.
Raman (wavenumber-assignment)	cm^{-1}/-	C–C – 1132; C=C – 1525	Blazevska-Gilev, J; Kupcik, J; Subrt, J; Vorlicek, V; Galikova, A; Pola, J, Polymer, 46, 8973-80, 2005.
NMR (chemical shifts)	ppm	H NMR: COOH – 8.76; OH – 5.61, 5.58	Chattopadhyay, S; Sivalingam, G; Madras, G, Polym. Deg. Stab., 80, 477-83, 2003.

PVAl poly(vinyl alcohol)

PARAMETER	UNIT	VALUE	REFERENCES
GENERAL			
Common name	-	poly(vinyl alcohol)	
Acronym	-	PVAl, PVOH	
CAS number	-	9002-89-5	
RTECS number	-	TR8100000, TR8101000	
Formula		$\left[\begin{array}{c}CH_2CH \\ | \\ OH\end{array}\right]_n$	
HISTORY			
Person to discover	-	Kuehn, E; Hopff, H	Kuehn, E; Hopff, H, US Patent 2,044,730, IG Farben, June 16, 1936.
Date	-	1936	
SYNTHESIS			
Monomer(s) structure	-	$H_2C{=}CHCOCH_3 \quad (H_2C{=}CHOH)$ with O double-bonded above C	
Monomer(s) CAS number(s)	-	108-05-4 (557-75-5)	
Monomer(s) molecular weight(s)	dalton, g/mol, amu	86.09 (44.053)	
Monomer ratio	-	100% (hydrolysis to a varying degree: 98.0-99.8 (fully hydrolyzed); 90-97 (intermediately hydrolyzed); 87-89 (partially hydrolyzed))	
Method of synthesis	-	because of instability of vinyl alcohol, poly(vinyl alcohol) is produced by polymerization of vinyl acetate (see more in PVAc) and its subsequent hydrolysis	
Typical impurities	ppm	sodium acetate, methanol and methyl acetate	
Number average molecular weight, M_n	dalton, g/mol, amu	7,000-101,000	
Mass average molecular weight, M_w	dalton, g/mol, amu	27,000-195,000	Leone, G, Consumi, M, Pepi, S, Magnani, A, J. Molec. Struct., 1202, 127264, 2020.
Polydispersity, M_w/M_n	-	1.2-1.5	Li, J, Liu, Y, Chen, Q, J. Chromatog. A, 1624, 461260, 2020.
Polymerization degree (number of monomer units)	-	150-2,200	
Molar volume at 298K	cm³ mol⁻¹	33.6 (crystalline)	
Van der Waals volume	cm³ mol⁻¹	25.1 (crystalline)	
Radius of gyration	nm	20-27	Li, J, Liu, Y, Chen, Q, J. Chromatog. A, 1624, 461260, 2020.
Chain-end groups	-	OH	
STRUCTURE			
Crystallinity	%	25-35 (syndiotactic); 30-60 (atactic); 18-24 (isotactic); 54.7 (electrospun fibers)	-; Zhang, Q, Li, Q, Zhang, L, Young, T M, Chem. Eng. J., 399, 125768, 2020.
Cell type (lattice)	-	monoclinic	
Cell dimensions	nm	a:b:c=0.781-0.784:0.543-0.552:0.252-0.253	

HANDBOOK OF POLYMERS 3rd Edition, Copyrights 2022; ChemTec Publishing

PVAl poly(vinyl alcohol)

PARAMETER	UNIT	VALUE	REFERENCES
Unit cell angles	degree	γ=91-93	
Crystallite size	nm	3.4-12.1; 3.3, 6.3 (electrospun nanofibers before and after heat treatment at 140ºC)	-; Zhang, Q, Li, Q, Zhang, L, Young, T M, Chem. Eng. J., 399, 125768, 2020.
Spacing between crystallites	nm	8.5-18.2	
Tacticity	%	78 (isotactic); 69.2 (syndiotactic)	Ohgi, H; Yang, H; Sato, T; Horii, F, Polymer, 48, 3850-57, 2007; Nagara, Y; Nakano, T; Okamoto, Y; Gotoh, Y; Nagura, M, Polymer, 42, 9679-86, 2001.
Avrami constants, k/n	-	n=4.54 (syndiotactic); n=1.48 (atactic)	Nagara, Y; Nakano, T; Okamoto, Y; Gotoh, Y; Nagura, M, Polymer, 42, 9679-86, 2001.

COMMERCIAL POLYMERS

PARAMETER	UNIT	VALUE	REFERENCES
Some manufacturers	-	Sekisui; Kuraray	
Trade names	-	Celvol; Elvanol	

PHYSICAL PROPERTIES

PARAMETER	UNIT	VALUE	REFERENCES
Density at 20ºC	g cm^{-3}	1.19-1.31	
Bulk density at 20ºC	g cm^{-3}	0.3-0.7	
Color	-	clear, white to yellow	
Refractive index, 20ºC	-	1.49-1.51	
Birefringence	-	0.0.55 (syndiotactic); 0.0353 (atactic)	Nagara, Y; Nakano, T; Okamoto, Y; Gotoh, Y; Nagura, M, Polymer, 42, 9679-86, 2001.
Odor	-	odorless	
Melting temperature, DSC	ºC	178; 230 (fully hydrolyzed), 180-190 (partially hydrolyzed)	
Decomposition temperature	ºC	100 (color change); 150 (rapid darkening); 200 (rapid decomposition)	Prosanov, I Y; Matvienko, A A, Phys. Solid State, 52, 10, 2203-6, 2010.
Thermal expansion coefficient, 23-80ºC	ºC^{-1}	0.7-1.2E-4	
Thermal conductivity	W m^{-1} K^{-1}	0.21	Yin, C-G, Liu, Z-J, Mo, R, Min, J-L, Polymer, 195, 122455, 2020.
Glass transition temperature	ºC	calc.=84; exp.=34-85	
Specific heat capacity	J K^{-1} kg^{-1}	1,500-1,650	
Heat of fusion	J g^{-1}	40.9-48.4	Zhang, W; Zhang, Z; Wang, X, J. Colloid Interface Sci., 333, 346-53, 2009.
Hansen solubility parameters, δ_D, δ_P, δ_H	MPa$^{0.5}$	17.0, 9.0, 18.0	Huang, Z, Ru, X-f, Zhu, Y-T, Teng, L-j, Chem. Eng. Res. Design, 144, 19-34, 2019.
Interaction radius		10.5	
Molar volume	kmol m^{-3}		
Hildebrand solubility parameter	MPa$^{0.5}$	21.7-25.78	
Surface tension	mN m^{-1}	calc.=37.0; 36.6-51.33	-; Alade, O S, Al Shehri, D, Mahmoud, M, Sasaki, K, Fuel, 285, 119128, 2021.
Dielectric constant at 100 Hz/1 MHz	-	2.6	
Volume resistivity	ohm-m	3.1-3.8E5	

PVAl poly(vinyl alcohol)

PARAMETER	UNIT	VALUE	REFERENCES
Permeability to oxygen, 25°C	cm^3 cm cm^{-2} s^{-1} Pa^{-1} x 10^{12}	0.000665	
Permeability to water vapor, 25°C	cm^3 cm cm^{-2} s^{-1} Pa^{-1} x 10^{12}	0.525	
Diffusion coefficient of water vapor	cm^2 s^{-1} x10^6	0.746	
Contact angle of water, 20°C	degree	51-72	Zhang, W; Zhang, Z; Wang, X, J. Colloid Interface Sci., 333, 346-53, 2009.
Surface free energy	mJ m^{-2}	44.2	

MECHANICAL & RHEOLOGICAL PROPERTIES

PARAMETER	UNIT	VALUE	REFERENCES
Tensile strength	MPa	23-55; 1,430 (drawn syndiotactic fiber); 1,490 (drawn atactic fiber)	Nagara, Y; Nakano, T; Okamoto, Y; Gotoh, Y; Nagura, M, Polymer, 42, 9679-86, 2001.
Young's modulus	MPa	38,100 (drawn syndiotactic fiber); 25,200 (drawn atactic fiber)	Nagara, Y; Nakano, T; Okamoto, Y; Gotoh, Y; Nagura, M, Polymer, 42, 9679-86, 2001.
Tenacity (fiber) (standard atmosphere)	cN tex^{-1} (daN mm^{-2})	20-65 (25-80)	Fourne, F, Synthetic Fibers. Machines and Equipment Manufacture, Properties. Carl Hanser Verlag, 1999.
Tenacity (wet fiber, as % of dry strength)	%	65-85	Fourne, F, Synthetic Fibers. Machines and Equipment Manufacture, Properties. Carl Hanser Verlag, 1999.
Fineness of fiber (titer)	dtex	1.5-10	Fourne, F, Synthetic Fibers. Machines and Equipment Manufacture, Properties. Carl Hanser Verlag, 1999.
Length (elemental fiber)	mm	38-200	Fourne, F, Synthetic Fibers. Machines and Equipment Manufacture, Properties. Carl Hanser Verlag, 1999.
Melt index, 190°C/21.6 kg	g/10 min	31-47	
Water absorption, 24h at 23°C	%	5	

CHEMICAL RESISTANCE

PARAMETER	UNIT	VALUE	REFERENCES
Acid dilute/concentrated	-	good-poor	
Alcohols	-	poor	
Alkalis	-	good	
Aromatic hydrocarbons	-	poor	
Esters	-	poor	
Halogenated hydrocarbons	-	poor	
Ketones	-	poor	
Θ solvent, Θ-temp.=25, 97°C	-	ethanol/water=41.5/58.5, water	
Good solvent	-	acetamide, DMF, DMSO, glycerol (hot), piperazine; hot water (fully hydrolized); cold water (partially hydrolyzed)	
Non-solvent	-	carboxylic acids, chlorinated hydrocarbons, esters, hydrocarbons, ketones, lower alcohols, THF	

PVAl poly(vinyl alcohol)

PARAMETER	UNIT	VALUE	REFERENCES
FLAMMABILITY			
Ignition temperature	ºC	79	
Autoignition temperature	ºC	450	
Limiting oxygen index	% O$_2$	20-22.5, 21.7	-; Serbezeanu, D, Vlad-Bubulac, T, Hamciuc, C, Enache, A A, Compos. Commun., 22, 100505, 2020
Char at 500ºC	%	3.3	Lyon, R E; Walters, R N, J. Anal. Appl. Pyrolysis, 71, 27-46, 2004.
Heat of combustion	J g^{-1}	23,310	Walters, R N; Hacket, S M; Lyon, R E, Fire Mater., 24, 5, 245-52, 2000.
Total heat release	kJ g^{-1}	20.75	Serbezeanu, D, Vlad-Bubulac, T, Hamciuc, C, Enache, A A, Compos. Commun., 22, 100505, 2020
Volatile products of combustion	-	CO$_2$, H$_2$O, CO, organic acids, aldehydes, alcohols	
WEATHER STABILITY			
Spectral sensitivity	nm	<280	
Activation wavelengths	nm	310, 326 (in the course of degradation)	
Transmittance	%	100 nm – 59; 300 nm – 20.6	
Stability to sunlight	-	excellent, addition of lignin gives 91% UVA absorption; lignin nanoparticles act as mechanical/thermal enhancer of poly(vinyl alcohol) film	Shikinaka, K, Nakamura, M, Otsuka, Y, Polymer, 190, 122254, 2020.
Important initiators and accelerators	-	products of thermal degradation, carbonyl groups, unsaturations, sensitizers (polynuclear aromatic compounds, benzophenones)	
Products of degradation	-	free radicals, unsaturations, carbonyl groups, hydroperoxides, chain scissions, water, polyenes	
BIODEGRADATION			
Colonized products		hydrogel, nanofiber, preparations	
Typical biodegradants	-	oxidase and dehydrogenase give β-hydroxylketone as well as 1,3-diketone moieties	
Stabilizers	-	Kathon LX; Dowicil 75, paraquat dichloride, quaternized chitosan	
TOXICITY			
NFPA: Health, Flammability, Reactivity rating	-	0-1/1/0	
Carcinogenic effect	-	not listed by ACGIH, NIOSH, NTP	
OSHA	mg m^{-3}	5 (respirable), 15 (total)	
Oral rat, LD$_{50}$	mg kg^{-1}	>20,000; 23,854	
ENVIRONMENTAL IMPACT			
Aquatic toxicity, *Daphnia magna*, LC$_{50}$, 48 h	mg l^{-1}	8.3	
Aquatic toxicity, *Bluegill sunfish*, LC$_{50}$, 48 h	mg l^{-1}	10	
Aquatic toxicity, *Fathead minnow*, LC$_{50}$, 48 h	mg l^{-1}	40	

PVAl poly(vinyl alcohol)

PARAMETER	UNIT	VALUE	REFERENCES
PROCESSING			
Typical processing methods	-	casting, coating, electrospinning, extrusion; molding, photopolymerization	Li, S, Zhou, D, Pei, M, Xiao, P, Eur. Polym. J., 134, 109854, 2020.
Processing temperature	°C	183-188 (extrusion)	
Additives used in final products	-	Fillers: aluminum oxide, calcium carbonate, clay, carbon black, ferrite, graphite, magnesium oxide, nanocellulose, sand, silica, titanium dioxide, zinc oxide, zirconia; Plasticizers: benzyl butyl phthalate, dipropylene glycol dibenzoate, glycerin, monostearyl citrate, polyethylene and polypropylene glycols, triacetin; Antistatics: alkyl aryl sulfonate, cadmium sulfide, ethoxylated fatty acid amine, tetraammonium salt; Antiblocking: talc; Release: silane modified PVOH; Slip: PTFE beads; Crosslinkers; Defoamers	
Applications	-	adhesives, belts, binders, cementitious laminate, coatings, controlled drug delivery, electroconductive film, endodontic applications, film, food, magnetic nanocomposite, membranes, optoelectronic applications, paper (with pigments and optical brighteners), packaging film, photographic papers, printing rolls, protective colloids, sanitary pads, seed tapes, sizing agents, toners, water-soluble laundry bags, warp sizing, wood glue, wound healing	
Outstanding properties	-	fibers, solvent-free protein cryopreservation	Fayter, A E R, Hasan, M, Congdon, T R, Gibson, M I, Eur. Polym. J., 140, 110036, 2020.
BLENDS			
Suitable polymers	-	chitosan, starch, NR, PAA, PCL, PEEK, PEG, PSI	
ANALYSIS			
FTIR (wavenumber-assignment)	cm^{-1}/-	O–H – 3550-3200; C–H – 2840-3000; C=O – 1750-1735; C–O – 1141; C–O–C – 1150-1085	Mansur, H S; Sadahira, C M; Souza, A N; Mansur, A A P, Mater. Sci. Eng., C28, 539-48, 2008.
Raman (wavenumber-assignment)	cm^{-1}/-	CH_2 – 2912; OH – 1440	Uddin, A J; Araki, J; Gotoh, Y, Composites, A42, 741-47, 2011.

PVB poly(vinyl butyrate)

PARAMETER	UNIT	VALUE	REFERENCES
GENERAL			
Common name	-	poly(vinyl butyrate)	
CAS name	-	ethenol, homopolymer, cyclic acetal with butanal	
Acronym	-	PVB	
CAS number	-	63148-65-2; 24991-31-9	
RTECS number	-	TR4955000	
Formula			

$$\left[\begin{array}{c} CH_2CH \\ | \\ O \\ | \\ CH_3(CH_2)_2-C=O \end{array} \right]_n$$

PARAMETER	UNIT	VALUE	REFERENCES
HISTORY			
Person to discover	-	Overholt, R L	Overholt, R L, US Patent 2,293,558, DuPont, Aug. 18, 1942.
Date	-	1942	
Details	-	use of PVB in coating composition	
SYNTHESIS			
Monomer(s) structure	-		

$$H_2C=CHOCCH_2CH_2CH_3 \quad (\overset{O}{\overset{\|}{})}$$

PARAMETER	UNIT	VALUE	REFERENCES
Monomer(s) CAS number(s)	-	123-20-6	
Monomer(s) molecular weight(s)	dalton, g/mol, amu	114.14	
Method of synthesis	-	PVB is prepared by reacting poly(vinyl alcohol) with butyraldehyde in the presence of an acid catalyst	Fernandez, M D; Fernandez, M J, Hoces, P, J. Appl. Polym. Sci., 102, 5007-17, 2006.
Temperature of polymerization	°C	30	
Time of polymerization	h	72	
Pressure of polymerization	Pa	atmospheric	
Catalyst	-	HCl	
Mass average molecular weight, M_w	dalton, g/mol, amu	40,000-120,000	
Chain-end groups	-	OH	
PHYSICAL PROPERTIES			
Density at 20°C	g cm^{-3}	1.07-1.1	
Color	-	white	
Refractive index, 20°C	-	1.47-1.50	
Odor		slightly pungent	
Melting temperature, DSC	°C	90-120	
Thermal conductivity, melt	W m^{-1} K^{-1}	0.236	
Glass transition temperature	°C	57-71	
Hansen solubility parameters, δ_D, δ_P, δ_H	MPa$^{0.5}$	18.6, 4.36, 13.03; 19.1, 9.5, 12.2	
Interaction radius		10.0	
Hildebrand solubility parameter	MPa$^{0.5}$	23.12; 24.6	

648   **HANDBOOK OF POLYMERS** 3rd Edition, Copyrights 2022 ChemTec Publishing

PVB poly(vinyl butyrate)

PARAMETER	UNIT	VALUE	REFERENCES
Surface tension	mN m⁻¹	calc.=38.0; exp.=28.9; 32 (245ºC); 26.4 (255ºC)	Morais, D; Valera, T S; Demarquette, N R, Macromol. Symp., 245-246, 208-14, 2006.
Dielectric constant at 100 Hz/1 MHz	-	2.6-3.2	
Dissipation factor at 100 Hz	E-4	0.0064-0.03	
Permeability to water vapor, 25ºC	cm³ cm cm⁻² s⁻¹ Pa⁻¹ x 10¹²	60.8	
Surface free energy	mJ m⁻²	38.0	
Speed of sound	m s⁻¹	39.2	
Acoustic impedance		2.60	

MECHANICAL & RHEOLOGICAL PROPERTIES

PARAMETER	UNIT	VALUE	REFERENCES
Tensile strength	MPa	22.2-23.0	Valera, T S; Demarquette, N R, Eur. Polym. J., 44, 755-68, 2008.
Tensile modulus	MPa	6.4	
Elongation	%	190-380	
Flexural modulus	MPa	14	
Young's modulus	MPa	100	Xu, J; Sun, Y; Liu, B; Zhu, M; Yao, X; Yan, Y; Li, Y; Chen, X, Eng. Failure Anal., in press, 2011.
Poisson's ratio	-	0.48-0.49	Xu, J; Sun, Y; Liu, B; Zhu, M; Yao, X; Yan, Y; Li, Y; Chen, X, Eng. Failure Anal., in press, 2011.
Shore A hardness	-	63-82	
Shore D hardness	-	27	
Melt index, 190ºC/2.16 kg	g/10 min	2.4-3	

CHEMICAL RESISTANCE

PARAMETER	UNIT	VALUE	REFERENCES
Alcohols	-	poor	
Aliphatic hydrocarbons	-	good	
Aromatic hydrocarbons	-	good	
Esters	-	good	
Halogenated hydrocarbons	-	poor	
Ketones	-	good	
Good solvent	-	alcohols, cyclohexanone, lower esters, methylene chloride	
Non-solvent	-	aliphatic ketones, hydrocarbons, MIBK	

FLAMMABILITY

PARAMETER	UNIT	VALUE	REFERENCES
Autoignition temperature	ºC	390	
Char at 500ºC	%	0.1	Lyon, R E; Walters, R N, J. Anal. Appl. Pyrolysis, 71, 27-46, 2004.

TOXICITY

PARAMETER	UNIT	VALUE	REFERENCES
NFPA: Health, Flammability, Reactivity rating	-	1/1/0	
Carcinogenic effect	-	not listed by ACGIH, NIOSH, NTP	
Oral rat, LD_{50}	mg kg⁻¹	10,000	

PVB poly(vinyl butyrate)

PARAMETER	UNIT	VALUE	REFERENCES
Skin rabbit, LD$_{50}$	mg kg^{-1}	7,940	

PROCESSING

PARAMETER	UNIT	VALUE	REFERENCES
Typical processing methods	-	compounding, electrospinning, extrusion, powder coating	
Processing temperature	°C	120 (bonding to glass)	
Processing pressure	bar	10 (bonding to glass)	
Additives used in final products	- `	Fillers: aluminum hydroxide, calcium carbonate, carbon black, graphite, rust protection fillers, zinc oxide; Plasticizers: biphenyl, dibutyl sebacate, diglycidyl ether of bisphenol A, dihexyl adipate, hexyl cyclohexyl adipate, polyethylene glycol, tetraethylene glycol di-n-heptanoate, triethylene glycol di-(2-ethylhexanoate); Antistatics: antimony-doped tin oxide, vanadium pentoxide; Antiblocking: silica; Release: liquid paraffin, n-butyl stearate, silicone	
Applications	-	adhesives and sealants, binders for rocket propellant, bullet-proof glass, ceramic binders, collapsible tubes, composite fiber binders, control of light, drum interiors, dry toners, heat and sound in construction glass, inks, magnetic tapes, nanofibers, paints, photoconductive papers, powder coating, safety glass interlayer (automotive windshields), wash primers, wood sealers and primers	

BLENDS

PARAMETER	UNIT	VALUE	REFERENCES
Suitable polymers	-	chitosan; PA6 (PVB is impact modifier), PP or PVC (tackifier)	
Compatibilizers	-	anhydride functionalized modifier	Hofmann, G H, Antec, 3241-45, 1999.

ANALYSIS

PARAMETER	UNIT	VALUE	REFERENCES
FTIR (wavenumber-assignment)	cm^{-1}/-	OH – 3489; C–H – 2970; C=O – 1730	Valera, T S; Demarquette, N R, Eur. Polym. J., 44, 755-68, 2008.
NMR (chemical shifts)	ppm	CH$_3$ – 13.7; CH$_2$ – 17.3, 36.7, 37.1, 41.5, 42.5; CH – 64.8, more	Fernandez, M D; Fernandez, M J, Hoces, P, J. Appl. Polym. Sci., 102, 5007-17, 2006.

PVC poly(vinyl chloride)

PARAMETER	UNIT	VALUE	REFERENCES
GENERAL			
Common name	-	poly(vinyl chloride)	
IUPAC name	-	poly(chloroethanediyl)	
CAS name	-	ethene, chloro-, homopolymer	
Acronym	-	PVC	
CAS number	-	9002-86-2	
EC number	-	208-750-2	
RTECS number	-	KV0350000	
Formula		$\left[CH_2CH\right]_n$ with Cl	
HISTORY			
Person to discover	-	Henri Victor Regnault (accidental polymerization), Fritz Klatte (technological developments), Waldo Semon (commercial applications	
Date	-	1835, beginning of 20th century, 1926	
Details	-	Henri Victor Regnault observed that vinyl monomer forms white solid material when exposed to sunlight; Klatte worked on processability; Semon continued Klatte efforts and succeeded in plasticization; extensive commercial applications had to wait on development of thermal stabilizers, which permitted industrial processing during Second World War in US	
SYNTHESIS			
Monomer(s) structure	-	$H_2C{=}CHCl$	
Monomer(s) CAS number(s)	-	75-01-4	
Monomer(s) molecular weight(s)	dalton, g/mol, amu	62.498820	
Monomer(s) expected purity(ies)	%	99.9	
Monomer ratio	-	100%	
Formulation example	-	suspension: water, suspending agent, initiator; emulsion: water, emulsifier, water-soluble initiator; microsuspension: water, emulsifier, oil-soluble initiator; bulk: initiator	
Common initiators		tert-octyl peroxyneodecanoate, dicyclohexyl peroxydicarbonate, tert-butyl peroxyneodecanoate, benzoyl peroxide, 2,2'-azobisbutylnitrile, tert-amyl peroxypivalate, dilauroyl peroxide	
Method of synthesis	-	suspension, microsuspension, emulsion, bulk, living radical polymerization	
Temperature of polymerization	°C	55-73	
Yield	%	80-90	
Activation energy of polymerization	$J\ g^{-1}$	398.4-1077.8	
Heat of polymerization	$kJ\ mol^{-1}$	-96 to -109	
Typical concentration of residual monomer	ppm	<1	
Mass average molecular weight, M_w	dalton, g/mol, amu	37,000-214,000	
Polydispersity, M_w/M_n	-	1.90-2.59 (suspension); 2.14-2.65 (emulsion); 2.00-2.06 (mass)	

HANDBOOK OF POLYMERS 3rd Edition, Copyrights 2022; ChemTec Publishing

PVC poly(vinyl chloride)

PARAMETER	UNIT	VALUE	REFERENCES
Polymerization degree (number of monomer units)	-	600-3,400	
K number		50-95 (suspension); 60-80 (emulsion); 58-69 (mass)	
Mean particle size	microm-eter	100-150 (suspension); 40-50 (general purpose emulsion); 2-25 (paste forming)	
Molar volume at 298K	cm^3 mol^{-1}	calc.=41.0 (crystalline); 45.1-58.4 (amorphous)	
Van der Waals volume	cm^3 mol^{-1}	calc.= 29.2 (crystalline); 29.2-38.0 (amorphous)	
Molecular cross-sectional area, calculated	cm^2 x 10^{-16}	18.5	
Radius of gyration	nm	5-10; 16.4-28.2	Wan, C; Qiao, X; Zhang, Y; Zhang, Y, Polym. Test., 22, 453-61, 2003; Mutin, P H; Guenet, J M, Polymer, 27, 7, 1098-1102, 1986.
Degree of branching	num-ber/1000 VC	3.3-4.8 (chloromethyl), 0.8 (short branches from backbiting, 0.1-0.2 (long branches), 0.9 (tertiary chlorines)	
Unsaturations	num-ber/1000 VC	0.1-0.3 (internal allylic chlorine), 0.1-0.6 (internal), 0.75-0.8 (end-group), 0.95-1.7 (total)	
Typical chain imperfections	num-ber/1000 VC	6-8 (head-to-head), 0.1-0.4 (initiator rests)	

STRUCTURE			
Crystallinity	%	4-10 (commercial); 5	-; Kiflemariam, B, Collin, D, Gavat, O, Carvalho, A, Moulin, E, Giuseppone, N, Guenet, J-M, Polymer, 207, 122814, 2020.
Crystalline structure	-	lamellar, fringed micelles	
Cell type (lattice)	-	orthorhombic	
Cell dimensions	nm	a:b:c=1.01-1.08:0.53-0.54:0.510-0.512	Natta, G; Corradini, P, J. Polym. Sci., 20, 251, 1956.
Unit cell angles	degree	$\alpha:\beta:\gamma=90:90:90$	
Number of chains per unit cell	-	2	Natta, G; Corradini, P, J. Polym. Sci., 20, 251, 1956.
Crystallite size	nm	0.7-15	
Spacing between crystallites	nm	0.26-20	
Tacticity	%	55-68 (syndiotactic dyads); typical: 27.6-44.0 (syndiotactic), 4.8-21.8 (isotactic), 30.5-52.0 (heterotactic)	
Chain conformation	-	planar zigzag	
Entanglement molecular weight	dalton, g/mol, amu	6,250	
Lamellae thickness	nm	2.5-6	Ballard, D G H; Burgess, A N; Deconinck, J W; Roberts, E A, Polymer, 28, 1, 3-9, 1987.

COMMERCIAL POLYMERS			
Some manufacturers	-	Geon, Vestolit/Orbia, Formosa Plastics, Shin-Etsu, Benvic, Solvay, Ineos, Inovyn	
Brand names	-	Geon, Vestolit, Formolon, Shintec, Benvic, SolVin, Norvinyl, Pevicon	

PVC poly(vinyl chloride)

PARAMETER	UNIT	VALUE	REFERENCES
PHYSICAL PROPERTIES			
Density at 20°C	g cm^{-3}	1.37-1.43; 1.53 (crystalline); 1.373 (amorphous)	
Bulk density at 20°C	g cm^{-3}	0.39-0.59	
Specific volume	10^{-4}m^{3}kg^{-1}	6.90-7.41	
Color	-	white	
Refractive index, 20°C	-	1.532-1.548	
Odor		odorless	
Melting temperature, DSC	°C	103-230; 400 (syndiotactic, estimate)	
Decomposition onset temperature	°C	200	Patel, P; Hull, T R; McCabe, R W; Flath, D; Grasmeder, J; Percy, M, Polym. Deg. Stab., 95, 709-18, 2010.
Fusion temperature	°C	185-195	
Thermal expansion coefficient, 23-80°C	°C^{-1}	3.5-7.1E-5	
Thermal conductivity, melt	W m^{-1} K^{-1}	0.13-0.17	
Thermal diffusivity at 20°C	10^{-4}cm^{2}s^{-1}	11.92	
Glass transition temperature	°C	calc.=81-82; exp.=82-87 (rigid); 66 (5 phr plasticizer); 13-52 (30 phr plasticizer); -52 to -82 (100 phr plasticizer); 90.2	Zhou, M-Y,, L-F,Sun, C-C, Chen, J-H, J. Membrane Sci., 572, 401-9, 2019.
Specific heat capacity	J K^{-1} kg^{-1}	900-970	
Heat deflection temperature at 1.8 MPa	°C	73-74	
Vicat temperature VST/A/50	°C	82-95	
Vicat temperature VST/B/50	°C	65-100	
Hansen solubility parameters, δ_D, δ_P, δ_H	MPa$^{0.5}$	18.82, 10.03, 3.07; 16.8, 8.9, 6.1; 18.4, 6.6, 8.0	
Interaction radius		8.5; 3.5; 3.0	
Molar volume	kmol m^{-3}	45.2	
Hildebrand solubility parameter	MPa$^{0.5}$	calc.=19.28-20.23; exp.=19.19-20.1; 19.4	Abreu,C M R, Fonseca, A C, Rocha, N M P, Coelho, J F J, Prog. Polym. Sci., 87, 34-69, 2018.
Surface tension	mN m^{-1}	32-46, 41.5	Wu, S, J. Adhesion, 5, 39, 1973; Negari, M S, Movahed, S O, Ahmadpour, A, Separation Purification Technol., 194, 368-76, 2018.
Dielectric constant at 1 kHz/1 MHz	-	3.39-3.5	Lu, Y, Khanal, S, Ahmed, S, Xu, S, Compos. Sci. Technol., 172, 29-35, 2019.
Dielectric loss factor at 1 kHz	-	0.81	
Relative permittivity at 100 Hz	-	0.009-0.017	
Volume resistivity	ohm-m	1E12 to 1E13	
Surface resistivity	ohm	1E11 to 1E12	
Arc resistance	s	60-80	
Coefficient of friction	-	0.35-0.8 (static), 0.72-0.93 (dynamic) on steel	DeCoste, J B, Antec, 232, 1969.
Permeability to nitrogen, 25°C	m^3 s^{-1} m^2 Pa^{-1} 10^{-9}	0.0089	
Permeability to oxygen, 25°C	m^3 s^{-1} m^2 Pa^{-1} 10^{-9}	0.034	

HANDBOOK OF POLYMERS 3rd Edition, Copyrights 2022; ChemTec Publishing

PVC poly(vinyl chloride)

PARAMETER	UNIT	VALUE	REFERENCES
Permeability to water vapor, 25°C	$m^3 s^{-1} m^2$ $Pa^{-1} 10^{-9}$	0.12	
Diffusion coefficient of nitrogen	$cm^2 s^{-1}$ $x10^6$	0.0038	
Diffusion coefficient of oxygen	$cm^2 s^{-1}$ $x10^6$	0.012	
Contact angle of water, 20°C	degree	80.4-91.9	Hammiche, D, Boukerrou, A, Guermazi, N, Arrakhiz, F E, Mater. Today: Proc.: in press, 2020
Surface free energy	$mJ m^{-2}$	40.1	
Surface tension	$mN m^{-1}$	32-46	
Speed of sound	$m s^{-1}$	39.7	
Acoustic impedance		3.27	
Attenuation	$dB cm^{-1}$, 5 MHz	11.2	

MECHANICAL & RHEOLOGICAL PROPERTIES

PARAMETER	UNIT	VALUE	REFERENCES
Tensile strength	MPa	7.1-68.9	
Tensile modulus	MPa	2,430-4,000	
Tensile stress at yield	MPa	39.2-88.3	
Elongation	%	3.3-430	
Flexural strength	MPa	67-107	
Flexural modulus	MPa	2,580-3,310	
Tear strength	MPa	2.1-7.9	
Izod impact strength, notched, 23°C	$J m^{-1}$	33-1302	
Tenacity (fiber) (standard atmosphere)	$cN tex^{-1}$ $(daN$ $mm^{-2})$	10-30	Fourne, F, Synthetic Fibers. Machines and Equipment Manufacture, Properties. Carl Hanser Verlag, 1999.
Tenacity (wet fiber, as % of dry strength)	%	100	Fourne, F, Synthetic Fibers. Machines and Equipment Manufacture, Properties. Carl Hanser Verlag, 1999.
Fineness of fiber (titer)	dtex	1.5-60	Fourne, F, Synthetic Fibers. Machines and Equipment Manufacture, Properties. Carl Hanser Verlag, 1999.
Length (elemental fiber)	mm	38-200	Fourne, F, Synthetic Fibers. Machines and Equipment Manufacture, Properties. Carl Hanser Verlag, 1999.
Poisson's ratio	-	0.380-0.385	
Shore A hardness	-	30-96	
Shore D hardness	-	22-25	
Rockwell hardness	-	M66-69	
Shrinkage	%	0.5-2.5	
Brittleness temperature (ASTM D746)	°C	-29 to -41	
Water absorption, equilibrium in water at 23°C	%	0.04-0.4	

PVC poly(vinyl chloride)

PARAMETER	UNIT	VALUE	REFERENCES
CHEMICAL RESISTANCE			
Acid dilute/concentrated	-	very good	
Alcohols	-	good	
Alkalis	-	very good	
Aliphatic hydrocarbons	-	good	
Aromatic hydrocarbons	-	fair-poor	
Esters	-	poor	
Greases & oils	-	good	
Halogenated hydrocarbons	-	poor	
Ketones	-	poor	
Q solvent, Θ-temp.=155.4, 22, 36.5, 84°C	-	benzyl alcohol, cyclohexanone, dimethylformamide, o-xylene	
Good solvent	-	chlorobenzene, cyclohexanone, DMF, DMSO, MEK, nitrobenzene, THF	
Non-solvent	-	acetone, non-oxidizing acids, alkalies, aniline, carbon disulfide, hydrocarbons, nitroparaffins	
Effect of EtOH sterilization (tensile strength retention)	%	113-115	Navarrete, L; Hermanson, N, Antec, 2807-18, 1996.
FLAMMABILITY			
Ignition temperature	°C	391	
Autoignition temperature	°C	435-454	
Limiting oxygen index	% O_2	37-49	
Heat release	kW m^{-2}	176	Yu, B; Liu, M; Lu, L; Dong, X; Gao, W; Tang, K, Fire Mater., 34, 251-61, 2010.
NBS smoke chamber	Ds	349-500	Padey, D; Walling, J; Wood A, Polymers in Defence and Aerospace 2007, Rapra, 2007, paper 15.
Char at 500°C	%	10.9-18.0	Lyon, R E; Walters, R N, J. Anal. Appl. Pyrolysis, 71, 27-46, 2004.
Heat of combustion	J g^{-1}	17,950	
Volatile products of combustion	-	CO, CO_2, H_2O, HCl; traces of benzene and phosgene	
CO yield	%	8 (with flame retardant)	
WEATHER STABILITY			
Spectral sensitivity	nm	310-370	Wypych, G, Handbook of Materials Weathering, 6th Ed., ChemTec Publishing, 2018.
Activation wavelengths	nm	310-325, 327, 364	
Excitation wavelengths	nm	284, 290	
Emission wavelengths	nm	350, 440	
Activation energy of photo-oxidation	kJ mol^{-1}	32.1 (nitrogen); 19.6 (air)	
Depth of UV penetration	μm	90; 150-200	

PVC poly(vinyl chloride)

PARAMETER	UNIT	VALUE	REFERENCES
Important initiators and accelerators	-	carbonyl groups, unsaturations, solvents forming hydroperoxides, sensitizing impurities (e.g., benzophenones), metaloorganics (copper-containing compounds, cadmium acetate, ferrocene, iron salts), metal chlorides produced from thermal stabilizers, products of degradation of some anti-oxidants, some pigments and fillers (containing cobalt, zinc, manganese, and lead), metal oxides (of titanium, zinc, and aluminum), hydrogen chloride (autocatalytic product of PVC degradation)	
Products of degradation	-	free radicals, unsaturations, carbonyl groups, hydroperoxides, chain scissions, crosslinks	
Stabilizers	-	UVA: 2-hydroxy-4-octyloxybenzophenone; 2-hydroxy-4-methoxybenzophenone; 2,2'-dihydroxy-4-methoxybenzophenone; 2-(2H-benzotriazol-2-yl)-p-cresol; 2-benzotriazol-2-yl-4,6-di-tert-butylphenol; 2-(2H-benzotriazole-2-yl)-4,6-di-tert-pentyl-phenol; 2-(2H-benzotriazole-2-yl)-4-(1,1,3,3-tetraethylbutyl) pheno; 2-(2H-benzotriazol-2-yl)-6-dodecyl-4-methylphenol, branched & linear; reaction product of methyl 3(3-(2H-benzotriazole-2-yl)-5-t-butyl-4-hydroxyphenyl propionate/ PEG 300; ethyl-2-cyano-3,3-diphenylacrylate; (2-ethylhexyl)-2-cyano-3,3-diphenylacrylate; N-(2-ethoxyphenyl)-N'-(2-ethylphenyl)oxamide; propanedioic acid, [(4-methoxyphenyl)-methylene]-dimethyl ester; Screener: carbon black, titanium dioxide, zinc oxide; Acid scavenger: hydrotalcite; HAS: 1,3,5-triazine-2,4,6-triamine, N,N'''[1,2-ethane-diyl-bis[[[4,6-bis[butyl(1,2,6,6-pentamethyl-4-piperidinyl) amino]-1,3,5-triazine-2-yl]imino]-3,1-propanediyl]bis[N',N''-dibutyl-N',N''-bis(1,2,2,6,6-pentamethyl-4-piperidinyl)-; bis(1,2,2,6,6-pentamethyl-4-piperidyl) sebacate and methyl 1,2,2,6,6-pentamethyl-4-piperidyl sebacate; bis(1,2,2,6,6-pentamethyl-4-piperidyl)sebacate + methyl-1,2,2,6,6-pentamethyl-4-piperidyl sebacate; bis(2,2,6,6-tetramethyl-4-piperidyl) sebacate; poly[[(6-[1,1,3,3-tetramethylbutyl)amino]-1,3,5-triazine-2,4-diyl][2,2,6,6-tetramethyl-4-piperidinyl)imino]-1,6-hexanediyl[2,2,6,6-tetramethyl-4-piperidinyl)imino]]; C20-24-. alpha.-, polymers with maleic anhydride, reaction products with 2,2,6,6-tetramethyl-4-piperidinamine; 1,6-hexanediamine, N,N'-bis(2,2,6,6-tetramethyl-4-piperidinyl)-, polymers with morpholine-2,4,6-trichloro-1,3,5-triazine reaction products, methylated; Phenolic antioxidants: ethylene-bis(oxyethylene)-bis(3-(5-tert-butyl-4-hydroxy-m-tolyl)-propionate); pentaerythritol tetrakis(3-(3,5-di-tert-butyl-4-hydroxyphenyl)propionate); 3,3',3',5,5',5'-hexa-tert-butyl-a,a',a'-(mesitylene-2,4,6-triyl) tri-p-cresol; 1,3,5-tris(3,5-di-tert-butyl-4-hydroxybenzyl)-1,3,5-triazine-2,4,6(1H,3H,5H)-trione; isotridecyl-3-(3,5-di-tert-butyl-4-hydroxyphenyl) propionate; 2,2'-ethylidenebis (4,6-di-tert-butylphenol); 3,5-bis(1,1-dimethyethyl)-4-hydroxy-benzenepropanoic acid, C13-15 alkyl esters; 1,1,3-tris(2'methyl-4'-hydroxy-5'tert-butylphenyl)butane	
BIODEGRADATION			
Colonized products	-	mattresses, plasticizers	
Typical biodegradants	-	phthalate esters are degraded by a wide range of bacteria and actinomycetes under both aerobic and anaerobic conditions	
Stabilizers	-	copper nanoparticle, 4,5-dichloro-2-n-octylisothiazolin-3-one, 2-n-octyl-isothiazolin-3-one, 10,10'-oxybisphenoxarsine, surface azidation,4 tebuconazole, 2,3,5,6-tetrachloro-4-(methylsulphonyl)pyridine, zeolite encapsulated 2-n-octyl-4-isothiazolin-3-one, zinc pyrithione	

PVC poly(vinyl chloride)

PARAMETER	UNIT	VALUE	REFERENCES
TOXICITY			
NFPA: Health, Flammability, Reactivity rating	-	1/1/0	
Carcinogenic effect	-	not listed by ACGIH, NIOSH, NTP	
TLV, ACGIH	mg m^{-3}	1 (respirable)	
OSHA	mg m^{-3}	5 (respirable); 15 (total)	
ENVIRONMENTAL IMPACT			
Aquatic toxicity, *Daphnia magna*, LC$_{50}$, 48 h	mg l^{-1}	800-8,000; 8,000-235,0000	Lithner, D; Damberg, J; Dave, G; Larsson, A, Chemosphere, 74, 1195-1200, 2009; Lithner, Ph D Thesis, Univresity of Gothenburg, 2011.
Cradle to grave non-renewable energy use	MJ/kg	53-55	
Cradle to pellet greenhouse gasses	kg CO$_2$ kg^{-1} resin	2.0-2.1	
PROCESSING			
Typical processing methods	-	blow molding, calendering, extrusion, injection molding, plastisol coating, rotational molding, thermoforming	
Additives used in final products	-	Fillers: aluminum fiber, aluminum hydroxide, antimony trioxide, calcium carbonate, carbon black, carbon fiber, clay, graphene, magnesium hydroxide, montmorillonite, sand, silica, talc, titanium dioxide, wood fiber; Plasticizers: adipates, azelates, benzoates, citrates, epoxidized soybean oil, ethylene inter-polymers, phosphates, phthalates, polyester-type polymeric plasticizers, sebacates; Antistatics: chlorinated polyethylene, carbon black, copper powder, ethoxylated fatty dimethyl ethylammoniumethosulfate, glycerol monostearate, graphite, polyethylene glycol monolaurate, propanesultone; Antiblocking: aluminosilicate, natural silica, synthetic silica; Release: ester wax, ethylene N,N'-bisstearamide, glyceryl monostearate; Slip: ethylene N,N'-bisoleamide, stearamide, zinc or calcium stearate or their mixture	
Applications	-	bottles, cables, coated fabrics, domestic appliances, drain pipes, film and sheet, fittings, flooring, foam backings of carpets, footwear, furniture trim, gloves, gutters, metal protection in automotive, office equipment, packaging, pipes, profiles, protective clothing, toys, tubing, siding, wallpaper, windows, and many more; ranking: high to low: pipe & fitting, window, rigid profile, wire and cable, flexible film, bottles, flooring, coating, flexible tube, roofing, medical, rigid sheet	
BLENDS			
Suitable polymers	-	CA, ENR, epoxy, EVA, NBR, NR, PANI, PMMA, PPy, PS, PUR, PVA, PVB, PVDF, SAN, SBR	
ANALYSIS			
FTIR (wavenumber-assignment)	cm^{-1}/-	1714, 1715, 1718, 1720, 1730 (carbonyl); 1785 (acid chloride); 1510 (carboxylate stabilizer); 3476-3420 (hydroperoxide); 3460 (hydroxyl); 1650 (isolated double bond); 1580 (conjugated double bond)	
Raman (wavenumber-assignment)	cm^{-1}/-	syndiotactic triads – 608, 630, 636; isotactic triads – 697	Dubault, A; Bokobza, L; Gandin, E; Halary, J L; polym. Int., 52, 7, 1108-18, 2003.

PVC poly(vinyl chloride)

PARAMETER	UNIT	VALUE	REFERENCES
NMR (chemical shifts)	ppm	C NMR: CH_2 – 46; CHCl – 58	Colombani, J; Labed, V; Joussot-Dubien, C; Perichaud, A; Raffi, J; Kister, J; Rossi, C, Nuclear Instruments Methods Phys. Rese., B265, 238-44, 2007.
x-ray diffraction peaks	degree	16-18, 25 (crystalline area)	

PVCA poly(vinyl chloride-co-vinyl acetate)

PARAMETER	UNIT	VALUE	REFERENCES
GENERAL			
Common name	-	poly(vinyl chloride-co-vinyl acetate)	
CAS name	-	acetic acid ethenyl ester, polymer with chloroethene	
Acronym	-	PVCA	
CAS number	-	9003-22-9; 9003-20-7	
RTECS number	-	AK0950000	
HISTORY			
Person to discover	-	Werntz, J H	Werntz, J H, US Patent 1,988,529, DuPont, Jan. 22, 1935.
Date	-	1935	
Details	-	polymerization of vinyl copolymers including PVCA	
SYNTHESIS			
Monomer(s) structure	-	$H_2C=CHOCCH_3$ (O) $H_2C=CHCl$	
Monomer(s) CAS number(s)	-	108-05-4; 75-01-4	
Monomer(s) molecular weight(s)	dalton, g/mol, amu	86.09; 62.50	
Vinyl chloride content	%	86-93 (SolVin); 61-89 (Vinnol)	
Method of synthesis	-	suspension polymerization	
Typical concentration of residual monomer	ppm	<1 (vinyl chloride)	
Number average molecular weight, M_n	dalton, g/mol, amu	44,000	
Mass average molecular weight, M_w	dalton, g/mol, amu	27,000 (SolVin); 40,000-80,000 (Vinnol)	
Polymerization degree (number of monomer units)	-	415	
K value		50-61 (SolVin); 41-51 (Vinnol)	
COMMERCIAL POLYMERS			
Some manufacturers	-	Solvay/Inovyn; Wacker	
Trade names	-	SolVin; Vinnol	
PHYSICAL PROPERTIES			
Density at 20°C	g cm^{-3}	1.30-1.37	
Bulk density at 20°C	g cm^{-3}	0.53 (SolVin); 0.7-0.75 (Vinnol)	
Color	-	white	
Refractive index, 20°C	-	1.45-1.47	
Odor	-	mild, pleasant	
Melting temperature, DSC	°C	100-110	
Softening point	°C	38	
Decomposition temperature	°C	100	
Glass transition temperature	°C	58-75	

PVCA poly(vinyl chloride-co-vinyl acetate)

PARAMETER	UNIT	VALUE	REFERENCES
MECHANICAL & RHEOLOGICAL PROPERTIES			
Tensile strength	MPa	3.27-4.86	Allie, L; Thorn, J; Aglan, H, Corrosion Sci., 50, 2189-96, 2008.
Tensile modulus	MPa	1,700	
Elongation	%	290	
CHEMICAL RESISTANCE			
Acid dilute/concentrated	-	good	
Alcohols	-	good	
Aromatic hydrocarbons	-	poor	
Ketones	-	poor	
Good solvent	-	acetone, MEK, THF	
FLAMMABILITY			
Autoignition temperature	°C	390; >450	
Volatile products of combustion	-	CO, CO_2, HCl	
TOXICITY			
HMIS: Health, Flammability, Reactivity rating	-	2/1/0	
Carcinogenic effect	-	not listed by ACGIH, NIOSH, NTP	
TLV, ACGIH	mg m^{-3}	3 (respirable), 10 (total)	
OSHA	mg m^{-3}	5 (respirable), 15 (total)	
Oral rat, LD_{50}	mg kg^{-1}	>25,000	
PROCESSING			
Typical processing methods	-	calendering, deep draw thermoforming, extrusion, solution	
Applications	-	aluminum foil coating, adhesives, blisters, credit cards, flooring, inks, mastics, membranes, paints, pharmaceutical packaging, printing inks, records, rigid sheets, sealers, strippable coatings, varnishes	
BLENDS			
Suitable polymers	-	CA, PDMS, PPy, PSA, SAN	
ANALYSIS			
FTIR (wavenumber-assignment)	cm^{-1}/-	C=O – 1740, C–O – 1237, 1025; C–Cl – 700, 639, 615; C–H – 1433, 1426, 1329, 1101, 966	Kupcik, J; Blazevska-Gilev, J; Subrt, J; Vorlicek, V; Pola, J, Polym. Deg. Stab., 91, 2560-66, 2006.
Raman (wavenumber-assignment)	cm^{-1}/-	C–C – 1128; C=C – 1518	Kupcik, J; Blazevska-Gilev, J; Subrt, J; Vorlicek, V; Pola, J, Polym. Deg. Stab., 91, 2560-66, 2006.

PVDC poly(vinylidene chloride)

PARAMETER	UNIT	VALUE	REFERENCES
GENERAL			
Common name	-	poly(vinylidene chloride)	
IUPAC name	-	poly(1,1-dichloroethene)	
Acronym	-	PVDC	
CAS number	-	9002-85-1	
Formula		$-\left[CH_2CCl_2\right]_n$	
HISTORY			
Person to discover	-	Henri Vicror Regnault; Ralph Wiley, Dow Chemical	
Date	-	1830; 1933; 1940; 1953	
Details	-	Regnault synthesized monomer and polymerized it by heating; Wiley discovered polymer by accident; produced in DOW in 2000 gal reactor; commercialized as Saran	Mounts, M L, Antec, 3849-53, 2003.
SYNTHESIS			
Monomer(s) structure	-	$H_2C{=}CCl_2$	
Monomer(s) CAS number(s)	-	75-35-4	
Monomer(s) molecular weight(s)	dalton, g/mol, amu	96.94	
Monomer ratio	-	100%	
Formulation example	-	vinylidene chloride, water, methyl hydroxypropyl cellulose, lauroyl peroxide	
Method of synthesis	-	free radical polymerization	
Temperature of polymerization	^{o}C	60	
Time of polymerization	h	30-60	
Yield	%	85-98	
Heat of polymerization	$J\ g^{-1}$	836-1104	Lebedev, B V; Kulagina, T G; Smirnova, N N, Vyssomol. Soed., A, 37, 1896, 1995.
Mass average molecular weight, M_w	dalton, g/mol, amu	105,000-125,000	
Polydispersity, M_w/M_n	-	1.5-2.0	
Molar volume at 298K	$cm^3\ mol^{-1}$	49.7 (crystalline)	
Van der Waals volume	$cm^3\ mol^{-1}$	38.0 (crystalline)	
CRYSTALLINE STRUCTURE			
Crystallinity	%	4-6	
Cell type (lattice)	-	monoclinic	
Cell dimensions	nm	a:b:c=0.672:1.252:0.468	
Unit cell angles	degree	γ=123	
Rapid crystallization temperature	^{o}C	140-150	
COMMERCIAL POLYMERS			
Some manufacturers	-	Dow	
Trade names	-	Saran	

PVDC poly(vinylidene chloride)

PARAMETER	UNIT	VALUE	REFERENCES
PHYSICAL PROPERTIES			
Density at 20°C	g cm^{-3}	1.70; 1.775 (amorphous); 1.97 (crystalline)	
Bulk density at 20°C	g cm^{-3}	0.8	
Refractive index, 20°C	-	1.60-1.63	
Odor	-	odorless	
Melting temperature, DSC	°C	158-205	
Softening point	°C	100-150	
Decomposition temperature	°C	120	
Glass transition temperature	°C	calc=-19; -17 to -20	
Heat of fusion	kJ mol^{-1}	5.62	
Hansen solubility parameters, δ_D, δ_P, δ_H	MPa$^{0.5}$	19.0, 9.6, 9.0	
Interaction radius		5.8	
Hildebrand solubility parameter	MPa$^{0.5}$	25.0; 23.1	
Surface tension	mN m^{-1}	calc.=45.4	Wu, S, J. Adhesion, 5, 39, 1973.
Permeability to nitrogen, 25°C	cm^3 cm cm^{-2} s^{-1} Pa^{-1} x 10^{12}	0.0000706	
Permeability to oxygen, 25°C	cm^3 cm cm^{-2} s^{-1} Pa^{-1} x 10^{12}	0.000383	
Permeability to water vapor, 25°C	cm^3 cm cm^{-2} s^{-1} Pa^{-1} x 10^{12}	0.7	
Contact angle of water, 20°C	degree	80	
Surface free energy	mJ m^{-2}	31.5	
MECHANICAL & RHEOLOGICAL PROPERTIES			
Tensile strength	MPa	24-69; 207-414 (oriented)	
Tensile stress at yield	MPa	19-26	
Elongation	%	10-20; 15-40 (oriented)	
Flexural strength	MPa	40	
Flexural modulus	MPa	500	
Compressive strength	MPa	55	
Rockwell hardness	-	R55	
Shrinkage	%	0.5-2.5	
Brittleness temperature (ASTM D746)	°C	-10 to 10	
Melt viscosity, shear rate=1000 s^{-1}	Pa s	750	
CHEMICAL RESISTANCE			
Acid dilute/concentrated	-	very good	
Alcohols	-	very good	

PVDC poly(vinylidene chloride)

PARAMETER	UNIT	VALUE	REFERENCES
Alkalis	-	good-poor	
Aliphatic hydrocarbons	-	good	
Aromatic hydrocarbons	-	fair-poor	
Esters	-	good-fair	
Greases & oils	-	good	
Halogenated hydrocarbons	-	fair-poor	
Ketones	-	fair	
Good solvent	-	benzonitrile, butyl acetate, cyclohexanone, 1,2-dichlorobenzene, dioxane, DMA, DMF, NMP, tetrahydrofurfuryl alcohol, tetralin (hot), THF (hot), trichloroethane	
Non-solvent	-	concentrated acids and alkalis (except ammonia), alcohols, carbon disulfide, chloroform, cyclohexanone, dioxane, hydrocarbons, phenols, THF	

FLAMMABILITY

PARAMETER	UNIT	VALUE	REFERENCES
Ignition temperature	°C	>530	
Autoignition temperature	°C	>530	
Limiting oxygen index	% O$_2$	60	
UL 94 rating	-	V-0	

WEATHER STABILITY

PARAMETER	UNIT	VALUE	REFERENCES
Important initiators and accelerators	-	Pd/AC (oxidation catalyst)	

PROCESSING

PARAMETER	UNIT	VALUE	REFERENCES
Typical processing methods	-	coating, extrusion, fiber spinning, molding	
Additives used in final products	-	Plasticizers: acetyl tri-n-butyl citrate, epoxidized soybean oil, polymeric condensation product of azelic acid and 1,3-butanediol, polymeric plasticizer of adipic acid and propylene glycol, Antistatics: ionic polymer, imidazoline/metal salt	
Applications	-	cling wrap (recently changed to LDPE), coatings, fiber, film layer reducing permeability of oxygen and flavors, filters, membranes, monofilaments, screens, shower curtains, stuffed animals, tape	

BLENDS

PARAMETER	UNIT	VALUE	REFERENCES
Suitable polymers	-	PBA, PBM	

PVDF poly(vinylidene fluoride)

PARAMETER	UNIT	VALUE	REFERENCES
GENERAL			
Common name	-	poly(vinylidene fluoride)	
IUPAC name	-	poly(1,1-difluoroethene)	
CAS name	-	ethene, 1,1-difluoro-, homopolymer	
Acronym	-	PVDF	
CAS number	-	24937-79-9	
Formula		$\left[CF_2CF_2\right]_n$	
HISTORY			
Person to discover	-	Ford, T A; Hanford, W E	Ebnesajjad, S, Fluoroplastics. Vol. 2. Melt Processible Fluoroplastics, William Andrew, 2003.
Date	-	1948 (patent), 1961 (commercialization)	
Details	-	DuPont scientists patented and commercialized PVDF	
SYNTHESIS			
Monomer(s) structure	-	$F_2C{=}CF_2$	
Monomer(s) CAS number(s)	-	75-38-7	
Monomer(s) molecular weight(s)	dalton, g/mol, amu	64.04	
Monomer(s) expected purity(ies)	%	97	
Monomer ratio	-	100%	
Formulation example	-	monomer 100, surfactant 0.1-0.2, initiator 0.05-0.6, paraffin wax 0.03-0.3, chain transfer agent 1.5-6.0	Ebnesajjad, S, Fluoroplastics. Vol. 2. Melt Processible Fluoroplastics, William Andrew, 2003.
Method of synthesis	-	emulsion and suspension polymerization	
Temperature of polymerization	°C	60-90	Ebnesajjad, S, Fluoroplastics. Vol. 2. Melt Processible Fluoroplastics, William Andrew, 2003.
Pressure of polymerization	MPa	2.8-4.8	Ebnesajjad, S, Fluoroplastics. Vol. 2. Melt Processible Fluoroplastics, William Andrew, 2003.
Number average molecular weight, M_n	dalton, g/mol, amu	64,000-380,000	
Mass average molecular weight, M_w	dalton, g/mol, amu	60,000-534,000	
Polydispersity, M_w/M_n	-	1.6-3.2	
Molar volume at 298K	cm^3 mol^{-1}	32.0 (crystalline)	
Van der Waals volume	cm^3 mol^{-1}	25.6 (crystalline)	
Radius of gyration	nm	14.8-26.5	Lutringer, G; Weill, G, Polym, 32, 5, 1896-1908, 1994.
STRUCTURE	-		
Crystallinity	%	32-76; 41-46 (DSC); 33-41 (WAXS)	Ebnesajjad, S, Fluoroplastics. Vol. 2. Melt Processible Fluoroplastics, William Andrew, 2003; Botelho, G; Silva, M M; Goncalves, A M; Sencadas, V; Serrado-Nunes, J; Lanceros-Mendez, S, Polym. Testing, 27, 818-22, 2008.

PVDF poly(vinylidene fluoride)

PARAMETER	UNIT	VALUE	REFERENCES
Cell type (lattice)	-	monoclinic, orthorhombic	
Cell dimensions	nm	a:b:c=0.496:0.964:0.462 (α); =0.848:0.491:0.256 (β)	Rutleedge, G C; Carbeck, J D; Lacks, D J, Antec, 2163-66, 1996.
Unit cell angles	degree	α:β:γ=90:90:90	
Number of chains per unit cell	-	2	
Crystallite size	nm	3,000-9,000 (spherulite radius)	
Thickness of layer	nm	6.86-6.95 (crystalline); 2.25-3.24	Linares, A; Nogales, A; Sanz, A; Ezquerra, T A; Peruccini, M, Phys. Rev. E, 82, 0.31802, 1-11, 2010.
Spacing between crystallites	nm	2.25-3.24	
Polymorphs	-	α, β, γ, δ	Rietveld, I B; Kobayashi, K; Honjo, T; Ishida, K; Yamada, H; Matsushige, K, J. Mater. Chem., 20, 8272-78, 2010.
Chain conformation	-	TGTG (α); TTTT (β), TTTG (γ); polar version of α (δ)	Li, W; Meng, Q; Zheng, Y; Zhang, Z; Xia, W; Xu, Z, Appl. Phys. Lett., 96, 192905,1-3, 2010.
Lamellae thickness	nm	8-10	Zhao, Z; Chu, J; Chen, X, Radiat. Phys. Chem., 43, 6, 523-26, 1994.
Heat of crystallization	kJ kg^{-1}	35-45	Gradys, A; Sajkiewicz, P; Adamovsky, S; Minakov, A; Schick, C, Thermochim. Acta, 461, 153-57, 2007.

COMMERCIAL POLYMERS

Some manufacturers	-	Arkema; Solvay; Ensinger	
Trade names	-	Kynar; Hylar, Solef, Tecaflon	

PHYSICAL PROPERTIES

Density at 20°C	g cm^{-3}	1.76-1.83; 1.68 (amorphous); 1.92-1.98 (crystalline)	
Color	-	white	
Refractive index, 20°C	-	1.42-1.49	
Transmittance	%	85-94	
Haze	%	3-13	
Gloss, 60°, Gardner (ASTM D523)	%	25	
Odor	-	odorless	
Melting temperature, DSC	$^{\circ}$C	158-200; 167-169 (main melting peak)	Linares, A; Nogales, A; Sanz, A; Ezquerra, T A; Peruccini, M, Phys. Rev. E, 82, 0.31802, 1-11, 2010.
Crystallization point	$^{\circ}$C	92-140	
Degradation temperature	$^{\circ}$C	375 (air), 410 (nitrogen)	
Thermal expansion coefficient, 23-80°C	$^{\circ}$C^{-1}	0.7-1.8E-4	
Thermal conductivity, 23°C	W m^{-1} K^{-1}	0.226	Song, Q, Zhu, W, Deng, Y, Hai, F, Guo, Z, Compos. Part A: Appl. Sci. Manuf., 127, 105654, 2019.
Glass transition temperature	$^{\circ}$C	calc.=-67; exp.=-29 to -57	
Specific heat capacity	J K^{-1} kg^{-1}	1,200-1,600 (23-100°C); 960 (melt)	
Heat of fusion	J g^{-1}	50.3-56.1	Mekhilef, N; Hedhli, L; Reynaud, S; Pasquariello, G O, Antec, 1133-38, 2007.
Heat of crystallization	kJ kg^{-1}	19-58	
Maximum service temperature	$^{\circ}$C	130-150	
Long term service temperature	$^{\circ}$C	150	

PVDF poly(vinylidene fluoride)

PARAMETER	UNIT	VALUE	REFERENCES
Heat deflection temperature at 0.45 MPa	°C	48-148	
Heat deflection temperature at 1.8 MPa	°C	36-115	
Vicat temperature VST/A/50	°C	110-172	
Enthalpy of melting	J g^{-1}	104.5; 53.5 (membrane)	Linares, A; Nogales, A; Sanz, A; Ezquerra, T A; Peruccini, M, Phys. Rev. E, 82, 0.31802, 1-11, 2010; Hejazi, M-A A,. Bamaga, O A, Al-Beirutty, M H, Abulkhair , H, Separation Purification Technol., 220, 300-8, 2019.
Fusion enthalpy	J g^{-1}	105 (fully crystalline PVDF)	Hejazi, M-A A,. Bamaga, O A, Al-Beirutty, M H, Abulkhair , H, Separation Purification Technol., 220, 300-8, 2019.
Acceptor number	-	17.0, 12.1, 10.2	
Interaction radius		4.1	
Hansen solubility parameters, δ_D, δ_P, δ_H	MPa$^{0.5}$	17.2, 12.5, 9.2	Yang, X, Zhu, H, Jiang, F, Zhou, X, J. Power Sources, 473, 228586, 2020.
Hildebrand solubility parameter	MPa$^{0.5}$	16.8-18.4	
Surface tension	mN m^{-1}	calc.=33.2; exp.=33.2	Wu, S, J. Adhesion, 5, 39, 1973.
Dielectric constant at 1000 Hz/1 MHz	-	9-10.5/7-9.9	
Dissipation factor at 100 Hz		0.03-0.05	
Dissipation factor at 1 MHz		0.03-0.05	
Volume resistivity	ohm-m	1.5-2.3E12; 2.5E12	-; Song, Q, Zhu, W, Deng, Y, Hai, F, Guo, Z, Compos. Part A: Appl. Sci. Manuf., 127, 105654, 2019.
Surface resistivity	ohm	1E13-1E14	
Electric strength K20/P50, d=0.60.8 mm	kV mm^{-1}	63-67	
Arc resistance	MV/m	700	
Coefficient of friction	-	0.3 (static), 0.24 (dynamic)	Tóth, L F, Sukumaran, J, Szebényi, G, De Baets, P, Wear, 440-441, 203083, 2019.
Diffusion coefficient of water vapor	cm^2 s^{-1} x10^7	8.71 (20°C); 5.57 (90°C)	Hansen, C M, Prog. Org. Coat., 42, 167-78, 2001.
Contact angle of water, 20°C	degree	79-93.4; 80 (adv) and 52 (rec); 77	Lee, S; Park, J-S; Lee, T R, Langmuir, 24, 4817-26, 2008; Abdu, B, Munirasu, S, Kallem, P, Banat, F, Separation Purification Technol., 252, 117416, 2020.
Surface free energy	mJ m^{-2}	31.5	

MECHANICAL & RHEOLOGICAL PROPERTIES

PARAMETER	UNIT	VALUE	REFERENCES
Tensile strength	MPa	14-60	
Tensile modulus	MPa	420-2,200	
Tensile stress at yield	MPa	14-60	
Elongation	%	20-600	
Tensile yield strain	%	3-12	
Flexural strength	MPa	8-78	
Flexural modulus	MPa	200-2,200	
Compressive strength	MPa	55-110	

PVDF poly(vinylidene fluoride)

PARAMETER	UNIT	VALUE	REFERENCES
Izod impact strength, notched, 23°C	J m⁻¹	50-1,000	
Poisson's ratio	-	0.383	
Shore D hardness	-	70-80	
Shrinkage	%	0.2-3	
Brittleness temperature (ASTM D746)	°C	-53 to 10	
Wear rate	m³ mm⁻¹	1.74E-14	Tóth, L F, Sukumaran, J, Szebényi, G, De Baets, P, Wear, 440-441, 203083, 2019.
Melt viscosity, shear rate=100 s⁻¹	Pa s	3,000	
Melt index, 230°C/2.16 kg	g/10 min	0.5-45	
Water absorption, equilibrium in water at 23°C	%	0.02-0.07	

CHEMICAL RESISTANCE

PARAMETER	UNIT	VALUE	REFERENCES
Acid dilute/concentrated	-	resistant	
Alcohols	-	resistant	
Alkalis	-	resistant (diluted)	
Aliphatic hydrocarbons	-	resistant	
Aromatic hydrocarbons	-	resistant	
Esters	-	non-resistant	
Halogenated hydrocarbons	-	resistant	
Ketones	-	non-resistant	
Good solvent	-	γ-butyrolactone, cyclohexanone, DMA, DMF, DMSO, ethylene, carbonate, NMP	
Non-solvent	-	acetone, alcohols, aliphatic and cycloaliphatic hydrocarbons, chlorinated solvents, MIBK	

FLAMMABILITY

PARAMETER	UNIT	VALUE	REFERENCES
Autoignition temperature	°C	268	
Limiting oxygen index	% O_2	44-75	
Char at 500°C	%	7	Lyon, R E; Walters, R N, J. Anal. Appl. Pyrolysis, 71, 27-46, 2004.
Heat of combustion	J g⁻¹	14,780	
UL 94 rating	-	V-0	

WEATHER STABILITY

PARAMETER	UNIT	VALUE	REFERENCES
Excitation wavelengths	nm	490	Martins, P; Serrado Nunes, J; Hungerford, G; Miranda, D; Ferreira, A; Sencadas, V; Lnaceros-Mendez, S, Phys. Lett., A373, 177-180, 2009.
Emission wavelengths	nm	580	Wypych, G, Handbook of Materials Weathering, 6th Edition, ChemTec Publishing, 2018.
Low earth orbit erosion yield	cm³ atom⁻¹ x 10⁻²⁴	1.29	Waters, D L; Banks, B A; De Groh, K K; Miller, S K R; Thorson, S D, High Performance Polym., 20, 512-22, 2008.

PVDF poly(vinylidene fluoride)

PARAMETER	UNIT	VALUE	REFERENCES
TOXICITY			
Carcinogenic effect	-	not listed by ACGIH, NIOSH, NTP	
PROCESSING			
Typical processing methods	-	blow molding, extrusion, injection molding, thermoforming	
Processing temperature	°C	195 (melt); 210 (extrusion)	
Processing pressure	MPa	13.8 (injection); 4.14 (back)	
Process time	min	2 (holding)	
Additives used in final products	-	Fillers: barium titanate, calcium carbonate, carbon black, carbon black coated with conductive polymer, copper powder, hafnium powder, lead zirconium titanate, silica, tantalum powder, titanium dioxide, zeolite, zinc sulfide; plasticizers: adipic polyester, dibutyl phthalate, dibutyl sebacate, glyceryl tributylate, tricresyl phosphate; Antistatics: carbon black, glycerol monooleate	
Applications	-	acid storage tanks, cables, capacitor films, filtration, fuel seals, heating cables for car seats, ignition cable, membranes, reactive liner of warheads, tubing, valves	
BLENDS			
Suitable polymers	-	PA-6, PA-11, PC, PEEK, PET, PES, PMMA, PS, PVCA	
ANALYSIS			
FTIR (wavenumber-assignment)	cm^{-1}/-	α-phase – 615, 766, 855; CF_2 – 615; CH_2 – 855, 766	Bao, S P; Liang, G D; Tjong, S C, Carbon, 49, 1758-68, 2011.
Raman (wavenumber-assignment)	cm^{-1}/-	γ-phase – 265, 434, 513, 811, 840, 883, 1234; α-phase – 287, 488, 613, 796, 875, 1200, 1429	Ince-Gunduz, B S; Alpern, R; Amare, D; Crawford, J; Dolan, B; Jones, S; Kobylarz, R; Reveley, M; Cebe, P, Polymer, 51, 1485-93, 2010.
NMR (chemical shifts)	ppm	H NMR: OCH_3 – 3.3; OCH_2 – 3.5; $COOCH_2$– – 4.1-4.2; H– – 0.7-2.3	Liu, F; Xu, Y-Y; Zhu, B-K; Zhang, F; Zhu, L-P, J. Membrane Sci., 345, 331-39, 2009.
x-ray diffraction peaks	degree	18.1, 18.8, 20.4 18.5 – monoclinic α-PVDF, 20–21° attributed to either α-, β-, or γ-PVDF	Huang, X; Jiang, P; Kim, C; Liu, F; Yin, Y, Eur. Polym. J., 45, 377-86, 2009; Huang, S, Hong, S, Su, Y, Zheng, X, Combustion and Flame, 219, 467-77, 2020.

PVDF-HFP poly(vinylidene fluoride-co-hexafluoropropylene)

PARAMETER	UNIT	VALUE	REFERENCES
GENERAL			
Common name	-	poly(vinylidene fluoride-co-hexafluoropropylene)	
CAS name	-	1-propene, 1,1,2,3,3,3-hexafluoro-, polymer with 1,1-difluoroethene	
Acronym	-	PVDF-HFP	
CAS number	-	9011-17-0	
HISTORY			
Person to discover	-	Moran, A L	Moran, A L, US Patent 2,951,832, DuPont, Sept. 6, 1960.
Date	-	1960	
SYNTHESIS			
Monomer(s) structure	-	$H_2C{=}CF_2$ $\quad$ $F_3CCF{=}CF_2$	
Monomer(s) CAS number(s)	-	75-38-7; 116-15-4	
Monomer(s) molecular weight(s)	dalton, g/mol, amu	64.04; 150.02	
Monomer(s) expected purity(ies)	%	99.998; 99.998	
Monomer ratio	-	3.5 (ranges from 11 mol% to 42 mol% of hexafluoropropylene)	
Fluorine content	%	<61-66	
Formulation example	-	in addition to monomer and water surfactant and initiator system (usually persulfate-sulfite) are used; copper salts are used as catalysts	Moore, A L, Fluoroelastomers Handbook. The Definitive User's Guide and Databook, William Andrew, 2006.
Method of synthesis	-	emulsion polymerization occurs by formation of monomer swollen polymer particles having size of 100-1000 nm	Moore, A L, Fluoroelastomers Handbook. The Definitive User's Guide and Databook, William Andrew, 2006.
Temperature of polymerization	°C	100-120	
Time of polymerization	h	2-4	
Pressure of polymerization	MPa	2-7	
Activation energy of polymerization	kJ mol^{-1}	-142	Moore, A L, Fluoroelastomers Handbook. The Definitive User's Guide and Databook, William Andrew, 2006.
Heat of polymerization	J g^{-1}	1340	Moore, A L, Fluoroelastomers Handbook. The Definitive User's Guide and Databook, William Andrew, 2006.
Number average molecular weight, M_n	dalton, g/mol, amu	110,000-380,000	
Mass average molecular weight, M_w	dalton, g/mol, amu	98,000-480,000	
Polydispersity, M_w/M_n	-	1.5-4.1	Sarno, M, Baldino, L, Scudieri, C, Reverchon, E, J. Phys. Chem. Solids, 136, 109132, 2020.
STRUCTURE			
Crystallinity	%	31, 27.45 (PVDF-HFP has relatively lower crystallinity than PVDF due to the incorporation of amorphous hexafluoropropylene phase, which has disrupted the regular arrangement of vinylidene fluoride)	Xu, P, Chen, H, Zhou, X, Xiang, H, J. Membrane Sci., 617, 118660, 2021.

PVDF-HFP poly(vinylidene fluoride-co-hexafluoropropylene)

PARAMETER	UNIT	VALUE	REFERENCES
Cell dimensions	nm	a:b:c=0.496:0.964:0.462 (α)	Abbrent, S; Plestil, J; Hlavata, D; Lindgren, J; Tegenfeldt, J; Wendsjo, A, Polymer, 42, 1407-16, 2001.
Polymorphs	-	$\alpha, \beta, \gamma, \delta$	Abbrent, S; Plestil, J; Hlavata, D; Lindgren, J; Tegenfeldt, J; Wendsjo, A, Polymer, 42, 1407-16, 2001.
Chain conformation	-	*trans–gauche–trans–gauche'* (α) (the most common phase); zig-zag, all *trans* (β)	Abbrent, S; Plestil, J; Hlavata, D; Lindgren, J; Tegenfeldt, J; Wendsjo, A, Polymer, 42, 1407-16, 2001.
Heat of crystallization	kJ kg^{-1}	28-37	
Rapid crystallization temperature	$^{\circ}$C	98-134	

COMMERCIAL POLYMERS

Some manufacturers	-	DuPont; Westlake Plastics	
Trade names	-	Viton; Kynar	

PHYSICAL PROPERTIES

PARAMETER	UNIT	VALUE	REFERENCES
Density at 20°C	g cm^{-3}	1.77-1.86	
Color	-	silver gray to amber	
Refractive index, 20°C	-	1.41	
Odor		odorless	
Melting temperature, DSC	$^{\circ}$C	125-164	
Decomposition temperature	$^{\circ}$C	>204; >330; 330-370 (TGA)	
Thermal expansion coefficient, 23-80°C	10^{-4} $^{\circ}$C^{-1}	1.5-1.7	
Thermal conductivity, 23°C	W m^{-1} K^{-1}	0.18	
Glass transition temperature	$^{\circ}$C	-5 to -40	
Specific heat capacity	J K^{-1} kg^{-1}	1,200	
Heat of fusion	kJ mol^{-1}	39	
Maximum service temperature	$^{\circ}$C	>200	
Long term service temperature	$^{\circ}$C	-40 to 200	
Heat deflection temperature at 0.45 MPa	$^{\circ}$C	100	
Heat deflection temperature at 1.8 MPa	$^{\circ}$C	39-52	
Vicat temperature VST/A/50	$^{\circ}$C	131-150	
Enthalpy of fusion	J g^{-1}	36.0-65.0	Mekhilef, N, Antec, 1821-26, 2000.
Hansen solubility parameters, δ_D, δ_P, δ_H	MPa$^{0.5}$	17.2, 12.5, 8.2	Wongchitphimon, S; Wang, R; Jiraratananon, R; Shi, L; Loh, C H, J. Membrane Sci., 369, 329-38, 2011.
Hildebrand solubility parameter	MPa$^{0.5}$	23.2	Wongchitphimon, S; Wang, R; Jiraratananon, R; Shi, L; Loh, C H, J. Membrane Sci., 369, 329-38, 2011.
Dielectric constant at 100 Hz/1 MHz	-	7.6-10.6; 11	-; Kishor K M J, Kalathi, J T, J. Alloys Compounds, 843, 155889, 2020.
Dissipation factor at 1 MHz	E-4	2000	
Volume resistivity	ohm-m	1E12	
Surface resistivity	ohm	1E14	
Coefficient of friction	-	0.25 (dynamic); 0.3 (static)	

PVDF-HFP poly(vinylidene fluoride-co-hexafluoropropylene)

PARAMETER	UNIT	VALUE	REFERENCES
Contact angle of water, 20°C	degree	138; 139.9	Wang, X, Xiao, C, Liu, H, Chen, M, Zhang, F, J. Membrane Sci., 596, 117583, 2020; Ponnamma, D, Aljarod, O, Parangusan, H, Al-Maadeed, M A A, Mater. Chem. Phys., 239, 122257, 2020.

MECHANICAL & RHEOLOGICAL PROPERTIES

Tensile strength	MPa	20-45	
Tensile modulus	MPa	500-1,100	
Tensile stress at yield	MPa	15-35	
Elongation	%	200-650	
Tensile yield strain	%	10-18	
Flexural strength	MPa	40	
Flexural modulus	MPa	400-1,000	
Abrasion resistance (ASTM D1044)	mg/1000 cycles	10	
Shore D hardness	-	65-72	
Shrinkage	%	2-3	
Brittleness temperature (ASTM D746)	°C	-17 to -62	
Intrinsic viscosity, 25°C	dl g^{-1}	1.0-1.7	
Melt index, 230°C/2.16 kg	g/10 min	1.3-8	
Water absorption, 24h at 23°C	%	0.04	

CHEMICAL RESISTANCE

Acid dilute/concentrated	-	resistant (only dilute)	
Alcohols	-	resistant	
Alkalis	-	non resistant	Mitra, S; Ghanbari-Siahkali, A; Kingshott, P; Almdal, K; Rehmeier, H K; Christensen, A G, Polym. Deg. Stab., 83, 195-206, 2004.
Aliphatic hydrocarbons	-	resistant	
Aromatic hydrocarbons	-	resistant	
Esters	-	non resistant	
Greases & oils	-	resistant	
Halogenated hydrocarbons	-	resistant	
Ketones	-	non resistant	
Good solvent	-	carbon dioxide, C_3F_3, $CClF_3$	

FLAMMABILITY

Ignition temperature	°C	>204	
Limiting oxygen index	% O_2	44-56	
Volatile products of combustion	-	CO, CO_2, HF, perfluoroolefins	
UL 94 rating	-	V-0	

WEATHER STABILITY

Excitation wavelengths	nm	510	Wypych, G, Handbook of Materials Weathering, 6th Ed., ChemTec Publishing, 2018.

PVDF-HFP poly(vinylidene fluoride-co-hexafluoropropylene)

PARAMETER	UNIT	VALUE	REFERENCES
Emission wavelengths	nm	647, 645	

BIODEGRADATION

PARAMETER	UNIT	VALUE	REFERENCES
Stabilizers	-	poly(4-vinyl-N-alkylpyridinium bromide)	Yao, C; Li, X; Neoh, K G; Shi, Z; Kang, E T, Appl. Surface Sci., 255, 3854-58, 2009.

TOXICITY

PARAMETER	UNIT	VALUE	REFERENCES
Carcinogenic effect	-	not listed by ACGIH, NIOSH, NTP	
OSHA	mg m^{-3}	5 (respirable), 15 (total)	
Oral rat, LD$_{50}$	mg kg^{-1}	>5,000	

ENVIRONMENTAL IMPACT

PARAMETER	UNIT	VALUE	REFERENCES
Aquatic toxicity, *Daphnia magna*, LC$_{50}$, 48 h	mg l^{-1}	>205	

PROCESSING

PARAMETER	UNIT	VALUE	REFERENCES
Typical processing methods	-	calendering, coating, compression molding, electrospinning, extrusion, injection molding, spinning, transfer molding	
Processing pressure	kPa	50 (spinning)	
Applications	-	aircraft, aerospace, chemical processing and transportation, food and pharmaceutical, oil and gas, petroleum refining; typical products: caulks, coatings, gaskets, membranes, o-rings, seals, vibration dampers, wire & cable	

BLENDS

PARAMETER	UNIT	VALUE	REFERENCES
Suitable polymers	-	PE	

ANALYSIS

PARAMETER	UNIT	VALUE	REFERENCES
FTIR (wavenumber-assignment)	cm^{-1}/-	C–F – 1287; CF$_2$ – 882; CH$_2$ – 842 peaks of α-phase, β-phase, and γ-phase of PVDF-HFP at around 615 cm−1, 730 cm−1 and 890 cm−1, respectively	Saikia, D; Wu, H-Y; Pan, Y-C; Lin, C-P; Huang, K-P; Chen, K-N; Fey, G T K; Kao, H-M, J. Power Sources, 196, 2826-34, 2011; Kishor K M J, Kalathi, J T, J. Alloys Compounds, 843, 155889, 2020.
NMR (chemical shifts)	ppm	CH$_2$ – 43.2; CF$_3$ – 164; CF$_2$ – 118.5	Saikia, D; Wu, H-Y; Pan, Y-C; Lin, C-P; Huang, K-P; Chen, K-N; Fey, G T K; Kao, H-M, J. Power Sources, 196, 2826-34, 2011.
x-ray diffraction peaks	degree	peak at 18.4° and 20.8° corresponds to alpha (α) and beta (β) phase, respectively, a broad peak at 26.6° and 38.8° corresponds to the gamma (γ) phase of PVDF-HFP	Kishor K M J, Kalathi, J T, J. Alloys Compounds, 843, 155889, 2020.

PVF poly(vinyl fluoride)

PARAMETER	UNIT	VALUE	REFERENCES
GENERAL			
Common name	-	poly(vinyl fluoride)	
IUPAC name	-	poly(vinyl fluoride)	
CAS name	-	ethene, fluoro-, homopolymer	
Acronym	-	PVF	
CAS number	-	24981-14-4	
Formula		$\left[CH_2CHF\right]_n$	
HISTORY			
Person to discover	-	Coffman, D D; Ford, T A	Coffman, D D; Ford, T A, US Patent 2,419,008 and 2,419,010, DuPont, Apr. 15, 1947.
Date	-	1947 (patent); 1961 (commercialization)	
SYNTHESIS			
Monomer(s) structure	-	$H_2C{=}CHF$	
Monomer(s) CAS number(s)	-	75-02-5	
Monomer(s) molecular weight(s)	dalton, g/mol, amu	46.04	
Monomer ratio	-	100%	
Formulation example	-	water 200, monomer 100, perfluorinated carboxylic acid 0.6, ammonium persulfate 0.2, water glass 3	Ebnesajjad, S, Fluoroplastics. Vol. 2. Melt Processible Fluoroplastics, William Andrew, 2003.
Method of synthesis	-	conventional PVF synthesis is accomplished by polymerization of vinyl fluoride in aqueous emulsion	Minnick, D L, Gilbert, W J R, Rocha, M A, Shiflett, M B, Fluid Phase Equilibria, 444, 61-8, 2017.
Temperature of polymerization	ºC	46; 50-250	-; Minnick, D L, Gilbert, W J R, Rocha, M A, Shiflett, M B, Fluid Phase Equilibria, 444, 61-8, 2017.
Time of polymerization	h	8	
Pressure of polymerization	MPa	4.3; 15-100	-; Minnick, D L, Gilbert, W J R, Rocha, M A, Shiflett, M B, Fluid Phase Equilibria, 444, 61-8, 2017.
Yield	%	95	
Number average molecular weight, M_n	dalton, g/mol, amu	80,000-790,000	Wang, J; Lu, Y; Li, H; Yuan, H, J. Appl. Polym. Sci., 102, 1780-86, 2006.
Mass average molecular weight, M_w	dalton, g/mol, amu	126,000-150,000	
Polydispersity, M_w/M_n	-	2.5-5.6	
Molar volume at 298K	cm^3 mol^{-1}	calc.=35.0; 32.0 (crystalline)	
Van der Waals volume	cm^3 mol^{-1}	23.12; 22.8 (crystalline)	
Concentration of head-to-head and tail-to-tail units	%	10-12	

PVF poly(vinyl fluoride)

PARAMETER	UNIT	VALUE	REFERENCES
Degree of branching	%	0.45-1.35	Aronson, M T; Bergeer, L L; Honsberg, U S, Polymer, 34, 12, 2546-53, 1993.

STRUCTURE

PARAMETER	UNIT	VALUE	REFERENCES
Crystallinity	%	52-68	Ebnesajjad, S, Fluoroplastics. Vol. 2. Melt Processible Fluoroplastics, William Andrew, 2003; Wang, J; Lu, Y; Li, H; Yuan, H, J. Appl. Polym. Sci., 102, 1780-86, 2006; Alchikh, M; Fond, C; Frere, Y, Polym. Deg. Stab., 95, 440-44, 2010.
Cell type (lattice)	-	hexagonal, orthorhombic, monoclinic	
Cell dimensions	nm	a:b:c=0.493:0.493:0.253 (hexagonal); 0.857:0.495:0.252 (orthorhombic); 0.494:0.494:0.252 (monoclinic)	
Unit cell angles	degree	2	
Crystallite size	nm	100,000	Alchikh, M; Fond, C; Frere, Y, Polym. Deg. Stab., 95, 440-44, 2010.
Polymorphs	-	α, β (trans), γ, δ	Alchikh, M; Fond, C; Frere, Y, Polym. Deg. Stab., 95, 440-44, 2010.
Tacticity	%	atactic	
Chain conformation	-	planar zig-zag (c-spacing=0.252 nm)	
Entanglement molecular weight	dalton, g/mol, amu	calc.=2,400; 2,546	

COMMERCIAL POLYMERS

PARAMETER	UNIT	VALUE	REFERENCES
Some manufacturers	-	DuPont, Daikin	
Trade names	-	Tedlar	

PHYSICAL PROPERTIES

PARAMETER	UNIT	VALUE	REFERENCES
Density at 20°C	g cm^{-3}	1.37-1.38	
Color	-	white	
Refractive index, 20°C	-	calc.=1.452-1.4926; exp.=1.45-1.46	
Transmittance	%	92	
Odor		odorless	
Melting temperature, DSC	°C	178-206	
Softening point	°C	125-130	
Decomposition temperature	°C	379-421	Wang, J; Lu, Y; Li, H; Yuan, H, J. Appl. Polym. Sci., 102, 1780-86, 2006.
Thermal expansion coefficient, 23-80°C	°C^{-1}	7.1-9E-5	
Thermal conductivity, melt	W m^{-1} K^{-1}	calc.=0.1566	
Glass transition temperature	°C	calc.=1-65; exp.=40-64	
Specific heat capacity	J K^{-1} kg^{-1}	1,000-1,760	
Maximum service temperature	°C	204	
Long term service temperature	°C	-73 to 107	
Hansen solubility parameters, δ_D, δ_P, δ_H	MPa$^{0.5}$	17.2, 12.3, 9.2	
Hildebrand solubility parameter	MPa$^{0.5}$	calc.=22.67; exp.=23.2	

674　　　**HANDBOOK OF POLYMERS** 3rd Edition, Copyrights 2022; ChemTec Publishing

PVF poly(vinyl fluoride)

PARAMETER	UNIT	VALUE	REFERENCES
Surface tension	mN m^{-1}	calc.=31.0-43.3; exp.=28-38.4	
Dielectric constant at 100 Hz/1 MHz	-	8.5/4.8	
Dissipation factor at 100 Hz	E-4	20	
Dissipation factor at 1 MHz	E-4	70	
Volume resistivity	ohm-m	6.9E11 to 1.8E12	
Surface resistivity	ohm	1.6-6.1E15	
Electric strength K20/P50, d=0.60.8 mm	kV mm^{-1}	120-140	
Coefficient of friction	-	0.13 (PVF/steel); 0.24 (PVF/PVF)	
Permeability to nitrogen, 25°C	cm^3 cm cm^{-2} s^{-1} Pa^{-1} x 10^{12}	0.00012-0.00167	
Permeability to oxygen, 25°C	cm^3 cm cm^{-2} s^{-1} Pa^{-1} x 10^{12}	0.00139-0.0062	
Permeability to water vapor, 25°C	cm^3 cm cm^{-2} s^{-1} Pa^{-1} x 10^{12}	1.01	
Contact angle of water, 20°C	degree	74-89	
Surface free energy	mJ m^{-2}	36.4	

MECHANICAL & RHEOLOGICAL PROPERTIES

PARAMETER	UNIT	VALUE	REFERENCES
Tensile strength	MPa	69-103	
Tensile modulus	MPa	2,075-2,138	
Elongation	%	95	
Tear strength	N m^{-1}	6,600-7,400	
Poisson's ratio	-	calc.=0.5	
Shrinkage	%	6 (150°C)	
Water absorption, equilibrium in water at 23°C	%	<0.5	

CHEMICAL RESISTANCE

PARAMETER	UNIT	VALUE	REFERENCES
Acid dilute/concentrated	-	resistant	
Alcohols	-	poor	Alchikh, M; Fond, C; Frere, Y, Polym. Deg. Stab., 95, 440-44, 2010.
Alkalis	-	resistant	
Aliphatic hydrocarbons	-	resistant	
Aromatic hydrocarbons	-	resistant	
Esters	-	resistant	
Greases & oils	-	resistant	
Halogenated hydrocarbons	-	resistant	
Ketones	-	resistant	
Good solvent	-	cyclohexanone (hot), dinitrile, DMA (hot), DMF, DMSO	
Non-solvent	-	aliphatic, cycloaliphatic, and aromatic hydrocarbons	

PVF poly(vinyl fluoride)

PARAMETER	UNIT	VALUE	REFERENCES
FLAMMABILITY			
Ignition temperature	°C	420	
Autoignition temperature	°C	480	
Limiting oxygen index	% O_2	22.6	
Volatile products of combustion	-	CO, HF	
WEATHER STABILITY			
Exposure result		5 years in Florida without change	
Important initiators and accelerators	-	aluminum	Drobny, J, Fluoroplastics, Rapra, 2006.
Stabilizers	-	UV absorbers in laminating films to protect substrate behind the film	
Results of exposure		82% tensile strength retention after 6 years of Florida exposure	
Low earth orbit erosion yield	cm^3 $atom^{-1}$ x 10^{-24}	3.19	Waters, D L; Banks, B A; De Groh, K K; Miller, S K R; Thorson, S D, High Performance Polym., 20, 512-22, 2008.
TOXICITY			
Carcinogenic effect	-	not listed by ACGIH, NIOSH, NTP	
PROCESSING			
Typical processing methods	-	coating, film extrusion, vacuum deposition	
Additives used in final products	-	Plasticizer: dimethyl phthalate; Stabilizer: pentaerythritol	Wang, J; Lu, Y; Yuan, H, Polym.-Plast. Technol. Eng., 46, 461-68, 2007.
Applications	-	aircraft interiors, architectural panels, awnings, coatings, glazing, molding, signs, wallcovering	
BLENDS			
Suitable polymers	-	PAC, PLA, PVDF	
ANALYSIS			
FTIR (wavenumber-assignment)	cm^{-1}/-	CH_2 – 2970, 2932, 2861, 1446, 1427, 1410, 1295, 831, 763; C–C – 1189, 1092; CF - 1033, 888, 465, 394	Aronson, M T; Bergeer, L L; Honsberg, U S, Polymer, 34, 12, 2546-53, 1993.
Raman (wavenumber-assignment)	cm^{-1}/-	CH_2 – 2932, 2859, 1436, 1302, 834; C–C – 1194, 1095; CF – 1150, 1032, 890, 454, 395	Aronson, M T; Bergeer, L L; Honsberg, U S, Polymer, 34, 12, 2546-53, 1993.
NMR (chemical shifts)	ppm	F NMR: head-to-tail monomer units – 178-182; head-to-head monomer units – 189, 197; CH_2F end groups – 220, 162, 147	Aronson, M T; Bergeer, L L; Honsberg, U S, Polymer, 34, 12, 2546-53, 1993.

PVK poly(N-vinyl carbazole)

PARAMETER	UNIT	VALUE	REFERENCES
GENERAL			
Common name	-	poly(N-vinyl carbazole)	
CAS name	-	9H-carbazole, 9-ethenyl-, homopolymer	
Acronym	-	PVK	
CAS number	-	25067-59-8	
RTECS number	-	FE6225480	
Formula			

PARAMETER	UNIT	VALUE	REFERENCES
HISTORY			
Person to discover	-	Reppe, W, Keyssner, E; Dorrer, E	Reppe, W, Keyssner, E; Dorrer, E, US Patent 2,072,465, IG Farben, Mar. 2, 1937.
Date	-	1937	
Details	-	production of polymeric N-vinyl compounds	
SYNTHESIS			
Monomer(s) structure	-		

PARAMETER	UNIT	VALUE	REFERENCES
Monomer(s) CAS number(s)	-	1484-13-5	
Monomer(s) molecular weight(s)	dalton, g/mol, amu	193.24	
Monomer(s) expected purity(ies)	%	less than 100 ppm impurities	
Monomer ratio	-	100% (can also be copolymerized with various monomers, see ref.)	Fink, J K, High Performance Polymers, William Andrew, 2008.
Method of synthesis	-	free radical polymerization, using AIBN as initiator	
Catalyst	-	Ziegler-Natta	
Activation energy of polymerization	J mol^{-1}	22.8-27.4 (propagation)	
Number average molecular weight, M_n	dalton, g/mol, amu	9,900-151,400	
Mass average molecular weight, M_w	dalton, g/mol, amu	40,000-3,230,000; 1,100,000	-; Menazea, A A, Ismail, A M, Elashmawi, I S, J. Mater. Res. Technol., 9, 3, 5689-98, 2020.
Polydispersity, M_w/M_n	-	2.06-3.30	
STRUCTURE			
Crystallinity	%	28-38	
Cell type (lattice)	-	hexagonal, orthorhombic	
Cell dimensions	nm	a:b:c=1.23:1.23:0.744 (hexagonal); 2.16:1.25:0.744 (orthorhombic)	

HANDBOOK OF POLYMERS 3rd Edition, Copyrights 2022; ChemTec Publishing

PVK poly(N-vinyl carbazole)

PARAMETER	UNIT	VALUE	REFERENCES
PHYSICAL PROPERTIES			
Density at 20°C	g cm^{-3}	1.18-1.20; 1.84 (amorphous)	
Color	-	white to off-white to yellow	
Refractive index, 20°C	-	1.683-1.696	
Odor	-	odorless	
Melting temperature, DSC	°C	>320	
Softening point	°C	>175	
Thermal expansion coefficient, 23-80°C	°C^{-1}	5E-5	
Glass transition temperature	°C	200-227; 126 (isotactic); 227 (amorphous); 276 (syndiotactic)	
Specific heat capacity	J K^{-1} kg^{-1}	1260	
MECHANICAL & RHEOLOGICAL PROPERTIES			
Tensile modulus	MPa	4,120; 4,000 (amorphous)	
Elongation	%	1.1	
Water absorption, equilibrium in water at 23°C	%	<0.1	
CHEMICAL RESISTANCE			
Alcohols	-	resistant	
Aliphatic hydrocarbons	-	resistant	
Aromatic hydrocarbons	-	non-resistant	
Esters	-	resistant	
Halogenated hydrocarbons	-	non-resistant	
Ketones	-	non-resistant	
Θ solvent, Θ-temp.=-37.5, -21.5, 37°C	-	chlorobenzene, nitrobenzene, toluene	
Good solvent	-	aromatic hydrocarbons, chloroform, chlorobenzene, methylene tetrachloride, THF	
Non-solvent	-	alcohols, aliphatic hydrocarbons, carbon tetrachloride, esters	
WEATHER STABILITY			
Absorption bands	nm	261, 295, 331, 344	Wu, H-X; Qiu, X-Q; Cai, R-F; Qian, S-X, Appl. Surface Sci., 253, 5122-28, 2007.
Excitation wavelengths	nm	256	Baibarac, M; Lira-Cantu, M; Sol, J O; Baltog, I; Casan-Pastor, N; Gomez-Romero, P, Composites Sci. Technol., 67, 2556-63, 2007.
TOXICITY			
NFPA: Health, Flammability, Reactivity rating	-	0/1/0	
Carcinogenic effect	-	not listed by ACGIH, NIOSH, NTP	
Oral rat, LD$_{50}$	mg kg^{-1}	>5,000	

PVK poly(N-vinyl carbazole)

PARAMETER	UNIT	VALUE	REFERENCES
PROCESSING			
Typical processing methods	-	spin coating, screen-printing, vacuum deposition	Fink, J K, High Performance Polymers, William Andrew, 2008.
Additives used in final products	-	Plasticizers: N-methylcarbazole, N-ethylcarbazole, N-butyl-carbazole, N-hexylcarbazole, N-phenylcarbazole, 1,3-biscar-bazolylpropane, o-nitroanisole, m-nitroanisole, p-nitroanisole, triphenylamine; Release: amorphous silica, PVDF; Fillers: functionalized carbon dots, fullerene, graphene oxide, zinc sulfide	
Applications	-	data storage, electrophotography, light emitting diodes, lithium-ion batteries, opto-electro transfer devices, photorefractive materials, photovoltaic devices	
BLENDS			
Suitable polymers	-	epoxy, PEO, PPy, PVDF	
ANALYSIS			
FTIR (wavenumber-assignment)	cm^{-1}/-	aromatic ring – 722; tail-to-tail – 744; aromatic ring – 1220; vinylidene group – 1320; vinyl carbazole – 1450	Baibarac, M; Lira-Cantu, M; Sol, J O; Baltog, I; Casan-Pastor, N; Gomez-Romero, P, Composites Sci. Technol., 67, 2556-63, 2007.
Raman (wavenumber-assignment)	cm^{-1}/-	C–H – 1128, 1156, 1316, 1451; C–C – 1388, 1618; benzene ring – 1514, 1570; C=C – 1594	Baibarac, M; Gomez-Romero, P; Lira-Cantu, M; Casan-Pastor, N; Mestres, N; Lefrant, S, Eur. Polym. J., 42, 2302-12, 2006.
x-ray diffraction peaks	degree	7.6 (intense, narrow), 20.4 (amorphous, broad, diffuse)	Menazea, A A, Ismail, A M, Elashmawi, I S, J. Mater. Res. Technol., 9, 3, 5689-98, 2020.

PVME poly(vinyl methyl ether)

PARAMETER	UNIT	VALUE	REFERENCES
GENERAL			
Common name	-	poly(vinyl methyl ether)	
CAS name	-	ethene, methoxy-, homopolymer	
Acronym	-	PVME	
CAS number	-	9003-09-2	
Formula		$\left[CH_2CH\,\big|\,OCH_3\right]_n$	
HISTORY			
Person to discover	-	Reppe, W; Schlichting, O	Reppe, W; Schlichting, O, US Patent 2,104,000, IG Farben, Dec. 28, 1937.
Date	-	1937	
Details	-	production of polymerization products from vinyl ethers	
SYNTHESIS			
Monomer(s) structure	-	$H_2C{=}CHOCH_3$	
Monomer(s) CAS number(s)	-	107-25-5	
Monomer(s) molecular weight(s)	dalton, g/mol, amu	70.09	
Number average molecular weight, M_n	dalton, g/mol, amu	10,000-18,100	
Mass average molecular weight, M_w	dalton, g/mol, amu	50,500-354,000	
Polydispersity, M_w/M_n	-	1.05-2.5	Morariu, S; Eckelt, J; Wolf, B A, Ind. Eng. Chem. Res., 48, 6943-48, 2009.
Molar volume at 298K	cm^3 mol^{-1}	calc.=50.0 (crystalline)	
Van der Waals volume	cm^3 mol^{-1}	calc.=34.4; 35.7 (crystalline)	
STRUCTURE			
Cell type (lattice)	-	trigonal	
Cell dimensions	nm	a:b:c=1.62:1.62:6.5	Bassi, I W; Atti. Accad. Nazl. Lincei, Cl. Sci. Fis., Mat. Nat., Rend., 29, 193, 1960.
Unit cell angles	degree	$\alpha{:}\beta{:}\gamma$=90:90:120	
Number of chains per unit cell	-	18	
Tacticity	%	59 (isotactic)	Hanykova, L; Labuta, J; Spevacek, J, Polymer, 47, 6107-16, 2006.
Chain conformation	-	helix 3/1	
Temperature of conformational change	°C	32-40	Narang, P, Venkatesu, P, Adv. Colloid Interface Sci., 274, 102042, 2019.
Avrami constants, k/n	-	n=1.0-1.24; k=0.83-7.72x104	Zhang, T; Li, T; Nies, E; Beghmans, H; Ge, L, Polymer, 50, 1206-13, 2009.

PVME poly(vinyl methyl ether)

PARAMETER	UNIT	VALUE	REFERENCES
COMMERCIAL POLYMERS			
Some manufacturers	-	BASF	
Trade names	-	Lutonal	
PHYSICAL PROPERTIES			
Density at 20°C	g cm^{-3}	1.03-1.05	
Bulk density at 20°C	g cm^{-3}	0.94-1.03	
Color	-	clear	
Refractive index, 20°C	-	1.467-1.478	
Melting temperature, DSC	°C	144	
Thermal expansion coefficient, 23-80°C	°C^{-1}	6.8E-4	
Glass transition temperature	°C	calc.=-21; exp.=-20 to -34	
Hildebrand solubility parameter	MPa$^{0.5}$	calc.=19.44	
Surface tension	mN m^{-1}	31.8	
MECHANICAL & RHEOLOGICAL PROPERTIES			
Tensile strength	MPa	11.65-14.00	
CHEMICAL RESISTANCE			
Acid dilute/concentrated	-	good/poor	
Alcohols	-	poor	
Alkalis	-	good	
Aromatic hydrocarbons	-	poor	
Esters	-	poor	
Ketones	-	poor	
Θ solvent, Θ-temp.=51°C	-	cyclohexane	
Good solvent	-	acetone, ethanol, ethyl acetate, methylene chloride, THF	
Non-solvent	-	diethyl ether, ethylene glycol, hexane	
FLAMMABILITY			
Ignition temperature	°C	230	
Autoignition temperature	°C	390	
Volatile products of combustion	-	CO, CO_2, hydrocarbons	
WEATHER STABILITY			
Products of degradation	-	tertiary peroxide, ketone groups, methanol, CO, CO_2, acetic acid, methyl acetate, formic acid, dimethyl malonate	
TOXICITY			
HMIS: Health, Flammability, Reactivity rating	-	1/1/0	
Carcinogenic effect	-	not listed by ACGIH, NIOSH, NTP	
Oral rat, LD$_{50}$	mg kg^{-1}	>5,000	

PVME poly(vinyl methyl ether)

PARAMETER	UNIT	VALUE	REFERENCES
ENVIRONMENTAL IMPACT			
Aquatic toxicity, *Bluegill sunfish*, LC$_{50}$, 48 h	mg l^{-1}	>500	
PROCESSING			
Applications	-	marine antifouling paints	
Outstanding properties	-	improves adhesion of liquid coatings, saponification resistant	
BLENDS			
Suitable polymers	-	PAA, PANI, PCL, PEG, iPS, sPS, PS, SAN	
ANALYSIS			
FTIR (wavenumber-assignment)	cm^{-1}/-	C–H – 2991, 2949, 2839, 1468, 1388; C–C – 1193; C–O – 1135, 1105, 1070	Maeda, Y, Langmuir, 17, 1737-42, 2001; Guo, Y; Peng, Y; Wu, P, J. Mol. Structure, 875, 486-92, 2008.

PVP poly(N-vinyl pyrrolidone)

PARAMETER	UNIT	VALUE	REFERENCES
GENERAL			
Common name	-	poly(N-vinyl pyrrolidone)	
Acronym	-	PVP	
CAS number	-	9003-39-8	
EC number	-	201-800-4	
RTECS number	-	TR8160000, TR8170000, TR8180000, TR8250000, TR8300000, TR8350000, TR8360000, TR8370000,	
Linear formula			
HISTORY			
Person to discover	-	Reppe, W; Schster, C; Hartmann, A	Reppe, W; Schster, C; Hartmann, A, US Patent 2,265,450, IG Farben, Dec. 9, 1941.
Date	-	1941	
Details	-	high pressure reactions catalyzed by heavy metal acetylides, especially copper acetylide, or metal carbonyls are called Reppe Chemistry, which is behind synthesis of PVP	
SYNTHESIS			
Monomer(s) structure	-		
Monomer(s) CAS number(s)	-	88-12-0	
Monomer(s) molecular weight(s)	dalton, g/mol, amu	111.14	
Monomer ratio	-	100% and copolymers	
Method of synthesis	-	monomer is thermally polymerized in the presence of hydrogen peroxide and ammonia	
Typical concentration of residual monomer	ppm	<100	
Number average molecular weight, M_n	dalton, g/mol, amu	2,000-400,000	
Mass average molecular weight, M_w	dalton, g/mol, amu	10,000-2,200,000	
Polydispersity, M_w/M_n	-	1.73-3.2; 1.6	-; Sevastos, A A, Thomson, N R, Lindsay, C, Khutoryanskiy, V V, Eur. Polym. J., 134, 109852, 2020.
End-to-end distance of unperturbed polymer chain	nm	2.2-7	
Radius of gyration	nm	53.3 (M_w=1,278,000)	Sütekin, S D, Güven, O, Appl. Radiat. Isotopes, 145, 161-9, 2019.
Hydrodynamic radius	nm	28-32	Sütekin, S D, Güven, O, Appl. Radiat. Isotopes, 145, 161-9, 2019.

PVP poly(N-vinyl pyrrolidone)

PARAMETER	UNIT	VALUE	REFERENCES
COMMERCIAL POLYMERS			
Some manufacturers	-	BASF	
Trade names	-	Kollidon, Luvitec, Luvicross (crosslinkable)	
PHYSICAL PROPERTIES			
Density at 20°C	g cm^{-3}	1.23-1.29	
Bulk density at 20°C	g cm^{-3}	0.2	
Color	-	white to off-white to creamy to yellow	
Refractive index, 20°C	-	1.5300	
Odor	-	faint, specific	
Melting temperature, DSC	°C	100-140	
Decomposition temperature	°C	130	
Glass transition temperature	°C	175-180, 102.3	-; Sevastos, A A, Thomson, N R, Lindsay, C, Khutoryanskiy, V V, Eur. Polym. J., 134, 109852, 2020.
Hildebrand solubility parameter	MPa$^{0.5}$	exp.=25.6; 24.3 (Fedor), 24.0 (van Krevelen)	-; Shan, X, Williams, A C, V.Khutoryanskiy, V, Int. J. Pharm., 590, 119884, 2020.
Surface free energy	mJ m^{-2}	46.0	
MECHANICAL & RHEOLOGICAL PROPERTIES			
Water absorption, equilibrium in water at 23°C	%	80	
Moisture absorption, equilibrium 23°C/50% RH	%	40	
CHEMICAL RESISTANCE			
Acid dilute/concentrated	-	non-resistant	
Alcohols	-	non-resistant	
Alkalis	-	non-resistant	
Aliphatic hydrocarbons	-	non-resistant	
Aromatic hydrocarbons	-	resistant	
Esters	-	resistant	
Halogenated hydrocarbons	-	non-resistant	
Ketones	-	resistant	
Θ solvent, Θ-temp.=10, 20, 20°C	-	dioxane, 2-propanol, water	
Good solvent	-	alcohols, amines, chlorinated hydrocarbons, glycols, water	
Non-solvent	-	hydrocarbons, ethers, esters, ketones, some chlorinated hydrocarbons	
FLAMMABILITY			
Ignition temperature	°C	>215	
Autoignition temperature	°C	420	
Volatile products of combustion	-	CO, CO_2, NO_x	

PVP poly(N-vinyl pyrrolidone)

PARAMETER	UNIT	VALUE	REFERENCES
TOXICITY			
NFPA: Health, Flammability, Reactivity rating	-	1/0-1/0-1	
Carcinogenic effect	-	not listed by ACGIH, NIOSH, NTP	
TLV, ACGIH	mg m^{-3}	3 (respirable), 10 (total)	
OSHA	mg m^{-3}	5 (respirable), 15 (total)	
Oral rat, LD$_{50}$	mg kg^{-1}	100,000; 1470	
Skin rabbit, LD$_{50}$	mg kg^{-1}	560	
ENVIRONMENTAL IMPACT			
Aquatic toxicity, *Daphnia magna*, LC$_{50}$, 48 h	mg l^{-1}	>100	
Aquatic toxicity, *Bluegill sunfish*, LC$_{50}$, 48 h	mg l^{-1}	>10,000	
WEATHER STABILITY			
Irradiation	-	At a mass fraction of 0.1% PVP and doses per pulse ranging from 7 to 117 Gy, the overall radical decay observed at 390 nm follows second order kinetics with rate constants on the order of 10^9 dm^3 mol^{-1} s^{-1}.	Dispenza, C, Antonietta Sabatino, M, Grimaldi, N, Jonsson, M, Radiat. Phys. Chem., 174, 108900, 2020.
PROCESSING			
Additives used in final products	-	Plasticizers: acetyl triethyl citrate, dioctyl adipate, polyethylene glycol, dipropylene glycol dibenzoate, glycerin, tributyl citrate, triethyl citrate	
Applications	-	additive in various polymers, agriculture (seed coating, antiseptic coatings, crop protection, fertilizer binder), batteries, coatings (antifogging, dispersant), cosmetics, emulsifier, glue sticks, hot-melt adhesives, medical devices, membranes, paper, photoresists, pressure-sensitive adhesives, printed circuit boards, sizing agent, tablet binder and excipient	
Outstanding properties	-	biodegradable, biocompatible, hygroscopic	
BLENDS			
Suitable polymers	-	CA, CAR, PA, PAA, PE, PC, PEG, PHEMA, PMMA, POM, PS, PSU, PU, PVC, PVOH	

PZ polyphosphazene

PARAMETER	UNIT	VALUE	REFERENCES
GENERAL			
Common name	-	polyphosphazene	
CAS name	-	poly[nitrilo[bis(2,2,2-trifluoroethoxy)phosphoranylidyne]]; poly[nitrilo(diphenoxyphosphoranylidyne)]	
Acronym	-	PZ	
CAS number	-	28212-50-2; 28212-48-8	
Formula			Allcock, H R, Current Opinion Solid State Mater. Sci., 10, 231-240, 2006.
HISTORY			
Person to discover	-	Allcock, H R	
Date	-	1965	
Details	-	Allcock's group published numerous papers and books on the subject	
SYNTHESIS			
Monomer(s) structure	-		
Monomer(s) CAS number(s)	-	940-71-6	
Monomer(s) molecular weight(s)	dalton, g/mol, amu	347.66	
Method of synthesis	-	thermal ring-opening polymerization of hexachlorophosphazene followed by esterification of the intermediate poly(dichlorophosphazene) with either amines or sodium salts of alcohols	
Temperature of polymerization	°C	250	
Time of polymerization	h	20	
Number average molecular weight, M_n	dalton, g/mol, amu	132,000-173,000	
Mass average molecular weight, M_w	dalton, g/mol, amu	10,000-5,000,000	Nakamura, H; Masuko, T; Kojima, M; Magill, J H, Macromol. Chem. Phys., 200, 2519-24, 1999.
STRUCTURE			
Crystallinity	%	19.96-52.35	Hazendonk, P; deDenus, C; Iuga, A; Cahoon, P; Nilsson, B; Iuga, D, J. Inorg. Organometal. Polym. Mater., 16, 4, 343-57, 2006.
Cell type (lattice)	-	orthorhombic (α and γ), monoclinic (β)	Hazendonk, P; deDenus, C; Iuga, A; Cahoon, P; Nilsson, B; Iuga, D, J. Inorg. Organometal. Polym. Mater., 16, 4, 343-57, 2006.

PZ polyphosphazene

PARAMETER	UNIT	VALUE	REFERENCES
Cell dimensions	nm	a:b:c=1.014:0.935:0.486 (α); a:b:c=2.06:0.94:0.486 (γ); a:b:c=1.003:0.937:0.486 (β); a:b:c=1.19:1.19:0.486 (hexagonal mesophase)	Hazendonk, P; deDenus, C; Iuga, A; Cahoon, P; Nilsson, B; Iuga, D, J. Inorg. Organometal. Polym. Mater., 16, 4, 343-57, 2006.
Polymorphs	-	α, β, γ	Hazendonk, P; deDenus, C; Iuga, A; Cahoon, P; Nilsson, B; Iuga, D, J. Inorg. Organometal. Polym. Mater., 16, 4, 343-57, 2006.

PHYSICAL PROPERTIES

PARAMETER	UNIT	VALUE	REFERENCES
Density at 20°C	g cm^{-3}	1.63-1.74	
Refractive index, 20°C	-	1.6-1.75	
Melting temperature, DSC	°C	110-249; 114 (poly(diaryloxy)phosphazene)	Zou, W, Basharat, M, Dar, S U, Zhang, T, Compos. Part A: Appl. SSci. Manuf., 119, 145-53, 2019.
Thermal expansion coefficient, 23-80°C	10^{-4} °C^{-1}	1.9-10.7	Nakamura, H; Masuko, T; Kojima, M; Magill, J H, Macromol. Chem. Phys., 200, 2519-24, 1999.
Glass transition temperature	°C	7 to -82	Hazendonk, P; deDenus, C; Iuga, A; Cahoon, P; Nilsson, B; Iuga, D, J. Inorg. Organometal. Polym. Mater., 16, 4, 343-57, 2006; Allegra, G; Meille, S V, Macromolecules, 37, 3487-96, 2004.
Surface free energy	mJ m^{-2}	15.8	

CHEMICAL RESISTANCE

PARAMETER	UNIT	VALUE	REFERENCES
Good solvent	-	water (some)	

FLAMMABILITY

PARAMETER	UNIT	VALUE	REFERENCES
Limiting oxygen index	% O$_2$	36.5 (poly(diaryloxy)phosphazene)	Zou, W, Basharat, M, Dar, S U, Zhang, T, Compos. Part A: Appl. SSci. Manuf., 119, 145-53, 2019.
UL rating	-	V-0 (poly(diaryloxy)phosphazene)	Zou, W, Basharat, M, Dar, S U, Zhang, T, Compos. Part A: Appl. SSci. Manuf., 119, 145-53, 2019.

WEATHER STABILITY

PARAMETER	UNIT	VALUE	REFERENCES
Spectral sensitivity	nm	translucent coating containing cyclomatrix-type polyphosphazene blocked 78.6% of UVA radiation and 99.8% of UVB radiation owing to the aminophenyl chromophores in structure.	Xu, H, Yan, C, Chen, X, Hu, J, Prog. Org. Coat., 149, 105933, 2020.

PROCESSING

PARAMETER	UNIT	VALUE	REFERENCES
Applications	-	biomedical applications, flame retardation, fuel cells, fuel hoses, medical (bone regeneration, nano-vesicles, stomach wall regeneration), membranes, optical film coatings, organic-inorganic hybrid optical coatings have potential applications in optical film and UV-shielding coatings, supercapacitors, waterproofing	Zhou, X, Qiu, S, Mu, X, Hu, Y, Compos. Part B: Eng., 202, 108397, 2020.
Outstanding properties	-	thermal stability	

BLENDS

PARAMETER	UNIT	VALUE	REFERENCES
Suitable polymers	-	epoxy, PDMS, PET, PI, PLA, PTFE (composite)	

PZ polyphosphazene

PARAMETER	UNIT	VALUE	REFERENCES
ANALYSIS			
NMR (chemical shifts)	ppm	C, H, F, and P data included in ref.	Hazendonk, P; deDenus, C; Iuga, A; Cahoon, P; Nilsson, B; Iuga, D, J. Inorg. Organometal. Polym. Mater., 16, 4, 343-57, 2006.
x-ray diffraction peaks	degree	diffraction data in ref.	Tang, H; Pintauro, P N, Eur. Polym. J., 35, 1023-35, 1999.

SAN poly(styrene-co-acrylonitrile)

PARAMETER	UNIT	VALUE	REFERENCES
GENERAL			
Common name	-	poly(styrene-co-acrylonitrile)	
CAS name	-	2-propenenitrile, polymer with ethenylbenzene	
Acronym	-	SAN	
CAS number	-	9003-54-7	
RTECS number	-	AT6978000	
HISTORY			
Person to discover	-	Fikentscher, H; Heuck, C	Fikentscher, H; Heuck, C, US Patent 2,140,048, IG Farben, Dec. 13, 1938.
Date	-	1938	
Details	-	production of polymerization products	
SYNTHESIS			
Monomer(s) structure	-	$H_2C{=}CHCN$	
Monomer(s) CAS number(s)	-	100-42-5; 107-13-1	
Monomer(s) molecular weight(s)	dalton, g/mol, amu	104.15; 53.06	
Acrylonitrile content	%	25-40	Liu, M; Zhang, X; Zammarano, M; Gilman, J W; Kashiwagi, T, Polym. Deg. Stab., 96, 1000-8, 2011.
Formulation example	-	water – 180-40, peroxidic catalyst – 0.1-0.5, styrene – 85-50, acrylonitrile – 15-50, emulsifying agent – 0.5-150, modifier – 0-1	Daly, L E, US Patent 2,439,202, US Rubber Company, Apr. 6, 1948.
Temperature of polymerization	ºC	60	Wang, W P; Pan, C Y, Wu, J S, J. Phys. Chem. Solids, 66, 1695-1700, 2005.
Time of polymerization	h	24	
Mass average molecular weight, M_w	dalton, g/mol, amu	165,000-185,000, 165,000	-; Ningaraju, S, Hegde, V N, Gnana Prakash, A P, Ravikumar, H B, Chem. Phys. Lett., 698, 24-35, 2018.
Polydispersity, M_w/M_n	-	2.08	Gao, C, Liu, P, Ding, Y, Li, T, Yang, M, Compos. Sci. Technol., 155, 41-9, 2018.
Van der Waals volume	cm^3 mol^{-1}	calc.=53.8	
Molecular cross-sectional area, calculated	cm2 x 10^{-1}6	33.8	
Free volume hole size	nm	11.062	Ningaraju, S, Hegde, V N, Gnana Prakash, A P, Ravikumar, H B, Chem. Phys. Lett., 698, 24-35, 2018.
STRUCTURE			
Crystallinity	%	26.35	Ningaraju, S, Hegde, V N, Gnana Prakash, A P, Ravikumar, H B, Chem. Phys. Lett., 698, 24-35, 2018.
Entanglement molecular weight	dalton, g/mol, amu	calc.=5030, 7005, 8716, 9154, 9536	

HANDBOOK OF POLYMERS 3rd Edition, Copyrights 2022;; ChemTec Publishing

SAN poly(styrene-co-acrylonitrile)

PARAMETER	UNIT	VALUE	REFERENCES
COMMERCIAL POLYMERS			
Some manufacturers	-	Ineos; Trinseo; Eni/Versalis	
Trade names	-	Luran; Tyril: Kostil	
PHYSICAL PROPERTIES			
Density at 20°C	g cm^{-3}	1.07-1.09; 0.97 (melt); 1.36 (35% glass fiber)	
Color	-	colorless	
Refractive index, 20°C	-	1.5700	Sultaanova, N G; Nikolov, I D; Ivanov, C D, Optical Quantum Electronics, 35, 21-34, 2003.
Transmittance	%	85-92	
Odor	-	odorless	
Decomposition temperature	°C	260	
Thermal expansion coefficient, 23-80°C	°C^{-1}	0.7E-4; 0.25E-4 (35% glass fiber)	
Thermal conductivity, melt	W m^{-1} K^{-1}	0.17 (melt); 0.21 (melt, 35% glass fiber)	
Glass transition temperature	°C	calc.=111; exp.=103-112	Silva, A L A; Takase, I; Pereira, R P; Rocco, A M, Eur. Polym. J., 44, 1462-74, 2008.
Specific heat capacity	J K^{-1} kg^{-1}	1,300 (25°C); 2,300-2,400 (melt); 1,900 (melt, 35% glass fiber)	
Heat deflection temperature at 0.45 MPa	°C	99-110; 108 (35% glass fiber)	
Heat deflection temperature at 1.8 MPa	°C	86-104; 104 (35% glass fiber)	
Vicat temperature VST/A/50	°C	105-120; 109 (35% glass fiber)	
Relative permittivity at 100 Hz	-	2.9-3; 3.5 (35% glass fiber)	
Relative permittivity at 1 MHz	-	2.7-2.8; 3.2 (35% glass fiber)	
Dissipation factor at 100 Hz	E-4	40-50; 70 (35% glass fiber)	
Dissipation factor at 1 MHz	E-4	70-80; 100 (35% glass fiber)	
Volume resistivity	ohm-m	>1E13	
Surface resistivity	ohm	>1E15	
Conductivity	S cm^{-1}	1E13	Oh, H, Kim, Y, Kim, J, Organic Electronics, 85, 105877, 2020.
AC conductivity	S cm^{-1}	2.16E-8	Ningaraju, S, Hegde, V N, Gnana Prakash, A P, Ravikumar, H B, Chem. Phys. Lett., 698, 24-35, 2018.
Comparative tracking index, CTI, test liquid A	-	400-550	
Speed of sound	m s^{-1}	41.8	
Acoustic impedance		2.68	
Attenuation	dB cm^{-1}, 5 MHz	5.1	
MECHANICAL & RHEOLOGICAL PROPERTIES			
Tensile strength	MPa	61-79; 110 (35% glass fiber)	
Tensile modulus	MPa	3,600-3,900; 12,000 (35% glass fiber)	
Tensile creep modulus, 1000 h, elongation 0.5 max	MPa	2,800; 7,500 (35% glass fiber)	

SAN poly(styrene-co-acrylonitrile)

PARAMETER	UNIT	VALUE	REFERENCES
Elongation	%	2.5-4	
Young's modulus	MPa	22-171 (core-shell particles)	Canche-Rscamilla, G; Duarte-Aranda, S; Rabelero_Velasco, M; Mendizabal-Mijares, E, Antec, 334-37, 2006.
Charpy impact strength, unnotched, 23°C	kJ m^{-2}	14-21; 17 (35% glass fiber)	
Charpy impact strength, unnotched, -30°C	kJ m^{-2}	16-21; 17 (35% glass fiber)	
Charpy impact strength, notched, 23°C	kJ m^{-2}	1.5-2.5; 4 (35% glass fiber)	
Poisson's ratio	-	0.366	
Shrinkage	%	0.3-0.7; 0.1-0.4 (glass fiber reinforced)	
Melt viscosity, shear rate=1000 s^{-1}	Pa s	120-200	
Melt volume flow rate (ISO 1133, procedure B), 220°C/10 kg	cm^3/10 min	7-23; 4 (35% glass fiber)	
Melt index, 230°C/3.8 kg	g/10 min	7.5	
Water absorption, equilibrium in water at 23°C	%	0.25-0.35	
Moisture absorption, equilibrium 23°C/50% RH	%	0.2-0.3	

CHEMICAL RESISTANCE			
Acid dilute/concentrated	-	good	
Alcohols	-	good	
Alkalis	-	very good	
Aliphatic hydrocarbons	-	good	
Aromatic hydrocarbons	-	poor	
Esters	-	poor	
Greases & oils	-	poor	
Halogenated hydrocarbons	-	poor	
Ketones	-	poor	
Θ solvent, Θ-temp.=43.2°C	-	ethyl acetate	
Good solvent	-	acetone, acetophenone, butyl acetate, chlorobenzene, cyclohexanone, diethyl ether, MEK, THF	
Non-solvent	-	acetamide, acetic acid, cetyl alcohol, diethylene glycol, formic acid	
Effect of EtOH sterilization (tensile strength retention)	%	63-88	Navarrete, L; Hermanson, N, Antec, 2807-18, 1996.
Chemicals causing environmental stress cracking	list	anionic and nonionic surfactants, sugar solution and fatty acids	Kawaguchi, T; Nishimura, H; Kasahara, K; Kuriyama, T; Narisawa, I, Polym. Eng. Sci., 43, 2, 419-30, 2003.

FLAMMABILITY			
Flammability according to UL-standard; thickness 1.6/0.8 mm	class	HB	
Ignition temperature	°C	380	

SAN poly(styrene-co-acrylonitrile)

PARAMETER	UNIT	VALUE	REFERENCES
Autoignition temperature	°C	450	
Limiting oxygen index	% O_2	18; 20 (35% glass fiber)	
WEATHER STABILITY			
Spectral sensitivity	nm	<360	
Activation wavelengths	nm	305	
Products of degradation	-	amide, aldehyde, acids	Mailhot, B; Gardette, J-L, Polym. Deg. Stab., 44, 237-47, 1994.
Activation energy	kJ mol^{-1}	19 (yellowing), 16 (gloss change)	Pickett, J E, Kuvshinnikova, O, Sung, L-P, Ermi, B D, Polym. Deg. Stab., 166, 135, 144, 2019.
TOXICITY			
HMIS: Health, Flammability, Reactivity rating	-	0/1/0	
Carcinogenic effect	-	not listed by ACGIH, NIOSH, NTP	
Oral rat, LD_{50}	mg kg^{-1}	1,800	
Skin rabbit, LD_{50}	mg kg^{-1}	2,000	
ENVIRONMENTAL IMPACT			
Aquatic toxicity, *Daphnia magna*, LC_{50}, 48 h	mg l^{-1}	13	
Aquatic toxicity, *Bluegill sunfish*, LC_{50}, 48 h	mg l^{-1}	28	
Aquatic toxicity, *Fathead minnow*, LC_{50}, 48 h	mg l^{-1}	10	
Cradle to grave non-renewable energy use	MJ/kg	90	
Cradle to pellet greenhouse gasses	kg CO_2 kg^{-1} resin	3.5	
PROCESSING			
Typical processing methods	-	blow molding, casting, extrusion, injection molding, thermoforming	
Preprocess drying: temperature/ time/residual moisture	°C/h/%	80/2-4/	
Processing temperature	°C	220-260 (injection molding); 220-240 (extrusion)	
Additives used in final products	-	Fillers: aluminum borate whiskers, barium sulfate, clay, glass fiber, montmorillonite, PTFE, reduced graphene oxide, zinc oxide, zirconium oxide	
Applications	-	appliances (housings, air conditioner parts, refrigerator shelves, blenders, lenses), automotive (dashboard, battery cases), bottles, high-density molding composition, housewares (eating utensils, beverage/food containers, display boxes), housings for electronic and electrical applications, instrument lenses packaging for high barrier properties	
Outstanding properties	-	outstanding transparency, good chemical resistance and stability in dishwashers together with high strength, rigidity, dimensional stability and thermal shock resistance	

SAN poly(styrene-co-acrylonitrile)

PARAMETER	UNIT	VALUE	REFERENCES
BLENDS			
Suitable polymers	-	EPDM, PC, PCL, PMMA, PPE, PS; incompatible with most thermoplastics; small amounts of admixtures lower mechanical properties and cause streaking	Luran Brochure KSEL 1001 BE, BASF, 2010.
ANALYSIS			
FTIR (wavenumber-assignment)	cm^{-1}/-	CN – 2220, C=O – 1725	Mailhot, B; Gardette, J-L, Polym. Deg. Stab., 44, 237-47, 1994.
x-ray diffraction peaks	degree	20,0 38.34, 44.6	Ningaraju, S, Hegde, V N, Gnana Prakash, A P, Ravikumar, H B, Chem. Phys. Lett., 698, 24-35, 2018.

SBC styrene-butadiene block copolymer

PARAMETER	UNIT	VALUE	REFERENCES
GENERAL			
Common name	-	styrene-butadiene block copolymer	
Acronym	-	SBC, S-TPE	
CAS number	-	9003-55-8	
HISTORY			
Person to discover	-	Hoeg, D F; Goldberg, E P; Pendleton, J F	Hoeg, D F; Goldberg, E P; Pendleton, J F, US Patent 3,598,886, Borg-Warner, Aug. 10, 1971.
Date	-	1971	
SYNTHESIS			
Monomer(s) structure	-	$H_2C=CHCH=CH_2$	
Monomer(s) CAS number(s)	-	100-42-5; 106-99-0	
Monomer(s) molecular weight(s)	dalton, g/mol, amu	104.15; 54.09	
Styrene content	%	60-82.6	
Method of synthesis	-	styrene, cyclohexane, and initiator are polymerized then butadiene is added and polymerized	Xiong, X; Eckelt, J; Wolf, B A; Zhang, Z; Zhang, L, J. Chromat., A1110, 53-60, 2006.
Temperature of polymerization	ºC	50	
Time of polymerization	h	5	
Pressure of polymerization	Pa	nitrogen atmosphere	
Number average molecular weight, M_n	dalton, g/mol, amu	87,000-109,000	
Mass average molecular weight, M_w	dalton, g/mol, amu	70,000-270,000	Nestle, N; Heckmann, W; Staininger, H; Knoll, K, Anal. Chim. Acta, 604, 54-61, 2007.
Polydispersity, M_w/M_n	-	$1.1^{-1}.74$	
STRUCTURE			
Cell dimensions	nm	a:b:c=0.885:0.908:0.479	Sakurai, K; Shrikawa, Y; Kashiwagi, T; Takahashi, T, Polymer, 35, 4238-9, 1994.
Unit cell angles	degree	β=113	Sakurai, K; Shrikawa, Y; Kashiwagi, T; Takahashi, T, Polymer, 35, 4238-9, 1994.
Tacticity	%	86 (*trans* in butadiene portion)	Sakurai, K; Shrikawa, Y; Kashiwagi, T; Takahashi, T, Polymer, 35, 4238-9, 1994.
Cis content	mol%	97 (butadiene units)	Zhu, H; Wu, Y-X; Zhao, J-W; Guo, Q-L; Huang, Q-g; Wu, G-Y, J. Appl. Polym. Sci., 106, 1, 103-9, 2007.
Lamellae thickness	nm	19	Balta-Calleja, F J; Cagiao, M E; Adhikari, R; Michler, G H, Polymer, 45, 247-54, 2004.
COMMERCIAL POLYMERS			
Some manufacturers	-	BASF; Chevron Phillips	
Trade names	-	Styroflex, Styrolux; K-Resin	

SBC styrene-butadiene block copolymer

PARAMETER	UNIT	VALUE	REFERENCES
PHYSICAL PROPERTIES			
Density at 20°C	g cm^{-3}	1.0-1.02; 0.91 (melt)	
Bulk density at 20°C	g cm^{-3}	0.55-0.65	
Color	-	colorless, clear to opaque	
Transmittance	%	90-92	
Haze	%	0.2-1.4	
Odor	-	faint, specific	
Softening point	°C	>35 to >90	
Decomposition temperature	°C	300	
Thermal expansion coefficient, 23-80°C	10^{-4} °C^{-1}	0.75E-4	
Thermal conductivity, melt	W m^{-1} K^{-1}	0.16	
Specific heat capacity	J K^{-1} kg^{-1}	2,300 (melt)	
Heat deflection temperature at 0.45 MPa	°C	62-75	
Heat deflection temperature at 1.8 MPa	°C	47-65	
Vicat temperature	°C	35-85	
Relative permittivity at 100 Hz	-	2.5	
Relative permittivity at 1 MHz	-	2.5	
Dissipation factor at 100 Hz	E-4	3	
Dissipation factor at 1 MHz	E-4	8	
Volume resistivity	ohm-m	>1E13	
Surface resistivity	ohm	1E15	
Comparative tracking index	-	600	
Permeability to oxygen, 23°C	cm^3 m^{-2} d^{-1} bar^{-1}	27.2	
Permeability to water vapor, 23°C	g m^{-2} d^{-1}	0.27	
MECHANICAL & RHEOLOGICAL PROPERTIES			
Tensile modulus	MPa	120 (Styroflex); 900-1800 (Styrolux)	
Tensile stress at yield	MPa	4 (Styroflex); 15-35 (Styrolux); 15-35 (K-Resin)	
Tensile creep modulus, 1000 h, elongation 0.5 max	MPa	490-1,050	
Elongation	%	20-360	
Tensile yield strain	%	2-5	
Flexural strength	MPa	24-37	
Flexural modulus	MPa	850-1,800	
Young's modulus	MPa	1,200	
Elmendorf tear resistance	N	650 (parallel); 800 (normal)	
Charpy impact strength, unnotched, 23°C	kJ m^{-2}	no break; 25 (Styrolux)	
Charpy impact strength, unnotched, -30°C	kJ m^{-2}	no break	
Charpy impact strength, notched, 23°C	kJ m^{-2}	no break; 2-85 (Styrolux)	

SBC styrene-butadiene block copolymer

PARAMETER	UNIT	VALUE	REFERENCES
Charpy impact strength, notched, -30°C	kJ m^{-2}	2	
Shore D hardness	-	56-63	
Melt volume flow rate (ISO 1133, procedure B), 200°C/5 kg	cm^3/10 min	11-16	
Melt index, 230°C/3.8 kg	g/10 min	7.5-15	
Water absorption, equilibrium in water at 23°C	%	0.07	
Moisture absorption, equilibrium 23°C/50% RH	%	0.07-0.09	
CHEMICAL RESISTANCE			
Acid dilute/concentrated	-	poor	
Alcohols	-	good	
Alkalis	-	very good	
Aliphatic hydrocarbons	-	not resistant	
Aromatic hydrocarbons	-	not resistant	
Esters	-	not resistant	
Greases & oils	-	not resistant	
Halogenated hydrocarbons	-	not resistant	
Ketones	-	not resistant	
FLAMMABILITY			
Flammability according to UL-94 standard; thickness 1.6/0.8 mm	class	HB	
Ignition temperature	°C	>288	
Autoignition temperature	°C	>400	
Volatile products of combustion	-	CO, CO$_2$, hydrocarbons	
WEATHER STABILITY			
Stabilizers	-	Antioxidant: p-phenylenediamine	Cibulkova, Z; Simon, P; Lehocky, P; Kosar, K; Chochulova, A, J. Therm. Anal. Calorim., 97, , 535-40, 2009.
TOXICITY			
HMIS: Health, Flammability, Re-activity rating	-	0-1/1/0-1	
ENVIRONMENTAL IMPACT			
Aquatic toxicity, *Bluegill sunfish*, LC$_{50}$, 48 h	mg l^{-1}	>100	
PROCESSING			
Typical processing methods	-	extrusion, injection molding, mixing	
Preprocess drying: temperature/ time/residual moisture	°C/h/%	60/1 (usually do not require drying)	
Processing temperature	°C	190-220 (Styroflex, injection molding); 170-190 (Styroflex, film extrusion); 170-240 (Styroflex, flat film extrusion); 180-250 (Styrolux, injection molding); 190-230 (Styrolux, extrusion)	

SBC styrene-butadiene block copolymer

PARAMETER	UNIT	VALUE	REFERENCES
Processing pressure	MPa	30-40 (clamping force)	
Additives used in final products	-	antiblock; slip; thermal stabilizer	
Applications	-	appliance housings, asphalt binder, blister packs, bottles, boxes, containers, cups, deli-trays, drink cups, flexible tubing, foam, hangers, lids, medical, modification of styrenic polymers (improves toughness and stress cracking resistance), packaging fresh meat, toys	
Outstanding properties	-	thermoplastic elastomer, transparency (sparkling clarity), surface gloss, impact strength, and toughness	

BLENDS

Suitable polymers	-	PS	

ANALYSIS

FTIR (wavenumber-assignment)	cm⁻¹/-	C=O – 1775, 1739, 1735, 1730, 1715, 1695, 1685, 1639	Allen, N S; Barcelona, A; Edge, M; Wilkinson, A; Merchan, C G; Sant Quiteria, V R, Polym. Deg. Stab., 86, 11-23, 2004.

SBR poly(styrene-co-butadiene)

PARAMETER	UNIT	VALUE	REFERENCES
GENERAL			
Common name	-	poly(styrene-co-butadiene)	
CAS name	-	benzene, ethenyl-, polymer with 1,3-butadiene	
Acronym	-	SBR	
CAS number	-	9003-55-8	
RTECS number	-	WL6478000	
HISTORY			
Date	-	1935; 1942	
Details	-	production in IG Farben, Germany; production in US	
SYNTHESIS			
Monomer(s) structure	-	$H_2C=CHCH=CH_2$	
Monomer(s) CAS number(s)	-	100-42-5; 106-99-0	
Monomer(s) molecular weight(s)	dalton, g/mol, amu	104.15; 54.09	
Styrene content	%	5-43; 25	Martinez-Barrera, G; Lopez, H; Castano, V M; Rodriguez, R, Radiat. Phys. Chem., 69, 155-62, 2004; Galimberti, M, Cipolletti, V, Cioppa, S, Conzatti, L, Appl. Clay Sci., 135, 168-75, 2017.
Formulation example	-	styrene – 25, butadiene – 75, water – 180, emulsifier – 5, dodecyl mercaptan – 0.2-0.8, cumene hydroperoxide – 0.17, $FeSO_4$ – 0.017, EDTA – 0.06	Maruyama, K; Kawaguchi, M; Kato, T, Colloids Surfaces, A189, 211-23, 2001.
Method of synthesis	-	emulsion polymerization in water medium initiated by peroxide or peroxydisulfate; also solution polymerized grades are available; typically produced by cold emulsion polymerization	Dube, M A; Li, L, Polym. Plast. Technol. Eng., 49, 648-56, 2010; Godoy, J L; Minari, R J; Vega, J R; Marchetti, J L, Chemometrics Intelligent Lab. Systems, in press, 2011.
Number average molecular weight, M_n	dalton, g/mol, amu	28,000-307,000	Rivera-Gastelum, M J; Puig, J E; Monroy, V M; Garcia Garduno, M; Castano, V M, Mater. Lett., 15, 253-59, 1992.
Mass average molecular weight, M_w	dalton, g/mol, amu	64,000-313,000	Rivera-Gastelum, M J; Puig, J E; Monroy, V M; Garcia Garduno, M; Castano, V M, Mater. Lett., 15, 253-59, 1992.
Polydispersity, M_w/M_n	-	1.04-3.31	Rivera-Gastelum, M J; Puig, J E; Monroy, V M; Garcia Garduno, M; Castano, V M, Mater. Lett., 15, 253-59, 1992.
End-to-end distance of the network chain	nm	4.83	Mondal, T, Bhowmick, A K, Ghosal, R, Mukhopadhyay, R, Polymer, 146, 31-41, 2018.
STRUCTURE			
Tacticity	%	*cis* – 9-38, *trans* – 53-75, vinyl – remainder	Martinez-Barrera, G; Lopez, H; Castano, V M; Rodriguez, R, Radiat. Phys. Chem., 69, 155-62, 2004.

SBR poly(styrene-co-butadiene)

PARAMETER	UNIT	VALUE	REFERENCES
COMMERCIAL POLYMERS			
Some manufacturers	-	Arlanxeo	
Trade names	-	Buna SE, SL, VSL	
PHYSICAL PROPERTIES			
Density at 20°C	g cm^{-3}	0.91-0.96	
Color	-	light amber to yellow to brown	
Refractive index, 20°C	-	1.53-1.56	
Odor	-	characteristic	
Decomposition temperature	°C	180	
Storage temperature	°C	0-35	
Thermal conductivity, melt	W m^{-1} K^{-1}	0.3-0.361	
Glass transition temperature	°C	-25 to -55	
Specific heat capacity	J K^{-1} kg^{-1}	1,895	Grieco, E; Bernardi, M; Baldi, G, J. Anal. Appl. Pyrolysis, 82, 304-11, 2008.
Maximum service temperature	°C	-40 to 100	
Long term service temperature	°C	65-70	
Vicat temperature VST/A/50	°C	92	
Hildebrand solubility parameter	MPa$^{0.5}$	17.4	
Coefficient of friction	-	1.5-2.5 (PE/SBR)	McNally, G M; Clarke, J L; Small, C M; Skelton, W J; Monroe, V, Antec, 2001.
MECHANICAL & RHEOLOGICAL PROPERTIES			
Tensile strength	MPa	13-28.5	
Tensile stress at yield	MPa	9.4-18	
Elongation	%	380-750; 268	-; Mondal, T, Bhowmick, A K, Ghosal, R, Mukhopadhyay, R, Polymer, 146, 31-41, 2018.
Young's modulus	MPa	2.-2.3	
Tear strength	kN m^{-1}	20.4-43	
Poisson's ratio	-	0.5 (or variable)	Starkova, O; Aniskevich, A, Polym. Test., 29, 310-18, 2010.
Compression set, 24h 105°C	%	34	Chakraborty, S; Kar, S; Dasgupta, S; Mukhopadhyay, R; Bandyopadhyay, S; Joshi, M; Ameta, S C, Polym. Test., 29, 679-84, 2010.
Shore A hardness	-	30-90	
Mooney viscosity	-	30-120	
Water absorption, equilibrium in water at 23°C	%	5	
CHEMICAL RESISTANCE			
Acid dilute/concentrated	-	poor	
Alcohols	-	good-fair	
Alkalis	-	poor	
Aliphatic hydrocarbons	-	poor	

SBR poly(styrene-co-butadiene)

PARAMETER	UNIT	VALUE	REFERENCES
Aromatic hydrocarbons	-	soluble	
Esters	-	poor	
Greases & oils	-	good-poor	
Halogenated hydrocarbons	-	poor	
Ketones	-	fair-poor	
Θ solvent, Θ-temp.=46, 21°C	-	methyl isobutyl ketone, methyl n-propyl ketone, n-octane	

FLAMMABILITY			
Autoignition temperature	^{o}C	320	
Volatile products of combustion	-	CO, CO_2, styrene, butadiene, aromatic tars	Grieco, E; Bernardi, M; Baldi, G, J. Anal. Appl. Pyrolysis, 82, 304-11, 2008.
UL 94 rating	-	HB	

WEATHER STABILITY			
Spectral sensitivity	nm	295-360	
Important initiators and accelerators	-	ozone	
Products of degradation	-	hydroperoxides, hydroxyl, carboxyl, ketone, and epoxy groups; butadiene is degraded	Arantes, T M; Leao, K V; Tavares, M I B; Ferreira, A G; Longo, E; Camargo, E R, Polym. Test., 28, 490-94, 2009.
Stabilizers	-	Screener: carbon black, zinc oxide, talc; Phenolic antioxidant: ethylene-bis(oxyethylene)-bis(3-(5-tert-butyl-4-hydroxy-m-tolyl)-propionate); 2,6,-di-tert-butyl-4-(4,6-bis(octylthio)-1,3,5,-triazine-2-ylamino) phenol; 2-(1,1-dimethylethyl)-6-[[3-(1,1-dimethylethyl)-2-hydroxy-5-methylphenyl] methyl-4-methylphenyl acrylate; 2',3-bis[[3-[3,5-di-tert-butyl-4-hydroxyphenyl]propionyl]]propionohydrazide; isotridecyl-3-(3,5-di-tert-butyl-4-hydroxyphenyl) propionate; 3,5-bis(1,1-dimethyethyl)-4-hydroxy-benzenepropanoic acid, C13-15 alkyl esters; 2,2'-isobutylidenebis(2,4-dimethylphenol); Phosphite: trinonylphenol phosphite; Thiosynergist: 4,6-bis(octylthiomethyl)-o-cresol; 4,6-bis(dodecylthiomethyl)-o-cresol; 2,2'-thiodiethylene bis[3-(3,5-ditert-butyl-4-hydroxyphenyl)propionate]; Amine: nonylated diphenylamine	

TOXICITY			
NFPA: Health, Flammability, Reactivity rating	-	1/1-3/0	
Carcinogenic effect	-	not listed by ACGIH, NIOSH, NTP	

ENVIRONMENTAL IMPACT			
Aquatic toxicity, *Daphnia magna*, LC_{50}, 48 h	mg l^{-1}	23	
Aquatic toxicity, *Bluegill sunfish*, LC_{50}, 48 h	mg l^{-1}	25.05	
Aquatic toxicity, *Fathead minnow*, LC_{50}, 48 h	mg l^{-1}	46.4-59.3	

PROCESSING			
Typical processing methods	-	calendering, coating, compression molding, mixing, vulcanization	

SBR poly(styrene-co-butadiene)

PARAMETER	UNIT	VALUE	REFERENCES
Additives used in final products	-	Fillers: barium sulfate, carbon black, carbon fiber, carbon nanotubes, clay, crosslinked PS beads, expanded graphite, lead oxide (g-radiation shields), kaolin, magnesium hydroxide, mica, rectorite, silica, sodium aluminum silicate; Plasticizers: aromatic mineral oil, paraffinic mineral oil, rosin esters, terpene resins; Antistatics: carbon black, steel fibers, trineoalkoxy amino and trineoalkoxy sulfonyl zirconate; Release: zinc stearate	
Applications	-	automotive goods, belts, caulking, coated fabrics, conveyor belts, flooring, gaskets, hollow fiber, lamination, mastics, pressure-sensitive adhesives, sealants, sheet, shoe products, sponge, sporting goods, tank and caterpillar tracks, tires, toys, tubing, wire & cable	

BLENDS

Suitable polymers	-	CMC, EVA, LDPE, NBR, NR, PE, PC, PMMA, PS, PVC, SAN, SBS	

ANALYSIS

FTIR (wavenumber-assignment)	cm^{-1}/-	aromatic hydrogen – 3025, 3061; CH_2 – 2917, 2847; *trans* C=C – 965; $C=CH_2$ – 913; phenyl – 699	Lei, Y; Tang, Z; Zhu, L; Guo, B; Jia, D, Polymer, 52, 1337-44, 2011.
Raman (wavenumber-assignment)	cm^{-1}/-	C–H – 2915, 1000; C=C – 1438, 620	Martinez-Barrera, G; Lopez, H; Castano, V M; Rodriguez, R, Radiat. Phys. Chem., 69, 155-62, 2004.
NMR (chemical shifts)	ppm	C NMR: C1 aromatic – 143.8, CH=CH vinyl – 141.2; C2 and C4 aromatic – 130.0; C3 – 128.6; CH=CH 1,4 – 126.7; C4 aromatic – 124.8; $=CH_2$ vinyl – 113.1; CH_2 styrene – 44.3, 41.9; CH_2 vinyl – 36.8, 31.4; CH_2 1,4 *trans* – 34.5, 26.2; CH_2 1,4 *cis* – 32.8, 28.9, 23.8	Arantes, T M; Leao, K V; Tavares, M I B; Ferreira, A G; Longo, E; Camargo, E R, Polym. Test., 28, 490-94, 2009.

SBS styrene-butadiene-styrene triblock copolymer

PARAMETER	UNIT	VALUE	REFERENCES
GENERAL			
Common name	-	styrene-butadiene-styrene triblock copolymer	
IUPAC name	-	buta-1,3-diene; styrene	
CAS name	-	benzene, ethenyl-, polymer with 1,3-butadiene, triblock	
Acronym	-	SBS	
CAS number	-	694491-73-1	
HISTORY			
Person to discover	-	Bailey, J T; Nyberg, D D	Bailey, J T; Nyberg, D D, US Patent 3,328,173, Shell, Mar. 1, 1966.
Date	-	1966	
Details	-	SBS block copolymers	
SYNTHESIS			
Monomer(s) structure	-	$H_2C=CHCH=CH_2$	
Monomer(s) CAS number(s)	-	100-42-5; 106-99-0	
Monomer(s) molecular weight(s)	dalton, g/mol, amu	104.15; 54.09	
Monomer ratio	-	18-70/30-82; 30/70 (block ratio)	Yang, X, Zhang, H, Chen, Z, Shi, C, Fuel, 286, 119314, 2021.
Block molecular weights	dalton, g/mol, amu	7,200, 34,000, 7,200; 10,600 (PS), 60,100 (PB) in Kraton D1102	Tzounis, L, Pegel, S, Zafeiropoulos, N E, Stamm, M, Polymer, 131, 1-9, 2017.
Star-shape - number of arms		1-4	Xiong, X; Zhang, L; Ma, Z; Li, Y, J. Appl. Polym. Sci., 95, 832-40, 2005.
Method of synthesis	-	copolymers are synthesized through anionic polymerization *via* either sequential or coupling methods. To produce SBS triblock copolymers, by both methods, the synthesis comprises initiation of styrene polymerization using a mono-anionic organolithium compound to form living polystyryl anion, followed by addition of a butadiene monomer to form living SB di-block. In the sequential method, a second quantity of styrene is added to the living SB di-block in order to complete the formation of SBS tri-block copolymer. The coupling process differs from the sequential one in that the tri-block copolymer is terminated by coupling two living SB di-blocks	Canto, L B; Mantovani, G L; deAzevedo, E R; Bonagamba, T J; Hage, E; Pessan, L A, Polym. Bull., 57, 513-24, 2006.
Catalyst	-	organolithium compound	
Number average molecular weight, M_n	dalton, g/mol, amu	65,000-381,000	
Mass average molecular weight, M_w	dalton, g/mol, amu	70,000-430,000	
Polydispersity, M_w/M_n	-	1.05-4.3	
STRUCTURE			
Tacticity	%	polybutadiene: 1,4-*trans* - 42-53, 1,4-*cis* - 36-49, 1,2-vinyl - 7-12	Lee, W-F; Lee, H-H, J. Elastomers Plast., 42, 1, 49-64, 2010; Canto, L B; Mantovani, G L; deAzevedo, E R; Bonagamba, T J; Hage, E; Pessan, L A, Polym. Bull., 57, 513-24, 2006.

SBS styrene-butadiene-styrene triblock copolymer

PARAMETER	UNIT	VALUE	REFERENCES
COMMERCIAL POLYMERS			
Some manufacturers	-	AlphaGary; Kraton Polymers	
Trade names	-	Evoprene; Kraton	
PHYSICAL PROPERTIES			
Density at 20°C	g cm^{-3}	0.91-1.03; 0.94	-; Gnanaseelan, M, Samanta, S, Pionteck, J, Voit, B, Mater. Chem. Phys., 229, 319-29, 2019.
Bulk density at 20°C	g cm^{-3}	0.3-0.4	
Color	-	white	
Refractive index, 20°C	-	1.520-1.560	
Transmittance	%	10-70	
Odor	-	odorless	
Glass transition temperature	°C	-60 to -95; -85 to -92 (butadiene block); 68-101 (styrene block)	Masson, J-F; Bundalo-Perc, S; Delgado, A, J. Polym. Sci. B, 43, 276-79, 2005; Wang, C, Macromolecules, 34, 9006-14, 2001; Adhikari, R; Michler, G H, Prog. Polym. Sci., 29, 949-86, 2004.
Long term service temperature	°C	-50 to 75	
Heat deflection temperature at 0.45 MPa	°C	90	
Heat deflection temperature at 1.8 MPa	°C	58-65	
Vicat temperature VST/A/50	°C	81-92	
Hansen solubility parameters, δ_D, δ_P, δ_H	MPa$^{0.5}$	17.37-17.41; 1.12-1.14; 2.60-2.64	Ovejero, G; Romero, M D; Diez, E; Diaz, I, Eur. Polym. J., 46, 2261-68, 2010.
Hildebrand solubility parameter	MPa$^{0.5}$	17.60-17.64	Ovejero, G; Romero, M D; Diez, E; Diaz, I, Eur. Polym. J., 46, 2261-68, 2010.
Contact angle of water, 20°C	degree	90	Lee, W-F; Lee, H-H, J. Elastomers Plast., 42, 1, 49-64, 2010.
Solvent solubility	-	diethyl ether, trichlorethylene	Yang, X, Zhang, H, Chen, Z, Shi, C, Fuel, 286, 119314, 2021.
Electrical conductivity	S cm^{-1}	1E-16	Gnanaseelan, M, Samanta, S, Pionteck, J, Voit, B, Mater. Chem. Phys., 229, 319-29, 2019.
Speed of sound	m s^{-1}	1,390-1,720	Adachi, K; North, A M; Pethrick, R A; Harrison, G; Lamb, J, Polymer, 23, 10, 1451-56, 1982.
MECHANICAL & RHEOLOGICAL PROPERTIES			
Tensile strength	MPa	20-39.8	
Tensile stress at yield	MPa	5-30	
Elongation	%	88-1610	
Tensile yield strain	%	15-51	
Flexural strength	MPa	23-50	
Flexural modulus	MPa	1,200-1,840	
Young's modulus	MPa	1,810	
Tear strength	kN m^{-1}	17-48	
Charpy impact strength, unnotched, 23°C	kJ m^{-2}	16 to no break	

SBS styrene-butadiene-styrene triblock copolymer

PARAMETER	UNIT	VALUE	REFERENCES
Charpy impact strength, notched, 23°C	kJ m^{-2}	1.3-2	
Izod impact strength, unnotched, 23°C	J m^{-1}	NB	
Izod impact strength, notched, 23°C	J m^{-1}	33 to NB	
Compression set, 24h 70°C	%	86	
Shore A hardness	-	40-94	
Shore D hardness	-	48-75	
Rockwell hardness	-	R10-88	
Shrinkage	%	0.2-1.8	
Mooney viscosity	-	47	
Melt index, 200°C/5 kg	g/10 min	4-60	
Water absorption, 24h at 23°C	%	0.06	
CHEMICAL RESISTANCE			
Acid dilute/concentrated	-	good	
Alcohols	-	good	
Aliphatic hydrocarbons	-	non-resistant	
Aromatic hydrocarbons	-	non-resistant	
Ketones	-	non-resistant	
FLAMMABILITY			
UL 94 rating	-	HB	
WEATHER STABILITY			
Spectral sensitivity	nm	<370	
Depth of UV penetration	μm	50-200 (degraded asphalt per month)	Wu, S; Pang, L; Zhu, G, Key Eng. Mater., 385-387, 481-84, 2008.
Products of degradation	-	hydroperoxides, hydroxyls, carboxylic, ketone, and epoxy groups	
Stabilizers	-	2,6-di-tert-butyl-4-methylphenol, 2-(29-hydroxy-59-methylphenyl)benzotriazole, tris(nonylphenyl) phosphite, 1,2,2,6,6-pentamethyl piperidinyl-4-acrylate; tocopherols	Singh, R P, Desai, S M; Solanky, S S; Thanki, P N, J. Appl. Polym. Sci., 75, 1103-14, 2000; Suffield, R M; Kiesser, J E; Dillman, S H, Antec, 3351-56, 2005.
TOXICITY			
NFPA: Health, Flammability, Reactivity rating	-	0/1/0	
Oral rat, LD$_{50}$	mg kg^{-1}	>2,000	
Skin rabbit, LD$_{50}$	mg kg^{-1}	>2,000	
ENVIRONMENTAL IMPACT			
Aquatic toxicity, *Daphnia magna*, LC$_{50}$, 48 h	mg l^{-1}	>1,000	
Aquatic toxicity, *Bluegill sunfish*, LC$_{50}$, 48 h	mg l^{-1}	>1,000	

SBS styrene-butadiene-styrene triblock copolymer

PARAMETER	UNIT	VALUE	REFERENCES
PROCESSING			
Typical processing methods	-	blow molding, calendering, coextrusion, compounding, extrusion, injection molding	
Processing temperature	°C	150-210 (injection molding, extrusion)	
Processing pressure	MPa	34.5-65.6 (injection)	
Additives used in final products	-	Plasticizers: asphalt, dibutyl or dioctyl phthalate, white mineral oil; Antistatics: carbon black, carbon nanotubes, polyaniline, quarternary ammonium compound; Antiblocking: behena-mide, oleyl palmitamide, stearyl erucamide; Release: stearyl erucamide; Antioxidants: pentaerythrityl terakis[3-(3,5-di-tert-butyl-4-hydroxyphenyl) propionate; Phosphites: 2,4-di-tert-bu-tylphenyl) phosphite; Filler: expanded vermiculite, multiwalled carbon nanotube, reduced graphene oxide, titanium dioxide, vanadium chloride, zinc oxide	
Applications	-	adhesives, asphalt modification, coatings, grommets, sealants, seals, toys	
BLENDS			
Suitable polymers	-	sPB, PEDOT, iPP, PP, PS, PVME	
ANALYSIS			
FTIR (wavenumber-assignment)	cm^{-1}/-	1,2-BD – 911; *cis*-1,4-BD – 737; *trans*-1,4-BD – 967; phenyl ring – 1596, 1491, 1449, 700; 966 - characteristic band of C=C	Lee, W-F; Lee, H-H, J. Elastomers Plast., 42, 1, 49-64, 2010; Yang, X, Zhang, H, Chen, Z, Shi, C, Fuel, 286, 119314, 2021.
NMR (chemical shifts)	ppm	1,2-vinyl – 40-37 and 120-110; 1,4-*cis* – 29-27; 1,4-*trans* – 34-32	Canto, L B; Mantovani, G L; deAzevedo, E R; Bonagamba, T J; Hage, E; Pessan, L A, Polym. Bull., 57, 513-24, 2006.

SEBS styrene-(ethylene-butylene)-styrene triblock copolymer

PARAMETER	UNIT	VALUE	REFERENCES
GENERAL			
Common name	-	styrene-(ethylene-butylene)-styrene triblock copolymer	
CAS name	-	styrene-butadiene rubber, hydrogenated, block, triblock	
Acronym	-	SEBS	
CAS number	-	308076-28-0	
SYNTHESIS			
Monomer(s) structure	-	$H_2C=CHCH=CH_2$	
Monomer(s) CAS number(s)	-	100-42-5; 74-85-1; 106-98-9	
Monomer(s) molecular weight(s)	dalton, g/mol, amu	104.15; 28.05; 56.11	
Polystyrene content	%	13-60; 30/70 (styrene/rubber ratio)	-; Ferreira Ribeiro, V, Naue Simões, D, Pittol, M, Campomanes Santana, R M, Polym. Testing, 63, 204-9, 2017.
Polystyrene block Mw		72,000	Chang, Y-W; Shin, J-Y; Ryu, S H, Polym. Int., 53, 1047-51, 2004.
Ethylene/butylene block Mw		37,500	
Number average molecular weight, M_n	dalton, g/mol, amu	50,000-154,000	Xu, W; Cheng, Z; Zhang, Z; Zhang, L; Zhu, X, Reactive Functional Polym., 71, 634-40, 2011.
Mass average molecular weight, M_w	dalton, g/mol, amu	57,000-183,000; 192,000 (Kraton)	Rungswang, W; Kotaki, M; Shimojima, T; Kimura, G; Sakurai, S; Chirachanchai, S, Polymer, 52, 844-53, 2011; Ferreira Ribeiro, V, Naue Simões, D, Pittol, M, Campomanes Santana, R M, Polym. Testing, 63, 204-9, 2017.
Polydispersity, M_w/M_n	-	1.05; 1.36	-; Ferreira Ribeiro, V, Naue Simões, D, Pittol, M, Campomanes Santana, R M, Polym. Testing, 63, 204-9, 2017.
STRUCTURE			
Lamellae thickness	nm	17.7-25.9	Rungswang, W; Kotaki, M; Shimojima, T; Kimura, G; Sakurai, S; Chirachanchai, S, Polymer, 52, 844-53, 2011.
COMMERCIAL POLYMERS			
Some manufacturers	-	Elasto; Kraton Polymers; Repsol	
Trade names	-	Dryflex; Kraton; Calprene	
PHYSICAL PROPERTIES			
Density at 20°C	g cm⁻³	0.88-0.95	
Bulk density at 20°C	g cm⁻³	0.3-0.4	
Glass transition temperature	°C	-53 to -60; -36 (EB); 100 (ST); -45	Chang, Y-W; Shin, J-Y; Ryu, S H, Polym. Int., 53, 1047-51, 2004; Teruel-Juanes, R, Pascual-Jose, B, del Río, C, Ribes-Greus, A, Reactive Functional Polymers, 155, 104715, 2020.
Long term service temperature	°C	-50 to 125	

706   **HANDBOOK OF POLYMERS** 3rd Edition, Copyrights 2022;; ChemTec Publishing

SEBS styrene-ethylene-butylene-styrene triblock copolymer

PARAMETER	UNIT	VALUE	REFERENCES
Diffusion coefficient of oxygen	cm^2 s^{-1} x10^6	0.1-0.12	
Contact angle of water, 20°C	degree	61	Peinado, C; Corrales, T; Catalina, F; Pedron, S; Quiteria, V R S; Parellada, M D; Barrio, J A; Olmos, D; Gonzalez-Benito, J, Polym. Deg. Stab., 95, 975-86, 2010.
activation energy of β dielectric relaxation	kJ mol^{-1}	23.2 (associated with intramolecular local mobility)	Teruel-Juanes, R, Pascual-Jose, B, del Río, C, Ribes-Greus, A, Reactive Functional Polymers, 155, 104715, 2020.

MECHANICAL & RHEOLOGICAL PROPERTIES

Tensile strength	MPa	3-40	
Tensile modulus	MPa	0.2-5.4	
Tensile stress at yield	MPa	5.6	
Elongation	%	470-880	
Flexural modulus	MPa	100-700	
Tear strength	kN m^{-1}	21-55	
Izod impact strength, notched, 23°C	J m^{-1}	NB	
Compression set, 72h 23°C	%	16-20	
Shore A hardness	-	30-95	
Shrinkage	%	1.2-3	
Brittleness temperature (ASTM D746)	°C	-60 to -21	
Melt index, 230°C/5 kg	g/10 min	1-22	

CHEMICAL RESISTANCE

Acid dilute/concentrated	-	good	
Alcohols	-	good	
Aliphatic hydrocarbons	-	poor	
Aromatic hydrocarbons	-	poor	
Greases & oils	-	good	
Halogenated hydrocarbons	-	poor	

WEATHER STABILITY

Excitation wavelengths	nm	290	Luengo, C; Allen, N S; Edge, M; Wilkinson, A; Parellada, M D; Barrio, J A; Santa, V R, Polym. Deg. Stab., 91, 947-56, 2006.
Emission wavelengths	nm	420, 450, 470	Luengo, C; Allen, N S; Edge, M; Wilkinson, A; Parellada, M D; Barrio, J A; Santa, V R, Polym. Deg. Stab., 91, 947-56, 2006.
Important initiators and accelerators	-	formation of OH was 3-4 greater at 55°C than 30°C and increased humidity also accelerated its formation	White, C C; Tan, K T; Huston, D L; Nguyen, T; Benatti, D J; Stanley, D; Chin, J W, Polym. Deg. Stab., 96, 1104-1110, 2011.
Important initiators and accelerators	-	titanium traces	Luengo, C; Allen, N S; Wilkinson, A; Edge, M; Parellada, M D; Barrio, J A; Santa, V R, J. Vinyl Addit. Technol., 12, 2-7, 2006.

SEBS styrene-(ethylene-butylene)-styrene triblock copolymer

PARAMETER	UNIT	VALUE	REFERENCES
Products of degradation	-	hydroperoxides, acetophenone, oxidation products, discoloration, chain scission	Luengo, C; Allen, N S; Wilkinson, A; Edge, M; Parellada, M D; Barrio, J A; Santa, V R, J. Vinyl Addit. Technol., 12, 2-7, 2006.
Stabilizers	-	hindered phenols and phosphites	Luengo, C; Allen, N S; Wilkinson, A; Edge, M; Parellada, M D; Barrio, J A; Santa, V R, J. Vinyl Addit. Technol., 12, 2-7, 2006.

TOXICITY

PARAMETER	UNIT	VALUE	REFERENCES
NFPA: Health, Flammability, Reactivity rating	-	0/1/0	
Carcinogenic effect	-	not listed by ACGIH, NIOSH, NTP	
Oral rat, LD_{50}	mg kg^{-1}	>2,000	
Skin rabbit, LD_{50}	mg kg^{-1}	>2,000	

ENVIRONMENTAL IMPACT

PARAMETER	UNIT	VALUE	REFERENCES
Aquatic toxicity, *Daphnia magna*, LC_{50}, 48 h	mg l^{-1}	>1,000	
Aquatic toxicity, *Bluegill sunfish*, LC_{50}, 48 h	mg l^{-1}	>1,000	

PROCESSING

PARAMETER	UNIT	VALUE	REFERENCES
Typical processing methods	-	compounding, injection molding	
Preprocess drying: temperature/time/residual moisture	°C/h/%	52-80/0-3/0.04	
Processing temperature	°C	170-230	
Processing pressure	MPa	4-53 (injection); 0.3-3.5 (back)	
Applications	-	appliances, adhesives, automotive (bumper, exterior parts and trim, interior parts), belts, cable jacketing, coatings, compatibilizer, electrical/electronic, footwear, gaskets, household goods, impact modifier, medical, modifier (plastics, bitumen), piping, sealants, sporting goods, tools, toys, wheels for office furniture	
Outstanding properties	-	chemical resistance, weather resistance	

BLENDS

PARAMETER	UNIT	VALUE	REFERENCES
Suitable polymers	-	EVA, HDPE, HIPS, PA6, PA66, PAN, PE, PP, PPS, PPy, sPS, PVDF, SIS	

ANALYSIS

PARAMETER	UNIT	VALUE	REFERENCES
FTIR (wavenumber-assignment)	cm^{-1}/-	CH_3 – 1379; CH_2 – 1460; C=O – 1711; ester – 1733; lactone – 1775; OH – 3150, 3600	White, C C; Tan, K T; Huston, D L; Nguyen, T; Benatti, D J; Stanley, D; Chin, J W, Polym. Deg. Stab., 96, 1104-1110, 2011.

SIS styrene-isoprene-styrene block copolymer

PARAMETER	UNIT	VALUE	REFERENCES
GENERAL			
Common name	-	styrene-isoprene-styrene block copolymer	
CAS name	-	isoprene-styrene rubber, block, triblock	
Acronym	-	SIS	
CAS number	-	308067-96-1	
HISTORY			
Person to discover	-	Bailey, J T; Nyberg, D D	Bailey, J T; Nyberg, D D, US Patent 3,328,173, Shell, Mar. 1, 1966.
Date	-	1966	
Details	-	SIS block copolymers	
SYNTHESIS			
Monomer(s) structure	-	$H_2C{=}CHC{=}CH_2$ with CH_3	
Monomer(s) CAS number(s)	-	100-42-5; 78-79-5	
Monomer(s) molecular weight(s)	dalton, g/mol, amu	104.15; 68.12	
Monomer ratio	-	14-30 (styrene)/70-86 (isoprene); 17 (styrene)	-; Chipara, M, Artiaga, R, Lau, K T, Hui, D, Compos. Commun., 3, 23-7, 2017.
Number average molecular weight, M_n	dalton, g/mol, amu	142,000	
Mass average molecular weight, M_w	dalton, g/mol, amu	156,000-237,000	
Polydispersity, M_w/M_n	-	1.10-1.49	
COMMERCIAL POLYMERS			
Some manufacturers	-	Kraton Polymers; Versalis	
Trade names	-	Kraton; Europrene	
PHYSICAL PROPERTIES			
Density at 20°C	g cm^{-3}	0.92-0.94	
Bulk density at 20°C	g cm^{-3}	0.35	
Color	-	clear	
Decomposition temperature	°C	190	Hacaloglu, J; Fares, M M; Suzer, S, Eur. Polym. J., 35, 939-44, 1999.
Glass transition temperature	°C	glass transition temperature of polyisoprene is ranging between -35 and -70°C, depending on the average molecular mass and on the microstructure (*cis, trans* composition) of the polymer. The glass transition temperature for soft segments (polyisoprene) is -55°C	Chipara, M, Artiaga, R, Lau, K T, Hui, D, Compos. Commun., 3, 23-7, 2017.
MECHANICAL & RHEOLOGICAL PROPERTIES			
Tensile strength	MPa	12-23.2	
Tensile stress at yield	MPa	6-32	
Elongation	%	750-1,500	
Tensile yield strain	%	1,300	

HANDBOOK OF POLYMERS 3rd Edition, Copyrights 2022;; ChemTec Publishing

SIS styrene-isoprene-styrene block copolymer

PARAMETER	UNIT	VALUE	REFERENCES
Shore A hardness	-	24-87	
Melt index, 190°C/5 kg	g/10 min	2-41	

CHEMICAL RESISTANCE			
Alcohols	-	good	
Aromatic hydrocarbons	-	poor	

FLAMMABILITY			
Volatile products of combustion	-	CO, CO_2, isoprene, styrene, benzene	Hacaloglu, J; Fares, M M; Suzer, S, Eur. Polym. J., 35, 939-44, 1999.

PROCESSING			
Typical processing methods	-	blow molding, coating, compression molding, extrusion, injection molding	
Preprocess drying: temperature/ time/residual moisture	°C/h/%	52/2-4/	
Processing temperature	°C	150-200	
Processing pressure	MPa	7-138 (injection); 0.35 (back)	
Additives used in final products	-	multiwalled carbon nanotubes	Chipara, M, Artiaga, R, Lau, K T, Hui, D, Compos. Commun., 3, 23-7, 2017.
Applications	-	adhesives, coatings, cosmetics, membranes, pharmaceuticals, plastics modification, sealants	

BLENDS			
Suitable polymers	-	PE, PPy, ULDPE	

SMA poly(styrene-co-maleic anhydride)

PARAMETER	UNIT	VALUE	REFERENCES
GENERAL			
Common name	-	poly(styrene-co-maleic anhydride)	
IUPAC name	-	poly(styrene-co-maleic anhydride)	
ACS name	-	2,5-furandione, polymer with ethenylbenzene	
Acronym	-	SMA	
CAS number	-	9011-13-6	
RTECS number	-	ON4240000	
HISTORY			
Person to discover	-	Wagner-Juaregg, T	Wagner-Juaregg, T, Chem. Ber., 63, 3213, 1930.
Date	-	1930	
SYNTHESIS			
Monomer(s) structure	-		
Monomer(s) CAS number(s)	-	100-42-5; 108-31-6	
Monomer(s) molecular weight(s)	dalton, g/mol, amu	104.15; 96.06	
Maleic anhydride content	%	7-35; 14 mol%	-, Dellacasa, E, Forouharshad, M, Rolandi, R, Pastorino, L, Monticelli, O, Mater. Lett., 220, 241-4, 2018.
Method of synthesis	-	precipitation polymerization	
Temperature of polymerization	ºC	60	
Time of polymerization	h	12	
Number average molecular weight, M_n	dalton, g/mol, amu	28,000-46,000	
Mass average molecular weight, M_w	dalton, g/mol, amu	5,000-224,000	
Polydispersity, M_w/M_n	-	2.5	
Polymerization degree (number of monomer units)	-	322-495	
Radius of gyration	nm	21	Krueger, S; Krahl, F; Arndt, K-F, Eur. Polym. J., 46, 1040-48, 2010.
STRUCTURE			
Entanglement molecular weight	dalton, g/mol, amu	calc.=14,522, 16,462,17,750	
Free volume	$cm^3 \, g^{-1}$	0.0668-0.0744	Kilburn, D; Dlubek, G; Pionteck, J; Bamford, D; Alam, M A, Polymer, 46, 869-76, 2005.
Hole size	$nm^3 \, K^{-1}$	0.075-0.102	Kilburn, D; Dlubek, G; Pionteck, J; Bamford, D; Alam, M A, Polymer, 46, 869-76, 2005.
COMMERCIAL POLYMERS			
Some manufacturers	-	Cray Valley, Ineos; Polyscope	
Trade names	-	SMA 2021, Lustran; Xiran	

SMA poly(styrene-co-maleic anhydride)

PARAMETER	UNIT	VALUE	REFERENCES
PHYSICAL PROPERTIES			
Density at 20°C	g cm^{-3}	1.05-1.08	
Bulk density at 20°C	g cm^{-3}	0.55	
Color	-	translucent, off-white	
Refractive index, 20°C	-	1.5640-1.577	
Transmittance	%	91	
Haze	%	2-2.5	
Odor	-	slight	
Melting temperature, DSC	°C	115	
Thermal expansion coefficient, 23-80°C	10^{-4} °C^{-1}	1	
Glass transition temperature	°C	118-176	
Heat deflection temperature at 0.45 MPa	°C	93-106	
Heat deflection temperature at 1.8 MPa	°C	80-92	
Vicat temperature VST/A/50	°C	104-129	
Dielectric constant at 100 Hz/1 MHz	-	2.65/2.73	
Dissipation factor at 100 Hz	E-4	62	
Dissipation factor at 1 MHz	E-4	38	
Volume resistivity	ohm-m	1.5E14	
Surface resistivity	ohm	4.3E14	
Electric strength K20/P50, d=0.60.8 mm	kV mm^{-1}	14	
Zeta potential	mV	-27.9	Dellacasa, E, Forouharshad, M, Rolandi, R, Pastorino, L, Monticelli, O, Mater. Lett., 220, 241-4, 2018.
MECHANICAL & RHEOLOGICAL PROPERTIES			
Tensile strength	MPa	29-43	
Tensile modulus	MPa	2,000-2,300	
Tensile stress at yield	MPa	30-44	
Elongation	%	3-10	
Flexural strength	MPa	57-72	
Flexural modulus	MPa	2,000-2,300	
Charpy impact strength, unnotched, 23°C	kJ m^{-2}	30 to NB	
Charpy impact strength, unnotched, -40°C	kJ m^{-2}	30-84	
Charpy impact strength, notched, 23°C	kJ m^{-2}	9-13	
Charpy impact strength, notched, -40°C	kJ m^{-2}	5-17	
Izod impact strength, notched, 23°C	J m^{-1}	180-210	
Izod impact strength, notched, -30°C	J m^{-1}	80-96	
Rockwell hardness	-	R98-100	

SMA poly(styrene-co-maleic anhydride)

PARAMETER	UNIT	VALUE	REFERENCES
Shrinkage	%	0.028-0.6	
Melt index, 220°C/10 kg	g/10 min	6-7	
Moisture absorption, equilibrium 23°C/50% RH	%	0.2	
CHEMICAL RESISTANCE			
Acid dilute/concentrated	-	very good	
Alcohols	-	very good	
Alkalis	-	poor	
Aliphatic hydrocarbons	-	good	
Aromatic hydrocarbons	-	poor	
Esters	-	poor	
Halogenated hydrocarbons	-	poor	
Ketones	-	poor	
Good solvent	-	toluene	
FLAMMABILITY			
Flammability according to UL-94 standard; thickness 1.6/0.8 mm	class	HB	
Ignition temperature	°C	343	
Autoignition temperature	°C	487	
Mass loss (194-476°C)	%	95.01	Sarı, A, Biçer, A, Alkan, C, Solar Energy Mater. Solar Cells, 161, 219-225, 2017.
WEATHER STABILITY			
Spectral sensitivity	nm	260, 300, 320	Holland, K A; Griesser, H J; Hawthorne, D G; Hodgkin, J H, Polym. Deg. Stab., 31, 269-89, 1991.
TOXICITY			
NFPA: Health, Flammability, Reactivity rating	-	1/1/0	
Carcinogenic effect	-	not listed by ACGIH, NIOSH, NTP	
TLV, ACGIH	mg m^{-3}	3 (respirable), 10 (total)	
OSHA	mg m^{-3}	5 (respirable), 15 (total)	
Oral rat, LD$_{50}$	mg kg^{-1}	21,000	
PROCESSING			
Typical processing methods	-	electrospinning, injection molding	
Preprocess drying: temperature/ time/residual moisture	°C/h/%	82-93/2-3/<0.1	
Processing temperature	°C	240-265	
Processing pressure	MPa	0.17-0.35 (back)	
Applications	-	adhesion promoter, antifog coatings, compatibilizer, fiber, ink additive, paper sizing, pigment binding, protein carrier, viscosity modifier	
Outstanding properties	-	heat resistance, impact resistance	

SMA poly(styrene-co-maleic anhydride)

PARAMETER	UNIT	VALUE	REFERENCES
BLENDS			
Suitable polymers	-	ABS, CA, epoxy, PA6, PCL, PET, PMMA, PS, PTMG, PVC, PVDF	
ANALYSIS			
FTIR (wavenumber-assignment)	cm^{-1}/-	C–H – 3100-3000 (aromatic) 2922, 2850 (aliphatic); C=C – 1601; aromatic ring – 1493	Ignatova, M; Stoilova, O; Manolova, N; Mita, D G; Diano, N; Nicolucci, C; Rashkov, I, Eur. Polym. J., 45, 2494-2504, 2009.
NMR (chemical shifts)	ppm	H NMR: CH$_3$ – 3.69; COOCH$_3$ – 7.13-7.22	Ignatova, M; Stoilova, O; Manolova, N; Mita, D G; Diano, N; Nicolucci, C; Rashkov, I, Eur. Polym. J., 45, 2494-2504, 2009.

SMAA poly(styrene-co-methylmethacrylate)

PARAMETER	UNIT	VALUE	REFERENCES
GENERAL			
Common name	-	poly(styrene-co-methylmethacrylate)	
ACS name	-	2-propenoic acid, 2-methyl-, methyl ester, polymer with ethenylbenzene	
Acronym	-	SMMA	
CAS number	-	25034-86-0	
SYNTHESIS			
Monomer(s) structure	-		
Monomer(s) CAS number(s)	-	100-42-5; 80-62-6	
Monomer(s) molecular weight(s)	dalton, g/mol, amu	104.15; 100.12	
Styrene content	mol%	40-59	Jiang, Z Y; Jiang, X Q; Huang, Y J; Lin, J; Li, S M; Li, S Z; Hsia, Y F, Nuclear Instruments Methods Phys. Res., B245, 491-94, 2006.
Method of synthesis	-	radical polymerization of styrene and methyl methacrylate; ultrasound assisted emulsion copolymerization	-; Buruga, K, Jagannathan, T K, Mater. Today: Proc., 4, 8, 7467-75, 2017.
Yield	%	98	Corona-Rivera, M A; Flores, J; Puig, J E; Mendizabal, E, Polym. Eng. Sci., 49, 2125-31, 2009.
Number average molecular weight, M_n	dalton, g/mol, amu	6,000-150,000	
Mass average molecular weight, M_w	dalton, g/mol, amu	217,000-315,000; 130,000	Zhu, S; Paul, D R, Polymer, 44, 3009-19, 2003; Buruga, K, T.Kalathi, J T, J. Alloys Compounds, 774, 370-7, 2019.
Polydispersity, M_w/M_n	-	1.9-2.3	
Molecular cross-sectional area, calculated	cm^2 x 10^{-16}	38.0	
STRUCTURE			
Entanglement molecular weight	dalton, g/mol, amu	calc.=7,624	
COMMERCIAL POLYMERS			
Some manufacturers	-	Ineos	
Trade names	-	Nas, Zylar	
PHYSICAL PROPERTIES			
Density at 20°C	g cm^{-3}	1.04-1.13	
Color	-	colorless to white	
Refractive index, 20°C	-	1.53-1.57	
Transmittance	%	88-91.3	
Haze	%	0.3-2	
Odor	-	odorless	
Softening point	°C	103	
Decomposition temperature	°C	260-280	

SMMA poly(styrene-co-methylmethacrylate)

PARAMETER	UNIT	VALUE	REFERENCES
Glass transition temperature	°C	86-118	
Heat deflection temperature at 1.8 MPa	°C	67-94	
Vicat temperature VST/B/50	°C	88-106	
Conductivity	mS cm^{-1}	0.119	Buruga, K, Jagannathan, T K, Mater. Today: Proc., 4, 8, 7467-75, 2017.
Zeta potential	mV	-123.2	Buruga, K, Jagannathan, T K, Mater. Today: Proc., 4, 8, 7467-75, 2017.

MECHANICAL & RHEOLOGICAL PROPERTIES

PARAMETER	UNIT	VALUE	REFERENCES
Tensile strength	MPa	19-52	
Tensile modulus	MPa	2,100-3,200	
Tensile stress at yield	MPa	29-62	
Elongation	%	2-140	
Tensile yield strain	%	4	
Flexural strength	MPa	46-100	
Flexural modulus	MPa	1,900-3,200	
Young's modulus	MPa	2500	Buruga, K, T.Kalathi, J T, J. Alloys Compounds, 774, 370-7, 2019.
Charpy impact strength, notched, 23°C	kJ m^{-2}	0.8	
Izod impact strength, unnotched, 23°C	J m^{-1}	160-260	
Izod impact strength, notched, 23°C	J m^{-1}	20-160	
Poisson's ratio	-	calc.=0.361	
Rockwell hardness	-	L95; M70-76	
Shrinkage	%	0.2-0.6	
Melt index, 230°C/5 kg	g/10 min	0.2-4.3 (190°C/5 kg); 0.13-5 (230°C/5 kg)	
Water absorption, equilibrium in water at 23°C	%	0.15-0.17	

CHEMICAL RESISTANCE

PARAMETER	UNIT	VALUE	REFERENCES
Alcohols	-	good	
Aliphatic hydrocarbons	-	good	
Aromatic hydrocarbons	-	poor	
Halogenated hydrocarbons	-	poor	
Ketones	-	poor	
Θ solvent, Θ-temp.=40-59, 61-68°C	-	2-ethoxy ethanol, cyclohexanol	
Good solvent	-	DMF, THF, toluene	

FLAMMABILITY

PARAMETER	UNIT	VALUE	REFERENCES
Flammability according to UL-standard; thickness 1.6/0.8 mm	class	HB	
Ignition temperature	°C	>250	
Autoignition temperature	°C	430	

SMAA poly(styrene-co-methylmethacrylate)

PARAMETER	UNIT	VALUE	REFERENCES
Volatile products of combustion	-	CO, CO_2	
WEATHER STABILITY			
Products of degradation	-	hydroperoxides, hydroxyl, radicals, chain scission	Torikai, A; Hozumi, A; Fueki, K, Polym. Deg. Stab., 16, 13-24, 1986.
TOXICITY			
NFPA: Health, Flammability, Reactivity rating	-	1/1/0	
Carcinogenic effect	-	not listed by ACGIH, NIOSH, NTP	
PROCESSING			
Typical processing methods	-	electrospinning, injection molding	
Preprocess drying: temperature/ time/residual moisture	$^oC/h/\%$	75-82/2/	
Processing temperature	^oC	182-243	
Processing pressure	MPa	0.7 (back)	
Applications	-	appliances, bathroom accessories, decorative displays, medical, nanofibers, toys	
Additives used in final products	-	haloysite nanotubes	
Outstanding properties	-	sterilizable (EtO, radiation), high clarity	
BLENDS			
Suitable polymers	-	PMMA, PS, SAN, SMA	
ANALYSIS			
FTIR (wavenumber-assignment)	$cm^{-1}/-$	C=O – 1730; phenyl – 1493; OH - 3400	Torikai, A; Hozumi, A; Fueki, K, Polym. Deg. Stab., 16, 13-24, 1986.
Raman (wavenumber-assignment)	$cm^{-1}/-$	vinyl – 1600-1675; C=O – 1675-1750	Corona-Rivera, M A; Flores, J; Puig, J E; Mendizabal, E, Polym. Eng. Sci., 49, 2125-31, 2009.

ST starch

PARAMETER	UNIT	VALUE	REFERENCES
GENERAL			
Common name	-	starch	
ACS name	-	starch	
Acronym	-	ST	
CAS number	-	9005-25-8	
EC number	-	232-679-6	
RTECS number	-	GM5090000	
HISTORY			
Person to discover	-	Beccari, J	
Date	-	1745	
Details	-	Prof. Beccari separated wheat flour to starch and protein; starch grains were ground on stone about 30,000 years ago in Europe. Egyptians are known to use wheat starch to stiffen cloth. Romans used it as thickening agent for sauces but also in cosmetics. Chinese used rice starch for smoothing paper.	
SYNTHESIS			
Monomer(s) structure	-	glucose	
Monomer(s) CAS number(s)	-	50-99-7	
Monomer(s) molecular weight(s)	dalton, g/mol, amu	180.16	
Amylose contents	%	1-30; 20-30 (typical; non-modified); 50-80 (modified starch, e. g., amylomaize); 26 (corn); 22 (potato); 3 (modified potato); 25.5-30.9 (wheat)	Lotti, C L; Corradini, E; de Medeiros, E S; Mattoso, L H C, Antec, 3994-98, 2002; arocas, A; Sanz, T; Hernando, M I; Fiszman, S M, Food Hydrocolloids, 25, 1554-62, 2011.
Number average molecular weight, M_n	dalton, g/mol, amu	1,800,000 (amylopectin in potato)	Bertoft, E; Blennow, A, Adv. Potato Chem. Technol., 83-98, 2009.
Mass average molecular weight, M_w	dalton, g/mol, amu	10,000,000-100,000,000 (amylopectin); 10,000-1,000,000 (amylose); 200,000-3,900,000 (amylose in potato starch); 60,900,000 (amylopectin in potato); 260,000,000-700,000,000 (amylopectin in wheat); 280,000,000 (amylopectin in corn); 340,000,000 (amylopectin in rice)	Mischnick, P; Momcilovic, D, Adv. Carbohydrate Chem. Biochem., 64, 117-210, 2010; Bertoft, E; Blennow, A, Adv. Potato Chem. Technol., 83-98, 2009; Maningat, C C; Seib, P A; Bassi, S D; Woo, K S; Lasater, G D, Starch, 3rd Ed., 441-510, Elsevier, 2009.
Polydispersity, M_w/M_n	-	1.29-6.9 (amylose in potato)	Bertoft, E; Blennow, A, Adv. Potato Chem. Technol., 83-98, 2009.
Polymerization degree (number of monomer units)	-	840-21,800 (amylose in potato); 11,200 (amylopectin in potato); 1,000-5,000 (amylose in wheat); 10,000 (amylopectin in wheat); 4,700-15,000 (amylopectin from various sources)	Bertoft, E; Blennow, A, Adv. Potato Chem. Technol., 83-98, 2009; Maningat, C C; Seib, P A; Bassi, S D; Woo, K S; Lasater, G D, Starch, 3rd Ed., 441-510, Elsevier, 2009.
Molar volume at 298K	cm^3 mol^{-1}	97.5 (wheat)	Habeych, E; Guo, X; van Soest, J; van der Goot, A J; Boom, R, Carbohydrate Polym., 77, 703-12, 2009.
Radius of gyration	nm	104-217; 244.3 (amylopectin in potato); 78.4-88.6 (tacca starch); 40.6-77.4	Tan, H-Z; Li, Z-G; Tan, B, Food Res. Int., 42, 551-76, 2009; Bertoft, E; Blennow, A, Adv. Potato Chem. Technol., 83-98, 2009; Nwokocha, L M; Senan, C; Williams, P A, Carbohydrate Polym., 86, 789-96, 2011; Li, B, Zhang, Y, Xu, F, Liu, A, Food Chem. 336, 127716, 2021.

ST starch

PARAMETER	UNIT	VALUE	REFERENCES
Degree of branching	%	25; 7.2-12.8	Kortstee, A J, Suurs, L C J M, Vermeesch, A M G, Visser, R G F, Carbohydrate Polym., 37, 2, 173-84, 1998; Yang, Y, Xu, X, Wang, Q, Carbohydrate Polym., 251, 117057, 2021.

STRUCTURE

PARAMETER	UNIT	VALUE	REFERENCES
Crystallinity	%	25-45; 32-36 (wheat starch); 57.6 (amorphous content in native maize starch); 100 (amorphous content in processed maize starch); 21.18-24.65 (starch with different amylose content)	Lotti, C L; Corradini, E; de Medeiros, E S; Mattoso, L H C, Antec, 3994-98, 2002; Du, X; Mac-Naughtan, B; Mitchell, J R, Food Chem., 127, 188-91, 2011; Zhang, B, Zhang, Q, Wu, H, Su, C, Li. W, LWT, 136, 2, 110380, 2021.
Cell type (lattice)	-	monoclinic; hexagonal	Perez, S; Baldwin, P M; Gallant, D J, Starch, 3rd Ed., 149-192, Elsevier, 2009.
Cell dimensions	nm	a:b:c= 2.124:1.172:1.069 (B2); a:b:c= 1.85:1.85:1.04 (B)	Perez, S; Baldwin, P M; Gallant, D J, Starch, 3rd Ed., 149-192, Elsevier, 2009.
Unit cell angles	degree	γ=123.5	Perez, S; Baldwin, P M; Gallant, D J, Starch, 3rd Ed., 149-192, Elsevier, 2009.
Crystallite size	nm	7-10 (wheat); 6.6-7.8 (retrograded starches)	Maningat, C C; Seib, P A; Bassi, S D; Woo, K S; Lasater, G D, Starch, 3rd Ed., 441-510, Elsevier, 2009; Villas-Boas, F, Facchinatto, W M, Colnago, L A, Franco, C M L, Int. J. Biol. Macromol., 163, 1333-43, 2020.
Spacing between crystallites	nm	3.5-3.7 (amylopectin in potato)	Bertoft, E; Blennow, A, Adv. Potato Chem. Technol., 83-98, 2009.
Polymorphs	-	A, B, V, I, II	Biliaderis, C G; Starch, 3rd Ed., 293-372, Elsevier, 2009.
Chain conformation	-	double helix (both amylose and amylopectin); diameter of 1 nm	Momany, F A; Willett, Antec, 1999.
Lamellae thickness	nm	5.3-5.8 (amylopectin in potato) In maize: thickness of amorphous layers: 1.97-2.14, thickness of crystalline layers: 7.72-7.89, Bragg lamellar repeat distance: 9.75-9.94	Bertoft, E; Blennow, A, Adv. Potato Chem. Technol., 83-98, 2009; Zhong, Y, Wu, Y, Blennow, A, Liu, X, LWT, 134, 11076, 2020.
Avrami constants, k/n	-	k=0.14-0.54 and n=0.47-0.96 (rice)	Hu, X; Xu, X; Jin, Z; Tian, Y; Bai, Y; Xie, Z, J. Food Eng., 106, 262-66, 2011.

COMMERCIAL POLYMERS

PARAMETER	UNIT	VALUE	REFERENCES
Some manufacturers	-	United Biopolymers, Novamont	
Trade names	-	Biopar	

PHYSICAL PROPERTIES

PARAMETER	UNIT	VALUE	REFERENCES
Density at 20°C	g cm^{-3}	1.34-1.65	
Color	-	white	
Birefringence	-	1.0131, 0.0139	
Odor	-	odorless	
Melting temperature, DSC	°C	decomposition; 240-250 (estimated above degradation temperature)	
Gelatinization temperature	°C	58-78	Biliaderis, C G; Starch, 3rd Ed., 293-372, Elsevier, 2009.

ST starch

PARAMETER	UNIT	VALUE	REFERENCES
Glass transition temperature	°C	-55 and 27-43 (two transitions)	Lotti, C L; Corradini, E; de Medeiros, E S; Mattoso, L H C, Antec, 3994-98, 2002.
Enthalpy of gelatinization	J g^{-1}	0.9-4.2 (wheat)	Maningat, C C; Seib, P A; Bassi, S D; Woo, K S; Lasater, G D, Starch, 3rd Ed., 441-510, Elsevier, 2009.
Surface tension	mN m^{-1}	45	Mora, C P, Martinez-Alejo, J M, Roman, L, Mora-Huertas, C E, Int. J. Pharm., 579, 119163, 2020.

MECHANICAL & RHEOLOGICAL PROPERTIES

Tensile strength	MPa	1.6-2.1	
Tensile modulus	MPa	1,020-1,140	
Tensile stress at yield	MPa	1.4-22	
Elongation	%	27-84	
Tensile yield strain	%	3-104	
Elastic modulus	MPa	9-38.7	
Intrinsic viscosity, 25°C	dl g^{-1}	118-384 (amylose); 116-171 (amylopectin)	
Water absorption, equilibrium in water at 23°C	%	22.5	
Moisture absorption, equilibrium 23°C/50% RH	%	10.2-13.3	

CHEMICAL RESISTANCE

Good solvent	-	liquid ammonium	
Non-solvent	-	alkalies, diethyl ether	

FLAMMABILITY

Ignition temperature	°C	>93.3	
Autoignition temperature	°C	>400	

WEATHER STABILITY

Spectral sensitivity	nm	644-662 (amylose, maximum absorption); 531-575 (amylopectin, maximum absorption)	Wypych, G, Handbook of Materials Weathering, 6th Ed., ChemTec Publishing, 2018.

BIODEGRADATION

Typical biodegradants	-	enzymolysis; biodegradation in composter	Jayasekara, R; Sheridan, S; Lourbakos, E; Beh, H; Christi, G B Y; Jenkins, M; Halley, P B; McGlashan, S; Lonergan, G T, Int. Biodeter. Biodeg., 51, 77-81, 2003.

TOXICITY

NFPA: Health, Flammability, Reactivity rating	-	1/1/0	
Carcinogenic effect	-	not listed by ACGIH, NIOSH, NTP	
Mutagenic effect	-	not known	
Teratogenic effect	-	not known	
Reproductive toxicity	-	not known	

ST starch

PARAMETER	UNIT	VALUE	REFERENCES
TLV, ACGIH	mg m^{-3}	10	
OSHA	mg m^{-3}	5 (respirable), 15 (total)	
Oral rat, LD$_{50}$	mg kg^{-1}	>5,000	
Skin rabbit, LD$_{50}$	mg kg^{-1}	>2,000	

ENVIRONMENTAL IMPACT

Biological oxygen demand, BOD$_5$	mg l^{-1}	1,100-3,900	
Chemical oxygen demand	mg l^{-1}	4,200-7,000	

PROCESSING

Typical processing methods	-	extrusion, injection molding	
Processing temperature	^{o}C	160 (extrusion); 180 (melt, injection)	
Processing pressure	MPa	5 (injection and holding)	
Additives used in final products	-	Plasticizers: diethylene glycol dibenzoate, dipropylene glycol dibenzoate, glycerin, glycerol esters, polyethylene and poly-propylene glycols, sorbitol, soybean oil, succinate polyester, sunflower oil, triacetin, tributyl acetyl citrate, vegetable oil; Antistatics: dicoconut alkyl dimethyl ammonium methyl sulfate, graft polymerized starch, polymeric systems based on poly-amide/polyether block amides; Release: magnesium stearate, polymethylhydrogensiloxane, potassium stearate, starch ester	
Applications	-	biodegradable plastics	
Outstanding properties	-	sustainable, biodegradable	

BLENDS

Suitable polymers	-	CA, chitosan, HDPE, LPDE, PCL, PEO, PLA, PR, PVOH	

ANALYSIS

FTIR (wavenumber-assignment)	cm^{-1}/-	starch conformation – 1045, 1022; COH – 1080, 1047, 1022, 995, 928; COC – 860	Wei, C; Qin, F; Zhou, W; Xu, B; Chen, C; Chen, Y; Wang, Y; Gu, M; Liu, Q, Food Chem., 128, 645-52, 2011; Mutungi, C; Onyango, C; Doert, T; Paasch, S; Thiele, S; Machill, S; Jaros, D; Rohm, H, Food Hydrocolloids, 25, 477-85, 2011.
Raman (wavenumber-assignment)	cm^{-1}/-	confocal Raman imaging	Wetzel, D L; Shi, Y-C; Schmidt, U, Vibrational Spect., 53, 173-77, 2010.
NMR (chemical shifts)	ppm	high-resolution solid-state NMR records characteristic spectra of ordered helices; C NMR permits determination of double helix contents	Lin, J-H; Singh, H; Wen, C-Y; Chang, Y-H, Cereal Sci., in press, 2011.
x-ray diffraction peaks	degree	15, 17, 18, 23 (A polymorph); 5, 17, 22, 24 (B polymorph)	Maningat, C C; Seib, P A; Bassi, S D; Woo, K S; Lasater, G D, Starch, 3rd Ed., 441-510, Elsevier, 2009.

TPU thermoplastic polyurethane

PARAMETER	UNIT	VALUE	REFERENCES
GENERAL			
Common name	-	thermoplastic polyurethane	
CAS number	-	75701-44-9; 9009-54-5, 9018-04-6; 9087-79-0	
Acronym	-	TPU	
HISTORY			
Person to discover	-	Charles Schollenberger	
Date	-	1959	
Details	-	after 10 years of experimental work, Charles Schollenberger developed technology of thermoplastic polyurethane for BFGoodrich, which resulted in fully automatized production of Estane	
SYNTHESIS			
Monomer(s) structure	-	polyols and isocyanates (see PU)	
Monomer ratio	-	usually stoichiometric with as small excess of polyol; example: MDI (aromatic isocyanate) – 10%, aliphatic isocyanate – 32%, aliphatic diol – 58% (Irogram PS455-203)	Lavall, R L; Ferrari, S; Tomasi, C; Marzantowicz, M; Quartarone, E; Magistris, A; Mustarelli, P; Lazzaroni, S; Fagnoni, M, J. Power Sources, 195, 5761-67, 2010.
Method of synthesis	-	polyol(s) are dried by azeotropic distillation (e.g., with toluene), catalyst is added followed by addition of isocyanate	Ojha, U; Kulkarni, P; Faust, R, Polymer, 50, 3448-57, 2009.
Temperature of polymerization	°C	80	
Time of polymerization	h	6	
Catalyst	-	most frequent: tin derivatives and amines	
Number average molecular weight, M_n	dalton, g/mol, amu	83,000-163,000, 48,000-157,000	Ojha, U; Kulkarni, P; Faust, R, Polymer, 50, 3448-57, 2009; Hu, S, Shou, T, Zhao, X, Zhang, L, Polymer, 205, 122764, 2020.
Mass average molecular weight, M_w	dalton, g/mol, amu	120,000;69,000-283,000	-; Hu, S, Shou, T, Zhao, X, Zhang, L, Polymer, 205, 122764, 2020.
Polydispersity, M_w/M_n	-	1.2-3.7; 1.32-1.79	Ojha, U; Kulkarni, P; Faust, R, Polymer, 50, 3448-57, 2009; Hu, S, Shou, T, Zhao, X, Zhang, L, Polymer, 205, 122764, 2020.
Surface roughness	nm	0.17-0.8	Xiao, S, Sue, H-J, Polymer, 169, 124-30, 2019.
STRUCTURE			
Crystallinity	%	45.7-47.8 (TSPU); 2.7-18.3 (TPU)	Chen, Y; Wang, R; Zhou, J; Fan, H; Shi, B, Polymer, 52, 1856-67, 2011; Buckley, C P; Prisacariu, C; Martin, C, Polymer, 51, 3213-24, 2010.
Crystallite size	nm	11.5	Deshmukh, Pranjali Khajanji, Swamini Chopra, D. R. Peshwe, D R, Mater. Today, 28, 2, 642-50, 2020.
COMMERCIAL POLYMERS			
Some manufacturers	-	Covestro; Huntsman; Lubrizol	
Trade names	-	Desmopan, Texin; Irogran; Estane	

TPU thermoplastic polyurethane

PARAMETER	UNIT	VALUE	REFERENCES
PHYSICAL PROPERTIES			
Density at 20°C	g cm^{-3}	1.02-1.12	
Melting temperature, DSC	°C	170-220	
Thermal expansion coefficient, 23-80°C	10^{-4} °C^{-1}	0.86	
Thermal conductivity	W m^{-1} K^{-1}	0.19	Ryu, S, Oh, H W, Kim, J, Mater. Chem. Phys., 223, 607-12, 2019.
Glass transition temperature	°C	-44 to -66	Ojha, U; Kulkarni, P; Faust, R, Polymer, 50, 3448-57, 2009.
Vicat temperature VST/A/50	°C	60.6-71.1	
Enthalpy melting	J g^{-1}	22	Loh, T W, Tran, P, Das, R, Ladani, R B, Orifici, A C, Compos. Commun., 22, 100465, 2020.
Hansen solubility parameters, δ_D, δ_P, δ_H	MPa$^{0.5}$	17.7, 5.9, 10.9 (MDI/BDO polymer), 19.1, 4.2, 6.7 (soft segment), 18.2, 8.2, 13.8 (hard segment)	Gallu, R, Méchin, F, Dalmas, F, Loup, F, Polymer, 207, 122882, 2020.
Surface tension	mN m^{-1}	41.69	Yao, N, Wang, H, Zhang, L, Tian, M, Appl. Surf. Sci., 530, 147124, 2020.
Dielectric constant at 60 Hz/1 MHz	-	8.39	Zhu, P, Weng, L, Zhang, X, Liu, L, Polym. Testing, 90, 106671, 2020.
Electric strength K20/P50, d=0.60.8 mm	kV mm^{-1}	15	
Coefficient of friction	-	0.31-0.32	Xiao, S, Sue, H-J, Polymer, 169, 124-30, 2019.
Permeability to nitrogen, 25°C	cm^3 cm cm^{-2} s^{-1} cmHg^{-1} x 10^{10}	1.4	Chen, Y; Wang, R; Zhou, J; Fan, H; Shi, B, Polymer, 52, 1856-67, 2011.
Permeability to oxygen, 25°C	cm^3 cm cm^{-2} s^{-1} cmHg^{-1} x 10^{10}	7.0	Chen, Y; Wang, R; Zhou, J; Fan, H; Shi, B, Polymer, 52, 1856-67, 2011.
Diffusion coefficient of nitrogen	cm^2 s^{-1} x10^7	2.2	Chen, Y; Wang, R; Zhou, J; Fan, H; Shi, B, Polymer, 52, 1856-67, 2011.
Diffusion coefficient of oxygen	cm^2 s^{-1} x10^7	3.5	Chen, Y; Wang, R; Zhou, J; Fan, H; Shi, B, Polymer, 52, 1856-67, 2011.
Contact angle of water, 20°C	degree	93	Goodwin, D G, Shen, S-J, Lyu, Y, Sung, L, Polym. Deg. Stab., 182, 109365, 2020.
MECHANICAL & RHEOLOGICAL PROPERTIES			
Tensile strength	MPa	17-66, 98.8-120.4	-; Xiao, S, Sue, H-J, Polymer, 169, 124-30, 2019.
Tensile modulus	MPa	120-330	
Tensile stress at yield	MPa	39-54.2	
Elongation	%	300-1500, 1369-1482	-; Xiao, S, Sue, H-J, Polymer, 169, 124-30, 2019.
Flexural strength	MPa	5.5-75.2	
Flexural modulus	MPa	17-1,990	
Young's modulus	MPa	33-72	Ojha, U; Kulkarni, P; Faust, R, Polymer, 50, 3448-57, 2009.
Tear strength	kN m^{-1}	33-256	

HANDBOOK OF POLYMERS 3rd Edition, Copyrights 2022;; ChemTec Publishing

TPU thermoplastic polyurethane

PARAMETER	UNIT	VALUE	REFERENCES
Compression set, 24h 70°C	%	11-87	
Shore A hardness	-	62-98	
Shore D hardness	-	28-73	
Shrinkage	%	0.3-0.8	

CHEMICAL RESISTANCE			
Alcohols	-	good	
Aliphatic hydrocarbons	-	good	
Aromatic hydrocarbons	-	poor	
Esters	-	poor	
Greases & oils	-	good	
Ketones	-	poor	
Good solvent	-	DMF, THF, MEK	

FLAMMABILITY			
Ignition temperature	°C	400	
Autoignition temperature	°C	>393	

WEATHER STABILITY			
Activation wavelengths	nm	313, 334, 365, 405, 435	
Excitation wavelengths	nm	320, 372	
Emission wavelengths	nm	420, 423, 455, 489	
Important initiators and accelerators	-	catalysts used in prepolymer synthesis, catalysts used in the curing process, heavy metals, peroxides in polyol, products of reaction of amine catalysts and polyols, nitrous oxide, acids and bases (hydrolysis), traces of solvents of types capable of producing hydroperoxides, products of thermooxidative degradation	
Products of degradation	-	photo-Fries rearrangement, yellowing, chains scission, hydroperoxides, carbonyls	
Stabilizers	-	UVA: 2,2'-dihydroxy-4-methoxybenzophenone; 2-(2H-benzotriazol-2-yl)-p-cresol; 2-benzotriazol-2-yl-4,6-di-tert-butylphenol; phenol, 2-(5-chloro-2H-benzotriazole-2-yl)-6-(1,1-dimethylethyl)-4-methyl-; 2-(2H-benzotriazole-2-yl)-4,6-di-tert-pentylphenol; 2-(2H-benzotriazole-2-yl)-4-(1,1,3,3-tetraethylbutyl)phenol; 2-(2H-benzotriazol-2-yl)-4,6-bis(1-methyl-1-phenylethyl)phenol; 2-(2H-benzotriazol-2-yl)-6-dodecyl-4-methylphenol, branched & linear; 2,4-di-tert-butyl-6-(5-chloro-2H-benzotriazole-2-yl)-phenol; 2-(3-sec-butyl-5-tert-butyl-2-hydroxyphenyl)benzotriazole; reaction product of methyl 3(3-(2H-benzotriazole-2-yl)-5-t-butyl-4-hydroxyphenyl propionate/PEG 300; ethyl-2-cyano-3,3-diphenylacrylate; (2-ethylhexyl)-2-cyano-3,3-diphenylacrylate; N-(2-ethoxyphenyl)-N'-(4-isododecylphenyl)oxamide; N-(2-ethoxyphenyl)-N'-(2-ethylphenyl)oxamide; benzoic acid, 4-[[(methylphenylamino)methylene]amino]-, ethyl ester;	Wypych, G, Handbook of Materials Weathering, 6th Ed., ChemTec Publishing, 2018.

TPU thermoplastic polyurethane

PARAMETER	UNIT	VALUE	REFERENCES
Stabilizers	-	Screeners: carbon black; HAS: 1,3,5-triazine-2,4,6-triamine, N,N'''[1,2-ethane-diyl-bis[[[4,6-bis[butyl(1,2,6,6-pentamethyl-4-piperidinyl)amino]-1,3,5-triazine-2-yl]imino]-3,1-propanediyl]bis[N',N''-dibutyl-N',N''-bis(1,2,2,6,6-pentamethyl-4-piperidinyl)-; 2,4-bis[N-butyl-N-(1-cyclohexyloxy-2,2,6,6-tetramethylpi-peridin-4-yl)amino]-6-(2-hydroxyethylamine)-1,3,5-triazine; bis(1,2,2,6,6-pentamethyl-4-piperidyl) sebacate and methyl 1,2,2,6,6-pentamethyl-4-piperidyl sebacate; bis(1,2,2,6,6-pentamethyl-4-piperidyl)sebacate + methyl-1,2,2,6,6-pentamethyl-4-piperidyl sebacate; bis(2,2,6,6-tetramethyl-4-piperidyl) sebacate; 2,2,6,6-tetramethyl-4-piperidinyl stearate; 2-dodecyl-N-(2,2,6,6-tetramethyl-4-piperidinyl)succinimide; poly[[(6-[1,1,3,3-tetramethylbutyl)amino]-1,3,5-triazine-2,4-diyl][2,2,6,6-tetramethyl-4-piperidinyl)imino]-1,6-hexanediyl[2,2,6,6-tetramethyl-4-piperidinyl)imino]]; butanedioic acid, dimethylester, polymer with 4-hydroxy-2,2,6,6-tetramethyl-1-piperidine ethanol; alkenes, C20-24-.alpha.-, polymers with maleic anhydride, 1, 6-hex-anediamine, N, N'-bis(2,2,6,6-tetramethyl-4-piperidinyl)-, polymers with 2,4-dichloro-6-(4-morpholinyl)-1,3,5-triazine; Phenolic antioxidants: ethylene-bis(oxyethylene)-bis(3-(5-tert-butyl-4-hydroxy-m-tolyl)-propionate); pentaerythritol tetrakis(3-(3,5-di-tert-butyl-4-hydroxyphenyl)propionate); octadecyl-3-(3,5-di-tert-butyl-4-hydroxyphenyl)-propionate; N,N'-hexane-1,6-diylbis(3-(3,5-di-tert-butyl-4-hydroxyphe-nylpropionamide); 3,4-dihydro-2,5,7,8-tetramethyl-2-(4,8,12-trimethyltridecyl)-2H-1-benzopyran-6-ol; isotridecyl-3-(3,5-di-tert-butyl-4-hydroxyphenyl) propionate; 2,2'-ethylidenebis(4,6-di-tert-butylphenol); 3,5-tris(4-tert-butyl-3-hydroxy-2,6-dimethyl benzyl)-1,3,5-triazine-2,4,6-(1H,3H,5H)-trione; 3,5-bis(1,1-dimethyethyl)-4-hydroxy-benzenepropanoic acid, C13-15 alkyl esters; Phosphite: isodecyl diphenyl phos-phite; Thiosynergist: 4,6-bis(dodecylthiomethyl)-o-cresol; 4,4'-thiobis(2-t-butyl-5-methylphenol); 2,2'-thiobis(6-tert-butyl-4-methylphenol); Amine: benzenamine, N-phenyl-, reaction products with 2,4,4-trimethylpentene; Optical brightener: 2,2'-(2,5-thiophenediyl)bis(5-tert-butylbenzoxazole)	

BIODEGRADATION

PARAMETER	UNIT	VALUE	REFERENCES
Biodegradable TPUs		biodegradable polyurethanes are typically prepared from poly-ester polyols, aliphatic diisocyanates and chain extenders	Tatai, L; Moore, T G; Adhikari, R; Malherbe, F; Jayasekara, R; Griffiths, I; Gunatillake, P A, Bioma-terials, 28, 5407-17, 2007.

PROCESSING

PARAMETER	UNIT	VALUE	REFERENCES
Typical processing methods	-	blow molding, calendering, coating, extrusion, injection mold-ing, slush molding	
Preprocess drying: temperature/time/residual moisture	°C/h/%	80-110/0.03	
Processing temperature	°C	185-240 (injection molding); 170-235 (extrusion)	
Applications	-	wearing parts (mine screens, strippers, cyclones); wheels and solid tires; rollers (transport, office equipment); seals (U and V packing rings, O-rings, cup seals); damping components (spacers, grippers, pickers, handles); drive elements (clutch components, timing belts, round belts, cog wheels); animal identification tags, automotive products, belts, cable sheath-ing, castors, coated fabrics, films, hoses, membranes, profiles, shoe soles	
Outstanding properties	-	abrasion resistance, resistance to oils and greases, rebound resilience, load-bearing capacity, damping properties	

TPU thermoplastic polyurethane

PARAMETER	UNIT	VALUE	REFERENCES
BLENDS			
Suitable polymers	-	CA, NR, PEO, phenoxy, POM, PP (functionalized), PR, PVB, PVC, PVDF, SBS	
ANALYSIS			
FTIR (wavenumber-assignment)	cm^{-1}/-	NH - 3455, 3320; amide – 1714, 1697, 1666; hydrogen bonded NH – 3316; hydrogen bonded C=O – 1708; C–N – 1525	Barick, A K; Tripathy, D K, Composites, A41, 1471-82, 2010.
Raman (wavenumber-assignment)	cm^{-1}/-	soft phase – 1115; hard phase – 1080; C=O – 1800, 1650	Ferry, A; Jacobsson, P; van Heumen, J D; Stevens, J R, Polymer, 37, 5, 737-44, 1996.
NMR (chemical shifts)	ppm	OH – 3.6; CH$_2$ – 4.1	Sonnenschein, M F; Guillaudeu, S J; Landes, B G; Wendt, B L, Polymer, 51, 3685-92, 2010.
x-ray diffraction peaks	degree	10, 11.2, 12	Buckley, C P; Prisacariu, C; Martin, C, Polymer, 51, 3213-24, 2010.

UF urea-formaldehyde resin

PARAMETER	UNIT	VALUE	REFERENCES
GENERAL			
Common name	-	urea-formaldehyde resin	
CAS name	-	urea, polymer with formaldehyde	
Acronym	-	UF	
CAS number	-	9011-05-6; 68611-64-3; 68071-45-4	
HISTORY			
Person to discover	-	Ellis, C	Ellis, C, US Patent 1,846,853, Feb. 23, 1932.
Date	-	1932 (filled 1924)	
Details	-	reaction between urea and formaldehyde in the presence of alkaline catalyst	
SYNTHESIS			
Monomer(s) structure	-		
Monomer(s) CAS number(s)	-	57-13-6; 50-00-0	
Monomer(s) molecular weight(s)	dalton, g/mol, amu	60.06; 30.03	
Monomer ratio	-	1-1.6 (F/U)	Park, B-D; Jeong, H-W; Int. J. Adhesion Adhesives, in press, 2011.
Method of synthesis	-	reaction of urea with formaldehyde dissolved in water (45-50% solution) to hydroxymethylated urea used for subsequent polycondensation	
Temperature of curing	ºC	90-120	Minopoulou, E; Dessipri, E; Chryssikos, G D; Gionis, V; Paipetis, A; Panayiotou, C, Int. J. Adhesion Adhesives, 23, 473-84, 2003.
Gelation time	s	51-201	Park, B-D; Jeong, H-W; Int. J. Adhesion Adhesives, in press, 2011.
Number average molecular weight, M_n	dalton, g/mol, amu	400-640 (water based dispersion)	Ferra, J M M; Mendes, A M; Costa, M R N; Carvalho, L H, J. Appl. Polym. Sci., 118, 1956-68, 2010.
Mass average molecular weight, M_w	dalton, g/mol, amu	2,500-500,000 (water based dispersion)	Gavrilovic-Grmusa, I; Neskovic, O; Diporovic-Momcilovic, M; Popovic, M, J. Serb. Chem. Soc., 75, 5, 689-701, 2010.
Polydispersity, M_w/M_n	-	5.2-7.3 (water based dispersion)	
STRUCTURE			
Crystallinity	%	74.9	Khan, T A, Gupta, A, Jamari, S S, Asim, M, Int. J. Adhesio Adhesives, 99, 102589, 2020.
COMMERCIAL POLYMERS			
Some manufacturers	-	Chemiplastica	
Trade names	-	Urochem	
PHYSICAL PROPERTIES			
Density at 20ºC	g cm⁻³	1.2-1.31	
Refractive index, 20ºC	-	1.43	

HANDBOOK OF POLYMERS 3rd Edition, Copyrights 2022; ChemTec Publishing

UF urea-formaldehyde resin

PARAMETER	UNIT	VALUE	REFERENCES
Melting temperature, DSC	°C	119	
Heat deflection temperature at 1.8 MPa	°C	130	
Hansen solubility parameters, δ_D, δ_P, δ_H	MPa$^{0.5}$	20.81, 8.29, 12.71	
Hildebrand solubility parameter	MPa$^{0.5}$	25.74	
Dielectric constant at 100 Hz/1 MHz	-	5	
Dissipation factor at 1000 Hz	E-4	1000	
Volume resistivity	ohm-m	1.1E9	
Surface resistivity	ohm	1.1E11	
MECHANICAL & RHEOLOGICAL PROPERTIES			
Tensile strength	MPa	55	
Tensile stress at yield	MPa	45-55	
Flexural strength	MPa	80-170	
Charpy impact strength, unnotched, 23°C	kJ m^{-2}	5-12	
Charpy impact strength, notched, 23°C	kJ m^{-2}	1.1-1.6	
Tenacity (fiber)	cN tex^{-1}	14	Rogers-Gentile, V; East, G C; McIntyre, J E; Snowden, P, J. Appl. Polym. Sci., 77, 64-74, 2000.
Shrinkage	%	0.8-1.4	
Melt viscosity, shear rate=1000 s^{-1}	mPa s	248-327; 150-350 (63-64% water emulsion)	Park, B-D; Jeong, H-W; Int. J. Adhesion Adhesives, in press, 2011.
Water absorption, equilibrium in water at 23°C	%	3	
FLAMMABILITY			
Flammability according to UL-94 standard; thickness 1.6/0.8 mm	class	V-0	
Ignition temperature	°C	>200	
Autoignition temperature	°C	393	
Limiting oxygen index	% O$_2$	30	
Char at 600°C	%	8.9	Khan, T A, Gupta, A, Jamari, S S, Asim, M, Int. J. Adhesion Adhesives, 99, 102589, 2020.
Volatile products of combustion	-	H_2O, CO_2, CO, NH_3, CH_4, HNCO, HCN	Jiang, X; Li, C; Chi, Y; Yan, J, J. Hazardous Mater., 173, 205-10, 2010.
TOXICITY			
Oral rat, LD$_{50}$	mg kg^{-1}	8,394	
Skin rabbit, LD$_{50}$	mg kg^{-1}	>2,000	
ENVIRONMENTAL IMPACT			
Aquatic toxicity, *Daphnia magna*, LC$_{50}$, 48 h	mg l^{-1}	>1,000	

UF urea-formaldehyde resin

PARAMETER	UNIT	VALUE	REFERENCES
Aquatic toxicity, *Bluegill sunfish*, LC_{50}, 48 h	mg l^{-1}	>1,000	
Cradle to grave non-renewable energy use	MJ/kg	85.9	Harding, K G; Dennis, J S; von Blottnitz, H; Harrison, S T L, J. Biotechnol., 130, 57-66, 2007.

PROCESSING			
Typical processing methods	-	injection molding	
Processing temperature	ºC	145-150 (mold); 95-115 (nozzle)	
Processing pressure	MPa	70-150 (injection); 30-80 (holding); 10-14 (back)	
Additives used in final products	-	Hardeners: ammonium chloride, ammonium sulfate, ammonium citrate, and zinc nitrate	
Applications	-	fiber, fireboard, particle board, plywood	Flores, J A; Pastor, J J; Martinez-Gabarron, A; Gimeno-Blanes, F J; Rodriguez-Guisado, I; Frutos, M J, Ind. Crops Prod., in press, 2011.
Outstanding properties	-	gloss, low water resistance, scratch resistance	

ANALYSIS			
FTIR (wavenumber-assignment)	cm^{-1}/-	N–H – 3350-3340, 900-650, 750-700; O–CH$_3$ – 2962-2960; C=O – 1654-46; C–N – 1560-50, 1260-1250, 1050-1030; C–H – 1465-1440, 1400-1380	Park, B-D; Kim, Y S; Singh, A P; Lim, K P, J. Appl. Polym. Sci., 88, 2677-87, 2003.
Raman (wavenumber-assignment)	cm^{-1}/-	N–H – 3300-3450; CH$_2$ – 2950-3020; C=O – 1650-1640; CN – 1180-1160, 1030-990, 920-890	Minopoulou, E; Dessipri, E; Chryssikos, G D; Gionis, V; Paipetis, A; Panayiotou, C, Int. J. Adhesion Adhesives, 23, 473-84, 2003.
NMR (chemical shifts)	ppm	mono-substituted amide (–CONH–) – 6.5-7.5; non-substituted amide (–CONH$_2$) – 5.5-6.5; hydroxyl (–OH) – 5.1-5.5; ether linkage –HNCH$_2$O– and methylol end groups –HNCH$_2$OH– – 4.40-4.70; methylene linkage –HNCH$_2$NH– – 4.20-4.40; OCH$_3$ and –OCH$_2$CH$_3$ – 3.0-3.3	Wibowo, E S, Lubis, M A R, Park, B-D, Causin, V, J. Ind. Eng. Chem., 87, 78-89, 2020.
x-ray diffraction peaks	degree	21.5, 25, 31, 40.5	Park, B-D; Jeong, H-W; Int. J. Adhesion Adhesives, in press, 2011.

UHMWPE ultrahigh molecular weight polyethylene

PARAMETER	UNIT	VALUE	REFERENCES
GENERAL			
Common name	-	ultrahigh molecular weight polyethylene	
IUPAC name	-	polyethylene	
Acronym	-	UHMWPE	
CAS number	-	9002-88-4	
Formula		$\left[CH_2CH_2\right]_n$	
SYNTHESIS			
Monomer(s) structure	-	$H_2C{=}CH_2$	
Monomer(s) CAS number(s)	-	74-85-1	
Monomer(s) molecular weight(s)	dalton, g/mol, amu	28.05	
Mass average molecular weight, M_w	dalton, g/mol, amu	1,600,000-9,200,000	Liu, H; Xie, D; Qian, L; Deng, X; Leng, Y X; Huamg, N, Surface Coat. Technol., 205, 2697-2701, 2011.
Polydispersity, M_w/M_n	-	12	
Polymerization degree (number of monomer units)	-	200,000	
STRUCTURE			
Crystallinity	%	41.2-91.1	Stephens, C P, Antec, 3433-37, 2003; Marcus, K; Allen, C, Wear, 178, 17-28, 1994.
Cell type (lattice)	-	orthorhombic	Sayyed, M I, Abdalsalam, A H, Taki, M M, Şakar, E, Radiat. Phys. Chem., 172, 108852, 2020.
Cell dimensions	nm	a:b:c=0748:0499:0.255	Marcus, K; Allen, C, Wear, 178, 17-28, 1994.
Crystallite size	nm	15-37	Marcus, K; Allen, C, Wear, 178, 17-28, 1994.
Spacing between crystallites	nm	13.6	Stephens, C P, Antec, 3433-37, 2003.
Chain length	μm	40-72	Marcus, K; Allen, C, Wear, 178, 17-28, 1994.
Entanglement molecular weight	dalton, g/mol, amu	1989	Xie, M; Li, H, Eur. Polym. J., 43, 3480-87, 2007.
Lamellae diameter	nm	13-18	Henry, C K, Sandoz-Rosado, E, Roenbeck, M R, Alvarez, N J, Polymer, 202, 122589, 2020.
Lamellae long spacing	nm	29-48	Henry, C K, Sandoz-Rosado, E, Roenbeck, M R, Alvarez, N J, Polymer, 202, 122589, 2020.
Crystallite size	nm	20-30	Zhang, H, Zhao, S, Xin, Z, Li, J, Chin. J. Chem. Eng., 28, 7, 1950-63, 2020.
Entanglement molecular weight	dalton, g/mol, amu	1144	Zhang, X, Zhao, S, Xin, Z, Polymer, 202, 122631, 2020.
Rapid crystallization temperature	°C	127-135	

730   **HANDBOOK OF POLYMERS** 3rd Edition, Copyrights 2022; ChemTec Publishing

UHMWPE ultrahigh molecular weight polyethylene

PARAMETER	UNIT	VALUE	REFERENCES
COMMERCIAL POLYMERS			
Some manufacturers	-	Mitsui Chemicals; Celanese	
Trade names	-	Mipelon; GUR	
PHYSICAL PROPERTIES			
Density at 20°C	g cm^{-3}	0.93-0.94	
Bulk density at 20°C	g cm^{-3}	0.3-0.45	
Melting temperature, DSC	°C	133-140; 124 (crystals)	Murase, H; Ohta, Y; Hashimoto, T, Polymer, 50, 4727-36, 2009.
Thermal expansion coefficient, 23-80°C	10^{-4} °C^{-1}	1.1-2	
Thermal conductivity, melt	W m^{-1} K^{-1}	0.39-0.42	
Glass transition temperature	°C	-118.9	Meng, Z, Wang, Y, Xin, X, Yan, F, Tribology Int., 153, 106628, 2021.
Specific heat capacity	J K^{-1} kg^{-1}	1,840-2,010	
Long term service temperature	°C	82	
Heat deflection temperature at 0.45 MPa	°C	95	
Heat deflection temperature at 1.8 MPa	°C	74-82	
Vicat temperature VST/A/50	°C	126	
Enthalpy of fusion	J g^{-1}	288; 121.6 (melting)	Stephens, C P, Antec, 3433-37, 2003.
Dielectric constant at 1000 Hz/1 MHz	-	2.3-2.35/2.3	
Dielectric loss factor at 1 kHz	-	2	
Relative permittivity at 100 Hz	-	2-2.4	
Dissipation factor at 1000 Hz	E-4	2-4	
Volume resistivity	ohm-m	1E13	
Surface resistivity	ohm	1E15	
Electric strength K20/P50, d=0.60.8 mm	kV mm^{-1}	28-39	
Arc resistance	s	250-350	
Coefficient of friction	-	0.077-0.15; 0.05-0.08 (wet); 0.11-0.22 (dry)	Marcus, K; Allen, C, Wear, 178, 17-28, 1994.
MECHANICAL & RHEOLOGICAL PROPERTIES			
Tensile strength	MPa	21-50.2; 150-250 (blown film)	
Tensile modulus	MPa	680-860; 1,800-3,300 (blown film)	
Tensile stress at yield	MPa	17-41	
Elongation	%	250-600; 50-124 (blown film)	
Tensile yield strain	%	11-20	
Flexural strength	MPa	20-26.5	
Flexural modulus	MPa	440-1,340	
Elastic modulus	MPa	700-800	
Compressive strength	MPa	14-23	
Young's modulus	MPa	1,800-3,300	Nakahara, T; Zenkoh, H; Yagi, K, Antec, 178-81, 2005.

UHMWPE ultrahigh molecular weight polyethylene

PARAMETER	UNIT	VALUE	REFERENCES
Charpy impact strength, notched, 23°C	kJ m^{-2}	NB	
Izod impact strength, unnotched, 23°C	J m^{-1}	NB	
Izod impact strength, notched, 23°C	J m^{-1}	NB	
Shear strength	MPa	3-5.5 (fiber)	Mead, J; Gabriel, R; Murray, T; Foley, G; McMorrow, J, Antec, 1722-27, 1997.
Abrasion resistance (ASTM D1044)	mg/1M cycles	2.4-31.8	Micheli, B R; Wannomae, K K; Lozynski, A J; Christensen, S D; Muratoglu, O K, J. Arthroplasty, in press, 2011; Fan, W; Song, H; Li, X; Liu, F; Wang, Q, J. Arthroplasty, 24, 4, 543-48, 2009.
Shore D hardness	-	57-68	
Brittleness temperature (ASTM D746)	°C	-70 to -84	
Intrinsic viscosity, 25°C	dl g^{-1}	7.5	
Melt index, 190°C/21.6 kg	g/10 min	5.5	
Water absorption, equilibrium in water at 23°C	%	0.01-0.02	
Moisture absorption, equilibrium 23°C/50% RH	%	0.01-0.012	
CHEMICAL RESISTANCE			
Acid dilute/concentrated	-	very good	
Alcohols	-	good	
Alkalis	-	very good	
Aliphatic hydrocarbons	-	poor	
Aromatic hydrocarbons	-	poor	
Esters	-	poor	
Greases & oils	-	good to poor	
Halogenated hydrocarbons	-	poor	
Ketones	-	poor	
Θ solvents	-	biphenyl, dibutyl phthalate, diphenyl ether, p-nonyl phenol	
Good solvent	-	1,2,4-trichlorobenzene, decalin, halogenated hydrocarbons, aliphatic ketones, xylene (all above 60°C)	
Non-solvent	-	most common solvents	
Environmental stress cracking resistance (Igepal)	h	>1,000	
FLAMMABILITY			
Flammability according to UL-94 standard; thickness 1.6/0.8 mm	class	HB	
Ignition temperature	°C	340-343	
Autoignition temperature	°C	350	
Heat of combustion	J g^{-1}	47,740	
Volatile products of combustion	-	CO, CO_2, aldehydes, benzene	

UHMWPE ultrahigh molecular weight polyethylene

PARAMETER	UNIT	VALUE	REFERENCES
WEATHER STABILITY			
Spectral sensitivity	nm	<300	
Activation wavelengths	nm	300, 330-360	
Excitation wavelengths	nm	230, 265, 275, 290, 292	
Emission wavelengths	nm	295, 312, 330, 344, 358, 450	
Important initiators and accelerators	-	unsaturations, aromatic carbonyl compounds (deoxyanisoin, dibenzocycloheptadienone, flavone, 4-methoxybenzophenone, 10-thioxanthone), hydrogen bound to tertiary carbon at branching points, aromatic amines, groups formed on oxidation (hydroperoxides, carbonyl, carboxyl, hydroxyl) substituted benzophenones, complexes with ground-state oxygen, quinones (anthraquinone, 2-chloroanthraquinone, 2-tert-butylathraquinone, 1-methoxyanthraquinone, 2-ethylanthraquinone, 2-methylanthraquinone), transition metal compounds (Ni < Zn < Fe < Co), ferrocene derivatives, titanium dioxide (anatase), ferric stearate, polynuclear aromatic compounds (anthracene, phenanthrene, pyrene, naphthalene	
Products of degradation	-	free radicals, hydroperoxides, carbonyl groups, chain scission, crosslinking	
Stabilizers	-	UVA: 2-hydroxy-4-octyloxybenzophenone; phenol, 2-(5-chloro-2H-benzotriazole-2-yl)-6-(1,1-dimethylethyl)-4-methyl-; 2,2'-methylenebis(6-(2H-benzotriazol-2-yl)-4-1,1,3,3-tetramethylbutyl)phenol; 2,4-di-tert-butyl-6-(5-chloro-2H-benzotriazole-2-yl)-phenol; reaction product of methyl 3(3-(2H-benzotriazole-2-yl)-5-t-butyl-4-hydroxyphenyl propionate/ PEG 300; 2-[4,6-bis(2,4-dimethylphenyl)-1,3,5-triazin-2-yl]-5-(octyloxy) phenol; Screener: titanium dioxide; zinc oxide; carbon black; Acid scavenger: hydrotalcite; Fiber: carbon nanotube; HAS: 1,3,5-triazine-2,4,6-triamine, N,N'''[1,2-ethane-diyl-bis[[[4,6-bis[butyl(1,2,6,6-pentamethyl-4-piperidinyl)amino]-1,3,5-triazine-2-yl]imino]-3,1-propanediyl] bis[N',N''-dibutyl-N',N''-bis(1,2,2,6,6-pentamethyl-4-piperidinyl)-; bis(1,2,2,6,6-pentamethyl-4-piperidyl)sebacate + methyl-1,2,2,6,6-pentamethyl-4-piperidyl sebacate; 2,2,6,6-tetramethyl-4-piperidinyl stearate; 1,6-hexanediamine- N,N'-bis(2,2,6,6-tetramethyl-4-piperidinyl)-polymer with 2,4,6-trichloro-1,3,5-triazine, reaction products with N-butyl-1-butanamine an N-butyl-2,2,6,6-tetramethyl-4-piperidinamine; butanedioic acid, dimethylester, polymer with 4-hydroxy-2,2,6,6-tetramethyl-1-piperidine ethanol; alkenes, C20-24-.alpha.-, polymers with maleic anhydride, reaction products with 2,2,6,6-tetramethyl-4-piperidinamine; 1,6-hexanediamine, N,N'-bis(2,2,6,6-tetramethyl-4-piperidinyl)-, polymers with morpholine-2,4,6-trichloro-1,3,5-triazine reaction products, methylated; Phenolic antioxidant: 2,6,-di-tert-butyl-4-(4,6-bis(octylthio)-1,3,5,-triazine-2-ylamino) phenol; pentaerythritol tetrakis(3-(3,5-di-tert-butyl-4-hydroxyphenyl)propionate); octadecyl-3-(3,5-di-tert-butyl-4-hydroxyphenyl)-propionate; 3,3',3',5,5',5'-hexa-tert-butyl-a,a',a'-(mesitylene-2,4,6-triyl) tri-p-cresol; 2-(1,1-dimethylethyl)-6-[[3-(1,1-dimethylethyl)-2-hydroxy-5-methylphenyl] methyl-4-methylphenyl acrylate; 2,2'-ethylidenebis (4,6-di-tert-butylphenol); ethylene bis[3,3-bis[3-(1,1-dimethylethyl)-4-hydroxyphenyl]butanoate]; 1,3,5-tris(4-tert-butyl-3-hydroxy-2,6-dimethyl benzyl)-1,3,5-triazine-2,4,6-(1H,3H,5H)-trione; 2,2'-methylenebis(4-methyl-6-tertbutylphenol); 3,5-bis(1,1-dimethyethyl)-4-hydroxy-benzenepropanoic acid, C13-15 alkyl esters; 2,2'-isobutylidenebis(2,4-dimethylphenol); 1,1,3-tris(2'methyl-4'-hydroxy-5'tert-butylphenyl)butane;	Wypych, G, Handbook of Materials Weathering, 6th Ed., ChemTec Publishing, 2018.

UHMWPE ultrahigh molecular weight polyethylene

PARAMETER	UNIT	VALUE	REFERENCES
Stabilizers	-	Phosphite: bis-(2,4-di-t-butylphenol) pentaerythritol diphosphite; tris (2,4-di-tert-butylphenyl)phosphite; trinonylphenol phosphite; distearyl pentaerythritol diphosphite; trilauryl trithiophosphite; Thiosynergist: didodecyl-3,3'-thiodipropionate; dioctadecyl 3,3'-thiodipropionate; 2,2'-thiodiethylene bis[3-(3,5-ditert-butyl-4-hydroxyphenyl)propionate]; 4,4'-thiobis(2-t-butyl-5-methylphenol); 2,2'-thiobis(6-tert-butyl-4-methylphenol); pentaerythritol tetrakis(b-laurylthiopropionate); Quencher: (2,2'-thiobis(4-tert-octyl-phenolato))-N-butylamine-nickel(II); Optical brightener: 2,2'-(2,5-thiophenediyl)bis(5-tert-butylbenzoxazole); Vitamin E in medical applications	Micheli, B R; Wannomae, K K; Lozynski, A J; Christensen, S D; Muratoglu, O K, J. Arthroplasty, in press, 2011.
TOXICITY			
NFPA: Health, Flammability, Reactivity rating	-	0/1/0	
Carcinogenic effect	-	not listed by ACGIH, NIOSH, NTP	
TLV, ACGIH	mg m^{-3}	3 (respirable), 10 (total)	
OSHA	mg m^{-3}	5 (respirable), 15 (total)	
Oral rat, LD$_{50}$	mg kg^{-1}	>7,950	
Skin rabbit, LD$_{50}$	mg kg^{-1}	>2,000	
ENVIRONMENTAL IMPACT			
Cradle to grave non-renewable energy use	MJ/kg	72-76	
Cradle to pellet greenhouse gasses	kg CO$_2$ kg^{-1} resin	1.5-2.0	
PROCESSING			
Typical processing methods	-	blow molding, blown film extrusion, cast film extrusion, extrusion, extrusion coating, injection molding, rotational molding, spinning	
Applications	-	corrugated pipe, drums, fiber, filters, implants, industrial tanks, modifier for resins and rubbers, total joint replacement prostheses	
Outstanding properties	-	abrasion resistance, chemical resistance, low friction coefficient, self-lubricating	
BLENDS			
Suitable polymers	-	LDPE, PMMA, PP, PVF	
ANALYSIS			
FTIR (wavenumber-assignment)	cm^{-1}/-	OH – 3300; C=O – 1750-1600; COO– – 1650-1560	Liu, H; Xie, D; Qian, L; Deng, X; Leng, Y X; Huamg, N, Surface Coat. Technol., 205, 2697-2701, 2011.
NMR (chemical shifts)	ppm	orthorhombic signal – 32.8	Tzou, D L; Schmidt-Rohr, K; Spiess, H W, Polymer, 35, 22, 4728-33, 1994.

ULDPE ultralow density polyethylene

PARAMETER	UNIT	VALUE	REFERENCES
GENERAL			
Common name	-	ultralow density polyethylene, ethene-1-octene copolymer	
Acronym	-	ULDPE	
CAS number	-	26221-73-8	
Formula		$\mathrm{-[CH_2CH_2CHCH_2-]_n}$ with $(CH_2)_5CH_3$ branch	
SYNTHESIS			
Monomer(s) structure	-	$H_2C=CH_2 \quad H_2C=CH(CH_2)_5CH_3$	
Monomer(s) CAS number(s)	-	74-85-1; 111-66-0	
Monomer(s) molecular weight(s)	dalton, g/mol, amu	28.05; 112.24	
Monomer(s) expected purity(ies)	%	99.0; 99.0	
Octene content	%	3.3-14.6	Haward, R N, Polymer, 40, 5821-32, 1999.
STRUCTURE			
Crystallinity	%	42.9	Woo, L; Westphal, S; Ling, T K, Thermochim. Acta, 226, 85-98, 1993.
COMMERCIAL POLYMERS			
Some manufacturers	-	DOW	
Trade names	-	Attane	
PHYSICAL PROPERTIES			
Density at 20°C	g cm^{-3}	0.865-0.912	
Color	-	white	
Transmittance	%	85-99	
Haze	%	0.6-8	
Gloss, 60°, Gardner (ASTM D523)	%	67-92	
Odor		odorless	
Melting temperature, DSC	°C	123-124	
Heat of fusion	J g^{-1}	125.6	
Vicat temperature VST/A/50	°C	71-93	
Seal initiation temperature	°C	84-97	
Permeability to carbon dioxide, 25°C	cm^3 mm m^{-2} atm^{-1} 24 h^{-1}	1,200-2,000	
Permeability to oxygen, 25°C	cm^3 mm m^{-2} atm^{-1} 24 h^{-1}	280-450	
Permeability to water vapor, 25°C	g mm m^{-2} atm^{-1} 24 h^{-1}	0.53-0.85	

ULDPE ultralow density polyethylene

PARAMETER	UNIT	VALUE	REFERENCES
MECHANICAL & RHEOLOGICAL PROPERTIES			
Tensile strength	MPa	29-53	
Tensile modulus	MPa	150	
Tensile stress at yield	MPa	4.9-9.8	
Elongation	%	450-660 (MD); 650-760 (TD)	
Dart drop impact	g	450 to >850 (0.02 mm thick film); 610-1,500 (0.051 mm)	
Film puncture resistance	J cm^{-3}	19-24 (0.02 mm thick film); 18-26 (0.051 mm)	
Elmendorf tear strength	g	260-330 (MD) and 450-530 (TD) (0.02 mm thick film); 550-1,000 (MD) and 870-1,200 (TD) (0.051 mm)	
Toughness	J cm^{-3}	1,050-1,280	
Melt index, 190°C/2.16 kg	g/10 min	0.5-4.0	
FLAMMABILITY			
Volatile products of combustion	-	CO, CO_2	
TOXICITY			
Carcinogenic effect	-	not listed by ACGIH, NIOSH, NTP	
Oral rat, LD$_{50}$	mg kg^{-1}	>5,000	
Skin rabbit, LD$_{50}$	mg kg^{-1}	>2,000	
PROCESSING			
Typical processing methods	-	blown film, cast film	
Processing temperature	°C	226-232	
Applications	-	food packaging	
Outstanding properties	-	abuse resistance, cling, optical properties, pinhole resistance, processability	
BLENDS			
Suitable polymers	-	PP, SIS	

UP unsaturated polyester

PARAMETER	UNIT	VALUE	REFERENCES
GENERAL			
Common name	-	unsaturated polyester	
CAS name	-	1,3-benzenedicarboxylic acid, polymer with 1,4-cyclohexanedimethanol, 2,2-dimethyl-1,3-propanediol and 2,5-furandione	
Acronym	-	UP	
CAS number	-	92230-55-2; 654641-87-9	
HISTORY			
Person to discover	-	Carleton Ellis	
Date	-	1936	
Details	-	discovered that product of reaction of glycol and maleic anhydride can be cured with peroxide	
SYNTHESIS			
Monomer(s) structure	-	neopentyl glycol; isophthalic acid; maleic anhydride; 1,4-cyclohexane-dimethanol	
Monomer(s) CAS number(s)	-	126-30-7; 121-91-5; 108-31-6; 105-08-8	
Monomer(s) molecular weight(s)	dalton, g/mol, amu	104.15; 166.13; 98.06; 144.24	
Method of synthesis	-	poly(butylene succinate-co-butylene itaconic), having binary acid composition, was synthesized by a two-step procedure, including direct esterification followed by polycondensation.	Gao, C, Wang, S, Zhang, J, Liu, Y, Wang, C, Polym. Deg. Stab., 181, 109336, 2020.
Catalyst	-	immobilized *Candida antarctica* lipase B, titanium(IV)butoxide, p-toluenesulfonic acid, sulfuric acid, 1,8-diazabicycloundec-7-ene, and 1,5,7-triazabicyclodec-5-ene	Gao, C, Wang, S, Zhang, J, Liu, Y, Wang, C, Polym. Deg. Stab., 181, 109336, 2020.
Number average molecular weight, M_n	dalton, g/mol, amu	2,446-4,543	Rajalakshmi, P, Marie, J M, Xavier, A J M, Poly. Deg. Stab., 170, 109016, 2019.
Mass average molecular weight, M_w	dalton, g/mol, amu	3,789-16,525	Rajalakshmi, P, Marie, J M, Xavier, A J M, Poly. Deg. Stab., 170, 109016, 2019.
Polydispersity, M_w/M_n	-	1.55-3.63	Rajalakshmi, P, Marie, J M, Xavier, A J M, Poly. Deg. Stab., 170, 109016, 2019.
STRUCTURE			
Crystallinity	%	33	Abral, H, Fajrul, R, Mahardika, M, Rosanti, S D, Polym. Testing, 81, 106193, 2020.
COMMERCIAL POLYMERS			
Some manufacturers	-	Reichhold; Ineos	
Trade names	-	Polylite; Aropol, Arotran, Envirez, Polaris	
PHYSICAL PROPERTIES			
Density at 20°C	g cm^{-3}	1.1-1.12	
Thermal expansion coefficient, 23-80°C	10^{-4} °C^{-1}	0.31	
Thermal conductivity, melt	W m^{-1} K^{-1}	0.17	

UP unsaturated polyester

PARAMETER	UNIT	VALUE	REFERENCES
Glass transition temperature	°C	-61 (before cure); 94-125 (after cure)	Wacker, M; Ehrenstein, G W, Antec, 836-41, 2001.
Maximum service temperature	°C	170	
Contact angle of water, 20°C	degree	62-98	Li, G; Wei, X; Wang, W; He, T; Li, X, Appl. Surf. Sci., 257, 290-95, 2010; Rajalakshmi, P, Marie, J M, Xavier, A J M, Poly. Deg. Stab., 170, 109016, 2019.
Surface free energy	mJ m^{-2}	47	Jia, Z; Li, Z; Zhao, Mater. Chem. Phys., 121, 193-97, 2010.

MECHANICAL & RHEOLOGICAL PROPERTIES

PARAMETER	UNIT	VALUE	REFERENCES
Tensile strength	MPa	22-85	
Tensile modulus	MPa	3,200-3,900	
Elongation	%	1.2-5.0	
Flexural strength	MPa	67-113	
Flexural modulus	MPa	3,500-4,600	
Compressive strength	MPa	104-131	
Young's modulus	MPa	1,970	Jasso-Gastinel, C F; Vivero-Marin, J M; Manero-Brito, O, Antec, 1615-19, 2005.
Water absorption, equilibrium in water at 23°C	%	0.2	

CHEMICAL RESISTANCE

PARAMETER	UNIT	VALUE	REFERENCES
Acid dilute/concentrated	-	good	
Alcohols	-	good	
Alkalis	-	good (dilute)	
Aliphatic hydrocarbons	-	poor	
Aromatic hydrocarbons	-	poor	
Greases & oils	-	poor	
Halogenated hydrocarbons	-	poor	

FLAMMABILITY

PARAMETER	UNIT	VALUE	REFERENCES
Limiting oxygen index	% O$_2$	22	Chen, X, Wan, M, Gao, M, Wang, Y, Yi, D, Chin. J. Chem. Eng., 28, 9, 2474-82, 2020.
Heat release	kW m^{-2}	202-720	Tibiletti, L; Longuet, C; Ferry, L; Coutelen, P; Mas, A; Robin, J-J; Lopez-Cuesta, J-M, Polym. Deg. Stab., 96, 67-75, 2011; Pereira, C M C; Herrero, M; Labajos, F M; Marques, A T; Rives, V, Polym. Deg. Stab., 94, 939-46, 2009.
Char at 500°C	%	1.5-4	Tibiletti, L; Longuet, C; Ferry, L; Coutelen, P; Mas, A; Robin, J-J; Lopez-Cuesta, J-M, Polym. Deg. Stab., 96, 67-75, 2011; Pereira, C M C; Herrero, M; Labajos, F M; Marques, A T; Rives, V, Polym. Deg. Stab., 94, 939-46, 2009.
Volatile products of combustion	-	CO, CO$_2$, styrene, phthalic anhydride	

UP unsaturated polyester

PARAMETER	UNIT	VALUE	REFERENCES
PROCESSING			
Typical processing methods	-	bulk molding, casting, compression molding, encapsulation, injection molding, printed circuit board, pultrusion, resin transfer molding, sheet molding	
Additives used in final products	-	Fillers: aluminum hydroxide, antimony trioxide, calcium carbonate, carbon black, chopped glass fiber, crashed marble, flyash, glass fiber, hollow glass spheres, kaolin, marble, montmorillonite, nano-TiO_2, polymeric bubbles, quartz, saw dust, silica, talc, wood flour	
Applications	-	automotive, boats, buttons, chairs, coatings, construction, ducts, electrical components, gel coats, marine laminates, pipes, sheet molding compounds, shower stalls, synthetic marble, tanks, wind turbine blades	
Outstanding properties	-	balance of toughness and other mechanical properties, renewable content (some), simple processing methods	
BLENDS			
Suitable polymers	-	EP, PCL, PEO, PLA, PMMA, PU	
ANALYSIS			
FTIR (wavenumber-assignment)	cm^{-1}/-	C–H – 2918, 1453; C=O – 1721; C–O–C – 1259, 1124, 1071	Zhao, Q; Jia, Z; Li, X; Ye, Z, Mater. Design, 31, 4457-60, 2010.
Raman (wavenumber-assignment)	cm^{-1}/-	vinyl – 1632, 1661; C=CH_2 – 1413; C=O – 1732	Cruz, J C; Osswald, T A; Kemper, M, Antec, 2828-32, 2006.
NMR (chemical shifts)	ppm	H NMR: CH_2–O – 3.8; CH_2–OH – 3.5; CH_2 – 4.3; aromatic proton – 7.3-7.8	Alemdar, N; Erciyes, A T; Bicak, N, Polymer, 51, 5044-50, 2010.

VE vinyl ester resin

PARAMETER	UNIT	VALUE	REFERENCES
GENERAL			
Common name	-	vinyl ester resin	
CAS name	-	vinyl ester resin; benzene, ethenyl-, polymer	
Acronym	-	VE	
CAS number	-	36425-15-7; 68002-44-8; 926021-66-1; 877997-30-3; 855482-93-8; 848485-23-4; 208520-04-1	
HISTORY			
Person to discover	-	Robertson, F	Robertson, F, US Patent 1,921,326, Carbide and Carbon Chemicals, Aug.8, 1933.
Date	-	1933	
SYNTHESIS			
Styrene content	%	27-41.5	
Method of synthesis	-	free radical polymerization	Yang, G; Liu, H; Bai, L; Jiang, M; Zhu, T, Microporous Mesoporous Mater., 112, 351-56, 2008.
Molecular weight between crosslinks	g mol^{-1}	300-900	La Scala, J J; Logan, M S; Sands, J M; Palmese, G R, Composite Sci. Technol., 68, 1869-76, 2008.
COMMERCIAL POLYMERS			
Some manufacturers	-	AOC; Ashland; Dow, Reichhold	
Trade names	-	Vipel; Hetron; Derakane; Dion	
PHYSICAL PROPERTIES			
Density at 20°C	g cm^{-3}	1.03-1.15	
Thermal expansion coefficient, 23-80°C	10^{-4} °C^{-1}	0.11	
Glass transition temperature	°C	55-145	Rosu, L; Cascaval, N; Rosu, D, Polym. Test., 28, 296-300, 2009; La Scala, J J; Logan, M S; Sands, J M; Palmese, G R, Composite Sci. Technol., 68, 1869-76, 2008.
Heat deflection temperature at 0.45 MPa	°C	108-115	
Heat deflection temperature at 1.8 MPa	°C	93-166	
Dielectric constant at 100 Hz/1 MHz	-	3.4-3.5/3.3-3.4	
Dissipation factor at 100 Hz	E-4	25-36	
Dissipation factor at 1 MHz	E-4	16-23	
Volume resistivity	ohm-m	1E14	
Surface resistivity	ohm	1E13	
Electric strength K20/P50, d=0.60.8 mm	kV mm^{-1}	120	
MECHANICAL & RHEOLOGICAL PROPERTIES			
Tensile strength	MPa	16-95; 80	Alia, C, Jofre-Reche, J A, Suárez, J C, Martín-Martínez, J M, Polym. Deg. Stab., 153, 88-99, 2018.

VE vinyl ester resin

PARAMETER	UNIT	VALUE	REFERENCES
Tensile modulus	MPa	3,000-3,800	
Tensile stress at yield	MPa	77-88	
Elongation	%	2.5-9	
Flexural strength	MPa	60-163	La Scala, J J; Logan, M S; Sands, J M; Palmese, G R, Composite Sci. Technol., 68, 1869-76, 2008.
Flexural modulus	MPa	3,200-4,200	La Scala, J J; Logan, M S; Sands, J M; Palmese, G R, Composite Sci. Technol., 68, 1869-76, 2008.
Compressive strength	MPa	82	
Abrasion resistance (ASTM D1044)	mg/1000 cycles	100	
Shrinkage	%	1.65	
Viscosity	mPa s	520-620	Alia, C, Jofre-Reche, J A, Suárez, J C, Martín-Martínez, J M, Polym. Deg. Stab., 153, 88-99, 2018.
Water absorption, equilibrium in water at 23°C	%	0.1	
CHEMICAL RESISTANCE			
Acid dilute/concentrated	-	good/poor	
Alcohols	-	good/poor	
Alkalis	-	good	
Aliphatic hydrocarbons	-	good	
Aromatic hydrocarbons	-	poor	
Esters	-	good	
Greases & oils	-	good	
Halogenated hydrocarbons	-	poor	
Ketones	-	poor	
FLAMMABILITY			
Ignition temperature	°C	35	
Limiting oxygen index	% O_2	19.5	Ji, S, Duan, H, Chen, Y, Ma, H, Polymer, 207, 122917, 2020.
Peak of heat release rate	kW m^{-2}	1157	Ji, S, Duan, H, Chen, Y, Ma, H, Polymer, 207, 122917, 2020.
Total heat release	MJ m^{-2}	94	Ji, S, Duan, H, Chen, Y, Ma, H, Polymer, 207, 122917, 2020.
Char at 800°C	%	2.0	Ji, S, Duan, H, Chen, Y, Ma, H, Polymer, 207, 122917, 2020.
TOXICITY			
Oral rat, LD$_{50}$	mg kg^{-1}	>4,000	
Skin rabbit, LD$_{50}$	mg kg^{-1}	>2,000	
PROCESSING			
Typical processing methods	-	lay-up, spray-up, filament winding, pultrusion, sheet and bulk molding, vacuum-assisted resin transfer molding	
Applications	-	aircraft, coatings, pipes, tanks, windmill blades	
Outstanding properties	-	chemical resistance, corrosion resistance	

VE vinyl ester resin

PARAMETER	UNIT	VALUE	REFERENCES
BLENDS			
Suitable polymers	-	PU	

XG xanthan gum

PARAMETER	UNIT	VALUE	REFERENCES
GENERAL			
Common name	-	xanthan gum, polysaccharide B-1459	Palaniraj, A; Jayaraman, V, J. Food Eng., 106, 1-12, 2011.
CAS name	-	xanthan gum	
Acronym	-	XG	
CAS number	-	11138-66-2	
EC number	-	234-394-2	
HISTORY			
Person to discover	-	Jeanes, A R	
Date	-	1961	
Details	-	developed process of XG biosynthesis	
SYNTHESIS			
Monomer ratio	-	glucose:mannose:glucuronic acid=2:2:1	
Acetate content	%	1.9-6.0	Garcia-Ochoa, F; Santos, V E; Casas, J A; Gomez, E, Biotech. Adv., 18, 7, 549-79, 2000.
Pyruvate content	%	1.0-5.7	Garcia-Ochoa, F; Santos, V E; Casas, J A; Gomez, E, Biotech. Adv., 18, 7, 549-79, 2000.
Chemical structure	-	xanthan is composed of the backbone and a side chain. The backbone is formed by the repeated glucose units, linked by the β-(1–4) glycosidic bonds. The side chain is a negatively charged trisaccharide [β-(1–3)-d-mannose-(1–4)-β-d-glucuronic acid-(1–2)-α-d-mannose]	Wang, L, Xiang, D, Li, C, Bai, X, Food Hydrocolloids, 112, 106352, 2021.
Method of synthesis	-	xanthan gum is produced by culturing *Xanthomonas campestris* on a well-aerated medium containing commercial glucose, organic nitrogen sources, dipotassium hydrogen phosphate and appropriate trace elements	Palaniraj, A; Jayaraman, V, J. Food Eng., 106, 1-12, 2011.
Temperature of biosynthesis	°C	28-30	
Time of polymerization	h	100	
Yield	g l^{-1}	11-15 (depending on carbon source; glucose is the most frequently used in commercial production); 50% sugar conversion	
Mass average molecular weight, M_w	dalton, g/mol, amu	2,000,000-20,000,000; 2,700,000	-; Nnyigide, O S, Nnyigide, T O, Hyun, K, Carbohydrate Polym., 251, 117061, 2021.
Polydispersity, M_w/M_n	-	1.014	Faria, S; de Oliveira Petkowicz, C L; de Morais, S A L; Terrones, M G H; de Resende, M M; de Franca, F P; Cardoso, V L, Carbohydrate Polym., in press, 2011.
STRUCTURE			
Chain conformation	-	right-handed, fivefold helix	right-handed, fivefold helix
COMMERCIAL POLYMERS			
Some manufacturers	-	Kelco, Cargill, Viachem	
Trade names	-	Keltrol	

XG xanthan gum

PARAMETER	UNIT	VALUE	REFERENCES
PHYSICAL PROPERTIES			
Color	-	white to light yellow to brown	
Odor	-	mild	
Initial decomposition temperature	°C	58 (dehydration, approx. 15%); 266.4 (weight loss exceeding 40%)	Zohuriaan, M J; Shokrolahi, F, Polym. Test., 23, 575-79, 2004.
MECHANICAL & RHEOLOGICAL PROPERTIES			
Water absorption, equilibrium in water at 23°C	%	8-15	
CHEMICAL RESISTANCE			
Good solvent	-	water (hot and cold)	
FLAMMABILITY			
Autoignition temperature	°C	>200	
Char at 600°C	%	27.5	Zohuriaan, M J; Shokrolahi, F, Polym. Test., 23, 575-79, 2004.
BIODEGRADATION			
Typical biodegradants	-	fungal cellulases catalyze cleavage of main chain	Katzbauer, B, Polym. Deg. Stab., 59, 1-3, 81-4, 1998.
TOXICITY			
NFPA: Health, Flammability, Reactivity rating	-	1/1/0	
Carcinogenic effect	-	not listed by ACGIH, NIOSH, NTP	
PROCESSING			
Applications	-	cosmetics (creams, lotions, shampoos, toothpaste), industrial (adhesives, agricultural chemicals, cleaners, drilling mud, paints, paper, textile and carpet printing), food thickening (bakery products, beverages, dairy products, salad dressings, sauces, soups), medical (anti-tumor activity), pharmaceutical (emulsions, suspensions, tablets)	Palaniraj, A; Jayaraman, V, J. Food Eng., 106, 1-12, 2011.
BLENDS			
Suitable polymers	-	chitosan, CMC, PEO, PR, starch	
ANALYSIS			
FTIR (wavenumber-assignment)	cm^{-1}/-	C=O – 1710-1730, 1530-1650; C–H – 1420-1430; C–O – 1050-1150	Faria, S; de Oliveira Petkowicz, C L; de Morais, S A L; Terrones, M G H; de Resende, M M; de Franca, F P; Cardoso, V L, Carbohydrate Polym., in press, 2011.
NMR (chemical shifts)	ppm	α-anomeric protons – 5.1 and 5.2; β carbons of pentoses and hexoses – 4.8 and 4.9; hydrogen near OH group – 4.0; uronic acid – 2.3	Faria, S; de Oliveira Petkowicz, C L; de Morais, S A L; Terrones, M G H; de Resende, M M; de Franca, F P; Cardoso, V L, Carbohydrate Polym., in press, 2011.

Printed in the United States
by Baker & Taylor Publisher Services